AF323769

Magnetic Micro-/Nano-Materials for Proteomics Analysis

Magnetic Micro-/Nano-Materials for Proteomics Analysis

Nianrong Sun • **Chunhui Deng**

Zhongshan Hospital of Fudan University, China

NEW JERSEY • LONDON • SINGAPORE • BEIJING • SHANGHAI • HONG KONG • TAIPEI • CHENNAI • TOKYO

Published by

World Scientific Publishing Co. Pte. Ltd.

5 Toh Tuck Link, Singapore 596224

USA office: 27 Warren Street, Suite 401-402, Hackensack, NJ 07601

UK office: 57 Shelton Street, Covent Garden, London WC2H 9HE

British Library Cataloguing-in-Publication Data
A catalogue record for this book is available from the British Library.

《磁性微纳米材料在蛋白质组学分析中的应用》
Originally published in Chinese by Fudan University Press
Copyright © Fudan University Press, 2017
This edition is distributed worldwide by World Scientific Publishing Co. Pte. Ltd., except China.

MAGNETIC MICRO-/NANO-MATERIALS FOR PROTEOMICS ANALYSIS

Copyright © 2021 by World Scientific Publishing Co. Pte. Ltd.

ISBN 978-981-123-054-7 (hardcover)
ISBN 978-981-123-055-4 (ebook for institutions)
ISBN 978-981-123-056-1 (ebook for individuals)

For any available supplementary material, please visit
https://www.worldscientific.com/worldscibooks/10.1142/12107#t=suppl

Typeset by Stallion Press
Email: enquiries@stallionpress.com

Contents

Chapter 1

Proteomics Analysis

1.1. Proteomics research

In 1994, the concept of "proteome" was firstly proposed at a science conference in Italy and this word was spliced together by the "prote" in protein and "ome" in genome. Originally, "proteome" was defined as "all proteins expressed by a genome". Although the definition reckons without the dynamic nature of proteomes, as well as the influence of surroundings on cells, tissues, or organisms which produce the proteins. But the proposition of proteome indicated that, along with the implementation and promotion of human genome project (HGP), it has been increasingly and clearly realized by scientists that the main feature of genome as the carrier of genetic information is identity. This can be reflected in the facts that the genome of a unicellular organism remains constant whatever growth conditions it faces, likewise for a multicellular organism, the genome of the individual is the same at whichever developmental stage or in whatever kinds of cells. Albeit all of human DNA base sequencing has been completed, there is a long way to recognize their function well in life activities, which is exactly done by proteins. The general consensus on proteome refers to all proteins in a known cell, including all subtypes and modified proteins, at a particular moment. Proteomics aims at analyzing the composition, expressed level, and modified status of proteins for dynamic changes within cells from a holistic level; understanding the protein–protein connection and interaction; suggesting protein functions and activity routines of cells. Currently, proteomics research, which takes large-scale analysis of proteins within cells or organisms as its main assignment, has accounted for primary position in the post-genomics era, whose importance and strategical significance to life science research are increasingly evident.

Although in the wake of the exponential growth of the publication on proteome research, proteomics has been rapidly accepted and valued by the scientific community, still it would be an uphill and long-term battle owing to its characteristics and difficulty. The characteristics of proteomes is as follows compared to that of genomes.

1.1.1. *Diversity*

Since protein is the chief executive of life activities, the constitution of proteome is diverse for different types of cells or the one cell in different active states. The currently accepted definition of proteome is "all proteins expressed by a genome", which is difficult to be employed in reality. In addition to the genes that have been known to have protein products, the rest are just open reading frame (ORF) which are determined by the initial and the terminal password, two specific sequences. ORF number is the number of genes in a genome. Obviously, the most convincing determination for ORF number should depend on the proteome. So there is a kind of circular definition between genome and proteome: proteome is constituted of expression product of all genes in a genome, while the determination of all genes must be affirmed by the proteome; on the other hand, even if the problem of ORF number can be solved, another problem still exists, that is determining whether there is one-to-one correspondence between the ORF and protein or not. Many pseudogenes have been discovered, but they never express though they have the same ORF with genuine gene. Moreover, it is difficult to find directly corresponding ORF for many proteins due to the presence of RNA editing and protein splicing.

1.1.2. *Dynamic nature*

Proteins in life activities can be classified into two categories: the first one is relatively stable, such as the proteins responsible for composition of various cell structures; the second one would appear as quantitative change or qualitative change along with change of cell status, such as proteins responsible for signal transduction of live cell regulation. Generally, functional proteomics refers to the research that focuses on protein changes. It aims at comparing the differential proteins of proteomes within one cell or tissue under different growth conditions or pathological conditions. The problem is that the stability of protein is quite inferior to nucleic acid, and the protein could be degraded or lost during the experimental process. In addition, taking the limitation of resolution and reproducibility of the present analytical technique into consideration, it is a tough task to realize the accurate measurement of the quantity difference of proteins.

1.1.3. *Spatiality*

Firstly, various proteins are distributed in different locations in a cell. It is necessary to know the spatial positions of proteins well for truly understanding their functions, since their functions are closely related with their spatial positions. Secondly, proteins are not stationary in cells, they usually play a role in cells by moving through different subcellular environments. Therefore, subcellular proteomic as a new research area closely associated with space is derived, and thereby a new challenge arises: how to obtain complete subcellular components by performing proper hierarchical separation of cells. During this process, it is vital to avoid the loss of constituent proteins which are to be studied and the contamination of non-constituent proteins simultaneously. Furthermore, in this area, another technological difficulty is how to distinguish subcellular constituent proteins from those moving within different regions just guided by physiological function.

1.1.4. *Group characteristic*

The interaction between protein–protein or protein–bio-macromolecule is the foundation of realizing protein functions. The new word interactome was thought up to describe the research on large-scale interaction of protein–protein. The research in this area can be classified into two categories. The first one is the research on the network of protein–protein interaction. Many activities within a cell are completed through a complex and extensive interaction network of protein–protein. The commonly used techniques for the research on large-scale interactions of protein–protein include large-scale dual-hybridization technology, phage-display technology, and protein-chip technology. Its problem is that there are more false signals. The second one is the analysis of the constitution of the protein complex. A protein complex can be divided into two categories. One is structured, such as core–pore complex, which is relatively stable. Classic research strategy on proteome such as electrophoresis and mass spectrometry (MS) could be adopted for this one after enrichment by hierarchical separation. The other is functionalized protein complex. This kind of complex aggregates together only when performing a function, so it is quite unstable. Currently, the prevalent method for this one is affinity tags technology. Namely, the plasmid of compound protein containing specific tag peptide is constructed first, which can form complexes with other kinds of proteins. Then affinity separation column is chosen to specifically extract the peptide with specific tag.

1.1.5. *Technology diversity*

All above characteristics of proteomes indicate that the required technology for proteome research is more than just one, and the technology difficulty is more than those for genome research. The research technology for proteomes can be simply classified into two categories. The first one is separation technology, such as 2D electrophoresis technology, hierarchical separation technology for cells, and affinity chromatography, etc. There are two problems in separation technology: one is "quantity", namely, there are tens of orders of magnitude of dynamic differences in protein quantity; the other is "quality", namely, the physical and chemical properties of different proteins are very different. The second one is identification technology, whose core is MS. There are also some other identification technologies such as protein chip technology, phage-display technology, and large-scale dual-hybridization method. One of the biggest problems of all identification technologies now is that the known sequences of proteins or genes are the basis for detection of proteins to be determined. In short, the development of proteomics is limited and driven by technology.

1.1.6. *Mutual complement and assistance with genomes*

Genomes need to use protein sequence to speculate and determine the full length of ORF. Proteomics research is closely associated with genome research, which can be noticed through the research on protein–protein interactions. For instance, dual-hybridization technology and phage-display technology and so forth take the genome data and technology as their basis. Moreover, there are much comparisons of gene data of known genomes during the process of protein identification by MS.

The above characteristics and difficulties manifest that the development of proteomes is long term and needs much efforts from scientists in many fields. In addition, these characteristics and difficulties are also the direction and trend of the development of proteomes. The intersection of proteomics and other large-scale scientific fields will be the most exciting new frontier for the life sciences in the future.

1.2. Proteomics research strategy

There are two research strategies along with the advent of proteomics. The first strategy is "exhaustion method", namely, high-throughput proteomics research technologies are adopted to analyze proteins in organisms as much as possible. This method emphasizes that proteomics analysis should be conducted from a large-scale and systemic perspective, which is more in line with the nature of proteomics. But it is an unattainable goal to analyze all proteins in organisms since protein expression is always varying with space and time. The second strategy is "function method", namely, studying the variation of cell protein composition in different periods, such as the differential expression of proteins in different environments, which regards the discovery of differential protein species as the main target. This method tends to take proteomics as the means and tools for studying biological phenomena. Correspondingly, proteomics research begins from two aspects. One aspect is constructing protein expression profiles by separation, recognition, quantification, orientation, and identification of post-translational modifications (PTMs), of all expressed proteins in a gene, tissue, cell, or organism. This one is called expression proteomics research. For instance, cluster regulation pattern, transcription pattern, and translation discrepancy regulation pattern of gene expression at the protein level can be discovered by studying the protein expression profile of organisms or a specific tissue/cell which has established a genomic database and gene expression profile. The other aspect is to take the specific cells and biological systems as objects to study protein–protein interactions, ascertain the role of proteins in specific channels and cellular structures, explore the interrelationships between protein structures and functions as well as protein expression differences under different physiological and pathological conditions. This is called functional proteomics. The research on functional proteomics can help us recognize the functions of genes, explore proteins related to the specific physiological and pathological conditions, build functional networks of proteins, and so on. The research scope of proteomics mainly refers to the protein expression profile in the early days, and is being improved and expanded with the development of the subject. The research of functional proteomics including low-abundance proteins with important functions, protein PTMs, and protein–protein interactions has increasingly become an important part of proteomics research.

The extraction, separation, and identification of samples are always the basis, whether for the early proteomics research on expression profile, or the prevailing protein PTMs research for the moment, or advanced structural parsing of proteins to be developed. At present, there are three research strategies in proteomics: the first is "bottom-up"; the second is "middle-down", the last is "top-down". The detailed technical route in proteomics is displayed in Figure 1.1. "Bottom-up" strategy is also called shotgun, which deals with protein samples by enzymolysis technique and then extracts hydrolyzed protein prior to tandem MS analysis and database search. "Middle-down" strategy refers to the process where the protein sample is digested into peptides after a series of pre-separations for MS analysis. "Top-down" strategy refers to the process where the protein sample is analyzed directly by MS after a series of pre-separations. Both "top-down" strategy and "middle-down" strategy purify holoproteins by making the most use of their physical and chemical properties, such as protein molecular weight, isoelectric point, hydrophobicity, acid alkalinity, and binding property, etc. One of the classical examples is two-dimensional gel electrophoresis (2DE). In recent years, "top-down" strategy based on liquid chromatography (LC), including capillary IEF-mass spectrometry (CIEF-MS), preparative-scale isoelectric focusing-size exclusion chromatography-mass spectrometry (IEF-SEC-MS), ion exchange chromatography-size exclusion chromatography-mass spectrometry (IEC-SEC-MS), size exclusion chromatography-reversed-phase liquid chromatography-mass

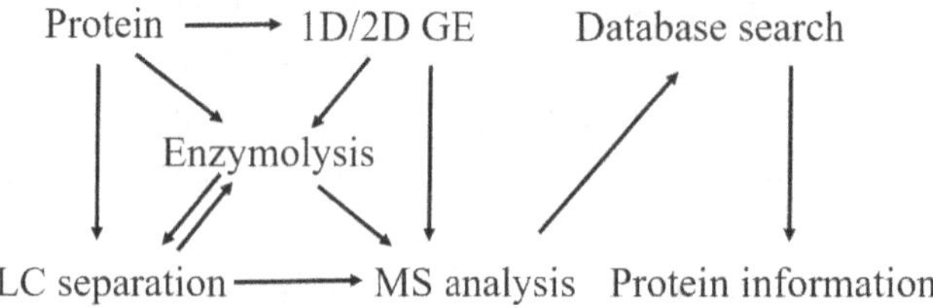

Figure 1.1: Technical route for proteomics.

spectrometry (SEC-RPLC-MS)., and so on, has been developed greatly, and is far better than the 2DE technique in the aspects of accuracy, precision analysis speed, and resolution.

1.3. Proteomics analysis technique

1.3.1. *Separation technique in traditional proteomics*

Electrophoresis and LC are two main techniques in traditional proteomics, which include common techniques such as 2DE, high performance liquid chromatography (HPLC), and 2D liquid chromatography (2D-LC). Polyacrylamide gel electrophoresis (PAGE), capillary electrophoresis (CE), chromatographic technique, and super-separation technique for protein purification, etc., are also the separation techniques in traditional proteomics.

1.3.1.1. *Two-dimensional gel electrophoresis*

2DE, which was developed in the mid-1970s, was one of the earliest supporting techniques and is still a powerful tool in proteomics research because of its high repeatability, maneuverability, and high resolution as well as the characteristics of proteins such as solubility and specificity, etc. At present, 2DE generally implies that the high-throughput separation of proteins rely on 2D treatment: the first dimension is based on isoelectric point (pI); the second dimension is based on the difference of relative molecular weight. Briefly, the proteins from prepared samples are firstly separated according to the different pI by isoelectric focusing (IEF), then the obtained gel strip from the first dimension is transferred to the second dimension, namely, proteins are separated again by PAGE according to their different relative molecular weights. All influences on IEF and gel electrophoresis should be taken into consideration during sample preparation. The selection of buffer solution, reductant and detergent, etc., in the process of sample preparation should depend on the differences of experimental requirement and sample particularity. None of the sample preparation methods is versatile, therefore, the optimized experimental conditions for each sample need to be explored by a large number of experiments. Effective and repeatable sample preparation method is the key of successful separation by using 2DE.

1.3.1.2. *High performance liquid chromatography*

HPLC technique features highly efficient separation, analysis quickness, high-test sensitivity, and a wide range of applications, and is widely used in the separation and analysis practice. According to the differences of the separation mechanism of components in the stationary phase and mobile phase, HPLC technique can fall into liquid–solid adsorption chromatography, liquid–liquid distribution chromatography, ion exchange chromatography, gel chromatography (exclusion chromatography), affinity chromatography, and so on. Further, liquid–liquid partition chromatography can fall into normal-phase chromatography (the polarity of the stationary phase is superior to that of the mobile phase) and reverse-phase chromatography (the polarity of the stationary phase is inferior to that of the mobile phase) according to the polarity difference between stationary phase and mobile phase. Reverse-phase chromatography is the most widely used and most effective method in current HPLC technique. Moreover, it is also the most widely used liquid-phase separation method for the separation of proteins and peptides.

1.3.1.3. *Two-dimensional or multi-dimensional LC*

Two-dimensional or multi-dimensional LC technique is a separation system that is composed of two chromatographic columns which possess different separation mechanisms and are mutually independent. The sample passes through the first dimensional chromatographic column into the interface, then is transferred into the second dimensional chromatographic column and detector after treatment by enrichment, capture, or cutting. The complex mixture (such as peptides) can be separated into single component by using the different sample characteristics such as molecular size, pI, hydrophilicity, charge, special molecular interaction (affinity), etc. The components which are not completely separated in the first dimensional separation system may obtain better separation result in the second dimensional system, therefore, both the separation capability and resolution are greatly improved. The peak capacity of fully orthogonal 2D-LC is the product of the peak capacity of two kinds of 1D separation modes.

The typical 2D-HPLC/MS method still has drawbacks, although there are many advantages, such as high peak capacity, wide range of application for proteins, and the whole process to be automated easily, when multi-dimensional

LC technique is applied to separate proteins and peptides. This is because peptide mixture is too complicated by direct enzymolysis of proteins, resulting in the suppression of low-abundance peptides during the separation process. Hence, it is unattainable to recognize and identify all proteins in a practical biological sample.

1.3.2. *Biological MS*

After the appearance of a series of new soft ionization techniques such as fast atomic bombardment ionization (FABI), matrix-assisted laser desorption ionization (MALDI), electrospray ionization (ESI), etc., biological MS has been developed rapidly since 1980s. MS technique has been the indispensable platform in proteomics research and beyond dispute because of the following merits: MS technique possesses the characteristics of high accuracy, high sensitivity and automation, and it can accurately measure the relative molecular weight of proteins and peptides and amino acid sequences, etc., so as to provide reliable basis for protein structure parsing. The information quality of MS data directly determines the reliability and identification number of protein.

So far, MALDI-MS and ESI-MS are two widely used biological MS. MALDI-MS features high sensitivity, strong operability, and good toleration to inorganic salt and buffer solution in biological sample. ESI-MS features excellent selectivity, wide analysis range of molecular weight, a little consumption of the sample, and is easy to be combined with various chromatography techniques. There are three main MS methods for proteomics analysis: peptide mass fingerprinting (PMF), tandem mass spectrometry (tandem MS, MS/MS), and ladder peptide sequencing (LPS).

Each protein can be digested into a group of exclusive and specific peptides with different lengths after treatment by specific enzymatic hydrolysis or chemical hydrolysis. PMF method refers to detecting the relative molecular weights of above obtained peptides by using MS for database search, so as to find the similar ones for depicting "PMF". In recent years, PMF technology has become the common identification method, which most closely matches with requirement of high throughput in proteomics research.

Tandem MS is equipped with multiple mass analyzer. Parent ion is separated by collision dissociation, and then the produced daughter ion is detected. Namely, metastable ions can be produced by molecules being tested in the process of ionization and flight, which are used for analyzing the mass discrepancy of peaks of adjacent group-type for the recognition of corresponding amino acid residues. Herein, the fragmentation of metastable ion includes self- and induced fragmentation. Furthermore, it is also possible to sequence the peptide by matrix-assisted laser desorption ionization time-of-flight mass spectrometry (MALDI-TOF-MS) with the function of post-source decay (PSD), however, the existence of parts of defects leads to an unsatisfied sequencing result. The peptide sequence diagram of tandem MS is more complex than that of PMF. When using tandem MS to identify proteins, it is necessary to combine the obtained parts of amine acid sequences with their fore and after ion mass as well as the parent mass of the peptide, which is called peptide sequence tag (PST).

LPS is usually used for the analysis of C-terminal sequence of amine acids. Namely, carboxypeptidase is used to cut the C-terminal amine acids of proteins and peptides to produce a serials of "ladder" peptides which contain the mass discrepancy of only one amino acid residue. The corresponding amino acid residue can be reckoned by the mass discrepancy of adjacent peptide peaks. However, enzymolysis is a dynamic process, which is multi-speed and easy to be interfered, therefore, the analysis result by LPS is also unsatisfied.

1.3.3. *Enzymatic hydrolysis technology*

Currently, biological MS is always critical for sample identification no matter whether from the "middle-down" or "bottom-up". The analysis of proteins by MS mainly deduces the "PMF" of the protein by taking advantage of the identification towards peptide. Enzymatic hydrolysis of protein is important for transferring large-size protein into small-size peptide. And that the completeness and specificity of enzymatic hydrolysis is vital to the determination of peptide sequence.

The traditional enzymatic hydrolysis method contains in-solution digestion and in-gel digestion according to the difference of separation method. As the name suggests, in-solution digestion is enzymatic hydrolysis within solution. The common proteases include trypsin with special effect on the carboxyl terminal of Arg and Lys (basic amino acids), α-chymotrypsin with special effect on the carboxyl terminal of aromatic amino acids, staphylococcal protease with special effect on the carboxyl terminal of Asp and Glu (acidic amino acid), thermolysin and subtilisin, etc. Currently, trypsin and chymotrypsin, which are commonly used in proteomics research, possess poor specificity since both of them are obtained from the pancreas and are difficult to separate completely. The TPCK-trypsin and TPCK-α-chymotrypsin are the best choices for guaranteeing the specificity of proteases.

Proteins are large molecules with hierarchical structure, of which the rupture of disulfide bond is usually regarded as the inducement of protein folding, and thereby leading it to be digested easily. This is a necessary step for the making of many peptide mappings. In addition, after the rupture of the disulfide bond, the peptide which binds to the disulfide bond can be removed from the peptide mixture to simplify the peptide mapping. The most common method for breaking disulfide bond is to reduce cystine to cysteine residue. However, because cysteine residues have high reactivity, the sequence determination would become complex, such as forming random disulfide bonds and being oxidized after removing the reductant. Therefore, cysteine residues need to be stabilized by a certain method such as alkylation. For S-carboxymethylation of the trace protein, dithioglycol (DTT) and iodoacetamide (IAA) are the commonly chosen reductant and alkylation reagent, respectively.

The mass ratio of substrate versus enzyme is a range from 200:1 to 50:1 in the enzymatic hydrolysis process of denatured proteins. Choosing a quite low concentration of enzymes is due to the fact that the free enzymes still remain in the sample solution after completion of enzymatic hydrolysis of denatured protein, and high concentration may interfere with subsequent chromatogram analysis or MS identification, even lead to redundant incubation time. Generally, the incubation lasts for 12–16 h at 37°C by using the low concentration of free enzyme, which is the primary defect of traditional in-solution digestion method. The other defect is that spontaneous degradation of protease left in the sample solution would generate serious signal suppression and interference for subsequent identification, especially for the identification of low-abundance sample. Moreover, online holoprotein separation-enzymatic hydrolysis-MS identification become a development trend of the proteomics technology with the rapid development of "middle-down", but the time-consuming digestion process affects the development flux of proteomics technology seriously. Besides, for complete digestion, some denaturants which are beneficial for unfolding the structure of proteins, such as carbamide, ethylene diamine tetraacetic acid (EDTA), and surfactant, etc., would be added in the process of traditional digestion. These additives pollute the digestion production, resulting in additional purification. All the above difficulties of traditional in-solution digestion method cannot be overcome. As a result, it has been an urgent issue to develop a highly efficienct and convenient digestion method for promoting development of proteomics technology.

1.3.4. *Low-abundance enrichment technology in proteomics*

Although MALDI-TOF-MS and LC-ESI-MS are very sensitive for the identification of trace proteins/peptides, it still shows defects in the aspect of identifying low-abundance peptides/proteins in practical biological samples. This is for two main reasons. On the one hand, the concentration of parts of peptides/proteins in practical biological samples is quite low (<1 $nmol \cdot L^{-1}$), their signals tend to be seriously interfered by high-abundance peptides/proteins and introduced impurity during the sample processing. On the other hand, the proteins associated with disease signaling such as transcription factors, growth factors, and protein kinases, etc., are mostly low abundance. The endogenous peptides associated with metabolism or life activities such as neuropeptides and hormones, etc., also tend to be suppressed by high-abundance proteins in the process of MS identification. Therefore, it is very significant that effective enrichment and separation of low-abundance peptides/proteins in complex bio-samples prior to MS analysis, which is one of the important conditions for realizing their accurate analysis and identification. In fact, many aspects are involved in effective enrichment of sample during the process of proteomics research. For instance, the volume of peptide extraction solution by in-gel digestion is too large to be analyzed by MS without aforehand enrichment.

1.3.4.1. *Solvent evaporation method*

Solvent evaporation method is one of the most commonly used methods for developing sample concentration at present, which is to put the sample into a lyophilizer to volatilize the solvent by vacuumizing to achieve sample concentration. This method is time-consuming and laborious. A large number of sample losses are unavoidable because of their adsorption on the surface of the container during the drying process, and the impurities such as inorganic salt can be concentrated at the same time. All the above seriously affect the sensitivity of MS identification.

1.3.4.2. *Chromatographic concentration method*

Chromatographic concentration method is another most commonly used method, which is taking advantage of the interaction between sample and adsorbent to concentrate the sample. And adsorbent is usually the silica gel which is modified by alkyl chain. This method can effectively concentrate the sample and remove salts and other impurities from the sample simultaneously. Especially for LC-MS, in order to prevent the impurities such as salts in samples to affect MS signal detection, a short reversed-phase column is usually connected before separation column, which can improve the effect of sample concentration and desalination.

Both commercial Zip-tip and Zip-plate depend on the principle of chromatographic concentration method for sample concentration and desalination, which fill a small amount of reverse-phase packing at pipette tip and the bottom of the 96-well plate via tedious operations, respectively. The sample amount that can be concentrated is limited because of the limited filler.

1.3.4.3. *On-plate enrichment method*

In recent years, the on-MALDI plate enrichment method has been developed with the wide application of MALDI-MS. Generally speaking, first is the modification of the plate, the proteins/peptides can be enriched selectively by various interactions between the modified surface and analytes, then the salts on the surface are washed away by deionized water. In this way, both enrichment and desalination can be carried out directly on the plate, eliminating sample loss that is caused by transfer process and embodying its advantage in the analysis and identification of trace proteins. Orland *et al.* have successively developed the selective on-plate enrichment method based on the specific interaction between immobilized antibody and antigen,[1] the on-plate desalination and enrichment method based on the hydrophobic interaction of immobilized monomolecular octadecanethiol,[2] as well as the on-plate enrichment method for proteins/peptides based on electrostatic interaction.[3] Liang *et al.*[4] immobilize the antibody on the target-plate by a layer of nitrocellulose membrane for enriching peptides via immune reaction. Schuerenberg *et al.*[5] have developed on-plate enrichment method based on hydrophilic interaction. More recently, Yang *et al.*[6] has immobilized the hydrophobic polymethyl methacrylate (PMMA) on plate and developed on-plate enrichment method for peptides, which realize the simultaneous enrichment and desalination. By the way, the on-plate enrichment and desalination method is more suitable for the analysis of small volume samples containing high concentrations of salts.

1.3.4.4. *Electroelution enrichment method*

Electroelution is an effective method for the MS identification and analysis of protein sample which is obtained by direct gel transfer after separation by 2DE. Traditional electroelution methods need large amounts of buffer solution, resulting in the dilution of the sample easily, which not only hinders the effective elution but also affects the subsequent MS detection. Clarke *et al.*[7] have designed a new electroelution system, which only needs a little buffer solution and contains the enrichment device for protein sample separation from gel. Briefly, polymeric styrene divinylbenzene membrane is firstly used to enrich protein. Then residues on the membrane such as salts and small hydrophilic organic molecules because of the non-specific enrichment are washed away by deionized water. Finally, the enriched proteins are eluted by using 80% methanol solution from the membrane for MALDI-MS analysis. The concentration of detected protein can be as low as 10^{-12} mol·L^{-1}.

1.3.4.5. *Nanomaterial-based solid-phase microextraction enrichment method*

Nanotechnology is praised as one of three technologies which will affect human life and economy most profoundly in the future. It helps recognize and remould the physical world at the level of atoms and molecules. With the appearance of microscopic characterization and manipulation techniques such as scanning tunneling microscope (STM), atomic force microscope (AFM), and scanning probe microscope (SPM), etc., nanotechnology gains rapid development and thereupon a series of new subjects emerge. The improvement of human survival and life is always the ultimate goal of scientific and technological development at every stage of history, the intersection of nanotechnology and bio-medicine is therefore an inevitable development trend, and nanobio-medicine and related technology are currently among the most active directions in nanotechnology field.

Nanoparticles (1–1,000 nm) for bio-medicine mainly include metal and alloy nanoparticles, metal oxide nanoparticles, organic polymer microspheres, and organo–inorganic composite microspheres. Compared with the conventional block materials, the nanoparticles possess many unusual characteristics, such as the special surface effect and some physical properties like bizarre light, heat, magnetism, etc. Generally speaking, specific surface area of a particle greatly increases with the decrease of the particle size, for instance, the specific surface area is 90 $m^2 \cdot g^{-1}$, 180 $m^2 \cdot g^{-1}$, and 4,500 $m^2 \cdot g^{-1}$, the corresponding particle size is 10 nm, 5 nm, and 2 nm, respectively. At the same time, due to its super-colossal specific surface area, the serious mismatching of bond state on the surface of a nanoparticle brings about many active centers, therefore resulting in very high surface activity of the nanoparticle. For magnetic nanomaterials, when magnetic nanoparticle size is small enough (namely, critical size), the material will show super-paramagnetism. For polymer particle, the easily functionalized surface coupled with its large specific surface area endows it with large functional group density, charge density, and good dispersion stability, making it especially suitable for bio-medicine applications because of the following advantages. First, reactive groups on the surface of polymer microsphere can be used to immobilize biological molecules. Second, charges on the surface of a polymer microsphere are beneficial to improve the stability of its synthesis, storage, and biological fluid (high particle intensity) hybrid process on the one hand, and to adsorb molecules with opposite charge (drug or biological macromolecule) by coulombic interaction on the other hand.

In recent years, with more and more attention being paid to proteomics, nanomaterials have developed rapidly and have been applied to proteomics analysis due to their great application potential. Early studies have adopted nanomaterials (such as nanodiamond, gold nanoparticles, etc.) to enrich biological macromolecules. And nanomaterial-based enrichment method has also become a recent research hotspot. The operation of traditional enrichment and desalination methods such as reversed-phase HPLC and size exclusion chromatography (SEC) is commonly complex, of which elution step may cause a lot of sample loss, thereby affecting sample detection. Therefore, the traditional method is not appropriate for trace protein/peptide. While nanomaterial-based solid-phase microextraction (SPME) enrichment method provides new ideas for solving the problem of complex operations, comparing to traditional methods, nanomaterial-based SPME method is directly used to enrich proteins/peptides by different interactions between the surface with diverse chemical modifications and samples, which is more suitable for detection and analysis of trace samples. Nanozeolite particles possess many merits, such as large specific surface area, variable pore shape, good thermal stability, and mechanical stability. Besides, they can combine with biological macromolecules by electrostatic and hydrophobic interaction, that's why they are widely used for enrichment and purification of biological macromolecules in recent years. On the one hand, many characteristics of nanozeolite particles, including the large specific surface area, high dispersibility in solution, and tunable surface properties (such as adjustable surface charge and hydrophilicity/hydrophobicity), enrich the protein through its interaction with the zeolite surface. Meanwhile salts in buffer solution are not enriched. Furthermore, the small size of nanozeolite particles make it possible for particles that have captured analytes to be directly coated on MALDI plate for MS detection, avoiding sample loss in the elution process. Yang *et al.* have developed a variety of methods for the enrichment of low-abundance proteins/peptides by nanozeolite particles,[8–10] for instance, nanozeolite containing Co^{2+} is fabricated, which provides a new way for the selective enrichment of low-abundance selenoproteins since this kind of nanozeolite material has rich ion-exchange property, rich silicon easily to be functionalized on the surface, good stability, strong rigidity, and uniform pore channel. The nanozeolite containing Co^{2+} is used to efficiently enrich selenoproteins by the strong chelation of Co^{2+} and the near histidine.[8] They also make use of the large specific surface area of nanozeolite material, its various interactions with peptide, as well as its specific size to realize high-efficient enrichment of trace proteins/peptides and the detection of peptides with low concentration as 20 ng·μL^{-1}, by directly coating nanozeolite with adsorbed sample on MALDI plate.[9] Furthermore, they fabricate Fe^{3+}-modified nanozeolite nanoparticles to effectively and selectively enrich phosphoproteins/peptides, which provides a new way for the research of protein PTMs.[10] In addition, they also develop polymer nanocomposite for trace proteins/peptides' enrichment.[11]

(1) Microbead material

Direct immobilization of antibody on microbead was the main method for proteins/peptides' enrichment at the early stage, which is generally limited by specificity. Later, Doucette *et al.*[12] developed chromatography microbead with hydrophobic polymer for pre-enrichment of proteins in the dilute solution, proteins from which were directly digested on microbead for MALDI-MS detection. Enrichment principle of this chromatography microbead is still based on the hydrophobic interaction between the packing of reversed-phase chromatography and the samples. The chromatography microbead with adsorbed peptides from protein digestion can be directly coated on the target plate for MS detection because of its small size. Therefore, this kind of method shortens the digestion time greatly, improves digestion efficiency, and avoids the interference from degradation products of trypsin, which can be used for high-contamination samples containing only 100 $nmol \cdot L^{-1}$ proteins.

(2) Magnetic bead material

The magnetic polymer microsphere is the most studied organic–inorganic composite particle. Girault *et al.*[13] chose streptavidin-coated magnetic beads to enrich biotinylated peptides. This kind of method, which is based on the connection of antigen–antibody interaction and magnetic bead has been increasingly commercialized and widely used. Lee *et al.*[14] fabricated aminophenylboronic acid (APBA)-immobilized magnetic bead for the effective enrichment of glycoproteins by the reversible reaction between boron hydroxyl groups and cis–diol structure. Zhang *et al.* have developed many magnetic bead nanomaterials for phosphoproteins' enrichment by the chelation and electrostatic interaction between phosphate group and metal ion/ metal oxide modified on the surface of the magnetic bead.[15,16] For instance,

Fe^{3+}-modified magnetic beads can realize the identification of phosphorylation site as well as the enrichment of mono-phosphopeptides and multi-phosphopeptides from 0.02 $nmol \cdot mL^{-1}$ β-casein digestion.[15] ZrO_2-modified magnetic bead can retain high selectivity when the molar ratio of β-casein versus bovine serum albumin (BSA) is 1:50 within 30 s.[16] Besides, their small size makes it possible for the magnetic beads with adsorbed phosphopeptides to be coated on MALDI plate without elution step. These ZrO_2-modified magnetic beads are successfully used to enrich and identify phosphoproteins in human serum. Also, the presence of magnetic beads simplifies the separation operation.

(3) Functionalized porous material-based SPME

Exploiting functionalized nanomaterials with different characteristics for interdisciplinary science has become an important content in current nanotechnology research along with rapid development of nanotechnology. In view of SEC enrichment principle, many studies have used porous materials as SPME adsorbent for protein/peptide enrichment. Molecular sieve is one of the most classic porous materials usually containing pore channel frame structure composed by ordered porous silicate or silica. The specificity of frame structure endows it the ability to allow the molecules with diameters smaller than the pore channel to go inside, while excluding the molecules with bigger diameters. Hence, this kind of molecular sieve can separate molecules with different sizes, different polarities, different boiling points, and even different saturation degrees. Mesoporous material is a new type of material with large specific surface area and 3D pore structure, with a pore size between micropore and macropore.

Mesoporous material can be divided into two major categories: silicon ones and non-silicon ones. Silicon-based mesoporous materials possess narrow pore size distribution and regular pore structure, which is usually used for catalysis, separation and purification, drug embedding and slow-release, gas sensors, etc. Besides, many properties of silicon-based material can be changed, such as stability change, hydrophilicity/hydrophobility change, and the change of catalytic activity, by introducing impure atoms to replace the silicon atoms, and thereby broadening the research field. Non-silicon mesoporous materials mainly include transition metal oxide, phosphate, sulfide, etc., which have been widely used for adsorption, catalyst loading, acid catalysis, oxidation catalyst, etc. For instance, activated carbon with a large specific surface area and high pore volume has become the main industrial adsorbent. The charge storage of electric double-layer capacitor (EDLC) material made of mesoporous carbon is higher than that of metal oxide particle assembly, and much higher than that of the marketed metal oxide EDLC. Titanium dioxide mesoporous material has the characteristics of strong photocatalytic activity and high catalyst loading capacity, there are many studies on its structural properties and characterization. In recent years, mesoporous material has been widely used in proteomics, and SiO_2 porous material has been successfully used in the enrichment of peptides in human plasma. However, these porous nanomaterials have irregular shapes and pore channels, which reduce the enrichment efficiency. Tian *et al.*[17] attempted to synthesize highly ordered mesoporous material with uniform pore size for the selective enrichment of peptides in human plasma. The large-size proteins were removed by the size-exclusion effect of pore size, which is similar to the mechanism of centrifugal ultrafiltration. Tian *et al.* found that MCM-41 with 2 nm of pore size owns the best enrichment efficiency for endogenous peptides in human plasma by contrasting with different pore sizes of mesoporous materials. Subsequently, they also studied the enrichment of endogenous peptides in complex biological samples using mesoporous materials with different surface functionalizations, which opens up a new direction for peptidomics analysis.

1.3.5. *Separation technology in PTM proteomics*

PTM refers to the process whereby the expressed proteins are chemically modified in an organism by the attachment of modification groups onto one or several amino acid residues of the peptide chain, or protein hydrolysis and shear. Protein PTM not only self-regulates its activity state, location, folding, and the interaction with other proteins but also plays a key role in cell biological process such as cell recognition, signal transduction, growth, and differentiation. However, the regularity of this key role during the formation of functional proteins cannot be explained at the gene level. Therefore, Protein PTM has become one of the important contents in proteome research. At present, there are more than 200 reported PTMs, including phosphorylation, glycosylation, methylation, acetylation, ubiquitination, etc., of which phosphorylation and glycosylation modification are most studied.

Although it is of great importance to study protein PTMs, large-scale studies on protein PTMs are still in the exploratory stage. The diversity of PTM is one reason. On the other hand, the spatial specificity of PTM make the modification type and degree of protein have great significance along with the change of the survival environment and inner state of the organism. Some PTMs are even fleeting. As a result, analysis technique and method face very high requirements. Currently, the pre-existing technologies, such as electrophoresis, HPLC, affinity technology, biological MS, and bio-informatics tool, are the mainstream methods for PTM proteomics research. The variety of PTM proteins is wide in bio-samples, while their abundance is quite low, leading to their suppression by some high-abundance proteins, besides, the existence of modified groups may affect their ionization efficiency in MS, therefore, the enrichment of PTM proteins and cleavage efficiency of tandem MS are critical to promote PTM proteomics research.

1.3.5.1. *Phosphoproteomics*

(1) Introduction of phosphoproteomics

Life science research has entered the post-genome era with the implementation and promotion of HGP. At present, the main research object of life science is functional genomics whose research strategy is from the perspective of mRNA in cells. But protein is the true executor of physiological function and the direct manifestation of life phenomenon. The implementation of HGP lays a solid foundation for determination of all gene sequences in the organism and the future life science research, but fails to provide direct molecular basis for understanding various life activities. Therefore, it is necessary to conduct research at the protein level to have a comprehensive and in-depth understanding of complex life activities and to elucidate the change mechanisms of life under physiological or pathological conditions. Moreover, the answers to the questions on the existing forms and the activities and their regularity of proteins, such as PTM, protein–protein interactions, and protein conformation, etc., also depend on the direct research on proteins. However, the traditional way for studying an individual protein cannot meet the requirement of the post-gene era, so proteomics has emerged as a new science.

The complexity of proteomics is not only embodied in the kinds and quantities of proteins but also in the diverse PTMs of proteins. Many proteins undergo different modifications in different locations. Reversible phosphorylation is the most important of all PTMs, and plays a key role in function regulation of organisms. In 1992, Edmond Fischer and Edwin Krebs were rewarded with the Nobel Prize in Physiology or Medicine for discovering reversible phosphorylation, which is a biological regulatory mechanism. As we know, many human diseases are caused by abnormal phosphorylation, and some phosphorylations are the products of some kinds of diseases. In view of the significance of phosphorylation in life activities, exploring the mechanism and function of phosphorylation has become a matter of great concern. The traditional method focuses on the research of individual proteins, while proteomics focuses on all types of proteins involved in specific physiological or pathological states and their relationship to the surrounding environment (molecules). The state and change of phosphorylation in cell or tissue can be observed from the overall study by using the philosophy and analysis method of proteomics. Thus, the new concept of "phosphoproteomics" has been derived.

(2) Reversible phosphorylation of protein

Reversible phosphorylation of protein is one of the most common PTMs and one of the most important bio-chemical processes in organisms. Genome prediction suggests that 2–3% of proteins encoded by genes in eukaryote are protein kinases involved in phosphorylation. The human genome sequence has predicted the existence of at least 100 phosphatases/phosphoesterases. It is estimated that about 50% of proteins have undergone phosphorylation during their life cycle, and there are at least 100,000 phosphorylation sites in human proteins. The importance of reversible phosphorylation of proteins can be seen directly.

The reversible process of phosphorylation and dephosphorylation of protein is controlled by the synergistic action between protein kinase and phosphatase. As seen in Figure 1.2, protein phosphorylation is catalyzed by protein kinase, for which the phosphate groups and energy are provided by adenosine triphosphate (ATP). While dephosphorylation is the hydrolysis reaction catalyzed by the phosphatase/phosphoesterase. Protein phosphorylation and dephosphorylation are vividly described as molecular switches in cell physio-

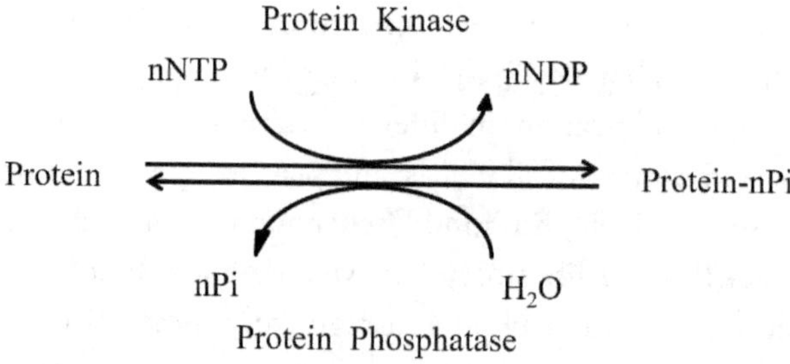

Figure 1.2: Phosphorylation and dephosphorylation of protein.

logical activities, since they can change the protein structure, activity, and the interaction for regulating almost all life activities, such as cell signal transduction, cell differentiation, growth, and apoptosis. The diversity of protein phosphorylation determines that they can regulate so many cellular functions. Protein phosphorylation is mainly divided into four types: (i) *O*-phosphate is the most common phosphorylation type, which mainly occurs on serine (Ser), threonine (Thr), and tyrosine residues; (ii) *N*-phosphate, which mainly occurs on arginine, lysine, and histidine residues; (iii) Acyl phosphate, which mainly occurs on aspartic acid and glutamic acid residues; (iv) S-phosphate, which mainly occurs on cysteine residues. Compared with the latter three, the form of the former is relatively stable, and is the most studied. Phosphorylation of Ser and Thr is unstable under alkaline condition, whose phosphate group will be decayed by β-elimination. For tyrosine phosphorylation, the phosphate group is connected to the benzene ring on tyrosine, therefore, it is relatively stable under alkaline condition and no β-elimination will occur. In eukaryote, Ser phosphorylation accounts for the highest proportion, and Thr phosphorylation accounts for the second, tyrosine phosphorylation accounts for the lowest. Their ratio is about 1800:200:1.

(3) Identification of phosphopeptide and phosphorylation sites

The flourishing of proteomics research mainly should be attributed to the rapid development of biological MS and the breakthrough of high-throughput separation technology. Especially biological MS has been an important means of detecting trace protein with the development of "soft ionization source" including ESI and MALDI. And is the key analytical technique to achieve high-throughput, high-efficiency, and high-accuracy proteomics research.

Most often, there are multiple phosphorylation sites on a phosphoprotein, but only one or several of them are phosphorylated. Therefore, the main problem in phosphoprotein research is how to accurately identify phospho-peptide sequence and determine the phosphorylation site at the same time. Nowadays, the determination of phosphopeptide sequence and phosphorylation site mainly depends on tandem MS. There have been many domestic and overseas reports about various tandem MS for phosphoprotein research. The following is a brief introduction according to the different methods of ion fragmentation.

(a) *Collision-induced dissociation*

The collision-induced dissociation (CID) technology brings in inert gas to break up the peptide ions, which means energy is introduced, resulting in the increase of energy level to force the peptide bond to be broken up. When using CID as fragmentation mode, a series of fragment ions are produced due to the break of peptide bonds after adsorbing collision energy by peptide ions. As seen in Figure 1.3, the N-terminal ion is b ion, and the C-terminal ion is y ion. Due to the stability of phosphorylation on tyrosine residue and the fragmentation mode of CID, tyrosine phosphopeptide can produce characteristic ion with m/z 216.043, which is the immonium ion of phosphotyrosine. Steen *et al.*[18] have applied the characteristic ion to identify tyrosine phosphopeptides. But for the phosphorylation on Ser or Thr residue, the phosphoester bond will be broken in the process of CID fragmentation, resulting in the neutral loss of phosphate group to produce dehydroalanine or dehydroaminobutyric acid. Therefore, neutral phosphate molecule as the characteristic ion has been applied to traditional triple quadrupole MS for the

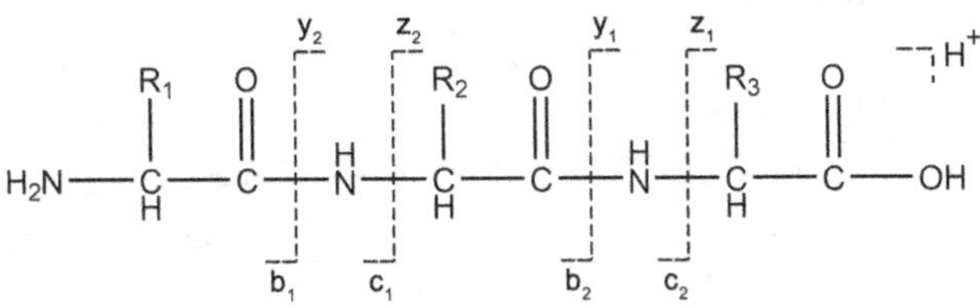

Figure 1.3: Peptide fragment ions from different fragmentation modes.

identification of Ser or Thr phosphorylation. The following are the two main scanning modes in MS identification.

Neutral loss scanning: In triple quadrupole MS, the first mass analyzer (Q1) and third (Q3) mass analyzer scan synchronously. But the scan range maintains a certain voltage difference, the m/z represented by the difference value is the mass of neutral phosphate molecule (m/z = 97.9 or m/z = 49). The ions scanned by Q1 enter into Q2 for splitting decomposition and then enter into Q3, and only those that have neutral loss in Q2 will be transmitted to the detector by Q3. The advantage of this method is that its scanning and analysis are conducted via positive ion mode, namely, the phosphorylation site and sequence can be analyzed directly once the phosphopeptide is discovered. However, the low specificity of this method generally leads to a false-positive result, which requires to be confirmed by tandem spectrogram.

Precursor ion scanning: In triple quadrupole MS, Q1 is scanned in succession. The ions scanned by Q1 enter into Q2 for splitting decomposition and then enter into Q3, which is set up to only transmit the daughter ion with a particular m/z such as m/z 79 (PO_3^-). So the obtained mass spectrum only shows the peak where neutral loss occurs. Since there is hardly any ion fragment of around 79 Da produced by the peptide, the specificity is very high by using the m/z to scan phosphopeptide precursor ion.

In the process of CID fragmentation, most of the collision energy absorbed by phosphopeptide ion is used for the fracture of its phosphoester bond, which reduces the fracture of its peptide bond. Therefore, the obtained number of b and y fragment ions is limited, which affects the accurate identification of the phosphopeptide sequence. The advance in biological MS, especially the emergence of liner ion trap, has provided a new way, namely three-stage tandem MS (MS/MS/MS, MS^3), for the identification of phosphopeptide and phosphorylation site.

As seen in Figure 1.4, in three-stage tandem MS, the first MS is used for full scanning within a certain mass range. Then the ions with strongest intensity are broken up by CID, from which the produced fragment ions are scanned by the secondary MS. If the shown m/z discrepancy of the fragment ion with neutral loss and parent ion is the mass of the neutral phosphate molecule (m/z = 97.9 or m/z = 49) in the secondary MS, the fragment ion will

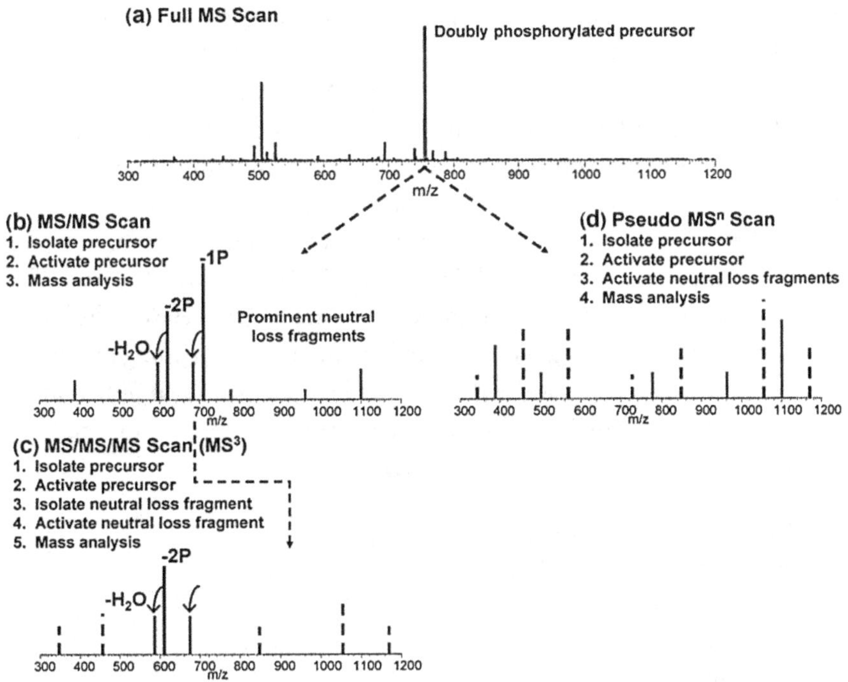

Figure 1.4: Process diagram of three-stage tandem MS and multi-stage activation MS (*Anal. Chem.* **2004**, 76, 3590–3598).

be broken up again by CID, then the new produced fragment ions will be scanned by the tertiary MS. Three-stage tandem MS can provide more information on phosphopeptide sequence and more accurate determination for phosphorylation site. Recently, multi-stage activation (MSA) derived from three-stage tandem MS has also been used to identify phosphopeptide. In multi-stage fragmentation technology, the collision fragmentation of parent ion and the ion with neutral loss occurs one after the other, the final tandem spectrogram is acquired by the superposition of the spectrogram from the secondary and the tertiary MS (Figure 1.4).

In 2005, Thermo introduced high-resolution MS, which is the tandem combination of linear ion trap and electrostatic field orbit trap (LTQ-Orbitrap), it can identify the composition in a complex mixture faster, more sensitively, and more reliably. Compared to other MS, LTQ-Orbitrap has the following significant advantages: (i) high precision; (ii) high resolution; (iii) wide dynamic range, and (iv) strong detection capability. The emergence of LTQ-Orbitrap makes high-resolution MS more widely applied and also provides new technology for phosphoprotein research. Villén *et al.*[19] reported that high-resolution MS can provide accurate MS information of parent ion, thus the secondary tandem MS can accurately identify phosphopeptide and determine phosphorylation site at the same time, while the three-stage tandem MS triggered by neutral loss can be omitted. In large-scale proteomics research, data acquisition rate of MS determines the identification degree of complex samples. Although MS/MS/MS and MSA can provide more effective information on fragment ions for identifying phosphopeptide, the longer time for the cycle-scanning means the reduction of cycle-scanning number in the same time, resulting in the reduction of identified peptide number on the whole. Therefore, MS/MS/MS and MSA are more suitable for studying phosphoprotein without high-resolution MS.

(b) *Electron capture dissociation*

In 1998, Zubarev *et al.*[20] developed a new fragmentation method, namely, electron capture dissociation (ECD), for multiple charged protein/peptide ions. Briefly, ECD produces ion fragments by the combination of low-energy electron and proteins/peptides with multiple positive charges, which is a non-ergodic process. The occurrence of cleavage reaction is based on the redistribution of vibrational energy within the molecule. Compared to other fragmentation modes, the fragmentation ion produced by ECD is quite different for the same parent ion. As seen in Figure 1.3, the cleavage of bond occurs between amide and α-carbon atom on peptide skeleton by using ECD fragmentation mode, and the produced fragment ion N-terminal is c ion and the C-terminal is z ion. The fragment ion produced by ECD is not affected by amino acid sequence of peptide. But ECD cannot produce fragment ion whose N-terminal is proline residue, due to the presence of the ring structure of the amino and α-carbon atom on proline. The advantage of ECD is that the rupture only occurs on skeleton structure of peptide, while the unstable phosphorylated group retains on the amino acid residue of c or z ion, which is beneficial for the accurate identification of phosphorylation site. To date, since the acquirement of low-energy electrons depends on the presence of a static magnetic field, the ECD is just implemented in the Fourier transform-ion cyclotron resonance mass spectrometry (FT-ICR MS). Furthermore, high price and maintenance cost of FT-ICR MS limit the wide application of ECD.

(c) *Electron transfer dissociation*

Electron transfer dissociation (ETD) is developed as an alternative technology for ECD, which can provide the same fragmentation mode as ECD. Syka *et al.* found that negative ion with low enough electron affinity (such as anthracene or azobenzene) is suitable as an electron donor which can be used on a large scale. This kind of electron donor can bind protein/peptide with multiple positive charges, then the peptide is broken up by the same non-ergodic process as ECD. Because there is no need for low-energy electrons, ETD technology has wider application range and has been applied to the research of phosphoproteins successfully. Molina *et al.*[21] reported that the number of phosphopeptide identified by ETD is 60% higher than that by CID, because the fragment ion information provided by ETD is 40% more than that by CID. At the same time, Molina *et al.* pointed out that the effective combination of CID and ETD can maximize the possibility of realizing whole phosphoproteomics analysis. Most peptides obtained by trypsin digestion have two positive charges after ionization, which is more suitable for CID fragmentation mode. While ECD and ETD fragmentation mode are suitable for peptide ions with more than three positive charges. Therefore, appropriate protease is the key to make the fragmentation more efficient. Intracellular protease is an appropriate choice, because it can only function at C-terminal lysine, thereby obtaining longer peptide which has more charge. According to the comparison result of trypsin and intracellular protease, Molina *et al.* found that the use of intracellular protease cannot improve the cleavage efficiency of ETD because the appearance of phosphate group affects the digestion efficiency of trypsin. Kjeldsen *et al.*[22] proposed a new method to increase charge number of peptide. Namely, adding 0.1% m-nitrobenzyl alcohol into mobile phase to increase the charge number of peptide from two to three or more, which can improve the fragmentation result of peptide in ETD.

(4) Separation and enrichment of phosphoprotein/peptide

Although the number of phosphoproteins is huge, only a small part of them can be phosphorylated in a certain time period. And phosphopeptide is low-abundance in the protein digestion product, resulting in its MS signal being inhibited by the high-abundance non-phosphopeptide or even cannot be detected. Therefore, there is only one way to make MS analysis feasible, that is to apply highly efficient separation and enrichment strategy to improve the relative content of phosphopeptides. This, how to easily and efficiently realize the enrichment of phosphoprotein/peptide, is a very important topic in phosphoproteomics research.

(a) *Immunoprecipitation*

Immunoprecipitation refers to utilizing antibodies to recognize and bind phosphoproteins. This is a common method to detect phosphoprotein by the immunoprecipitation reaction between anti-phosphoamino acid antibody and phosphoprotein. Commercial anti-tyrosine phosphate antibody has been developed for phosphoprotein research. Rush *et al.*[23] utilized tyrosine phosphorylated antibody to identify a total of more than 600 tyrosine phosphopeptides in four types of cells, of which include at least 559 phosphorylation sites. The antigenic determinants of phosphorylated Ser and Thr are relatively small, resulting in the steric hindrance of the binding site of antigen–antibody, therefore, the binding force and specificity of this antibody are not good enough for enriching Ser and Thr phosphoprotein effectively.

(b) *Chemical modification*

Chemical modification is also a new method for the separation and enrichment of phosphopeptide. The main idea is to use different chemical methods to treat phosphopeptide and mark specific group on the phosphorylation site, so as to realize the selective enrichment of the phosphopeptide. Chemical modification method can be mainly divided into two categories: (i) based on β-elimination; (ii) based on 1-(3-dimethylaminopropyl)-3-ethylcarbodiimide hydrochloride (EDC) catalysis.

(i) **Based on β-elimination:** Phosphorylation on Ser and Thr residue is not stable and easily forms dehydroalanine or dehydroaminobutyric acid due to the occurrence of β-elimination under strong alkaline condition. As seen in Figure 1.5, the produced double bond after elimination of phosphate group becomes the receptor of Michael addition reaction. Several methods have been reported to chemically label phosphopeptides by means of β-elimination for enriching them selectively.

Oda *et al.*[24] pointed out that the β-elimination of phosphopeptide occured firstly under strong alkaline condition, dithioglycol or dimercaptopropane as nucleophile would then be added on Ser or Thr residue by Michael addition reaction, and the obtained mercapto phosphopeptide was finally

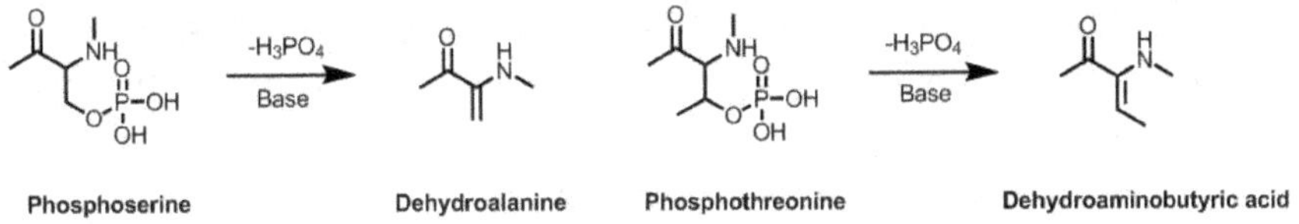

Figure 1.5: β-elimination on phosphoserine/threonine.

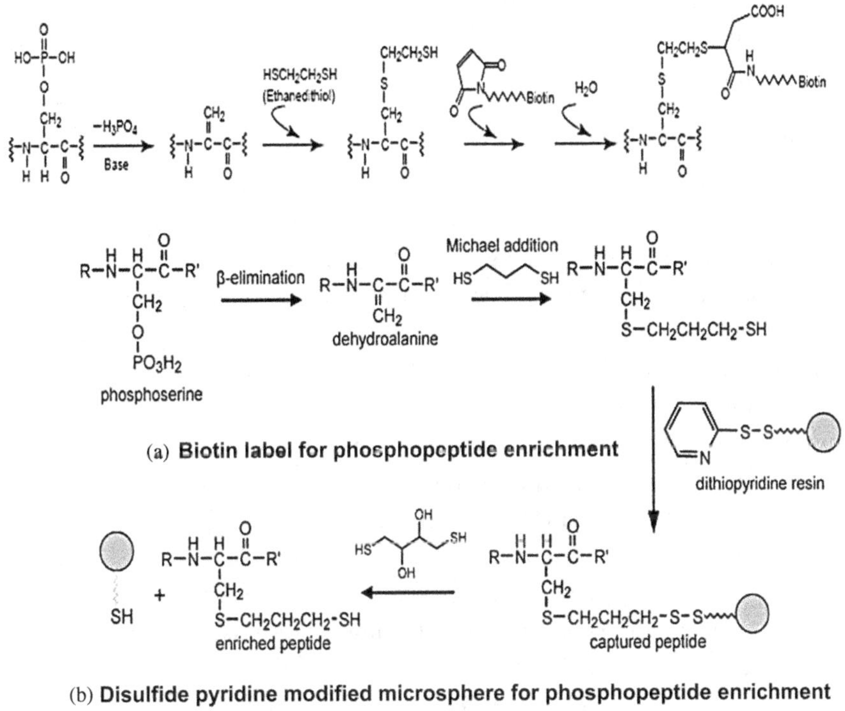

Figure 1.6: Phosphopeptide enrichment based on β-elimination.

labeled with a biotin tag. In this way, phosphopeptide can be enriched selectively by the specificity of biotin. The detailed process is displayed in Figure 1.6(a). However, this method has some shortcomings such as low sensitivity and easy loss of low-abundance protein, etc.

Thaler *et al.*[25] provided a more simple method. They modified the sulfydryl on phosphopeptide by the same method as Oda's, then the obtained mercapto phosphopeptide was immobilized on the surface of resin microsphere with disulfide pyridine modification, and thereby realizing selective separation of phosphopeptides. The detailed process is displayed in Figure 1.6(b). Tseng *et al.*[26] depended on the above principle to make sulfydryl covalently connect with iodoacetamide that was also connected to the resin by the other end, in this way, phosphopeptide can be captured on the resin by solid-phase Michael addition. Chemical modification based on β-elimination is simple and feasible, but the presence of cysteine and methionine in protein would result in side reactions. Therefore, cysteine and methionine residues need to be oxidized by using performic acid before the modification of sulfhydryl. In addition, this method is not suitable for phosphorylation on tyrosine residue.

(ii) **Based on EDC catalysis:** Zhou *et al.*[27] developed a multistep solid-phase enrichment technique for the separation of phosphopeptide. Firstly, tert-butyl dicarbonic acid (tBoc) is used to protect the amino group of peptides. Then carboxyl and phosphate groups of peptides react with ethanolamine under the catalysis of EDC to transform to amide and phosphamide groups, and the phosphamide group can be hydrolyzed to phosphate group in 10% trifluoroacetic acid (TFA) solution, which can protect the carboxyl group. Finally, under the catalysis of EDC, regained phosphate group is labeled with cystine on which sulfydryl reacts with iodine acetyl-modified glass bead by dithiothreitol catalysis. So the separation of phosphopeptide can be realized. The detailed process is displayed in Figure 1.7(a). Bodenmiller *et al.*[28] adopted the similar method to immobilize the phosphopeptide labeled with cystine on maleimide-modified porous glass, and realized the selective enrichment of phosphopeptide. Tao *et al.*[29] developed a more convenient method based on EDC catalysis for the enrichment of phosphopeptide. Firstly, carboxyl group of phosphopeptide is protected through the methyl esterification, then phosphonate group directly reacts with amino-terminated dendrimer under the catalysis of EDC, thus realizing the separation of phosphopeptide. Since amino-terminated dendrimer has a large number of amino groups, its emergence prevents phosphate group of peptide to react with the amino group on itself or other peptides. Accordingly, the protection step for amino groups can be omitted, which simplifies reaction steps and improves reaction efficiency. The detailed process is displayed in Figure 1.7(b). This method is suitable for various phosphorylations, but it involves too much chemical reaction, leading to sample loss.

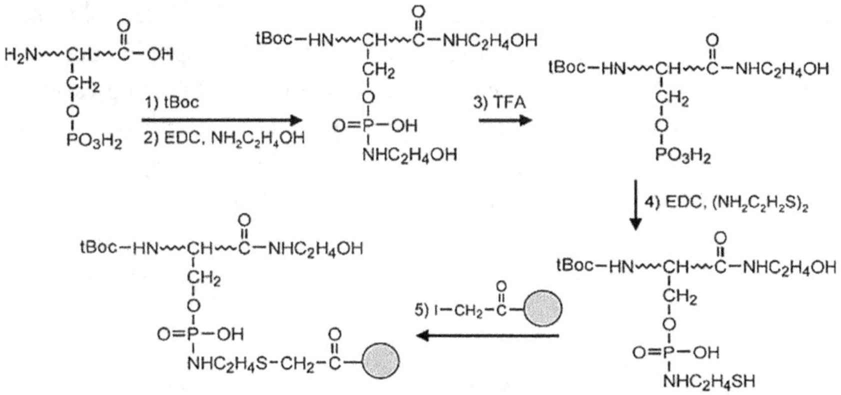

(a) **Iodine acetyl-modified glass bead for phosphopeptide enrichment**

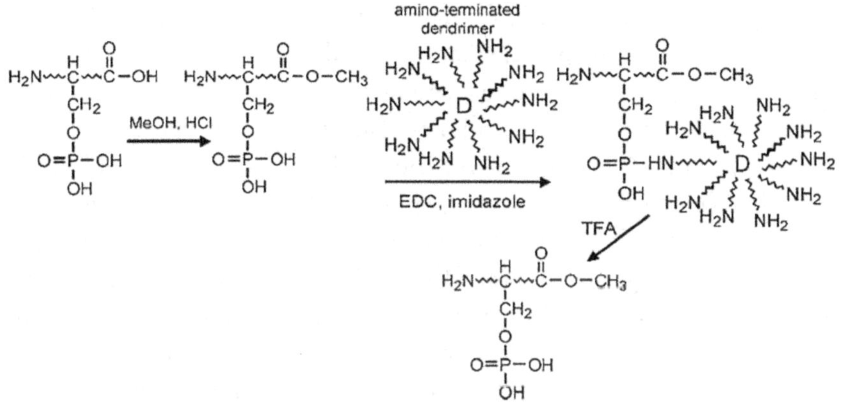

(b) **amino-terminated dendrimer for phosphopeptide enrichment**

Figure 1.7: Phosphopeptide enrichment based on EDC catalysis.

There are some new methods derived from these two methods. For instance, Lansdell *et al.*[30] designed and synthesized diazo-modified solid-phase carrier for directly reacting with phosphate group, thus realizing the separation of phosphopeptide and non-phosphopeptide. Knight *et al.*[31] reported a specific enzyme-cutting technique in allusion to the phosphorylation site. But this technique is still at the stage of method optimization.

(c) *Liquid chromatography*

(i) **Reversed-phase liquid chromatography (RPLC):** RPLC is a common and important method for the separation of peptide mixture, which has good reproducibility and does not require special equipment. For RPLC, phospho-peptides can be separated according to their different hydrophobicities. Very few phosphopeptides can be separated at low flow rate by using capillary column. Multi-phosphopeptide with relatively high hydrophilicity may not be retained in RPLC but directly go through the chromatographic column. Only at the highest gradient can the peptide with high hydrophobicity be eluted. Thus, some multi-phosphopeptides cannot be detected. Nevertheless, RPLC has still been widely used in the analysis of phosphopeptide because it is easy to connect with MS.

(ii) **Strong cation-exchange chromatography (SCX):** Ion exchange chromatography achieves the goal of separation and purification by taking advantage of the interaction difference between the charged part of analyte and ion exchanger with opposite charge. According to the report of Beausoleil et al.,[32] at a pH of 2.7, N-terminal amino of peptide by trypsin digestion as well as the amino on lysine and arginine residue are protonated, which equip the peptide with +2 of static charge. But under the above same pH condition, phosphate group of phosphopeptide is charged negatively with one because its phosphate group does not undergo protonation, resulting in phosphopeptide equipment with +1 of static charge. The detailed information is displayed in Figure 1.8. SCX can be used to separate phosphopeptide from non-phosphopeptide depending on their charge difference.

Beausoleil et al. has carried out analysis for phosphoproteome in HeLa cell. Briefly, nuclear protein is pre-separated by sodium dodecyl sulfate polyacrylamide gel electrophoresis (SDS-PAGE), then treated by trypsin and separated again by SCX at a pH of 2.7, a total of 967 phosphoproteins and 2002 phosphorylation sites are finally identified by LC-MS. Ballif et al.[33] identified more than 500 phosphopeptides in the research of phosphoproteomics of fetal rat brain by using the same technique route. Sui et al.[34] used SCX to study phosphoproteomics of yeast bacteria. Their results suggest that the optimized SCX separation system can be used as a convenient enrichment method for the separation of phosphopeptides. Especially for complex practical samples, SCX can improve the relative abundance of phosphopeptides in a very short time, so as to facilitate the detection of phosphopeptides. The shortcoming of SCX is that it is only applicable to peptides with trypsin digestion. Moreover, if there is histidine in the phosphopeptide sequence, the phosphopeptide will

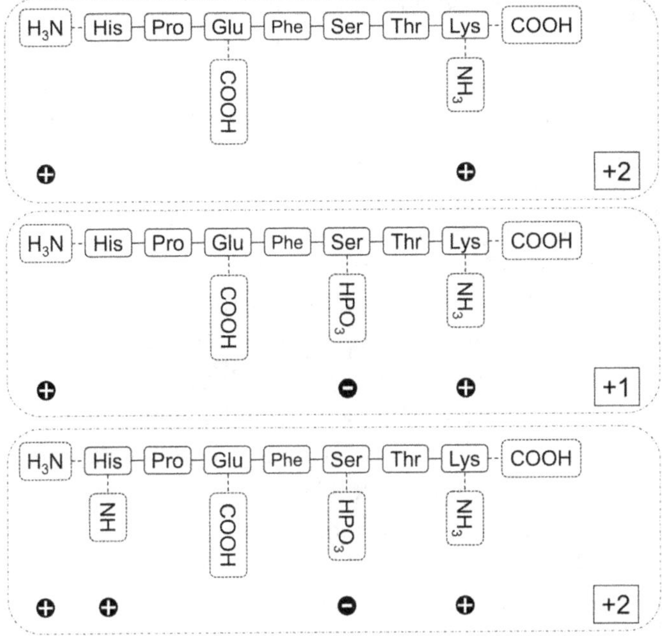

Figure 1.8: The static charge of trypsin digestion peptide at pH of 2.7.

be equipped with +2 of static charge, resulting in its separation difficulty from non-phosphopeptide. The combination of SCX and other enrichment methods will increase the number of identified phosphopeptides. Trinidad *et al.*[35] made a comparison about the results from combined usage of SCX and immobilized metal ion affinity chromatography (IMAC) and respective usage of SCX and IMAC. The number of identified phosphopeptides by combination usage increases three times compared with that by respective usage. Besides, their results show that only 35% of peptides with trypsin digestion are equipped with +2 of static charge, most peptides have higher static charge, which lengthens the elution time of phosphopeptide. In other words, the elution of phosphopeptide runs through the whole SCX fraction rather than only in the earlier fraction. Olsen *et al.*[36] identified 6,600 of phosphorylation sites on 2,400 of phosphoproteins by using SCX and TiO_2, which not only greatly improves the identification level of phosphorylation site but also provides a safeguard for large-scale phosphoproteome research.

(iii) **Strong anion-exchange chromatography (SAX):** The principle of SAX is same as that of SCX for phosphopeptide separation. Another drawback of SCX is that it cannot realize the identification of multi-phosphopeptides well. The multiple negative charges of multi-phosphopeptide will lead it to leave the chromatographic column without reservation in SCX separation. While SAX can avoid such a problem. Han *et al.*[37] identified 274 phosphorylation sites on 168 phosphoproteins in phosphoproteome of human liver by using SAX. Liver proteins/peptides were directly analyzed by MS after SAX separation. The combination of SAX and IMAC can not only achieve the goal of enrichment, but also realize grading separation of phosphopeptide by a certain gradient elution.

The identification of phosphoprotein by single SAX or SCX is limited. The combination of SAX and SCX can increase the coverage of phosphopeptide significantly. Dai *et al.*[38] reported an anion–cation multi-dimensional LC-MS system for separation of peptide mixture, which is the combination of SCX, SAX, and RPLC. The method has high-resolution and easy to operate and needs low-dosage samples. 14,105 different peptides including 13,256 non-phosphopeptides and 849 phosphopeptides containing 809 phosphorylation sites have been identified from 1 mg rat liver sample by using this method.

(iv) **Hydrophilic interaction chromatography (HILIC):** The stationary phase in HILIC is a kind of material with strong polarity, and the mobile phase is 10–40% water. The separation is based on the interaction between the hydrophilic and charged group of compounds and the hydrogen bond and ionic bond of stationary phase. The stronger the polarity of the compound, the stronger the function of adsorption. HILIC is mainly used for the separation of small molecular compounds, such as carbohydrate, saponin, sugar, and pharmaceutical metabolite, as well as relatively less used for the separation of peptide/protein. The latest application of HILIC is for the separation and enrichment of phosphopeptide. The phosphonate group of phosphopeptide endows it with hydrophilicity and charge, which gives it a stronger binding capacity than non-phosphopeptide in HILIC, so it can be separated from the non-phosphopeptide.

McNulty *et al.*[39] used the combination of HILIC, RPLC, and MS to treat enzymolysis products of Hela cell line. 1,000 non-redundant phosphopeptides, of which more than 700 were identified for the first time, amplify the phosphoprotein database in Hela further. At the same time, their results show that HILIC and RPLC have good orthogonality, and the repetition probability of phosphopeptide in the two non-adjacent components of HILIC is very small. Compared with SCX method, separation and enrichment method based on HILIC has higher sensitivity and can identify more phosphopeptides. Of more importance is that the method without desalination is simple and reduces the sample loss. Albuquerque *et al.*[40] also used HILIC to construct a new multi-dimensional chromatography enrichment strategy. Namely, they used the combination of IMAC, HILIC, and RPLC to study DNA damage of saccharomyces cerevisiae phosphoproteomics. 8,764 non-redundant phospho-peptides belonging to 2,278 phosphoproteins were identified. This analysis strategy shows unique advantage in large-scale and in-depth analysis of phosphoproteomics.

(v) **Electrostatic repulsion hydrophilic interaction chromatography (ERLIC):** In ion-exchange chromatography (IEC), when the mobile phase is organic solvent principally, even if the analyte has the same charge as the stationary phase, they can still be combined by hydrophilic interaction, which is the basic principle of ERLIC. ERLIC is the latest method for phosphopeptide enrichment. In ERLIC, glutamic acid, aspartic acid, and C-terminal carboxyl of peptide are in the protonation state when pH $\leq$ 2. The positive charges of N-terminal, lysine, and arginine of peptides make the electrostatic repulsion existing between peptide and packing of chromatographic column when the peptide is going through a weak anion exchange column. While the negative charge of the phosphate group produces electrostatic attraction between phosphopeptide and packing of chromatographic column, resulting in the separation of phosphopeptide and non-phosphopeptide. The hydrophilic interaction of phosphate group can be improved by adding an appropriate amount of organic solvents into the mobile phase, and thereby enhancing the retention ability of chromatographic column for phosphopeptides. Different from other enrichment methods, this method has realized enrichment and grading separation of phosphopeptide in a one-step experiment.

Gan *et al.*[41] evaluated the effectiveness of ERLIC in the separation and enrichment of phosphopeptides. Briefly, ERLIC and SCX-IMAC are used to study phosphoproteomics in A431 cells, respectively. The result shows that 17,311 and 4,850 phosphopeptides are identified by ERLIC and SCX-IMAC, respectively, moreover, 926 and 1,315 non-redundant phosphopep-tides are also identified by ERLIC and SCX-IMAC, respectively. But the identification overlap of these two methods is only 12%, which proves the high complementarity of these two methods. Zhang *et al.*[42] used ERLIC to enrich glycopeptide and phosphopeptide in membrane protein of rat brain simultaneously. A total of 942 glycosylation sites on 519 non-redundant glycoproteins are identified, and 823 phosphorylation sites on 337 non-redundant phosphoproteins are identified. With the further optimization of ERLIC method, it will play a more important role in the study of large-scale phosphoproteome.

(d) *Immobilized metal ion affinity chromatography*

In the 1970s, Porath *et al.*[43] reported a new type of chromatography, which was originally named metal chelate chromatography, later renamed IMAC, which has been one of the common methods for phosphopeptide enrichment so far. In 1986, Andersson *et al.*[44] fabricated Fe^{3+}-immobilized IMAC for selective enrichment of phosphoprotein/peptide. The enrichment principle of IMAC is that phosphate group and metal ion can combine by the electrostatic effect, which is affected by pH and ionic strength and the organic phase in the solution. Therefore, phosphopeptide can be eluted via destruction of the combination using alkaline conditions or phosphate.

The stationary phase in IMAC consists of carrier, chelator, and metal ion. The carrier is solid for carrying chelator, which immobilizes metal ions by forming coordination compounds. Metal ions are used for selective enrichment of phosphopeptides. The crosslinking ability of the chelator to different metal ions is varied due to the difference in the number of polar atoms in the chelator. According to the theory of coordination compound, the chelator and metal ion tend to form a five or six-member ring, which is more stable than the simple coordination compound; the more the number of coordination atoms of the chelator, the more stable the formed coordination compound, of which the more the number of rings, the more stable the structure. Nitrile triacetic acid (NTA) and imino diacetic acid (IDA) are the most commonly used chelators in IMAC. As seen in Figure 1.9, NTA has one nitrogen and three carboxyl oxygen atoms, a metal ion, and three carboxyl oxygen atoms to form a four-dentate structure through nitrogen. For IDA, metal ion chelates two carboxyl oxygen atoms through nitrogen to form tri-dentate structure. Ficarro *et al.*[45] made a comparison about the influence of NTA and IDA on selectivity of phosphopeptide enrichment. The result shows

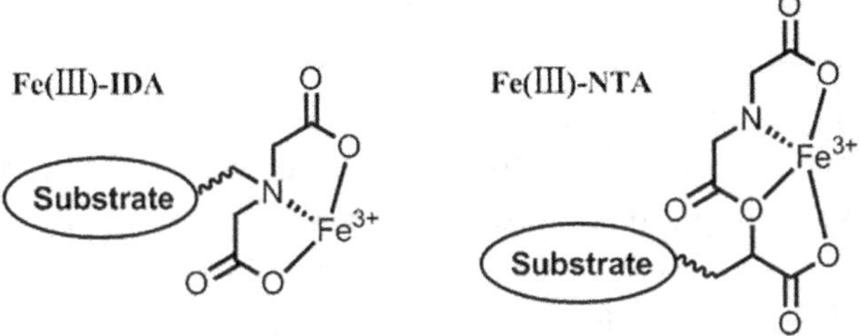

Figure 1.9: The coordination compounds formed by Fe^{3+}, IDA, and NTA.

that NTA as chelator can provide higher selectivity than IDA. Since the property of different metal ions is different, many other metal ions such as Zr (IV) and Al^{3+} are also applied to the enrichment of phosphopeptide. According to the report of Ficarro *et al.*, the enrichment performance of Fe^{3+} and Ga^{3+} is better. At present, IMAC with NTA as chelator is most widely used in enrichment of phosphopeptides. Recently, a new IMAC method with phosphate group as chelator for Zr (IV) immobilization has also been used for the enrichment of phosphopeptides.[46] The IMAC method has the characteristics of being fast and direct. The samples enriched by IMAC can be directly used for MS analysis after desalination treatment. However, it also has many limitations.

First, loss of metal ion may occur during the process of enrichment and elution because the metal ion combines with the carrier by non-covalent bond, leading to the loss of some phosphopeptides or unnecessary pollution. These problems can be avoided by thoroughly cleaning the IMAC column before use and choosing the appropriate chelator. For example, IDA is a tridentate chelator, while NTA is a four-dentate chelator, so the stability of NTA for immobilizing metal ion is better. Second, peptide containing aspartic acid and glutamic acid residue can bind to metal ion non-specifically, thus reducing the selectivity of IMAC enrichment method. Moreover, the non-specific binding will seriously affect the subsequent MS detection since the ionization efficiency of the non-phosphopeptide is higher than that of the phosphopeptide. According to a report,[47] electrostatic interaction between acid residue and metal ion can be decreased by improving ionic strength in the sample solution and the washing buffer. While another report shows that the selectivity of IMAC method cannot be improved by adding salt in the process of enrichment.[48] Ficarro *et al.*[47] converted the carboxyl group on aspartic acid and glutamic acid residue into carboxymethyl ester using acetyl chloride before the enrichment of the phosphopeptide, thus inhibiting the non-specific binding of the acid peptide. A total of 216 phosphopeptides in 171 phosphoproteins as well as 383 phosphorylation sites are identified from yeast by using this method. In order to improve the efficiency of methyl esterification, Moser *et al.*[49] used sulfur dichloride instead of acetyl chloride to esterify carboxyl group. The drawback of this method is that incomplete methyl esterification will increase the complexity of the sample, and the lyophilization step during the reaction will result in serious sample loss.

Bodenmiller *et al.*[50] studied the effect of methyl esterification on phosphopeptide enrichment by IMAC. The result shows that a total of 199 phosphopeptides are identified if methyl esterification occurs before IMAC enrichment, and 193 phosphopeptides are identified if methyl esterification occurs after IMAC enrichment. The enrichment efficiency of the two methods is not significantly different, but it is worth noting that only 30% of phosphopeptides are identified by both of them. To reduce the non-specific binding of aspartic acid and glutamic acid residue to metal ion, Seeley *et al.*[51] used Glu-C instead of trypsin for enzymolysis. Since Glu-C cuts peptide at the C-terminal of aspartic acid and glutamic acid, most peptides contain only one acid residue, which can reduce the non-specific binding of acid peptides. The enrichment selectivity of IMAC to phosphopeptide increases from 40% to 70% after using Glu-C. In addition, acidification of sample before IMAC enrichment can protonate the carboxyl group on acidic residue, thus reducing non-specific binding. The third limitation of IMAC is that its enrichment efficiency for multi-phosphopeptide is higher than that of mono-phosphopeptide, because the binding force between multi-phosphopeptide and metal ions is stronger.

(e) *Metal oxide affinity chromatography*

In recent years, metal oxide affinity chromatography (MOAC) technology develops rapidly in the application of phosphopeptide enrichment. Metal oxides can be shown as Lewis acid or Lewis base at different pH conditions. Under acidic conditions, metal atom with positive charge is shown as Lewis acid, which can combine with anion. Under alkaline conditions, it is shown as Lewis base, which can combine with cation. Therefore, the phosphonate group of the phosphopeptide can bind to metal atom under acidic condition, and the binding can be broken under alkaline condition. Many different metal oxides, such as TiO_2, ZrO_2, Al_2O_3, and Nb_2O_5, have been used in the enrichment of phosphopeptides.

Nowadays, TiO_2 is the most commonly used metal oxide in the enrichment of phosphopeptide. In 2004, Pinkse et al.[52] provided an online 2D-LC method for phosphopeptide research. The first dimension is MOAC filled with spherical TiO_2, the second dimension is RPLC. Firstly, phosphopeptides in peptide mixture are enriched by TiO_2 under the condition of 0.1 mol·L^{-1} of acetic acid (pH = 2.7), the unenriched non-phosphopeptides are collected by RPLC and identified by MS. Then, the phosphopeptides are eluted from TiO_2 under alkaline condition (pH = 9.0), and subsequently concentrated by RPLC for MS detection. Pinkse et al. evaluate the enrichment efficiency of this method by using guanosine cyclophosphate-dependent protein kinase. A total of eight phosphopeptides are identified, two of which were discovered earlier. However, the results also show that there is non-selective binding between the acidic non-phosphopeptide and TiO_2, which has a great influence on the enrichment selectivity. They point out that methyl esterification of acid peptide before enrichment can help improve enrichment selectivity. Another way to improve the selectivity of TiO_2 is to choose the appropriate sample solution, washing buffer, and eluent. According to the report of Larsen et al.,[53] when the sample solution is changed from 0.1 mol·L^{-1} acetic acid to 0.1% TFA, at the same time, an appropriate amount of 2,5-dihydroxybenzoic acid (DHB) is added, the enrichment selectivity can be greatly improved for TiO_2. The pH values of 0.1 mol·L^{-1} acetic acid and 0.1% TFA are 2.7 and 1.9, respectively, so 0.1% TFA can protonate carboxyl group on acid residue more effectively, thus improving the selectivity of TiO_2 enrichment. It is worth noting that when DHB is added to the sample solution and washing buffer, a total of 20 phosphopeptides can be identified from 500 fmol α-casein digestion, and there is no non-phosphopeptide. Larsen et al. also test the effect of other substituted aromatic carboxylic acids (such as DHB) on enrichment selectivity. They think due to the different spatial structure, substituted aromatic carboxylic acid and titanium atom are bidentate chelating, and phosphate group and titanium atom are bridged bidentate chelating, therefore, the binding force of DHB and titanium atom is stronger than aliphatic carboxylic acid containing only one carboxyl group, but weaker than phosphate group. The existence of such competitive combinations is exactly what enhances the selectivity of enrichment. Moreover, the authors change the pH values of eluent from the previous 9.0 to 10.5, another four phosphopeptides are identified from α-casein digestion. This result suggests that increase of pH values of eluent can improve recovery rate of phosphopeptide more effectively, thus improving sensitivity of the method.

There are also other metal oxides that have been used for the enrichment of phosphopeptides. Kweon *et al.*[54] indicated that both pH value as the main factor and ionic strength affect the binding ability of ZrO_2 to phosphopeptide. They test the performance of TiO_2 and ZrO_2 by using α-casein and β-casein as standard samples. The result shows that TiO_2 tends to enrich multi-phosphopeptide, while ZrO_2 tends to enrich mono-phosphopeptide. In order to further improve the detection level of phosphorylation, the combination of TiO_2 and ZrO_2 can significantly increase the coverage of identified phosphopeptide and reduce sample loss. In order to reduce non-specific enrichment of acidic amino acid, the authors investigate different enrichment effects by using Glu-C and trypsin as enzyme-cutting agentia. They find that the use of Glu-C cannot obviously decrease non-specific enrichment of acidic amino acid either by ZrO_2 or by TiO_2. Li *et al.*[55] designed Al_2O_3-modified Fe_2O_3 for phosphopeptide enrichment, which has high capacity of enrichment and easily separates phosphopeptides from non-phosphopeptides. Ficarro *et al.*[56] used Nb_2O_5 to enrich phosphopeptides. The result shows that DHB can also improve the enrichment selectivity of Nb_2O_5, and there is 30% enrichment difference between Nb_2O_5 and TiO_2.

Compared with IMAC, MOAC possesses higher selectivity and is more resistant to low pH, detergent, salt, and other low molecular pollutants. The report of Larsen's group[53] indicates that the enrichment efficiency of metal oxide for mono-phosphopeptide is higher than that for multi-phosphopeptide, which has been demonstrated by Ficarro's group.[56] But in fact, Larsen *et al.* thought that the enrichment efficiency of metal oxide for multi-phosphopeptide and mono-phosphopeptide is the same, the enrichment effect for multi-phosphopeptide is not obvious because it is more difficult to be eluted from metal oxide. As mentioned earlier, the enrichment effect of IMAC is exactly opposite to that of MOAC, so there is a great complementarity between them. In order to combine the characteristics of the two kinds of methods, Thingholm *et al.*[57] examined the relationship between phosphopeptide eluted from IMAC and pH value of eluent, the result shows that all mono-phosphopeptides are eluted at a pH of 1.0, while multi-phosphopeptides only can be eluted in alkaline condition. In view of the characteristics of IMAC, Thingholm *et al.* developed a new enrichment method, namely sequential elution from IMAC (SIMAC). Firstly, the phosphopeptides in peptide mixture are enriched by IMAC column, the effluent of which is then treated by TiO_2. Then mono-phosphopeptides enriched by IMAC are eluted under acidic condition (pH = 1.0), meanwhile, TiO_2 is used for secondary enrichment to improve the selectivity of enrichment. Finally, the multi-phosphopeptides enriched by IMAC are eluted under alkaline conditions (pH = 11.3). The flow diagram is shown in Figure 1.10. By using SIMAC, a total of 492 phosphopeptides including 186 multi-phosphopeptides are identified. However, if only TiO_2 is used, only 286 phosphopeptides including 54 multi-phosphopeptides are identified. The biggest advantage of SIMAC is that mono-phosphopeptides can be separated from multi-phosphopeptides before MS analysis, which greatly improves the identification efficiency of multi-phosphopeptides.

(f) *Other method*

Several common and significant enrichment methods in phosphoproteomics are described above. But in consideration of the importance of protein phosphorylation, scientists have also developed many other methods. The following is a brief introduction.

(i) **Calcium phosphate precipitation:** In 1994, Reynolds *et al.*[58] used calcium ion and 50% ethanol to precipitate peptide containing multiple Ser phosphorylation from casein digestion. In 2007, Zhang *et al.*[59] proposed the use of calcium phosphate precipitate method for enrichment of phosphopeptide. Briefly, sodium hydrogen phosphate and ammonium hydroxide are added into peptide mixture, then excessive calcium chloride is added to precipitate the phosphopeptide. Finally, the precipitate is dissolved by formic acid for MS analysis. The effectiveness of this method is validated by using α-casein and β-casein as standard samples. Moreover, the authors make a comparison about the enrichment effect among calcium phosphate precipitation, IMAC and TiO_2 are compared, which indicates there is good

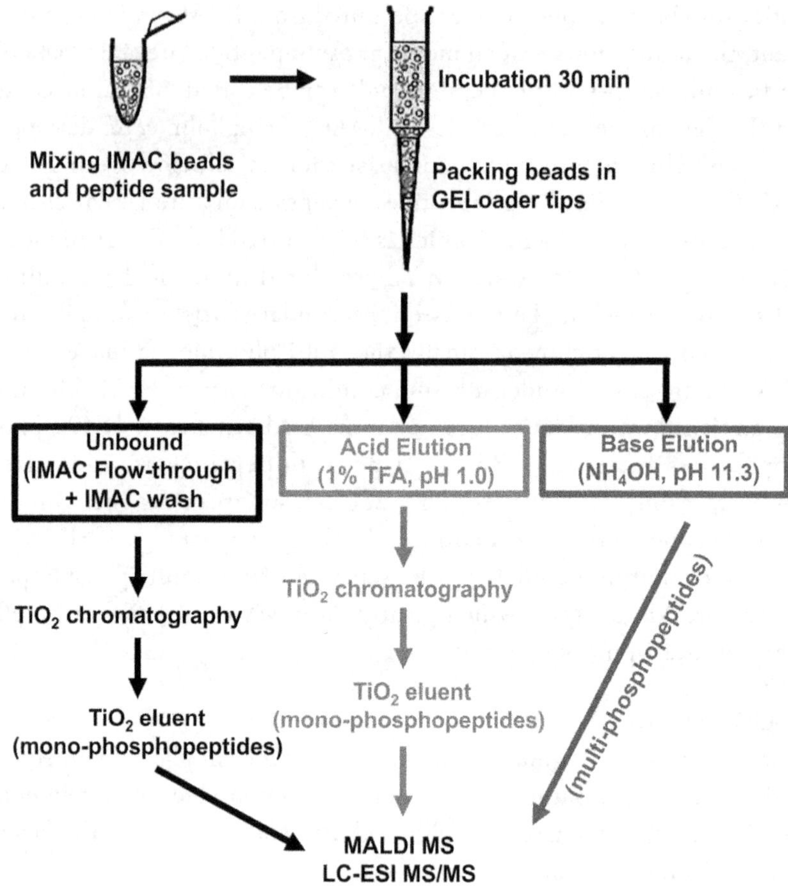

Figure 1.10: SIMAC strategy for phosphopeptide enrichment.

enrichment complementarity between calcium phosphate precipitation and IMAC. Therefore, they combine the two methods further for phosphoprot-eomics research in rice, a total of 227 non-redundant phosphorylation sites, 213 of which are on Ser residue and 14 on Thr residue. The phosphorylation of tyrosine residue is not identified in the result, which was attributed to its low phosphorylation.

(ii) **Polyarginine for multi-phosphopeptide enrichment:** Although the enrichment method of phosphopeptide has made great progress in recent

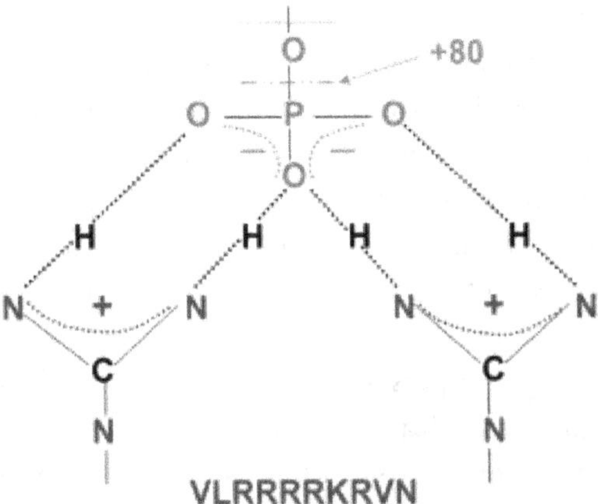

Figure 1.11: Cartoon showing the electrostatic interaction between the guanidinium group of Arg (blue) and the phosphate group of the phosphorylated serine (red).

years, the analysis of multi-phosphopeptide still faces many problems in the research of large-scale phosphoproteomics. The low ionization efficiency and low abundance of multi-phosphopeptides are not conducive for MS analysis. In 2008, Chang *et al.*[60] developed a new method for selective enrichment of multi-phosphopeptides. Guanidine group on arginine can specifically bind to phosphate group, and the adjacent arginine residue and phosphate group can form a stable compound by some interactions similar to covalent binding,[61] as shown in Figure 1.11. Therefore, Chang *et al.* synthesized polyarginine modified solid-phase microsphere for multi-phosphopeptide enrichment. This method possesses high selectivity and can be run under extreme conditions, but it has not been used in the research of large-scale phosphoproteomics.

1.3.5.2. *Glycoproteomics*

(1) Introduction of glycoproteomics

Glycosylation is one of the most important PTMs in an organism. According to the record and prediction of swiss-prot database, more than half of human proteins are glycosylated. More and more data show that glycosylation as a major PTM has an important effect on protein function.

For example, glycosylation plays an important role in protein folding, transport, and localization; almost all key molecules in the immune system are glycoproteins; glycosylation degree and abnormal change of glycan structure are often signs of cancer and other diseases. Generally speaking, glycosylation exists in those proteins within extracellular environment, such as membrane proteins, secreted proteins, and many proteins in body fluids, which are most likely to be contacted in the process of diagnosis and treatment. Many diagnostic markers and therapeutic targets, such as Her2/neu of breast cancer, prostate specific antigen of prostate cancer, and CA125 of ovarian cancer, are glycoproteins. The theoretical and practical significance of glycosylation research attracts more and more scientists to engage in this field. Glycobiology has also become an important subject in current proteome research.

At present, the research content of glycoproteomics mainly includes four aspects: (i) separation and identification of glycoprotein; (ii) determination of glycosylation site; (iii) the analysis of the composition, connection sequence, and mode of glycan, namely identification of glycan structure; (iv) the influence of glycosylation on protein function. It is necessary to recognize, detect, and identify the expression and change of glycoproteins on a large scale in whichever aspect.

(2) The structure and types of glycoproteins

The basic component of glycoprotein includes glycan and peptide chain. Research suggests that there are 41 connection modes which are derived from the different combinations of 13 monosaccharides and eight amino acids co-existing in glycoproteins. In eukaryotic cell, however, usually only 11 kinds of monosaccharides serve as glycan constitution of glycoprotein, including β-D-glucose (Glc), α-D-mannose (Man), α-D-galactose (Gal), α-D-xylose (Xyl), α-D-arabinose (Ara), α-L-fucose (Fuc), glucuronic acid (GlcuA), iduronic acid (IduA), N-acetyl glucosamine (GlcNAc), N-acetylgalactosamine (GalNAc), N-acetylneuraminic acid (NeuNAc), and sialic acid (Sia).

Glycoprotein is mainly divided into four types: N-glycoprotein, O-glycoprotein, GPI-anchored glycoprotein, and C-glycoprotein. The glycan and peptide chain are linked by GlcNAc and GalNAc in most glycoproteins. β-hydroxyl of GlcNAc and amide group of asparagine (Asn) are combined to form N-glycosidic bond, then glycan is linked to the amine acid by N-glycosidic bond for forming N-glycoprotein. α-hydroxyl of GalNAc and the hydroxyl of Ser and Thr are combined to form O-glycosidic bond, then glycan is linked to the amine acid by O-glycosidic bond for forming O-glycoprotein. GPI-anchored glycoprotein refers to that amide group of glycosylphosphatidylinositol (GPI) linked to carboxyl terminal of protein so as to make protein anchored on the membrane. C-glycoprotein refers to that glycan that is linked to tryptophan (Trp) residue on peptide chain by C–C bond. In addition, there are also several rare glycosidic bond types, such as β-xylosyltaxol-Ser, β-galactosyl-hydroxylysine, and α-L-arabinose-hydroxyproline. At present, N-glycoprotein is the most studied, followed by O-glycoprotein, and there are few studies on GPI-anchored glycoprotein and C-glycoprotein. In glycoprotein, Asn for the formation of N-glycosidic bond always exists in particular amino acid sequence of Asn-X-Ser/Thr, where X can be any amino acid except proline. However, the amino acid sequence of O-glycosylation has no particular rule to follow. In addition to the difference of glycosylation site, the glycan composition of N-glycoprotein and O-glycoprotein is also different. N-glycan can be mainly divided into three categories: (i) high-mannose type, consisting of GlcNAc and mannose; (ii) compound type: containing fructose, galactose, and sialic acid in addition to GlcNAc and mannose; (iii) hybrid type, containing two types of glycans in (iv) and (v). All N-glycans are based on the common core pentasaccharide: Man3GlcNAc2. There is no common oligosaccharide structure in O-glycosylation, but the derivations of O-glycan are usually based on four core structures, namely, Galβ1–3GalNAc1–, GlcNAcβ1–6(Galβ1–3)GalNAc1–, Glcβ1–3GalNAc1–, and GlcNAcβ1–6(Glcβ1–3)GalNAcα1–.

(3) Separation and analysis technology for glycoprotein

There are two problems in glycoproteomics research, one is the separation and enrichment of glycoprotein, the other is low ionization efficiency of glycopeptide due to the existence of glycan, which interferes with MS detection. Due to the relatively low abundance of glycoprotein and the complex microheterogeneity of glycosylation, the priority is to separate and purify glycoprotein from complex mixture in order to study it more clearly and reliably.

Currently, there are many enrichment methods for glycoprotein/peptide that have been developed according to the special physical and chemical property of glycan on glycoprotein, which promotes the advance of glycosylation research. The separation and enrichment technology for glycoprotein is as follows.

(a) *Lectin affinity technology*

Lectin affinity technology is the most widely used separation and enrichment technology in glycoproteinomics. In 1888, Lectin first appeared in the literature where Stillmark described its interaction with glycoprotein, glycan, and glycolipid on cell membrane. Lectin is a kind of glycan-binding protein secreted by both animals and plants, it can exclusively recognize a certain specific sequence which exists in monosaccharide/glycan with particular structure, then binds to the specific sequence. This kind of binding is non-covalent and reversible. Lectin, which is a kind of protein with specific affinity to different saccharides, has been used to study saccharides for a long time. Now there are a variety of lectins with different sources and affinities that have been successfully isolated and commercialized.

In practice, lectin can be immobilized on chromatographic packing such as agarose bead. Lectin-immobilized bead can be used directly, or loaded into a chromatographic column. Glycoproteins/peptides are usually eluted by using specific monosaccharides to compete with them, after they are captured by lectin. Different lectins have affinity to different types of oligosaccharide chains. For example, leukoagglutinin in sambucus nigra and hemagglutinin in maackia amurensis are respectively used for specific recognition of α_{2-6} and α_{2-3} sialic acid residues in glycan on the terminal of glycoproteins. Jacalin is commonly used to specifically recognize oligosaccharide chain containing galactose and is often used to study *O*-glycoprotein. Concanavalin A (Con A) is commonly used to isolate glycoprotein/peptide which is rich in high mannose and high glucose. Wheat germ agglutinin (WGA) can selectively bind GalNAc and NeuNAc. Con A and WGA are commonly used in the study of *N*-glycoprotein.

Con A and WGA are the most commonly used lectins for glycoprotein enrichment in various biological samples. It is feasible to use one particular lectin to enrich a certain type of glycoprotein, or to use lectin affinity purification technology in series/array, namely, use two or more lectins to enrich different types of glycoproteins. Yang *et al.*[62] applied multi-lectin affinity chromatography to enrich glycoproteins in serum/plasma, and made a comparison about identification result. Bunkenborg *et al.*[63] applied Con A and WGA in the double enrichment of *N*-glycoproteins in human serum proteomic, 86 glycosylation sites on 77 glycoproteins were finally identified by LC-MS/MS.

Lectin-modified solid-phase material has been commercialized, which is very convenient to purchase and apply. Lectin is a kind of protein. To date, much research about protein immobilization techniques based on various substrates have been carried out. Therefore, now both commercialized lectin-modified bead and lectin protein can be bought. According to the requirement of different experiments, lectin can be immobilized on different substrates by various methods for meeting requirements of special experimental design or developing some new technology. Zhou *et al.*[64] prepared a glycoproteomics chip reactor which was connected to Con A affinity column for acidification, glycan take-off, and enzymatic hydrolysis of glycoprotein online as well as the SCX separation of peptide mixture, that whole process is quick and easy. 82 glycopeptides belonging to 41 glycoproteins are identified from 5 μL human plasma by using this method. Feng *et al.*[65] prepared a nanoliter-level capillary monolithic column, that is, immobilized a layer of Con A within the capillary chromatographic column by sandwich chelating of IDA-Cu^{2+}-Con A, which is suitable for finite samples in glycoproteomics research. Taking the interaction between dopamine and protein into consideration, Tarlov *et al.*[66] used dopamine to immobilize Con A onto sheet metal, such as gold, indium, and iridium, to construct lectin chip for monitoring different glycan forms on glycan.

One point to note is that the interaction between lectin affinity and oligosaccharide chain is relative, not absolute, therefore lectin can be used for glycoprotein enrichment, which is not suitable for the identification of glycan structure.

(b) *Boric acid affinity technology*

Boric acid affinity technology is also one of the common methods for the separation and enrichment of glycoproteins. Boric acid group can interact with oligosaccharide chain under a certain condition, the schematic diagram is shown in Figure 1.12. In alkaline aqueous solution, the hydroxylation of boric acid group transforms its molecular conformation from plane triangle into tetrahedron anion, which can react with cis–diol structure to generate a

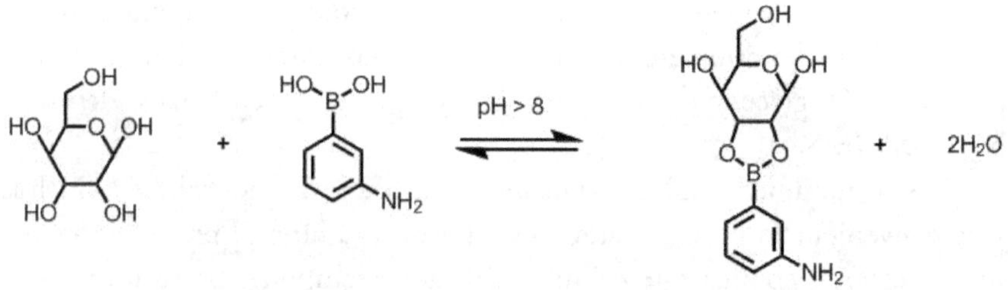

Figure 1.12:   Schematic diagram of the boric acid affinity method.

five-member ring diester and release two water molecules. Under acidic condition, the reaction is reversed, namely, ring diester is hydrolyzed. The essential condition in which boric acid group reacts with oligosaccharide chain is the existence of cis–diol structure on glycan. Since the monosaccharide unit on glycan generally has cis–diol structure, boric acid affinity technology can enrich glycoprotein/peptide.

In principle, there should rarely be discrimination effect when affinity interaction between boric acid derivatives and glycoprotein/peptide occurs. As long as the glycan structure has cis–diol structure, no matter for N-type or for O-type, it can be captured.

Practically, however, whether using boric acid modified material or boric acid affinity chromatography, there are also many secondary interactions that occur between sample and substrate such as hydrophobic interaction, ionic interaction, hydrogen bond interaction and complexation, etc., except primary interaction between boric acid group and cis–diol. Those secondary actions will influence this enrichment method to a certain degree. The selectivity of boric acid affinity technology is not outstanding in several common enrichment technologies for glycoprotein, resulting in its limited application for large-scale glycoproteomics in complex biological samples. Zhang *et al.*[67] first used boric acid affinity chromatography to enrich glycoprotein in serum that was then digested into peptide mixture, from which glycopeptide was enriched subsequently. Finally, ETD MS result showed that a higher proportion of glycopeptide (87.5%) can be identified by the two-step enrichment. Boric acid compounds are common small molecules, which are easy to derive. It is very suitable for exploring new technology and methods. Lu *et al.* modified boric acid group on Fe_3O_4 (see Ref. [68]) and porous material[69] by different methods for the enrichment of standard glycoprotein/peptide. Li *et al.*[70] synthesized nucleotide modified with boric acid group for the synthesis and screening of DNA aptamer, which can selectively bind glycoprotein.

(c) *Hydrophilic interaction chromatography*

HILIC is a chromatographic technique that uses polar stationary phase and non-polar mobile phase, and has been previously used to analyze small polar molecules. Since monosaccharide molecules such as glucose, galactose, and mannose contain a large number of hydroxyl groups, making glycan with strong hydrophilicity, the hydrophobicity of non-glycoprotein/peptide is relatively strong. Therefore, according to the hydrophilicity difference, hydrophilic chromatography media such as cellulose and sepharose, etc., can be used to enrich glycoprotein/peptide. These media can be used either directly or as stationary phase in HILIC for automatic operation. HILIC method is easy to operate and can non-selectively enrich various types of glycopeptide, but its specificity is not high. Especially when non-glycopeptides are rich in hydrophilic amino acids, they are susceptible to non-specific enrichment in samples.

HILIC is commonly used in glycopeptide enrichment. Hägglund *et al.*[71] identified four glycopeptides from fetuin digestion by HILIC and successfully combined this method with lectin affinity technology. Wada *et al.*[72] enriched glycopeptide from standard glycoprotein such as IgG by directly using the hydrophilic binding force of sepharose CL-4B and oligosaccharide, and parsed their glycan structure at the same time. Picariello *et al.*[73] identified 32 glycoproteins including 63 *N*-glycosylation sites from human milk digestion by HILIC and MS.

(d) *Hydrazine chemical reaction*

Like boric acid method, hydrazine chemical reaction method also uses the cis–diol structure of glycan on glycoprotein/peptide for enrichment. But it has a different chemical reaction principle. Hydrazine chemical reaction method mainly includes four steps: (i) oxidation, the cis–diol on glycan was oxidized to aldehyde by periodate; (ii) coupling, aldehyde reacts with solid phase support such as the hydrazine group on the resin to form hydrazone bond, resulting in the capture of glycoprotein/peptide; (iii) cleanout, the removal of unbound protein/peptide in reaction system, for solid-phase glycoprotein on resin, it should be directly treated with protease and then washed again for removing non-glycopeptide; (iv) peptide release for MS analysis. There is no discrimination effect when hydrazine chemical method is used for glycan enrichment. This method has high efficiency and can easily identify peptides and recognize glycosylation sites, even make quantitative analysis by isotope or other reagent tags. Now it has more and more applications in glycoproteomics research. But hydrazine chemical method also has some disadvantages, like it requires multi-step derivatization and complicated operations, moreover, glycan is retained on the stationary phase after peptide release, which limits the application of this method in the parsing of glycan structure.

Hydrazine chemistry is a popular method in glycoproteomics research, which can be used either alone or in combination with other glycoprotein enrichment techniques or with the quantitative technique. Hydrazine chemistry has been used to study biological samples such as plasma, membrane protein, and saliva. Whelan *et al.*[74] used hydrazine chemistry to study glycoproteins in three types of breast cancer tumor cells that may be cancer markers, and identified 27 glycosylation sites on 25 glycoproteins. Blake *et al.*[75] combined hydrazine chemistry method with HILIC for studying glycosylation site and glycan structure of erythrocyte lectin in H_5N_1 influenza virus. Beuerman *et al.*[76] combined hydrazine chemistry with iTRAQ quantitative technique to qualitatively and quantitatively study the glycoprotein in tears of seasonal keratitis patients. Chen *et al.*[77] treated human liver tissue protein with trypsin, pepsin, and thermolysin firstly, then enriched glycopeptide by hydrazine chemical method, therewith cut glycan, finally 939 glycosylation sites on 523 glycoproteins were identified by RPLC-MS/MS.

(e) *β-elimination-Mie addition method*

In current proteomics research, the development mainstream of separation technology is based on *N*-glycosylation, since there is a lack of universal enzymes like PNGase F for *O*-glycosylation research. *β*-elimination-Mie addition method is widely used in *O*-glycosylation research. Actually, *β*-elimination-Mie addition is used in phosphopeptide enrichment at first, Wells *et al.*[78] drew lessons from this method to make *O*-glycopeptide produce unsaturated double bond by *β*-elimination in alkaline condition, then made the Mie addition reaction happen by adding dithiothreitol (DTT) or biotinpentylamine (BAP), and finally used thiol affinity column or biotin affinity column to capture *O*-glycopeptide for MS identification and glycosylation site analysis. Furthermore, Cai *et al.*[79] tested the reaction effects of three nucleophiles including ammonium hydroxide, methylamine, and dimethylamine on the Mie addition, and identified *O*-glycosylation site by LC-MS. Hanisch *et al.*[80] developed a method for the on-column degradation of *O*-glycan, which can degrade glycan in moderate conditions and avoid peptide degradation or dehydration. Besides, the preliminarily step-by-step degradation of *O*-glycan is beneficial to analyze its fine structure.

(4) Biological MS for glycoprotein

Biological MS in proteomics research mainly faces two problems. One is the identification of peptide in glycopeptide. Compared to other peptides, glycopeptide is not easy to be used in the ionization in MS. And its abundance is low in biological samples, which generally accounts for only 2–5% of protein digestion peptides. The other problem is that complexity of glycan structure makes it difficult to analyze glycoproteins. With the development of MS, the study of glycoproteins has made a breakthrough in recent years. The further combination with advanced dissociation technology such as CID, ECD, and ETD on the basis of MALDI and ESI, superadding the detection by multistage MS with high sensitivity and high resolution, make it possible to realize PTM protein with more sophisticated MS identification and structural analysis. In glycoproteomics research, the general analytical procedure is to first isolate glycoprotein/peptide and then conduct MS analysis and identification, or the glycan is cut from the peptide by glycosidase or glycan degradation, then with peptide together or respectively processed for MS analysis. In the latter way, the glycosylation site can be determined by comparing digestion peptide mapping of glycoprotein before and after treatment with glycosidase, and glycan structure on glycopeptide can be analyzed by tandem MS.

The determination of glycosylation site is the main content of glycoproteomics. For *N*-glycoprotein, the *N*-glycopeptide is usually released by glycosidase such as PNGase F and so on, but the glycan with core structure of chitobiose cannot be cut by PNGase F. *O*-glycan can be cut by β-elimination. Generally speaking, when using PNGase F for *N*-glycopeptide release, the Asn on glycosylation site is transformed into aspartic acid, generating about 1 Da of mass

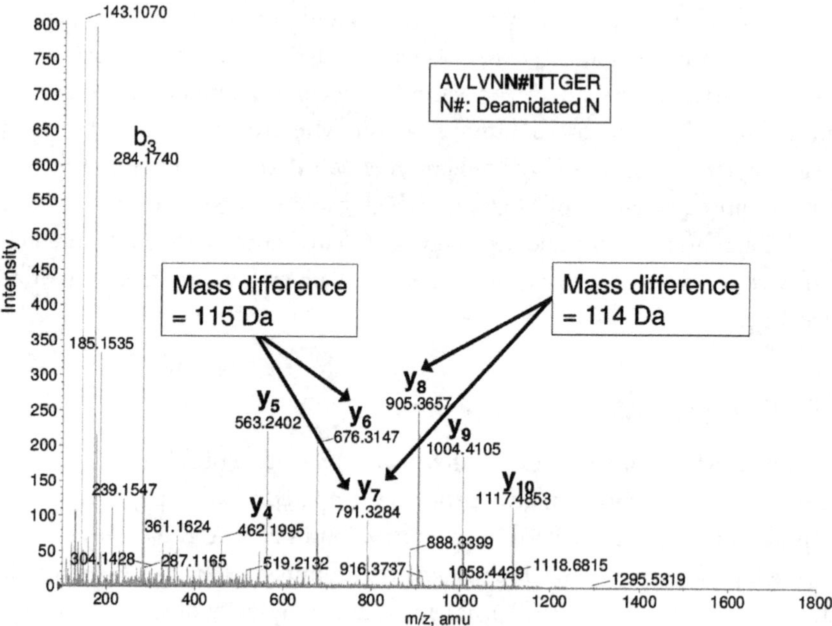

Figure 1.13: MS/MS spectrum of *N*-linked glycopeptide fragment AVLVNN#ITTGER (*m/z* = 644.31, charge = 2+) originated from Golgi membrane protein 1. (#) Represents the *N*-glycosylation site and the consensus motif of *N*-glycosylation is in bold.[76]

discrepancy, which can serve as mass marker for the identification of glycosylation site (Figure 1.13). If the resolution of MS is not high enough, the addition of $H_2{}^{18}O$ during PNGase F enzymolysis can make Asn transform to aspartic acid labeled with ^{18}O, then 3 Da of mass discrepancy will be generated, which can not only improve the accuracy of the mass marker but also make for quantitative analysis. *O*-glycosylation site can be identified by adding nucleophile such as alkylamine as mass marker. Moreover, glycopeptide can also be determined by monitoring the signal of characteristic fragment ion. Some signals of characteristic glycan fragment are as follows: *m/z* 163 (Hex), *m/z* 204 (HexNAc), *m/z* 292 (sialic acid), *m/z* 366 (HexHexNAc). The identification of glycan structure includes not only the identification of types and quantity of monosaccharide that make up the glycan, but also the analysis of its configuration. For single glycan analysis, the first is to get to know if every monosaccharide is cyclized to either pyran ring or furan ring, the isomer after cyclization is either α-configuration or

β-configuration, and, to understand the connection mode among each monosaccharide.

It is more complex that the branch structure of a glycan may appear due to the presence of multiple active hydroxyls on every monosaccharide, or even that those reduction terminals undergo various modifications and post-processing such as phosphorylation, sulfating, dehydration of hydroxyls in the same glycosyl to form inner ether, or with other compound containing hydroxyl to form acetal, etc. Some of these complexities are unique to glycan, leading to latent isomer number of oligosaccharide chains much more than that of peptide or polynucleotide chains as far as they contain the same number of monosaccharides. Fortunately, the monosaccharide composition and structure of glycan on glycoprotein is not arbitrary, there are certain rules to follow, such as the core pentasaccharide of N-glycan, which reduces the complexity of glycan structure greatly. Even so, the glycan structure still is more complicated than the peptide or polynucleotide chain, which results in its identification becoming more difficult.

The combination of chemical derivatization and multi-stage MS is an effective method to analyze the fine structure of unknown glycan. It is important to note that the cutting method of N-glycan and O-glycan is different, so it is vital to minimize the impact of cutting process on glycan for simultaneously identifying both N-glycan and O-glycan in biological samples. Glycosidase is commonly used for N-glycan cutting. The degradation of O-glycan usually adopts chemical reactions whose condition must be as mild as possible. When parsing some oligosaccharides with low complexity, such as disaccharide, trisaccharide, chemical derivatization for the hydroxyls on monosaccharide is a good choice. The most common is methylated derivative, which can increase the mass discrepancy of fragment ion signals, thereby parsing the glycan structure. Both glucosidic bond cleavage and cross-ring cleavage of the cyclic monosaccharide can occur in the fragmentation mode of the glycan, generating some characteristic fragmentation ion signals. For example, the presence of core pentasaccharide in N-glycan make glycopeptide have at least 892 Da increment; GlcNAc directly connected to the Asn (glycosylation site) can produce three consecutive peaks with a certain mass discrepancy (83 Da and 120 Da) due to cross-ring cleavage in tandem MS analysis. Glycosylation site can be determined according to the characteristic mass discrepancy. Chen et al.[81] has parsed the glycopeptide structure of horseradish peroxidase (HRP) including glycosylation site and glycan structure by combining MALDI-MS with ESI-MS, indicating the analytical ability of multi-stage MS to parse glycan structure.

One of the difficulties in glycoproteomics analysis is to consider gycopeptides as a whole, which can be blamed on the subsistent limitations such as separation method and enrichment specificity and the response of glycopeptide to MS. There are a few reports but they are only limited to standard glycoprotein or a certain specific glycoprotein in complex samples. However, with glycan cutting first, the technological difficulties in the high-throughput identification of glycosylation sites have been overcome through the combination of glycosidase and high-precision multi-stage MS. Besides, there are a lot of research results on the use of biological MS to target a certain glycan pattern in biological samples, or reflect change trends of some glycan patterns in the entire system. Therefore, there is no doubt that the development of biological MS will be an important force for advancing glycopeoteomics, although many difficulties still exist in glycopeoteomics research, especially in the high-throughput parsing of glycopeptide.

1.3.6. *Peptidomics and peptidome research technique*

1.3.6.1. *Peptidome and peptidomics*

In biology, "peptide" usually refers to polypeptide with a maximum molecular weight of about 10,000 Da. "Peptide" is derived from "peptos", which is a Greek word and means digestible, indicating peptides are products of proteolysis that also are regarded often as biological junk. Peptidome refers to all proteins with low molecular weight, like endogenous peptides existed in complex biological samples such as cell lysate, tissue extraction, and body fluid, etc. In recent years, it has been found that endogenous peptide usually plays as regulator and indicator in biological process with the development of MS-based bio-informatics, for instance, bio-active peptide hormone and somatotropin have important influences on growth, health, and disease. Therefore, endogenous peptide contains many bio-markers that may record human physiological and pathological states. These bio-markers may have higher clinical sensitivity and specificity than conventional markers. Endogenous peptide has turned itself into something gorgeous, from "biological junk" to "undeveloped resource for disease-specific diagnostic information". Peptidomics research refers to the global, qualitative, and quantitative analysis of endogenous peptide and small molecular proteins in biological samples at a particular time and regional system. Generally speaking, the peptide in the peptidome can be divided into two categories: one is the bio-active peptide that plays a vital role in biological process, such as hormone and somatotropin, etc.; the other is degradation fragment of protein that reflects the proteolysis mode and biological state of the individual. Today, peptidome research attracts more and more attention in both academia and industry due to its simpler separation than proteome research (no digestion process), as well as its record of the present physiological state in an organism. Circulatory protein fragments produced in bodily fluid or tissue may reflect major biological events and provide salient information for clinical diagnosis. Petricoin once pointed out[82] that peptidome can in various forms include or represent all of the analyte in the organism. In the meantime, Petricoin also said that peptidome is a combination of those relatively undeveloped information continents, and it may be one of the richest analyte archives, the analytes of which are sensitive and specific to ongoing disease process. Whatever, the real function of peptidomes, they need to be further studied and testified.

1.3.6.2. *Peptidome research technique*

Peptidomics analysis process (separation and analysis of endogenous peptide) mainly includes sample pretreatment, separation, detection, quantification, and data analysis (Figure 1.14).[83] The technology focuses on sample separation and MS detection.

Many new techniques have been applied for the separation and enrichment of endogenous peptides. Ultrafiltration is the most widely used method to separate peptides from biological samples with simultaneous removal of proteins. For example, Hancock *et al.*[84] used centrifugal ultrafiltration to remove molecule (>10 kDa) in human serum first, and then identified endogenous peptide in the filtrate successfully by MALDI-TOF-MS, LC-MS, and LTQ-FT-MS. However, the time consumption of centrifugal ultrafiltration increases sharply with the increase of sample dosage, moreover, other small molecules such as salt and so on are concentrated together with endogenous peptide, which will affect the subsequent

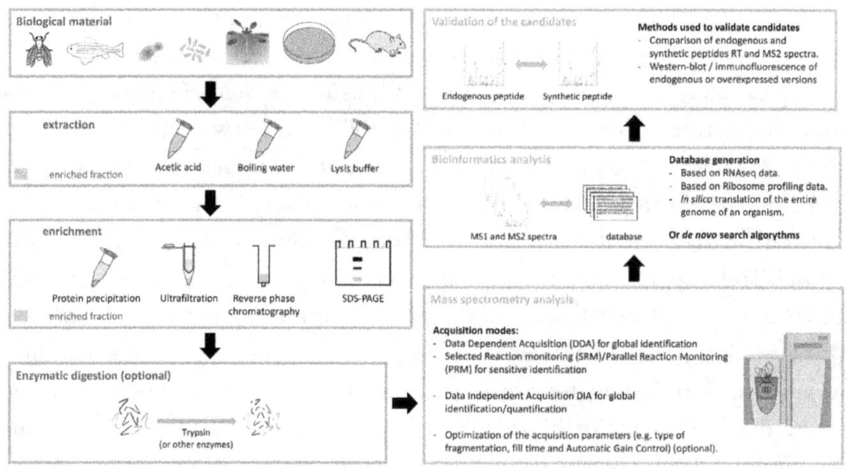

Figure 1.14:   Method for the analysis of peptidome.[83]

MS identification. Organic solvent precipitation method adds organic solvent into biological samples for protein precipitation, while small peptides dissolve in organic solvent, and finally protein is removed through centrifugal operation. Chromatographic column separation method separates endogenous peptide by filling chromatographic column with special packing. This column possesses the dual function to exclude high molecular weight protein and enrich low molecular weight protein/peptide. If this column can maintain the dual function of special packing when original biological samples are added continuously, then it is very suitable for online separation and analysis. With the development of nanoscience, another bio-sample pretreatment method is to use functionalized material as SPME adsorbent to separate and enrich endogenous peptides. This method often uses inert material as a carrier, such as organic polymer magnetic bead and inorganic polymer particle, etc., then adsorbs endogenous peptide by taking advantage of its special surface structure and property, or by the functional group modified on its surface. For instance, C8, Cu^{2+}, or weak cation modified chromatogram magnetic bead (about 0.1–10 μm of particle diameter) generally takes organic polymer magnetic bead as the core; C8, PMMA, and Cu^{2+} modified nanomaterials adsorb peptide by their large specific surface area and the interaction between surface group and peptide. In addition, there are multi-dimensional LC separation methods.

In summary, previous reports indicate that most of the work is based on centrifugation and affinity chromatography in the enrichment and separation of endogenous peptides. However, repeated centrifugations in these enrichment processes lead to complex operations, and sample loss may occur in the process of extraction and enrichment, moreover, impurities may be concentrated together with peptide. All the above set barriers to rapid and high-throughput detection of the endogenous peptides. Therefore, it is urgent to develop fast, efficient, and convenient extraction and enrichment technologies for endogenous peptides.

The analysis and detection of peptidome also depends on biological MS. Non-volatile salts are often needed in the treating process of bio-samples, which will increase noise and cause signal suppression during MS analysis. In addition, endogenous peptide associated with disease and signaling is generally low-abundance in complex tissues, cells, and body fluid, which leads to its difficult identification by MS. Therefore, pre-enrichment technology is the bottleneck for the successful analysis and identification of endogenous peptides and detection of disease-related markers in peptidomics research.

1.4. Magnetic micro-/nanomaterial for proteomics

Magnetic polymer microsphere has attracted more and more attention since the 1970s. It is a functional microsphere which combines inorganic magnetic particles with organic/inorganic polymer. Besides those properties of conventional polymer microsphere (e.g., easily functionalized on surface, adjustable particle size, controllable electrical property of surface, etc.), it can also respond to external magnetic fields, namely, it can be quickly separated from media solution, which greatly simplifies the separation process. Recently, magnetic polymer microspheres have been widely used in many fields such as targeted drug delivery, cell separation, catalyst carrier, bio-analysis, and proteomics. Especially in proteomics research, magnetic material has been successfully used for rapid enzymatic hydrolysis, separation, and enrichment of functional protein and peptidomics.

1.4.1. *Preparation of magnetic micro-/nanomaterial*

Magnetic micro-/nanomaterials include magnetic inorganic materials, magnetic organic materials, and magnetic composites. There are many kinds of inorganic magnetic nanoparticles, such as metal alloys (Fe, Co, Ni), iron oxide (γ-Fe_2O_3, Fe_3O_4), ferrite ($CoFe_2O_4$, $BaFe_{12}O_{19}$, $BaFeO_4$), iron nitride (Fe_4N), and chromium dioxide (CrO_2), etc. The toxicity of cobalt and nickel strictly restricts their application in biology and medicine, etc. While Fe_3O_4 has been widely used as an inorganic magnetic material due to its low toxicity and easy preparation, etc. Fe_3O_4 can be prepared by the co-precipitation of ferrous ion and iron ion in alkaline aqueous solution, or the oxidative precipitation of ferrous particles. And its size, shape, and composition can be controlled by adjusting reaction conditions. The common functional materials can be mainly divided into natural biological macromolecules, synthetic macromolecules, and inorganic substances. Different kinds of magnetic polymer microspheres have different preparation methods.

1.4.2. *Magnetic micro-/nanomaterial for drug delivery*

The rapid development of life science, information science, and material science and so forth brings new ideas to pharmacy. The research of new drug formulation has been developed from previous velocity controlled-release and directivity controlled-release to time controlled-release, self-regulating release, and individualized administration. Thereinto, directivity controlled-release also known as targeted drug delivery system (TDDS) refers to a drug delivery system in which the drug is released into the target tissue through carrier conjugate or being buried in the carrier. With the development of new drug formulations in the lasy 10 years, the research on targeting preparation has become one of the hot spots, domestically and overseas.

Magnetic pharmaceutical preparation is commonly used in the magnetic targeted drugs delivery system (MTDDS) which refers to the process of packing drugs and magnetic material in polymer carrier, then leading drugs to directional movement and positioning deposition in the body under the influence of external magnetic field, and then releasing the drug in the magnetic field for targeted local concentration or targeted intercepting. Currently, the magnetic pharmaceutical preparations are mainly magnetic particles (MPs), including magnetic microspheres (MMSs), and magnetic nanoparticles (MNPs). Compared with other targeted preparations, magnetic drug-loading particles are not only able to effectively reduce the capture of reticuloendothelial system, but also have unique properties: a certain slow-release effect for reducing dosage; increasing drug concentration of the target area under external magnetic field for improving the curative effect; reducing the toxic and side effects of drugs on other organs and normal tissues; absorbing energy under the action of alternating magnetic field to generate heat for heat therapy.

The research on anticancer drug keeps digging deeply with the augmenting of cancer incidence, while anticancer drugs kill cancer cells and poison normal cells at the same time through conventional conveying. The emergence of magnetic drug-loading particles that can locate accurately solves the above problem and processes.

The biggest drawback of chemotherapy for most cancers is non-specificity. Drugs used for therapy are usually injected intravenously and then dispersed throughout the body. The non-specificity results in the side effects of chemotherapeutic cytotoxin, in which the drug kills some normal cells while attacking the target tumor cells. However, the targeting ability of micro-/nanoparticles can reduce the risk that drugs spread in the whole body, thereby reducing or eliminating side effects of chemotherapy, and reducing compound dose with cell toxicity, besides, can realize drug administration more precisely.

MNPs have been used as drug carriers for targeting drug sites in the body since the 1970s. Widder *et al.* find that MNPs can bind to cytotoxic drugs, and MNP carriers loaded with drugs can be injected into the human body through veins or arteries. Magnetic fields are used to guide the drug to the target tumor site and to concentrate the drug.

Therapeutic agents are released from magnetic carriers due to enzymatic hydrolysis or changes of other physiological conditions (such as pH, temperature, osmotic pressure, etc.) once magnetic carriers are concentrated inside the tumor or other target site, increasing the uptake of drugs by tumor cells at targeted sites.

Besides the above described unique advantages, drug-loading magnetic particles also face some challenges on clinical application: (i) the selection of external magnetic field including magnetic field intensity, gradient, time consumption, and stereo-location, etc. will affect the staying stability of magnetic drugs in target organs; (ii) some properties of magnetic particles need to be improved further, such as the uniformity and controllability of particle size and surface property, etc. for increasing load rate of carrier and enhancing drug slow-release property, stability, and active targeting property of magnetic particles; (iii) to study the effect of internal environment on magnetic drug targeting.

1.4.3. *Magnetic micro-/nanomaterial for rapid digestion*

The application of magnetic nanomaterials immobilized with enzymes in protein proteolysis not only make the separation of enzymes from the substrate easier but also improve the thermal stability and operating stability of the enzymes. The immobilized enzyme possesses good reproducibility and high efficiency, therefore, it can be used in continuous production and reduce the production cost. In addition, the microenzyme reactor made of magnetic materials is simple to fill and easy to regenerate and operate. Bíková *et al.*[85] prepared a microenzyme reactor by immobilizing trypsin onto magnetic microspheres with carboxyl groups and then filling it into microchip channel under external magnetic field. The products of standard proteins by this microenzyme reactor are separated parallelly by RP-HPLC and HPCE, and the collecting fractions are identified by MS. They also test the effects of different buffer and denaturation conditions on integrity of enzymatic hydrolysis by using SDS-PAGE. Chen *et al.*[86] found that microwave assisted enzymatic hydrolysis can be completed within 30 s in the presence of magnetic microspheres, because magnetic microspheres are excellent microwave radiation absorbers. This discovery leads to a rapid change in the field of microwave-assisted enzymatic hydrolysis. Trypsin can be immobilized on the surface of magnetic microspheres prepared via hydrothermal synthesis by choosing proper coupling agent, for large-dose in-solution digestion, microchip digestion, nanoliter-level on-plate digestion, and microwave-assisted fast digestion, which can shorten digestion time compared to the traditional method by about 12 h to 15 s.[87]

1.4.4. *Magnetic micro-/nanomaterial for PTM proteomics*

It is very common to identify phosphoprotein by PMF and tandem MS with the development of MS. However, compared to other proteins, the low abundance of phosphoprotein suppresses its ion signals by non-phosphorylated ones under the common positive ion mode of MS. Therefore, the enrichment of phosphoprotein/peptide is the premise for their effective identification by the MS. The common enrichment methods for phosphoprotein/peptide include solid-phase metal affinity chromatography, chemical modification, metal oxide/metal hydroxide affinity chromatography, etc. In addition, magnetic micro-/nanomaterials modified with boric acid and lectin, etc., are also used in magnetic separation and enrichment of trace glycoprotein/peptide.

1.4.5. *Magnetic micro-/nanomaterial for separation and enrichment of low-abundance protein/peptide*

The separation and enrichment of low-abundance protein/peptide is an important link in proteomics and peptidomics. Many virtues of magnetic polymer microspheres, including easily modified surface, good dispersion, and sensitive response to magnetic field, make it possible for the separation and enrichment of trace peptides in proteomics analysis.

Currently, there are many applications of commercial C8-magnetic materials prepared by using organic polymer as linker in laboratory. But this kind of magnetic material has weak magnetic responsiveness, large particle size (generally micron-size), relatively small specific surface area, poor biocompatibility, and poor compatibility with MALDI-MS. Deng's group[88] prepared novel functionalized magnetic material by making some coating (such as SiO_2) and succedent chemical modifications on the surface of Fe_3O_4 with particle size around 200 nm. The novel functionalized magnetic materials are successfully used for the separation and enrichment of trace peptide and the analysis of peptide in human serum by MALDI-TOF MS directly, which simplifies the analysis and determination procedure of low-abundance peptide and solves the analysis difficulty of trace sample, as well as opens up a new way for the application of magnetic polymer microsphere.

References

1. Brockman A. H., Orlando R. Probe Immobilized Affinity-Chromatography Mass-Spectrometry. *Analytical Chemistry*, 1995, 67: 4581–5.
2. Brockman A. H., Dodd B. S., Orlando N. A Desalting Approach for MALDI-MS Using on-Probe Hydrophobic Self Assembled Monolayers. *Analytical Chemistry*, 1997, 69: 4716–20.

3. Warren M. E., Brockman A. H., Orlando R. On-Probe Solid-Phase Extraction MALDI-MS Using Ion-Pairing Interactions for the Cleanup of Peptides and Proteins. *Analytical Chemistry*, 1998, 70: 3757–61.

4. Liang X. L., Lubman D. M., Rossi D. T., Nordblom G. D., Barksdale C. M. On Probe Immunoaffinity Extraction by Matrix-Assisted Laser Desorption/Ionization Mass Spectrometry. *Analytical Chemistry*, 1998, 70: 498–503.

5. Schuerenbeg M., Luebbert C., Eickhoff H., Kalkum M., Lehrach H., Nordhoff E. Prestructured MALDI-MS Sample Supports. *Analytical Chemistry*, 2000, 72: 3436–42.

6. Jia W., Wu H., Lui H., Li N., Zhang Y., Cai R., Yang P. Rapid and Automatic on-Plate Desalting Protocol for MALDI-MS: Using Imprinted Hydrophobic Polymer Template. *Proteomics*, 2007, 7: 2497–506.

7. Clarke N. J., Li F., Tomlinson A. J., Naylor S. One Step Microelectroelution Concentration Method for Efficient Coupling of Sodium Dodecylsulfate Gel Electrophoresis and Matrix-Assisted Laser Desorption Time-of-Flight Mass Spectrometry for Protein Analysis. *Journal of the American Society for Mass Spectrometry*, 1998, 9: 88–91.

8. Xu F., Wang Y. J., Wang X. D., Zhang Y. H., Tang Y., Yang P. Y. A Novel Hierarchical Nanozeolite Composite as Sorbent for Protein Separation in Immobilized Metal-Ion Affinity Chromatography. *Advanced Materials*, 2003, 15: 1751–3.

9. Zhang Y. H., Wang X. Y., Shan W., Wu B. Y., Fan H. Z., Yu X. J., Tang Y., Yang P. Y. Enrichment of Low-Abundance Peptides and Proteins on Zeolite Nanocrystals for Direct Maldi-Tof Ms Analysis. *Angewandte Chemie-International Edition*, 2005, 44: 615–7.

10. Zhang Y. H., Yu X. J., Wang X. Y., Shan W., Yang P. Y., Tang Y. Zeolite Nanoparticles with Immobilized Metal Ions: Isolation and MALDI-TOF-MS/MS Identification of Phosphopeptides. *Chemical Communications*, 2004, 2882–3.

11. Jia W., Chen X., Lu H., Yang P. $CaCO_3$-Poly(Methyl Methacrylate) Nanoparticles for Fast Enrichment of Low-Abundance Peptides Followed by $CaCO_3$-Core Removal for MALDI-TOF-MS Analysis. *Angewandte Chemie-International Edition*, 2006, 45: 3345–9.

12. Doucette A., Craft D., Li L. Protein Concentration and Enzyme Digestion on Microbeads for Maldi-Tof Peptide Mass Mapping of Proteins from Dilute Solutions. *Analytical Chemistry*, 2000, 72: 3355–62.

13. Girault S., Chassaing G., Blais J. C., Brunot A., Bolbach G. Coupling of MALDI-TOF Mass Analysis to the Separation of Biotinylated Peptides by Magnetic Streptavidin Beads. *Analytical Chemistry*, 1996, 68: 2122–6.

14. Lee J. H., Kim Y. S., Ha M. Y., Lee E. K., Choo J. B. Immobilization of Aminophenylboronic Acid on Magnetic Beads for the Direct Determination of

Glycoproteins by Matrix Assisted Laser Desorption Ionization Mass Spectrometry. *Journal of the American Society for Mass Spectrometry*, 2005, 16: 1456–60.

15. Xu X., Deng C., Gao M., Yu W., Yang P., Zhang X. Synthesis of Magnetic Microspheres with Immobilized Metal Ions for Enrichment and Direct Determination of Phosphopeptides by Matrix-Assisted Laser Desorption Ionization Mass Spectrometry. *Advanced Materials*, 2006, 18: 3289–93.

16. Li Y., Leng T., Lin H., Deng C., Xu X., Yao N., Yang P., Zhang X. Preparation of Fe_3O_4@ZrO_2 Core-Shell Microspheres as Affinity Probes for Selective Enrichment and Direct Determination of Phosphopeptides Using Matrix-Assisted Laser Desorption Ionization Mass Spectrometry. *Journal of Proteome Research*, 2007, 6: 4498–510.

17. Tian R., Zhang H., Ye M., Jiang X., Hu L., Li X., Bao X., Zou H. Selective Extraction of Peptides from Human Plasma by Highly Ordered Mesoporous Silica Particles for Peptidome Analysis. *Angewandte Chemie-International Edition*, 2007, 46: 962–5.

18. Steen H., Kuster B., Mann M. Quadrupole Time-of-Flight Versus Triple-Quadrupole Mass Spectrometry for the Determination of Phosphopeptides by Precursor Ion Scanning. *Journal of Mass Spectrometry*, 2001, 36: 782–90.

19. Villen J., Beausoleil S. A., Gygi S. P. Evaluation of the Utility of Neutral-Loss-Dependent Ms3 Strategies in Large-Scale Phosphorylation Analysis. *Proteomics*, 2008, 8: 4444–52.

20. Zubarev R. A., Kelleher N. L., McLafferty F. W. Electron Capture Dissociation of Multiply Charged Protein Cations. A Nonergodic Process. *Journal of the American Chemical Society*, 1998, 120: 3265–6.

21. Molina H., Horn D. M., Tang N., Mathivanan S., Pandey A. Global Proteomic Profiling of Phosphopeptides Using Electron Transfer Dissociation Tandem Mass Spectrometry. *Proceedings of the National Academy of Sciences of the United States of America*, 2007, 104: 2199–204.

22. Kjeldsen F., Giessing A. M. B., Ingrell C. R., Jensen O. N. Peptide Sequencing and Characterization of Post-Translational Modifications by Enhanced Ion-Charging and Liquid Chromatography Electron-Transfer Dissociation Tandem Mass Spectrometry. *Analytical Chemistry*, 2007, 79: 9243–52.

23. Rush J., Moritz A., Lee K. A., Guo A., Goss V. L., Spek E. J., Zhang H., Zha X. M., Polakiewicz R. D., Comb M. J. Immunoaffinity Profiling of Tyrosine Phosphorylation in Cancer Cells. *Nature Biotechnology*, 2005, 23: 94–101.

24. Oda Y., Nagasu T., Chait B. T. Enrichment Analysis of Phosphorylated Proteins as a Tool for Probing the Phosphoproteome. *Nature Biotechnology*, 2001, 19: 379–82.

25. Thaler F., Valsasina B., Baldi R., Jin X., Stewart A., Isacchi A., Kalisz H. M., Rusconi L. A New Approach to Phosphoserine and Phosphothreonine Analysis

in Peptides and Proteins: Chemical Modification, Enrichment Via Solid-Phase Reversible Binding, and Analysis by Mass Spectrometry. *Analytical and Bioanalytical Chemistry*, 2003, 376: 366–73.

26. Tseng H. C., Ovaa H., Wei N. J. C., Ploegh H., Tsai L. H. Phosphoproteomic Analysis with a Solid-Phase Capture-Release-Tag Approach. *Chemistry & Biology*, 2005, 12: 769–77.

27. Zhou H. L., Watts J. D., Aebersold R. A Systematic Approach to the Analysis of Protein Phosphorylation. *Nature Biotechnology*, 2001, 19: 375–8.

28. Bodenmiller B., Mueller L. N., Pedrioli P. G. A., Pflieger D., Juenger M. A., Eng J. K., Aebersold R., Tao W. A. An Integrated Chemical, Mass Spectrometric and Computational Strategy for (Quantitative) Phosphoproteomics: Application to Drosophila Melanogaster Kc167 Cells. *Molecular Biosystems*, 2007, 3: 275–86.

29. Tao W. A., Wollscheid B., O'Brien R., Eng J. K., Li X. J., Bodenmiller B., Watts J. D., Hood L., Aebersold R. Quantitative Phosphoproteome Analysis Using a Dendrimer Conjugation Chemistry and Tandem Mass Spectrometry. *Nature Methods*, 2005, 2: 591–8.

30. Lansdell T. A., Tepe J. J. Isolation of Phosphopeptides Using Solid Phase Enrichment. *Tetrahedron Letters*, 2004, 45: 91–3.

31. Knight Z. A., Schilling B., Row R. H., Kenski D. M., Gibson B. W., Shokat K. M. Phosphospecific Proteolysis for Mapping Sites of Protein Phosphorylation. *Nature Biotechnology*, 2003, 21: 1047–54.

32. Beausoleil S. A., Jedrychowski M., Schwartz D., Elias J. E., Villen J., Li J. X., Cohn M. A., Cantley L. C., Gygi S. P. Large-Scale Characterization of Hela Cell Nuclear Phosphoproteins. *Proceedings of the National Academy of Sciences of the United States of America*, 2004, 101: 12130–5.

33. Ballif B. A., Villen J., Beausoleil S. A., Schwartz D., Gygi S. P. Phosphoproteomic Analysis of the Developing Mouse Brain. *Molecular & Cellular Proteomics*, 2004, 3: 1093–101.

34. Sui S., Wang J., Lu Z., Cai Y., Zhang Y., Yu W., Qian X. Phosphopeptide Enrichment Strategy Based on Strong Cation Exchange Chromatography. *Chinese Journal of Chromatography*, 2008, 26: 195–9.

35. Trinidad J. C., Specht C. G., Thalhammer A., Schoepfer R., Burlingame A. L. Comprehensive Identification of Phosphorylation Sites in Postsynaptic Density Preparations. *Molecular & Cellular Proteomics*, 2006, 5: 914–22.

36. Olsen J. V., Blagoev B., Gnad F., Macek B., Kumar C., Mortensen P., Mann M. Global, in Vivo, and Site-Specific Phosphorylation Dynamics in Signaling Networks. *Cell*, 2006, 127: 635–48.

37. Han G., Ye M., Zhou H., Jiang X., Feng S., Jiang X., Tian R., Wan D., Zou H., Gu J. Large-Scale Phosphoproteome Analysis of Human Liver Tissue by

Enrichment and Fractionation of Phosphopeptides with Strong Anion Exchange Chromatography. *Proteomics*, 2008, 8: 1346–61.

38. Dai J., Jin W.-H., Sheng Q.-H., Shieh C.-H., Wu J.-R., Zeng R. Protein Phosphorylation and Expression Profiling by Yin-Yang Multidimensional Liquid Chromatography (Yin-Yang Mdlc) Mass Spectrometry. *Journal of Proteome Research*, 2007, 6: 250–62.

39. McNulty D. E., Annan R. S. Hydrophilic Interaction Chromatography Reduces the Complexity of the Phosphoproteome and Improves Global Phosphopeptide Isolation and Detection. *Molecular & Cellular Proteomics*, 2008, 7: 971–80.

40. Albuquerque C. P., Smolka M. B., Payne S. H., Bafna V., Eng J., Zhou H. A Multidimensional Chromatography Technology for in-Depth Phosphoproteome Analysis. *Molecular & Cellular Proteomics*, 2008, 7: 1389–96.

41. Gan C. S., Guo T., Zhang H., Lim S. K., Sze S. K. A Comparative Study of Electrostatic Repulsion-Hydrophilic Interaction Chromatography (Erlic) Versus Scx-Imac-Based Methods for Phosphopeptide Isolation/Enrichment. *Journal of Proteome Research*, 2008, 7: 4869–77.

42. Zhang H., Guo T., Li X., Datta A., Park J. E., Yang J., Lim S. K., Tam J. P., Sze S. K. Simultaneous Characterization of Glyco- and Phosphoproteomes of Mouse Brain Membrane Proteome with Electrostatic Repulsion Hydrophilic Interaction Chromatography. *Molecular & Cellular Proteomics*, 2010, 9: 635–47.

43. Porath J., Carlsson J., Olsson I., Belfrage G. Metal Chelate Affinity Chromatography, a New Approach to Protein Fractionation. *Nature*, 1975, 258: 598–9.

44. Andersson L., Porath J. Isolation of Phosphoproteins by Immobilized Metal (Fe-3+) Affinity-Chromatography. *Analytical Biochemistry*, 1986, 154: 250–4.

45. Ficarro S. B., Adelmant G., Tomar M. N., Zhang Y., Cheng V. J., Marto J. A. Magnetic Bead Processor for Rapid Evaluation and Optimization of Parameters for Phosphopeptide Enrichment. *Analytical Chemistry*, 2009, 81: 4566–75.

46. Zhou H., Xu S., Ye M., Feng S., Pan C., Jiang X., Li X., Han G., Fu Y., Zou H. Zirconium Phosphonate-Modified Porous Silicon for Highly Specific Capture of Phosphopeptides and MALSI-TOF MS Analysis. *Journal of Proteome Research*, 2006, 5: 2431–7.

47. Ficarro S. B., McCleland M. L., Stukenberg P. T., Burke D. J., Ross M. M., Shabanowitz J., Hunt D. F., White F. M. Phosphoproteome Analysis by Mass Spectrometry and Its Application to Saccharomyces Cerevisiae. *Nature Biotechnology*, 2002, 20: 301–5.

48. Ndassa Y. M., Orsi C., Marto J. A., Chen S., Ross M. M. Improved Immobilized Metal Affinity Chromatography for Large-Scale Phosphoproteomics Applications. *Journal of Proteome Research*, 2006, 5: 2789–99.

49. Moser K., White F. M. Phosphoproteomic Analysis of Rat Liver by High Capacity Imac and Lc-Ms/Ms. *Journal of Proteome Research*, 2006, 5: 98–104.

50. Bodenmiller B., Mueller L. N., Mueller M., Domon B., Aebersold R. Reproducible Isolation of Distinct, Overlapping Segments of the Phosphoproteome. *Nature Methods*, 2007, 4: 231–7.

51. Seeley E. H., Riggs L. D., Regnier F. E. Reduction of Non-Specific Binding in Ga(Iii) Immobilized Metal Affinity Chromatography for Phosphopeptides by Using Endoproteinase Glu-C as the Digestive Enzyme. *Journal of Chromatography B-Analytical Technologies in the Biomedical and Life Sciences*, 2005, 817: 81–8.

52. Pinkse M. W. H., Uitto P. M., Hilhorst M. J., Ooms B., Heck A. J. R. Selective Isolation at the Femtomole Level of Phosphopeptides from Proteolytic Digests Using 2D-NanoLC-ESI-MS/MS and Titanium Oxide Precolumns. *Analytical Chemistry*, 2004, 76: 3935–43.

53. Larsen M. R., Thingholm T. E., Jensen O. N., Roepstorff P., Jorgensen T. J. D. Highly Selective Enrichment of Phosphorylated Peptides from Peptide Mixtures Using Titanium Dioxide Microcolumns. *Molecular & Cellular Proteomics*, 2005, 4: 873–86.

54. Kweon H. K., Hakansson K. Selective Zirconium Dioxide-Based Enrichment of Phosphorylated Peptides for Mass Spectrometric Analysis. *Analytical Chemistry*, 2006, 78: 1743–9.

55. Li Y., Liu Y., Tang J., Lin H., Yao N., Shen X., Deng C., Yang P., Zhang X. $Fe_3O_4@Al_2O_3$ Magnetic Core-Shell Microspheres for Rapid and Highly Specific Capture of Phosphopeptides with Mass Spectrometry Analysis. *Journal of Chromatography A*, 2007, 1172: 57–71.

56. Ficarro S. B., Parikh J. R., Blank N. C., Marto J. A. Niobium(V) Oxide (Nb_2O_5): Application to Phosphoproteomics. *Analytical Chemistry*, 2008, 80: 4606–13.

57. Thingholm T. E., Jensen O. N., Robinson P. J., Larsen M. R. SIMAC (Sequential Elution from IMAC), a Phosphoproteomics Strategy for the Rapid Separation of Monophosphorylated from Multiply Phosphorylated Peptides. *Molecular & Cellular Proteomics*, 2008, 7: 661–71.

58. Reynolds E. C., Riley P. F., Adamson N. J. A Selective Precipitation Purification Procedure for Multiple Phosphoseryl-Containing Peptides and Methods for Their Identification. *Analytical Biochemistry*, 1994, 217: 277–84.

59. Zhang X., Ye J., Jensen O. N., Roepstorff P. Highly Efficient Phosphopeptide Enrichment by Calcium Phosphate Precipitation Combined with Subsequent IMAC Enrichment. *Molecular & Cellular Proteomics*, 2007, 6: 2032–42.

60. Chang C.-K., Wu C.-C., Wang Y.-S., Chang H.-C. Selective Extraction and Enrichment of Multiphosphorylated Peptides Using Polyarginine-Coated Diamond Nanoparticles. *Analytical Chemistry*, 2008, 80: 3791–7.

61. Woods A. S., Ferre S. Amazing Stability of the Arginine-Phosphate Electrostatic Interaction. *Journal of Proteome Research*, 2005, 4: 1397–402.

62. Yang Z. P., Hancock W. S., Chew T. R., Bonilla L. A Study of Glycoproteins in Human Serum and Plasma Reference Standards (Hupo) Using Multilectin Affinity Chromatography Coupled with RPLC-MS/MS. *Proteomics*, 2005, 5: 3353–66.

63. Bunkenborg J., Pilch B. J., Podtelejnikov A. V., Wisniewski J. R. Screening for N-Glycosylated Proteins by Liquid Chromatography Mass Spectrometry. *Proteomics*, 2004, 4: 454–65.

64. Zhou H., Hou W., Denis N. J., Zhou H., Vasilescu J., Zou H., Figeys D. Glycoproteomic Reactor for Human Plasma. *Journal of Proteome Research*, 2009, 8: 556–66.

65. Feng S., Yang N., Pennathur S., Goodison S., Lubman D. M. Enrichment of Glycoproteins Using Nanoscale Chelating Concanavalin a Monolithic Capillary Chromatography. *Analytical Chemistry*, 2009, 81: 3776–83.

66. Morris T. A., Peterson A. W., Tarlov M. J. Selective Binding of Rnase B Glycoforms by Polydopamine-Immobilized Concanavalin A. *Analytical Chemistry*, 2009, 81: 5413–20.

67. Zhang Q., Tang N., Brock J. W. C., Mottaz H. M., Ames J. M., Baynes J. W., Smith R. D., Metz T. O. Enrichment and Analysis of Nonenzymatically Glycated Peptides: Boronate Affinity Chromatography Coupled with Electron-Transfer Dissociation Mass Spectrometry. *Journal of Proteome Research*, 2007, 6: 2323–30.

68. Zhang L., Xu Y., Yao H., Xie L., Yao J., Lu H., Yang P. Boronic Acid Functionalized Core-Satellite Composite Nanoparticles for Advanced Enrichment of Glycopeptides and Glycoproteins. *Chemistry-a European Journal*, 2009, 15: 10158–66.

69. Xu Y., Wu Z., Zhang L., Lu H., Yang P., Webley P. A., Zhao D. Highly Specific Enrichment of Glycopeptides Using Boronic Acid-Functionalized Mesoporous Silica. *Analytical Chemistry*, 2009, 81: 503–8.

70. Li M., Lin N., Huang Z., Du L., Altier C., Fang H., Wang B. Selecting Aptamers for a Glycoprotein through the Incorporation of the Boronic Acid Moiety. *Journal of the American Chemical Society*, 2008, 130: 12636–8.

71. Hagglund P., Bunkenborg J., Elortza F., Jensen O. N., Roepstorff P. A New Strategy for Identification of N-Glycosylated Proteins and Unambiguous Assignment of Their Glycosylation Sites Using Hilic Enrichment and Partial Deglycosylation. *Journal of Proteome Research*, 2004, 3: 556–66.

72. Wada Y., Tajiri M., Yoshida S. Hydrophilic Affinity Isolation and Maldi Multiple-Stage Tandem Mass Spectrometry of Glycopeptides for Glycoproteomics. *Analytical Chemistry*, 2004, 76: 6560–5.

73. Picariello G., Ferranti P., Mamone G., Roepstorff P., Addeo F. Identification of N-Linked Glycoproteins in Human Milk by Hydrophilic Interaction Liquid Chromatography and Mass Spectrometry. *Proteomics*, 2008, 8: 3833–47.

74. Whelan S. A., Lu M., He J., Yan W., Saxton R. E., Faull K. F., Whitelegge J. P., Chang H. R. Mass Spectrometry (LC-MS/MS) Site-Mapping of N-Glycosylated Membrane Proteins for Breast Cancer Biomarkers. *Journal of Proteome Research*, 2009, 8: 4151–60.

75. Blake T. A., Williams T. L., Pirkle J. L., Barr J. R. Targeted N-Linked Glycosylation Analysis of H5N1 Influenza Hemagglutinin by Selective Sample Preparation and Liquid Chromatography/Tandem Mass Spectrometry. *Analytical Chemistry*, 2009, 81: 3109–18.

76. Zhou L., Beuerman R. W., Chew A. P., Koh S. K., Cafaro T. A., Urrets-Zavalia E. A., Urrets-Zavalia J. A., Li S. F. Y., Serra H. M. Quantitative Analysis of N-Linked Glycoproteins in Tear Fluid of Climatic Droplet Keratopathy by Glycopeptide Capture and ITRAQ. *Journal of Proteome Research*, 2009, 8: 1992–2003.

77. Chen R., Jiang X., Sun D., Han G., Wang F., Ye M., Wang L., Zou H. Glycoproteomics Analysis of Human Liver Tissue by Combination of Multiple Enzyme Digestion and Hydrazide Chemistry. *Journal of Proteome Research*, 2009, 8: 651–61.

78. Wells L., Vosseller K., Cole R. N., Cronshaw J. M., Matunis M. J., Hart G. W. Mapping Sites of O-Glcnac Modification Using Affinity Tags for Serine and Threonine Post-Translational Modifications. *Molecular & Cellular Proteomics*, 2002, 1: 791–804.

79. Zheng Y., Guo Z., Cai Z. Combination of Beta-Elimination and Liquid Chromatography/Quadrupole Time-of-Flight Mass Spectrometry for the Determination of O-Glycosylation Sites. *Talanta*, 2009, 78: 358–63.

80. Hanisch F.-G., Teitz S., Schwientek T., Mueller S. Chemical De-O-Glycosylation of Glycoproteins for Application in LC-Based Proteomics. *Proteomics*, 2009, 9: 710–9.

81. Chen Y., Yan G., Zhou X., Yang P. Combination of Matrix-Assisted Laser Desorption Ionization and Electrospray Ionization Mass Spectrometry for the Analysis of Intact Glycopeptides from Horseradish Peroxidase. *Chinese Journal of Chromatography*, 2010, 28: 135–9.

82. Petricoin E. F., Belluco C., Araujo R. P., Liotta L. A. The Blood Peptidome: A Higher Dimension of Information Content for Cancer Biomarker Discovery. *Nature Reviews Cancer*, 2006, 6: 961–7.

83. Fabre B., Combier J. P., Plaza S. Recent Advances in Mass Spectrometry–based Peptidomics Workflows to Identify Short-open-readingframe-encoded

Peptides and Explore Their Functions. *Current Opinion in Chemical Biology*, 2021, 60: 122–130.

84. Zheng X., Baker H., Hancock W. S. Analysis of the Low Molecular Weight Serum Peptidome Using Ultrafiltration and a Hybrid Ion Trap-Fourier Transform Mass Spectrometer. *Journal of Chromatography A*, 2006, 1120: 173–84.

85. Bilkova Z., Slovakova M., Minc N., Futterer C., Cecal R., Horak D., Benes M., le Potier I., Krenkova J., Przybylski M., Viovy J.-L. Functionalized Magnetic Micro- and Nanoparticles: Optimization and Application to Mu-Chip Tryptic Digestion. *Electrophoresis*, 2006, 27: 1811–24.

86. Chen W.-Y., Chen Y.-C. Acceleration of Microwave-Assisted Enzymatic Digestion Reactions by Magnetite Beads. *Analytical Chemistry*, 2007, 79: 2394–401.

87. Lin S., Yao G., Qi D., Li Y., Deng C., Yang P., Zhang X. Fast and Efficient Proteolysis by Microwave-Assisted Protein Digestion Using Trypsin-Immobilized Magnetic Silica Microspheres. *Analytical Chemistry*, 2008, 80: 3655–65.

88. Chen H., Xu X., Yao N., Deng C., Yang P., Zhang X. Facile Synthesis of C-8-Functionalized Magnetic Silica Microspheres for Enrichment of Low-Concentration Peptides for Direct MALDI-TOF MS Analysis. *Proteomics*, 2008, 8: 2778–84.

Chapter 2

Magnetic Micro-/Nanomaterial: Synthesis and Characterization

2.1. Introduction of magnetic nanomaterial

The research on magnetism of matter has a long history. The invention of compass more than 4,000 years ago marked that magnetism had entered into people's lives. The advancement of modern science and technology, as well as several investigations and practices during the past decades, eloquently indicate that magnetism as a common property of all matter leads to the ubiquity of the magnetic phenomenon, and that magnetic field is a kind of physical field which is present in any space. Metal magnetism reigned supreme before 1950s, following that was the golden age of ferrite which dominated absolute advantage in different application areas except electric power industry, while magnetic nanomaterials began to be produced, developed, and expanded gradually since the 1980s. The progress of nanotechnology brings major opportunities and challenges, leapfrogging the development of traditional magnetic industry, rejuvenating old magnetics and thereby exhibiting attractive and widespread prospects, as well as making it one of the most striking and energetic research areas in the current world.

2.1.1. *Magnetic nanomaterial*

Magnetic material refers to that with available magnetic property.[1] Nanomaterial refers to that with at least one dimension in the 3D space belonging to the nanoscale, or that consists of structural units at the nanoscale. Therefore, magnetic nanomaterial can be classified into three categories according to the number of dimensions: (i) zero-dimension, in which the three spacial dimensions of the material are all at the nanoscale, such as a magnetic nanoparticle; (ii) one-dimension, in which two spacial dimensions of the material are at the nanoscale, such as magnetic nanowire, nanorod, etc.; and (iii) two-dimension, in which only one spacial dimension is at the nanoscale, such as magnetic ultrathin membrane, multi-layer film, etc.

Magnetic materials can be divided into several categories according to their functions: (i) magnetic core material, which is easily magnetized by the external magnetic field; (ii) permanent magnetic material, which is capable of producing a sustained magnetic field; (iii) magnetic recording material, which is capable of conducting an information note by changing the direction of magnetization; (iv) photomagnetic recording material, which is capable of taking a note and regeneration by using light or heat to change magnetization; (v) magneto-resistance material, the resistance of which will change under the action of magnetic field; (vi) magneto-striction material, the size of which will change due to magnetization; (vii) magnetic fluid, the shape of which can change freely. By taking advantage of the above functions, a variety of magnetic devices can be prepared, such as transformer, motor, loudspeaker, magneto-strictive oscillator, recording medium, sensor, damper, printer, magnetic field generator, electromagnetic absorber, etc. Furthermore, the above devices can be used to fabricate various facilities, including those industrial machineries such as robots, computers, and mother machines, besides the endless list of application machines, such as automobiles, sound equipment, televisions, video/audio recorders, telephones, washing machines, dust collectors, electronic clocks, refrigerators, air-conditioners, electric cookers, etc.

In recent years, new materials with amorphous state, permanent magnet of rare earth, ultra-magneto-striction, and giant magneto-resistance, etc., have been successively discovered, meanwhile, the performance of magnetic materials has also been improved significantly owing to the development of new technologies such as micronization, crystallographic azimuth control, thin-film, superlattice, etc. Currently, magnetic materials have become one of the key materials for supporting and promoting social development, and have been the significant pillar and foundation for the national economy and national defence industry, as well as an integral part of different areas such as information storage, processing, and transmission. The general development trend of information technology has been the pursuit of small, light, thin, and multi-functioning, resulting in the requirement that the progress direction of magnetic materials be after high performance and new functions. Magnetic materials have gone through the development stages of crystalline, amorphous, nanocrystalline, nanoparticle, and nanostructure. Nanomaterial is the most important and fundamental component of nanotechnology. When the characteristic size of material is at the nanoscale (1–100 nm), the electron fluctuation and the interaction between atoms will show strong size-dependent property, therefore, many properties of nanomaterials, such as melting point, mechanical property, magnetic property, electrical property, thermal property, optical, chemical property, etc., are obviously distinct from those of conventional materials.

2.1.2. *Magnetism resource of magnetic nanomaterial*

The magnetism of the matter comes from the magnetism of its constituent atoms, the orbital motion of an electron in an atom is equivalent to a closing current, which produces magnetic moment. Besides, the spin of an electron also produces spin magnetic moment. Therefore, the magnetism of a material mainly derives from the orbital magnetic moment and spin magnetic moment. For the multi-electron atom, magnetism is provided by the orbital magnetic moment and spin magnetic moment of electrons in an incomplete shell, since the sum of the orbital magnetic moment and spin magnetic moment of electron is zero when the shell is complete, which makes no contribution to atom magnetic moment. For instance, atom magnetic moment of iron family element including *Fe, Ni,* and *Co* and so on results from 3D electron in incomplete shell. In fact, magnetism of the matter can be divided into five categories according to the arrangement of magnetic dipole within the matter in the presence or absence of a magnetic field: diamagnetism, paramagnetism, ferromagnetism, ferrimagnetism, and anti-ferromagnetism.[1] For diamagnetic material, there is no magnetic dipole in the absence of magnetic field, while there is a very weak induced magnetic dipole in the presence of magnetic field. The magnetization direction of diamagnetic material is opposite to that of the external field. For paramagnetic material, with an external magnetic field, the disordered magnetic dipole in the absence of magnetic field will be arranged in order along the direction of the external magnetic field, in other words, the magnetization direction of paramagnetic material is consistent with the external field. For ferromagnetic material, whether there is an external magnetic field or not, magnetic dipole will be arranged in the same direction and displays long-term orderly arrangement, therefore, ferromagnetic material shows the permanent magnetic moment at the macro level. For ferromagnetic material, each magnetic dipole always has an adjacent one which possesses opposite magnetic moment orientation and relatively weak magnetism. For anti-ferromagnetic material, the two adjacent magnetic dipoles own opposite orientations but equal size, resulting in reciprocal offset.

2.1.3. *Main magnetic characteristic of magnetic nanomaterial*

2.1.3.1. *Single magnetic domain structure*

Multiple magnetic domain structures exist in large magnetic particles.[1] The balance between magneto-static energy (ΔE_{MS}) and domain wall energy (ΔE_{dw}) leads to the formation of domain wall, where the magneto-static energy is proportional to the volume of the matter, and the domain wall energy is proportional to the interface between the walls. The domain wall energy will be greater than the magneto-static energy of single domain particle when the size of the matter is low to a certain critical dimension, which leads to the appearance of the single magnetic domain particle.

2.1.3.2. *Superparamagnetism*

Presumably, a magnetic material is regarded as an aggregation of single magnetic domain particles. For each particle, magnetic moments are aligned in parallel, the direction of which is in line with that of easy magnetization due to the strong exchange effect between magnetic atoms or ions. But the directions of magnetic moments between particles are diverse due to their different directions of easy magnetization. Since the total magneto-crystalline anisotropy energy and thermal perturbation energy are, respectively, in direct proportion to $K_{eff}V$ and $k_B T$ (K_{eff} is magneto-crystalline anisotropy constant, V is the particle volume, k_B is boltzmann constant, T is the absolute temperature of sample), the direction of magnetic moment within particles may keep parallel integrally and change repeatedly among different directions of easy magnetization, therefore, when particle size is further reduced to a certain value, the thermal perturbation energy will be equivalent to the total magneto-crystalline anisotropy energy. From the aggregation of single domain particle, it can be seen that the magnetic moment orientation of different particles is changing all the time, which is similar to what occurs in the paramagnetic materials. One of the differences between them is, for paramagnetic material, the magnetic moment of each atom or ion has only a few Bohr magnetons in spherical particle aggregation with 5 nm of diameter, the total magnetic moment of which may be more than 10,000 Bohr magnetons. This kind of magnetic property is called superparamagnetism, which has two significant characteristics: firstly, taking magnetization intensity M as vertical coordinate and H/T as horizontal coordinate (H is the intensity of the applied magnetic field, and T is the absolute temperature), within the certain temperature range where the aggregation of single domain particle shows superparamagnetism, the measured magnetization curves must be coincident; secondly, no magnetic hysteresis occurs, namely, the remanence and coercivity of the aggregation are nought.

There are two physical quantities for describing the aggregation of single domain magnetic particle. One is superparamagnetic critical dimension, r_0. Only when particle size is less than or equal to r_0 ($r \leq r_0$) can superparamagnetism appear on the premise of remaining constant temperature of a particle system. The superparamagnetic critical dimension of various kinds of magnetic nanoparticles is not the same. For instance, the superparamagnetic critical dimensions of α-Fe, Fe_3O_4 and α-Fe_2O_3 are 5 nm, 30 nm, and 20 nm, respectively. The other one is cut-off temperature, T_B. There is generally a characteristic temperature T_B for magnetic nanoparticles with sufficiently small size. When T is less than T_B ($T < T_B$), particles exhibit strong magnetism (ferromagnetism or ferrimagnetism); when T is more than or equal to T_B ($T \geq T_B$), particles exhibit superparamagnetism.

2.1.3.3. *Coercivity*

The coercivity of magnetic nanoparticle is related to its size. Large-size magnetic particles tend to form multi-domain structures, so nucleation and domain wall motion lead to magnetization reversal. Along with the decrease of particle size, the magnetic domain structure of strong magnetic particles will change from multiple to single, so that the magnetic reversal pattern will change from domain wall displacement to magnetic domain rotation, thereby showing the trend that coercivity increases significantly until it reaches a maximum value at the critical dimension of single magnetic domain (r_c). Then coercivity decreases with the further decrease of the size of magnetic particles, and when particle size decreases to the critical dimension of super-paramagnetism (r_0), coercivity will approach zero.

2.1.3.4. *Curie temperature*

Curie temperature (T_c), as one of the significant parameters of material magnetism, is usually proportional to the exchange integral (J). For those magnetic nanoparticles, the existence of small size and surface effects generally leads to the change of their intrinsic magnetism, and thereby possessing a moderately low T_c. For instance, the T_c of Ni nanoparticle with 85 nm of diameter is measured as around 350°C, which is slightly lower than block Ni $(T_c \approx 385°C)$. In addition, its surface effect becomes more obvious along with the decrease in the size of magnetic nanoparticles, that is, the lesser the coordination number of the internal atoms compared to that of atoms on the surface of the nanoparticles, the more remarkable the decrease in the average coordination number of atoms in the nanoparticle, which is accompanied by the decrease of J_e with dependencies on the number of nearest atoms, finally resulting in the decrease of T_c.

2.1.3.5. *Magnetizability*

Magnetizability refers to the magnetization intensity of a matter induced by the external magnetic field, H. The magnetism of nanoparticle is closely related to the parity of the total number of electrons it contains. The electrons in each particle can be regarded as a system, and the parity of the number of electrons can be either odd or even. The magnetism of a particle which contains odd or even electrons exhibits different temperature feature.

2.1.3.6. *Magnetic parameters of magnetic nanomaterial*

For ferromagnetic or anti-ferromagnetic material, its magnetism can be represented by the basic parameters on the magnetic curve.[2] Figure 2.1 shows a typical magnetic curve which presents the change relation between magnetization intensity of sample and external magnetic field. *OB* represents

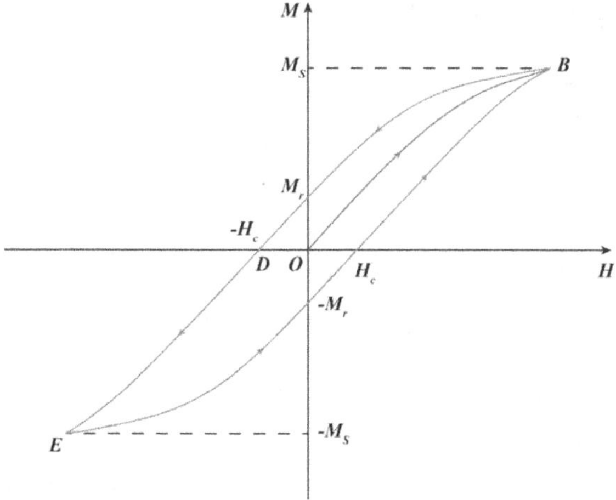

Figure 2.1: Schematic diagram of typical magnetic hysteresis loop.

the increase in the magnetization intensity of samples continuously along with the strengthening of external magnetic field for the unmagnetized sample. And the magnetization intensity of sample achieves saturation when the intensity of external magnetic field attains point B, which is called saturation magnetic value (M_s). If the external magnetic field is decreased at this moment, the magnetization intensity does not reversibly go down along the original magnetic curve, but varies along the curve BD in Figure 2.1. The magnetization intensity that does not disappear when the intensity of external magnetic field decreases to nought is called residual magnetization (M_r). Only when the external magnetic field extends to point H_c (H_c is called coercivity) in the opposite direction, can the magnetization intensity become nought. The magnetization intensity achieves reversed saturation if further extending to point E in the opposite direction, from where the external magnetic field increases to point B, the curve will be seen as in Figure 2.1. The particular curve is called magnetic hysteresis loop. Actually, the magnetic hysteresis is produced by the difficulty of restoring the magnetic domain to its original shape when removing the external magnetic field.

As seen above, many magnetic materials show mesoscopic magnetism, which is different from the common material when the grain size is reduced to nanoscale, the main feature of which includes quantum size effect, macroscopic quantum tunneling effect, small-size effect of magnetic ordered particle, specifically apparent magnetic property and even magnetic phase transition at some time, etc.

2.1.4. *Ferrite nanomaterial*

Ferrite is a typical and most widely used magnetic nanomaterial, the ferrimagnetism of which derives from two inside opposing magnetic moments without canceling out each other. The detailed explanation is as follows: there are two or more kinds of cations in ferrite, these ions have different magnetic moments (some of them are not magnetic at all), plus the number of ions occupying A or B positions is different, so material remains part of magnetism, which is called ferrimagnetism due to the offset effect caused by the anti-parallel orientation of the magnetic moments.

Generally speaking, the size of the material affects its magnetism significantly. When the particle size of a ferromagnetic or ferrimagnetic material is reduced to a sort of critical value (r_c), the magnetic particle will change from multi-domain to single-domain status. If the particle size continues to decrease to r_0, the magnetic anisotropy energy will decrease to be comparative to the energy of thermal motion, moreover, the magnetic dipole and magnetization direction will vary randomly, the small particle therefore does not have permanent magnetic moment without external magnetic field, this feature is called superparamagnetism. The hysteresis loop of a superparamagnetic particle shows no hysteresis. It is worth mentioning that the critical sizes of different types of magnetic materials are diverse. For example, the critical size of α-Fe is 16 nm, while for Fe_3O_4 it is 20 nm.[3] Superparamagnetism can prevent particles from accumulating in solution caused by magnetic interaction, resulting in wide application of magnetic particles in the field of bio-medicine. Also, ferrite particle of spinel type can exhibit superparamagnetism when it decreases to a certain size. According to Herndon's report, MFe_2O_4 (M = Co, Ni, Zn) particle with size of 30 nm can show superparamagnetism.[4] Besides, the magnetism of a ferrite particle is also affected by other factors such as composition and morphology. Among the ferrite materials, Fe_3O_4 nanoparticle is most widely used due to its diverse synthetic methods, low toxicity, and small side effects on the human body, high biological safety, and superparamagnetism.

2.1.4.1. *The structure of Fe_3O_4*

Fe_3O_4 is a kind of ionic crystal consisting of Fe^{2+}, Fe^{3+}, and O^{2-} via ionic bonds.[5] The black magnetite of Fe_3O_4 is an anti-spinel structure and a cubic crystal system, in which the arrangement mode of ions is similar to the configuration of spinel. For crystallographic protocell of Fe_3O_4, a crystal cell contains 56 ions, which include 24 metal ions and 32 O^{2-} ions. Since O^{2-} ions are relatively larger than the metal ions, the arrangement of O^{2-} ions forms a close-packed structure, resulting in the formation of many voids. And metal ions fill these voids. As seen in Figure 2.2, the close-packed face-centered cubic lattice formed by O^{2-} ions contains relatively smaller tetrahedral voids (t) and relatively larger octahedral voids (o), both of which are occupied by Fe ions. Among much else, 8 of Fe^{2+} ions and 8 of Fe^{3+} ions occupy the octahedron voids, 8 of Fe^{3+} ions occupy the tetrahedral voids. However, since the arrangement order of these two ions is messy and electrons can move between them, Fe_3O_4 is not a simple mixture of FeO and Fe_2O_3 but an oxide with mixed valence. A part of the crystal cell is shown in Figure 2.3.

In Fe_3O_4, electrons are easily transferred from one atom with valence state to another owing to the existence of mixed valence of Fe element, leading to inter-valency transition and inter-valency absorption. The absorption band produced by inter-valency absorption between Fe^{2+} and Fe^{3+} is in the visible light region. Therefore, all visible lights are absorbed so that Fe_3O_4 appears black. In the structure of Fe_3O_4, the octahedral units of $[FeO_6]$ of edge-shared connection form infinite long chains (as shown in Figure 2.4), in which electrons are easy to transfer due to the alternate arrangement of Fe^{2+} and Fe^{3+}, that's why the Fe_3O_4 possesses high electrical conductivity and strong

Figure 2.2: Schematic diagram of tetrahedron (t) and octahedron (o) voids in a dense double layer of Fe_3O_4.

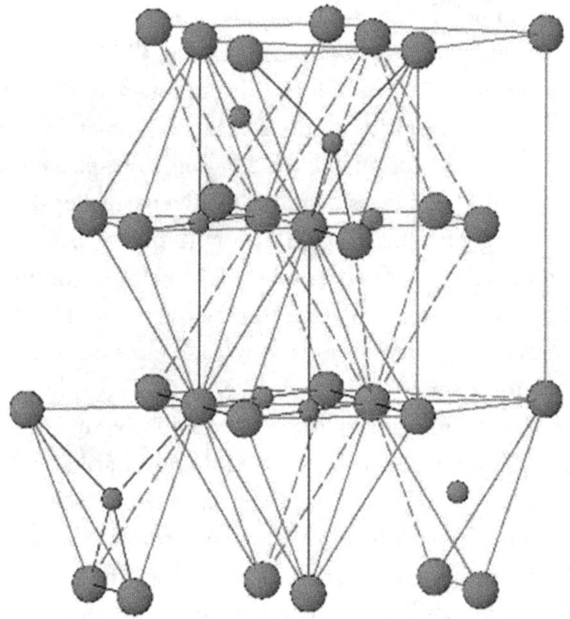

Figure 2.3: Schematic diagram of a part of Fe_3O_4 crystal cell.

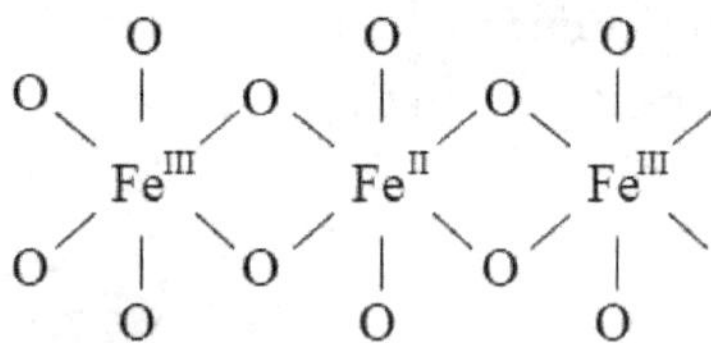

Figure 2.4: Schematic diagram of chain-like structure of Fe_3O_4.

magnetism. Fe_3O_4 is a semimetal with 0.3 eV of small band gap (E_g), and the saturation magnetic value (M_s) of which reaches 92 emu·g^{-1} at room temperature.

2.1.4.2. *Synthetic methods of Fe$_3$O$_4$*

The synthesis of Fe_3O_4 nanoparticles has been widely studied in recent decades.[5] The reported synthetic methods include coprecipitation method, high-temperature decomposition method, microemulsion method, water (solvent) thermal method, sol–gel method, ultrasonic chemical method, and so on.

(1) Coprecipitation method

Coprecipitation that is characterized by simplicity is the most commonly used method at present, the principle of which can be expressed in equation (2.1).

$$Fe^{2+} + 2\,Fe^{3+} + 8\,OH^- \rightarrow Fe_3O_4 + 4\,H_2O \qquad (2.1)$$

In general, ferric salt and ferrous salt are mixed on the basis of the ratio of 2:1 (or more) under the protection of inert gas, then the excessive NH_4OH or NaOH is added under a certain temperature to adjust the pH value to be in the range of 8–14. The coprecipitation reaction is carried out with high-speed stirring. The Fe_3O_4 powder can be obtained through a series of steps including washing, filtering, and drying after the coprecipitation reaction. The size and shape of Fe_3O_4 nanoparticles are related to the type of used ferric salts (such as chloride, sulfate, or nitrate), the ratio of Fe^{2+}/Fe^{3+}, reaction temperature, pH value, and ionic strength of the solution. The mass of the resultant Fe_3O_4 nanoparticles is repeatable by using the same synthetic conditions. The M_s of Fe_3O_4 nanoparticles synthesized by coprecipitation method is generally smaller than that of bulk materials (92 $emu \cdot g^{-1}$). The biggest problem of coprecipitation method is to acquire monodisperse Fe_3O_4 nanoparticles. In order to solve this problem, many researchers try to add surfactant to cover the surface of Fe_3O_4 to reduce the agglomeration of nanoparticles. By contrast, the addition of organic additives as stabilizers or reductants before the formation of Fe_3O_4 nanoparticles is a better improvement method for coprecipitation, which can modify Fe_3O_4 nanoparticles on their surface at the same time with their nucleation and growth. With this improved coprecipitation method, the resultant Fe_3O_4 nanoparticles can not only disperse evenly in solution, but also be stable for a long time. The control of particle size of Fe_3O_4 nanoparticles by organic additive ions is achieved through two aspects of competition. On the one hand, organic ions and ferric ions can form complexes to prevent the formation of nuclei, which is conducive to the growth of nanoparticles, and thereby obtaining relatively large size of nanoparticles. On the other hand, the adsorption of organic additives on the surface can prevent the further growth of nanoparticles, which is conducive to the synthesis of nanoparticles with relatively small size. The competition between the two aspects can lead to the dynamic balance so as to regulate the size distribution of Fe_3O_4 nanoparticles. In short, the selection of suitable organic additives is very important to control the size of Fe_3O_4 nanoparticles.

(2) High-temperature decomposition method

High-temperature decomposition method is to heat and decompose organic metal compounds, such as $Fe(Cup)_3$, $Fe(CO)_5$, $Fe(acac)_3$ and so forth, in high boiling organic solvents to obtain Fe_3O_4 nanoparticles which have high monodispersity and narrow size distribution. The stable dispersity of resultant nanoparticles in solution is attributed to the adsorbed surfactant on the surface of Fe_3O_4 during the synthetic process. The size and morphology of Fe_3O_4 nanoparticles can be controlled by changing reaction time, temperature, concentration of reactant, properties of solvent, etc. Compared to coprecipitation method, high-temperature decomposition method has obvious advantages, for instance, the obtained superparamagnetic nanoparticles have better monodispersity, narrower size distribution, higher crystallinity, higher M_s, and better steerability of composition. The main reasons are as follows: firstly, high reaction temperature (200–350°C) is adopted for high-temperature decomposition method, which is conducive to control of nucleation and growth of magnetic nanoparticles; secondly, non-aqueous solvent is used to avoid coordination reaction between water and ferric ions. In addition, the high-temperature decomposition method has many disadvantages, such as high cost for synthesis, low yield, safety risks caused by high temperature, and the difficulty to achieve industrialized production.

(3) Microemulsion method

Microemulsion method, also known as reverse-phase micelle method, is an isotropic thermodynamic stable system composed of two mutually dissoluble liquids and surfactants. Interfacial films of surfactants can stabilize microemulsion beads in one or two phases. In the oil-in-water microemulsion system, the oil phase is a continuous phase, and the surfactant separates the water phase into microemulsion beads of the size of 1–50 nm. In this sense, microemulsion beads can provide a nanoreactor for the formation of nanoparticles. So the agglomeration of nanoparticles can be effectively avoided, and the resultant magnetic nanoparticles have narrow size distribution and good dispersity. Moreover, most of the nanoparticles present spherical morphology that is similar to the shape of the nanoreactor. The monodisperse Fe_3O_4 nanoparticles with different sizes can be acquired by easily adjusting the size of microemulsion beads, which depends on the molar ratio of water and surfactant. Although microemulsion method has certain advantages in terms of controlling the size distribution and morphology of nanoparticles, its shortcomings are obvious, such as the narrow range of controlled synthesis conditions, the relatively low output of nanoparticles, large consumption of raw materials including oil phase, surfactant and assistant surfactant, long production cycle, troublesome product postprocessing as well as high cost of industrialized production.

(4) Hydrothermal/solvothermal method

Hydrothermal/solvothermal method is to make insoluble substances dissolve, react and recrystallize, using aqueous solution (or organic solvent) as reaction medium, to obtain the ideal product in tailor-made hermetic reaction vessel (autoclave) under the condition of high temperature and high pressure. All of the reaction conditions such as solvent, reaction temperature, and reaction time in the synthetic process of Fe_3O_4 nanoparticles by hydrothermal/solvothermal method have great influence on the synthesis reaction. In general, the size of Fe_3O_4 nanoparticles increases along with the extension of reaction time.

Hydrothermal/solvothermal method has two advantages: firstly, the relatively high temperature (130–250°C) is conducive to the improvement of magnetic properties; secondly, hydrothermal/solvothermal method is carried out in a hermetic reaction vessel, thereby causing relatively high pressure (0.3–4 MPa), which avoids the volatilization of components and is conducive to improving purity of products and protecting the environment. The disadvantage is that the equipment requirements are high because the reaction conditions need high temperature and pressure.

(5) Sol–gel method

Sol–gel method is a widely used wet-chemical method for the synthesis of nanometal oxides. Usually, the Fe^{2+} and Fe^{3+} solutions with molar ratio of 1:2 are firstly mixed, then a certain amount of organic acids is added to adjust the pH value to obtain nanoparticle sol of ferric oxides, and then the solvent is evaporated slowly to form a 3D-netty gel, the product finally is obtained by removing organic residues via high-temperature. During the process of high-temperature treatment, the escape of organic groups leaves a large number of interconnected micropores in the material, which is beneficial to improve the specific surface area.

The main process of preparing iron oxide nanoparticles by sol–gel method is to add a certain amount of crystal growth enhancer of OH^- ion into the solution containing Fe^{3+} to prepare $Fe(OH)_3$ suspension which is then heated up, filtered, washed with deionized water several times, and finally dried to produce α-Fe_2O_3 nanoparticles. The nanoparticles prepared by sol–gel method have the characteristics of high purity, good chemical uniformity, small size, narrow size distribution, and low reaction temperature. However, sol–gel method also has disadvantages such as poor sintering property, large drying shrinkage property, and long preparation cycle.

(6) Ultrasonic chemical method

Fe_3O_4 nanoparticles can also be synthesized by ultrasonic chemistry method. Ultrasonic chemistry method is to convert ferric salt into Fe_3O_4 nanoparticles under the extreme conditions of high temperature, high pressure, and extremely high cooling rate and so on, which are generated by the cavitation of ultrasonic wave.

Apart from the above-mentioned methods, superparamagnetic nanoparticles can also be prepared by microwave heating method. In general, the synthesis and application of nanoparticles should be considered together to maximize the benefits, as it does not make much industrial significance to consider the synthesis method alone.

2.1.5. *Magnetic micro-/nanomaterial for proteomic separation and analysis*

There are many kinds of magnetic micro-/nanomaterials with different properties for proteomics separation and analysis. According to surface properties of the material, they can be divided into hydrophilic surface, hydrophobic surface, metal affinity surface, mesoporous surface, and so on. According to the purpose, they can be divided into magnetic material for enzymatic hydrolysis and magnetic material for enrichment. Among much else the magnetic material for enzymatic hydrolysis can be divided into adsorption type, covalent type, ion chelation type, etc.; the magnetic material for enrichment can be divided into low abundance enrichment, phosphoprotein enrichment, glycoprotein enrichment, endogenous peptide enrichment, etc., in accordance with the different enrichment targets.

In this chapter, the synthesis of magnetic nanoparticles in proteomics separation and analysis will be described according to their structures and properties.

2.2. Synthesis of unmodified Fe_3O_4 nanoparticle

Fe_3O_4 nanoparticles are most widely used in proteomics analysis because of diverse synthetic methods, low toxicity, and small side effects on human body, high biological safety, and superparamagnetism at the nanometer scale.[6] Fe_3O_4 nanoparticles are also the source of magnetism for most of the magnetic micro-/nanomaterials described in this book. This section describes the detailed synthesis process of Fe_3O_4 nanoparticles.

Hydrothermal method is commonly adopted for the synthesis of Fe_3O_4 nanoparticles. Firstly, 2.70 g $FeCl_3 \cdot 6H_2O$ is dissolved in 100 mL ethylene glycol, and the yellow transparent solution can be obtained after magnetic stirring for 0.5 h. Then 7.20 g anhydrous NaAc is added into the aforementioned yellow transparent solution, and kept under continuous magnetic stirring for another 0.5 h. Next, the resulting solution is transferred to a 200-mL Teflon-lined stainless steel reactor, which is further placed in an oven at 200°C for 8 h, 16 h, and 24 h, respectively. After the reaction is finished, the Teflon-lined stainless-steel reactor is cooled to room temperature. Finally, the obtained materials are washed with ethanol 3 times (3 × 30 mL) and ionized water 5 times (2 × 30 mL) to remove water-soluble impurities such as sodium acetate and ethylene glycol, for vacuum drying at 60°C for 12 h. The size of Fe_3O_4 nanoparticles could be adjusted by changing the amount of $FeCl_3 \cdot 6H_2O$, the reaction temperature, and time.

Figure 2.5 shows the transmission electron microscope (TEM) image of Fe_3O_4 nanoparticles, showing that the average size of Fe_3O_4 nanoparticles is about 250 nm. Figure 2.6 is Fourier transform infrared spectroscopy (FTIR) spectrum of Fe_3O_4 nanoparticles, the strong absorption peak around 576 cm^{-1} is assigned to Fe–O vibration, indicating the formation of oxide crystal of iron.

There is no special functional group on the surface of Fe_3O_4 nanoparticles when using the above synthetic method. Therefore, this kind of Fe_3O_4 nanoparticles is usually used as a core for further modification instead of applied directly in proteomics separation and analysis.

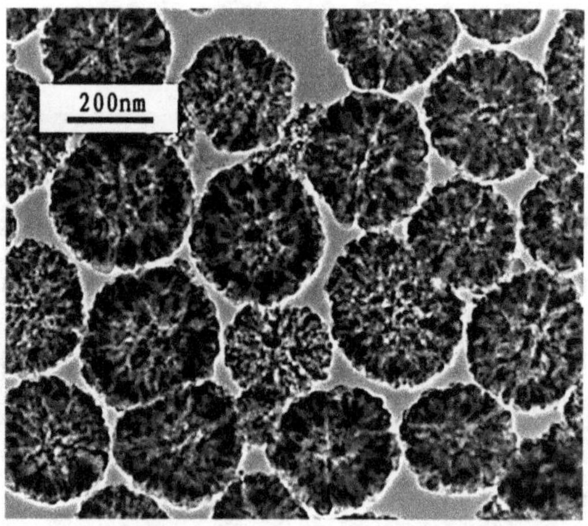

Figure 2.5: TEM image of Fe_3O_4.

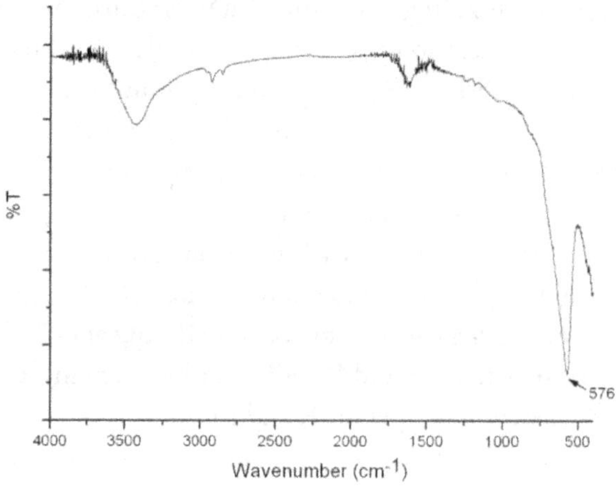

Figure 2.6: FTIR spectrum of Fe_3O_4.

2.3. One-pot synthesis of Fe_3O_4 nanoparticle with functional group

Fe_3O_4 nanoparticles with functional groups on surface can be obtained by adding some chemical reagents directly in the synthesis process.

2.3.1. *Synthesis of amino-modified Fe₃O₄ nanoparticle (Fe₃O₄–NH₂)*

Amino-modified Fe_3O_4 nanoparticles are synthesized by hydrothermal method.[7] Similarly, 1.0 g $FeCl_3·6H_2O$ is first dissolved in 30 mL ethylene glycol, and the yellow transparent solution can be obtained after magnetic stirring for 0.5 h. Then 4.0 g anhydrous NaAc is added into the aforementioned yellow transparent solution, and kept under continuous magnetic stirring for another 0.5 h. Next, 3.6 g 1,6-hexanediamine is added to the above solution, with continued magnetic stirring for another 0.5 h. The resulting brown yellow transparent solution is transferred to a 200 mL Teflon-lined stainless-steel reactor, which is further placed in an oven at 200°C for 6–18 h. After the reaction is finished, the Teflon-lined stainless-steel reactor is cooled to room temperature. The obtained materials are washed with hot water 6 times, and then washed repeatedly with ethanol and deionized water to fully remove the remaining water-soluble impurities such as hexanediamine and ethylene glycol, for vacuum drying at 50°C

From the scanning electron microscope (SEM) image in Figure 2.7, it can be seen that the amino-modified magnetic nanoparticles with smooth surface are nearly spherical. The good uniformity and small size of amino-modified magnetic nanoparticles result in a large specific surface area, which can provide more sites for enzyme immobilization and surface chemical modification. In addition, the amino-modified magnetic nanoparticles can disperse well in

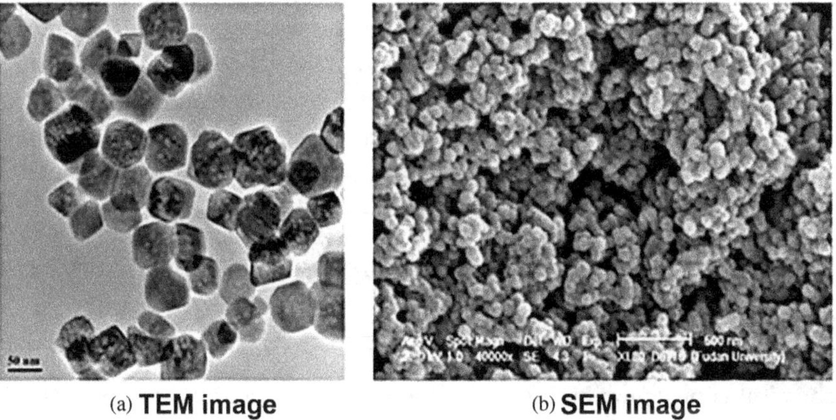

(a) **TEM image**　　　　　(b) **SEM image**

Figure 2.7: TEM image and SEM image of Fe_3O_4–NH_2 by optimized hydrothermal method.

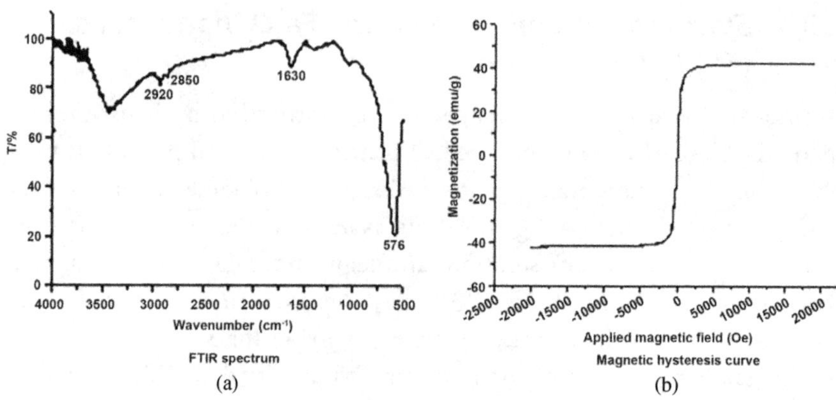

Figure 2.8: FTIR spectrum and hysteresis curve of Fe_3O_4–NH_2.

aqueous solution and organic phase and have good magnetic responsiveness. All of these characteristics are beneficial to their surface modification and practical application. In the FTIR spectrum of amino-modified magnetic nanoparticles of Figure 2.8(a), the absorption peak of 576 cm^{-1} corresponds to the vibration of Fe–O, the absorption peak of 1,630 cm^{-1} corresponds to the vibration of amino group of 1,6-hexanediamine, and the absorption peak of about 2,850–2,920 cm^{-1} corresponds to the stretching vibration of C–H in the alkyl chain, indicating the presence of free amino groups on the surface of amino-modified magnetic nanoparticles. From the hysteresis curve in Figure 2.8(b), the M_s of amino-modified magnetic nanoparticles is about 45.0 $emu·g^{-1}$, indicating the strong magnetic responsiveness. The amino-modified magnetic nanoparticles have good physical and chemical properties such as large specific surface area, superparamagnetism, and large number of modifiable functional groups on the surface, which makes it possible to separate and enrich the biological macromolecules rapidly and effectively.

2.3.2. *Synthesis of oleic acid-modified Fe_3O_4 nanoparticle (OA–Fe_3O_4)*

The synthesis of oleic acid-modified Fe_3O_4 nanoparticles adopts the simple chemical coprecipitation method.[8] The typical preparation process is as follows: first, 2.70 g $FeCl_3·6H_2O$ and 0.64 g $FeCl_2$ are dissolved in 100 mL water in a three-necked bottle, then N_2 is passed through for 2 h at 30°C. Secondly, the three-necked bottle is immersed in an oil bath of 50°C and then 8 $mol·L^{-1}$

NaOH aqueous solution (20 mL) is added dropwise within 30 min under rapid mechanical stirring of 600 rpm, the resulting black dispersion liquid is stirred continuously for 1 h under 50°C. Thirdly, 10 mL oleic acid is quickly injected under mechanical stirring, and the reaction lasts for 1 h. Then the temperature is raised to 80°C, with continuous stirring for 2 h. Finally, the dispersion liquid is cooled to room temperature under mechanical stirring, and the black pulp product is washed with ethanol several times for vacuum drying at 40°C.

From the TEM image of OA–Fe_3O_4 in Figure 2.9(a), it can be seen that the average size of nanoparticles is about 15 nm. From the high-resolution TEM of OA–Fe_3O_4 in Figure 2.9(b), the ordered lattice edge in line with plane 311 of Fe_3O_4 nanocrystal can be observed clearly, moreover, the non-nanocrystal layer can be seen at the edge of nanocrystal, which might be caused by oleic acid molecules on the surface. In general, the carboxyl group of oleic acid can bind to iron atoms of Fe_3O_4 nanoparticles through complexation since oleic acid can form carboxylate under alkaline condition, which can inhibit the growth of magnetite nanocrystals and obtain high crystallization of OA–Fe_3O_4 (as seen in the inset of Figure 2.9(a)). The resulting OA–Fe_3O_4 nanoparticles are stable in both air and water, and can retain initial properties for months. In the FTIR spectrum of OA–Fe_3O_4 nanoparticles of Figure 2.10, the 2,921 cm^{-1} and 2,854 cm^{-1} correspond to the asymmetric stretching vibration and symmetric stretching vibration of CH_2, respectively. The 1,540 cm^{-1} and 1,635 cm^{-1} correspond to the oscillating absorption peaks of typical carboxylate, confirming the chelation of oleic acid with iron atoms.

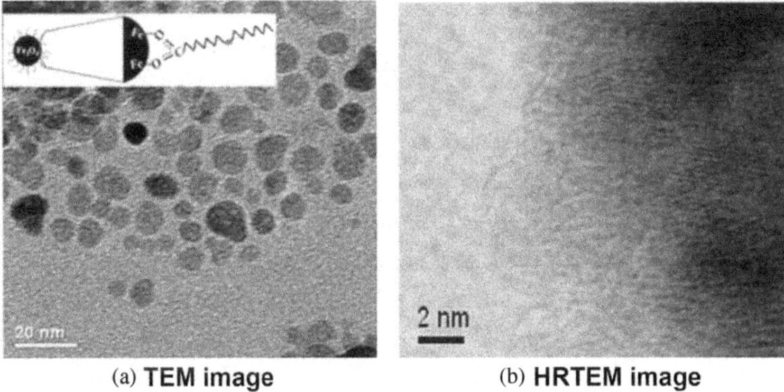

| (a) **TEM image** | (b) **HRTEM image** |

Figure 2.9: TEM image and high-resolution TEM image of OA–Fe_3O_4 (the inset represents the sketch map of the composition of OA–Fe_3O_4).

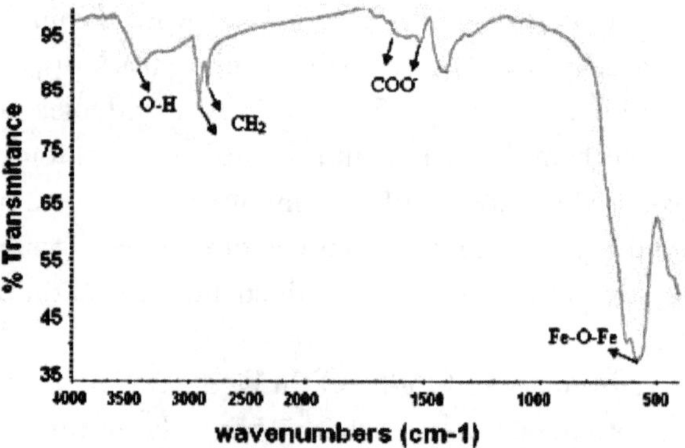

Figure 2.10: FTIR spectrum of OA–Fe_3O_4.

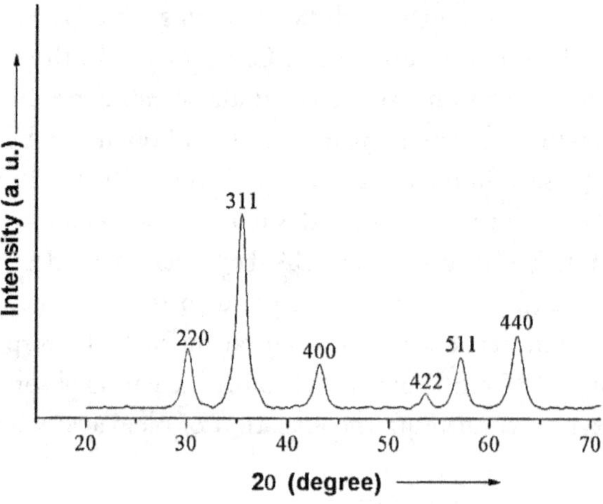

Figure 2.11: XRD pattern of OA–Fe_3O_4.

According to thermogravimetric (TGA) analysis, the amount of oleic acid in OA–Fe_3O_4 nanoparticles is about 9.7%, indicating that the OA–Fe_3O_4 is a material with high-content hydrophobic organics.

X-ray diffraction (XRD) measurement shows the characteristic diffraction pattern of Fe_3O_4 (as seen in Figure 2.11), all diffraction peaks are widened because the size of magnetite nanoparticle is small. According to the Debye–Scherer formula and the strongest 311 diffraction peak, the size of OA–Fe_3O_4 nanoparticle is calculated to be about 14.4 nm, which is very close to the previous evaluation based on TEM image (as seen in Figure 2.9). Magnetic characterization at 300 K shows that the M_s of OA–Fe_3O_4 nanoparticle is as high as 34.3 emu·g^{-1}, and there is no remanence, indicating the superparamagnetism of OA–Fe_3O_4. Shortly, the OA–Fe_3O_4 can not only respond quickly to external magnetic field because of its high magnetism and superparamagnetism, it can also disperse well in the solution after removing the external magnetic field.

The special functional groups on surface make it possible to use these magnetic micro-/nanomaterials directly for proteomics separation and analysis. Besides, the special functional groups such as amino groups also make it very convenient to conduct further surface modification.

2.4. Synthesis of inorganic magnetic micro-/ nanomaterial with core–shell structure

Compared with single-component magnetic nanoparticles, magnetic nanoparticles with surface modification can effectively change the surface charge, functional properties and reactivity, and further enhance stability and dispersity of nanoparticles. This surface modification also has a certain impact on the magnetic properties of the material. For instance, if ferromagnetic particles are embedded in a non-magnetic matrix, the generated giant magneto-resistance phenomenon can be used for magnetic recording. The dielectric or conductive properties of the matrix also affect the magnetic properties of the nanocomposites. For core–shell nanoparticles, nanocomposites rely on surface spin and interactions of matrix to effectively shield and protect magnetic components. The design and construction of core–shell magnetic nanomaterials with special properties and stability have been long sought. Magnetic nanocomposite can be classified into three categories: magnetic-inorganic core–shell material, magnetic-organic core–shell material, and magnetic-inorganic–organic core–shell materials. In this section, we will introduce the magnetic-inorganic core–shell materials in detail.

2.4.1. *Silica-coated magnetic core–shell material* *(Fe_3O_4@SiO_2)*

Fe_3O_4@SiO_2 microspheres are most widely used in studies because of the following three reasons.[6] Firstly, as is well known, silica is a bioinert material with good bio-compatibility, which provides an inert surface for the application of $Fe_3O_4@SiO_2$ microspheres in biological systems. Therefore, $Fe_3O_4@SiO_2$ microspheres can be used in the field of bio-medicine. Secondly, the silica layer shields the magnetic dipole attraction between the magnetic microspheres effectively, which facilitates their dispersion in liquid media and protects them from corrosion and dissolution in acidic environments.

Thirdly, the surface of silica prepared by sol–gel method is rich in hydroxyl groups. This can not only improve the hydrophilicity of magnetic nanoparticles, but also facilitate the surface functionalization through chemical modification from reacting with other compounds, resulting in that different functional groups are bonded to meet the requirements of different practical applications.

The synthetic process of $Fe_3O_4@SiO_2$ microspheres is elaborated as follows:

- Firstly, prepare Fe_3O_4 microspheres by hydrothermal method.
- Secondly, conduct the surface treatment of Fe_3O_4 microspheres. 3.0 g Fe_3O_4 microspheres are added into 2 $mol \cdot L^{-1}$ HCl solution (5 mL) in a centrifuge tube for 5–10 min of ultrasonic treatment. The green supernatant caused by the dissolved surface part of Fe_3O_4 microspheres is separated and removed using a magnet. Then the Fe_3O_4 microspheres are washed by 10 mL deionized water. Afterwards, 5 mL 10% of trisodium citrate solution is used to ultrasonically disperse Fe_3O_4 microspheres after standing for about 1–2 h. After removing the supernatant, 5 mL deionized water is used to disperse the microspheres (denoted as solution B) again for later use.
- Finally, $Fe_3O_4@SiO_2$ microspheres are prepared. 1.0 g solution B is diluted by 5 mL deionized water and 30 mL ethanol and then 0.5 mL ammonium hydroxide is added. After ultrasonic dispersion of 5–10 min, 1.0 mL tetraethyl orthosilicate (TEOS) is added under mechanical stirring. After 12 h of stirring at room temperature, the microspheres are collected by magnet. The experiment is optimized by changing the ratio between Fe_3O_4 and TEOS such as adjusting the amount of TEOS as 0.5 mL, 0.75 mL, and 1.25 mL. From TEM images of Fe_3O_4 and $Fe_3O_4@SiO_2$ microspheres in Figure 2.12, it can be seen that an even layer of SiO_2 has been coated on the surface of Fe_3O_4.

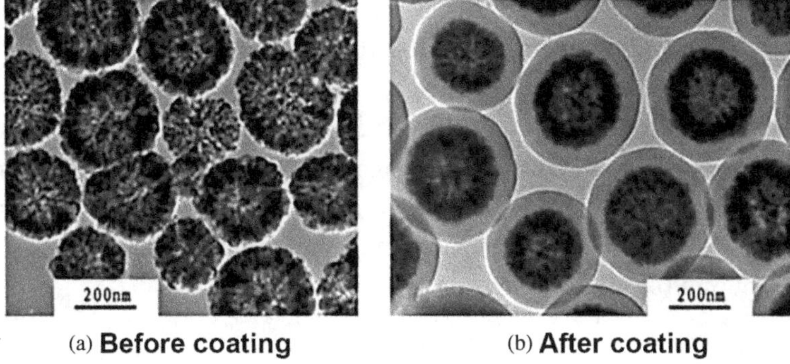

Figure 2.12: TEM images of Fe_3O_4 before and after coating of SiO_2 layer (Fe_3O_4 is prepared by hydrothermal method).

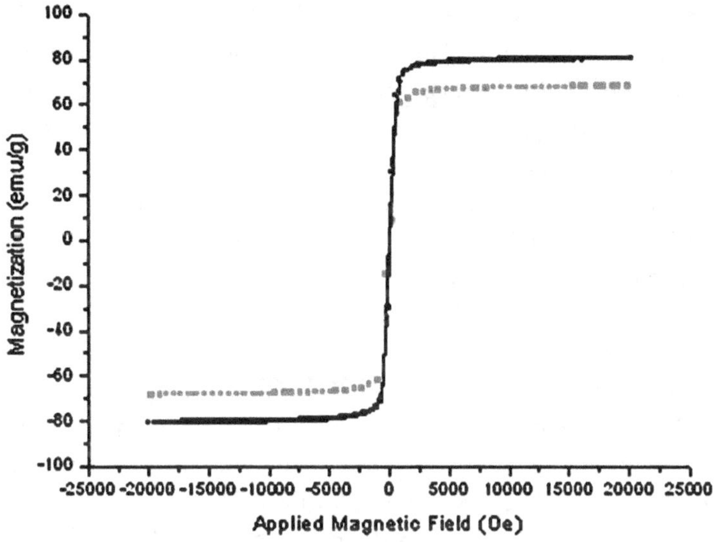

Figure 2.13: Hysteresis curve of Fe_3O_4 before (continuous line) and after (dotted line) coating SiO_2 layer.

The results of Figure 2.13 indicate that both Fe_3O_4 before and after coating SiO_2 layer possess good superparamagnetism, which means the magnetism will vanish once the external magnetic field disappears. The saturation magnetic values of magnetic microsphere and magnetic silica are 80.7 emu·g^{-1} and 68.2 emu·g^{-1}, indicating the embedding of silica has no significant impact on the superparamagnetism of microspheres. This is mainly attributed to the less than 30 nm size of magnetic nanoparticles inside magnetic microspheres. It is well known that superparamagnetism of magnetic microspheres is crucial for their applications in bio-medical and bio-engineering fields. When the magnetic field disappears, superparamagnetism prevents them from coming together and quickly dispersing. Due to superparamagnetism and shielding effect of silica layer, magnetic silica can be stably dispersed in water with solid capacity up to 0.1% (mass ratio).

2.4.2. *Carbonaceous polysaccharide-coated magnetic core–shell material (Fe$_3$O$_4$@CP)*

Firstly, Fe$_3$O$_4$ is synthesized by hydrothermal method.[9] Secondly, 0.5 g Fe$_3$O$_4$ nanoparticles are dispersed in 0.1 mol·L^{-1} HNO$_3$ aqueous solution (80 mL). After 5 min of ultrasound treatment, the pale yellow supernatant (caused by the dissolution of the surface part of Fe$_3$O$_4$) is removed by magnetic separation, then the process is repeated twice including the dispersion of nanoparticles in deionized water and their magnetic separation. Thirdly, the nanoparticles are further dispersed into 0.5 mol·L^{-1} glucose aqueous solution (80 mL) for 180°C high-temperature reaction for about 4 h. The reaction kettle is cooled naturally in the air. Finally, the products are collected with magnets, washed with ethanol and water repeatedly to remove impurities, and then dried in vacuum for 24 h at 60°C to obtain carbonaceous polysaccharide (CP)-coated magnetic microspheres (Fe$_3$O$_4$@CP).

The TEM image in Figure 2.14(a) shows that the average size of Fe$_3$O$_4$ is around 250 nm. As seen from Figure 2.14(b), the thickness of CP layer is around 20 nm, and Fe$_3$O$_4$@CP with 290 nm of particle size presents clear core–shell structure. Figure 2.15 is the FTIR spectra of Fe$_3$O$_4$ nanoparticles and Fe$_3$O$_4$@CP microspheres. The peaks at ~3,500 cm^{-1} and ~1,630 cm^{-1} correspond to the absorption of water on particles. From the spectrum of Fe$_3$O$_4$@CP microspheres in Figure 2.15(b), the peaks at ~1,700 cm^{-1} and ~1,625 cm^{-1} are assigned to the swing absorption of C=O and C=C, respectively, which is because of the dehydration and aromatization of glucose during the hydrothermal reaction. The peaks in the range of 1,200–1,400 cm^{-1} are assigned to the stretching and bending vibration of –OH, indicating that a large number of hydroxyl groups remain after carbonization and dehydration of glucose. These large covalently bonded –OH and –C(H)=O groups on the

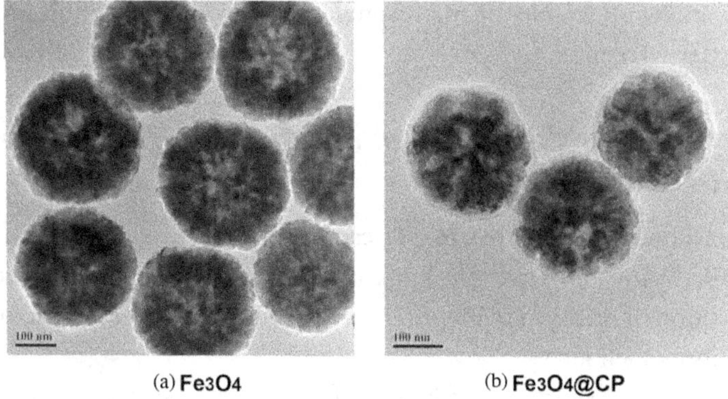

Figure 2.14: TEM images of Fe$_3$O$_4$ and Fe$_3$O$_4$@CP.

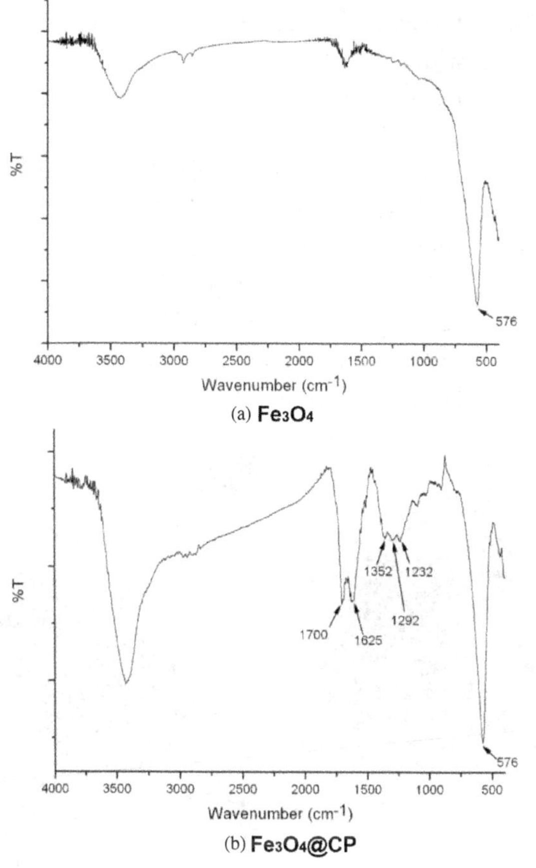

Figure 2.15: FTIR spectra of Fe$_3$O$_4$ and Fe$_3$O$_4$@CP.

carbon framework enable the microspheres to be both hydrophilic and stable in the water system.

2.4.3. *Metal oxide-coated magnetic core–shell material $(Fe_3O_4@M_xO_y)$*

The corresponding metal hydroxide can be immobilized on the surface of $Fe_3O_4@CP$ by means of hydrolysis of metal alkoxide, and then calcined under the protection of nitrogen to obtain the core–shell $Fe_3O_4@M_xO_y$ microspheres with uniform shell layer.[10]

Let's take the synthesis of $Fe_3O_4@TiO_2$ as an example. First, 100 mg $Fe_3O_4@CP$ microspheres are dissolved in mixed solution consisting of 5 mL butyl titanate and 35 mL ethanol for 5 min of ultrasound treatment. Then the mixture of deionized water and ethanol at volume ratio of 1:5 is added drop-wise under stirring, with stirring continued for 1 h. Finally, the microspheres are separated by a magnet and washed with ethanol 5 times, the obtained solid products are calcined at 500°C under the protection of nitrogen.

According to the above steps, $Fe_3O_4@TiO_2$, $Fe_3O_4@ZrO_2$, $Fe_3O_4@Al_2O_3$, $Fe_3O_4@Ga_2O_3$, $Fe_3O_4@CeO_2$, $Fe_3O_4@InO_2$, $Fe_3O_4@Ta_2O_5$, and $Fe_3O_4@SnO_2$ can be prepared by using 5 mL butyl titanate, 0.05 g isopropanol zirconium, 0.5 g isopropanol aluminum, 0.05 g isopropanol gallium, 0.05 g pentmethoxy tantalum, and 113 mg potassium stannate (0.75 g urea), respectively.

As shown in Figure 2.16, the carbon layer on the surface of Fe_3O_4 micro-spheres that is coated by taking advantage of hydrothermal reaction of glucose

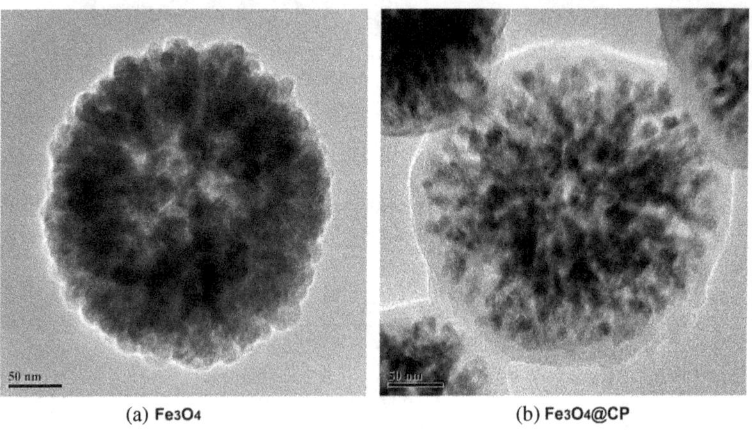

Figure 2.16: TEM images of different materials.

aqueous solution is about 20 nm. Figure 2.17 displays a series of SEM images of $Fe_3O_4@M_xO_y$, from which it can be observed the particle sizes of $Fe_3O_4@M_xO_y$ are uniform. Figure 2.18 displays corresponding TEM images. Compared with Figure 2.16(b), the translucent carbon layer on Fe_3O_4 micro-

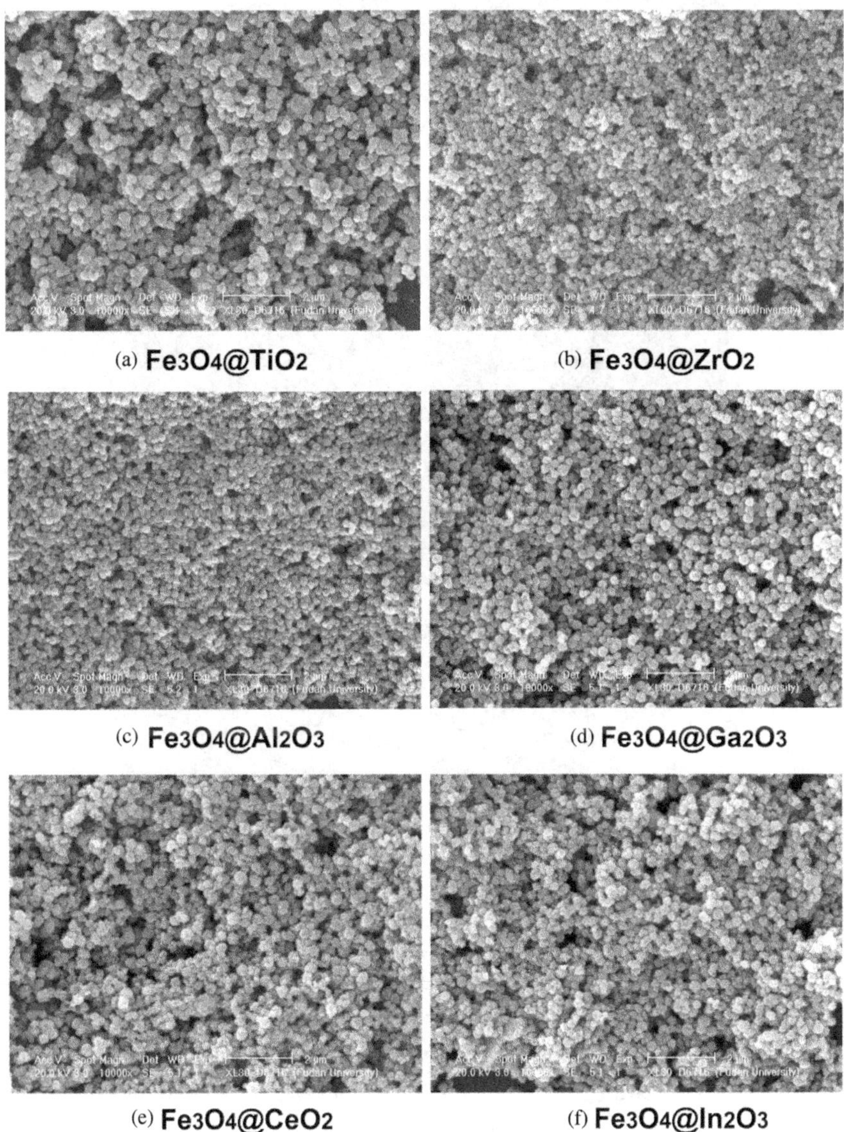

(a) **Fe3O4@TiO2**

(b) **Fe3O4@ZrO2**

(c) **Fe3O4@Al2O3**

(d) **Fe3O4@Ga2O3**

(e) **Fe3O4@CeO2**

(f) **Fe3O4@In2O3**

Figure 2.17: SEM images of $Fe_3O_4@M_xO_y$ microspheres.

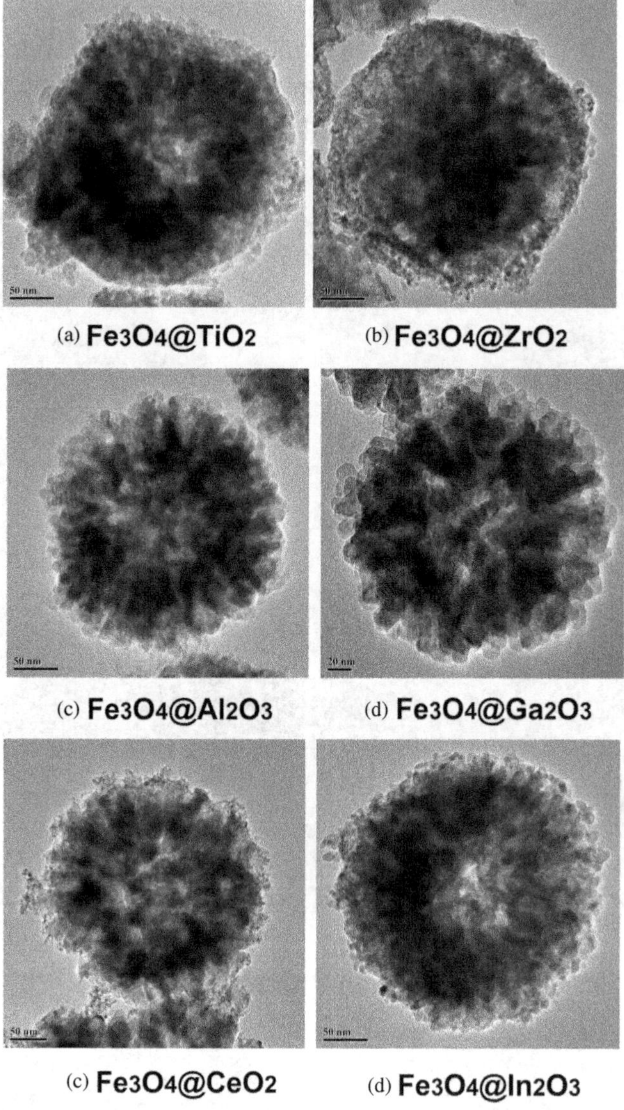

Figure 2.18: TEM images of $Fe_3O_4@M_xO_y$ microspheres.

spheres disappears and is replaced by a dense layer of nanoparticles about 20 nm in diameter. The results of energy dispersive X-ray analysis (EDX) in Figure 2.19 indicate the designed metal oxides are indeed coated on Fe_3O_4 microspheres. Table 2.1 displays the content of metal elements immobilized

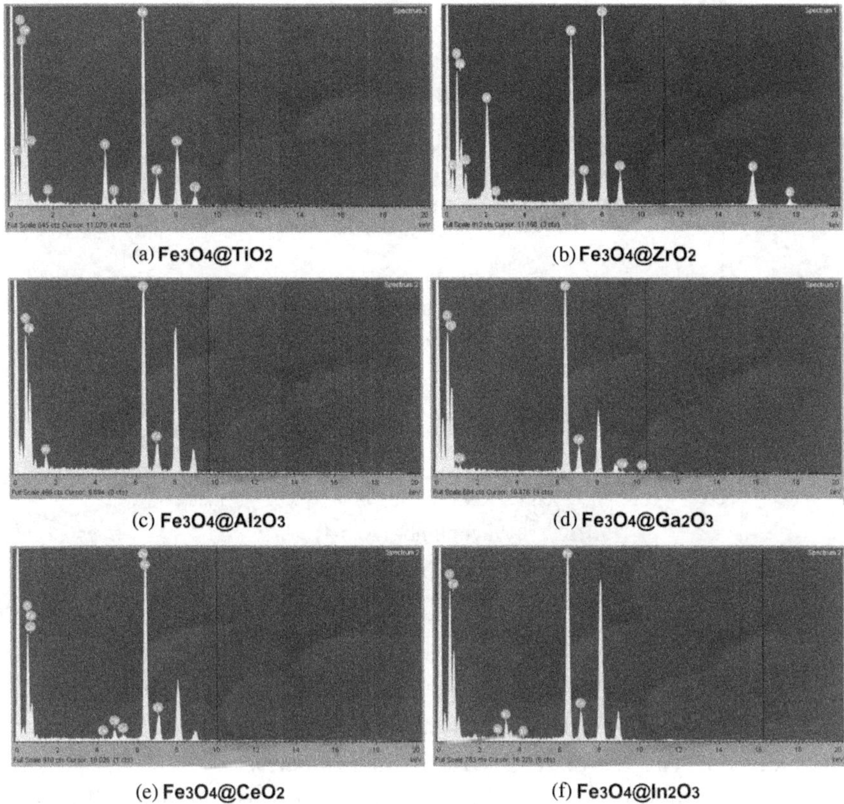

Figure 2.19: EDX analysis of $Fe_3O_4@M_xO_y$ microspheres.

Table 2.1: The content of metal elements on the surface of Fe_3O_4 microspheres.

	Corresponding Element Content
$Fe_3O_4@TiO_2$	5.96%
$Fe_3O_4@ZrO_2$	9.70%
$Fe_3O_4@Al_2O_3$	2.01%
$Fe_3O_4@Ga_2O_3$	1.22%
$Fe_3O_4@CeO_2$	2.26%
$Fe_3O_4@InO_2$	2.58%

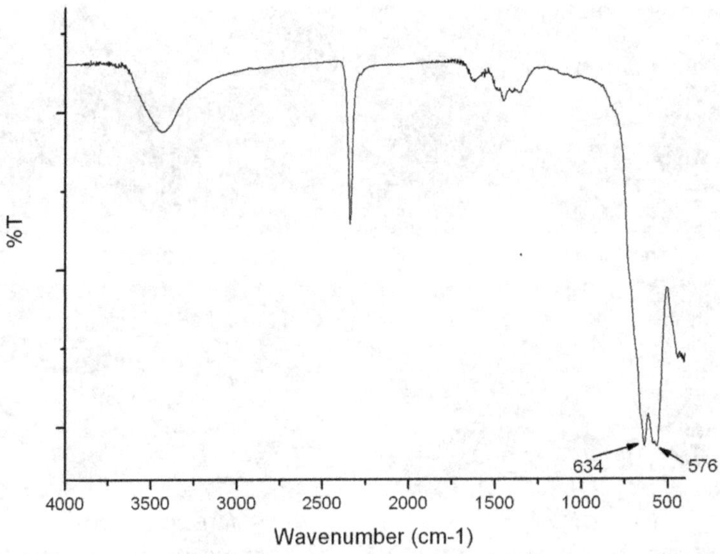

Figure 2.20: FTIR spectrum of Fe_3O_4@ZrO_2 microspheres.

on the surface of Fe_3O_4 microspheres. Although the element content obtained by EDX analysis is only semi-quantitative data, it is enough to demonstrate the successful synthesis of Fe_3O_4@M_xO_y microspheres.

FTIR is used for further characterization to verify that metal oxides have been successfully coated on the surface of Fe_3O_4 microspheres. Figure 2.15(b) indicates the successful synthesis of Fe_3O_4@CP microspheres and there are a large number of hydrophilic groups on their surface. The presence of these hydrophilic groups not only endows Fe_3O_4@CP microspheres with better hydrophilicity and stability, but also enhances the affinity between Fe_3O_4@CP microspheres and metal hydroxide oligomers obtained after hydrolysis of metal alkoxide, which is conducive to the adsorption of metal hydroxide to the surface of Fe_3O_4@CP. Taking Fe_3O_4@ZrO_2 microspheres as example, as shown in Figure 2.20, the characteristic peak of Zr–O around 634 cm^{-1} can be observed in addition to Fe–O peak at around 576 cm^{-1} in the low wavenumber region, indicating the successful coating of metal oxide further.

2.5. Organic magnetic material with core–shell structure

2.5.1. *Polydopamine-coated magnetic core–shell material (Fe$_3$O$_4$@PDA)*

Polydopamine (PDA) can be coated on Fe$_3$O$_4$ nanoparticles according to the following method: 10 mg Fe$_3$O$_4$ microspheres are firstly dispersed in 10 mmol·L^{-1} Tris aqueous solution (10 mL, pH=8.5), and then 20 mL ethanol is added for ultrasonic mixing.[11] Then 40 mg dopamine hydrochloride is dissolved in 15 mL deionized water, and the dissolved dopamine hydrochloride solution is added to the Tris ethanol mixture solution containing Fe$_3$O$_4$ microspheres for 20 h of stirring at room temperature. The resulting product is washed with deionized water and ethanol several times and dried to obtain PDA-coated magnetic microspheres.

2.5.2. *Polymethylacrylic acid-coated magnetic core–shell material (Fe$_3$O$_4$@PMAA)*

The synthetic process is as follows:[12] Fe$_3$O$_4$@PMAA microsphere is firstly prepared through distillation precipitation polymerization by using methacryloxypropyltrimethoxysilane (MPS) modified magnetic microspheres as a seed, methacrylic acid (MAA) as monomer, 2,2′-azobis(2-methylpropionitrile (AIBN) as initiator, *N,N′*-methylenebis acrylamide (MBA) as cross-linking agent. It is worth mentioning that the modification of MPS is found to be essential for the preparation of Fe$_3$O$_4$@PMAA microsphere. The shell thickness and cross-linking degree of Fe$_3$O$_4$@PMAA microsphere can be well regulated by changing the experimental conditions of distillation precipitation polymerization, such as the concentration of polymerization reaction and the amount of MBA.

2.5.3. *Chitosan-coated magnetic core–shell material (Fe$_3$O$_4$@CS)*

As shown in Figure 2.21,[13] the preparation of Fe$_3$O$_4$@CS microspheres is as follows: 2.163 g FeCl$_3$·6H$_2$O is dissolved in 70 mL ethylene glycol for 15 min of ultrasonic dispersion, and then 3.7 g NH$_4$Ac and 1.0 g chitosan (CS) are added. The above mixture is reacted at 160°C for 2 h under the

Figure 2.21: Schematic diagram of preparation of Fe$_3$O$_4$@CS microspheres.

(a) **TEM of Fe3O4@CS**   (b) **TGA curves**

(c) **Zeta potential**   (d) **Magnetic hysteresis curve**

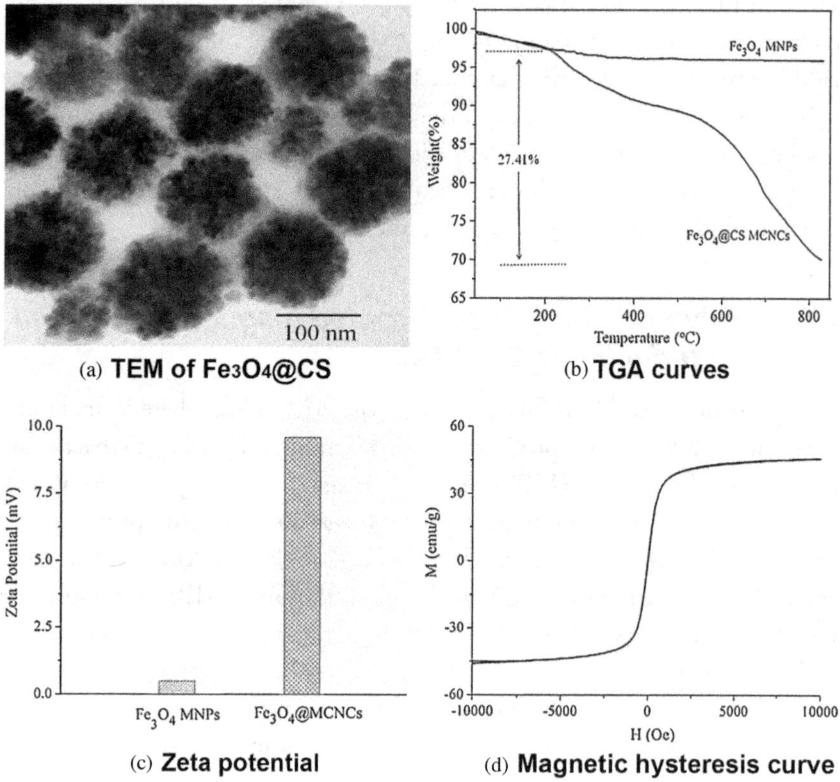

Figure 2.22: Characterizations of Fe$_3$O$_4$@CS microspheres.

protection of nitrogen. Then reactants are transferred to 100 mL Teflon-lined stainless steel reactor for reaction for 16 h at 200°C. The Fe$_3$O$_4$@CS microspheres are washed with deionized water and ethanol, and dried in vacuum for later use.

Figure 2.22(a) shows the TEM image of Fe$_3$O$_4$@CS microsphere, from which its average size is estimated to be about 140 nm. TGA analysis is used to estimate the mass ratio of inorganic and organic components in Fe_3O_4@CS microsphere. As shown in Figure 2.22(b), the mass loss of Fe_3O_4@CS microspheres is about 27.41% when exceeding 200°C. Moreover, after modification with chitosan, the zeta potential on the surface of magnetic microspheres changes to be 9.6 mV (Figure 2.22(c)). All the above results indicate that the content of CS in Fe_3O_4@CS microspheres is very high. In addition, the M_s of Fe_3O_4@CS microspheres is measured to be 46.1 emu·g^{-1} (Figure 2.22(d)), indicating that it has good magnetic responsiveness.

2.5.4. *Immobilization of Con A on amino-phenylboronic acid modified magnetic microsphere (Fe$_3$O$_4$@ APBA–sugar–Con A)*

2.5.4.1. *The preparation of Fe$_3$O$_4$@APBA-sugar–Con A*

The preparation process is as seen in Figure 2.23.[14]

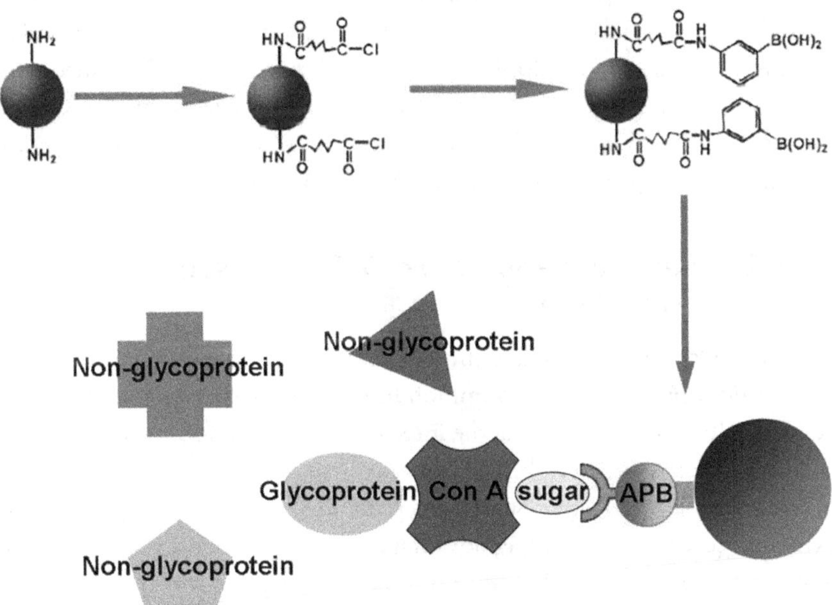

Figure 2.23: Schematic diagram of preparation of Fe_3O_4@APBA–sugar–Con A.

(1) Synthesis of Fe_3O_4@APBA microsphere

Firstly, the amino-modified magnetic microspheres are synthesized as described in section 2.3.

Then, 0.1 g amino-modified magnetic microspheres are dispersed in 30 mL anhydrous chloroform, and 0.2 mL of adipoyl chloride is slowly added under of ultrasonic treatment and mechanical stirring. The reaction lasts for 4 h. The obtained solid material is cleaned for 3 times with 10 mL anhydrous chloroform, and then the solid material is dispersed in 30 mL anhydrous chloroform.

Finally, 0.2 g of 3-aminophenylboronic acid is added, and the resulting solid product is washed with anhydrous chloroform for vacuum drying for later use.

(2) Synthesis of Fe_3O_4@APBA–sugar–Con A microsphere

After washing with 50 mmol·L^{-1} NH_4HCO_3 solution, the Fe_3O_4@APBA microspheres are dispersed in 10% of methyl alpha-D-pyranomannan solution (dissolved in 50 mmol·L^{-1} NH_4HCO_3 solution) for oscillatory reaction at room temperature for 30 min. The obtained solid product is washed with 50 mmol·L^{-1} NH_4HCO_3 solution, and then mixed with 2 mg·L^{-1} Con A solution (dissolved in phosphate buffer saline, PBS) for oscillatory reaction for 1 h. The final product is washed with PBS solution.

2.5.4.2. Characterization of Fe_3O_4@APBA–sugar–Con A microsphere

Two methods have been used to immobilize Con A on the surface of Fe_3O_4@APBA microspheres. One is to immobilize Con A directly on the surface of Fe_3O_4@APBA microspheres via the interaction between boric acid and sugar, and the other is to immobilize Con A on the surface of Fe_3O_4@APBA microspheres by using the "sandwich" structure of APBA-methyl-alpha-D-pyrmannose-Con A. By taking advantage of ultraviolet and visible spectrophotometer measurement, the immobilized protein content on Fe_3O_4@APBA microspheres by the former method is about 10 μg·mg^{-1}, and that on

Fe$_3$O$_4$@APBA microspheres by latter method is about 40 μg·mg^{-1}, by ultraviolet and visible spectrophotometer measurement.

2.6. Magnetic-inorganic–organic material with core–shell structure

2.6.1. *Polymethyl methacrylate-modified magnetic silica (Fe$_3$O$_4$@SiO$_2$@PMMA)*

Figure 2.24 presents the schematic diagram of preparation of Fe$_3$O$_4$@SiO$_2$@ PMMA microsphere.

Firstly, the dense SiO$_2$-coated magnetic core (Fe$_3$O$_4$@SiO$_2$) is synthesized, on the surface of which contains a large number of Si-O-H groups.[15]

Then the obtained Fe$_3$O$_4$@SiO$_2$ microspheres are mixed with 80 mL ethanol and 1.5 mL MPS for 48 h of mechanical stirring at 30°C. The resulting Fe$_3$O$_4$@SiO$_2$–MPS microspheres are washed with ethanol many times and dispersed in 50 mL ethanol for later use.

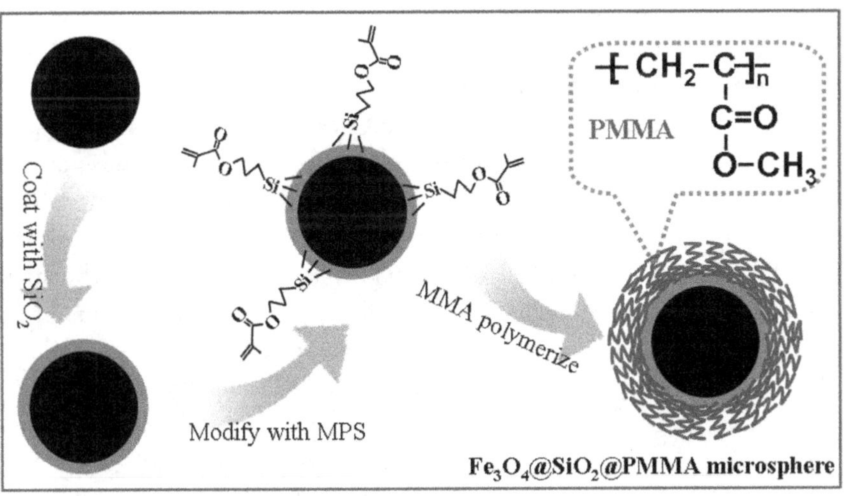

Figure 2.24: Schematic diagram of preparation of Fe$_3$O$_4$@SiO$_2$@PMMA.

Finally, the Fe_3O_4@SiO_2@PMMA microspheres with sandwich structure are prepared by free radical polymerization in aqueous phase. A mechanical stirrer, return line and nitrogen inlet are installed on a 100 mL three-necked bottle. 10 mL ethanol dispersion solution of Fe_3O_4@SiO_2–MPS microspheres is added into a 100 mL three-necked bottle, then 50 mL aqueous solution of 5.0 mg sodium dodecyl benzene sulfonate is added into the three-necked bottle under the condition of stirring. After stirring about 1 h, 1.0 g methyl methacrylate (MMA) monomer is added, and the mixture solution is treated by bubbling nitrogen for 30 min. Next, 40 mg potassium persulfate is added as initiator. The reaction lasts for 24 h with 70°C oil bath and the stirring speed of 200 rpm. The resulting Fe_3O_4@SiO_2@PMMA microspheres are washed with ethanol for cleaning residual impurities and further dried at 40°C for later use.

TEM image in Figure 2.25(a) shows that the magnetic nanoparticles of Fe_3O_4 are nearly spherical in shape with an average size of about 170 nm. The selected area electron diffraction (SAED) image that records the single particle shows a multi-layer diffraction ring (the inset in Figure 2.25(a)), which corresponds to the polycrystalline structure characteristic of magnetite. The TEM image of Fe_3O_4@SiO_2 microspheres in Figure 2.25(b) shows that each black magnetic core is uniformly coated by a gray SiO_2 layer with a thickness of about 35 nm. The thickness of the SiO_2 layer can be adjusted from tens to hundreds of nanometers through sol–gel process. The TEM image in Figure 2.25(c) shows that Fe_3O_4@SiO_2@PMMA microspheres disperse well and have an average size of about 270 nm. The high-powered TEM image in Figure 2.25(d) shows Fe_3O_4@SiO_2@PMMA microspheres have a sandwich-like structure, namely, a black Fe_3O_4 core in the center, a gray SiO_2 layer in the middle, and a thin gray layer at the outside. The thickness of PMMA layer is estimated to be 20 nm. The unique structure of the microspheres can be clearly observed because of the apparent quantitative distribution of the three components. The SEM image in Figure 2.26(a) shows that Fe_3O_4@SiO_2 is a nearly spherical particle with smooth surface and particle size of about 240 nm (with deviation of less than 12%). After covering of PMMA layer, the radius increases slightly and the morphology is more uniform as seen in Figure 2.26(b), which are consistent with the previous TEM images.

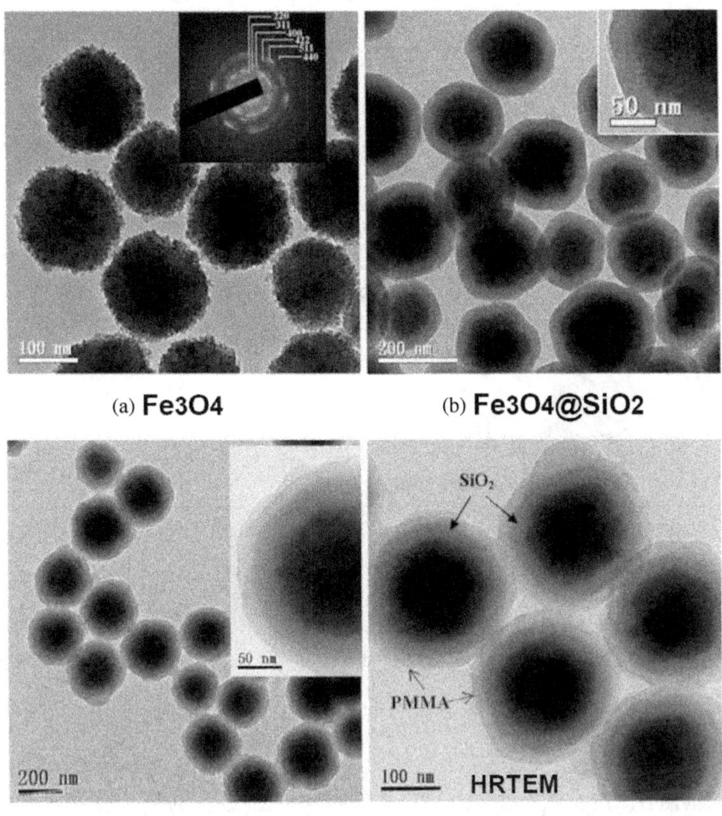

(a) **Fe3O4** (b) **Fe3O4@SiO2**

(c) **Fe3O4@SiO2@PMMA** (d) **Fe3O4@SiO2@PMMA**

Figure 2.25: TEM images of microspheres.

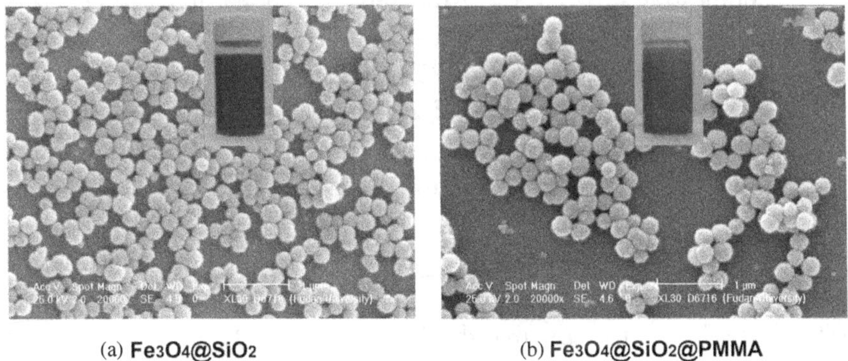

(a) **Fe3O4@SiO2** (b) **Fe3O4@SiO2@PMMA**

Figure 2.26: SEM images of $Fe_3O_4@SiO_2$ and $Fe_3O_4@SiO_2@PMMA$ microspheres (the inset is the dispersion photo of microsphere in water).

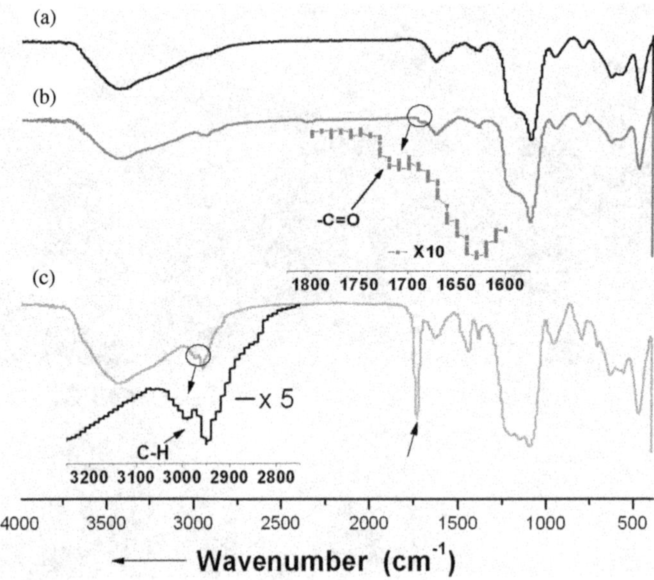

Figure 2.27:   FTIR images of three microspheres (curve (a) Fe_3O_4@SiO_2 microsphere; curve (b) Fe_3O_4@SiO_2–MPS microsphere; curve (c) Fe_3O_4@SiO_2@PMMA microsphere; the inset is the enlarged part of FTIR images).

The surface modification of Fe_3O_4@SiO_2 and the composition of Fe_3O_4@SiO_2@PMMA microspheres are further characterized by FTIR. As shown in Figure 2.27, the peak of Fe_3O_4@SiO_2 microspheres at 1,090 cm^{-1} corresponds to the vibration absorption of Si–O–Si, while those at about 1,630 cm^{-1} and 3,400 cm^{-1} correspond to the absorption peaks of water and hydroxyl groups. After modification of MPS, the extra absorption peaks at 1,720 cm^{-1} and 2,940 cm^{-1} are caused by C=O and C–H in MPS, respectively. After coating of PMMA, the FTIR spectrum of Fe_3O_4@SiO_2@PMMA microspheres shows strong absorption at 1,730 cm^{-1} and 2,950 cm^{-1}, respectively, representing C=O and C–H, this is because there is a large amount of PMMA on the microsphere surface, which further proves that PMMA has been successfully modified on the microsphere surface.

The magnetic properties of Fe_3O_4@SiO_2 and Fe_3O_4@SiO_2@PMMA are measured at 300 K by superconducting quantum interferometer device (SQUID). The results show that the M_s of Fe_3O_4@SiO_2 and Fe_3O_4@SiO_2@PMMA are 49.5 emu·g^{-1} and 36.7 emu·g^{-1}, respectively. Figure 2.28 shows

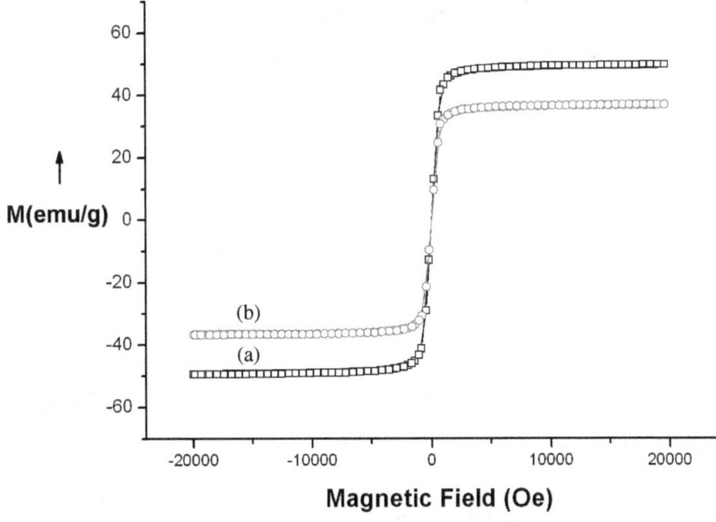

Figure 2.28: Magnetic hysteresis curves of two microspheres (curve (a): Fe_3O_4@SiO_2 microsphere; curve (b): Fe_3O_4@SiO_2@PMMA microsphere).

the hysteresis curves of these two microspheres. It indicates both the microspheres exhibits superparamagnetic properties, this is because the magnetic core of the microspheres is composed of magnetic particles at nanoscale. Similar to Fe_3O_4@SiO_2, the dispersion of Fe_3O_4@SiO_2@PMMA in aqueous solution is very good, and no precipitation can be seen within 8 h after dispersion. The inset in Figure 2.26 is the photo of their uniform dispersion in water.

The Fe_3O_4@SiO_2@PMMA microspheres are negatively charged, so they have good dispersion, which also enables the material to have a high adsorption capacity when applied. Besides, the high magnetic intensity of Fe_3O_4@SiO_2@PMMA microspheres leads to a very good magnetic response. The Fe_3O_4@SiO_2@PMMA microspheres can be rapidly separated from their dispersion solution (~1 wt.%) within 0.5 min by a magnet (1,000 Oe). These unique properties of Fe_3O_4@SiO_2@PMMA microspheres ensure the rapid separation–redispersion process, thus achieving rapid and effective magnet-assisted enrichment or separation.

2.6.2. *Zwitterionic polymer-modified magnetic silica (Fe$_3$O$_4$@SiO$_2$@PMSA)(poly(2-(methacryloyloxy) ethyl)dimethyl-(3-sul-fopropyl)ammonium hydroxide, PMSA)*

As described above, Fe$_3$O$_4$ particles are firstly synthesized and then coated with SiO$_2$ layer, and then MPS is modified on the outside of SiO$_2$ layer to obtain Fe$_3$O$_4$@SiO$_2$–MPS microspheres.[16] Next, a zwitterionic polymer layer of PMSA is coated on the outside of SiO$_2$ layer through MPS. The detailed synthetic process is as follows: 50 mg Fe$_3$O$_4$@SiO$_2$–MPS microspheres are ultrasonically dispersed in 80 mL ACN/H$_2$O (v/v, 3/1). Then 200 mg MSA, 50 μL MAA, 200 mg MBA and 6 mg AIBN are added for 20 min of ultrasonic treatment, and then the reaction temperature rises to 90°C within 30 min for keeping micro-boiling condition about 1.5 h. The product is cleaned by ethanol and dried by vacuum for later use.

In this work example, the synthesis of Fe$_3$O$_4$@SiO$_2$@PMSA microspheres is carried out according to the route 1 in Figure 2.29. The final core–shell

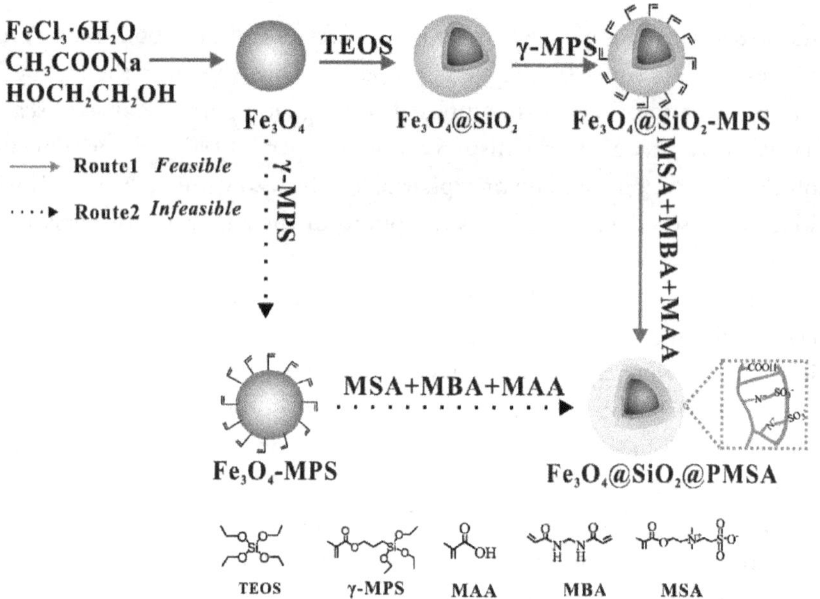

Figure 2.29: Schematic diagram of synthesis of Fe$_3$O$_4$@SiO$_2$@PMSA.

product cannot be obtained according to the route 2 in Figure 2.29. The morphology and size of Fe_3O_4 microspheres are relatively uniform as shown in Figures 2.30(a) and 2.30(b), the average size of which is about 280 nm. After coating a silica layer, the surface is smoother than that of Fe_3O_4 microspheres and the size increases slightly by about 10 nm as shown in Figures 2.30(c) and 2.30(d). From the TEM image of $Fe_3O_4@SiO_2@PMSA$ microspheres in Figure 2.30(e), it can be seen the PMSA polymer layer is successfully coated on the surface of the magnetic silica with a thickness of about 60 nm. Figure 2.30(f) is the image of $Fe_3O_4@SiO_2@PMSA$ microspheres by field emission scanning electron microscopy (FESEM).

Figure 2.31(a) shows the FTIR spectra of (I) $Fe_3O_4@SiO_2$ microspheres, (II) $Fe_3O_4@SiO_2$–MPS microspheres, and (III) $Fe_3O_4@SiO_2@PMSA$ microspheres. Compared with the FTIR spectrum of $Fe_3O_4@SiO_2$ (581 cm^{-1}, $v_{Fe-O-Fe}$; 1,091 cm^{-1}, $v_{Si-O-Si}$), an absorption peak at 1,730 cm^{-1} appears in the FTIR spectrum of $Fe_3O_4@SiO_2$–MPS, which is assigned to the C=O, indicating the successful modification of MPS. In the FTIR spectrum of $Fe_3O_4@SiO_2@PMSA$, the newly appeared absorption peaks at 1,040 cm^{-1} and 1,190 cm^{-1} are attributed to the stretching vibration of O=S=O in $-SO_3^-$, the absorption peak at 1,550 cm^{-1} is attributed to the bending vibration of N–H in MBA, and the absorption peak at 1,730 cm^{-1} is enhanced. All the above results indicate the successful modification of PMSA.

Figure 2.31(b) exhibits the EDX analysis of (I) $Fe_3O_4@SiO_2$ microspheres and (III) $Fe_3O_4@SiO_2@PMSA$ microspheres. The N and S elements which newly appear in the EDX spectrum of $Fe_3O_4@SiO_2@PMSA$ microspheres indicate that PMSA is successfully modified on the surface of $Fe_3O_4@SiO_2$–MPS microspheres.

Figure 2.31© shows the TGA curves of (I) $Fe_3O_4@SiO_2$ microspheres and (III) $Fe_3O_4@SiO_2@PMSA$ microspheres. The 6.7% mass loss of $Fe_3O_4@SiO_2$ microspheres is because the loss of water absorbed by the microspheres. It is calculated that the mass loss of $Fe_3O_4@SiO_2@PMSA$ microspheres is about 35.5%, indicating that there is a large amount of PMSA polymers modified on the surface.

Figure 2.31(d) shows the magnetic hysteresis curves of (IV) Fe_3O_4 microspheres, (I) $Fe_3O_4@SiO_2$ microspheres, and (III) $Fe_3O_4@SiO_2@PMSA$ microspheres. Their saturation magnetic values are estimated to be 64.3 emu·g^{-1},

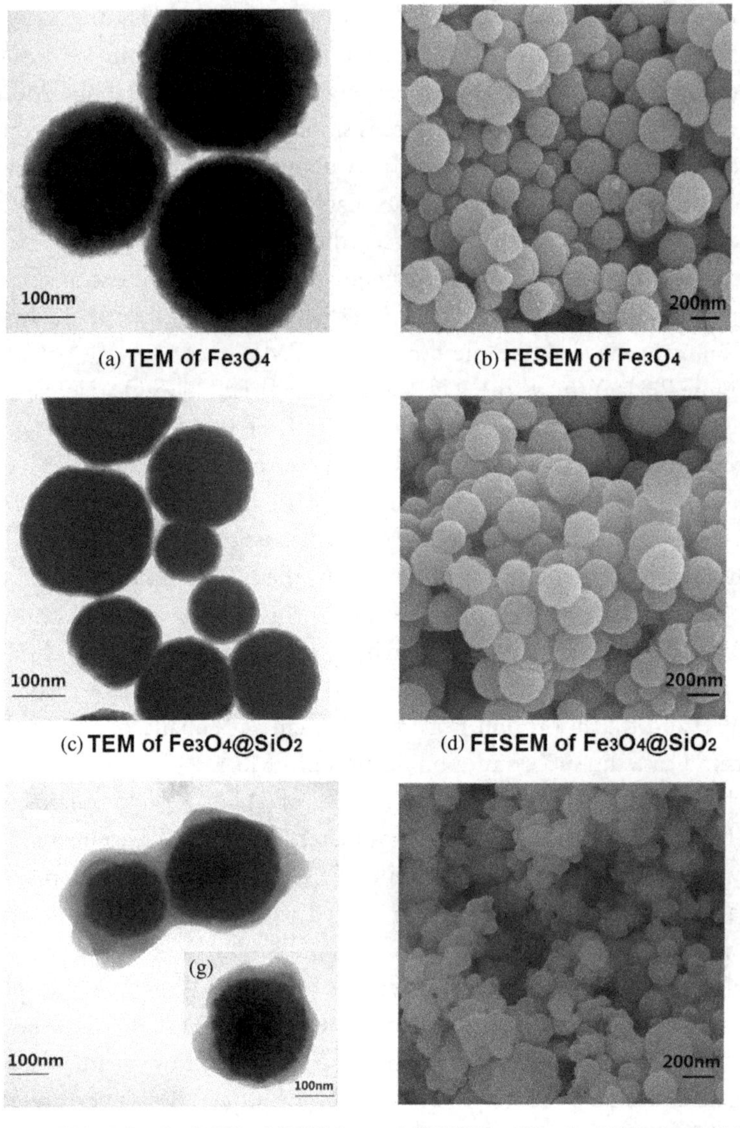

Figure 2.30: TEM and FESEM images of three different microspheres.

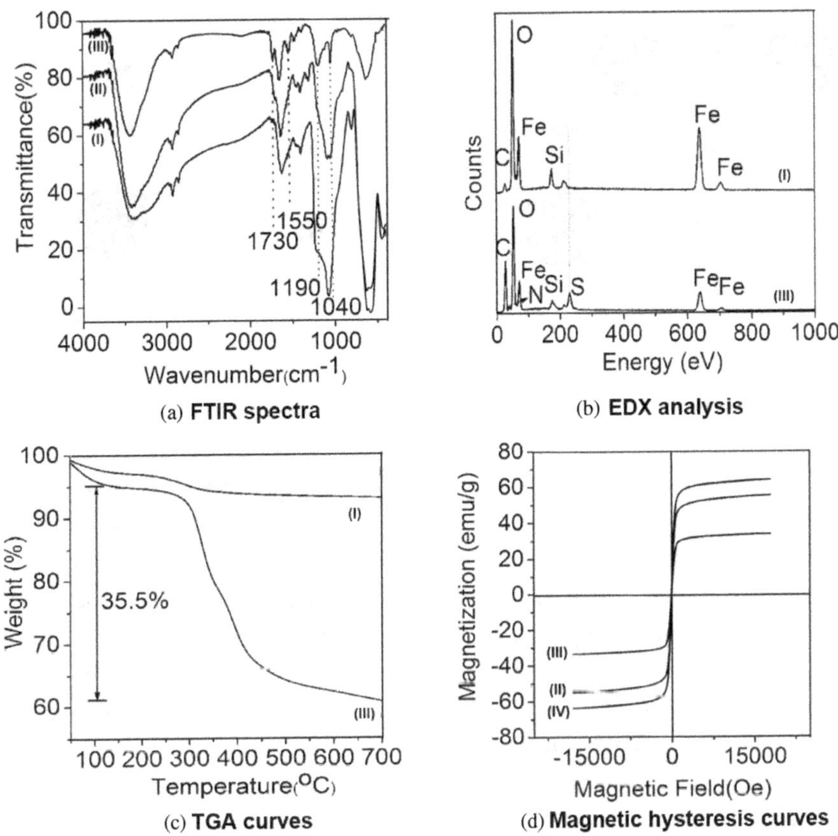

Figure 2.31: FTIR spectra, EDX analysis, TGA and magnetic hysteresis curves of (I) $Fe_3O_4@SiO_2$, (II) $Fe_3O_4@SiO_2$–MPS, and (III) $Fe_3O_4@SiO_2@PMSA$.

55.5 emu·g^{-1}, and 33.6 emu·g^{-1}, respectively, indicating all three materials have good magnetic response ability.

2.6.3. *Magnetic silica co-modified by maltose and PEG ($Fe_3O_4@SiO_2@PEG$-maltose)*

The synthetic process of $Fe_3O_4@SiO_2@PEG$-maltose is shown in Figure 2.32, and described as follows[17]:

Firstly, $Fe_3O_4@SiO_2$ microspheres are synthesized as described before. 400 mg $Fe_3O_4@SiO_2$ microspheres are rinsed by ethanol and then dispersed

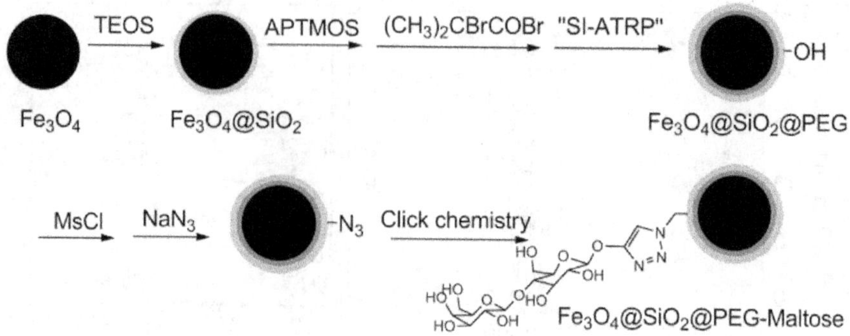

Figure 2.32: Schematic diagram of synthesis of Fe$_3$O$_4$@SiO$_2$@PEG–maltose.

in 60 mL isopropanol. Then 1 mL 3-aminopropyltrimethoxysilane (APTMOS) is added dropwise. Then the reaction lasts for 24 h under mechanical stirring. The resulting products are dispersed in 50 mL anhydrous dichloromethane after washed by isopropanol and dichloromethane. With the protection of argon gas and the condition of ice bath, 0.711 mL distilled triethylamine (5.0 mmol) and 0.528 mL 2-bromoisobutyryl bromide (4.2 mmol) are added in order. After continuous stirring for 2 h, the reaction then lasts for 16 h at room temperature. The resulting Fe$_3$O$_4$@SiO$_2$–Br are washed by dichloromethane, ethanol and deionized water and dried in vacuum for later use.

Polyethylene glycol (PEG) polymers are then grafted onto the surface of magnetic silica by surface-initiated atom transfer radical polymerization (SI-ATRP). And then hydroxyl groups on PEG branches are converted into azide groups that are used to immobilize maltose by click chemistry.

As shown in Figures 2.33(a) and 2.33(b), Fe$_3$O$_4$@SiO$_2$@PEG–maltose microspheres have a smooth and spherical surface and are sandwich shaped. The magnetic microsphere as core is about 240 nm, the silica layer as intermediate layer is about 6 nm, and the polymer layer as outermost layer is about 20 nm. The magnetic hysteresis curve of Fe$_3$O$_4$@SiO$_2$@PEG–maltose microspheres is shown in Figure 2.33(c), and it is estimated that its M_s is about 35.9 emu·g^{-1}. In the FTIR spectrum of Fe$_3$O$_4$@SiO$_2$@PEG–N$_3$ in Figure 2.33(d), the absorption peaks at 1,640 cm^{-1}, 2,855 cm^{-1}, and 1,450 cm^{-1} are attributed to the stretching vibration of C=O, the

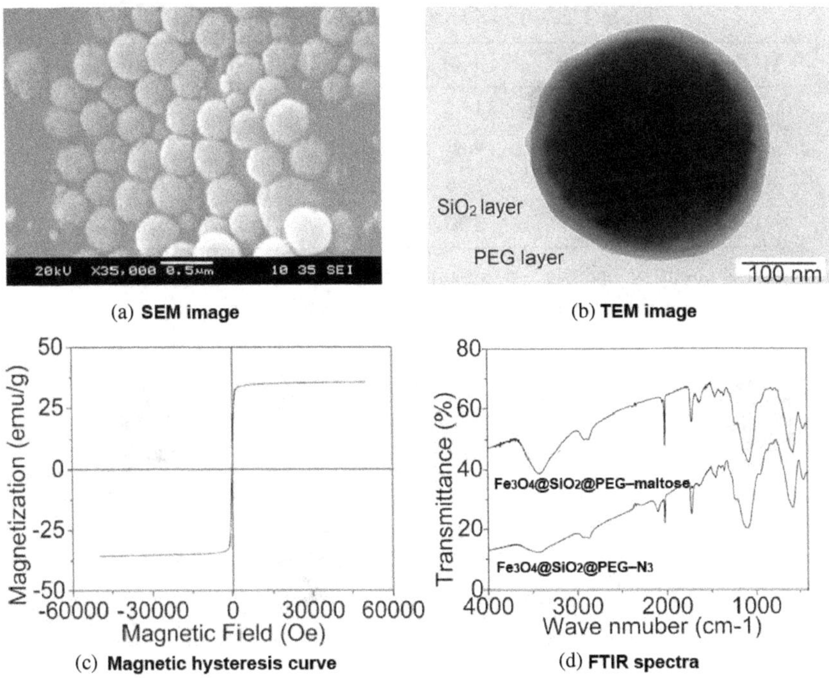

Figure 2.33: SEM image, TEM image, magnetic hysteresis curve of Fe_3O_4@SiO_2@PEG–maltose and FTIR spectra of different materials.

stretching vibration of C–H and the bending vibration of C–H, respectively, indicating that the PEG polymer layer is successfully modified on the surface of Fe_3O_4@SiO_2–Br microspheres. The absorption peak at 2,104 cm^{-1} is attributed to the symmetric stretching vibration of N=N=N. In the FTIR spectrum of Fe_3O_4@SiO_2@PEG–maltose, the characteristic peak of azide group at 2,104 cm^{-1} disappears, suggesting that maltose is successfully modified. According to EDX analysis in Table 2.2, the content of maltose in Fe_3O_4@SiO_2@PEG–maltose is 88.56 $\mu mol\cdot m^{-2}$. However, if maltose is modified directly on the surface of silica (Fe_3O_4@SiO_2–maltose, the schematic diagram of which is shown in Figure 2.34), the content of maltose is only 5.58 $\mu mol\cdot m^{-2}$. This result indicates that PEG plays an important role in maltose grafting. Moreover, PEG enhances the hydrophilicity of the material.

Table 2.2: EDX analysis of different materials.

MNP	C(%)	N(%)	H(%)	$X^{a}_{maltaose}$/μmol m^{-2}
Fe$_3$O$_4$@SiO$_2$@PEG-N$_3$	11.90	1.24	1.72	—
Fe$_3$O$_4$@SiO$_2$@PEG-Maltose	14.87	1.05	2.22	88.56
Fe$_3$O$_4$@SiO$_2$-N$_3$	2.10	<0.3	0.42	—
Fe$_3$O$_4$@SiO$_2$-Maltose	2.30	<0.3	0.45	5.58

[a]The amount of maltose was obtained *via* calculation based on the carbon content from the elemental analysis.

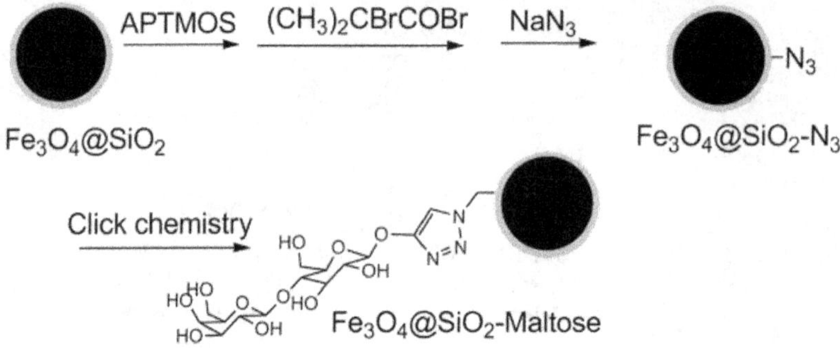

Figure 2.34: Schematic diagram of synthesis of Fe$_3$O$_4$@SiO$_2$–maltose.

2.6.4. *Magnetic silica co-modified by chitosan and hyaluronic acid (MNPs–(HA–CS)$_n$)*

As shown in Figure 2.35, Fe$_3$O$_4$@SiO$_2$ microspheres are synthesized.[18] Then 200 mg Fe$_3$O$_4$@SiO$_2$ microspheres are dispersed in 50 mL isopropanol, and 1 mL APTEOS is added dropwise. Then the reaction lasts for 24 h under mechanical stirring at room temperature. The resulting product (MNPs–NH$_2$ microsphere) is dried in vacuum for later use. Next, 50 mg MNPs–NH$_2$ microspheres are dispersed in hyaluronic acid (HA) solution (1 mg·mL^{-1}, 0.135 mol·L^{-1} NaCl, pH = 5) after washed by ethanol and deionized water. Then after stirring for 20 min, the resulting MNPs–HA are washed by deionized water, and then dispersed in chitosan (CS) solution (1 mg·mL^{-1}, 0.135 mol·L^{-1} NaCl, pH = 5) for another 20 min of stirring. If n = 10, the product is MNPs–(HA–CS)$_{10}$. The MNPs–(HA–CS)$_n$ is dispersed in PBS

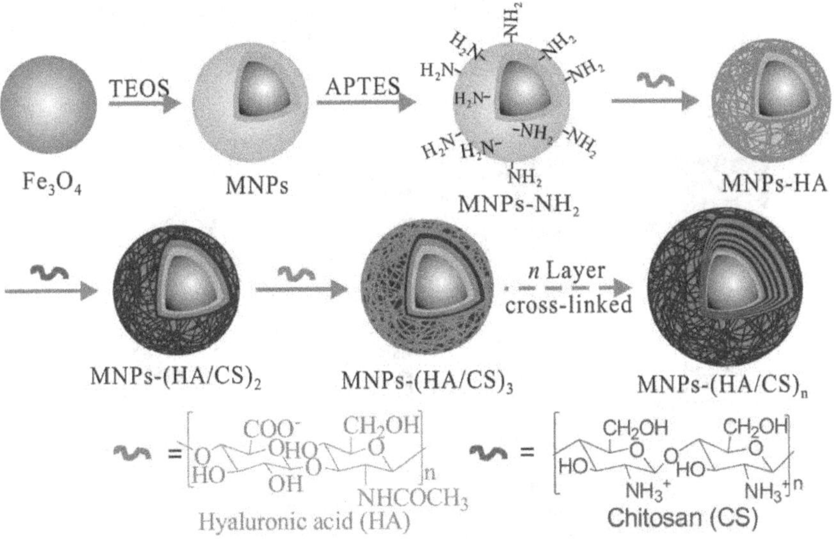

Figure 2.35: Schematic diagram of synthesis of MNPs–(HA–CS)$_n$.

solution (10 mmol·L^{-1}, pH = 5.5, containing 2 mg·mL^{-1} EDC and 2 mg·mL^{-1} NHS) for overnight. The obtained product is washed by deionized water, ethanol and acetonitrile to obtain the MNPs–(HA–CS)$_n$.

As shown in Figure 2.36(a), MNPs–NH$_2$ microspheres with core–shell structure contain magnetic microsphere as core and SiO$_2$ as outer layer. The particle size of core is about 220 nm and the thickness of SiO$_2$ layer is about 3 nm. When n = 10, the thickness of polymer layer is about 9 nm (as seen in Figure 2.36(b)), it can be inferred that the thickness of (HA–CS)$_n$ is proportional to the number of polymer layers by comparing to Figures 2.36(c) and 2.36(d) further. The average thickness of the bilayer is about 1.8 nm. In addition, the saturation magnetic value of MNPs–(HA–CS)$_{10}$ microspheres is calculated to be 58.8 emu·g^{-1}. Table 2.3 shows the elemental analysis results of different materials. According to the data in the table, the content of HA–CS deposited on the surface of the micro-sphere is basically equivalent every time. The disaccharide content of MNPs–(HA–CS)$_{10}$ microspheres reaches 164.33 μmol·m^{-2}, which greatly enhances the hydrophilicity of the material.

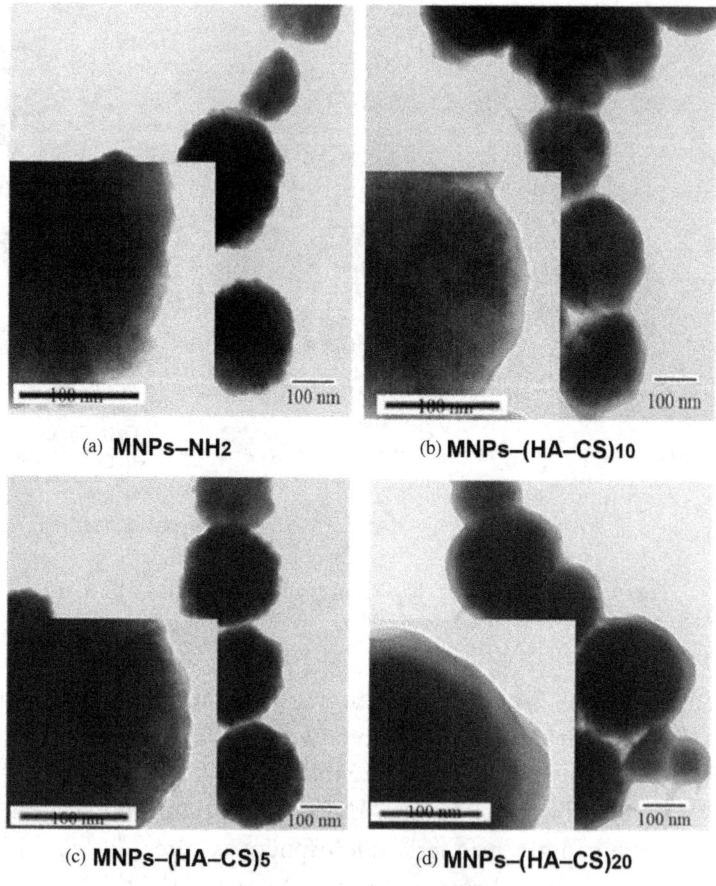

Figure 2.36:   TEM images of different materials.

Table 2.3:   EDX analysis of different materials.

MNP	C(%)	N(%)	H(%)	$X^{a}_{\text{disaccharide unit}}$ [μmol m^{-2}]
MNPs-NH$_2$	2.61	1.01	<0.3	—
MNPs-HA	3.28	1.14	<0.3	15.05
MNPs-(HA/CS)$_5$	5.13	1.00	0.64	79.34
MNPs-(HA/CS)$_{10}$	7.54	1.33	1.25	164.33
MNPs-(HA/CS)$_{20}$	10.74	2.10	2.84	299.87

[a]The amount of disaccharide units was obtained *via* a calculation based on the carbon content from the elemental analysis.

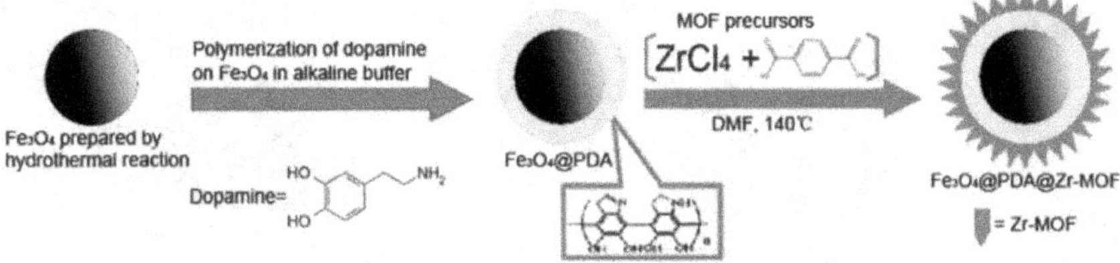

Figure 2.37: Schematic diagram of synthesis of Fe_3O_4@PDA@Zr–MOF.

2.6.5. *Polydopamine-modified magnetic microsphere for immobilization of metal organic framework (Fe_3O_4@PDA@Zr–MOF)*

As shown in Figure 2.37, the magnetic Fe_3O_4 is firstly synthesized, and then a layer of PDA is coated on the surface of Fe_3O_4.[19] Next, 78 mg $ZrCl_4$ and 10 mg terephthalic Fe_3O_4@PDA microspheres are added into the mixed dispersion solution. Finally, the whole mixed solution reacts at 140°C for 20 min, and the resulting product is washed with ethanol.

TEM images of Fe_3O_4@PDA microspheres are shown in Figures 2.38(a) and 2.38(c), and the thickness of the PDA layer is about 40 nm. The TEM images of Fe_3O_4@PDA@Zr–MOF in Figures 2.38(b) and 2.38(d) show that the boundary between the PDA layer and the Zr–MOF layer is not very clear, but the diameter of the whole microsphere increases by 47 nm after coating the Zr–MOF layer on the surface of the Fe_3O_4@PDA microsphere. It can be seen from the EDX spectrum of Fe_3O_4@PDA@Zr–MOF (Figure 2.39) that Zr element do exist in the material, indicating that Zr–MOF is successfully coated on the surface of Fe_3O_4@PDA microspheres. This is confirmed further by FTIR and XRD characterization.

Figure 2.40 shows the FTIR spectra of Fe_3O_4@PDA@Zr–MOF and a series of intermediate products. In the spectrum of Fe_3O_4 microspheres, the absorption peaks at 571 cm^{-1} and 3,396 cm^{-1} are attributed to the vibration of Fe–O–Fe and the stretching vibration of O–H on the surface of Fe_3O_4 microsphere, respectively. In the spectrum of Fe_3O_4@PDA microspheres, new absorption peak at 1,645 cm^{-1} can be attributable to bending vibration of N–H, the absorption peaks at 1,616 cm^{-1} and 1,434 cm^{-1}, 1,498 cm^{-1} belong to stretching vibration of C=C, the absorption peak at 1,292 cm^{-1} belongs to

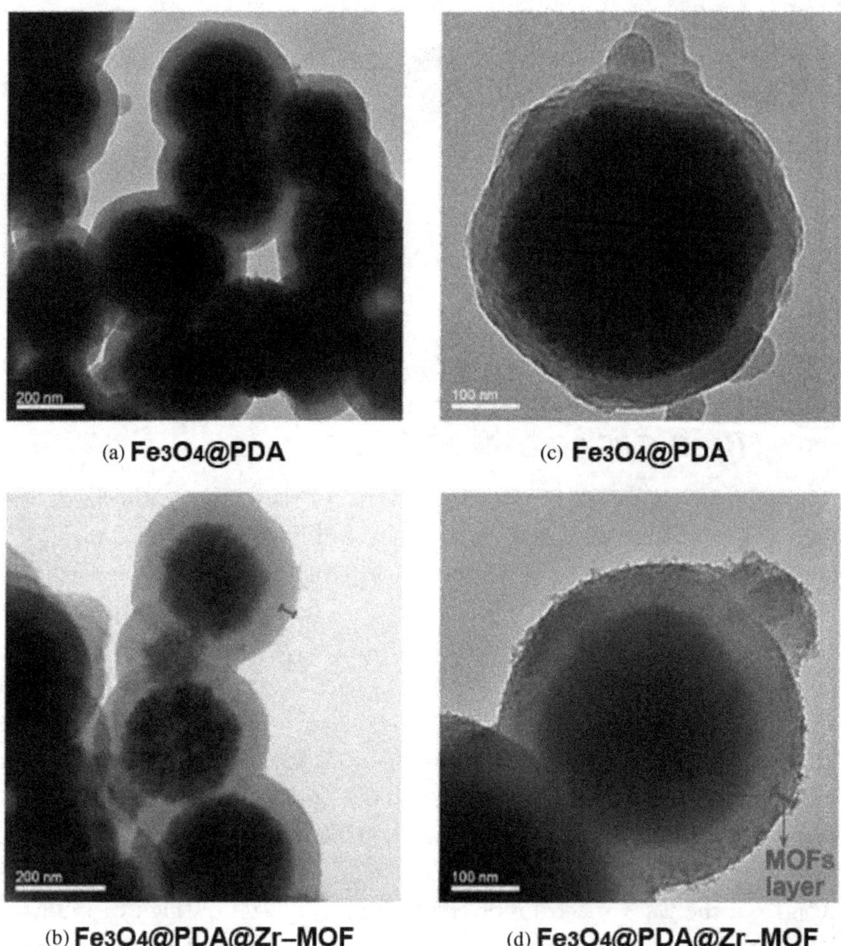

Figure 2.38: TEM images of Fe$_3$O$_4$@PDA and Fe$_3$O$_4$@PDA@Zr–MOF.

stretching vibration of C–O, the absorption peaks in the range from 1,200 cm^{-1} to 1.400 cm^{-1} belong to the bending vibration of –CH$_2$, indicating the successful coating of PDA layer. In the spectrum of Fe$_3$O$_4$@PDA@Zr–MOF, the absorption peak at 1,740 cm^{-1} is attributed to the stretching vibration of C=O in the ligand of terephthalic, and the absorption peak at 3,438 cm^{-1} is the stretching vibration of O–H in the carboxyl groups. Further, the XRD pattern of Fe$_3$O$_4$@PDA@Zr–MOF in Figure 2.41 indicates the success-

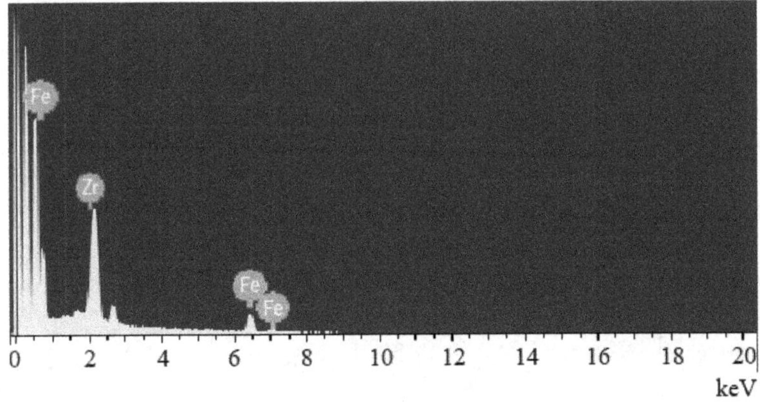

Figure 2.39: EDX analysis of Fe_3O_4@PDA@Zr–MOF.

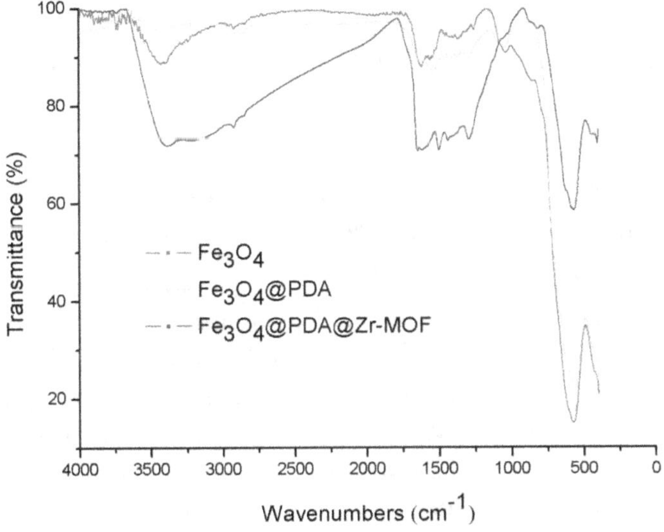

Figure 2.40: FTIR spectra of different materials.

ful coating of Zr–MOF layer. In addition, the BET specific surface area of Fe_3O_4@PDA@Zr–MOF is measured as 216.14 $m^2 \cdot g^{-1}$ according to the N_2 adsorption–desorption isotherms.

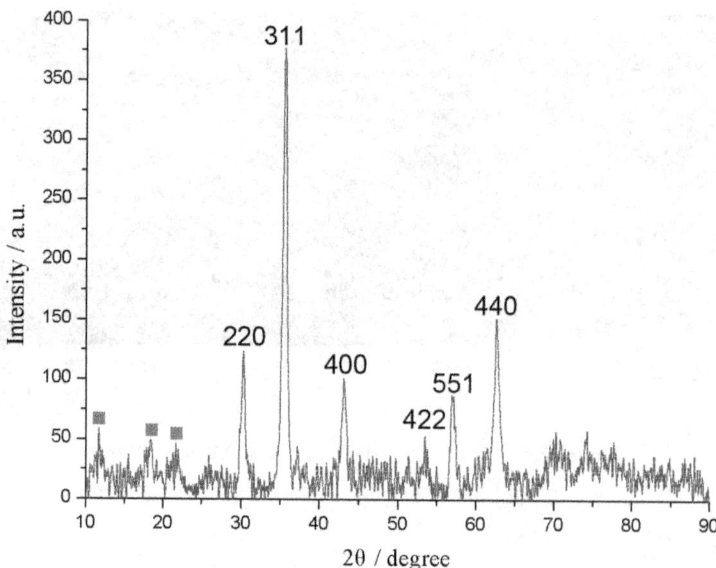

Figure 2.41: XRD pattern of Fe_3O_4@PDA@Zr–MOF (▪ represents the characteristic peak of Zr–MOF).

2.6.6. *Mercaptoacetic acid-modified magnetic microsphere for immobilization of MOF (Fe_3O_4@MIL–100(Fe))*

As shown in Figure 2.42, the magnetic Fe_3O_4 is first synthesized, and then the mercaptoacetic acid (MerA) is modified on the surface of Fe_3O_4 as follows: 350 mg Fe_3O_4 microspheres are washed with ethanol 3 times, and then dissolved in 0.58 mmol·L^{-1} MerA solution (75 mL).[20] This is then kept under stirring gently for 24 h under protection of nitrogen to obtain Fe_3O_4–MerA which should be washed with ethanol several times until there is no irritating smell. Next, Fe_3O_4@MIL–100(Fe) is prepared according to the following process: 100 mg Fe_3O_4–MerA microspheres are dispersed in 5 mL $FeCl_3$ solution (10 mmol·L^{-1}), then left standing for 15 min after blending. The resulting product is dispersed in ethanol solution of 5 mL trimesic acid (H_3btc, 10 mmol·L^{-1}) after it is washed with ethanol. This self-assembly process is repeated 31 times according to the schematic diagram of synthesis in Figure 2.42 to obtain the final product that is washed with ethanol and dried in vacuum for later use.

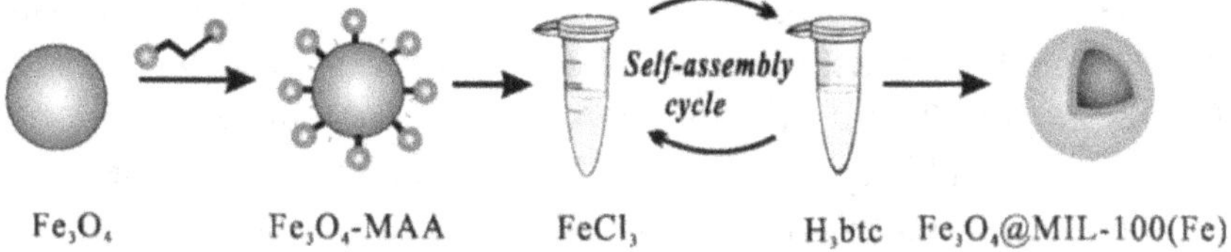

Figure 2.42: Schematic diagram of synthesis of Fe_3O_4@MIL–100(Fe).

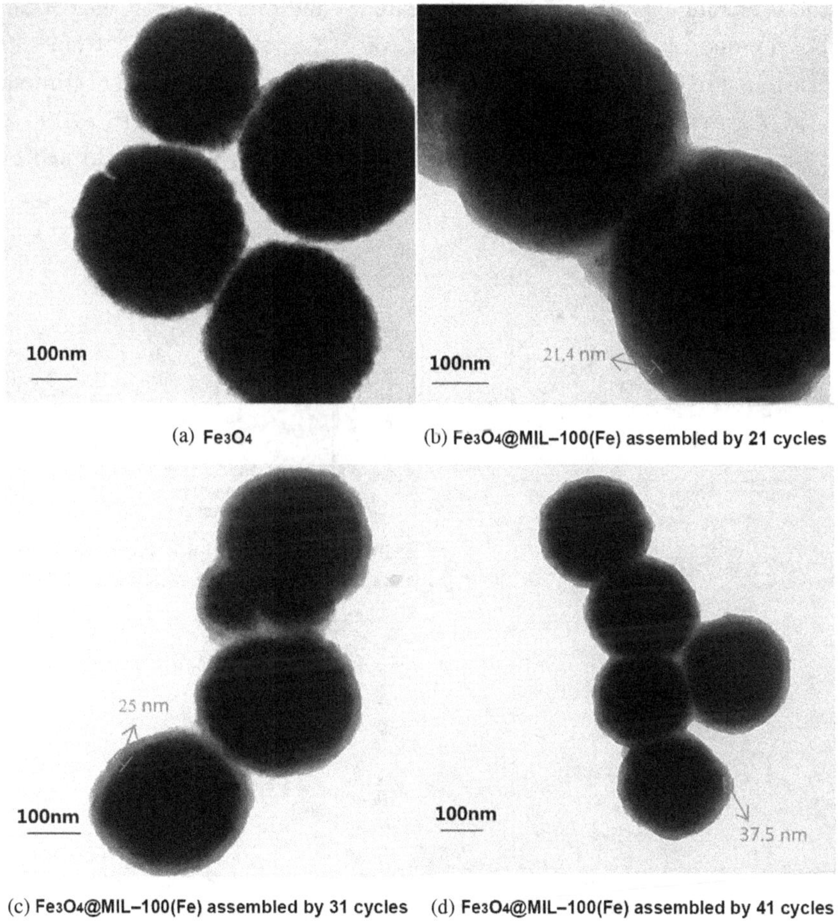

Figure 2.43: TEM images of Fe_3O_4 and Fe_3O_4@MIL–100(Fe) produced by different cycles of self-assembly.

As shown in Figure 2.43(a), the diameter of Fe_3O_4 microspheres is about 345 nm and they disperse uniformly without agglomeration. Compared with Figures 2.43(b)–2.43(d), the thickness of MOF layer increases with the number of assembly cycles in the synthetic process. Compared the FTIR spectra of Fe_3O_4 microspheres with Fe_3O_4@MIL–100(Fe), the absorption peak at 1,715 cm^{-1} in Figure 2.44(a) is attributed to the stretching vibration of C=O in the carboxyl group, which is a new peak in the spectrum of Fe_3O_4@MIL–100(Fe), moreover, the absorption peak at 1,375 cm^{-1} is attributed to the stretching vibration of C–O, these jointly indicate the emergence of carboxyl groups. Further, the absorption peaks at 1,566 cm^{-1} and 1,450 cm^{-1} are attributed to the stretching vibration of benzene ring in the ligand of trimesic acid. All the above results indicate the successful coating of MIL–100(Fe) layer on the surface of magnetic microspheres, which is also confirmed by

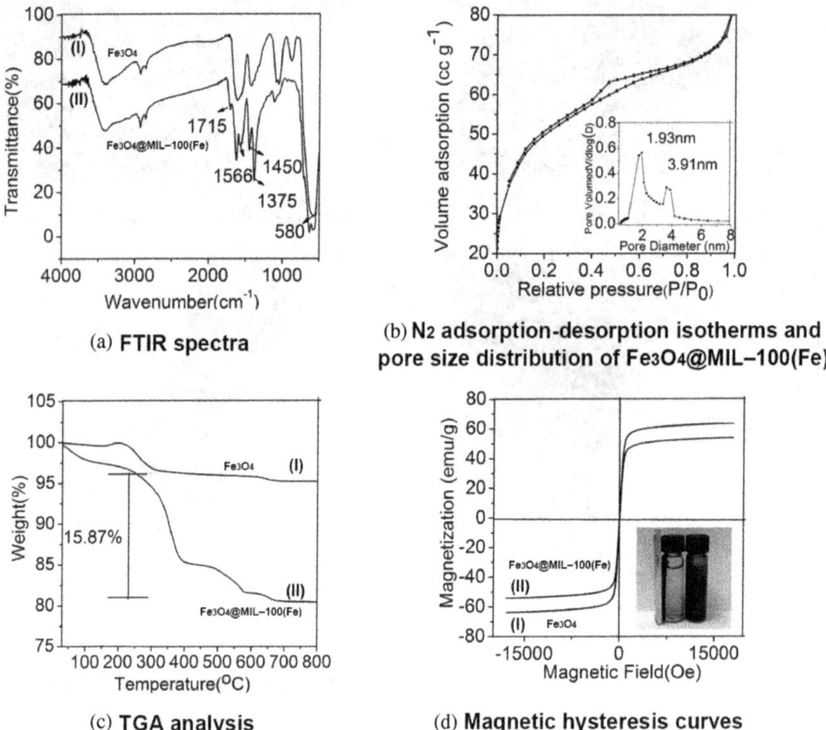

Figure 2.44: FTIR spectra, N_2 adsorption–desorption isotherms, TGA analysis and magnetic analysis of two kinds of microspheres.

TGA analysis in Figure 2.44(c). As shown in Figure 2.44(b), Fe_3O_4@MIL–100(Fe) possesses two main pore sizes of 1.93 nm and 3.91 nm, indicating its porous structure. From Figure 2.44(d), the saturation magnetic values of Fe_3O_4 and Fe_3O_4@MIL–100(Fe) are 63.7 emu·g^{-1} and 53.1 emu·g^{-1}, respectively.

2.6.7. *Fe_3O_4@SiO$_2$@PSV (Using 3-methacryloxypropyl trimethoxysilane, MPS, as linker)*

As shown in Figure 2.45, Fe_3O_4@SiO$_2$ microspheres are firstly synthesized, and then polystyrene-vinylphenylboronic acid (PSV) is coated on the surface of Fe_3O_4@SiO$_2$ microspheres according to the following process: 1 mL MPS and 50 mL ethanol solution of Fe_3O_4@SiO$_2$ microspheres are mixed for mechanical stirring about 6 h.[21] The resulting solid products are dispersed in 200 mL sodium dodecyl sulfate (SDS, 0.01 g) aqueous solution, then 0.5 mL styrene and 0.038 g 4-vinylphenyl boronic acid (VPBA) are added. Under the protection of nitrogen, aqueous solution containing 0.03 g potassium persulfate is added, and then the mixture is mechanically stirred at 75°C for 8 h. The reaction is terminated by passing into air. The obtained Fe_3O_4@SiO$_2$@PSV are cleaned with deionized water and ethanol, and dried in vacuum for later use.

As shown in Figures 2.46(a) and 2.47(a), Fe_3O_4@SiO$_2$@PSV microspheres have a core–shell–shell structure, in which the core of Fe_3O_4 microspheres is about 120 nm in diameter, the middle SiO$_2$ layer is about 50 nm,

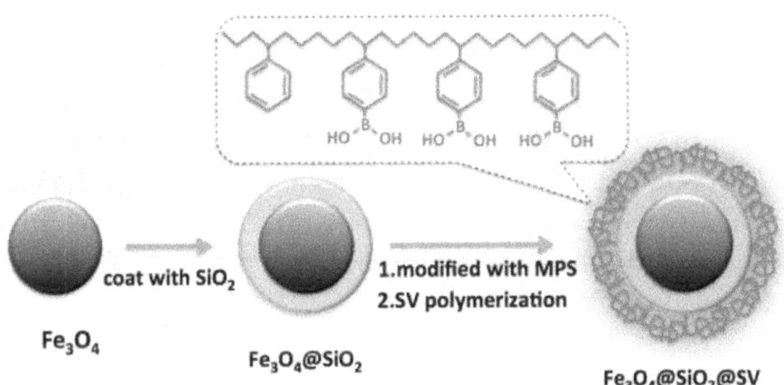

Figure 2.45: Schematic diagram of synthesis of Fe_3O_4@SiO$_2$@PSV.

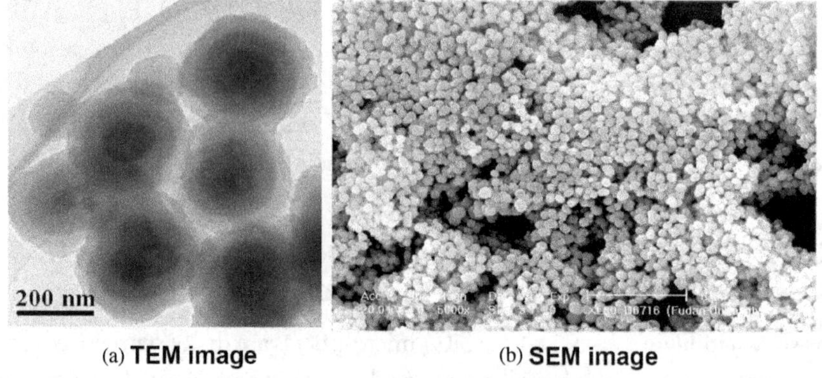

(a) TEM image   (b) SEM image

Figure 2.46:   TEM and SEM image of $Fe_3O_4@SiO_2@PSV$.

and the outermost PSV copolymer layer is about 50 nm. It can be observed from Figure 2.46(b) that $Fe_3O_4@SiO_2@PSV$ microspheres are spherical and have uniform size. The EDX mapping of $Fe_3O_4@SiO_2@PSV$ microspheres are shown in Figures 2.47(c) and 2.47(d). The distribution of Fe, Si, and B further indicates that $Fe_3O_4@SiO_2@PSV$ microspheres are composed of Fe_3O_4 core, intermediate SiO_2 layer and outermost PSV copolymer layer.

Figure 2.48(a) shows the FTIR spectrum of $Fe_3O_4@SiO_2$ microspheres. The absorption peaks at 1,089 cm^{-1} is attributed to the stretching vibration of Si–O–Si, and those at 1,634 cm^{-1} and 3,428 cm^{-1} are attributed to the absorbed water and the hydroxyl groups on the surface of $Fe_3O_4@SiO_2$ microspheres, indicating that the SiO_2 layer is successfully coated on the surface of magnetic microspheres. Figure 2.48(b) shows the FTIR spectrum of $Fe_3O_4@SiO_2$–MPS microspheres. The new absorption peak at 1,713 cm^{-1} is attributed to the C=O of MPS. The 1,377 cm^{-1} in Figure 2.48(c) belongs to the vibration characteristic peak of B–O, and the absorption peaks at 1,602 cm^{-1} and 909 cm^{-1} are attributed to the C=C and H–C (=C) in styrene and VPBA, respectively. The absorption peaks of benzene ring at 1,453 cm^{-1}, 1,493 cm^{-1}, and 1,602 cm^{-1} are attributed to the twisted vibration of m-benzene ring. All the above results indicate the successful synthesis of $Fe_3O_4@SiO_2@PSV$ microspheres.

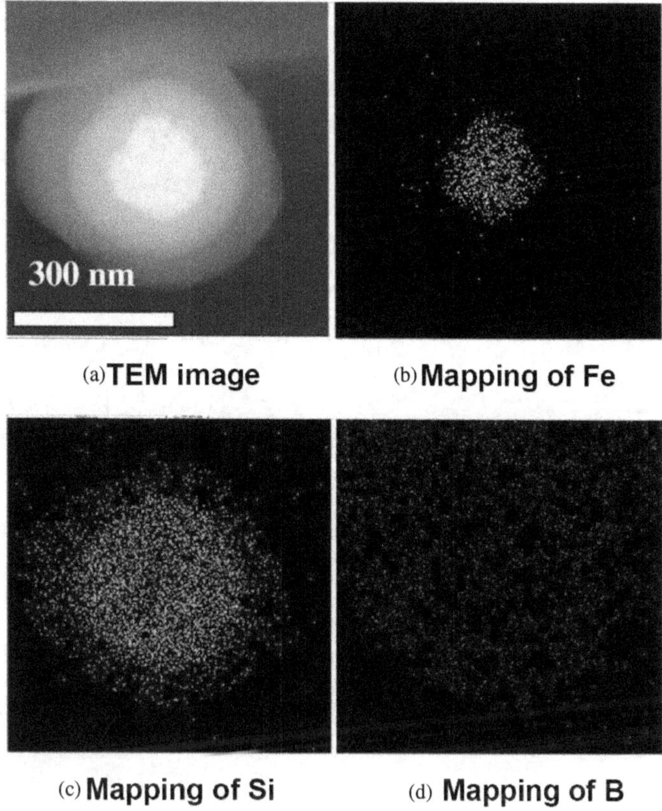

Figure 2.47: TEM image and EDX mapping of $Fe_3O_4@SiO_2@PSV$.

In addition, the mass loss caused by the copolymer layer of MPS and PSV on $Fe_3O_4@SiO_2@PSV$ microspheres can be calculated as 30.03 wt.% by TGA analysis. It can be seen from Figure 2.49 that the mass loss of $Fe_3O_4@SiO_2–$MPS is 2.78 wt.% within the same temperature range, indicating that the mass percentage of PSV is 27.25 wt.%, namely, there is a large amount of boric acid in $Fe_3O_4@SiO_2@PSV$ microspheres.

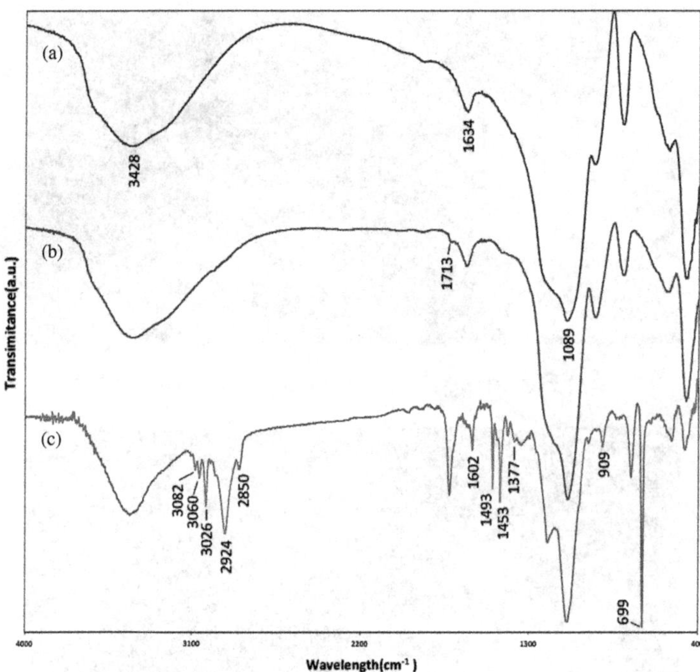

Figure 2.48: FTIR spectra of different materials (curve (a) $Fe_3O_4@SiO_2$; curve (b) $Fe_3O_4@SiO_2$–MPS; curve (c) $Fe_3O_4@SiO_2@PSV$).

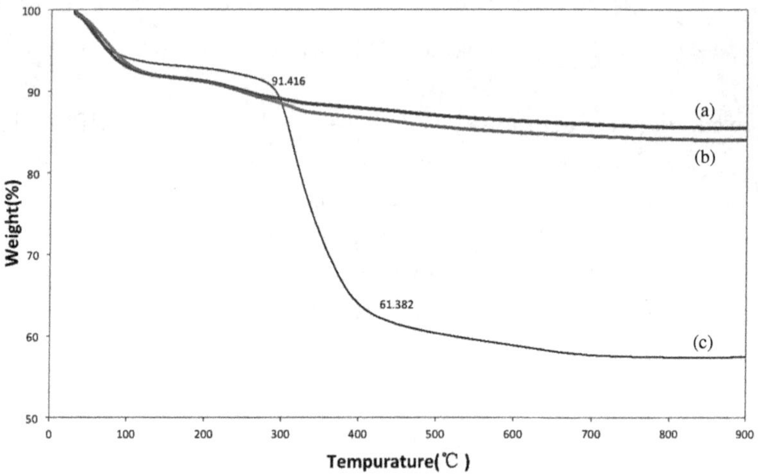

Figure 2.49: TGA analysis of different materials (curve (a) $Fe_3O_4@SiO_2$; curve (b) $Fe_3O_4@SiO_2$–MPS; curve (c) $Fe_3O_4@SiO_2@PSV$).

2.6.8. *Iminodiacetic acid-modified magnetic microsphere for immobilization of metal ion*

2.6.8.1. *Preparation of Fe_3O_4@SiO_2@Glymo–IDA–Fe^{3+}*

The synthetic process of Fe_3O_4@SiO_2@Glymo–IDA–Fe^{3+} is shown in Figure 2.50.[22]

(1) Synthesis of Fe_3O_4@SiO_2

The detailed synthetic process can be found in Section 2.4.

(2) Synthesis of Glymo–IDA

Firstly, 4.20 g iminodiacetic acid (IDA) is dispersed in 50 mL Na_2CO_3 solution (2 mol·L^{-1}), the pH value of which is adjusted to be 11 by using 10 mol·L^{-1} NaOH solution. Then the entire reaction system is subjected to an ice bath and stirred magnetically for 1 h, in the meanwhile, 1.5 g 3-glycidoxypropyltrimethoxysilane (Glymo) is added dropwise within 0.5 h. The reaction system is then heated to 65°C for 6 h, and the reaction system is then cooled to 0°C. The above steps are repeated twice. Finally, the pH value of the mixture solution is adjusted to 6 by concentrated hydrochloric acid for next step.

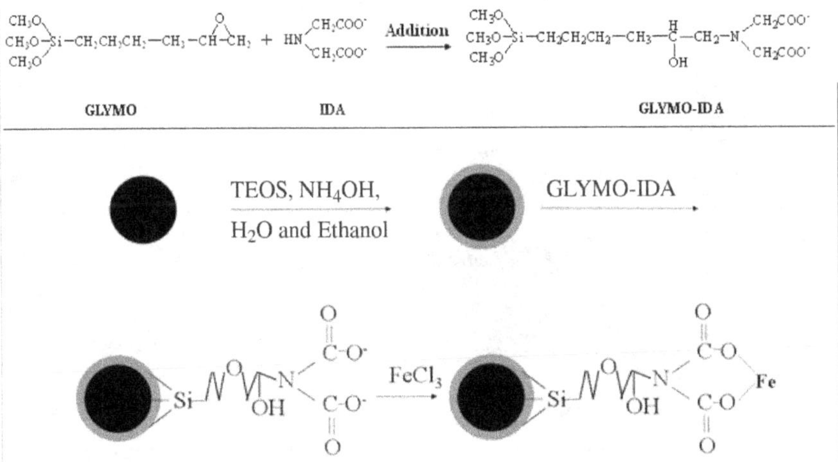

Figure 2.50: Schematic diagram of synthesis of Fe_3O_4@SiO_2@Glymo–IDA–Fe^{3+}.

(3) Synthesis of Fe_3O_4@SiO_2@Glymo–IDA

Firstly, 0.02 g Fe_3O_4@SiO_2 microspheres are dispersed in 50 mL anhydrous ethanol, and 10.0 mL the above Glymo–IDA solution is added to maintain reaction for 24 h at 40°C. The solid product Fe_3O_4@SiO_2@Glymo–IDA microspheres are cleaned with ethanol.

(4) Synthesis of Fe_3O_4@SiO_2@Glymo–IDA–Fe^{3+}

Fe_3O_4@SiO_2@Glymo–IDA is dispersed in 20 mL $FeCl_3$ solution (0.2 mol·L^{-1}), after oscillated and dispersed for 2 h, the obtained Fe_3O_4@SiO_2@ Glymo–IDA–Fe^{3+} microspheres are cleaned with deionized water and dried in vacuum for later use.

Fe_3O_4@SiO_2@Glymo–IDA can be dispersed in other metal salt solutions to prepare materials that immobilize other metal ions on the surface such as Cu^{2+}, TiO^{2+}, etc.

2.6.8.2. Characterization of Fe_3O_4@SiO_2@Glymo–IDA–Fe^{3+}

From the SEM images of Fe_3O_4 and Fe_3O_4@SiO_2@Glymo–IDA–Fe^{3+} in Figure 2.51, it can be seen that magnetic silica microspheres have good dispersity and morphological uniformity before and after chemical bonding.

The FTIR spectra of Fe_3O_4@SiO_2 and Fe_3O_4@SiO_2@Glymo–IDA–Fe^{3+} microspheres are shown in Figure 2.52. The stretching vibration absorption

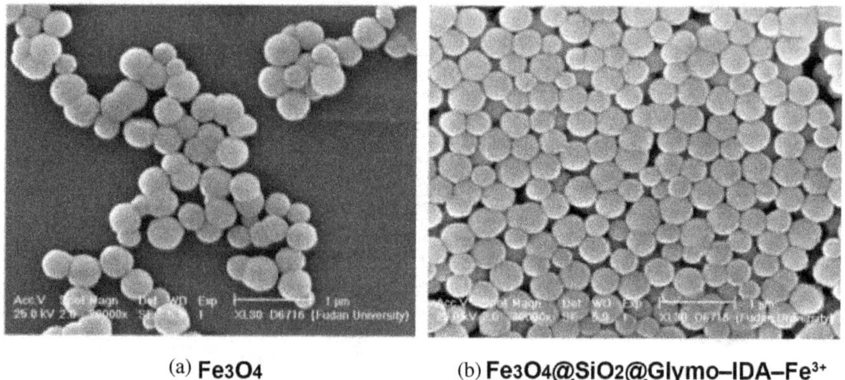

(a) **Fe3O4** (b) **Fe3O4@SiO2@Glymo–IDA–Fe³⁺**

Figure 2.51: SEM images of two kinds of microspheres.

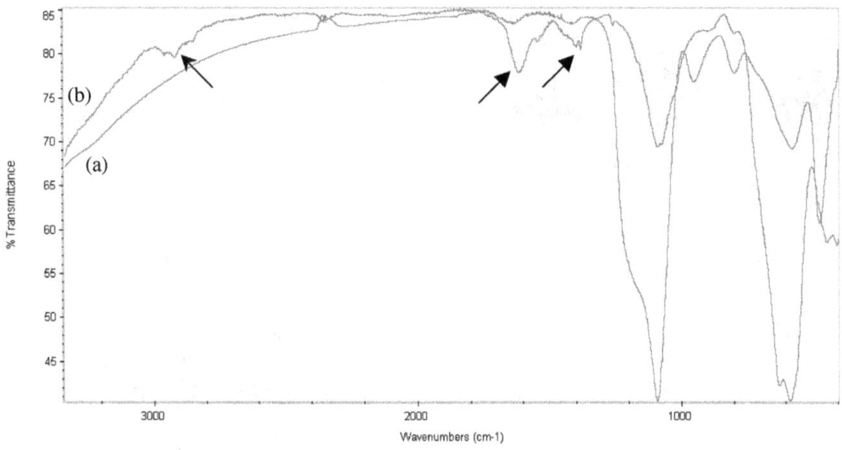

Figure 2.52: FTIR spectra of different microspheres (curve (a) $Fe_3O_4@SiO_2$; curve (b) $Fe_3O_4@SiO_2@Glymo–IDA–Fe^{3+}$).

peaks of Si–O–Si can be found in both Figures 2.52(a) and 2.52(b), indicating that SiO_2 is successfully coated on the surface of Fe_3O_4 microspheres. The absorption peak around 2,900 cm^{-1} could be attributed to CH_2 in the silane coupling reagent, and those at 1,610 cm^{-1} and 1,400 cm^{-1} are attributed to carboxyl groups, indicating the successful modification of Glymo–IDA on the surface of silica.

2.6.9. *Aminophenylboronic acid-modified magnetic silica ($Fe_3O_4@SiO_2$–APBA, using Glymo as linker)*

2.6.9.1. *Preparation of $Fe_3O_4@SiO_2$-APBA*

The synthetic process of $Fe_3O_4@SiO_2$–APBA is shown in Figure 2.53.[23]

(1) Synthesis of $Fe_3O_4@SiO_2$

The detailed synthetic process can be found in Section 2.4.

(2) Synthesis of $Fe_3O_4@SiO_2$–APBA

Firstly, 50 mg $Fe_3O_4@SiO_2$ microspheres are distributed in 40 mL toluene containing 400 μL Glymo for reflux reaction for 12 h at 80°C. Then 50 mg

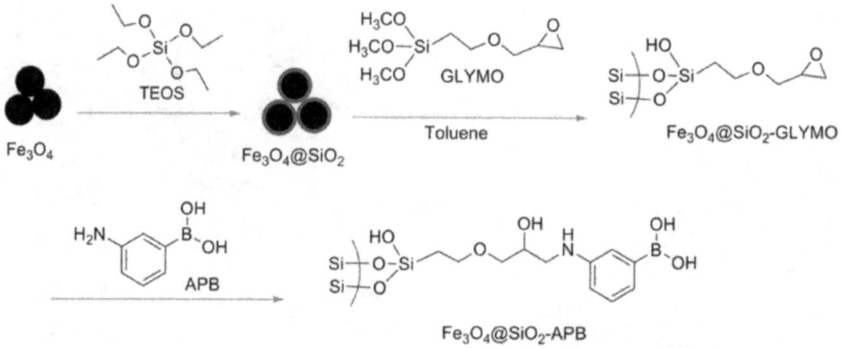

Figure 2.53: Schematic diagram of synthesis of Fe_3O_4@SiO_2–APBA.

3-aminophenylboronic acid (APBA) is ultrasonically dispersed in 60 mL NH4HCO3 solution (50 mmol·L^{-1}, pH > 8), 20 mL from which is mixed with Fe_3O_4@SiO_2–Glymo for mechanical stirring for 3 h at 65°C. The obtained solid product by magnetic separation is redispersed in 20 mL APBA solution. Repeat the process 3 times to obtain the Fe_3O_4@SiO_2–APBA.

2.6.9.2. *Characterization of Fe_3O_4@SiO_2–APBA*

Figure 2.54 shows the FTIR spectra of Fe_3O_4 microspheres, Fe_3O_4@SiO_2 microspheres and Fe_3O_4@SiO_2–APBA microspheres. For curve (a), the absorption peak at 575 cm^{-1} is attributed to the bending vibration of Fe–O. For curve (b), the absorption peak at 1,088 cm^{-1} is attributed to the stretching vibration of Si–O–Si, proving the successful coating of SiO_2 on the surface of Fe_3O_4 microspheres. For curve (c), the absorption peaks at 2,865 cm^{-1} and 2,923 cm^{-1} correspond to the CH_2 in Glymo, the absorption peaks at 1,622 cm^{-1}, 1,437 cm^{-1}, and 796 cm^{-1} are attributed to the stretching vibration and bending vibration of benzene ring, the absorption peak of B–O is at 1,380 cm^{-1}.

Figure 2.55 shows the magnetic hysteresis curves of Fe_3O_4 microspheres, Fe_3O_4@SiO_2 microspheres and Fe_3O_4@SiO_2–APBA microspheres. It can be calculated that the saturation magnetic values of these three materials are 94.46 emu·g^{-1}, 52.71 emu·g^{-1}, and 49.70 emu·g^{-1}, respectively. The decrease of M_s indirectly indicates that the intermediate layer and the outermost layer of the microsphere are successfully coated with silica and boric acid, respectively. Moreover, the content of B elements on the surface of Fe_3O_4@

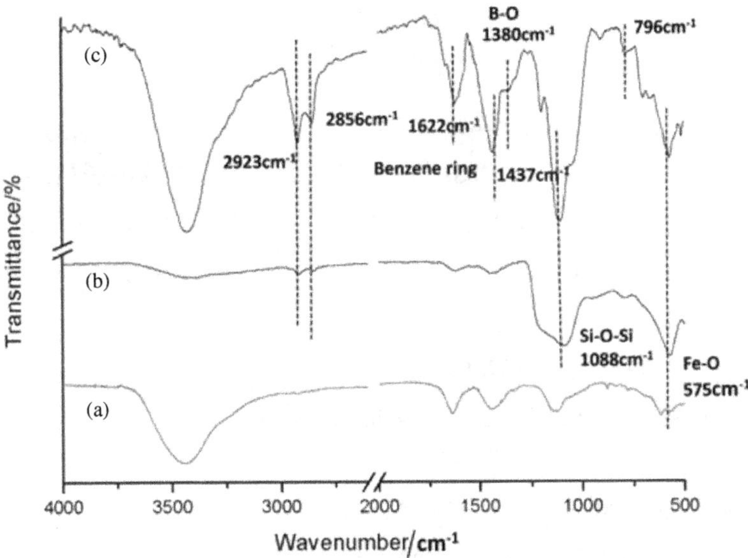

Figure 2.54: FTIR spectra of different materials (curve (a) Fe_3O_4; curve (b) $Fe_3O_4@SiO_2$; curve (c) $Fe_3O_4@SiO_2$–APBA).

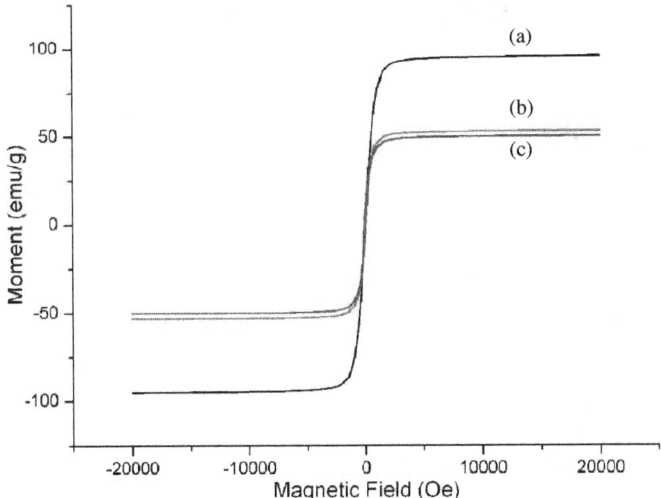

Figure 2.55: Magnetic hysteresis curves of different materials (curve (a) Fe_3O_4; curve (b) $Fe_3O_4@SiO_2$; curve (c) $Fe_3O_4@SiO_2$–APBA).

SiO$_2$–APBA microspheres is about 15 mmol·g^{-1} by inductively coupled plasma-atomicemission spectroscopy (ICP-AES) analysis.

2.6.10. *Aminophenylboronic acid-modified magnetic Au–silica microsphere (Fe$_3$O$_4$@SiO$_2$@Au–APBA)*

2.6.10.1. *Preparation of Fe$_3$O$_4$@SiO$_2$@Au–APBA*

The synthetic process of Fe$_3$O$_4$@SiO$_2$@Au–APBA is shown in Figure 2.56.[24]

(1) Synthesis of Fe$_3$O$_4$@SiO$_2$

The detailed synthetic process can be found in Section 2.4.

(2) Synthesis of Fe$_3$O$_4$@SiO$_2$@Au

Synthesis of Au nanoparticles: 500 mL HAuCl$_4$ solution (1 mmol·L^{-1}) is stirred at 100°C for reflux reaction, and 50 mL 38.8 mmol·L^{-1} sodium citrate

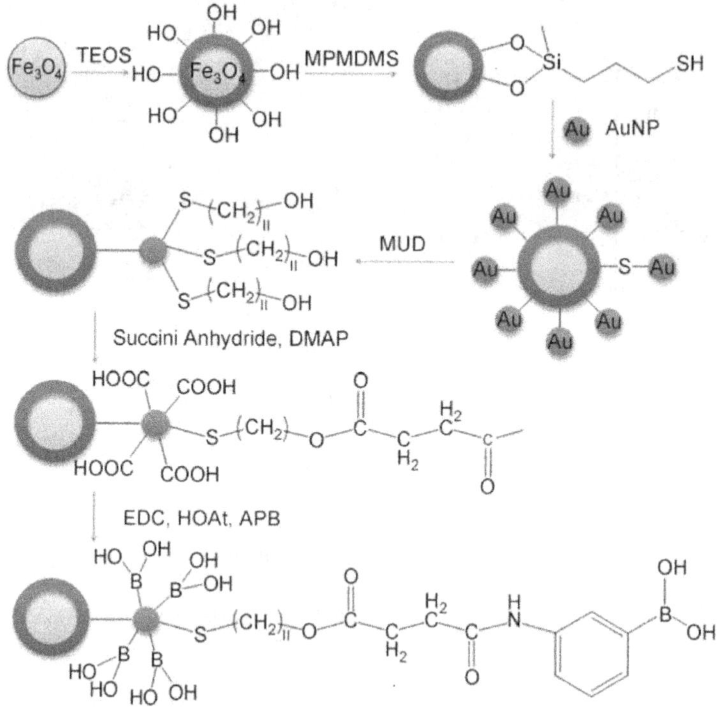

Figure 2.56: Schematic diagram of synthesis of Fe$_3$O$_4$@SiO$_2$@Au–APBA.

solution is quickly added during the stirring process. The reflux is continued for 10 min after the reaction system turns wine red. Then stirring is continued for another 15 min at room temperature.

Synthesis of $Fe_3O_4@SiO_2@Au$: the ethanol solution of $Fe_3O_4@SiO_2$ microspheres is treated by ultrasonic for 30 min, and then 600 μL mercaptopropyl methyldimethoxysilane (MPMDMS) is added for stirring overnight at room temperature. The solid product is fully cleaned with anhydrous ethanol and then dispersed in 50 mL anhydrous ethanol for 30 min of ultrasonic treatment. Then 100 mL dispersion solution of Au nanoparticles is added. This is then stirred overnight at room temperature. The resulting solid product is then dispersed in 100 mL DMF for later use.

(3) Synthesis of $Fe_3O_4@SiO_2@Au–APBA$

First, 15 mg 11-mercapto-1-undecanol (MUD) is dispersed in the DMF solution of $Fe_3O_4@SiO_2@Au$. It is then continuously stirred overnight at room temperature. The solid product is washed with DMF and then dispersed in 100 mL DMF solution (denoted as dispersion solution A).

Secondly, 100 mg succinic anhydride and 200 mg 4-dimethylaminopyridine (DMAP) are dissolved in the dispersion solution A. Then, the reaction system is placed under 55°C and mechanically stirred for 3 h. The resulting solid products are cleaned with DMF and then dispersed in 50 mL anhydrous ethanol (denoted as dispersion solution B).

Thirdly, 100 mg 1-[3-(dimethylamino)propyl]-3-ethylcarbodiimide hydrochloride (EDC) and 150 mg 1-hydroxy-7-azabenzotriazde (HOAt), and 60 mg 3-aminophenylboronic acid-hydrate (APBA) are added into the above dispersion solution B. The whole reaction system is stirred mechanically for 1 h at room temperature. The resulting solid product is washed fully by anhydrous ethanol, and then dispersed in 10 mL anhydrous ethanol (denoted as dispersion solution C), namely, ethanol dispersion solution of $Fe_3O_4@SiO_2@Au–APBA$.

2.6.10.2. *Characterization of $Fe_3O_4@SiO_2@Au–APBA$*

As shown in Figures 2.57(a) and 2.57(b), Fe_3O_4 microspheres with an average particle size of about 200 nm have good dispersity and uniform size. Figure 2.57(c) shows the TEM image of $Fe_3O_4@SiO_2@Au–APBA$ microsphere, from which it can be seen that there are a large number of Au nanoparticles on the surface of the magnetic microsphere.

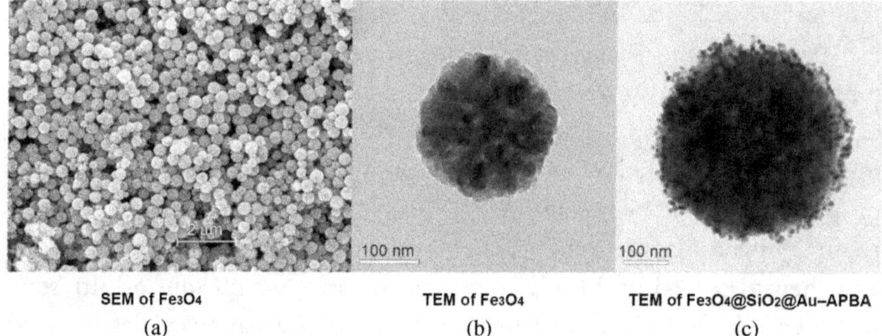

Figure 2.57: SEM and TEM images of Fe$_3$O$_4$, as well as TEM image of Fe$_3$O$_4$@SiO$_2$@Au–APBA.

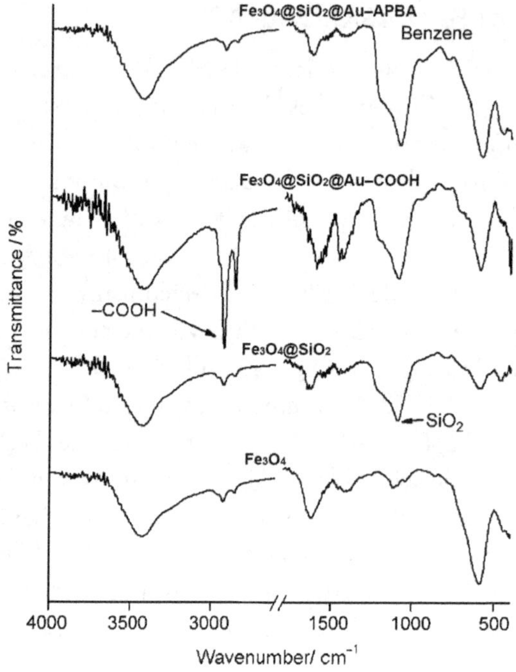

Figure 2.58: FTIR spectra of different materials.

In order to prove the feasibility of synthetic route of Fe$_3$O$_4$@SiO$_2$@Au–APBA microspheres, FTIR is performed for characterizing different intermediates. As shown in Figure 2.58, the absorption peak at 1,089 cm^{-1} in the FTIR spectrum of f Fe$_3$O$_4$@SiO$_2$ microspheres is attributed to the stretching vibration of Si–O–Si, indicating that the SiO$_2$ layer is successfully coated on the surface of the magnetic microspheres. In the FTIR spectrum of Fe$_3$O$_4$@SiO$_2$@Au–COOH, the absorption peak at 3,045 cm^{-1} is attributed to the stretching vibration of O–H, while in the FTIR spectrum of f Fe$_3$O$_4$@SiO$_2$@Au–APBA microspheres, this characteristic peak disappears and the vibration peak of hydrogen belonging to benzene ring appears near 700–800 cm^{-1}. These results indicate that boric acid groups are successfully modified on the surface of microspheres.

2.6.11. *Mercaptophenylboronic acid-modified magnetic Au–carbon microsphere (Fe$_3$O$_4$@CP@Au–MPBA)*

2.6.11.1. *Preparation of Fe$_3$O$_4$@CP@Au–MPBA*

The synthetic process of Fe$_3$O$_4$@CP@Au–MPBA is shown in Figure 2.59.[25]

(1) Synthesis of Fe$_3$O$_4$@CP

The detailed synthetic process can be found in Section 2.4.

(2) Synthesis of Fe$_3$O$_4$@CP@Au

Synthesis of Au nanoparticles: under stirring condition, 0.6 mL NaBH$_4$ aqueous solution (0.01 mol·L^{-1}) is slowly added to 20 mL mixture aqueous solution containing 0.25 mmol·L^{-1} HAuCl$_4$ and 0.25 mmol·L^{-1} trisodium citrate. Stirring is continued for another 30 s. When the solution turns wine

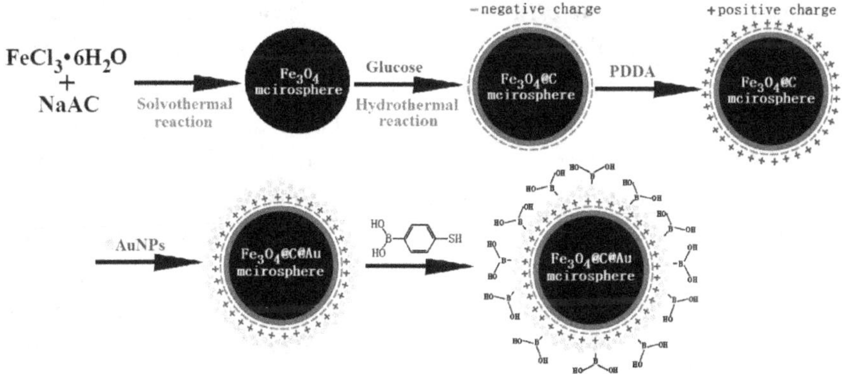

Figure 2.59: Schematic diagram of synthesis of Fe$_3$O$_4$@CP@Au-MPBA.

red, Au nanoparticles are successfully synthesized. The synthesized Au nano-particles are used within 24 h.

Synthesis of Fe_3O_4@CP@Au: 200 mg Fe_3O_4@CP microspheres are dispersed in 20 mL mixture aqueous solution containing 0.20% poly dimethyl diallyl ammonium chloride (PDDA), 20×10^{-3} mol·L^{-1} Tris and 20×10^{-3} mol·L^{-1} NaCl. After stirring for 20 min, the obtained solid product is fully washed by deionized water. Then 40 mg the PDDA-modified products are dispersed in 60 mL the above dispersion solution of Au nanoparticles for stirring for another 20 min. The obtained Fe_3O_4@CP@Au microspheres are washed by deionized water and dried in vacuum for later use.

50 mg Fe_3O_4@CP@Au microspheres are dispersed in 20 mL 1 mg·mL^{-1} ethanol solution of 4-mercaptophenylboronic acid (MPBA), and continuously stirred for 2 h. The obtained Fe_3O_4@CP@Au–MPBA are washed by deionized water and ethanol in order for later use.

2.6.11.2. *Characterization of Fe_3O_4@CP@Au–MPBA*

It can be seen from TEM image of Fe_3O_4@CP microspheres in Figure 2.60(a) that the diameter of Fe_3O_4 microspheres is about 240 nm and the thickness of carbon layer on surface is about 20 nm. As shown in Figure 2.60(b), the peak at 1,700 cm^{-1} is attributed to the vibration absorption of the C=O, and the peak at 1,620 cm^{-1} is attributed to the vibration absorption of the C=C, indicating that glucose molecules have been successfully carbonized. The FTIR absorption peaks in the range from 1,000 cm^{-1} to 1,300 cm^{-1} are attributed to the stretching vibration of C–O and the bending vibration of O–H, indicating the presence of a large number of hydroxyl groups on the surface of the microspheres, as well as incomplete carbonization of glucose. The zeta potential of Fe_3O_4@CP microspheres in aqueous solution is −59.3 mv, that is, the surface of Fe_3O_4@CP microspheres is negatively charged. Meanwhile, when the positively charged PDDA is modified on the surface of Fe_3O_4@CP microspheres, the zeta potential of Fe_3O_4@CP–PDDA in aqueous solution is + 2.33 mv, indicating that PDDA is successfully modified on the surface of Fe_3O_4@CP microspheres through electrostatic adsorption.

From the TEM image of Fe_3O_4@CP@Au microspheres in Figure 2.60(c), it can be seen that Au nanoparticles with a diameter of about 3 nm scatter densely on the surface of the carbon layer. The presence of Au nanoparticles

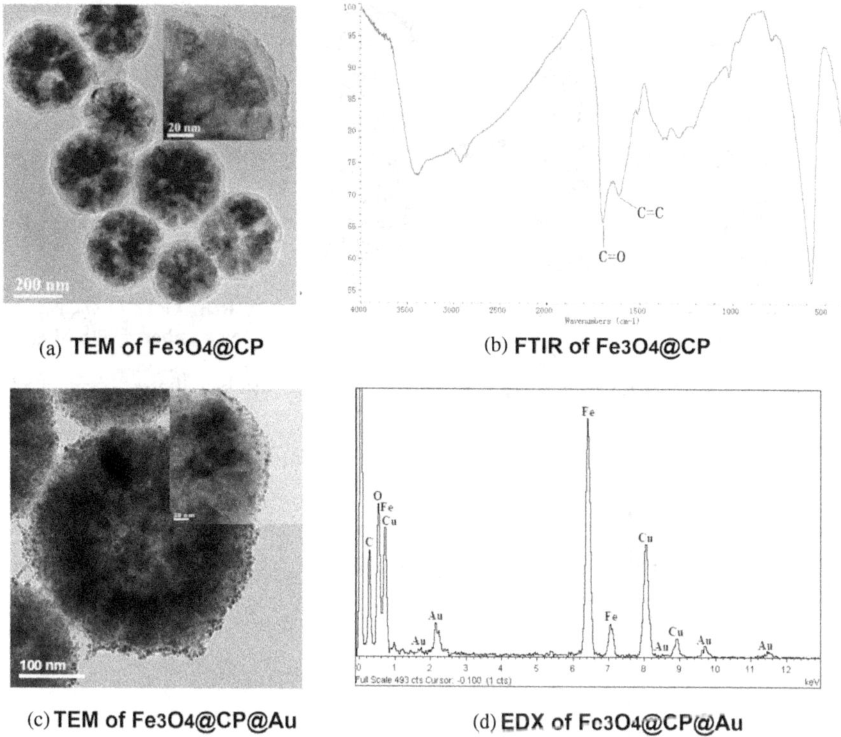

(a) **TEM of Fe3O4@CP**

(b) **FTIR of Fe3O4@CP**

(c) **TEM of Fe3O4@CP@Au**

(d) **EDX of Fc3O4@CP@Au**

Figure 2.60: TEM images, EDX analysis, and FTIR spectra of different materials.

is further confirmed by EDX analysis of Fe_3O_4@CP@Au microspheres as seen in Figure 2.60(d).

It is noteworthy that Au nanoparticles are prepared through the reduction of $HAuCl_4$ by $NaBH_4$. As shown in Figure 2.61(a), and the dispersion solution of Au nanoparticles presents wine red. As shown in Figure 2.61(d), the dispersion solution of Au nanoparticles became colorless after adding Fe_3O_4@CP–PDDA microspheres, indicating that Au nanoparticles are all adsorbed on the surface of Fe_3O_4@CP microspheres.

The immobilization of the boric acid group is based on the strong interaction between the mercapto group and Au. The immobilized amount of MPBA on Fe_3O_4@CP@Au microspheres is determined by RPLC, as shown in Figure 2.62. The immobilized amount of MPBA reaches equilibrium after reaction for 1 h. The immobilized content of MPBA on surface of Fe_3O_4@ CP@Au is 50 $\mu g \cdot mg^{-1}$ through analysis and parallel calculation.

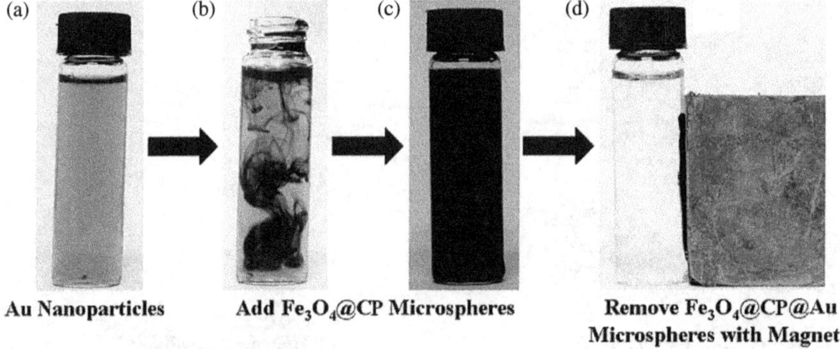

Figure 2.61: Schematic diagram of synthesis of Fe_3O_4@CP@Au by self-assembly method.

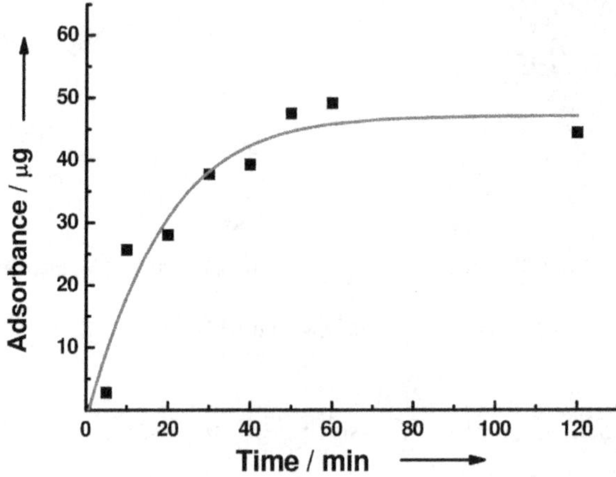

Figure 2.62: Schematic diagram of the change of immobilized amount of MPBA on the surface of Fe_3O_4@CP@Au over time.

2.6.12. *Phosphate group-modified magnetic microsphere for immobilization of Zr(IV) metal ions (Fe_3O_4@Phosph–Zr(IV))*

2.6.12.1. *Preparation of Fe_3O_4@Phosph–Zr(IV)*

The synthetic process of Fe_3O_4@Phosph–Zr(IV) is shown in Figure 2.63.[26]

Firstly, Fe_3O_4@CP microspheres are prepared according to the synthetic method in Section 2.4. Then 80 mg Fe_3O_4@CP microspheres and

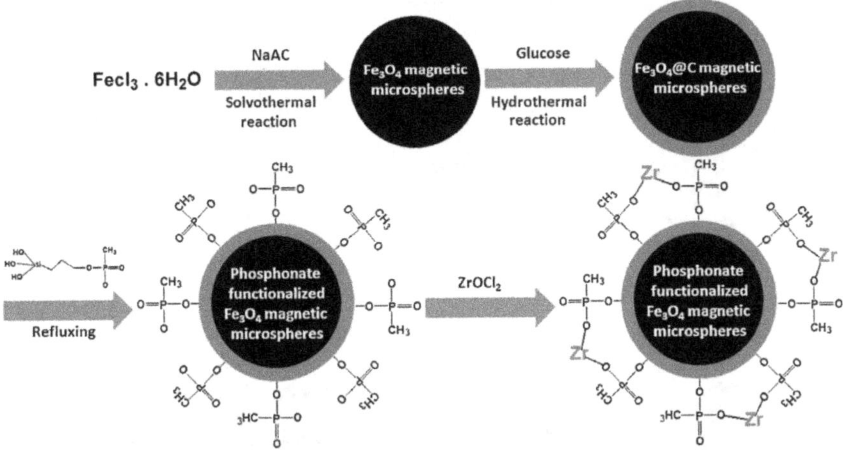

Figure 2.63: Schematic diagram of synthesis of Fe$_3$O$_4$@Phosph–Zr(IV).

0.5 mL trihydroxy silicomethyl methyl phosphate are dispersed in 20 mL toluene for reflux reaction for 12 h at 80°C. The obtained solid product is washed by deionized water and ethanol, respectively, and dried in vacuum for later use.

Next, the phosphate group-modified magnetic microspheres (Fe$_3$O$_4$@Phosph) are dispersed in 0.2 mol·L^{-1} ZrOCl$_2$ aqueous solution for stirring overnight to immobilize Zr(IV) metal ions on the surface. Finally, the prepared magnetic microspheres immobilized with Zr(IV) ions are washed by deionized water and ethanol, respectively, and dried in vacuum for later use.

2.6.12.2. *Characterization of Fe$_3$O$_4$@Phosph–Zr(IV)*

FTIR is used to characterize the functional groups on the surface of Fe$_3$O$_4$@CP microspheres. As shown in the solid line in Figure 2.64, the peak at 1,700 cm^{-1} is attributed to the vibration absorption of C=O, and the peak at 1,620 cm^{-1} is attributed to the vibration absorption of the C=C. The presence of the two bonds of C=O and C=C confirms the carbonization reaction between glucose molecules. The peaks in the range of 1,000–1,300 cm^{-1} are attributed to the stretching vibration absorption of C–OH and the bending vibration absorption of O–H, indicating that a

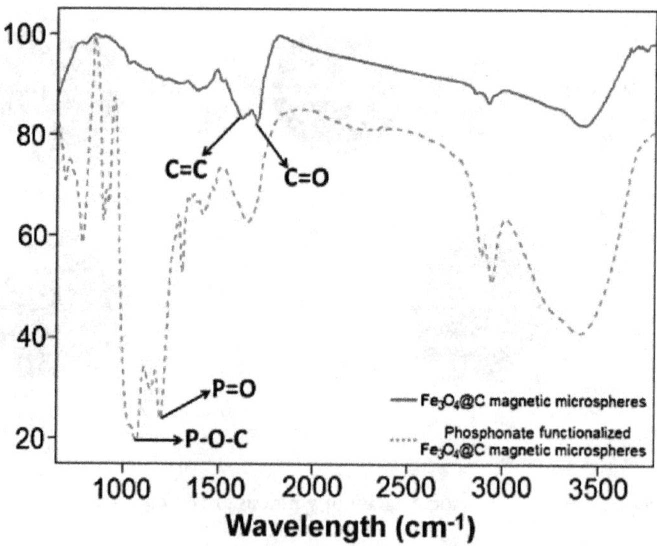

Figure 2.64: FTIR spectra of different materials.

large number of hydroxyl groups exist on the surface of Fe_3O_4@CP microspheres. The presence of plenty of hydroxyl groups on the surface can not only greatly improve the dispersity and stability of Fe_3O_4@CP microspheres in aqueous solution, but also provide enough chemical groups for the subsequent surface chemical modification. The phosphate groups are modified on the surface of Fe_3O_4@CP microspheres by a simple silanization reaction. The FTIR spectrum of magnetic microspheres modified with phosphate groups is shown in the dotted line in Figure 2.64. The peak at 1,050 cm^{-1} is attributed to the vibration absorption of P–O–C, and the peak at 1,275 cm^{-1} is attributed to the vibration absorption of P=O, indicating that phosphate groups have been successfully modified on the surface of Fe_3O_4@CP microspheres.

The presence of zirconium is confirmed by the EDX analysis of Fe_3O_4@Phosph–Zr(IV) microspheres. From Figure 2.65, Zr(IV) ions are successfully immobilized on the surface of Fe_3O_4@Phosph microspheres. The above results indicate that the new method proposed in this work example can synthesize phosphate group-modified magnetic microsphere effectively and rapidly. And the phosphate group-modified magnetic

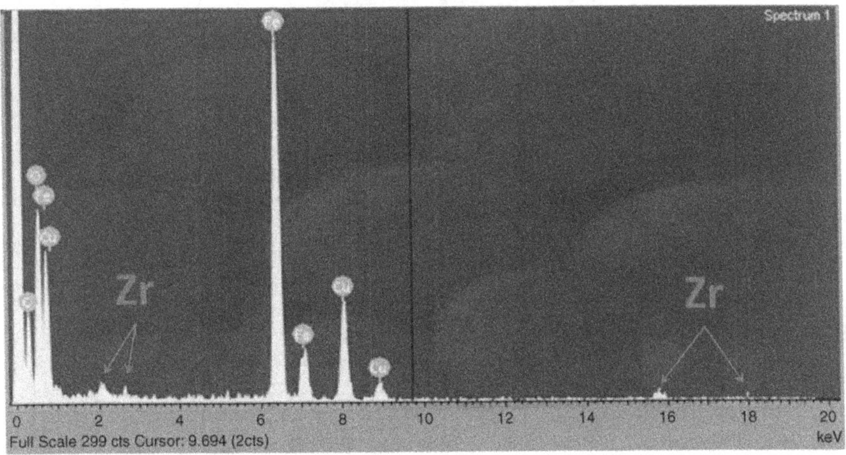

Figure 2.65: EDX analysis of Fe_3O_4@Phosph–Zr(IV).

microsphere is also very effective for fixing metal ions. In this method, modification of phosphate groups is achieved by a simple silanization reaction.

2.6.13. *Tetraazacyclododecane (DOTA)-modified magnetic silica microsphere for immobilization of rare earth ions (Fe_3O_4@TCPP–DOTA–M^{3+})*

2.6.13.1. *Preparation of Fe_3O_4@TCPP–DOTA–M^{3+}*

The synthetic process of Fe_3O_4@TCPP–DOTA–M^{3+} is shown in Figure 2.66.[27,28]

Firstly, Fe_3O_4@SiO_2 microspheres are prepared according to the synthetic method in Section 2.4.

Secondly, Fe_3O_4@SiO_2–NH_2 microspheres are synthesized as follows: Fe_3O_4@SiO_2 microspheres are washed successively with 40 mL 1 mol·L^{-1} HCl, H_2O, 40 mL 20% HNO_3 and H_2O. Then the dry Fe_3O_4@SiO_2 microspheres are dispersed in 60 mL ethanol and then degassed under stirring (using nitrogen to eliminate air). Then 6 mL 3-aminopropyl-3-ethoxy silane is added to react for 12 h at 60°C. The resulting product is washed 3 times with ethanol and acetone, respectively, and then re-dispersed in 40 mL ethanol.

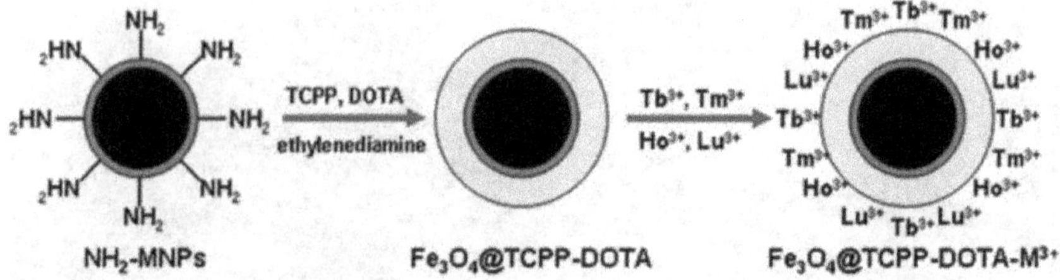

Figure 2.66: Schematic diagram of synthesis of Fe_3O_4@TCPP–DOTA–M^{3+}.

Finally, Fe_3O_4@TCPP–DOTA–M^{3+} microspheres are synthesized as follows: 10 mg tetrakis (4-carboxyphenyl) porphyrin (TCPP) is dissolved in ethanol for ultrasound about 1 h. Then the mixture containing EDC and *N*-hydroxysuccinimide (NHS) with molar ratio of 10/3 is added to react for 30 min. Then 4 mL dispersion solution of Fe_3O_4@SiO_2–NH_2 microspheres is added, the obtained product is washed by deionized water. Next, the mixture of EDC and NHS with molar ratio of 2/1 is added again for further reaction for 30 min, then 6 μL ethylenediamine is added, and then the product is washed by deionized water and dispersed in solution containing 20 mg 1,4,7,10-tetraazocyclododecane (DOTA) for reaction for 4 h. after washed by deionized water and ethanol, the microspheres are dispersed in mixture solution of 4 mmol·L^{-1} $TbCl_3$, $TmCl_3$, $HoCl_3$ and $LuCl_3$ (molar ratio of 1:1:1:1) to react for 6 h at 70°C. The final product is washed by deionized water and stored at 4°C for later use.

2.6.13.2. *Characterization of Fe_3O_4@TCPP–DOTA–M^{3+}*

The FTIR spectra of Fe_3O_4 microspheres and Fe_3O_4@TCPP–DOTA–M^{3+} microspheres are shown in Figure 2.67. The absorption peak at 590 cm^{-1} is attributed to the Fe–O bond in Fe_3O_4 microspheres, and the absorption peak at 1,076 cm^{-1} in Figure 2.67(b) is attributed to the Si–O bond, indicating the successful preparation of Fe_3O_4@SiO_2 microspheres. The enhanced absorption peak at 1,629 cm^{-1} is attributed to the vibration of N–H bond, indicating the successful modification of amino linker on the surface of magnetic silica. The EDX analysis of Fe_3O_4@TCPP–DOTA–M^{3+} microspheres is shown in Figure 2.68, Fe, C, O, Si, Tb, Tm, Ho, and Lu elements are observed in the spectra, indicating the successful immobilization of rare earth ions.

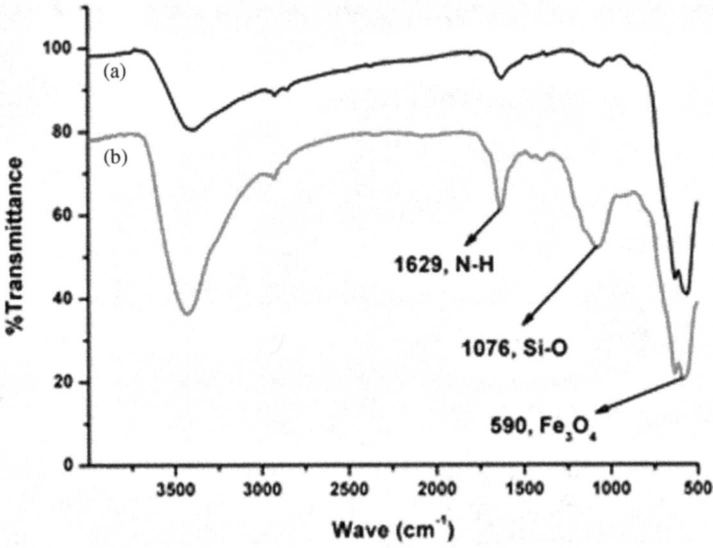

Figure 2.67: FTIR spectra of different materials (curve (a) Fe_3O_4 microsphere; curve (b) Fe_3O_4@TCPP–DOTA–M^{3+} microsphere).

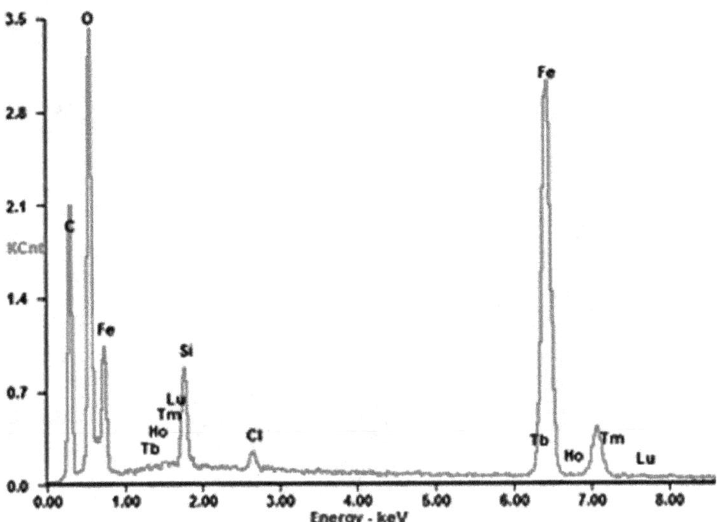

Figure 2.68: EDX analysis of Fe_3O_4@TCPP–DOTA–M^{3+} microsphere.

2.6.14. *Glycopeptide polymer-modified magnetic microsphere (dM-MNPs)*

2.6.14.1. *Preparation of dM-MNPs*

The synthetic process of dM-MNPs is shown in Figure 2.69.[29]

(1) Synthesis of MNPs–NH$_2$

MNPs–NH$_2$ microspheres are prepared according to the synthetic method of MNPs–(HA–CS)$_n$ in Section 2.6.4.

(2) Synthesis of MNPs–dN$_3$

20 mg azido-modified peptide is dissolved in PBS solution (10 mmol·L^{-1}, pH = 5.5), then 50 mg EDC and 50 mg NHS are added. Oscillation is maintained for 30 min. Next, 20 mg MNPs–NH$_2$ microspheres are added for

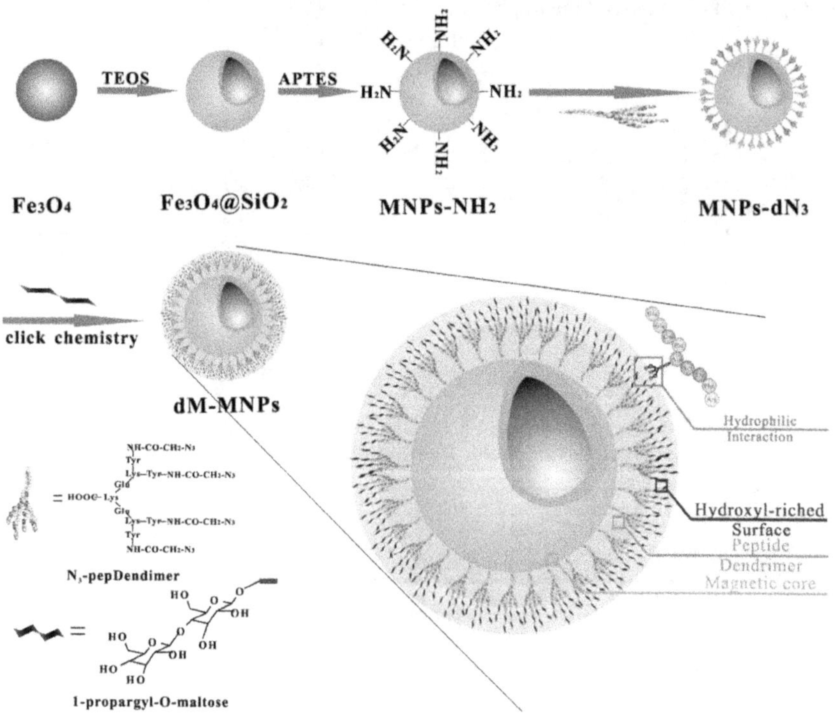

Figure 2.69: Schematic diagram of synthesis of dM-MNPs.

ultrasonic treatment for 5 min, and then kept oscillating for 36 h at room temperature, among much else 50 mg EDC and 50 mg NHS are added every 4 h during the process. The obtained MNPs-dN$_3$ microspheres are washed with ethanol and deionized water and then dispersed in 4 mL mixture solution of methanol and water with volume ratio of 50:50 for later use.

(3) Synthesis of maltose with terminal alkynyl modification

Based on Ref. [29].

(4) Synthesis of dM-MNPs

4 mL dispersion solution of MNPs–dN$_3$ microspheres prepared in the above step 2 is ultrasonic for 30 min, then 30 μL catalyst solution (ascorbic acid/copper sulfate, mmol·L^{-1}/mmol·L^{-1}: 200/100) is added. Finally, 5 mg maltose with terminal alkynyl modification is added and continuously oscillated for 12 h. The obtained dM-MNPs are washed fully with methanol, ethanol, and deionized water, and dried in vacuum for later use.

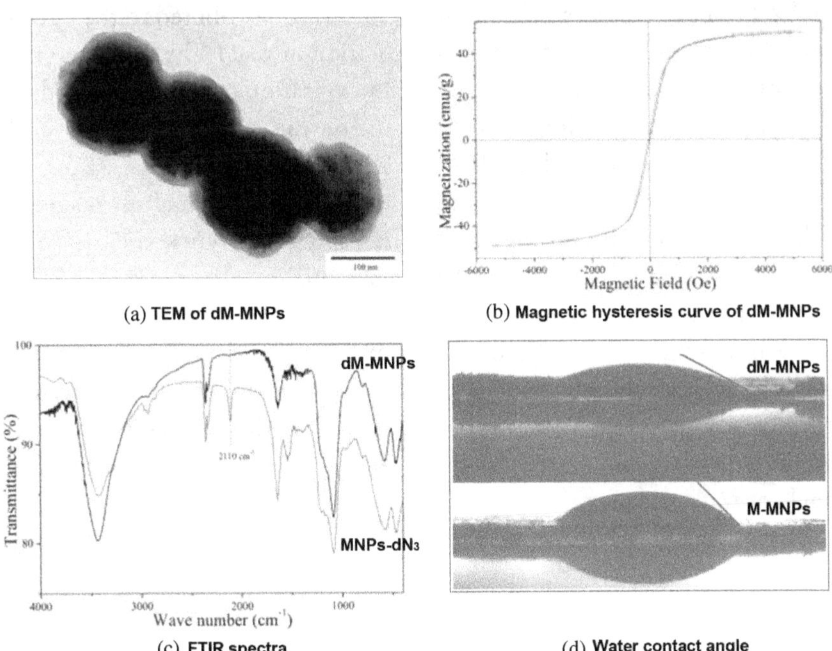

(a) TEM of dM-MNPs

(b) Magnetic hysteresis curve of dM-MNPs

(c) FTIR spectra

(d) Water contact angle

Figure 2.70: Characterization of dM-MNPs microspheres.

Table 2.4: EDX analysis of different materials.

MNP	C(%)	N(%)	H(%)	$\chi^{a}_{maltose}$/**nmol mg^{-1}**
MNPs-dN$_3$	13.44	2.957	1.152	—
dM-MNPs	14.95	3.424	2.049	89.88
MNPs-N$_3$	3.920	0.953	0.721	—
M-MNPs	4.016	0.326	0.551	5.71

[a]The amount of maltose was calculation based on the carbon content from elemental analysis.

2.6.14.2. *Characterization of dM-MNPs*

As shown in Figure 2.70(a), dM-MNPs microspheres present core–shell structure and the size of magnetic core is about 170 nm. The saturation magnetic value of the dM-MNPs microspheres is estimated to be about 49.7 emu·g^{-1} in Figure 2.70(b), indicating that dM-MNPs microspheres possess strong magnetic response ability. In the FTIR spectrum of the MNPs-dN$_3$ microspheres, the peaks at 1,640 cm^{-1}, 2,855 cm^{-1}, 1,450 cm^{-1}, and 2,110 cm^{-1} in Figure 2.70(c) are, respectively, attributed to the stretching vibration of C=O, the stretching vibration of C–H, the bending vibration of C–H and the symmetric stretching vibration of $^{+}$N=N=N^{-}, indicating the successful grafting of azido on the on the surface of silica layer. In the FTIR spectrum of dM-MNPs microspheres in Figure 2.70(c), the absorption peak at 2,110 cm^{-1} disappears, indicating that maltose is successfully modified on the surface of the microspheres via click chemical reaction. According to EDX analysis in Table 2.4, maltose content in dM-MNPs microspheres is 89 nmol·mg^{-1}, and the content of M-MNPs microspheres is 5.7 nmol·mg^{-1}. The maltose content of former is about 15 times that of maltose in later, indicating that dM-MNPs microspheres have extremely strong hydrophilicity. As shown in Figure 2.70(d), dM-MNPs microspheres are more hydrophilic than M-MNPs microspheres by analysis of contact angle of water balance.

2.6.15. *Guanidine silane-modified magnetic silica (Fe$_3$O$_4$@SiO$_2$@GDN)*

2.6.15.1. *Preparation of Fe$_3$O$_4$@SiO$_2$@GDN*

The synthetic process of Fe$_3$O$_4$@SiO$_2$@GDN is shown in Figure 2.71.[30]

Firstly, guanidine silane monomer is prepared according to the following steps: 3.5 g 2-ethyl-thiopseudourea hydrobromide is dissolved in a mixture solution containing 3.0 mL dimethyl sulfoxide (DMSO) and 3.5 mL tetrahydrofuran (THF) and kept under magnetic stirring for 20 min. Then 4.45 mL 3-aminopropyltriethoxysilane (APTEOS) is slowly added at 0°C, then the temperature rises to 25°C as the stirring reaction is continued for 48 h.

Secondly, Fe$_3$O$_4$@SiO$_2$ microspheres are prepared according to the synthetic method in Section 2.4.

Finally, Fe$_3$O$_4$@SiO$_2$@GDN microspheres are prepared according to the following steps: 400 mg Fe$_3$O$_4$@SiO$_2$ microspheres are dispersed in 80.0 mL *Tris*–HCl buffer (pH = 8.21, 0.1 mol·L^{-1}), and 1.2 mL (3.17 mmol) guanidine silane monomer and 0.6 mL (2.69 mmol) TEOS are added for reaction at stirring speed of 150 rpm for 16 h. The resulting product is washed and continue to react at 35°C for 24 h.

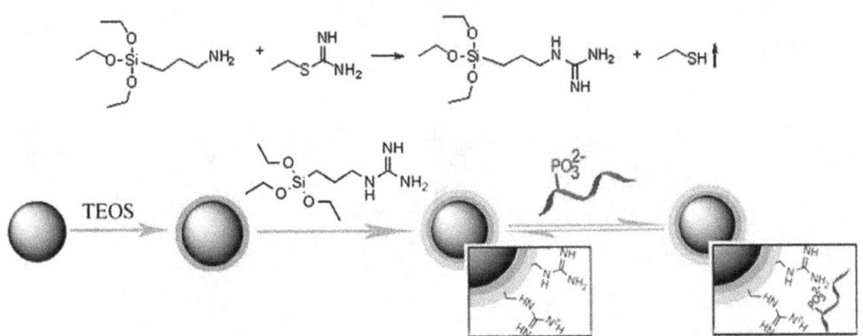

Figure 2.71:   Schematic diagram of synthesis of Fe$_3$O$_4$@SiO$_2$@GDN.

2.6.15.2. *Characterization of $Fe_3O_4@SiO_2@GDN$*

TEM and SEM are used to observe the morphology of $Fe_3O_4@SiO_2$ and $Fe_3O_4@SiO_2@GDN$ to ensure the successful synthesis of $Fe_3O_4@SiO_2@GDN$ microspheres. As shown in Figure 2.72, the size of Fe_3O_4 microsphere is about 200 nm, the silica layer is about 8.5 nm, and the silica layer containing guanidine is about 20 nm.

According to X-ray photoelectron spectroscopy (XPS) analysis in Figure 2.73(a), the surface components of $Fe_3O_4@SiO_2@GDN$ microspheres include N, C, O, and Si elements. As shown in Figure 2.73(b), the characteristic peaks of guanidine at 3,430 cm^{-1} and 1,635 cm^{-1} can be seen from the

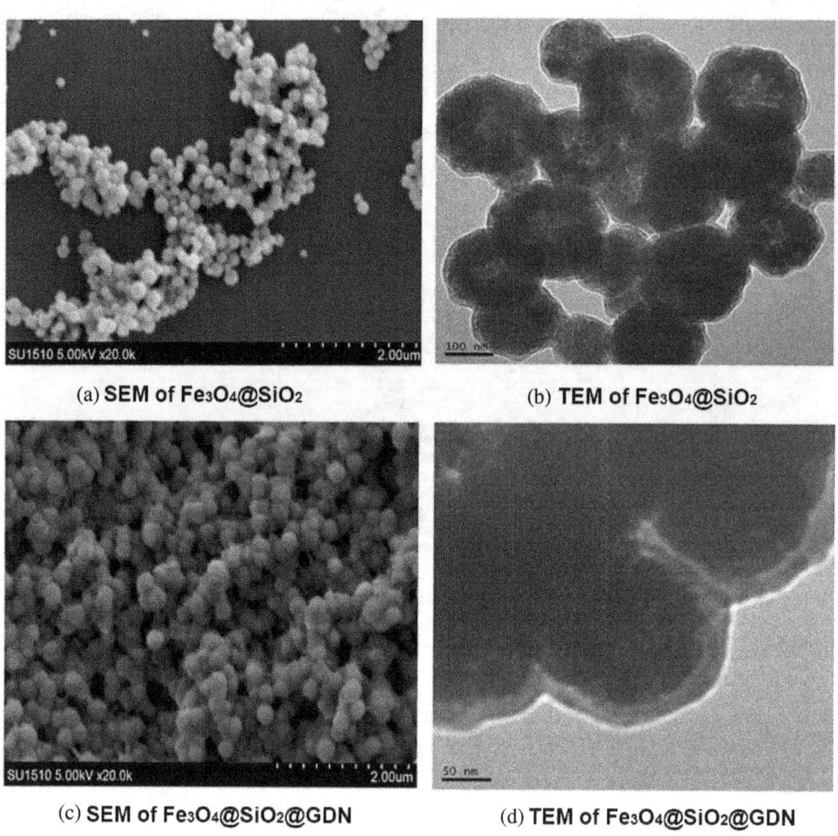

(a) **SEM of Fe₃O₄@SiO₂**   (b) **TEM of Fe₃O₄@SiO₂**

(c) **SEM of Fe₃O₄@SiO₂@GDN**   (d) **TEM of Fe₃O₄@SiO₂@GDN**

Figure 2.72: SEM and TEM images of different materials.

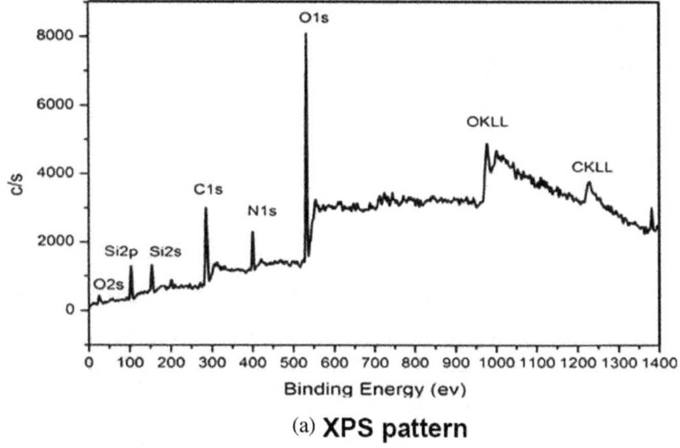

(a) **XPS pattern**

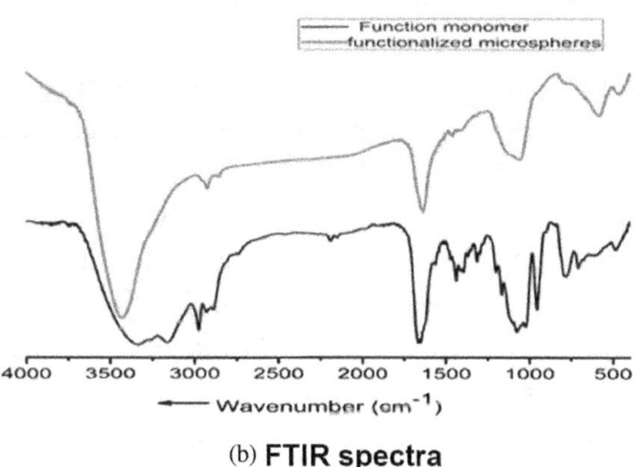

(b) **FTIR spectra**

Figure 2.73: XPS analysis and FTIR spectra of Fe_3O_4@SiO_2@GDN.

FTIR spectra, indicating that a large number of guanidine groups exist in Fe_3O_4@SiO_2@GDN microspheres.

2.6.16. *Polyethyleneimine-modified magnetic silica (Fe_3O_4@SiO_2@PEI)*

2.6.16.1. *Preparation of Fe_3O_4@SiO_2@PEI*

Firstly, Fe_3O_4 microspheres are synthesized according to the following steps: 2 g $FeCl_2 \cdot 4H_2O$ and 6 g $FeCl_3 \cdot 6H_2O$ are dispersed in 15 mL 2 mol·L^{-1}

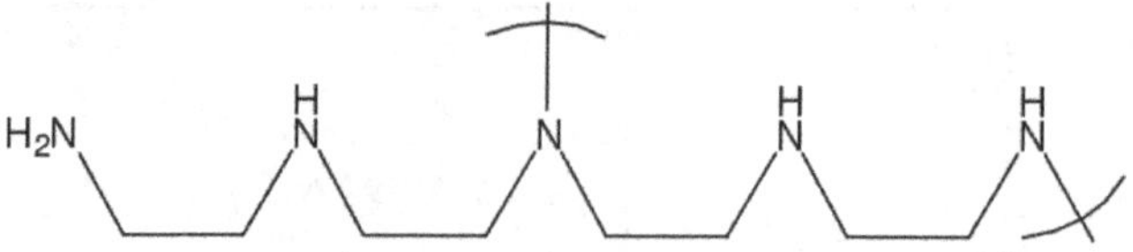

Figure 2.74: Schematic diagram of structure of PEI.

hydrochloric acid which has been deoxygenated by nitrogen, then 30 mL 33% (v/v) ammonia water is added.[31] The above mixture solution is vigorously stirred for 30 min under the protection of nitrogen, and then vigorously stirred for 1 h at room temperature. The obtained magnetic microspheres are washed with deionized water for 3 times, and finally dispersed in 50 mL ethanol for 30 min with ultrasound to suspend them.

The structure of polyethyleneimine (PEI) is shown in Figure 2.74. $Fe_3O_4@SiO_2@PEI$ microspheres are synthesized according to the following steps: 50 mL the above suspension solution of the magnetic microspheres is mixed with 9 mL ammonia water, 0.15 mL TEOS and 7.5 mL H_2O. The mixture solution is vigorously stirred for 2 h at 40°C. The obtained product is dispersed in 50 mL H_2O after washed by deionized water, and then 120 mg PEI is added for stirring vigorously for 12 h. The final product is dispersed ultrasonically in 50 mL H_2O after washed by deionized water.

2.6.16.2. *Characterization of $Fe_3O_4@SiO_2@PEI$*

Figure 2.75(a) shows the TEM image of $Fe_3O_4@SiO_2@PEI$ microsphere. It can be observed that the average size is about 25 nm. Figure 2.75(b) is the zeta potential analysis of $Fe_3O_4@SiO_2$ microsphere and $Fe_3O_4@SiO_2@PEI$ microsphere. It can be seen that $Fe_3O_4@SiO_2$ microsphere is negatively charged. PEI is modified on the surface of $Fe_3O_4@SiO_2$ microsphere via extremely strong electrostatic interaction, resulting in that the zeta potential of $Fe_3O_4@SiO_2$ increases from −32 mV to 34 mV, and the surface of the whole microsphere is positively charged. Then the influence of pH on surface zeta potential of $Fe_3O_4@SiO_2@PEI$ microsphere is also tested. As shown in Figure 2.75(c), when the pH value of the solution reaches 11, the zeta potential of $Fe_3O_4@SiO_2@PEI$ microsphere begins to decrease, and the $Fe_3O_4@SiO_2@PEI$ microsphere maintains a positive charge in the range of pH value from 3 to 10.

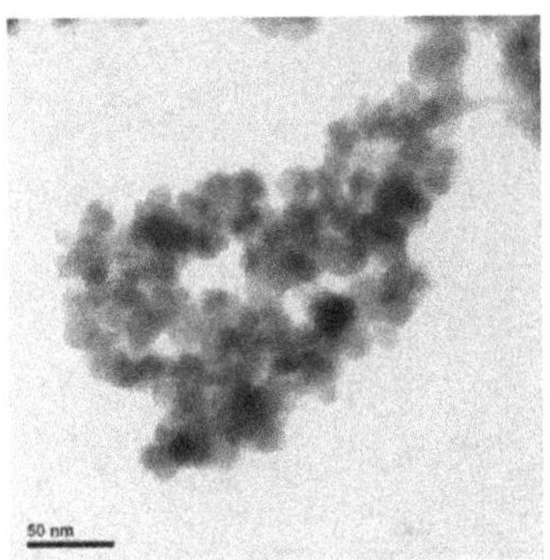

(a) TEM of Fe₃O₄@SiO₂@PEI

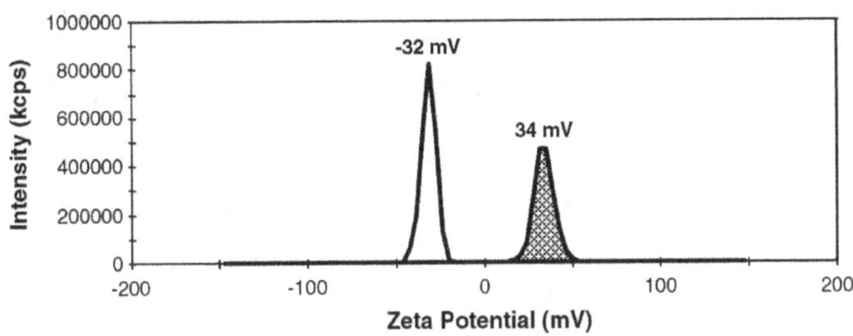

(b) Zeta potential analysis of Fe₃O₄@SiO₂ (blank) and Fe₃O₄@SiO₂@PEI (shadow)

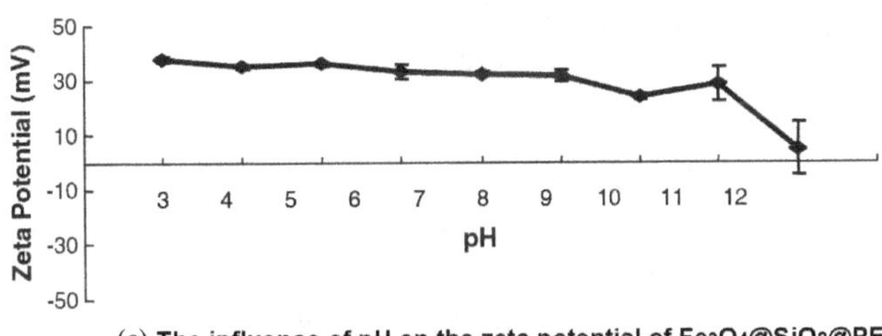

(c) The influence of pH on the zeta potential of Fe₃O₄@SiO₂@PEI

Figure 2.75: TEM image of Fe_3O_4@SiO_2@PEI, zeta potential analysis of Fe_3O_4@SiO_2 and Fe_3O_4@SiO_2@PEI, and the influence of pH on the zeta potential.

2.6.17. *Immobilization of boronic acid groups on Fe₃O₄@pVBC@APBA by Click chemistry*

2.6.17.1. *Preparation of Fe₃O₄@pVBC@APBA*

The synthetic process of Fe_3O_4@pVBC@APBA is shown in Figure 2.76.[32]

Firstly, Fe_3O_4 microspheres are synthesized according to the following steps: 0.675 g $FeCl_3 \cdot 6H_2O$, 0.2 g sodium citrate and 1.927 g NH_4Ac are

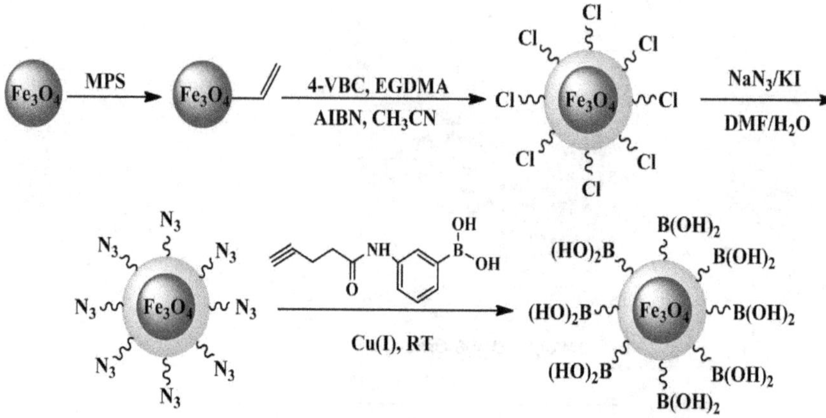

Figure 2.76: Schematic diagram of synthesis of Fe_3O_4@pVBC@APBA.

dispersed in 35 mL polyethylene glycol and vigorously stirred to mix the solution evenly. Then the mixture solution is transferred to the reaction kettle for reaction at 200°C for 16 h. The Fe_3O_4 microspheres are cleaned with ethanol and dried in vacuum for later use.

Secondly, Fe_3O_4@MPS microspheres are synthesized according to the following steps: 0.2 g Fe_3O_4 microspheres are dispersed in 50 mL solution of H_2O/ethanol (v/v, 1/4), then 1.5 mL $NH_3 \cdot H_2O$ and 0.4 mL MPS are added, and the whole reaction system is vigorously stirred at 60°C for 24 h. The Fe_3O_4@MPS microspheres are cleaned with ethanol and dried in vacuum for later use.

Next, Fe_3O_4@pVBC microspheres are synthesized according to the following steps: 50 mg Fe_3O_4@MPS microspheres are ultrasonically dispersed in 40 mL ACN, then 200 μL 2-(4-chlorphenyl) propylene (4-VBC), ethylene glycol methacrylate (EGDMA, 200/400/600/800/1000 μL), and 10 mg azodiisobutyronitrile (AIBN) are added. Then the reaction system is placed in a heated oil bath, and the temperature of the reaction system is heated to the oil bath temperature within 30 min. When half of the solvent in the reaction system is distilled, the reaction is regarded as the end (within 60 min). The obtained Fe_3O_4@pVBC microspheres are cleaned with ethanol and dried in vacuum for later use.

Finally, Fe_3O_4@pVBC@APBA microspheres are synthesized according to the following steps: 100 mg Fe_3O_4@pVBC microspheres are dispersed in 30 mL solution of DMF/H2O (v/v, 2/1), then 1.25 g NaN_3 and 300 mg KI are added and mechanically stirred at 50°C for 24 h under nitrogen protection. The obtained $Fe_3O_4@pVBC@N_3$ microspheres are cleaned with ethanol and then mixed with 30 mg APBA in 20 mL solution of $MeOH/H_2O$ (v/v, 1/1). $CuSO_4 \cdot 5H_2O$ and L-sodium ascorbate are added to make the final concentration 0.5 $mmol \cdot L^{-1}$ and 2.5 $mmol \cdot L^{-1}$, respectively. Then the whole reaction system is mechanically stirred for 24 h at 25°C. The product is cleaned with deionized water and ethanol and dried in vacuum for later use.

2.6.17.2. *Characterization of Fe₃O₄@pVBC@APBA*

In the synthesis process of $Fe_3O_4@pVBC@APBA$ microspheres, the appropriate molar ratio of 4-VBC and EGDMA is crucial to the immobilized amount of Cl element. Generally, the more the immobilized amount of Cl element, the more conducive to the modification of boric acid. Therefore, the Cl content on the surface of microspheres prepared by 4-VBC and EGDMA with different molar ratios is determined by ion chromatography. As shown in Table 2.5, when the molar ratio of 4-VBC and EGDMA increases from 1.00:0.75 to 1.00:2.25, the content of immobilized Cl elements shows an increasing trend. When the molar ratio increases from 1.00:2.25 to 1.00:3.75, the immobilized content shows a decreasing trend, so the molar ratio of 4-VBC to EGDMA is finally determined to be 1.00:2.25.

Figure 2.77(a) shows the TEM image of Fe_3O_4 microspheres, from which it can be seen that the magnetic microspheres are uniform in size. The size is about 400 nm. Figure 2.77(b) is a TEM image of $Fe_3O_4@pVBC@APBA$ microspheres, and it is observed that the outer layer of the core–shell structure is about 50 nm.

Table 2.5: The Cl content corresponding to different molar ratios of 4-VBC and EGDMA.

Molar ratio of 4-VBC to EGDMA	Content of Cl (%)
1.00 : 0.75	2.81
1.00 : 1.50	3.96
1.00 : 2.25	5.36
1.00 : 3.00	4.41
1.00 : 3.75	2.57

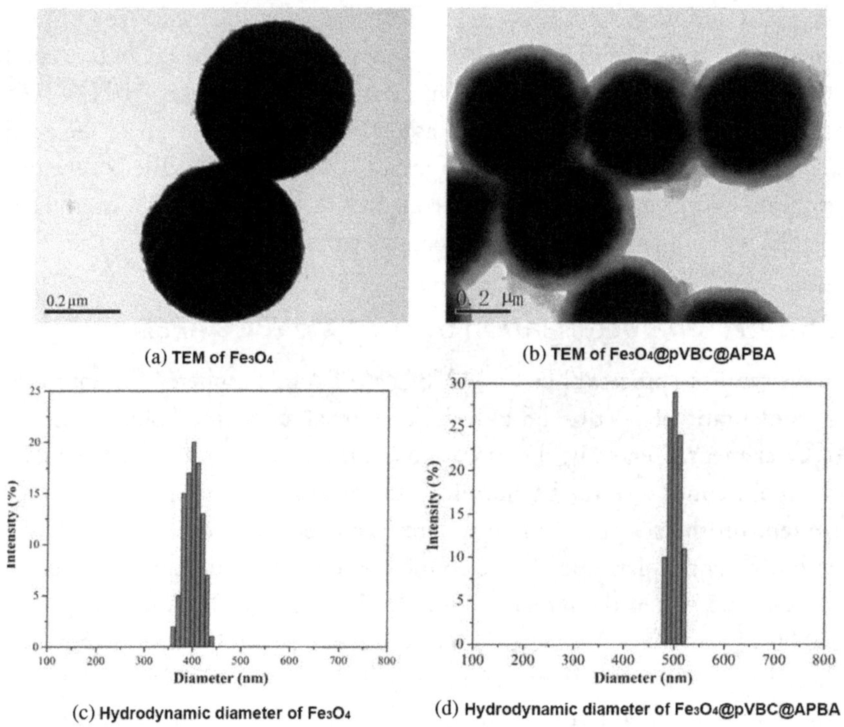

Figure 2.77: TEM images and hydrodynamic diameters of two different microspheres.

As shown in Figures 2.77(c) and 2.77(d), the hydrodynamic diameters of Fe_3O_4 microspheres and Fe_3O_4@pVBC@APBA microspheres are 392 ± 40 nm and 500 ± 20 nm, respectively, which are basically close to those observed in TEM images.

The absorption peaks at $1,610$ cm^{-1} and $1,401$ cm^{-1} in Figure 2.78(a) are attributed to carbonyl groups in stabilizer of sodium citrate, and that at 580 cm^{-1} is attributed to Fe–O. The absorption peaks at $1,159$ cm^{-1} and $1,228$ cm^{-1} in Figure 2.78(b) are attributed to the presence of Si–O–Si, indicating that the silane reagent of MPS is successfully modified. The absorption peak at $1,632$ cm^{-1} is attributed to the C=C bond in MPS. The absorption peak at $1,730$ cm^{-1} in Figure 2.78(c) is attributed to the stretching vibration of C=O, indicating that pVBC shell is successfully coated on the surface of Fe_3O_4@MPS. The absorption peak at $2,940$ cm^{-1} is attributed

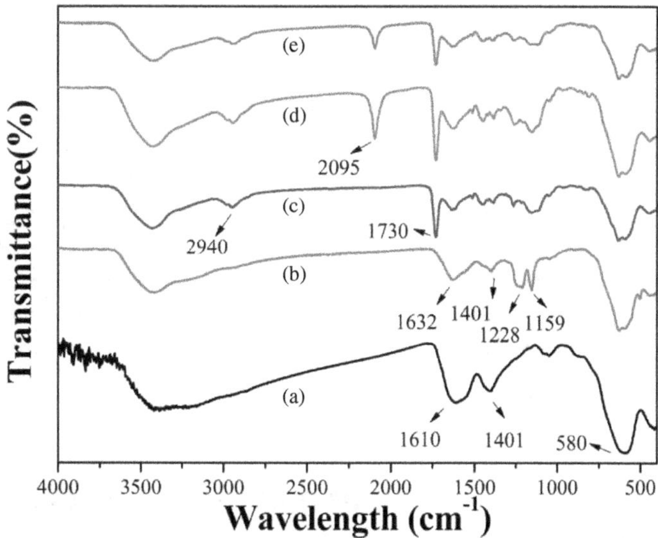

Figure 2.78: FTIR spectra of different materials (curve (a) Fe_3O_4; curve (b) Fe_3O_4@MPS; curve (c) Fe_3O_4@pVBC; curve (d) Fe_3O_4@pVBC@N_3; curve (e) Fe_3O_4@pVBC@APBA).

to the stretching vibration of CH_2 in pVBC shell. The absorption peak at 2,095 cm^{-1} in Figure 2.78(d) is attributed to the symmetric stretching vibration of azide group. In Figure 2.78(e), the characteristic absorption peak of azide group disappears, which indirectly indicates that click chemical reaction has been successfully carried out.

As shown in Figure 2.79(a), the mass loss of Fe_3O_4 microspheres is about 12% due to the presence of the stabilizer of sodium citrate on the surface. The mass loss of Fe_3O_4@MPS microspheres is about 15% due to the presence of the stabilizer of sodium citrate and the silane reagent of MPS. Fe_3O_4@pVBC microspheres show a very sharp curve of mass loss between 310°C and 430°C, indicating that the Fe_3O_4@pVBC microspheres contain 60% p(VBC-EGDMA). The Fe_3O_4@pVBC@APBA microspheres show more mass loss due to the immobilization of boric acid ligands on the microsphere surface by a click chemical reaction. The content of boric acid is estimated to be about 0.3 mmol·g^{-1}, and its presence in Fe_3O_4@pVBC@APBA is further confirmed by XPS analysis as shown in Figure 2.79(b).

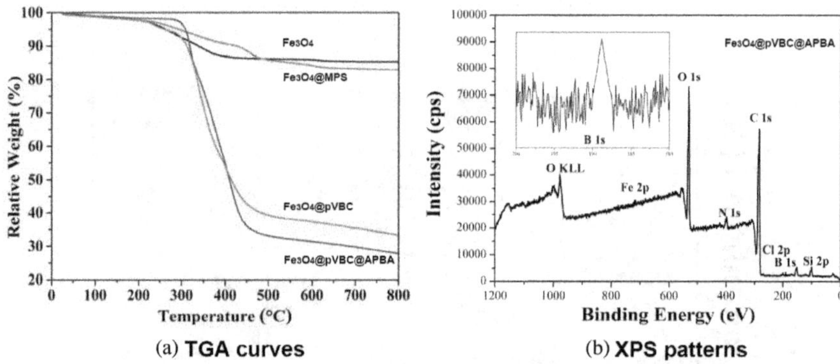

Figure 2.79: TGA and XPS analysis of different materials.

2.7. Magnetic mesoporous material with core–shell structure

Mesoporous material is a new type of material with large specific surface area and 3D porous structure, the pore size of which is between micropore and macropore. It also has the function of screening molecules based on size.

2.7.1. *Magnetic mesoporous silica microsphere with core–shell–shell structure ($Fe_3O_4@nSiO_2@mSiO_2$)*

$Fe_3O_4@nSiO_2@mSiO_2$ microspheres are prepared by using $Fe_3O_4@nSiO_2$ as the core and cetyl trimethyl ammonium bromide (CTAB) as the template to coat mesoporous silica on the surface of $Fe_3O_4@nSiO_2$ as seen in Figure 2.80.[33] Firstly, hydrothermal method is used to prepare Fe_3O_4 with uniform particle size and superparamagnetism. Secondly, the $Fe_3O_4@nSiO_2$ microspheres with the core of Fe_3O_4 and the shell of non-porous SiO_2 are obtained by sol–gel method. Then, the $CTAB/SiO_2$ composite is grown outside the $Fe_3O_4@nSiO_2$ microspheres by using CTAB as template, via sol–gel method. Finally, the CTAB is removed by acetone extraction. The detailed synthetic process of the outermost mesoporous layer is as follows:

Prepare CTAB–80 mL ethanol and 60 mL H_2O are mixed in a beaker, then 0.37 g CTAB is added under stirring condition, and continuously stirred until completely dissolved to obtain dispersion solution of CTAB for later use.

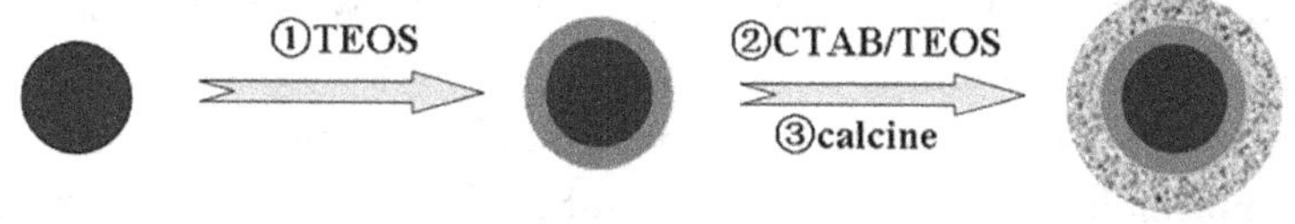

Figure 2.80: Schematic diagram of synthesis of Fe$_3$O$_4$@nSiO$_2$@mSiO$_2$.

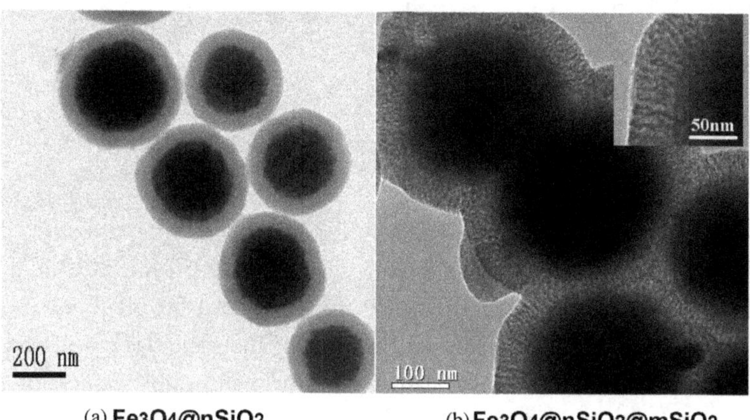

(a) **Fe3O4@nSiO2**   (b) **Fe3O4@nSiO2@mSiO2**

Figure 2.81: TEM images (apparatus: JEM-2100F, JEOL).

Immobilize CTAB–50 mg Fe$_3$O$_4$@nSiO$_2$ and 60 mL H$_2$O are added in a proper three-necked bottle for ultrasonic dispersion for 2–3 min. Then 1.0 mL ammonia water and the dispersion solution of CTAB are added at the same time and mechanically stirred at room temperature. After mixing well, stirring is continued for another 0.5–1 h at room temperature.

Synthesize Fe$_3$O$_4$@nSiO$_2$@mSiO$_2$–1 mL TEOS is injected into the above three-necked bottle with a syringe and stirred quickly for 5 min to make it even. Then the stirring speed is reduced slightly and the reaction is maintained at room temperature for 12 h. The product is then washed with a small amount of ethanol and acetone 3 times, respectively, to remove the unreacted CTAB and other impurities, and dispersed in 50 mL acetone in a clean flask. The acetone is replaced after reflux at 70°C for 8 h for another reflux of another 8 h. The process is repeated twice to remove CTAB from the microspheres. The resulting product is washed with ethanol several times, and dried in vacuum at 60°C for 12 h.

The obtained $Fe_3O_4@nSiO_2@mSiO_2$ microspheres with core–shell–shell structure have a large number of hydroxyl groups on the surface, which can be used for further modification or surface modification by calcining.

The morphology and surface coating of $Fe_3O_4@nSiO_2@mSiO_2$ microspheres are observed by TEM. As shown in Figure 2.81(a), $Fe_3O_4@nSiO_2$ has a Fe_3O_4 core with a diameter of about 250 nm, and a dense and nonporous SiO_2 layer with a thickness of about 20 nm. Compared with $Fe_3O_4@nSiO_2$, the $Fe_3O_4@nSiO_2@mSiO_2$ microspheres in Figure 2.81(b) have an additional mesoporous layer of about 70 nm.

2.7.2. *Magnetic mesoporous silica microsphere ($Fe_3O_4@mSiO_2$)*

Mesoporous silica can also be coated on the surface of Fe_3O_4 directly.[34]

50 mg Fe_3O_4 and 500 mg CTAB are dissolved in 50 mL H_2O in a flask for ultrasound for 30 min. Then 400 mL H_2O and 50 mL 0.01 $mol\cdot L^{-1}$ NaOH are added and shaken well for 5 min. Then the flask is placed in a water bath of 60°C for 30 min to form a uniform and stable dispersion solution. Next, 2.5 mL mixture solution of TEOS/ethanol (volume ratio 1:4) is slowly added to the flask and then continuously stirred for 12 h at 60°C. The resulting product is collected by magnetic separation and dried, and then re-dispersed in acetone to remove the surfactant in a water bath of 80°C. After 5 times of extraction, the $Fe_3O_4@mSiO_2$ microspheres are washed with deionized water over 5 times and dried in vacuum at 50°C for later use.

2.7.3. *Glucose-modified magnetic mesoporous silica microsphere ($Fe_3O_4@mSiO_2$–glucose)*

2.7.3.1. *Preparation of $Fe_3O_4@mSiO_2$–glucose*

The synthetic process of $Fe_3O_4@mSiO_2$–glucose is shown in Figure 2.82.[35]

Firstly, Fe_3O_4 microspheres are synthesized according to the following steps: 6.8 g $FeCl_3$, 12.0 g NaAc and 2 g $Na_3Cit\cdot2H_2O$ are dissolved in 200 mL ethylene glycol, and transferred to Teflon-lined stainless steel reactor for reaction at 200°C for 7 h. After the reactor is cooled to room temperature, the product is washed with deionized water and ethanol, and dried in vacuum for later use.

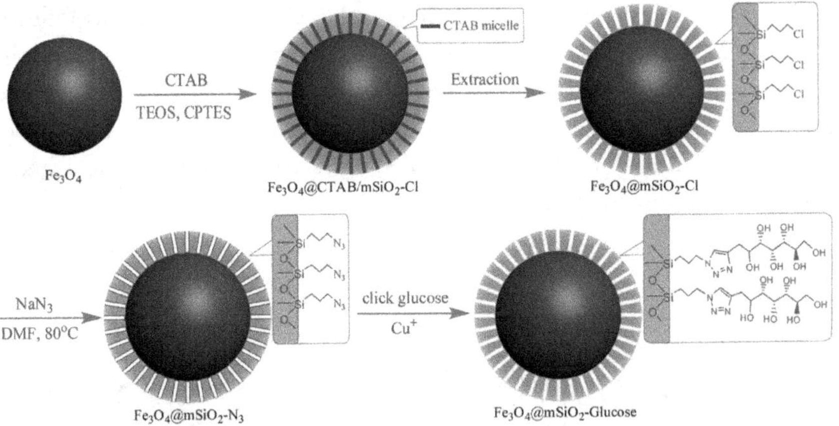

Figure 2.82: Schematic diagram of synthesis of $Fe_3O_4@mSiO_2$–glucose.

Secondly, $Fe_3O_4@mSiO_2$–Cl microspheres are synthesized according to the following steps: 0.5 g Fe_3O_4 microspheres and 1.0 g CTAB are dissolved in 500 mL H_2O for ultrasonic dispersion for 1 h. Then 80 mL 7.5 mmol·L^{-1} NaOH aqueous solution is added for reaction for 30 min at 60°C. Then 20 mL mixture solution of TEOS/3-chloropropyltriethoxy silane(CPTES)/ethanol (v/v/v, 2/1/2) is added for reaction for another 12 h at 60°C. The obtained product is cleaned with deionized water and ethanol, and then dispersed in 60 mL ethanol for reflux reaction at 90°C to remove CTAB. The reflux process is repeated 5 times. The resulting product is dried in vacuum for later use.

Thirdly, the $Fe_3O_4@mSiO_2$–N_3 microspheres are synthesized according to the following steps: 0.5 g $Fe_3O_4@mSiO_2$–Cl microspheres are dispersed in 100 mL DMF saturation solution of NaN_3 for stirring for 24 h at 80°C. The obtained $Fe_3O_4@mSiO_2$–N_3 microspheres are cleaned with ethanol and dried at room temperature for later use.

Next, propargyl glucose is prepared as follows: 0.90 g D-glucose, 1.6 mL 80 wt.% propargyl bromotoluene solution, and 100 mL THF solution containing 2.43 g $FeCl_3$ are mixed, and the mixture is stirred vigorously for 10 min at 10–15°C. Then 0.975 g zinc powder is added in batches at 8 min intervals, and the mixture is stirred at room temperature for 12 h. The obtained product is dissolved in 100 mL ether and 50 mL H_2O for 10 min, respectively, and then filtered. The filtrate is treated with ammonium bicarbonate and then filtered again, the final filtrate is applied directly to the next step.

Finally, $Fe_3O_4@mSiO_2$–glucose microspheres are synthesized as follows: 0.5 g $Fe_3O_4@mSiO_2$–N_3 microspheres are dispersed in 100 mL H_2O, and the above-mentioned propargyl glucose solution, 372 mg $CuSO_4 \cdot 5H_2O$, and 892 mg sodium ascorbate are added. Reaction is maintained at room temperature for 12 h. The product is cleaned with deionized water and ethanol and dried at room temperature for later use.

(a) **SEM of Fe3O4@mSiO2–glucose**

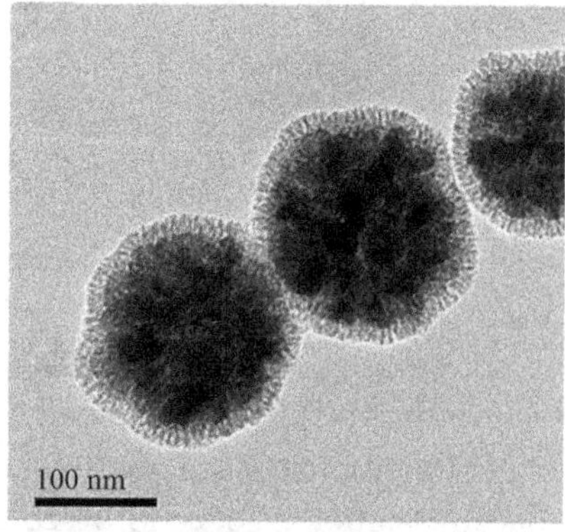

(b) **TEM of Fe3O4@mSiO2–glucose**

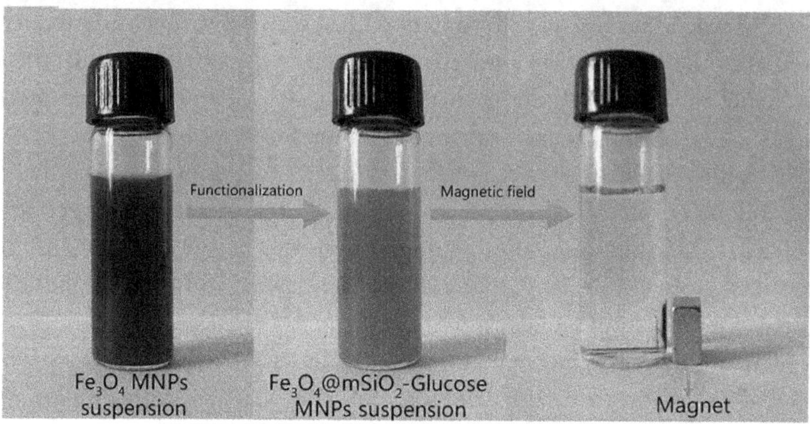

(c) **Fe3O4 disperses in water**   (d) **Fe3O4@mSiO2–glucose disperses in water**   (e) **Magnetic separation of Fe3O4@mSiO2–glucose**

Figure 2.83: SEM and TEM images of $Fe_3O_4@mSiO_2$–glucose, and the dispersion diagram of two different microspheres in water, and separation diagram by magnet.

2.7.3.2. *Characterization of $Fe_3O_4@mSiO_2$-glucose*

As shown in Figures 2.83(a) and 2.83(b), SEM and TEM images show that $Fe_3O_4@mSiO_2$-glucose microspheres are spherical and core–shell structure, the diameter of Fe_3O_4 microspheres is about 200 nm, and the thin layer of ordered mesoporous silica is about 12 nm. As shown in Figure 2.83(c), the Fe_3O_4 microspheres can be uniformly dispersed in water, and so do $Fe_3O_4@mSiO_2$-glucose microspheres in Figure 2.83(d). Besides, Figure 2.83(e) shows a excellent magnetic responsiveness of $Fe_3O_4@mSiO_2$-glucose.

Figure 2.84(a) shows the N_2 adsorption–desorption isotherms of $Fe_3O_4@mSiO_2$-glucose and its pore size distribution curve. The specific surface area of $Fe_3O_4@mSiO_2$-glucose microspheres is about 324 $m^2 \cdot g^{-1}$, and the total pore volume is 0.25 $cm^3 \cdot g^{-1}$, and the pore size distribution is mainly concentrated at 2.2 nm. This indicates that the mesoporous silica layer is successfully coated on the surface of magnetic microspheres, and the $Fe_3O_4@mSiO_2$-glucose has a large specific surface area.

As shown in Figure 2.84(b), in the FTIR spectrum of $Fe_3O_4@SiO_2$ microspheres, the absorption peaks at 584 cm^{-1} and 1,080 cm^{-1} are attributed to the vibration of Fe–O–Fe and Si–O–Si, respectively. The strong absorption peaks at 2,925 cm^{-1} and 2,850 cm^{-1} are attributed to $-CH_2$, indicating the successful coating of silica layer and CTAB template. In the FTIR spectrum of $Fe_3O_4@SiO_2$-Cl microspheres, the characteristic absorption peak of $-CH_2$ is obviously very weak, indicating that CTAB template is successfully removed. Compared with the FTIR spectrum of $Fe_3O_4@SiO_2$-Cl microspheres, the absorption peak at 2,104 cm^{-1} is the stretching vibration of azide group, which disappears in the FTIR spectrum of $Fe_3O_4@mSiO_2$-glucose, while the characteristic absorption peaks of $-CH_2$ are enhanced at 2,945 cm^{-1} and 2,875 cm^{-1}, indicating that the click chemical reaction between azide group and propargyl glucose has occurred, namely, glucose has been successfully modified on magnetic mesoporous silica microsphere.

Figure 2.84(c) shows the wide-angle XRD patterns of Fe_3O_4 (I) and $Fe_3O_4@mSiO_2$-glucose (II), the characteristic diffraction peaks indicate the surface modifications including the coating of mesoporous silica and grafting of glucose have no influence on the crystalline nature of Fe_3O_4.

Figure 2.84(d) shows that the saturation magnetic values of Fe_3O_4 and $Fe_3O_4@mSiO_2$-glucose are 79.8 $emu \cdot g^{-1}$ and 69.1 $emu \cdot g^{-1}$, respectively, indicating the good magnetic responsiveness to external magnetic field.

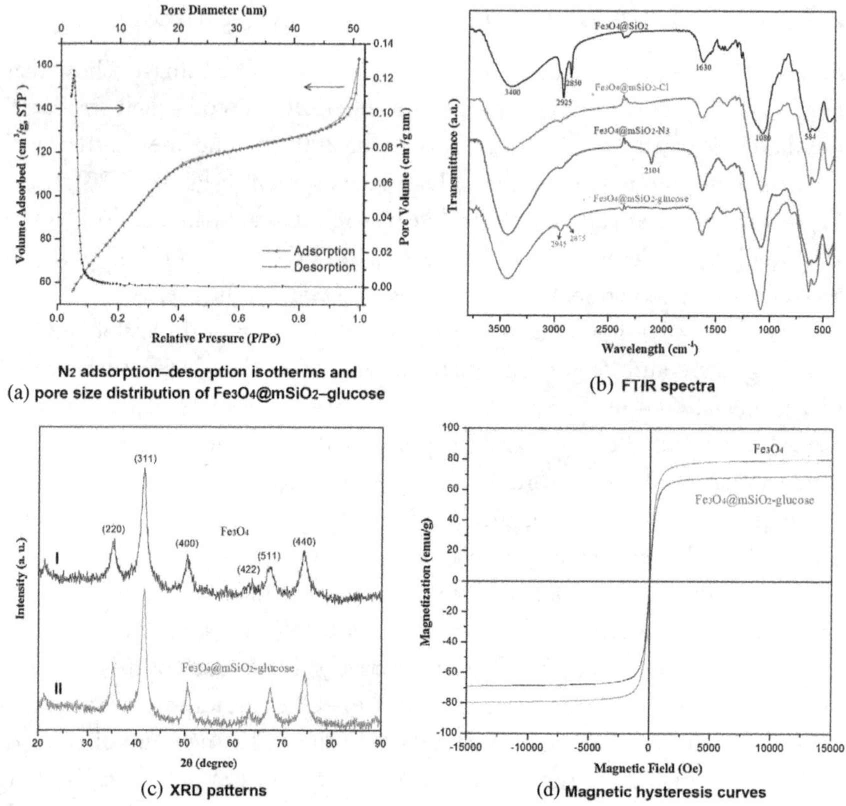

N2 adsorption–desorption isotherms and (a) pore size distribution of Fe₃O₄@mSiO₂–glucose

(b) FTIR spectra

(c) XRD patterns

(d) Magnetic hysteresis curves

Figure 2.84: Characterization diagrams of different materials.

2.7.4. *Magnetic mesoporous metal oxide microsphere (Fe₃O₄@mTiO₂)*

Magnetic mesoporous TiO_2 ($Fe_3O_4@mTiO_2$) is used as an example for introducing mesoporous metal oxide microspheres.[36]

2.7.4.1. *Preparation of Fe₃O₄@mTiO₂*

The synthetic process of $Fe_3O_4@mTiO_2$ is shown in Figure 2.85.

Fe₃O₄ microspheres are synthesized firstly and then the precursor of $Fe_3O_4@mTiO_2$ is prepared through hydrolyzing and condensation of tetrabutyl orthotitanate (TBOT) as follows: 150 mg Fe₃O₄ microspheres are dissolved in 200 mL ethanol, then 0.9 mL concentrated ammonia (28 wt.%) are

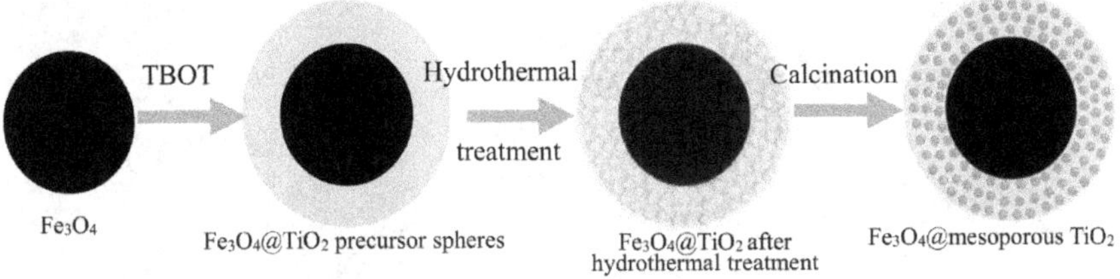

Figure 2.85: Schematic diagram of synthesis of $Fe_3O_4@m\text{TiO}_2$.

added for ultrasonic dispersion for 15 min. Then 2.0 mL TBOT is added drop by drop within 5 min under mechanical stirring, and then stirring is continued at 45°C for 24 h. The obtained solid product is cleaned by deionized water and ethanol, respectively.

Finally, the $Fe_3O_4@m\text{TiO}_2$ microspheres are prepared as follows: 0.5 g the aforementioned the precursor of $Fe_3O_4@m\text{TiO}_2$ is dispersed in 20 mL H_2O, and then the mixture is transferred to 30 mL Teflon-lined stainless steel reactor for 24 h at 160°C. After the reactor temperature drops to room temperature, the solid product is cleaned with deionized water and ethanol, respectively. The dry solid product is calcined at 400°C for 2 h under the protection of nitrogen.

2.7.4.2. *Characterization of $Fe_3O_4@m\text{TiO}_2$*

SEM and TEM images of $Fe_3O_4@m\text{TiO}_2$ microspheres are shown in Figure 2.86. According to Figure 2.86(a), $Fe_3O_4@m\text{TiO}_2$ microspheres are indeed spherical in shape, uniform in size and evenly dispersed. According to Figure 2.86(b), the diameter of $Fe_3O_4@m\text{TiO}_2$ microspheres is about 600 nm, and the thickness of TiO_2 layer is about 100 nm.

As shown in Figure 2.87, it can be seen from the wide-angle XRD patterns that the TiO_2 layer is amorphous before hydrothermal and calcining treatment. After these treatments, TiO_2 layer is transformed into crystallized particles as seen in Figure 2.87(b). The spectrum of $Fe_3O_4@m\text{TiO}_2$ microspheres is in good agreement with the standard spectrum of Fe_3O_4 and anatase, indicating the successful synthesis of $Fe_3O_4@m\text{TiO}_2$ microspheres.

Figure 2.88 shows the N_2 adsorption–desorption isotherms of $Fe_3O_4@m\text{TiO}_2$ microspheres and their precursors and corresponding pore size distri-

(a) **SEM image**   (b) **TEM image**

Figure 2.86: SEM and TEM images of $Fe_3O_4@mTiO_2$.

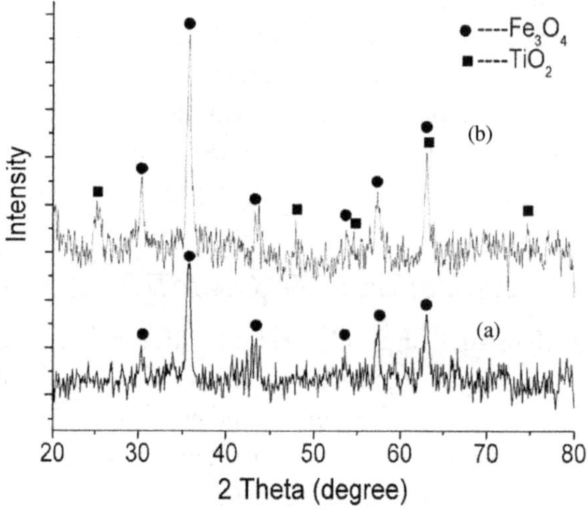

Figure 2.87: XRD patterns of different materials (curve (a) precursor of $Fe_3O_4@mTiO_2$; curve (b) $Fe_3O_4@mTiO_2$).

bution curves. The BET specific surface areas of the precursors and $Fe_3O_4@mTiO_2$ microspheres are $139.8 \ m^2 \cdot g^{-1}$ and $162.6 \ m^2 \cdot g^{-1}$, respectively. $Fe_3O_4@mTiO_2$ microspheres with larger specific surface area indicate the necessity of hydrothermal and calcining treatment.

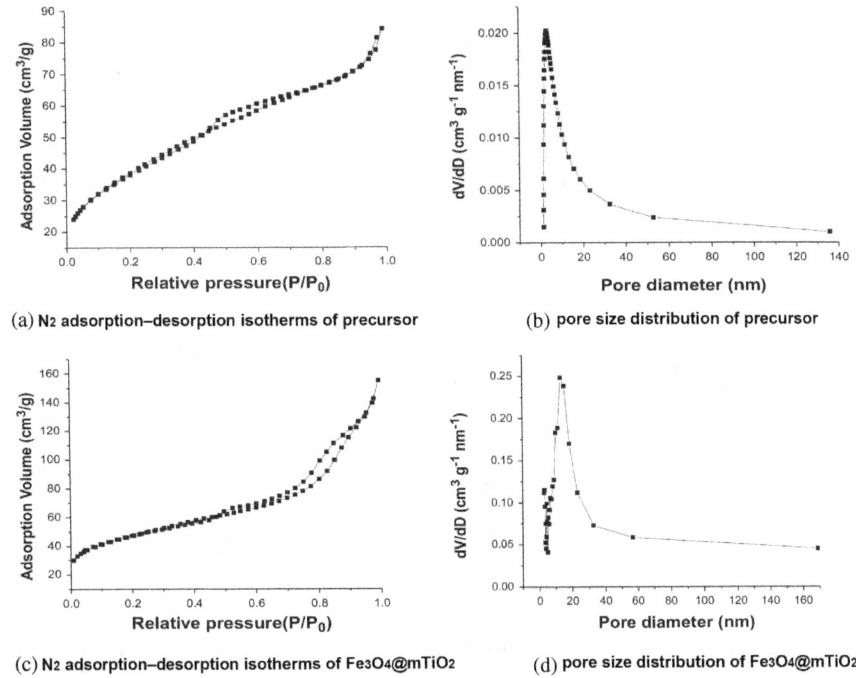

(a) N2 adsorption–desorption isotherms of precursor

(b) pore size distribution of precursor

(c) N2 adsorption–desorption isotherms of Fe3O4@mTiO2

(d) pore size distribution of Fe3O4@mTiO2

Figure 2.88: N_2 adsorption–desorption isotherms and pore size distribution curves of two different materials.

2.7.5. *Magnetic hybrid mesoporous microsphere ($Fe_3O_4@mTiO_2@mSiO_2$)*

2.7.5.1. *Preparation of $Fe_3O_4@mTiO_2@mSiO_2$*

As shown in Figure 2.89, Fe_3O_4 microspheres are synthesized firstly and then the precursor of $Fe_3O_4@mTiO_2$ is prepared through hydrolyzing and condensation of TBOT.[37] Next, the precursor is converted into $Fe_3O_4@mTiO_2$ microspheres by hydrothermal reaction. Finally, the mesoporous silica of outer layer is generated as follows: the $Fe_3O_4@mTiO_2$ microspheres are treated with polyvinyl pyrrolidone (PVP) overnight to obtain $Fe_3O_4@mTiO_2$–PVP, then the $Fe_3O_4@mTiO_2$–PVP are mixed with 23 mL ethanol, 4.3 mL H_2O, 0.62 mL ammonia water and 0.86 mL TEOS for reaction about

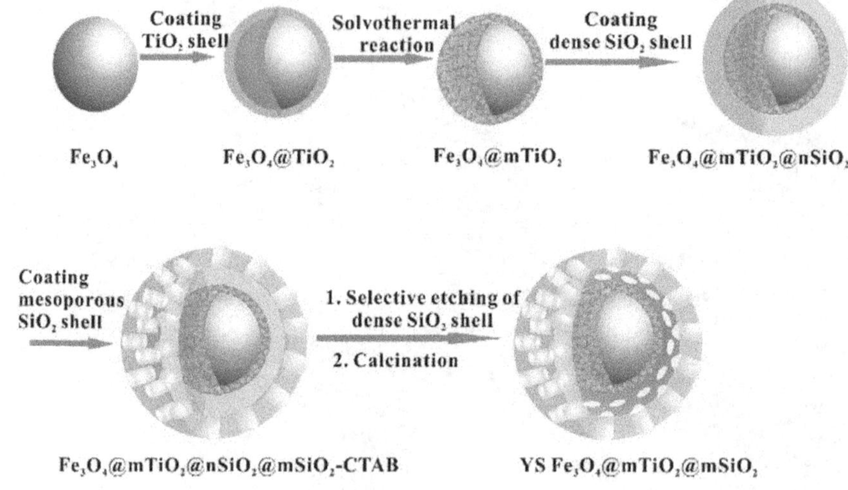

Figure 2.89: Schematic diagram of synthesis of $Fe_3O_4@mTiO_2@mSiO_2$.

6 h at room temperature to obtain $Fe_3O_4@mTiO_2@nSiO_2$ microspheres. Then the $Fe_3O_4@mTiO_2@nSiO_2$ microspheres are mixed with CTAB, 25 mL ethanol, 15 mL H_2O, and 0.275 mL ammonia water for reaction about 30 min under mechanical stirring. Then 0.125 mL TEOS is added drop by drop for another 6 h of reaction, the obtained product is to react with 10 mL 21.2 mg·L^{-1} Na_2CO_3 for 10 h at 50°C, and then calcined for 6 h at 450°C.

2.7.5.2. Characterization of $Fe_3O_4@mTiO_2@mSiO_2$

As shown in Figure 2.90(c), the diameter of $Fe_3O_4@mTiO_2$ is about 200–300 nm, and the core–shell structure is obvious after the coating of SiO_2 layer (Figure 2.90(d)). After coating the outermost layer of mesoporous silica by using CTAB as the guiding agent, the size of the entire microsphere increases to 500–600 nm (Figure 2.90(e)).

The wide-angle XRD patterns in Figure 2.91 shows that the spectrum of $Fe_3O_4@mTiO_2@mSiO_2$ is in good agreement with the standard spectrum of Fe_3O_4 and anatase. In addition, the EDX element mapping further confirms the successful synthesis of $Fe_3O_4@mTiO_2@mSiO_2$ microspheres (Figure 2.92).

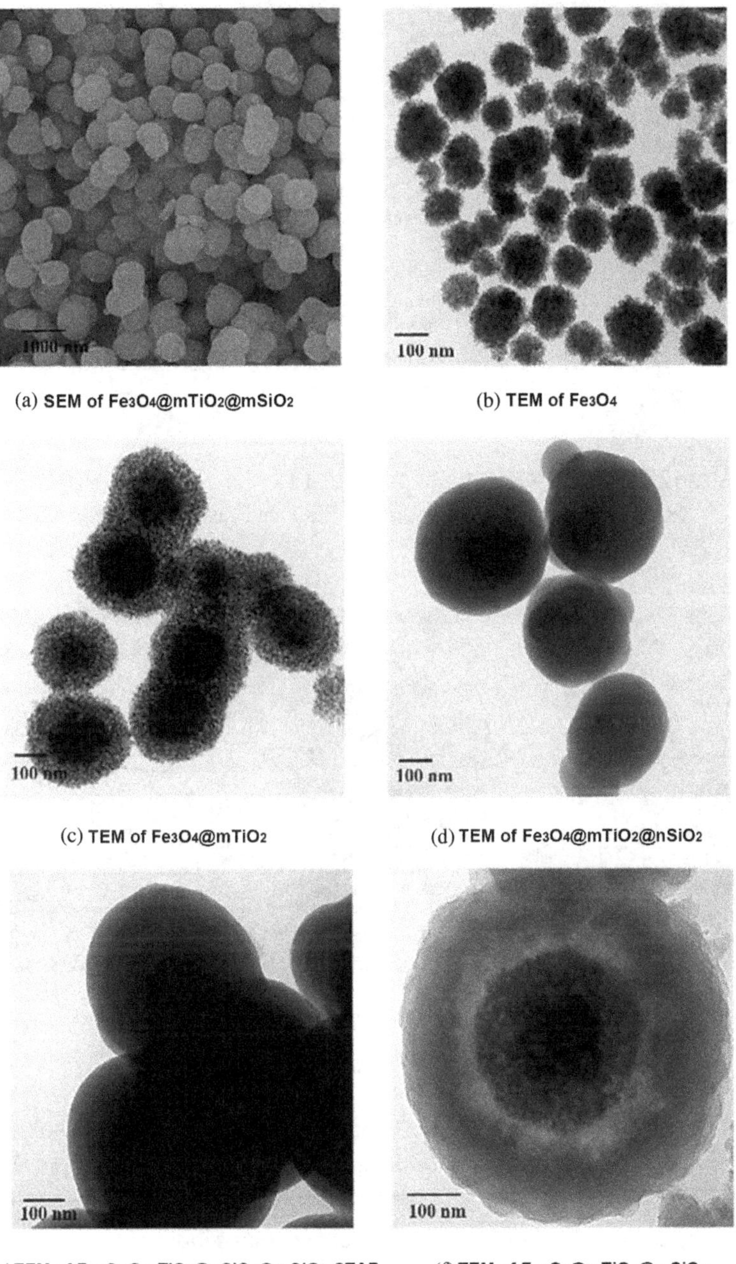

(a) SEM of Fe₃O₄@mTiO₂@mSiO₂ — (b) TEM of Fe₃O₄

(c) TEM of Fe₃O₄@mTiO₂ — (d) TEM of Fe₃O₄@mTiO₂@nSiO₂

(e) TEM of Fe₃O₄@mTiO₂@nSiO₂@mSiO₂-CTAB — (f) TEM of Fe₃O₄@mTiO₂@mSiO₂

Figure 2.90: SEM and TEM images of different materials.

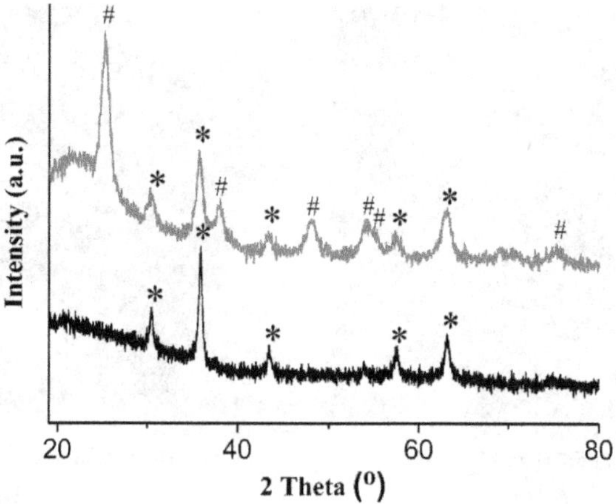

Figure 2.91: XRD patterns of Fe₃O₄ (black) and Fe₃O₄@*m*TiO₂@*m*SiO₂ (red) (* and # represent XRD characteristic peaks of magnetite Fe₃O₄ and anatase TiO₂, respectively).

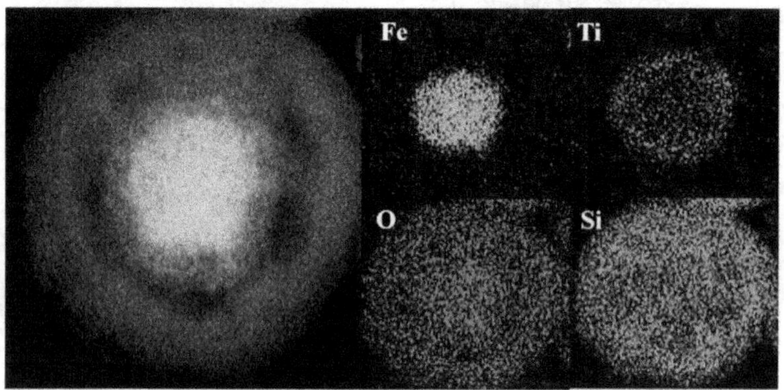

Figure 2.92: EDX element mapping of Fe₃O₄@*m*TiO₂@*m*SiO₂.

As shown in Figure 2.93(b), Fe₃O₄@*m*TiO₂@*m*SiO₂ microspheres are mainly concentrated in three pore sizes, namely, 2.7 nm, 3.9 nm, and 9.1 nm. In Figure 2.93(d), the pore sizes of Fe₃O₄@*m*TiO₂ microspheres are mainly concentrated at 3.7 nm and 9.5 nm. The above results indicate that both *m*TiO₂ and *m*SiO₂ are stable in the core–shell structure. In addition, the specific surface area of Fe₃O₄@*m*TiO₂ microspheres is measured as 169.79 m²·g⁻¹.

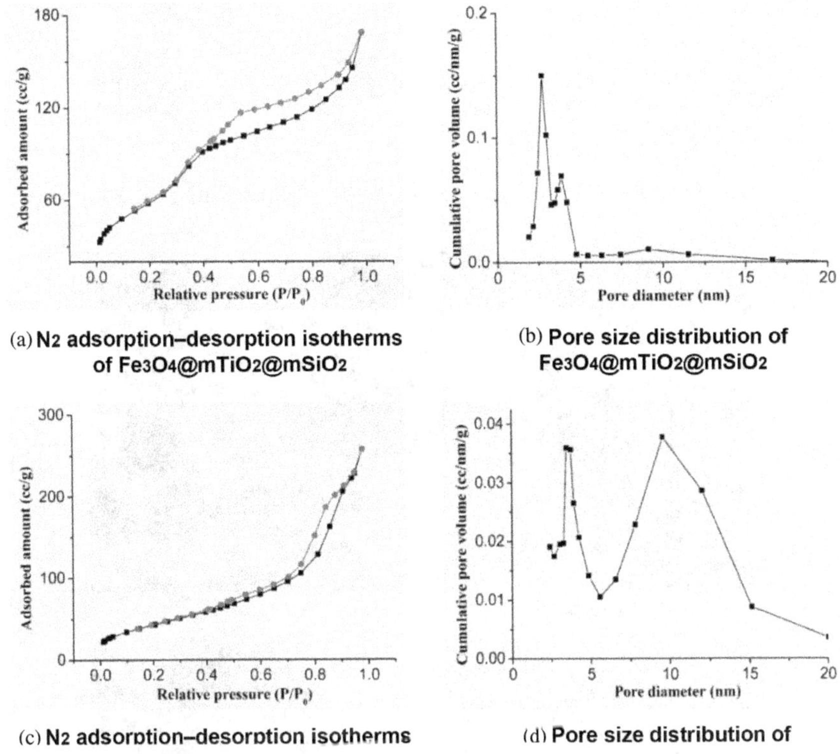

(a) **N2 adsorption–desorption isotherms of Fe3O4@mTiO2@mSiO2**

(b) **Pore size distribution of Fe3O4@mTiO2@mSiO2**

(c) **N2 adsorption–desorption isotherms**

(d) **Pore size distribution of**

Figure 2.93: N_2 adsorption–desorption isotherms and pore size distribution curves of two different microspheres.

2.7.6. *Mesoporous rare earth oxide-coated magnetic silica microsphere (Fe₃O₄@SiO₂@mCeO₂)*

2.7.6.1. *Preparation of Fe₃O₄@SiO₂@mCeO₂*

As shown in Figure 2.94, Fe_3O_4 microspheres are synthesized firstly and then coated with SiO_2 to obtain $Fe_3O_4@SiO_2$ microspheres.[38] Then $Fe_3O_4@SiO_2$ microspheres and 50 mg $Ce(NO_3)_3 \cdot 6H_2O$ are dispersed in 30 mL ethanol for 15 min by ultrasound. Then 20 mL 0.01 $g \cdot mL^{-1}$ cyclohexylenetetramine is added for another 15 min by ultrasound, and the whole reaction solution is reacted at 70°C for 2 h. The product is cleaned by deionized water and ethanol, and dried in vacuum for calcination for about 2 h at 400°C further.

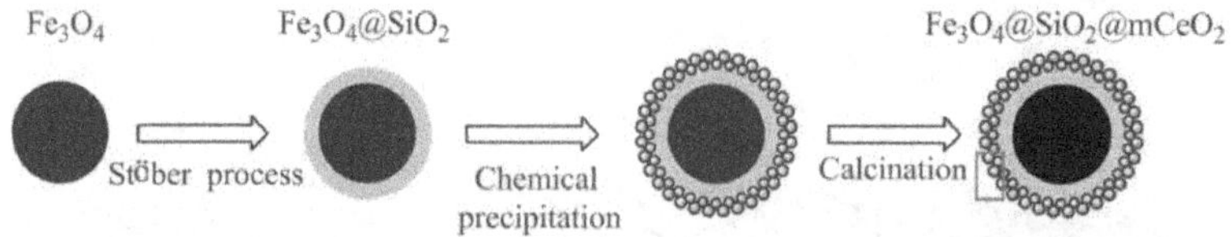

Figure 2.94: Schematic diagram of synthesis of Fe$_3$O$_4$@SiO$_2$@mCeO$_2$.

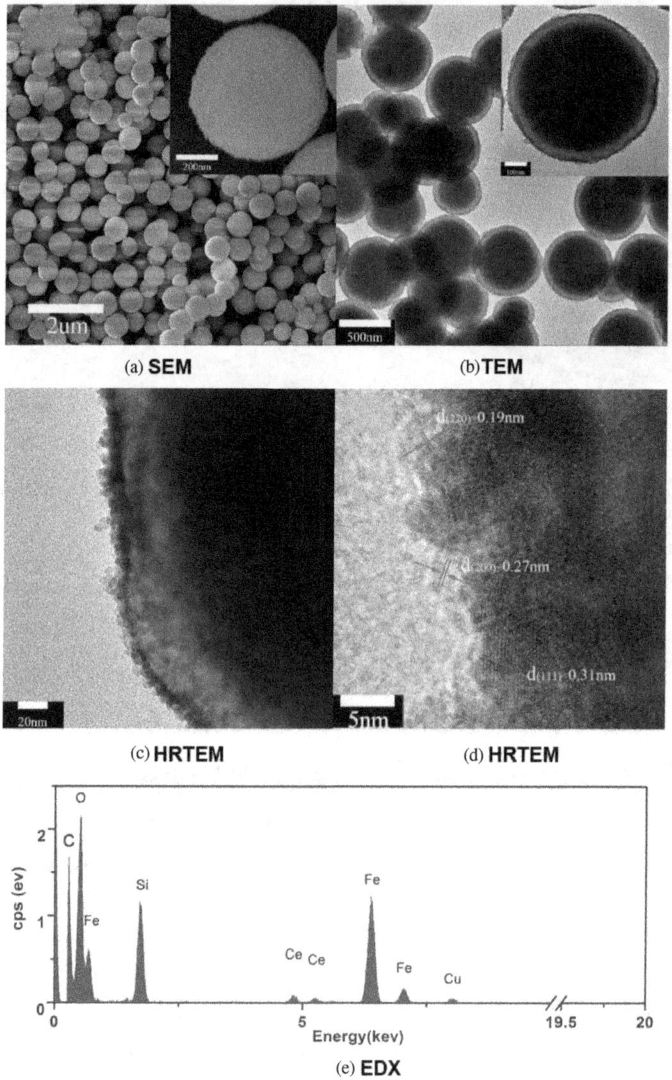

Figure 2.95: SEM, TEM, and HRTEM images, and EDX analysis of Fe$_3$O$_4$@SiO$_2$@mCeO$_2$.

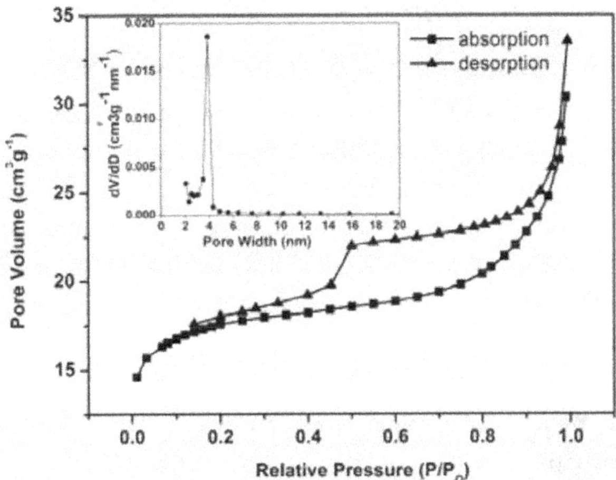

(a) N2 adsorption–desorption isotherms and pore size distribution of Fe3O4@SiO2@mCeO2

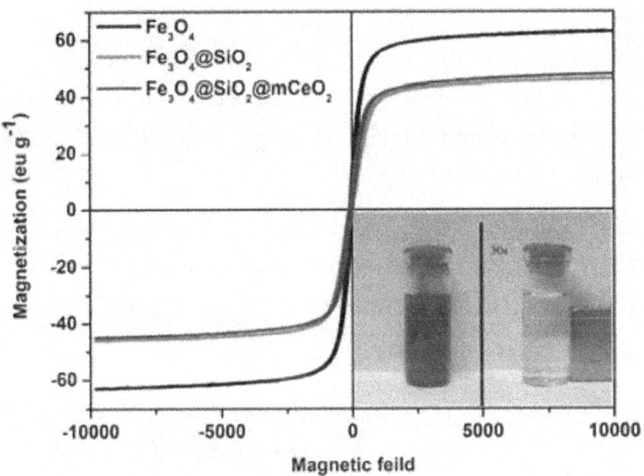

(b) magnetic hysteresis curves of different materials and magnetic separation of Fe3O4@SiO2@mCeO2

Figure 2.96: Characterizations of Fe$_3$O$_4$@SiO$_2$@mCeO$_2$ microspheres.

2.7.6.2. Characterization of $Fe_3O_4@SiO_2@mCeO_2$

The SEM and TEM images of $Fe_3O_4@SiO_2@mCeO_2$ microspheres are shown in Figures 2.95(a) and 2.95(b). It can be seen that the $Fe_3O_4@SiO_2@mCeO_2$ microspheres have a rough surface and a sandwich structure. The thickness of the core layer is about 500 nm, the middle layer is about 70 nm, and the outermost layer is about 20 nm. It can be observed from Figures 2.95(c) and 2.95(d) that the outermost layer is porous, and composed of numerous CeO_2 nanocrystals of about 8 nm in size. The surface spacing of CeO_2 single crystal labeled in Figure 2.95(d) of 0.19 nm, 0.27 nm, and 0.31 nm corresponds to the (220), (200), and (111) surfaces of CeO_2, respectively. The existence of the Ce element is further confirmed by EDX analysis (Figure 2.95(e)).

The N_2 adsorption–desorption isotherms of $Fe_3O_4@SiO_2@mCeO_2$ microspheres and the pore size distribution curve are shown in Figure 2.96(a). The BET specific surface area, average pore size, and total pore volume are 53.98 $m^2 \cdot g^{-1}$, 3.08 nm, and 0.032 $cm^3 \cdot g^{-1}$, respectively. In addition, the magnetic hysteresis curves of Fe_3O_4, $Fe_3O_4@SiO_2$, and $Fe_3O_4@SiO_2@mCeO_2$ are shown in Figure 2.96(b), and their saturation magnetic values are 62.8 $emu \cdot g^{-1}$, 46.0 $emu \cdot g^{-1}$, and 47.7 $emu \cdot g^{-1}$, respectively, indicating that $Fe_3O_4@SiO_2@mCeO_2$ microspheres have good magnetic separation ability and can hasten the whole enrichment process.

2.7.7. Lanthanum silicate-modified magnetic microsphere based on mesoporous silica ($Fe_3O_4@La_xSi_yO_5$)

2.7.7.1. Preparation of $Fe_3O_4@La_xSi_yO_5$

As shown in Figure 2.97, Fe_3O_4 microspheres are synthesized first and then coated with $mSiO_2$ to obtain $Fe_3O_4@mSiO_2$ microspheres.[39] Then the $Fe_3O_4@mSiO_2$ and 50 mg $La(NO_3)_3 \cdot 6H_2O$ are dissolved in 30 mL ethanol to be dispersed ultrasonically for 30 min. Next, 20 mL 0.01 $g \cdot mL^{-1}$ cyclohexylenetetramine is added for another 60 min of ultrasound. Finally, the whole reaction solution is

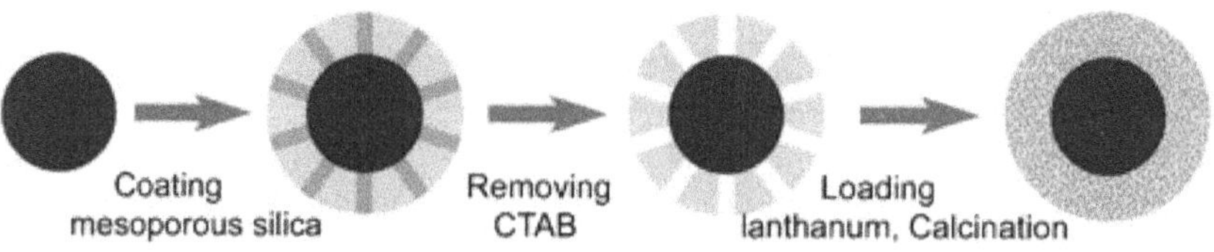

Figure 2.97: Schematic diagram of synthesis of $Fe_3O_4@La_xSi_yO_5$.

reacted at 70°C for 1 h. The product is cleaned with deionized water and ethanol, and dried in vacuum for calcination for about 6 h at 400°C further.

2.7.7.2. Characterization of $Fe_3O_4@La_xSi_yO_5$

In this work example, lanthanum silicate-modified non-porous silica is first analyzed to verify the importance of mesoporous silica for the formation of lanthanum silicate shells. As shown in Figures 2.98(e) and 2.98(f), lanthanum

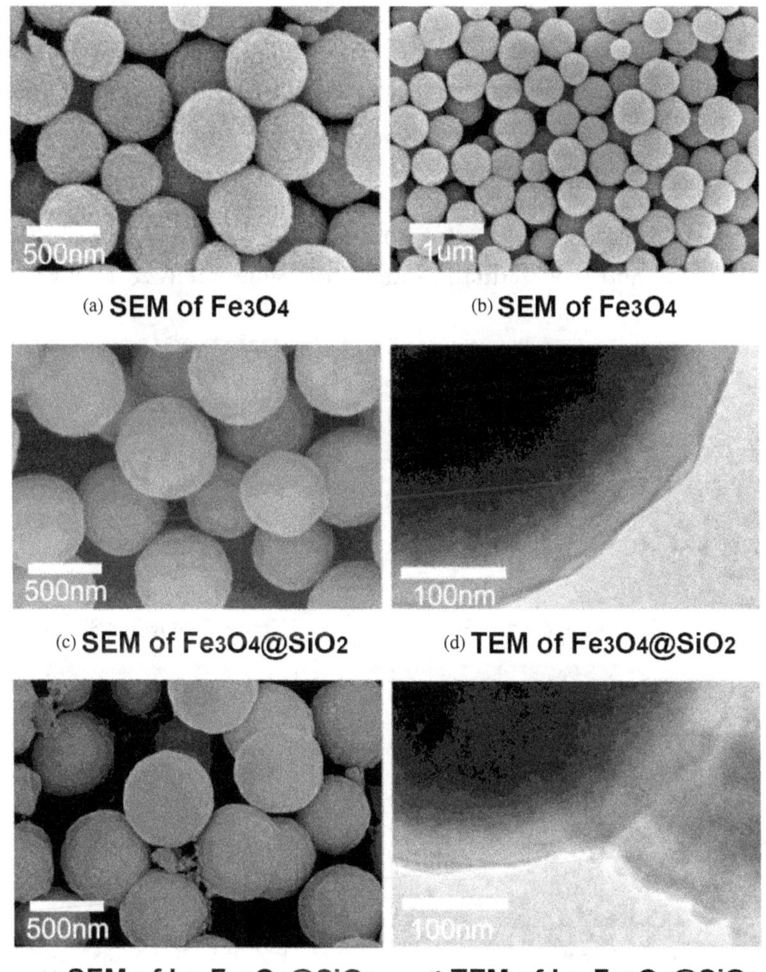

(a) **SEM of Fe3O4**   (b) **SEM of Fe3O4**

(c) **SEM of Fe3O4@SiO2**   (d) **TEM of Fe3O4@SiO2**

(e) **SEM of La-Fe3O4@SiO2**   (f) **TEM of La-Fe3O4@SiO2**

Figure 2.98: SEM and TEM images of three different materials.

nanoparticles are randomly scattered on the surface of non-porous silica shells instead of forming lanthanum silicate. Figures 2.98(a) and 2.98(b) show that Fe_3O_4 microspheres have rough surface and uniform size of about 450 nm. Figure 2.99(a) shows the SEM image of $Fe_3O_4@mSiO_2$ microspheres, from which it can be seen that the size of the whole microsphere increases and the surface becomes smooth after the coating of mesoporous silica. Figure 2.99(b) shows the distribution of ordered mesoporous channels and the obvious core–shell structure. As seen from Figures 2.99(c) and 2.99(d), the size of the whole microsphere decreases after lanthanum silicate is formed, and the mesoporous channels on the surface of the microsphere disappear.

FTIR is further used to characterize $Fe_3O_4@La_xSi_yO_5$ microspheres (Figure 2.100), and the absorption peak at 971 cm^{-1} is attributed to Si–O–La, indicating the formation of lanthanum silicate shells. This is further confirmed by EDX analysis (Figure 2.101).

In addition, the magnetic hysteresis curves of Fe_3O_4 microspheres, $Fe_3O_4@mSiO_2$ microspheres and $Fe_3O_4@La_xSi_yO_5$ microspheres are shown in Figure 2.102, and their saturation magnetic values are 62.8 emu·g^{-1}, 38.8

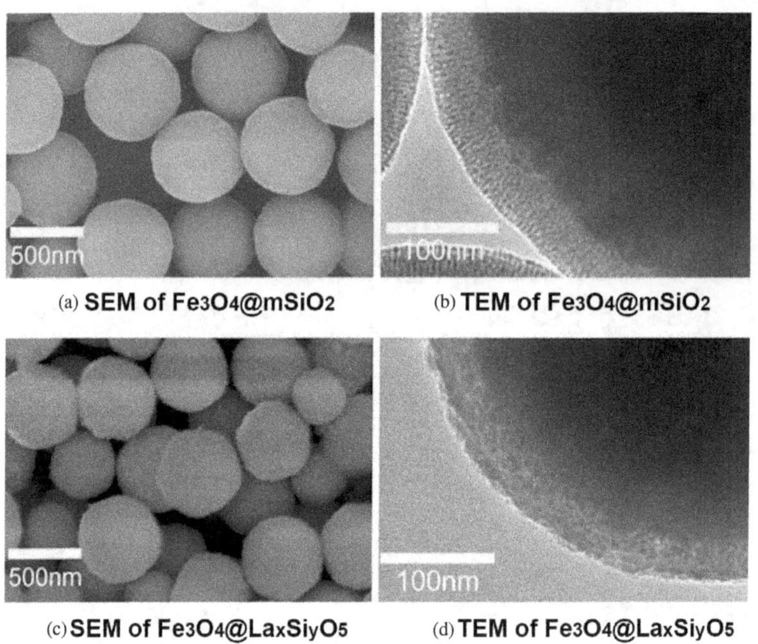

(a) SEM of Fe3O4@mSiO2   (b) TEM of Fe3O4@mSiO2

(c) SEM of Fe3O4@LaxSiyO5   (d) TEM of Fe3O4@LaxSiyO5

Figure 2.99: SEM and TEM images of two different materials.

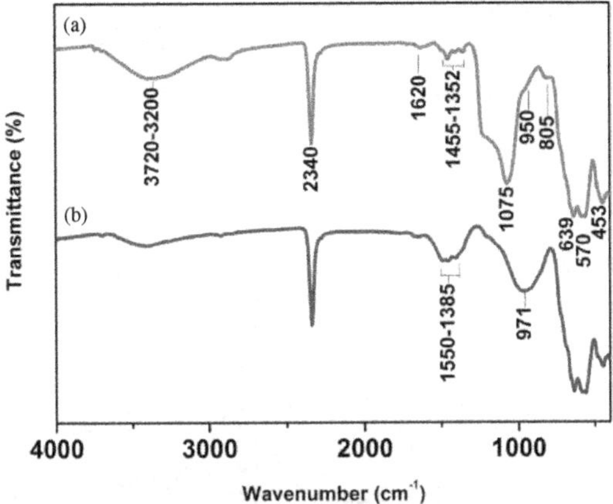

Figure 2.100: FTIR spectra of different materials (curve (a) $Fe_3O_4@mSiO_2$; curve (b) $Fe_3O_4@La_xSi_yO_5$).

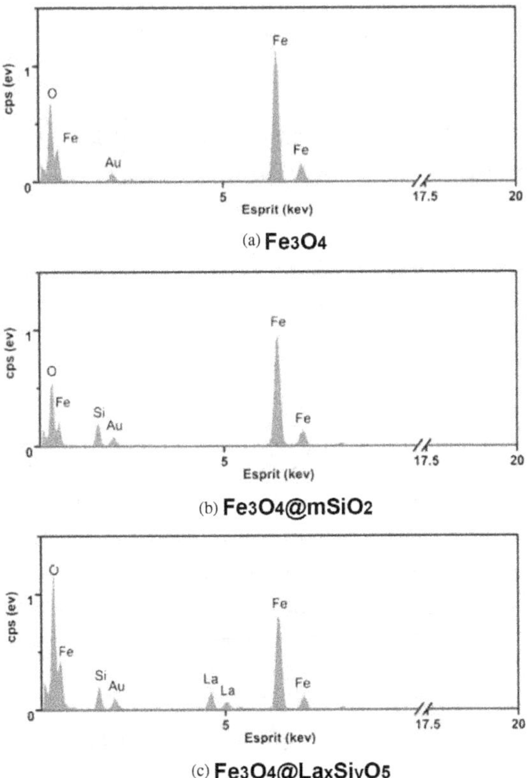

Figure 2.101. EDX analysis of different materials.

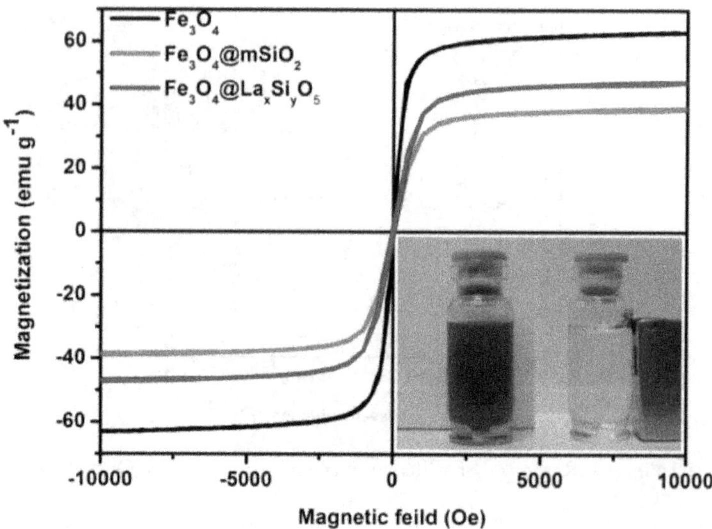

Figure 2.102: Magnetic hysteresis curves of different materials and the separation process of Fe$_3$O$_4$@La$_x$Si$_y$O$_5$ under external magnetic field.

emu·g^{-1}, and 47.2 emu·g^{-1}, respectively, indicating that Fe$_3$O$_4$@La$_x$Si$_y$O$_5$ microspheres have good magnetic separation ability.

2.8. Magnetic carbon nanotube and magnetic graphene material

2.8.1. *Magnetic carbon nanotube*

2.8.1.1. *Acidification of carbon nanotube*

Multi-walled carbon nanotubes (MWCNTs) are firstly acidified with nitric acid.[40] A certain amount of MWCNTs is dispersed in concentrated nitric acid, and then continuously stirred for 6 h at 60°C. The obtained black dispersion solution is treated with ultrasound for 30 min, and then washed by deionized water 5 times to neutral pH. After centrifugal separation, the acidified MWCNTs are dried in vacuum at 50°C.

2.8.1.2. *Co-assembly of Fe$_3$O$_4$ and MWCNTs*

The co-assembly of Fe$_3$O$_4$ and MWCNTs is carried out by hydrothermal synthesis.

About 0.81 g (3 mmol) $FeCl_3 \cdot 6H_2O$ is dissolved in 40 mL ethylene glycol, and then continuously stirred until the solution is clarified. 200 mg acidified MWCNTs are added, which are dispersed evenly by ultrasound for 3 h. Next, 3.6 g NaAc and 1.0 g polyethylene glycol are added and continuously stirred for about 30 min. The mixture solution is then transferred into a 200 mL Teflon-lined stainless steel reactor for reaction for 12 h at 200°C. After the reactor is cooled to room temperature, the product is washed by ethanol 5 times, and dried in vacuum at 60°C for later use.

2.8.2. Magnetic graphene

2.8.2.1. Acidification of graphene

The graphene slices are fully soaked in a three-necked flask containing a certain volume of concentrated nitric acid, and continuously stirred in 60°C of water bath for 6 h.[41] Then NaOH solution is gradually dropped to neutralize the excess nitric acid. Next, the solid product is washed repeatedly with a large amount of deionized water to neutral pH. The acidified graphene is collected by centrifuge at high speed, and dried in vacuum at 60°C.

2.8.2.2. Combination of Fe_3O_4 and graphene

About 800 mg solid yellow particles of $FeCl_3 \cdot 6H_2O$ are dissolved in ethylene glycol with the help of magnetic stirring, and then 150 mg sodium citrate and 150 mg acidified graphene are added. The mixture is placed in ultrasonic apparatus for nearly 2 h to disperse evenly. Then 3.5 g sodium acetate and 2.0 g PEG-20000 are added. The above mixture is then transferred to Teflon-lined stainless steel reactor for reaction for 12 h at 200°C. After the reactor temperature drops to room temperature, the solid product is cleaned with deionized water and ethanol, and dried in vacuum at 60°C for later use.

2.8.3. Ordered mesoporous silica-coated magnetic MWCNTs (MWCNTs/Fe_3O_4-@mSiO_2)

2.8.3.1. Preparation of ordered mesoporous silica-coated magnetic MWCNTs

As shown in Figure 2.103, magnetic carbon nanotubes (MWCNTs/Fe_3O_4) are firstly synthesized.[40] Then the mesoporous silica is coated on MWCNTs/

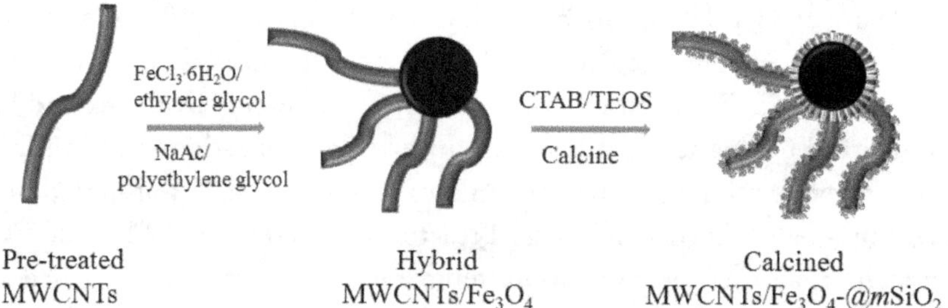

Figure 2.103: Schematic diagram of synthesis of calcined MWCNTs/Fe$_3$O$_4$–@mSiO$_2$.

Fe$_3$O$_4$ through a one-step sol–gel reaction under the template action of the positively charged surfactant CTAB as follows: MWCNTs/Fe$_3$O$_4$, CTAB, and H$_2$O are mixed at a ratio of 50 mg/500 mg/50 mL and treated ultrasonically for 30 min to achieve good dispersion. Then 400 mL H$_2$O and 50 mL 0.01 mol·L^{-1} NaOH solution are added for ultrasound for 5 min to form a homogeneous and stable dispersion solution, and the solution is placed in the 60°C of water bath. After reaction for 30 min, the mixture of 2.50 mL TEOS/ethanol (v/v, 1/4) is added slowly while stirring, and the reaction lasts for 12 h at 60°C. After magnetic separation, washing by ethanol, and drying in vacuum, the obtained product is dispersed in 50 mL acetone and reflux in 50°C of water bath to remove CTAB. In order to remove the surfactant CTAB completely, the reflux process is repeated 5 times, and the resulting product is washed with ethanol and dried in vacuum at 50°C. Finally, the product is placed in a tubular furnace, which is heated up to 300°C at the rate of 2°C/min, under the protection of nitrogen. And the temperature is maintained at 300°C for 1 h. After naturally cooling to room temperature, the final product could be used for the following enrichment experiment.

2.8.3.2. *Characterization of calcinated MWCNTs/Fe$_3$O$_4$–@ mSiO$_2$*

SEM image in Figure 2.104(a) shows that several MWCNTs are extended on Fe$_3$O$_4$, and the calcinated MWCNTs/Fe$_3$O$_4$–@mSiO$_2$ is a kind of tadpole-like material. TEM is used to observe the tadpole structure of the calcinated MWCNTs/Fe$_3$O$_4$–@mSiO$_2$. From the TEM image in Figure 2.104(b), the gray mesoporous silica layer can be clearly seen on the outer layer of calcinated

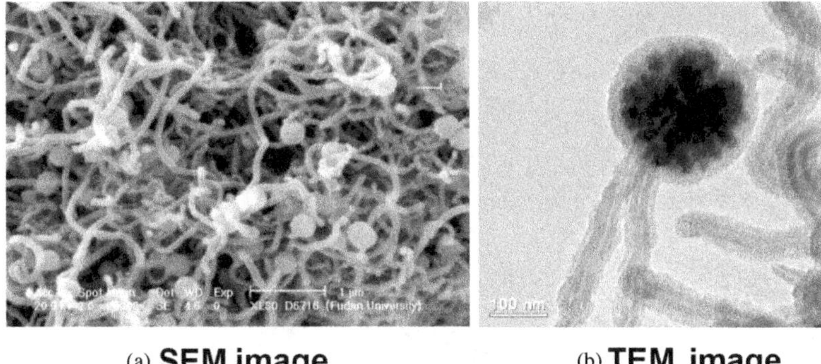

(a) SEM image　　　　(b) TEM image

Figure 2.104: SEM and TEM images of octopus-like calcined MWCNTs/Fe$_3$O$_4$–@mSiO$_2$.

MWCNTs/Fe$_3$O$_4$–@mSiO$_2$, in which the diameter of Fe$_3$O$_4$ is about 250 nm, and the diameter of cylinder MWCNTs is about 50 nm, moreover, the mesoporous silica layer (gray part) with a thickness of about 50 nm can be clearly observed in TEM image. In addition, it can be found that the porous channels are vertically arranged on the surface of silica through the co-assembly reaction of surfactant CTAB and silica oligomer from the image of high-power microscope. The special channel orientation increases the surface area and the possibility of guest molecules entering the channels, and lays a certain foundation for adsorption and enrichment.

FTIR is used to characterize the surface groups of materials as shown in Figure 2.105. The FTIR analysis of MWCNTs/Fe$_3$O$_4$–@mSiO$_2$ and MWCNTs/Fe$_3$O$_4$ is carried out at the same time, both of which show the absorption peak at 581 cm^{-1} that corresponds to the typical stretching vibration of Fe–O–Fe. By comparison, the absorption peaks at 1,087 cm^{-1}, 1,697 cm^{-1}, and ~3,400 cm^{-1} only appear in the FTIR spectrum of MWCNTs/Fe$_3$O$_4$–@mSiO$_2$, indicating that these three absorption peaks may be unique absorption peaks of mesoporous silica. The absorption peak at 1,087 cm^{-1} is attributed to the stretching vibration of Si–O–Si, the absorption peaks at 3,400 cm^{-1} and 1,697 cm^{-1} are attributed to silicon hydroxyl (Si–OH) in the outer layer of silica. The above results indicate that the silica layer is successfully coated on the outer surface of MWCNTs/Fe$_3$O$_4$.

As shown in Figure 2.106, the peak at 2θ = 26.2° in Figure 2.106(a) perhaps is the diffraction peak of MWCNTs in phase 002. By comparison, it

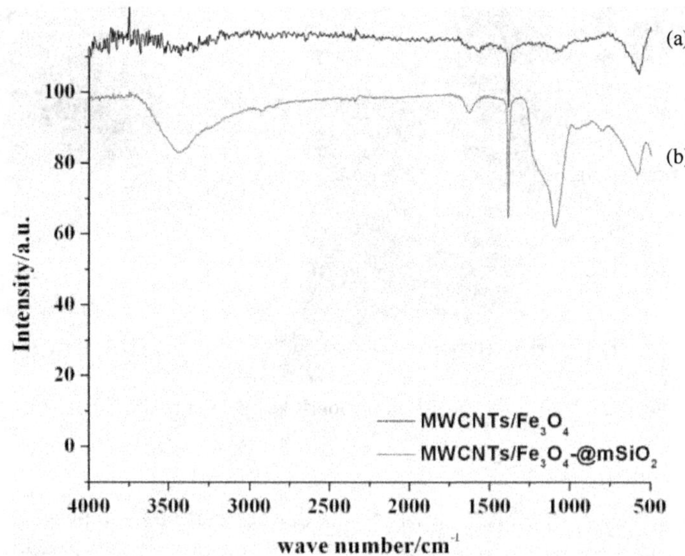

Figure 2.105: FTIR spectra of MWCNTs/Fe$_3$O$_4$ before and after coating with mesoporous silica (curve (a): MWCNTs/Fe$_3$O$_4$; curve (b): MWCNTs/Fe$_3$O$_4$–@mSiO$_2$).

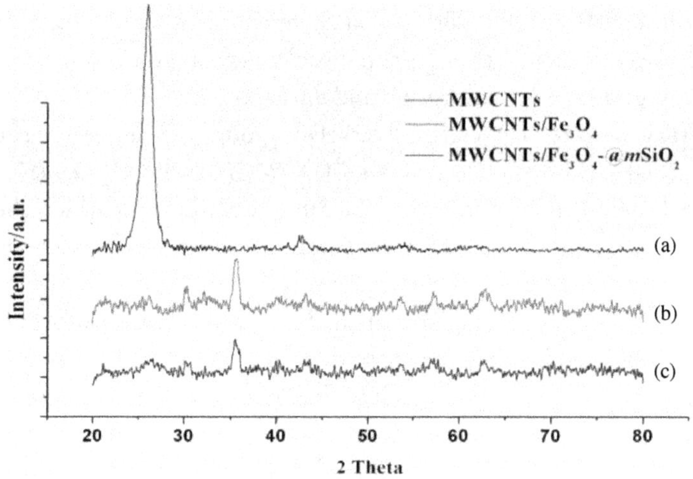

Figure 2.106: The wide-angle XRD patterns of three different materials (curve a: MWCNTs; curve b: MWCNTs/Fe$_3$O$_4$; curve c: MWCNTs/Fe$_3$O$_4$–@mSiO$_2$).

is found that MWCNTs/Fe$_3$O$_4$ and calcinated MWCNTs/Fe$_3$O$_4$–@mSiO$_2$ have the phase states of magnetite and MWCNTs. The diffraction peaks of MWCNTs/Fe$_3$O$_4$ and calcinated MWCNTs/Fe$_3$O$_4$–@mSiO$_2$ are in the same position, indicating that the sol–gel process in the synthetic process has no influence on the the basic skeleton and phase state of MWCNTs/Fe$_3$O$_4$.

The N$_2$ adsorption–desorption measurement is used to characterize the porous property of calcinated MWCNTs/Fe$_3$O$_4$–@mSiO$_2$. As shown in Figure 2.107, the N$_2$ adsorption–desorption isotherms of calcinated MWCNTs/Fe$_3$O$_4$–@mSiO$_2$ present an IV-type curve with hysteresis ring. The pore size distribution of calcinated MWCNTs/Fe$_3$O$_4$–@mSiO$_2$ is obtained by calculating the adsorption branch by BJH method, and its pore size is concentrated at 3.69 nm, indicating that the porous size on surface is relatively uniform. The BET surface area is 161.1 m^2·g^{-1} and the total pore volume is 0.277 cm^3·g^{-1}. The dense porosity and narrow pore size distribution provide the possibility of selective enrichment.

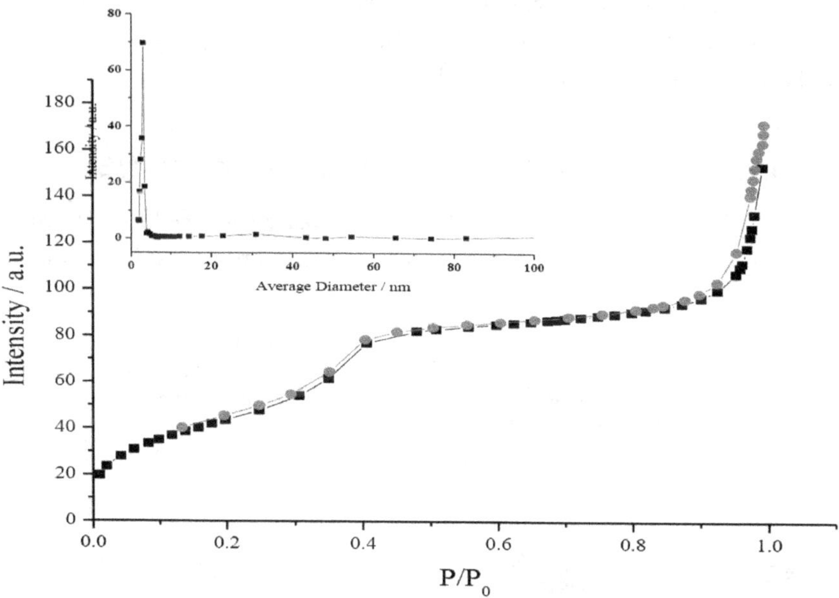

Figure 2.107: N$_2$ adsorption-desorption isotherms and pore size distribution curve of calcined MWCNTs/Fe$_3$O$_4$–@mSiO$_2$.

2.8.4. *Double-sided magnetic mesoporous graphene (Fe₃O₄-graphene@mSiO₂)*

2.8.4.1. *Preparation of Fe₃O₄–graphene@mSiO₂*

As shown in Figure 2.108, magnetic graphene is first synthesized and then coated with a layer of mesoporous silica.[41] The coating of mesoporous silica is performed as follows: 50 mg hydrophilic magnetic graphene and 500 mg CTAB are dispersed in 50 mL deionized water for ultrasonic dispersion about 30 min. Then 400 deionized water and 50 mL 0.01 mol·L⁻¹ NaOH are added for ultrasonic dispersion about 10 min. After the solution is evenly mixed and preheated to 60°C, the mixture of 2.5 mL TEOS/ethanol (v/v, 1/4) is added drop by drop using a medical syringe under mechanical stirring. The mechanical stirring reaction is continued for 12 h in 60°C of water bath. The product is separated by magnet and washed with deionized water several times. After reflux by 60 mL acetone for 2 h, the product is washed with deionized water several times and dried in vacuum at 60°C for 12 h. Finally, the magnetic mesoporous graphene is calcined for 6 h under the protection of nitrogen at 550°C to remove hydroxyl and other polar groups on its surface and turn its surface into hydrophobic.

2.8.4.2. *Characterization of Fe₃O₄–graphene@mSiO₂*

As seen in Figures 2.109(a) and 2.109(b), SEM images show Fe₃O₄–graphene@ *m*SiO₂ presents unique lamellar structure and there are several Fe₃O₄ micro-

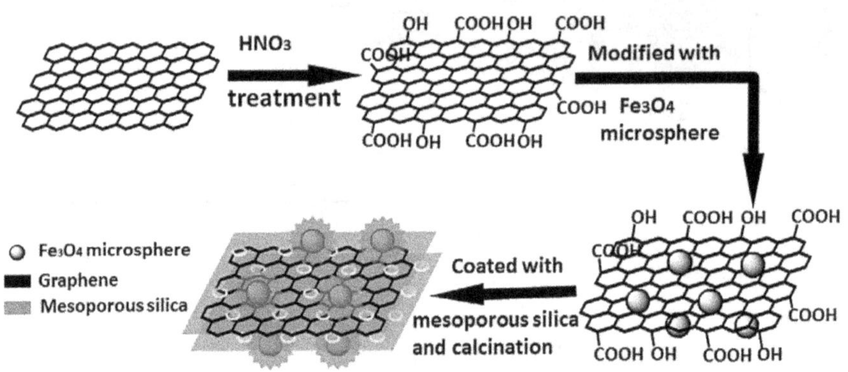

Figure 2.108: Schematic diagram of synthesis of Fe₃O₄–graphene@*m*SiO₂.

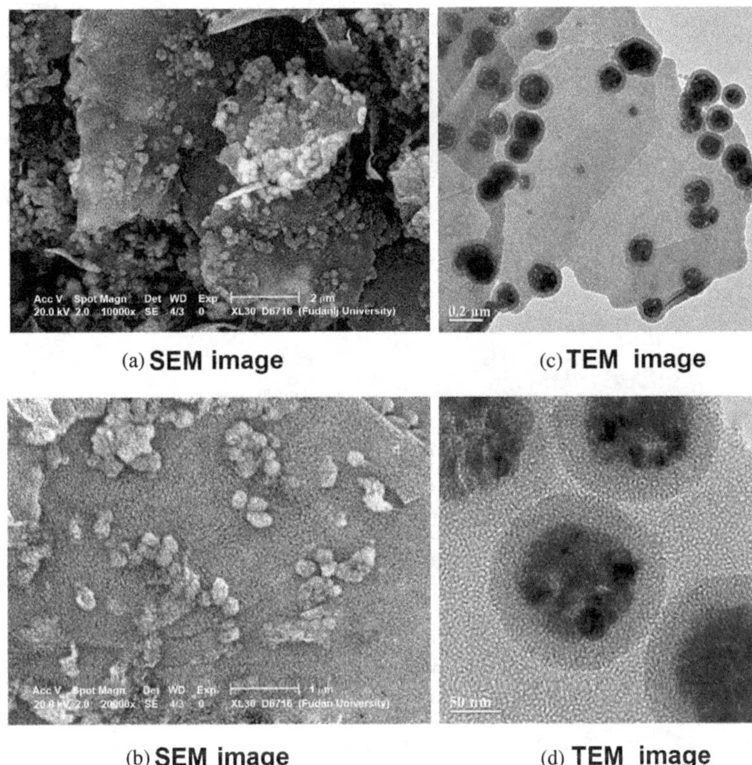

<table>
<tr><td align="center">(a) SEM image</td><td align="center">(c) TEM image</td></tr>
<tr><td align="center">(b) SEM image</td><td align="center">(d) TEM image</td></tr>
</table>

Figure 2.109: SEM and TEM images of Fe_3O_4-graphene@mSiO$_2$.

spheres on its surface, with the diameter of the lamellar structure measured to be about 4 μm. Similarly, the lamellar structure and the magnetic microspheres are also observed from the TEM images in Figures 2.109(c) and 2.109(d), and it can be seen that a mesoporous film is covered on the surface of the whole nanocomposite. The pore size of mesoporous film is about 3.2 nm, the thickness of silica film on both sides of the lamellar structure is about 50 nm.

As shown in Figure 2.110, the wide-angle XRD pattern shows that the strong diffraction peak appears around 26°, which is a characteristic diffraction peak, indicating that there are a number of graphene flakes. At the same time, there are also some obvious diffraction peaks around 30°, 36°, 46°, 58°, and 62°, which are characteristic diffraction peaks of Fe_3O_4 microspheres according to relevant literature and data analysis.

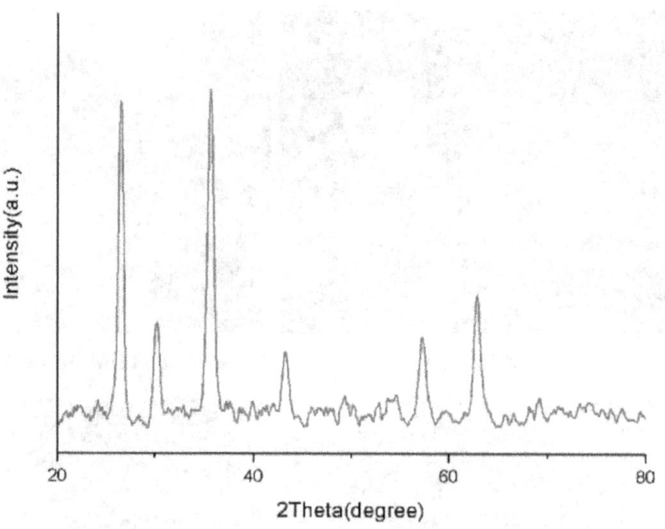

Figure 2.110: Wide-angle XRD pattern of Fe_3O_4–graphene@mSiO$_2$.

The N_2 adsorption–desorption isotherms are measured under the condition of 77 K, as shown in Figure 2.111, the IV-type adsorption–desorption characteristic curves appear in the data graph, and a sharp capillary condensation process can be observed at around the relative pressure of $0.4p/p_0$, indicating Fe_3O_4–graphene@mSiO$_2$ possesses a typical ordered mesoporous structure. The pore size of the double-sided magnetic mesoporous graphene is calculated to be about 3.2 nm, BET surface area and total pore volume are $167.8 \ m^2 \cdot g^{-1}$ and $0.2 \ cm^3 \cdot g^{-1}$, respectively, proving that this mesoporous material has characteristics of large surface area and large pore volume.

The absorption peak at $3,500 \ cm^{-1}$ can be clearly found in Figure 2.112(a), which is the vibration peak of O–H in hydroxyl and carboxyl groups on the surface of acidified graphene. The absorption peaks around $1,715 \ cm^{-1}$ and $1,600 \ cm^{-1}$ are attributed to the vibration of C=O in carboxyl groups on the surface of acidified graphene and the vibration of C=C in the graphene framework. These results indicate that the introduction of carboxyl group and carbonyl group causes no damage to the graphene skeleton during the acidification process of graphene, and the successful introduction of carboxyl group and carbonyl group also contributes to the further modification. As shown in Figure 2.112(b), in the FTIR spectrum of calcinated Fe_3O_4–

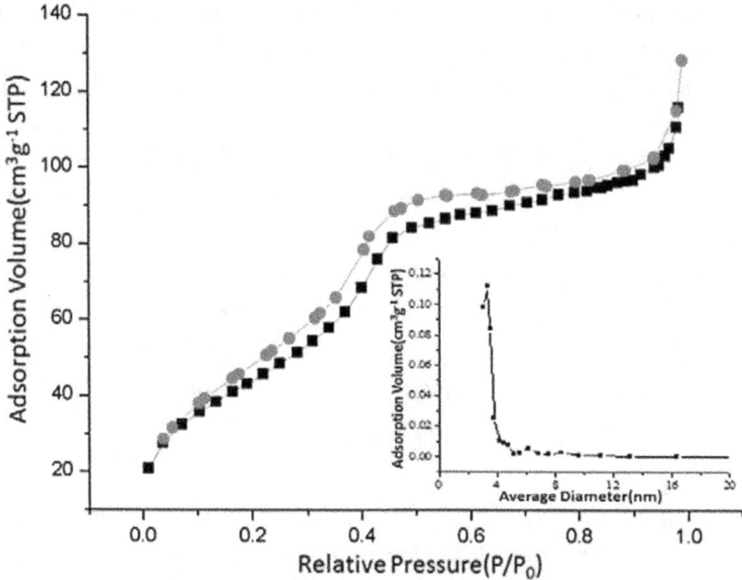

Figure 2.111: N_2 adsorption–desorption isotherms and pore size distribution curve of Fe_3O_4–graphene@mSiO$_2$.

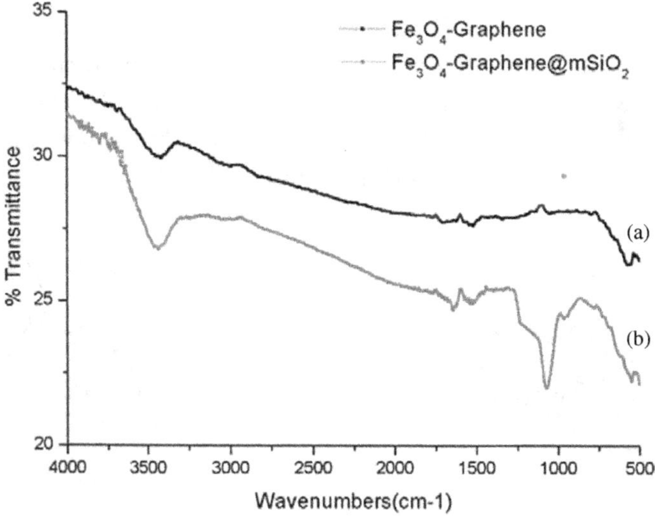

Figure 2.112: FTIR spectra of graphene materials (curve (a) acidified graphene; curve (b) Fe_3O_4–graphene@mSiO$_2$).

graphene@mSiO$_2$, the absorption peak at 1,080 cm^{-1} is attributed to the vibration of Si–O–Si of silica film on surface. The almost exactly same absorption peak at 1,600 cm^{-1} corresponding to the C=C of graphene skeleton also appears, confirming that the high strength of graphene maintains during the modification process. Thanks to the high-temperature anaerobic calcining process, compared to the FTIR spectrum of acidified graphene, the intensity of vibration peak of O–H at 3,500 cm^{-1} is significantly reduced in that of calcinated Fe$_3$O$_4$–graphene@mSiO$_2$, indicating an improved hydrophobicity. From the above, the excellent physical and chemical properties of Fe$_3$O$_4$–graphene@mSiO$_2$ make it expected to perform well in the process of separation and selective enrichment of endogenous peptides from complex biological samples.

2.8.5. *Mag GO@(Ti–Sn)O$_4$ hybrid material (Hybrid binary metal oxide-grafted magnetic graphene)*

2.8.5.1. *Preparation of Mag GO@(Ti–Sn)O$_4$*

Mag GO (Fe$_3$O$_4$–graphene) are synthesized according to the aforementioned method.[42]

The synthesis of Mag GO@(Ti–Sn)O$_4$ hybrid material is shown in Figure 2.113. 0.5 mL TBOT and 0.3 g SnCl$_4$·5H$_2$O are dispersed in 50 mL anhydrous ethanol for ultrasonic treatment for 30 min. Then 15 mg Mag GO is added, and ultrasound is performed for 30 min. Then 60 mL mixed solution of ethanol/water (v/v, 5/1) is added drop by drop within 30 min under mechanical stirring, and then continued to be stirred for 8 h. The

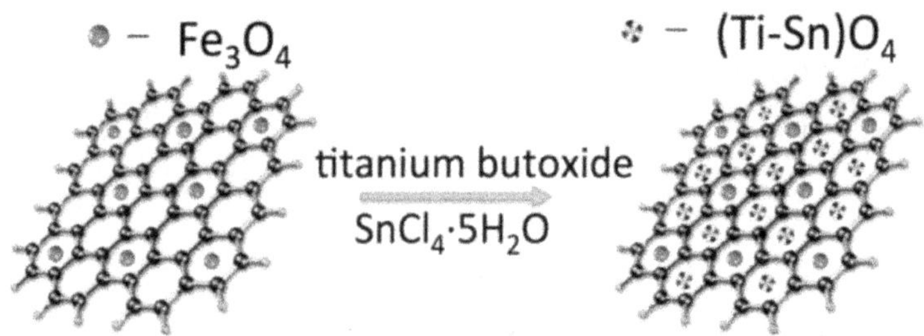

Figure 2.113: Schematic diagram of synthesis of Mag GO@(Ti–Sn)O$_4$.

solid product is cleaned with deionized water and ethanol, respectively, and dried in vacuum and calcined at 400°C for 2 h under the protection of nitrogen.

2.8.5.2. *Characterization of Mag GO@(Ti–Sn)O$_4$*

Figure 2.114(a) shows the SEM image of Mag GO@(Ti–Sn)O$_4$ hybrid material from which it can be seen that graphene is transparent and the diameter of magnetic microsphere is about 400 nm. As shown in Figures 2.114(b) and 2.114(c), the surface of nanospheres on Mag GO@(Ti–Sn)O$_4$ is rough and granular. At the same time, HRTEM image shows that the binary metal oxide microspheres of (Ti–Sn)O$_4$ on the hybrid material have lattice spacing of 0.25 nm and 0.26 nm, respectively, corresponding to the crystal surface of rutile TiO$_2$ (101) and the crystal surface of square SnO$_2$ (101), which indicate that (Ti–Sn)O$_4$ formed by TiO$_2$ and SnO$_2$ is a binary metal hybrid oxide microsphere at the atomic level (Figure 2.114(d)). As

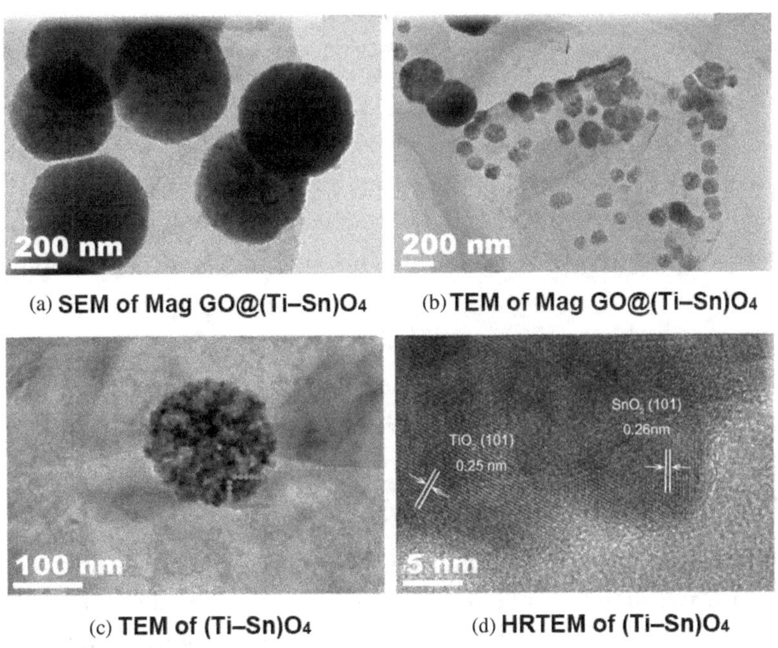

(a) **SEM of Mag GO@(Ti–Sn)O₄**

(b) **TEM of Mag GO@(Ti–Sn)O₄**

(c) **TEM of (Ti–Sn)O₄**

(d) **HRTEM of (Ti–Sn)O₄**

Figure 2.114: SEM, TEM, and HRTEM images of two hybrid materials.

Table 2.6: EDX analysis of Mag GO@(Ti–Sn)O$_4$.

Element	Wt. %	Mole %
C	68.25	83.21
O	13.13	12.02
Ti	0.4	0.12
Fe	17.32	4.54
Sn	0.91	0.11
Total	100	100

shown in Table 2.6, the atomic molar ratio of Ti and Sn is 0.12/0.11, which means that Ti and Sn form binary metal hybrid oxide microspheres in equal proportions. The specific surface area of Mag GO@(Ti–Sn)O$_4$ hybrid material is tested to be 361.5 m^2·g^{-1}, while that of Mag GO@TiO$_2$ composite is only 191.9 m^2·g^{-1}.

2.8.6. *Synthesis of Mag GO@PDA (polydopamine-coated magnetic graphene)*

80 mg dopamine hydrochloride is dissolved in 80 mL (10 mmol·L^{-1}) Tris buffer for ultrasound about 5 min.[43] Then 20 mg Mag GO is added and kept under mechanical stirring for 10 h at room temperature. The solid product is cleaned by deionized water and ethanol, respectively, and dried in vacuum for later use.

2.8.7. *Mag GO@PDA@(Zr–Ti)O$_4$ hybrid material (hybrid binary metal oxide grafted on dopamine-coated magnetic graphene)*

2.8.7.1. *Preparation of Mag GO@PDA@(Zr–Ti)O$_4$*

0.5 mL titanium isopropanol and 0.98 g zirconium isopropanol are dispersed in 50 mL anhydrous ethanol for ultrasound for about 30 min.[43] Then 15 mg Mag GO@PD is added and kept in ultrasound for another 30 min. Then,

60 mL mixture solution of ethanol/water (v/v, 5/1) is added drop by drop within 30 min under mechanical stirring, and continued to be stirred for 8 h. The solid product is cleaned with deionized water and ethanol, respectively, and dried in vacuum and calcined at 400°C for 2 h under the protection of nitrogen.

In addition, Mag GO@PDA@ZrO$_2$ and Mag GO@PD@TiO$_2$ are synthesized according to the above method.

2.8.7.2. *Characterization of Mag GO@PDA@(Zr–Ti)O$_4$*

Figure 2.115 shows the TEM image of Mag GO@PDA@(Zr–Ti)O$_4$ hybrid material, from which it can be seen that the surface of graphene is rough. On the one hand, this is because the surface of graphene is coated with dopamine; and on the other hand, the (Zr–Ti)O$_4$ does not have a spherical structure, but uniformly covers the surface of PDA-coated magnetic graphene. As shown in Figure 2.116, the distribution of Zi and Ti shows that they overlap and hybridize to form hybrid binary metal oxide.

Similar to Mag GO@PDA@(Ti–Sn)O$_4$ hybrid material, the atomic molar ratio of Zr and Ti is 15.54/14.569 (Table 2.7), that is, Zr and Ti also form hybrid binary metal oxide microspheres in equal proportions.

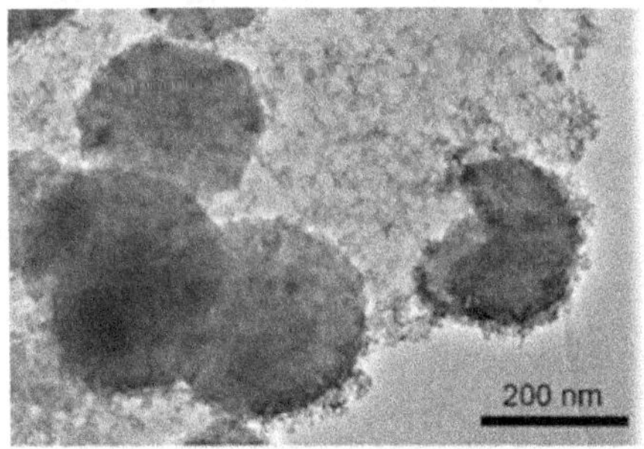

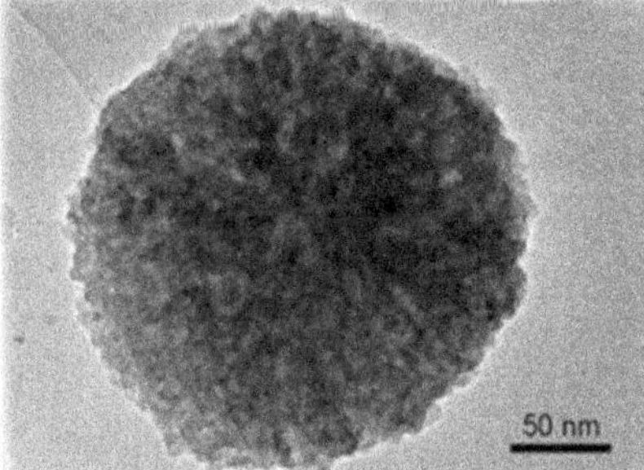

(a) Mag GO@PDA@(Zr–Ti)O₄   (b) Fe₃O₄ of Mag GO@PDA@(Zr–Ti)O₄

Figure 2.115: TEM images of two different materials.

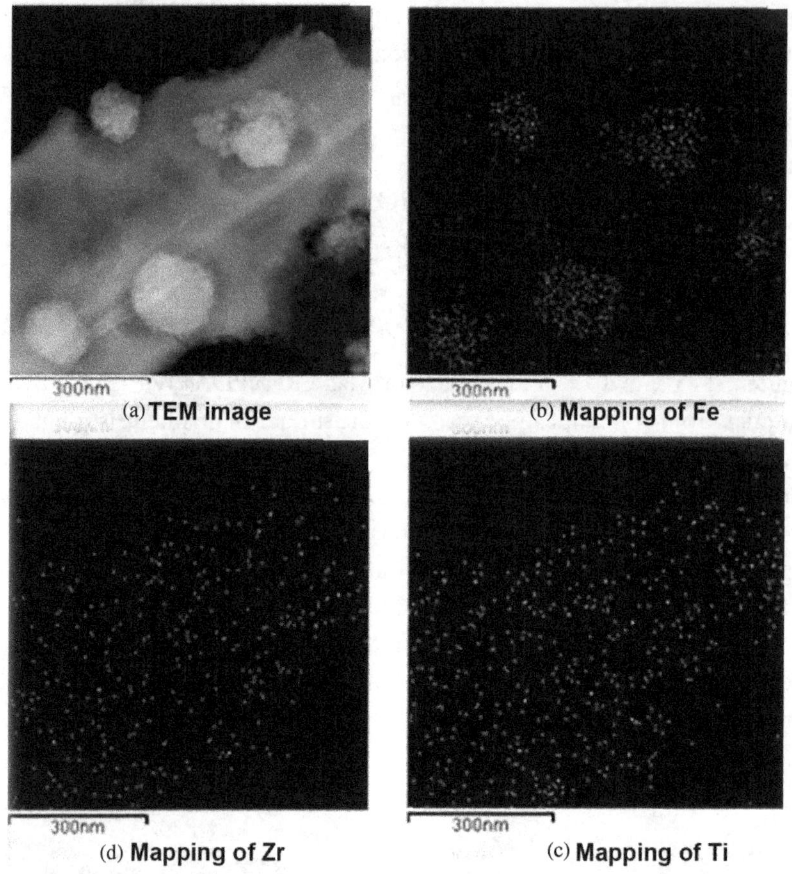

Figure 2.116: TEM image and EDX metal element mapping of Mag GO@PDA@ $(Zr–Ti)O_4$.

Table 2.7: EDX analysis of Mag GO@PDA@$(Zr-Ti)O_4$.

Element	Wt.%	Mole %
Ti	11.56	14.56
Fe	64.87	69.90
Zr	23.57	14.54

2.8.8. GO/Fe₃O₄/Au/PEG (polyethylene glycol-modified magnetic graphene)

2.8.8.1. Preparation of GO/Fe₃O₄/Au/PEG

The synthetic process of GO/Fe$_3$O$_4$/Au/PEG is shown in Figure 2.117.[44]

(1) Synthesis of GO/Fe₃O₄

10 mg graphene oxide (GO) is dissolved in 40 mL 2-morpholine ethanesulfonic acid (MES, 0.1 mol·L^{-1}, pH = 5.6) for ultrasound for about 3 h. Then 95.5 mg EDC and 57.5 mg NHS are added for ultrasound for another 2 h.

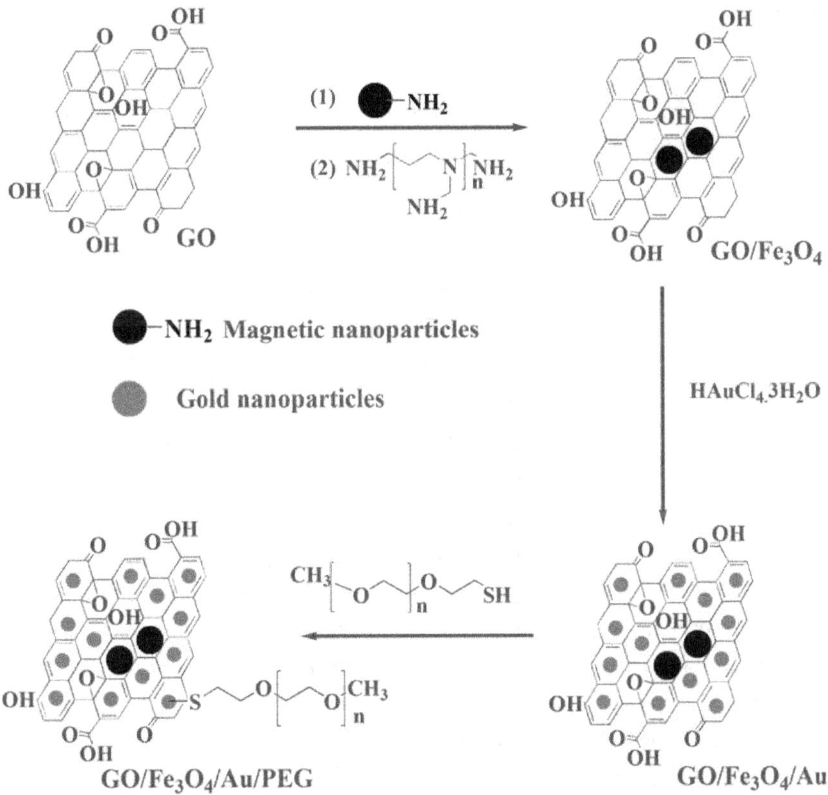

Figure 2.117: Schematic diagram of synthesis of GO/Fe$_3$O$_4$/Au/PEG.

Then 20 mg amino-functionalized magnetic microspheres are added for ultrasound for 1 h. The product is cleaned by deionized water and ethanol and dried in vacuum.

(2) Synthesis of GO/Fe$_3$O$_4$/Au

30 mg PEI is dissolved in 1 mL 1.5 mg·mL^{-1} GO/Fe$_3$O$_4$ dispersion solution for continuous and vigorous stirring for 1 h. The product is separated and dissolved in 1 mL H$_2$O, then 4 μL 100 mg·mL^{-1} HAuCl$_4$·3H$_2$O solution is added, and the reaction is conducted at 70°C for 1 h. The product is washed and dried in vacuum for later use.

(3) Synthesis of GO/Fe$_3$O$_4$/Au/PEG

2 mg GO/Fe$_3$O$_4$/Au and 20 mg SH-PEG are dispersed in 1 mL H$_2$O for continuous stirring at room temperature for 24 h. The product is washed and dried in vacuum for later use.

2.8.8.2. Characterization of GO/Fe$_3$O$_4$/Au/PEG

The characteristic peaks of C 1s, O 1s, N 1s, Si 2p, Fe 2p, and Au 4f can be observed from the XPS pattern of GO/Fe$_3$O$_4$/Au in Figure 2.118(a), and as seen in Figure 2.118(b), the binding energy of 84.1 eV and 87.8 eV in the Au 4f, respectively, belongs to the Au 4f 7/2 and Au 4f 5/2 of Au°. These results suggest that Au nanoparticles are successfully fixed on the surface of the magnetic graphene oxide. Figures 2.118(c)–2.118(e) show the TEM images of GO/Fe$_3$O$_4$, GO/Fe$_3$O$_4$/Au, and GO/Fe$_3$O$_4$/Au/PEG, respectively, from which it can be seen that magnetic nanoparticles are scattered on the surface of GO, and Au nanoparticles are scattered on the surface of magnetic microspheres and GO. The unique structure of GO provides sufficient support for the SH-PEG modification to ensure the high hydrophilicity of the material.

The TGA curves of GO/Fe$_3$O$_4$, GO/Fe$_3$O$_4$/Au, and GO/Fe$_3$O$_4$/Au/PEG are shown in Figure 2.119. At 900°C, GO/Fe$_3$O$_4$/Au/PEG produces a small amount of mass loss compared with GO/Fe$_3$O$_4$/Au. The content of SH-PEG in GO/Fe$_3$O$_4$/Au is estimated to be about 9.28 wt.%.

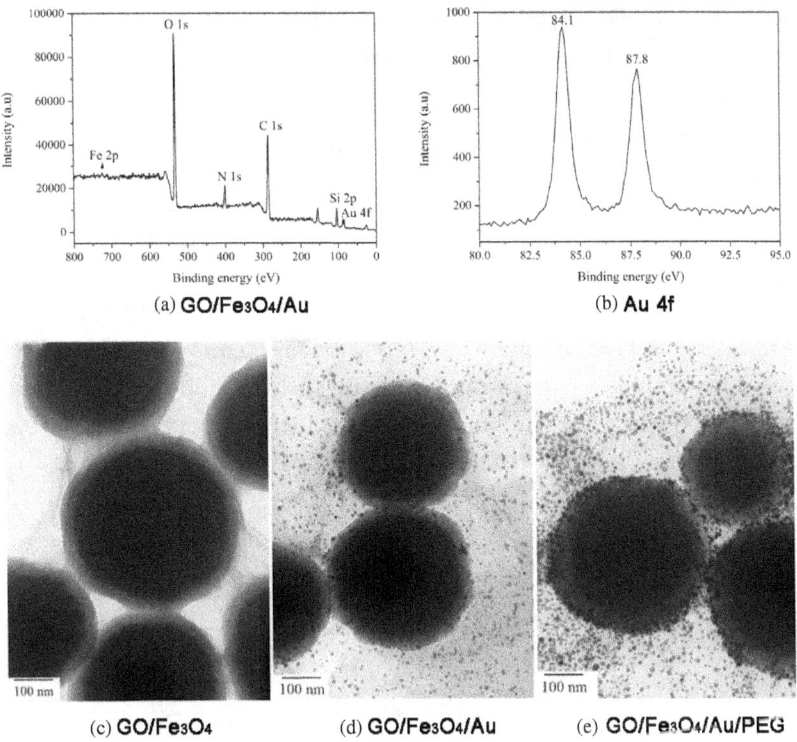

(a) GO/Fe₃O₄ (b) Au 4f

(c) GO/Fe₃O₄ (d) GO/Fe₃O₄/Au (e) GO/Fe₃O₄/Au/PEG

Figure 2.118: XPS patterns and TEM images of different materials.

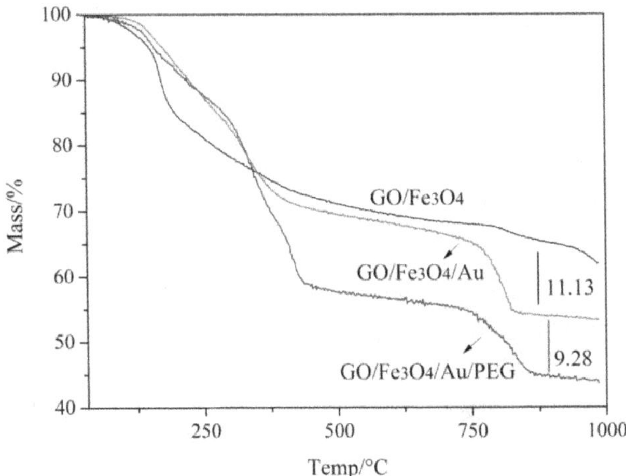

Figure 2.119: The TGA curves of different materials.

2.8.9. Fe_3O_4–GO@nSiO$_2$–PAMAM–Au–maltose (magnetic graphene co-modified by maltose and polyamidoamine dendrimers)

2.8.9.1. Preparation of Fe_3O_4–GO@nSiO$_2$–PAMAM–Au–maltose

The synthetic process of Fe_3O_4–GO@nSiO$_2$–PAMAM–Au–maltose is shown in Figure 2.120.[45]

The magnetic GO is first synthesized by hydrothermal method, and then the magnetic GO is coated with a layer of silica for bonding to PAMAM polymer, which provide a large number of active sites for modification of hydrophilic components. The Au nanoparticles are then attached to the surface of Fe_3O_4–GO@nSiO$_2$–PAMAM as a linker to immobilize SH-maltose.

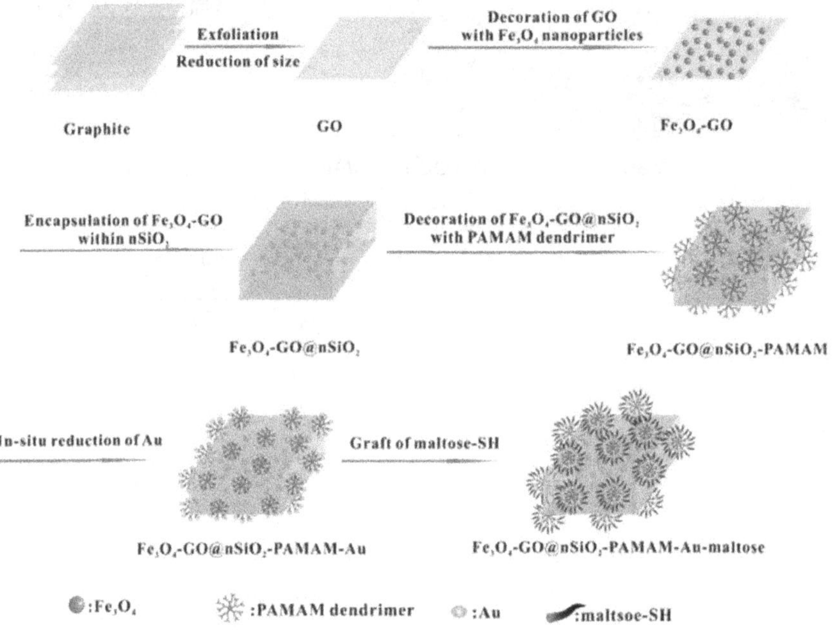

Figure 2.120: Schematic diagram of synthesis of Fe_3O_4–GO@nSiO$_2$–PAMAM–Au–maltose.

2.8.9.2. Characterization of Fe$_3$O$_4$–GO@nSiO$_2$–PAMAM–Au–maltose

Figure 2.121 is the TEM image of Fe$_3$O$_4$–GO@nSiO$_2$–PAMAM–Au–maltose and a series of intermediates during the synthetic process, which show that Fe$_3$O$_4$–GO@nSiO$_2$–PAMAM–Au–maltose is a sandwich structure.

In the FTIR spectrum of Fe$_3$O$_4$–GO in Figure 2.122(a), there are only the vibration absorption peak of Fe–O bond at 580 cm^{-1} and the fingerprint spectrum peaks of GO can be observed. However, the vibration absorption peaks of Si–O–Si at 1,084 cm^{-1}, 953 cm^{-1}, and 459 cm^{-1} appear in the FTIR spectrum of Fe$_3$O$_4$–GO@nSiO$_2$ after the coating of silica layer. The characteristic absorption peak of amide appears in the FTIR spectrum of Fe$_3$O$_4$–GO@nSiO$_2$–PAMAM–Au–maltose, indicating that the PAMAM polymer is successfully coated. The XRD in Figure 2.122(b) and EDX analysis in

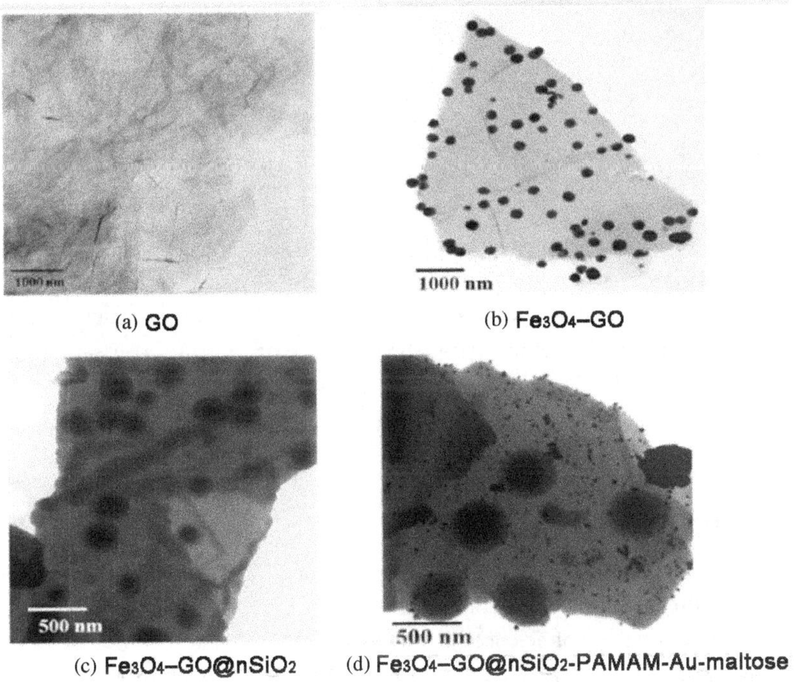

Figure 2.121: TEM images of different materials.

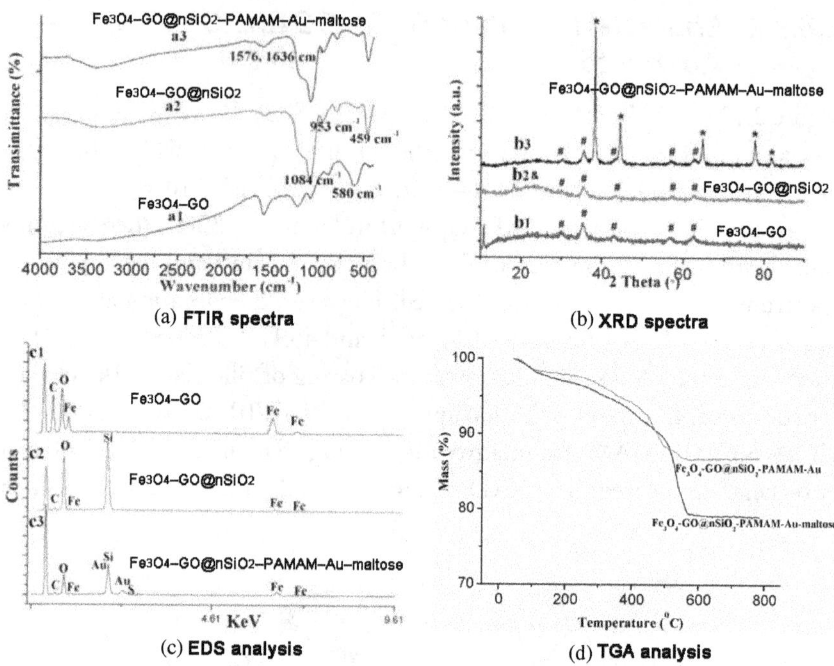

Figure 2.122: FTIR spectra, XRD patterns, EDX analysis, and TGA curves of different materials (&, #, * represent the characteristic peak of silica, Fe_3O_4, and Au nanoparticles, respectively).

Figure 2.122(c) further indicate the presence of Au nanoparticles. The TGA curve in Figure 2.122(d) indicates the presence of maltose. It is seen that the mass loss of the Fe_3O_4–GO@nSiO$_2$–PAMAM–Au–maltose is about 8.03% at 600°C compared with Fe_3O_4–GO@nSiO$_2$–PAMAM–Au.

References

1. Zhu L.-P. Design, Synthesis and Magnetic Properties of Some Micro- and Nano-Structured Magnetic Materials, Technical Institute of Physics and Chemistry, Chinese Academy of Sciences (2008).

2. Wang H. Synthesis, Assembly and Application of Magnetic Fe_3O_4/C Core–Shell Nanoparticles. University of Science and Technology of China (2011).

3. Zhang L.-d. Nanomaterial and Nanostructure. *Bulletin of the Chinese Academy of Sciences* (Beijing Science Press, 2001).

4. Herndon M. K., Collins R. T., Hollingsworth R. E., Larson P. R., Johnson M. B. Near-Field Scanning Optical Nanolithography Using Amorphous Silicon Photoresists. *Applied Physics Letters*, 1999, 74: 141–3.

5. Zhou C.-J. Synthesis and Application of Surface Coating Superparamagnetic Fe_3O_4 Nanoparticles, Hunan University (2010).

6. Xu X.-Q. Development of Novel Isolation and Identification Approaches in Proteomics Analysis Using Functional Magnetic Technologies, Fudan University (2007).

7. Lin S., Yun D., Qi D., Deng C., Li Y., Zhang X. Novel Microwave-Assisted Digestion by Trypsin-Immobilized Magnetic Nanoparticles for Proteomic Analysis. *Journal of Proteome Research*, 2008, 7: 1297–307.

8. Chen H., Liu S., Li Y., Deng C., Zhang X., Yang P. Development of Oleic Acid-Functionalized Magnetite Nanoparticles as Hydrophobic Probes for Concentrating Peptides with MALDI-TOF-MS Analysis. *Proteomics*, 2011, 11: 890–7.

9. Chen H., Deng C., Li Y., Dai Y., Yang P., Zhang X. A Facile Synthesis Approach to C-8-Functionalized Magnetic Carbonaceous Polysaccharide Microspheres for the Highly Efficient and Rapid Enrichment of Peptides and Direct Maldi-Tof-Ms Analysis. *Advanced Materials*, 2009, 21: 2200–5.

10. Li Y., Zhang X., Deng C. Application of Functionalized Magnetic Microspheres in Proteomics Analysis, Fudan University (2008).

11. Yan Y. Study on Phosphorylation and Glycosylation of Proteins Based on Functional Hybrid Nanomaterials, Fudan University (2014).

12. Ma W., Zhang Y., Li L., Zhang Y., Yu M., Guo J., Lu H., Wang C. Ti^{4+}-Immobilized Magnetic Composite Microspheres for Highly Selective Enrichment of Phosphopeptides. *Advanced Functional Materials*, 2013, 23: 107–15.

13. Fang C., Xiong Z., Qin H., Huang G., Liu J., Ye M., Feng S., Zou H. One-Pot Synthesis of Magnetic Colloidal Nanocrystal Clusters Coated with Chitosan for Selective Enrichment of Glycopeptides. *Analytica Chimica Acta*, 2014, 841: 99–105.

14. Tang J., Liu Y., Yin P., Yao G., Yan G., Deng C., Zhang X. Concanavalin a-Immobilized Magnetic Nanoparticles for Selective Enrichment of Glycoproteins and Application to Glycoproteomics in Hepatocelluar Carcinoma Cell Line. *Proteomics*, 2010, 10: 2000–14.

15. Chen H., Deng C., Zhang X. Synthesis of Fe_3O_4@SO_2@PMMA Core–Shell–Shell Magnetic Microspheres for Highly Efficient Enrichment of Peptides and Proteins for MALDI-TOF MS Analysis. *Angewandte Chemie-International Edition*, 2010, 49: 607–11.

16. Chen Y., Xiong Z., Zhang L., Zhao J., Zhang Q., Peng L., Zhang W., Ye M., Zou H. Facile Synthesis of Zwitterionic Polymer-Coated Core–Shell

Magnetic Nanoparticles for Highly Specific Capture of N-Linked Glycopeptides. *Nanoscale*, 2015, 7: 3100–8.

17. Xiong Z., Zhao L., Wang F., Zhu J., Qin H., Wu R. a., Zhang W., Zou H. Synthesis of Branched PEG Brushes Hybrid Hydrophilic Magnetic Nanoparticles for the Selective Enrichment of N-Linked Glycopeptides. *Chemical Communications*, 2012, 48: 8138–40.

18. Xiong Z., Qin H., Wan H., Huang G., Zhang Z., Dong J., Zhang L., Zhang W., Zou H. Layer-by-Layer Assembly of Multilayer Polysaccharide Coated Magnetic Nanoparticles for the Selective Enrichment of Glycopeptides. *Chemical Communications*, 2013, 49: 9284–6.

19. Zhao M., Deng C., Zhang X. The Design and Synthesis of a Hydrophilic Core–Shell–Shell Structured Magnetic Metal-Organic Framework as a Novel Immobilized Metal Ion Affinity Platform for Phosphoproteome Research. *Chemical Communications*, 2014, 50: 6228–31.

20. Chen Y., Xiong Z., Peng L., Gan Y., Zhao Y., Shen J., Qian J., Zhang L., Zhang W. Facile Preparation of Core–Shell Magnetic Metal Organic Framework Nanoparticles for the Selective Capture of Phosphopeptides. *ACS Applied Materials & Interfaces*, 2015, 7: 16338–47.

21. Wang M., Zhang X., Deng C. Facile Synthesis of Magnetic Poly(Styrene-Co-4-Vinylbenzene-Boronic Acid) Microspheres for Selective Enrichment of Glycopeptides. *Proteomics*, 2015, 15: 2158–65.

22. Xu X., Deng C., Gao M., Yu W., Yang P., Zhang X. Synthesis of Magnetic Microspheres with Immobilized Metal Ions for Enrichment and Direct Determination of Phosphopeptides by Matrix-Assisted Laser Desorption Ionization Mass Spectrometry. *Advanced Materials*, 2006, 18: 3289–93.

23. Wang Y., Liu M., Xie L., Fang C., Xiong H., Lu H. Highly Efficient Enrichment Method for Glycopeptide Analyses: Using Specific and Nonspecific Nanoparticles Synergistically. *Analytical Chemistry*, 2014, 86: 2057–64.

24. Zhang L., Xu Y., Yao H., Xie L., Yao J., Lu H., Yang P. Boronic Acid Functionalized Core–Satellite Composite Nanoparticles for Advanced Enrichment of Glycopeptides and Glycoproteins. *Chemistry — A European Journal*, 2009, 15: 10158–66.

25. Qi D., Zhang H., Tang J., Deng C., Zhang X. Facile Synthesis of Mercaptophenylboronic Acid-Functionalized Core–Shell Structure Fe_3O_4@C@Au Magnetic Microspheres for Selective Enrichment of Glycopeptides and Glycoproteins. *Journal of Physical Chemistry C*, 2010, 114: 9221–6.

26. Qi D., Mao Y., Lu J., Deng C., Zhang X. Phosphate-Functionalized Magnetic Microspheres for Immobilization of Zr^{4+} Ions for Selective Enrichment of the Phosphopeptides. *Journal of Chromatography A*, 2010, 1217: 2606–17.

27. Zhai R., Jiao F., Feng D., Hao F., Li J., Li N., Yan H., Wang H., Jin Z., Zhang Y., Qian X. Preparation of Mixed Lanthanides-Immobilized Magnetic Nanoparticles for Selective Enrichment and Identification of Phosphopeptides by Ms. *Electrophoresis*, 2014, 35: 3470–8.

28. Wei J., Zhang Y., Wang J., Tan F., Liu J., Cai Y., Qian X. Highly Efficient Enrichment of Phosphopeptides by Magnetic Nanoparticles Coated with Zirconium Phosphonate for Phosphoproteome Analysis. *Rapid Communications in Mass Spectrometry*, 2008, 22: 1069–80.

29. Li J., Wang F., Liu J., Xiong Z., Huang G., Wan H., Liu Z., Cheng K., Zou H. Functionalizing with Glycopeptide Dendrimers Significantly Enhances the Hydrophilicity of the Magnetic Nanoparticles. *Chemical Communications*, 2015, 51: 4093–6.

30. Deng Q., Wu J., Chen Y., Zhang Z., Wang Y., Fang G., Wang S., Zhang Y. Guanidinium Functionalized Superparamagnetic Silica Spheres for Selective Enrichment of Phosphopeptides and Intact Phosphoproteins from Complex Mixtures. *Journal of Materials Chemistry B*, 2014, 2: 1048–58.

31. Chen C.-T., Wang L.-Y., Ho Y.-P. Use of Polyethylenimine-Modified Magnetic Nanoparticles for Highly Specific Enrichment of Phosphopeptides for Mass Spectrometric Analysis. *Analytical and Bioanalytical Chemistry*, 2011, 399: 2795–806.

32. Zhang X., He X., Chen L., Zhang Y. A Combination of Distillation-Precipitation Polymerization and Click Chemistry: Fabrication of Boronic Acid Functionalized Fe_3O_4 Hybrid Composites for Enrichment of Glycoproteins. *Journal of Materials Chemistry B*, 2014, 2: 3254–62.

33. Chen H., Liu S., Yang H., Mao Y., Deng C., Zhang X., Yang P. Selective Separation and Enrichment of Peptides for Ms Analysis Using the Microspheres Composed of Fe_3O_4@$nSiO_2$ Core and Perpendicularly Aligned Mesoporous SiO_2 Shell. *Proteomics*, 2010, 10: 930–9.

34. Liu S., Chen H., Lu X., Deng C., Zhang X., Yang P. Facile Synthesis of Copper(II) Immobilized on Magnetic Mesoporous Silica Microspheres for Selective Enrichment of Peptides for Mass Spectrometry Analysis. *Angewandte Chemie-International Edition*, 2010, 49: 7557–61.

35. Zheng J., Xiao Y., Wang L., Lin Z., Yang H., Zhang L., Chen G. Click Synthesis of Glucose-Functionalized Hydrophilic Magnetic Mesoporous Nanoparticles for Highly Selective, Enrichment of Glycopeptides and Glycans. *Journal of Chromatography A*, 2014, 1358: 29–38.

36. Lu J., Wang M., Deng C., Zhang X. Facile Synthesis of Fe_3O_4@Mesoporous TiO_2 Microspheres for Selective Enrichment of Phosphopeptides for Phosphoproteomics Analysis. *Talanta*, 2013, 105: 20–7.

37. Wan H., Li J., Yu W., Liu Z., Zhang Q., Zhang W., Zou H. Fabrication of a Novel Magnetic Yolk-Shell $Fe_3O_4@mTiO_2@mSiO_2$ Nanocomposite for Selective Enrichment of Endogenous Phosphopeptides from a Complex Sample. *RSC Advances*, 2014, 4: 45804–8.

38. Cheng G., Zhang J.-L., Liu Y.-L., Sun D.-H., Ni J.-Z. Synthesis of Novel $Fe_3O_4@SiO_2@CeO_2$ Microspheres with Mesoporous Shell for Phosphopeptide Capturing and Labeling. *Chemical Communications*, 2011, 47: 5732–4.

39. Cheng G., Liu Y.-L., Zhang J.-L., Sun D.-H., Ni J.-Z. Lanthanum Silicate Coated Magnetic Microspheres as a Promising Affinity Material for Phosphopeptide Enrichment and Identification. *Analytical and Bioanalytical Chemistry*, 2012, 404: 763–70.

40. Liu S. New Methods for Peptidomics Separation and Analysis Based on Functionalized Magnetic Mesoporous Materials, Fudan Univerity (2011).

41. Yin P., Sun N., Deng C., Li Y., Zhang X., Yang P. Facile Preparation of Magnetic Graphene Double-Sided Mesoporous Composites for the Selective Enrichment and Analysis of Endogenous Peptides. *Proteomics*, 2013, 13: 2243–50.

42. Wang M., Deng C., Li Y., Zhang X. Magnetic Binary Metal Oxides Affinity Probe for Highly Selective Enrichment of Phosphopeptides. *ACS Applied Materials & Interfaces*, 2014, 6: 11775–82.

43. Wang M., Sun X., Li Y., Deng C. Design and Synthesis of Magnetic Binary Metal Oxides Nanocomposites through Dopamine Chemistry for Highly Selective Enrichment of Phosphopeptides. *Proteomics*, 2016, 16: 915–9.

44. Jiang B., Wu Q., Deng N., Chen Y., Zhang L., Liang Z., Zhang Y. Hydrophilic $Go/Fe_3O_4/Au/PEG$ Nanocomposites for Highly Selective Enrichment of Glycopeptides. *Nanoscale*, 2016, 8: 4894–7.

45. Wan H., Huang J., Liu Z., Li J., Zhang W., Zou H. A Dendrimer-Assisted Magnetic Graphene-Silica Hydrophilic Composite for Efficient and Selective Enrichment of Glycopeptides from the Complex Sample. *Chemical Communications*, 2015, 51: 9391–4.

Chapter 3

Enzymatic Hydrolysis Technology Based on Magnetic Micro-/Nanomaterial in Proteomics

3.1. Basic principle of protein enzymolysis based on magnetic micro-/nanomaterial

3.1.1. *Protein enzymolysis based on magnetic micro-/nanomaterial*

The study of protein enzymolysis based on magnetic microspheres began in the 1980s.[1] Currently, there are two widely used methods for protein enzymolysis. One is to immobilize the target protein on magnetic micro-/nanomaterial, followed by enzymatic hydrolysis in solution. The other is to immobilize the enzyme on magnetic micro-/nanomaterial first and then add enzyme-modified material into the protein solution for enzymolysis. The former is mainly suitable for some special proteins (like phosphorylated protein) which can be selectively adsorbed onto the material via physical interaction, and then separated and digested. The latter can greatly enhance the enzymolysis efficiency and preservative stability of the enzyme through the immobilization of the enzyme on appropriate material, although the immobilized process and operation of enzyme are relatively complex and time-consuming.[2] Thus, the latter is more extensively applied in proteomics and will be mainly discussed in this chapter.

3.1.2. *Commercial magnetic bead for enzyme immobilization*

Immobilizing enzymes on magnetic bead is a process that restricts free enzyme in a limited space totally or mostly. The basic principle is as follows: the magnetic microsphere is directly put into the mixed solution containing a certain amount of enzymes, so that the enzyme and the active group on the surface of the microsphere are fully cross-linked. After the immobilization reaction, the external magnetic field is applied to separate the microsphere from the solution. The methods of immobilizing enzymes on magnetic microspheres include physical adsorption, cross-linking, metal ion chelation, and covalent bonding.[3] The manner of enzyme immobilization, the capacity of the immobilized enzyme, stability, and enzyme activity are the key factors for the research of protein enzymolysis based on magnetic material.

Slovakova[4] and Bilkova[5] successfully use commercial magnetic beads (particle size is about several micrometers) to perform enzymatic hydrolysis of protein in microfluidic microchips. The commercial magnetic beads are prepared by suspension polymerization, that is, the polymer microspheres containing magnetic nanoparticles are obtained by adding initiators and stabilizers into the reaction system in the presence of magnetic nanoparticles and organic monomers with functional groups. Due to the weak affinity between inorganic magnetic nanoparticles and organic monomers, the amount of organic copolymers coated on the surface of magnetic nanoparticles is greatly affected. Therefore, the magnetic content of the commercial magnetic beads is low and the magnetic response is not strong enough. In Slovakova's study, the saturation magnetic value of the polystyrene-grafted magnetic bead with –COOH on its surface is only 46 emu·g^{-1}. The insufficient magnetic response prolongs the time of immobilizing magnetic beads in the chip channel, the preparation of microarray enzyme reactor requires 20 min. In Bilkova's work[5] it needs more time (60 min). This is clearly not sufficient for high-throughput proteomic analysis.

3.1.3. *Advantage of magnetic micro-/nanomaterial for enzyme immobilization*

Compared with common commercial magnetic microsphere, magnetic nanomaterial has good surface effect and volume effect, with the following advantages. (i) It has a large specific surface area, high density of functional groups, strong selective adsorption ability, and short time of reaching adsorption equilibrium.[6] (ii) It has good selective magnetic response. It has superparamagnetism when the size of magnetic ferric oxide is small to a certain size, so as to avoid the occurrence of magnetic agglomeration between particles when used. (iii) It has stable physical and chemical properties, and a certain mechanical strength. Also, it can withstand a certain concentration of acid and alkali solution and microbial degradation. In addition, the embedded magnetic material is not easy to oxidize. At the same time, it has certain biocompatibility and will not cause obvious harm to the organism. (iv) The surface of magnetic microspheres has or is endowed with a variety of functional groups by surface modification. These functional groups can connect to biological active substances (such as nucleic acids, enzymes, etc.), and can also couple specific molecules (such as specific ligands, antibodies, antigens, etc.) for specific separation of biological macromolecules. More importantly, due to the good magnetic responsiveness, the positioning of the enzyme-immobilized magnetic microsphere in the capillary/chip channel can be easily realized by external magnetic field, thus eliminating the tedious steps of making plug. In addition, when the external magnetic field is removed, the magnetic microsphere can be easily derived from the microchannel, and the capillary/chip enzyme reactor can be regenerated by introducing new enzyme-immobilized magnetic microspheres. Based on the above advantages, magnetic nanomaterials have a good prospect in the preparation of capillary/chip enzyme reactor.[7,8]

3.1.4. *Immobilization method of enzyme on magnetic micro-/nanomaterial*

According to the interaction between protease molecule and carrier, the immobilization method of enzyme can be divided into four categories: adsorption, cross-linking, covalent bonding, and embedding. The classification is based on whether the realization of enzyme immobilization is mainly embedded in a limited space or adsorbed on the support or carrier.

3.1.4.1. *Covalent bonding for enzyme immobilization*

Covalent bonding is a method that combines enzymes with functional groups on the surface of magnetic micro-/nanomaterials in the form of covalent bonds. This method is relatively mature. Figure 3.1 presents the schematic

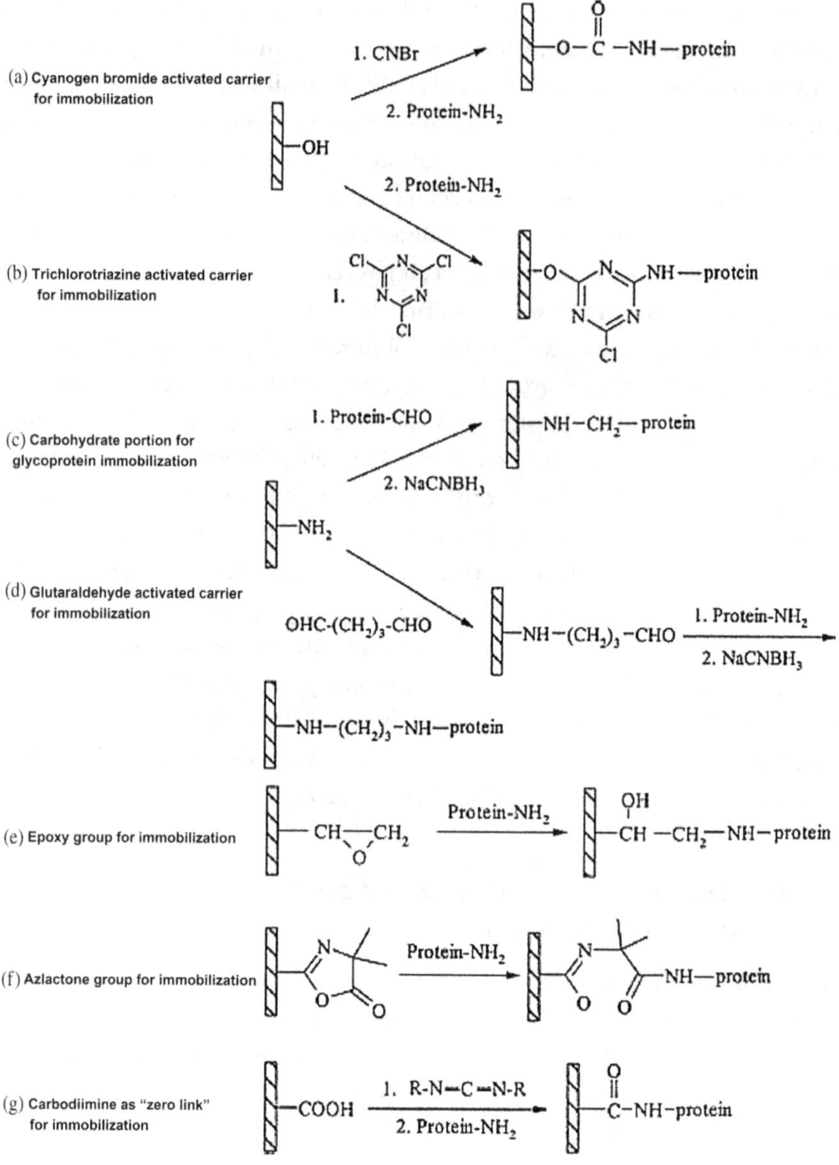

Figure 3.1: Schematic diagram of covalent cross-linking of molecular enzymes.[10]

diagram of seven reactions of covalent cross-linking of enzyme molecules. One of the advantages of this method is that the the binding of enzyme and the carrier is firm, and the enzyme does easily fall off even when the solution contains high ionic strength. Moreover, this method can reduce the possibility of enzyme activity caused by enzyme self-degradation and prolong the life of the immobilized enzyme reactor. In addition, it also improves the thermal stability of the enzyme, which is because the binding of the enzyme to the material matrix leads to the relative stabilization of the protein structure, thus limiting the thermal movement of the protein at high temperature.[9]

Many functional groups of enzymes are suitable for covalent bonding, including (i) α-amino, and ε-amino of lysine and arginine at the terminal of the chain, (ii) α-carboxyl, and β- and γ-amino acid of glutamic acid and aspartic acid at the terminal of the chain; (iii) phenol ring of tyrosine, (iv) sulfhydryl group of cysteine, (v) hydroxyl group of serine and threonine, imidazole group of histidine, and (vi) indolyl group of tryptophan.[10]

Among all the bonding methods, the most widely used one is immobilizing enzyme on active amino-modified magnetic micro-/nanomaterial through glutaraldehyde bonding. Glutaraldehyde reacts with the amino group on the surface to form an imine bond. The aldehyde group at the other terminal of glutaraldehyde can bond to Σ-amino of lysine to achieve the goal of immobilizing the enzyme. The reaction conditions are mild and the catalytic activity and stability of the enzyme are well preserved. The research groups in Fudan university synthesize amino-modified magnetic microsphere (Fe_3O_4–NH_2) with high paramagnetism and good dispersity by one-step synthetic method, and then develop the technology of enzyme immobilization based on magnetic material through the modification of plenty of glutaraldehyde on the surface of Fe_3O_4–NH_2. With this technology, the enzymatic hydrolysis of the whole protein can be completed within 5 min, leaving out any complicated reduction alkylation (as seen in Section 3.2).

Another preferred covalent bonding method for enzyme immobilization is to use the epoxy group on the surface of magnetic material. Deng's group[11,12] concluded that Glymo is a very good cross-linking agent, the ethylene oxide group of which has a stronger recognition ability to the molecular structure of enzymes and stronger reaction ability, compared to other modification groups such as hydroxyl groups, aldehyde groups, and sulfonyl groups, etc. Cross-linking agent of Glymo can be modified on the magnetic microsphere by one-step reaction, which greatly simplifies the reaction process. Moreover, no organic reagents are introduced in this process, and the obtained magnetic microsphere has a good magnetic response to the external magnetic field. This is very important for the application of magnetic materials in the field of biomedicine.

There have been many reports of enzyme immobilization on carboxy-modified magnetic microspheres. The surface of magnetic materials with different properties obtained by different modification methods have different effects on the performance of enzymes fixed on the surface of the materials. For the covalent bonding methods, the capacity of enzyme immobilized by materials is different. The largest amount of enzymes immobilized by material per unit mass is 78–88 $\mu g \cdot mg^{-1}$, which is realized by the enzyme immobilization method based on active amino-modified magnetic micro-/nanomaterial via glutaraldehyde bonding. Followed by 10.8–37.2 $\mu g \cdot mg^{-1}$ via the bonding of magnetic micro-/nanomaterial with amino group in trypsin using Glymo. The lowest loading is 14.8–21 $\mu g \cdot mg^{-1}$ through the reaction between carboxyl-modified magnetic material and NHS-EDAC or EDC.

3.1.4.2. *Physical adsorption for enzyme immobilization*

Adsorption methods can be divided into physical adsorption method and electrostatic adsorption method (Figure 3.2).[3]

The physical adsorption refers to using the carrier to adsorb an enzyme for realizing its immobilization. The commonly used inorganic carriers include active carbon, porous ceramics, acid clay, calcium phosphate, metal oxide, etc. The organic carriers include starch, gluten, cellulose and its deriva-

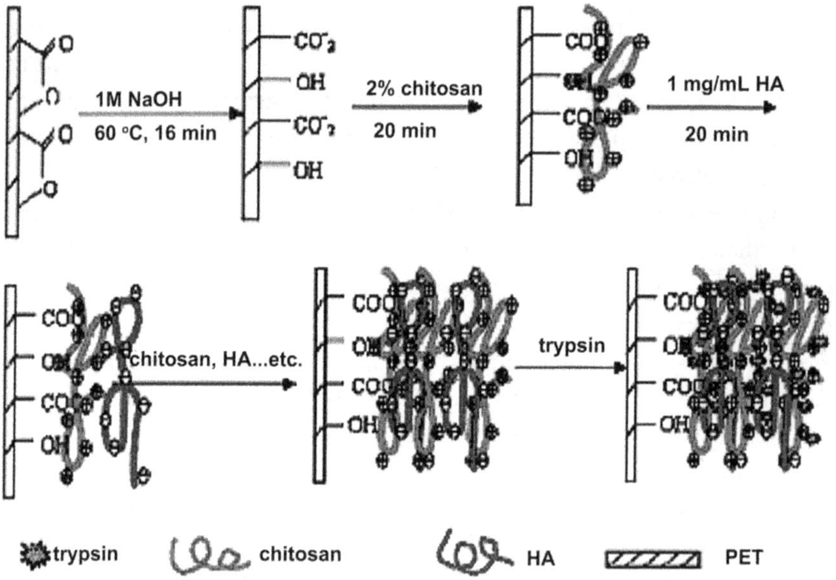

Figure 3.2: Schematic diagram of the reaction of immobilizing protease molecules according to electrostatic adsorption.

tives, and chitin and its derivatives. This method has the characteristic that the active site and the conformation of the enzyme are not easy to be destroyed, but the enzyme is attached to the carrier and easily falls off. Enzymes immobilized by this method include α-amylase, saccharification enzyme, glucose oxidase, etc.

The immobilization method that combines the enzyme with the water-insoluble carrier containing the ion exchange group by the electrostatic force is called the ion adsorption method. The carriers used in this method are polysaccharide ion exchanger and synthetic polymer ion exchange resin. For example, DEAE-cellulose, TEAE-cellulose, CM-cellulose, etc. The ion adsorption method is simple to operate. The treatment conditions are mild and the amino acid residues in the active sites are not easily destroyed. However, the binding force between the carrier and the enzyme is relatively weak, which is easily affected by the type of buffer or pH. When the reaction is carried out under the condition of high ionic strength, the enzyme will often fall off the carrier. The enzymes immobilized by this method include glucose isomerase, saccharification enzyme, β-amylase, cellulase, etc.

Zhu *et al.*[13] prepared porous alkaline silk fibroin by treating degummed silk with diluted alkali solution, and further immobilized α-amylase by physical adsorption method. The total activity of the immobilized enzyme per gram of alkaline silk fibroin was 439.81 U, the recovery ratio of the activity was 48.33%, and the performance ratio of the activity was 74.18%.

Yang *et al.*[14] immobilized pig pancreatic lipase by adsorption method using cellulose acetate as the carrier, which was used to catalyze the glycerolysis of lard to synthesize monoglyceride, and the maximum yield of monoglyceride was 50.05%.

3.1.4.3. *Cross-linking method for enzyme immobilization*

The cross-linking method uses a bifunctional/multifunctional reagent to cross-link enzyme molecules. It forms covalent bond between an enzyme and bifunctional/multifunctional reagent, so as to obtain a three-way cross-linked structure. In addition to the cross-linking among enzyme molecules, there are also some intramolecular cross-links. The immobilized enzymes with different physical properties can also be produced according to different conditions of use and adding of materials. The commonly used cross-linking agents are glutaraldehyde, diazobenzidine-2, 2-disulfonic acid, etc.

3.1.4.4. *Metal ion chelating for enzyme immobilization*

Metal ion chelation is a kind of physical adsorption method that mainly uses ion chelating reagent (such as iminodiacetic acid, IDA) on the surface of the solid-phase carrier to immobilize metal ion, and then realizes the immobilization of the enzyme through the Lewis acid–base interaction between metal ion and protease molecule. It is commonly seen in the interaction of transition metal ion with *N*-atom in the amino group of protein.

3.1.4.5. *Embedding method for enzyme immobilization*

Embedding method, as the name suggests, involves burying an entire molecule in a medium, and the resulting mixture can be shaped as needed.[3] The embedding method can be divided into grid type and microcapsule type. The former means that the enzyme is embedded in the macromolecular gel fine grid as shown in Figure 3.3. The latter is to embed the enzyme in the macromolecular semi-permeable membrane. Generally, the embedding method omits the binding reaction with the amino acid residue of the enzyme, rarely changes the spatial conformation of the enzyme, and can obtain high recovery ratio of the enzyme. Therefore, it can be applied to the immobilization of many enzymes. Grid type of gel embedding is commonly used, featuring mild conditions, and will not result in chemical modification of amino acid residue of the enzyme or change the structure of the enzyme. But the activity of enzyme immobilized by this method largely depends on the interaction between the embedding material and the enzyme, as well as microenvironment, resulting in the poor repeatability in different batches. Hydrophilic inorganic gel system is often used as an embedding material. SiO_2 gel with

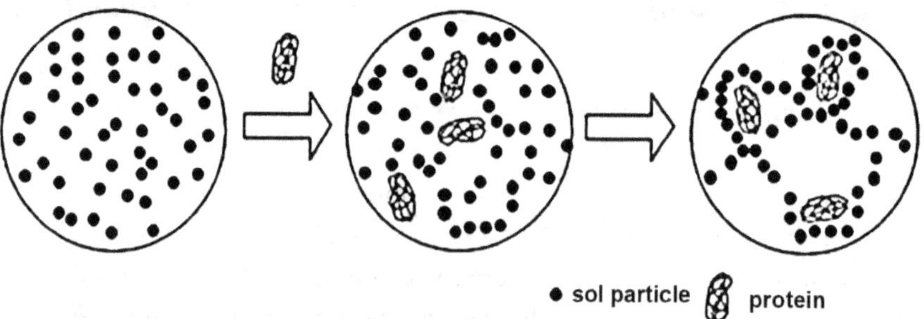

Figure 3.3: Schematic diagram of the process of embedding trypsin by sol–gel method.

grid structure of O–Si–O prepared by using silane oxide reagent as raw material for protein immobilization has been applied in affinity chromatography and enzymatic catalysis.[15,16] One of the outstanding advantages of the gel embedding method for protein immobilization is that the gel membrane can be prepared into various shapes such as film, ball, or particle. However, the embedding method is only suitable for those enzymes that act on small molecular substrates and products, and it is not suitable for those that act on large molecular substrates and products, because only small molecules can diffuse through the mesh of the macromolecular gel, and this diffusion resistance will also lead to changes in the kinetic behavior of the immobilized enzyme and reduce the enzyme activity.

3.1.5. *Immobilized enzyme reactor*

A variety of immobilized enzyme reactors can be made through the immobilization of enzyme using the aforementioned methods, for realizing fast enzymatic hydrolysis offline or online. Online means to combine the immobilized enzyme reactor with other means of separation and analysis, and offline is the contrary. For offline or online enzymatic hydrolysis technologies, immobilized enzyme reactors are widely accepted for their convenience, speed, high efficiency and reusability.[17]

Zou *et al.*[18] prepared a nanoscale enzyme reactor, and the peptide mapping is obtained by MALDI-TOF-MS. The experimental results show that the sample amount of 10–13 mmol or even 10–15 mmol can meet the analytical requirement using the capillary as reactor, which greatly reduces the required sample amount for peptide mapping analysis. Kuhr *et al.*[19] prepared a capillary wall-based trypsin reactor by making full use of the specific interaction between biotin and anti-biotin protein. In other words, they first activated the Si–OH on the capillary inwall with (3-aminopropyl)-triethoxysilane, and then fixed the biotin-labeled trypsin to the capillary inwall through the action of biotin–antibiotin, protein–biotin. An enzyme reactor with high stability and catalytic activity can be obtained by this method (as seen in Figure 3.4).

Michaels first proposed the concept of enzyme membrane reactor.[20] The enzyme membrane reactor can combine the catalytic property of the enzyme with the excellent separation property of the membrane material, so that the enzymatic hydrolysis reaction and product separation can be carried out simultaneously. This effectively accelerates the reaction, breaks through the

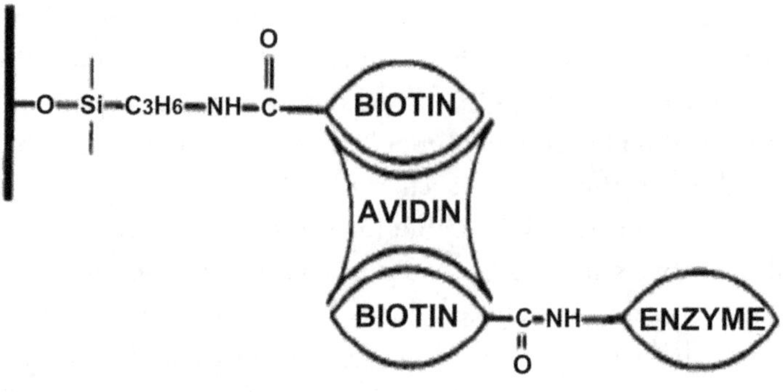

Figure 3.4: Immobilization of biotin-labeled trypsin to the inner wall of capillary by the action of avidin–biotin.

limit of chemical equilibrium, and improves the conversion rate. Zou *et al.*[21] prepared the immobilized trypsin microliter reactor and analyzed the peptide mapping by combining with MALDI-TOF-MS. The results showed that there were few unhydrolyzed proteins in the samples after an hour's reaction. Subsequently, they[22] investigated the properties of this trypsin reactor using an HPLC system.

More and more attention has been paid to the integration of trypsin reactor into micro total analysis systems such as microfluidic microarray. Gao *et al.*[23] established a miniaturized membrane reactor and an online analysis platform of instantaneous isokinetic electrophoresis-capillary zone electro-phoresis-mass spectrometry (CITP/CZE-ESI-MS). They used polyvinylidene fluoride (PVDF) to strongly interact with the protein molecules to adsorb trypsin onto the porous PVDF membrane, which in turn was connected to the microfluidic channels of the PDMS matrix. The sample was hydrolyzed by trypsin as it was pushed across the membrane by syringe pump, and the obtained peptide was focused by CITP and then separated by CZE, and finally analyzed by ESI-MS. The protein can be identified in minutes using this device, and the amount of consumed protein is at the nanogram level (Figure 3.5). Cooper *et al.*[24] prepared a micro-sized trypsin PVDF membrane reactor (2 mm × 5 mm) by winding it around a polymer sheath (a capillary component). The reactor can complete enzymatic hydrolysis of proteins in a few seconds, consuming less than 5 fmol of protein, and the capillary inter-face makes it easy to be used in conjunction with various separation methods and MS detectors (Figure 3.6).

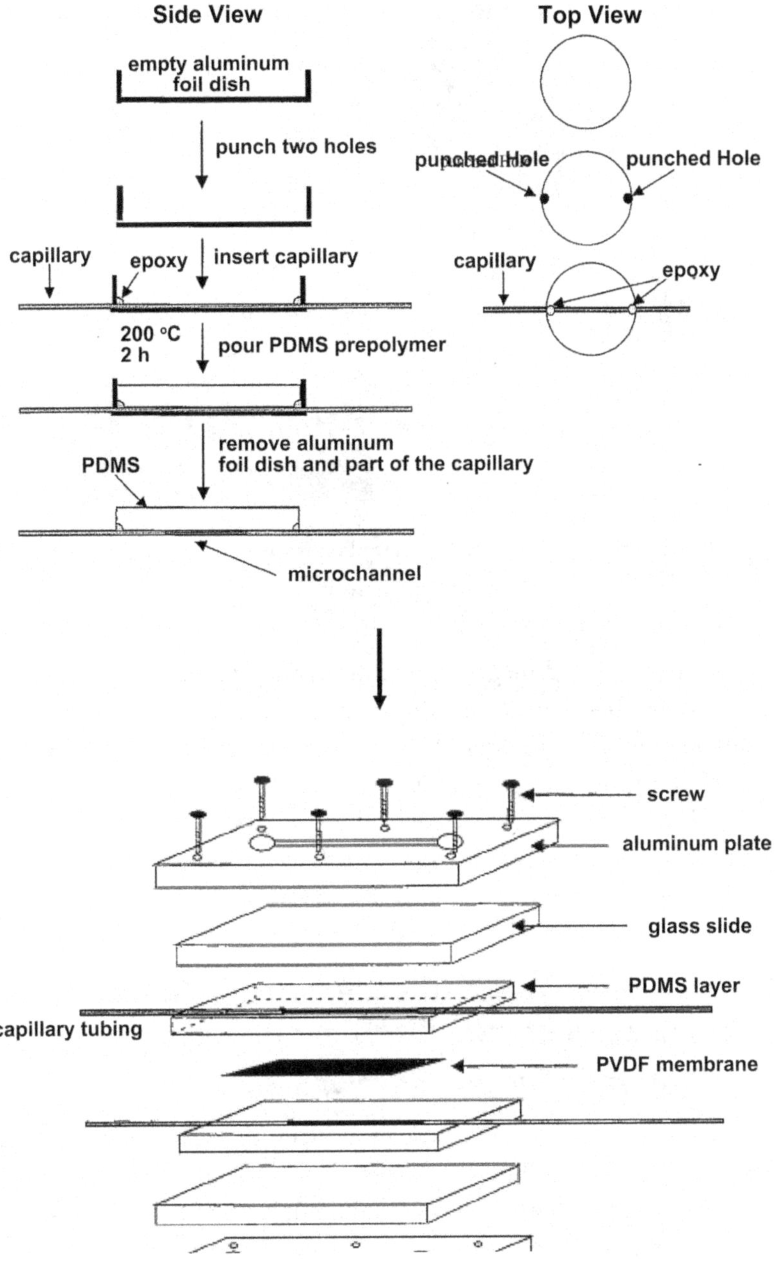

Figure 3.5: Processing of enzyme reactor on microfluidic chip and composition of microfluidic chip.

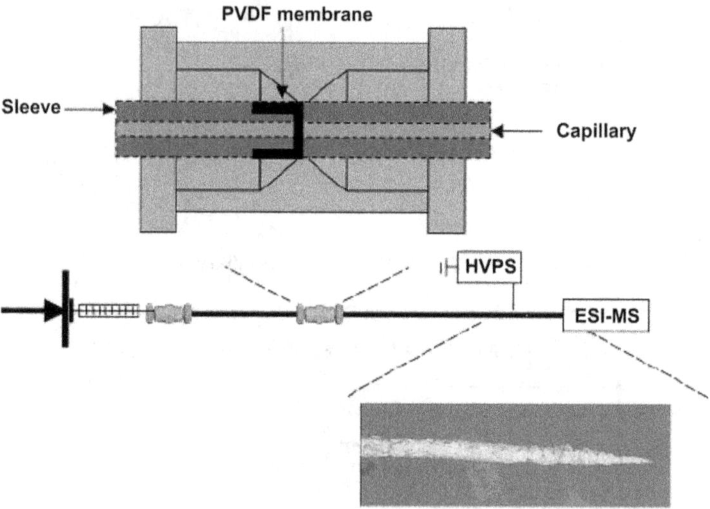

Figure 3.6: Schematic diagram of PVDF membrane-based trypsin reactor and online connection of ESI-MS.

Bikova *et al.*[5] used magnetic microspheres with carboxyl groups to fix trypsin through covalent cross-linking, and filled it into the microchip channel by using an external magnetic field. The standard protein products of enzymatic hydrolysis in the microenzyme reactor were separated in parallel by RP-HPLC and HPCE, and the finally collected fraction was identified by MS. SDS-PAGE was also used to investigate the effects of different buffers and denaturation conditions on the completeness of enzymatic hydrolysis (Figure 3.7).

3.2. Microwave-assisted enzymatic hydrolysis based on enzyme-immobilized magnetic material

3.2.1. *Basic principle and development status*

Microwave-assisted enzymatic hydrolysis is another development trend of enzymatic hydrolysis technology.[9] Microwave radiation is an effective heat source. The reaction that requires several hours under the conventional heat source can be completed within a few minutes under the heating of microwave source, and the yield and selectivity of the reaction are very high.[25] In recent years, several research groups have applied microwave-assisted proteolysis to reduce the time of enzymatic hydrolysis from a few hours to a few

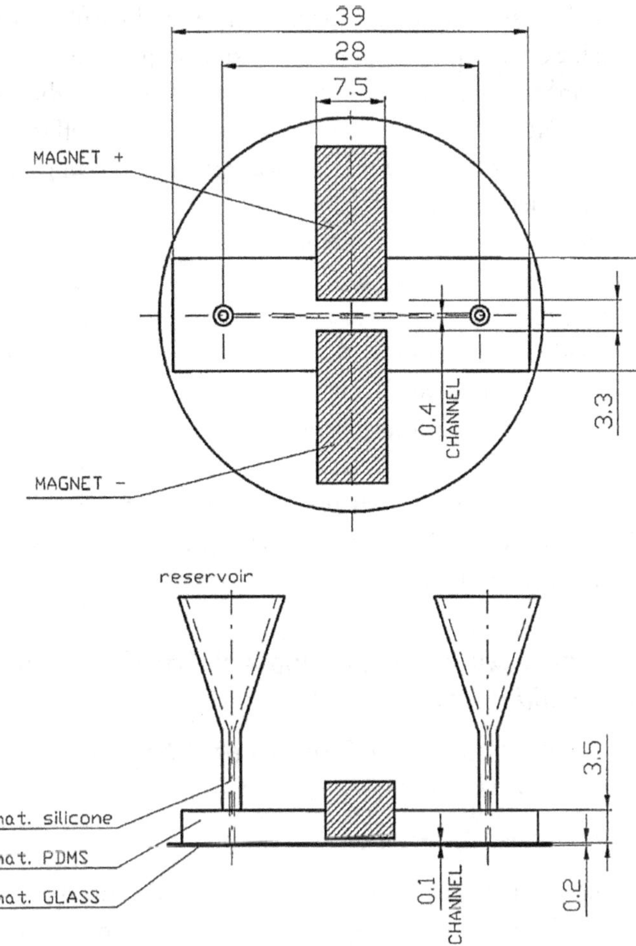

Figure 3.7: Magnetic micro-enzyme reactor etched on microfluidic chip and its processing diagram.

minutes. Pramanik *et al.*[26] applied microwave technology to the enzymatic hydrolysis of standard proteins in solution or on glue. The whole process took only 10 min, including bovine ubiquine. Juan et al. applied the microwave-assited enzymatic hydrolysis technology to the digestion of known proteins on the glue, and achieved a good effect in 5 minutes.[27] In Lin's study,[28] the effects of several organic solvents on the microwave-assisted enzymatic hydrolysis were investigated. They found that methanol and acetonitrile at a certain concentration had promoting effects on the microwave-assisted enzymatic hydrolysis. Recently, Sun *et al.*[29] successfully applied microwave-assisted enzymatic hydrolysis technology to the enzymatic hydrolysis of complex samples in solution and gel. The obtained results in just 6 min were comparable to standard overnight enzymatic hydrolysis. Moreover, they found that microwave irradiation had a good effect on the extraction efficiency of peptide. Other research groups studied acid cleavage during enzymatic hydrolysis under microwave radiation.[30–32]

It has long been recognized that magnetic materials absorb microwave radiation well. Kirschvink $et\ al.$[33] found that even a small amount of biomagnetic substances in the body can cause serious damage to animal organs under microwave irradiation. Before that, Walkeiwicz $et\ al.$ found that magnetic materials are good microwave absorbers.[34] These findings suggested the use of magnetic materials in microwave-assisted enzymatic hydrolysis. Recently, Chen $et\ al.$[35] found that in the presence of magnetic nanospheres, the microwave-assisted enzymatic hydrolysis process could be completed within 30 s with significant effect, which lead to rapid changes in the field of microwave-assisted enzymatic hydrolysis.

3.2.2. Synthesis of magnetic material and trypsin immobilization

3.2.2.1. Synthesis of Fe_3O_4–Glymo and trypsin immobilization

(1) Synthesis of Fe_3O_4–Glymo

First, Fe_3O_4 magnetic microsphere is synthesized as described in Section 2.4 of Chapter 2. Then, Glymo is grafted on the surface of Fe_3O_4 magnetic microsphere according to the following method: 0.05 g Fe_3O_4 microsphere is added into 0.1 mol·L^{-1} HNO$_3$ solution for being activated by 10 min of ultrasonication, and then the solid product is washed with deionized water 3 times and dried in vacuum oven for 12 h under 60°C, and then dispersed in toluene solution containing 0.35 mL Glymo that is continously stirred for about 0.5 h. In the end, the reaction lasts 6 h of reflux at 65°C.

(2) Immobilization of trypsin on Fe_3O_4–Glymo

5 mg Fe_3O_4–Glymo microspheres are added into 1.5 mL EP tube, and then 500 μL 2 μg·L^{-1} tosyl-phenylalanine chloromethyl-kelone (TPCK) trypsin is added. The mixture is exposed to ultrasonic irradiation for 1 min to form a

unifor m suspension. Then it is vibrated at 37°C for 3 h on a shaker. After the reaction is completed, the supernatant is removed with the help of a magnet, and the magnetic microsphere is washed with deionized water 3 times and then re-dispersed in 500 μL 25 mmol·L^{-1} NH$_4$HCO$_3$ (pH ~8.3).

(3) Measurement of loading capacity of Fe$_3$O$_4$–Glymo towards trypsin

The trypsin solutions before and after reaction are diluted 3 times for ultraviolet spectrophotometry measurement, the loading capacity of Fe$_3$O$_4$–Glymo towards trypsin is calculated as 8.7–11.2 μg·mg^{-1} according to the absorbancy at 280 nm.

3.2.2.2. *Synthesis of Fe$_3$O$_4$@SiO$_2$–Glymo and trypsin immobilization*

(1) Synthesis of Fe$_3$O$_4$@SiO$_2$–Glymo

Synthesis of Fe$_3$O$_4$@SiO$_2$–Glymo and trypsin immobilization are shown in Figure 3.8. First, Fe$_3$O$_4$@SiO$_2$ is synthesized as described in Section 2.4 of Chapter 2. Then Glymo is grafted on the surface of Fe$_3$O$_4$@SiO$_2$ as follows: 0.05 g Fe$_3$O$_4$@SiO$_2$ is dispersed and washed ultrasonically by ethanol, and then dried at 37°C. Next, the dried Fe$_3$O$_4$@SiO$_2$ is dispersed in 20 mL toluene solution, and 0.35 mL Glymo is added. After uniform dispersion by vibration, the mixture keeps 12 h of circulation reflux at 80°C. The obtained solid product is washed with ethanol 4 times, and then stored in ethanol for later use.

Figure 3.9 shows the FTIR spectrum of Fe$_3$O$_4$@SiO$_2$–Glymo. There are three characteristic peaks in total for ternary ring ether. The first is the absorption peak in the range of 3,050–2,990 cm^{-1} which is attributed to the stretching vibration

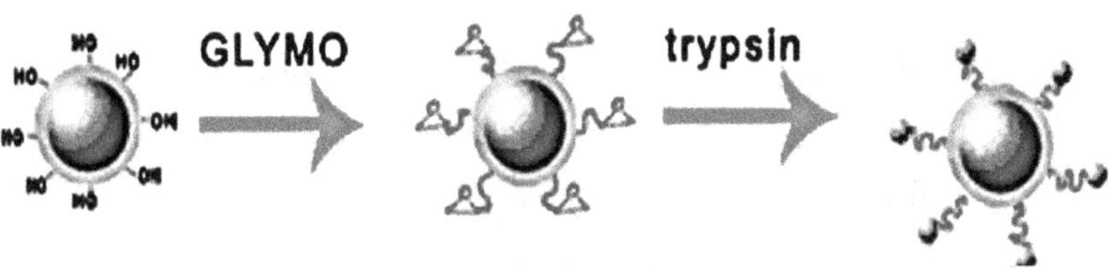

Figure 3.8: Schematic diagram of preparation of Fe$_3$O$_4$@SiO$_2$–Glymo–trypsin.

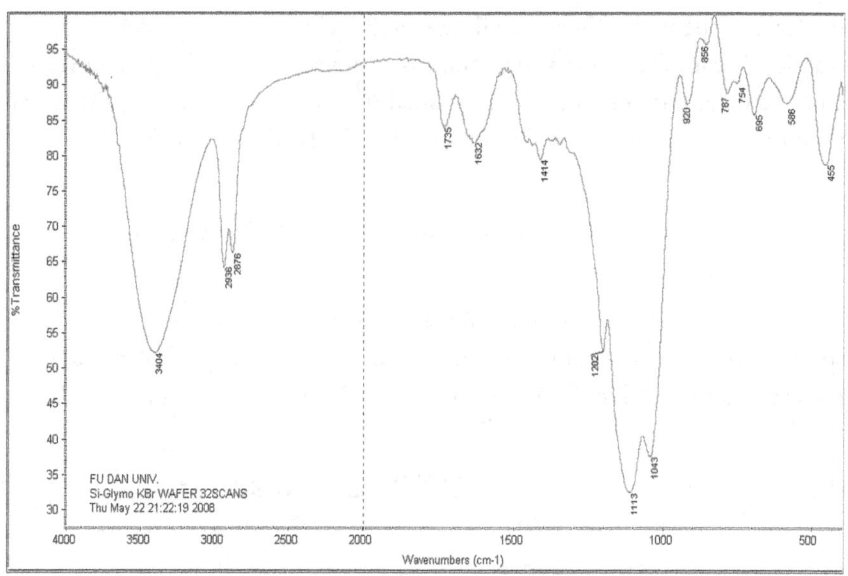

Figure 3.9: FTIR spectrum of $Fe_3O_4@SiO_2$–Glymo.

of C–H on the epoxy carbon of ternary ring ether. The second is the absorption peak in the range of $1{,}280$–$1{,}240$ cm^{-1} which is attributed to the stretching vibration of C–O–C bond of ring ether, this absorption peak is strong in small-molecule ring ether, but relatively weak in large-molecule ring ether. The third is the two medium-strong peaks at 950–863 cm^{-1} and 865–786 cm^{-1} generated by the skeleton vibration absorption of ternary ether ring, which is the main basis to speculate whether there is a ternary ring ether in the molecule. The peaks in the $3{,}050$–$2{,}990$ cm^{-1} are covered by the strong absorption peak of O–H, while there is a moderate strong absorption at 950–863 cm^{-1} and 865–786 cm^{-1}. Therefore, it can be confirmed that there is an epoxy group on the surface of Fe_3O_4.

(2) Immobilization of trypsin on $Fe_3O_4@SiO_2$–Glymo

The immobilization of trypsin is as follows: 5 mg $Fe_3O_4@SiO_2$–Glymo is added into 1.5 mL EP tube and then 500 μL 2 μg·μL^{-1} trypsin solution (buffers are H_2O and 25 mmol·L^{-1} NH_4HCO_3, respectively) is added. The mixture is exposed to ultrasonic irradiation for 1 min to form a uniform suspension, and then vibrated at 37°C for 3 h on a shaker. After the reaction is completed,

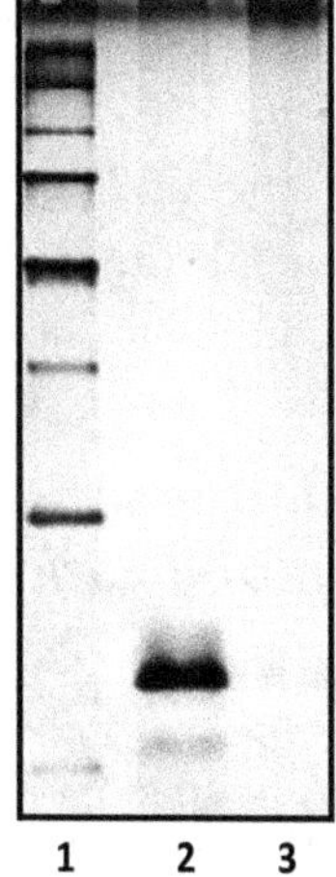

Figure 3.10: Silver-stained SDS-PAGE to verify the covalent interaction between magnetic microsphere and enzyme (12% separation gel, Band 1: Marked protein; Band 2: Trypsin; Band 3: Fe_3O_4@SiO_2–Glymo–trypsin).

the supernatant is removed with the help of an external magnet, and the magnetic product is washed 3 times with deionized water before being dispersed in the buffer again.

The suspension of Fe_3O_4@SiO_2–Glymo–trypsin is analyzed by SDS-PAGE to verify the covalent effect of immobilizing the enzyme. Because a series of denaturants are added and heating is carried out before SDS-PAGE analysis, the author thought that the trypsin which was immobilized on surface of Fe_3O_4@SiO_2–Glymo particles through the non-covalent effect would be separated from the surface driven by electric current under the strong denaturation condition of SDS-PAGE, and thereby generating bands at the position of corresponding molecular marker. Figure 3.10 shows the SDS-PAGE analysis of silver-stained Fe_3O_4@SiO_2–Glymo–trypsin. Band 1 is the Marker band, band 2 is the trypsin band. A clear band is shown at 24 KDa with 2 μg sample, band 3 corresponds to the enzyme immobilized magnetic microsphere and has no obvious band. The results show that: (i) protease has been fixed on the surface of magnetic silica by strong covalent effect; (ii) cleaning multiple times after the synthesis of Fe_3O_4@SiO_2–Glymo–trypsin has effectively removed the molecules that are adsorbed via non-covalent effect.

(3) Measurement of loading capacity of $Fe_3O_4@SiO_2$–Glymo towards trypsin

The trypsin solutions before and after reaction are diluted 3 times for ultraviolet spectrophotometry measurement, the loading capacity of Fe_3O_4–Glymo towards trypsin is calculated according to the absorbancy at 280 nm.

3.2.2.3. *Synthesis of amino-magnetic nanoparticle and enzyme immobilization*

(1) Synthesis and characterization of amino-magnetic nanoparticle

The amino-magnetic nanoparticle is synthesized according to a previous report,[36] and the detailed description can be found in Section 2.3 of Chapter 2.

(2) Glutaraldehyde modification on amino-magnetic nanoparticle

Compared with microscale, amino-magnetic nanoparticles have many advantages. (i) The average particle size is about 50 nm, with a higher specific surface area and more reactive sites. (ii) There is no need for coating silica on the surface. High content of Fe_3O_4 endows it with good magnetic responsiveness and superparamagnetism, both of which are extremely important for its application in practical analysis. (iii) Due to its small particle size and a large number of hydrophilic groups (amino groups) on the surface, it has a good dispersion in aqueous solution. More importantly, the amino-magnetic nanoparticles prepared in this study are synthesized by one-step hydrothermal method. After the activation of glutaraldehyde, the enzyme could be fixed on its surface. Therefore, the cumbersome modification steps before trypsin immobilization are greatly simplified, as shown in Figure 3.11.

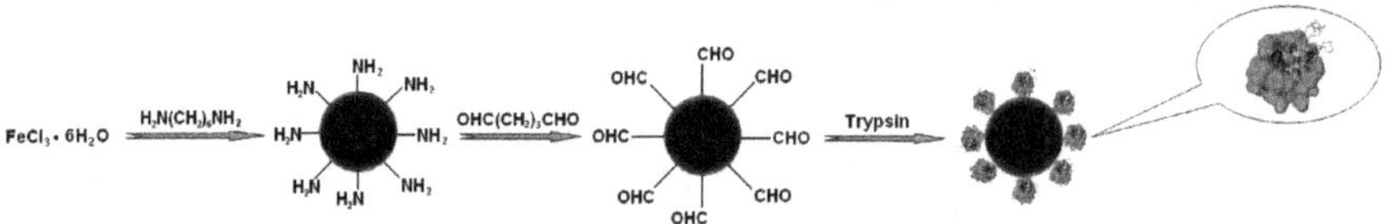

Figure 3.11: Schematic diagram of synthesis of amino-magnetic nanoparticle and trypsin immobilization.[3]

The trypsin immobilization method is as follows: 3 mg amino-magnetic nanoparticles are washed 3 times with 200 μL coupling buffer (CB buffer: 50 mmol·L^{-1} NH$_4$OAC, 1 mmol·L^{-1} CaCl$_2$, 1 mmol·L^{-1} MnCl$_2$). And then the obtained solid product is dispersed in 200 μL CB solution containing 5% glutaraldehyde for maintaining the rea ction at 650 rpm for 1.5 h on the shaker. The final product is washed by CB buffer 4 times.

(3) Immobilization of trypsin on amino-magnetic nanoparticle

The amino-magnetic nanoparticle is added into CB solution containing 1% NaCNBH$_3$ and 2 mg·mL^{-1} trypsin, the pH value of which is adjusted to 8–8.5 using ammonium hydroxide for keeping reaction at 37°C about 3 h. The solid product is dispersed in 400 μL CB solution containing 0.75% glycine and 1% NaCNBH$_3$ for standing for 1 h. Then the final product is washed with CB solution for 4 times.

(4) Measurement of loading capacity of amino-magnetic nanoparticle towards trypsin

The amount of trypsin immobilized on the surface of amino-magnetic nanoparticle is determined by ultraviolet spectrophotometry measurement. A buffer solution of alkaline pH is used in the reaction of enzyme immobilization. The results show that the loading capacity towards trypsin is 62–80 μg·mg^{-1}. This indicates that the specific surface area of magnetic nanoparticle is greatly increased, and more free amino groups exist on the surface, resulting in more sites for trypsin immobilization, compared with magnetic microspheres. Besides, amino magnetic spheres have greater advantages in immobilizing protein molecules.

3.2.3. *Enzyme immobilized magnetic material for microwave-assisted enzymatic hydrolysis*

The process of microwave-assisted enzymatic hydrolysis is shown in Figure 3.12. Compared with the magnetic silica in Chen's work,[35] the microwave absorption ability of naked Fe$_3$O$_4$ is stronger.

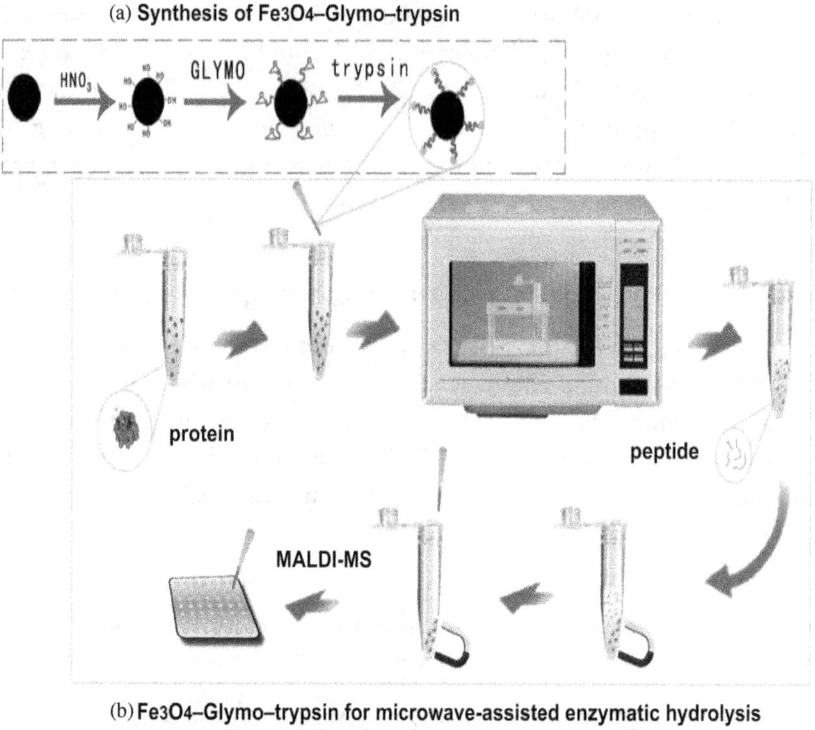

(b) Fe3O4–Glymo–trypsin for microwave-assisted enzymatic hydrolysis

Figure 3.12: Schematic diagram of microwave-assisted enzymatic hydrolysis.

3.2.3.1. *Application of Fe_3O_4-Glymo-trypsin in microwave-assisted enzymatic hydrolysis*

(1) Fe_3O_4–Glymo–trypsin for microwave-assisted enzymatic hydrolysis of Cytochrome C

The conditions of microwave-assisted enzymatic hydrolysis are optimized by using Cytochrome C (Cyc), and its feasibility in proteomics is verified by practical samples.

Figures 3.13(a)–3.13(c) tests the effect of the dosage of Fe_3O_4–Glymo–trypsin on the microwave-assisted enzymatic hydrolysis. 50 μg, 200 μg, and 500 μg are used, respectively, and heated by microwave for 15 s. The completeness of enzymatic hydrolysis is observed by using the linear model of MALDI-TOF MS. It can be seen that when the dosage increases from 50 μg to 200 μg, the protein peak disappears, which can be considered as complete

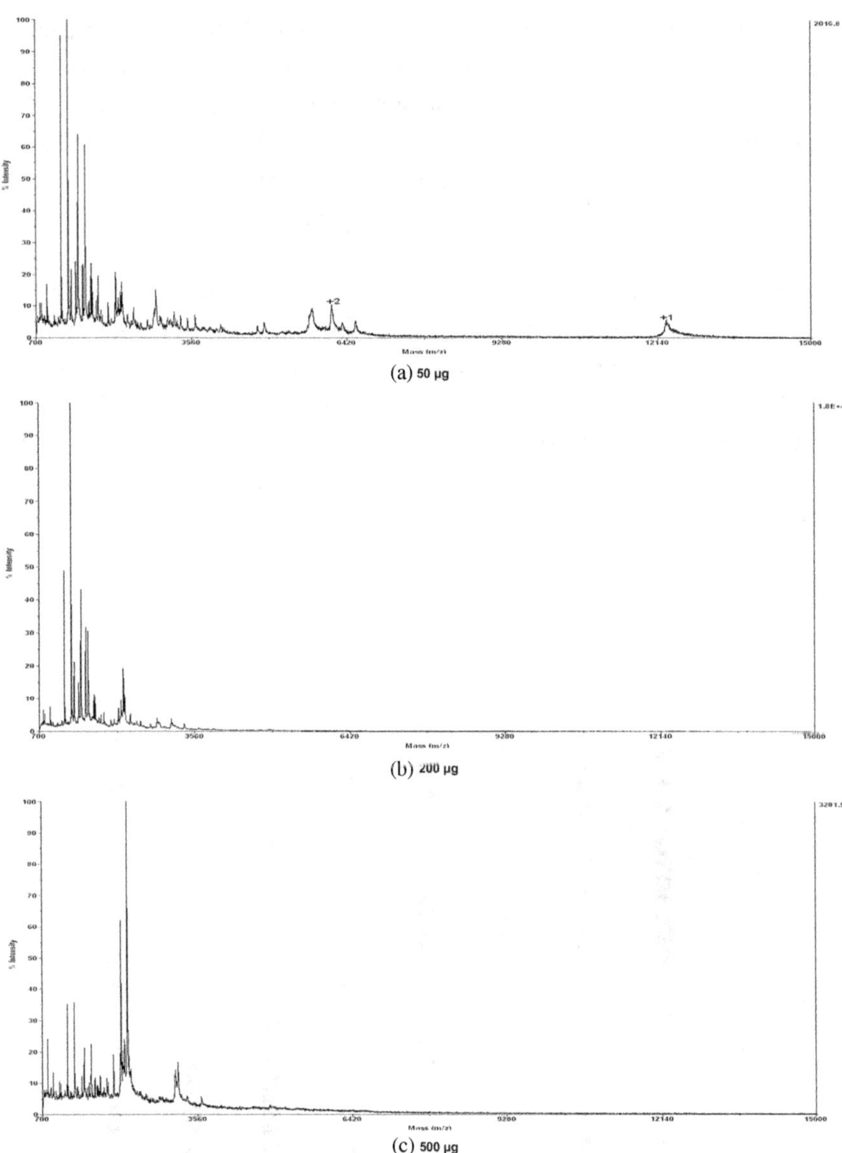

Figure 3.13: The effect of dosage of Fe_3O_4–Glymo–trypsin on microwave-enzymatic hydrolysis. Experimental conditions: Cyc (50 μL, 50 $\mu g \cdot \mu L^{-1}$); microwave (output power 700 W, time 15 s).

enzymatic hydrolysis. When the dosage increases to 500 μg, more fragment peaks appear at $m/z < 2500$. When using the reflection mode of MALDI-TOF MS, the fragments of enzymatic hydrolysis of Cyc are accurately identified. After 15 s of microwave-assisted enzymatic hydrolysis, a total of 17 peptides are identified, reaching a sequence coverage of 92%, which is better than that of the 12 peptides and 78% of sequence coverage through 12 h of traditional in-solution digestion.

Because trypsin is immobilized on the magnetic sphere, it can be recovered and reused after enzymatic hydrolysis. After each application, the supernatant is removed by external magnetic field, and the Fe_3O_4–Glymo–trypsin is washed twice with 25 mmol·L^{-1} NH_4HCO_3 before further use. It is found that Fe_3O_4–Glymo–trypsin can still maintain high activity after 7 applications. Figure 3.14 shows the results of its repeatability.

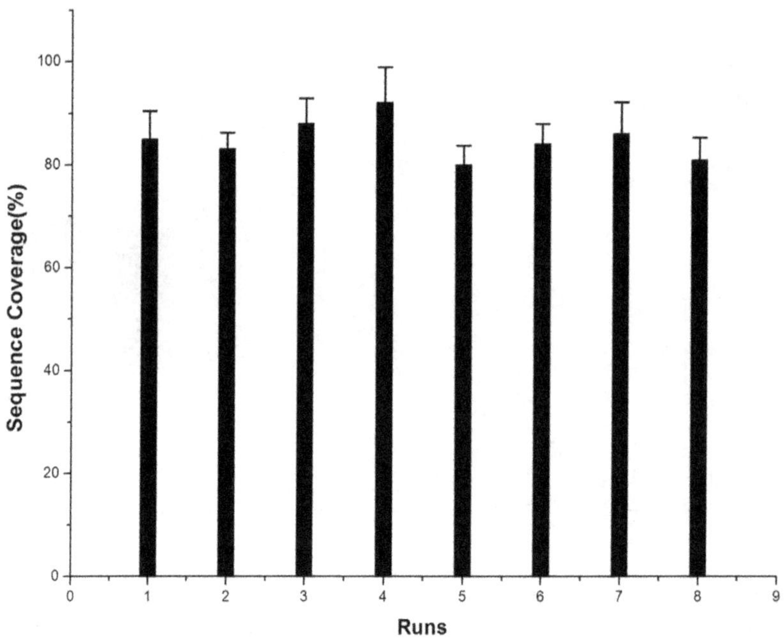

Figure 3.14: Peptide coverage histogram with repeated use of Fe_3O_4–Glymo–trypsin for Cyc digestion 8 times (MALDI-TOF MS analysis).

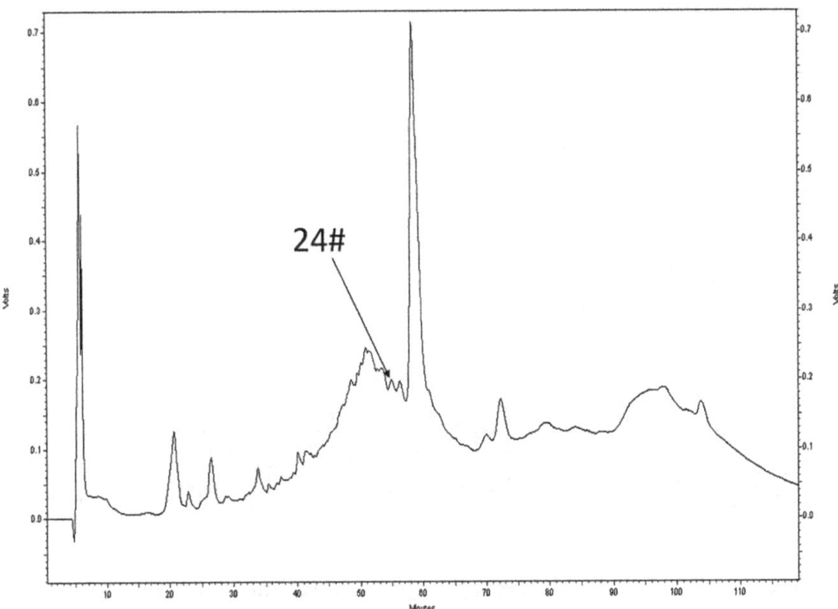

Figure 3.15: RP-HPLC separation of water-soluble total proteins in mouse liver. Flow rate 0.7 mL·min^{-1}; Capillary packed column: 250 × 4.6 mm, 5 μm, 300 Å; UV: λ = 215 nm. The 24# fraction is indicated by an arrow.

(2) Fe$_3$O$_4$–Glymo–trypsin for microwave-assisted enzymatic hydrolysis of mouse liver protein

The technical route of middle-down has incomparable advantages over the "shot-gun" method in providing protein information and post-modification of protein.[37,38] However, the separation methods based on various physical and chemical properties of proteins (such as isoelectric point, molecular weight, hydrophobicity, etc.) make the whole experiment process much longer. Moreover, the time for enzymatic hydrolysis of each fraction is as long as 12–16 h, which severely limits the throughput analysis of proteomics. Therefore, the microwave-assisted enzymatic hydrolysis method based on the enzyme immobilized magnetic material will greatly shorten the process and make rapid middle-down analysis possible.

Figure 3.15 shows the RP-HPLC separation process of water-soluble total proteins extracted from mouse liver. A fraction is collected at 2-min

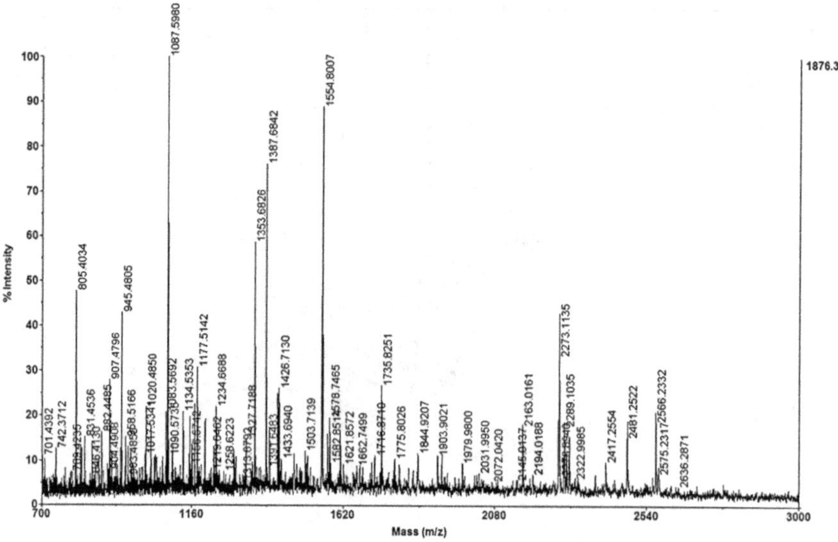

Figure 3.16: MALDI-TOF MS spectrum of the 24# fraction that is digested by microwave-assisted Fe_3O_4–Glymo–trypsin.

Table 3.1: Protein information identified by MALDI-TOF MS/MS after 24# fraction is digested by microwave-assisted Fe_3O_4–Glymo–trypsin.

Name of Protein	Accession	Molecular Weight (MW)	Isoelectric Point (PI)	Peptide Count	Score
Hemoglobin alpha 2 chain [*Rattus norvegicus*]	gi|60678292	15274.78	8.45	5	142
Glial fibrillary acidic protein [*Rattus norvegicus*]	gi|8393431	49912.57	5.35	14	65
Carbamoyl-phosphate synthetase 1, mitochondrial [*Rattus norvegicus*]	gi|8393186	164475.5	6.33	12	61
Glial fibrillary acidic protein, astrocyte (GFAP)	gi|115311597	49926.59	5.35	13	58

intervals from 5 min to 15 min and from 19.5 min to 109.5 min, and a total of 50 fractions are collected. All the fractions are collected, freeze-dried, and redissolved in 50 μL 25 mmol·L^{-1} NH$_4$HCO$_3$ (pH = 8.3) solution for microwave enzymatic hydrolysis. The 24# fraction is identified by MALDI-TOF MS/MS. The mass spectrum is shown in Figure 3.16, and the identified protein information is shown in Table 3.1.

3.2.3.2. *Application of Fe$_3$O$_4$@SiO$_2$–Glymo–trypsin in microwave-assisted enzymatic hydrolysis*

(1) Fe$_3$O$_4$@SiO$_2$–Glymo–trypsin for microwave-assisted enzymatic hydrolysis of standard protein

Since the agglomeration of magnetic spheres is significantly improved after coating silica and the amount of immobilized enzyme is increased, the concentration of standard protein is therefore increased to 0.5 μg·μL^{-1} for further study. First, with 15 s of microwave time, different amounts of Fe$_3$O$_4$@SiO$_2$–Glymo–trypsin are compared to explore the impact on completeness of enzymatic hydrolysis. The results show that the peaks of both BSA and myoglobin (MYO) are obviously reduced, while the small molecular weight peptides increase gradually, with the increase of dosage of Fe$_3$O$_4$@SiO$_2$–Glymo–trypsin. Then 500 μg Fe$_3$O$_4$@SiO$_2$–Glymo–trypsin is used to investigate the influence of different microwave times on the efficiency of enzymatic hydrolysis. What's worth mentioning is that the aim of developing this rapid enzymatic hydrolysis technology is to maximize the time-effectiveness of the entire proteomics technology route. Therefore, the effect of microwave time from 15 s to 120 s on the sequence coverage of the generated peptide is investigated under the condition that the amount of protein and Fe$_3$O$_4$@SiO$_2$–Glymo–trypsin remain unchanged. As shown in Figure 3.17, the sequence coverage of these two proteins does not increase significantly from 15 s to 120 s. Considering the requirement of rapid time-effectiveness, the experiment time is not prolonged.

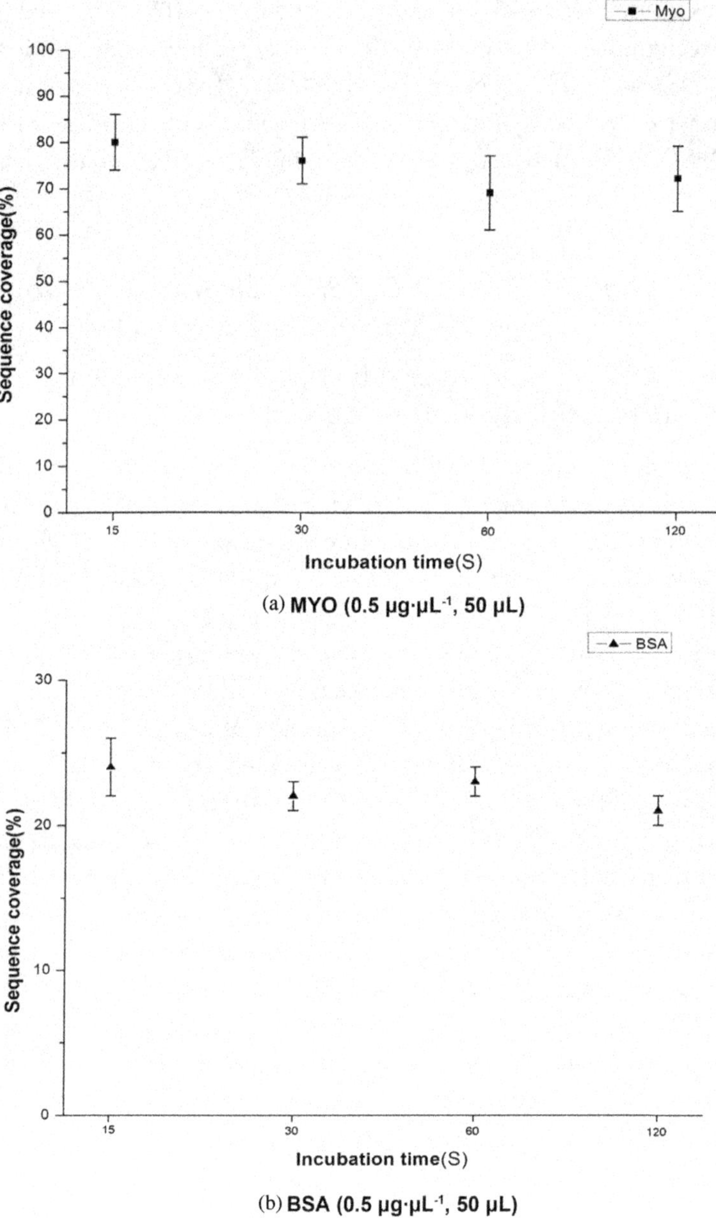

Figure 3.17: The effect of microwave time on efficiency of enzymatic hydrolysis of $Fe_3O_4@SiO_2$–Glymo–trypsin (500 μg).

(2) Microwave-assisted enzymatic hydrolysis of $Fe_3O_4@SiO_2$– Glymo–trypsin applied in "bottom-up" technology route

The pure "bottom-up" technique (also known as "shot-gun") represented by Yates[39] advocates treating biological samples with enzymes directly, followed by separation and analysis of the generated peptides. In terms of the analysis of proteins with different physical and chemical properties, this method loses a lot of information of the proteins, however, it also greatly reduces the difficulties faced by the analysis method, hence it has been a common analysis method in the field of proteomics for a long time. If the new technology of microwave-assisted enzymatic hydrolysis is combined with the "bottom-up" technology, the whole analytical process will be greatly shortened, resulting in a broad prospect in medical diagnosis and clinical application.

After the reductive alkylation reaction, the freshly extracted mouse liver protein is divided into two parts. One part is mixed with $Fe_3O_4@SiO_2$– Glymo–trypsin for 15 s of microwave-assisted enzymatic hydrolysis, the other is digested according to 12 h of conventional in-solution enzymatic hydrolysis. Both of which are analyzed by LC-ESI-MS/MS under the same conditions.

The reverse-phase separation system sets the separation time for 50 min, and including the processing time of microwave-assiated enzymatic hydrolysis, the whole process takes about an hour. Figure 3.18 shows the total ion current (TIC) chromatograms of the peptides produced by the microwave assisted enzymatic hydrolysis of $Fe_3O_4@SiO_2$–Glymo–trypsin and the conventional in-solution digestion.

According to the principle of SEQUEST algorithm, the MS data is strictly checked. After the searches of positive and negative library, the data is superimposed and simulated, and the corresponding X_{corr}/(charge number) values are obtained, respectively, under the credibility standards of 95% and 99%, as shown in Figure 3.19.

Finally, a total of 318 proteins are identified using microwave-assisted enzymatic hydrolysis at confidence coefficient of 99%. A total of 364 proteins are identified by traditional in-solution enzymatic hydrolysis. The results fully show that 15 s of microwave-assisted enzymatic hydrolysis of $Fe_3O_4@SiO_2$– Glymo–trypsin equals 12 h of conventional in-solution enzymatic hydrolysis. Also, the combination of this enzymolysis technology and proteomics technique shows obvious time effectiveness, and can greatly shorten the separation and identification process, which has great potential for clinical diagnosis and statistical analysis of disease signaling proteins.

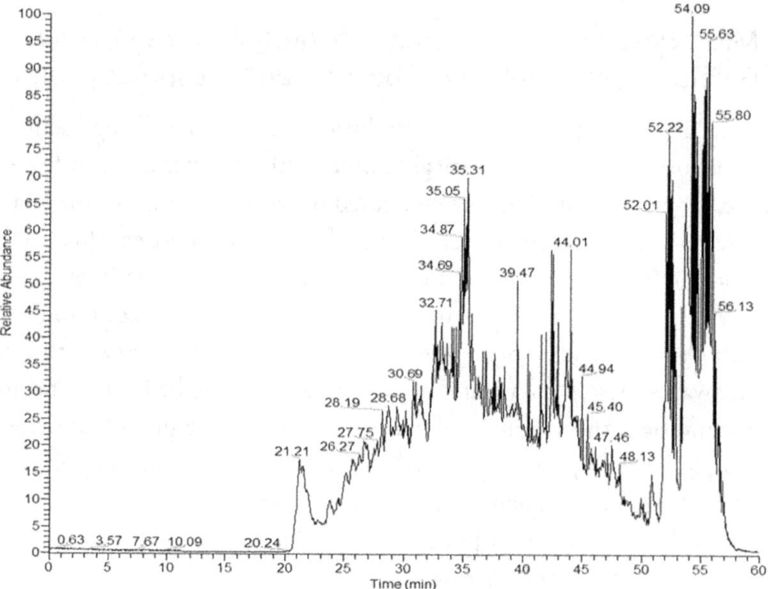

(a) TIC chromatograms of LC-ESI-MS/MS by microwave-assisted enzymatic hydrolysis

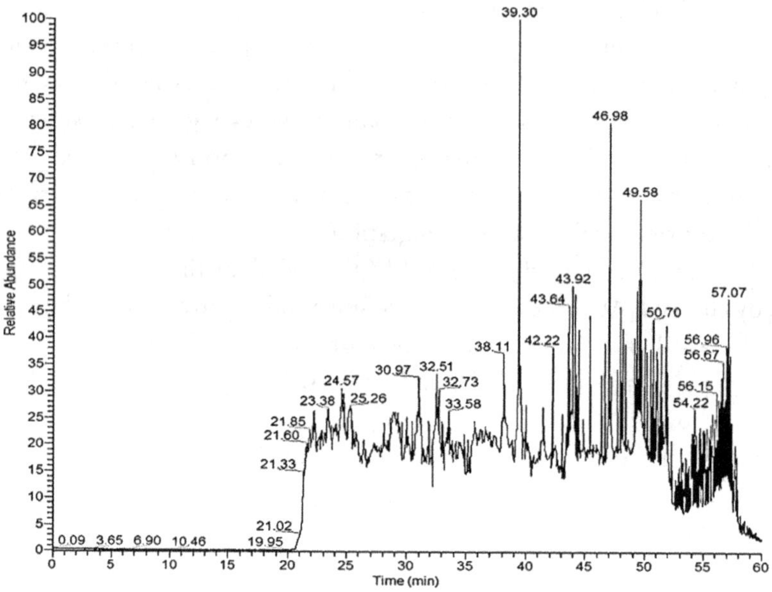

(b) TIC chromatograms of LC-ESI-MS/MS by conventional in-solution digestion

Figure 3.18: The TIC chromatograms of the peptides produced by the microwave-assisted enzymatic hydrolysis of $Fe_3O_4@SiO_2$–Glymo–trypsin and the conventional in-solution digestion.

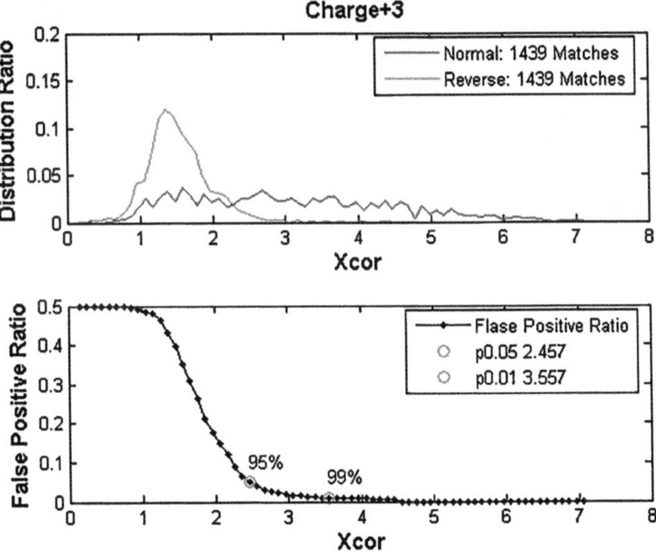

Figure 3.19: Calculation process of X_{corr}.

3.2.3.3. *Application of trypsin immobilized amino-magnetic nanoparticle in microwave-assisted enzymatic hydrolysis*

(1) Trypsin immobilized amino-magnetic nanoparticle for microwave-assisted enzymatic hydrolysis of standard protein

Due to the increase of immobilized amount of enzyme and good magnetic responsiveness, the concentration of standard protein is increased to detect the efficiency of microwave-assisted enzymatic hydrolysis of trypsin immobilized amino-magnetic nanoparticle. First, the effect of different amounts of nanoparticles on the completeness of enzymatic hydrolysis is investigated in a 15-s microwave time. Then the effect of microwave time on the efficiency of enzymatic hydrolysis is investigated under the condition of complete enzymatic hydrolysis.

As seen in Figure 3.20, the peaks of proteins gradually disappear along with the increase of dosage of trypsin immobilized amino-magnetic nanoparticles. It can be considered that complete enzymatic hydrolysis has been

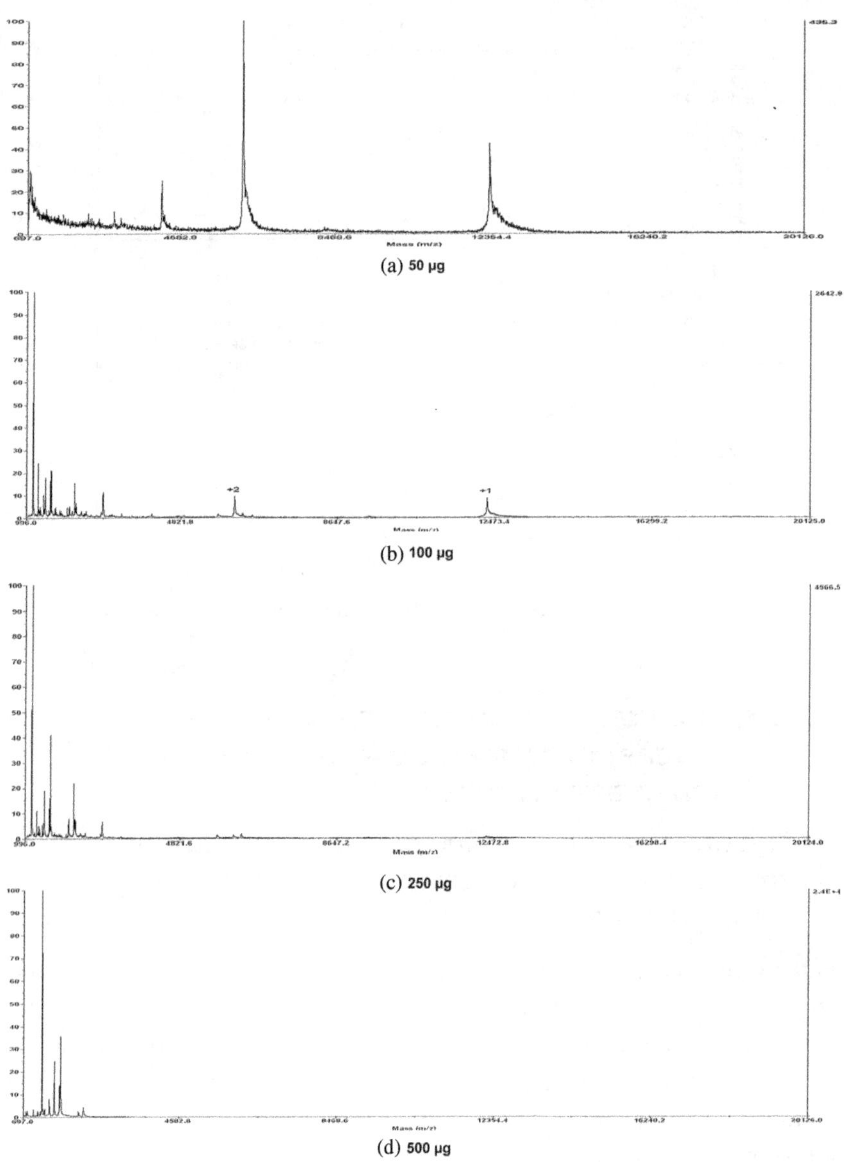

Figure 3.20: MALDI-TOF-MS linear pattern spectra of peptides from Cyc (1 mg·mL^{-1}, 200 µL) through microwave-assisted enzymatic hydrolysis of different amounts of trypsin immobilized amino-magnetic nanoparticles.

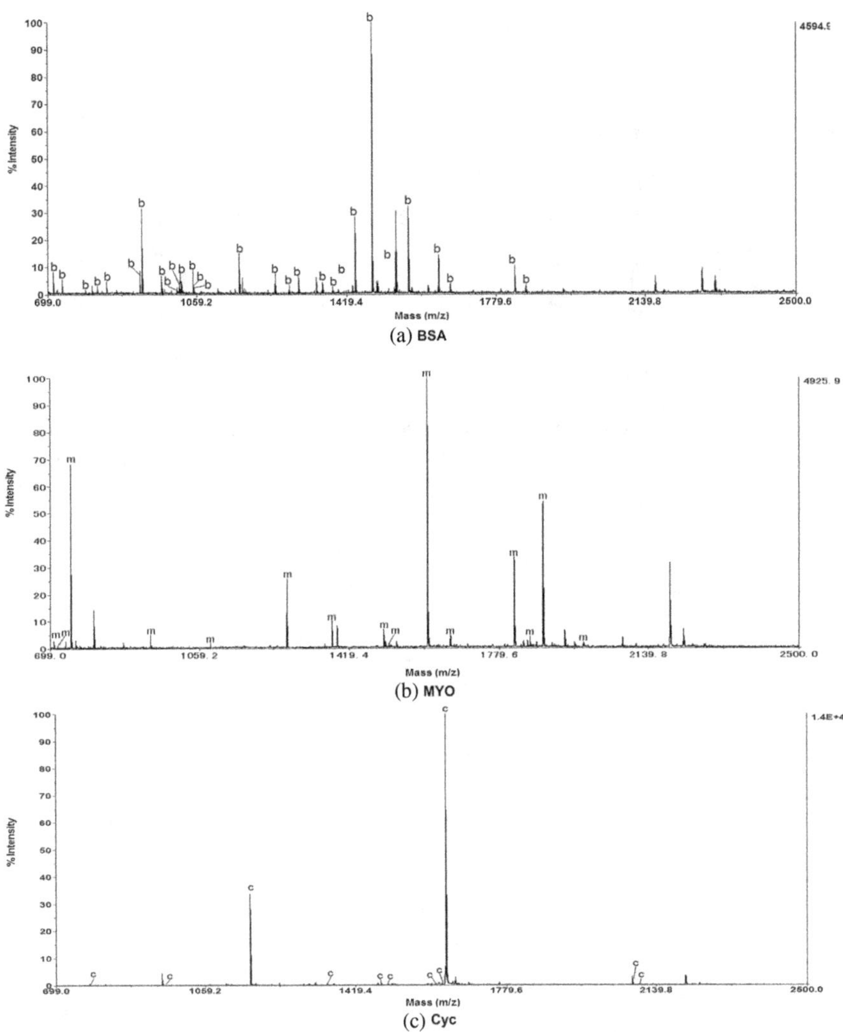

Figure 3.21: MALDI MS spectra of peptides from three proteins through microwave-assisted enzymatic hydrolysis of trypsin immobilized amino-magnetic nanoparticle (250 μg). Conditions: 200 μL protein solution (1 mg·mL⁻¹); microwave time: 15 s, power: 850 W.

achieved when the dosage is 250 μg (Figure 3.21). Then the effects of microwave time and microwave power on the efficiency of enzymatic hydrolysis are investigated in detail. The results show that the peptide sequence coverage has no obvious regular change with the change of these two factors (Figure 3.22).

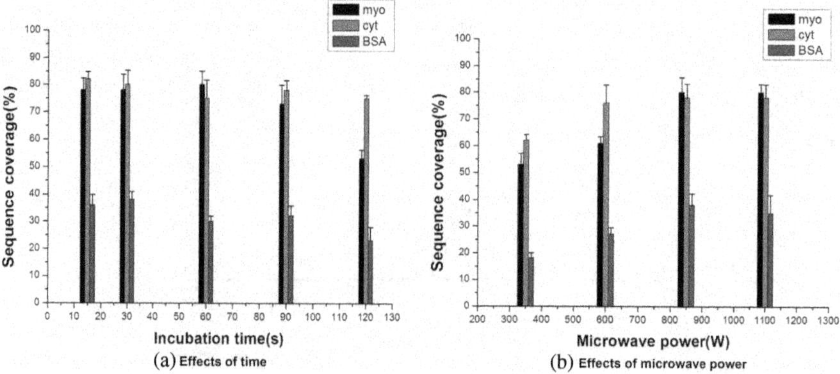

Figure 3.22: Effects of microwave time and microwave power on protein sequence coverage. Enzymatic conditions: 200 μL 1 mg·mL^{-1}of protein dosage, 250 μg of trypsin immobilized amino-magnetic nanoparticle.

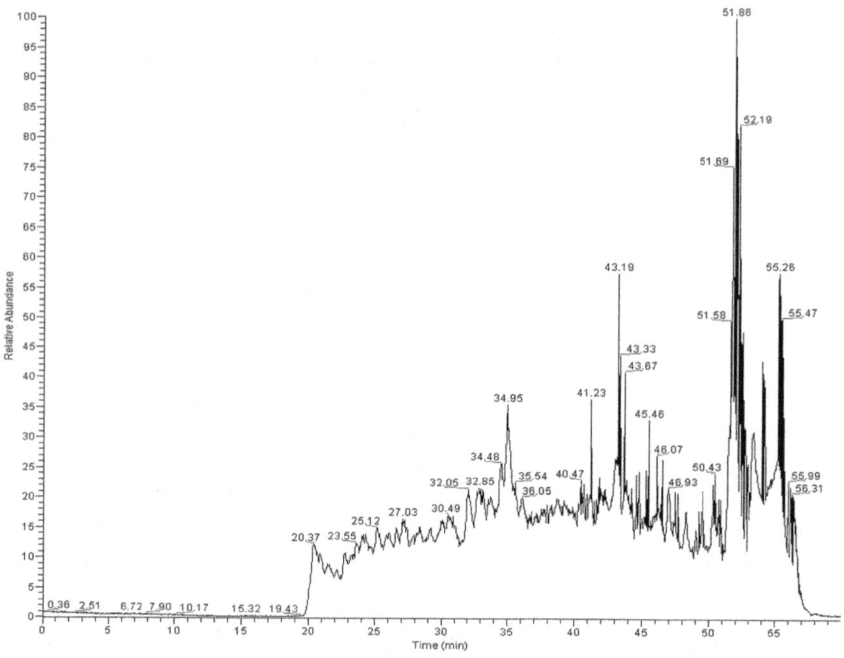

Figure 3.23: The TIC chromatogram of peptides from the mouse liver protein through microwave-assisted enzymatic hydrolysis of trypsin immobilized amino-magnetic nanoparticle.

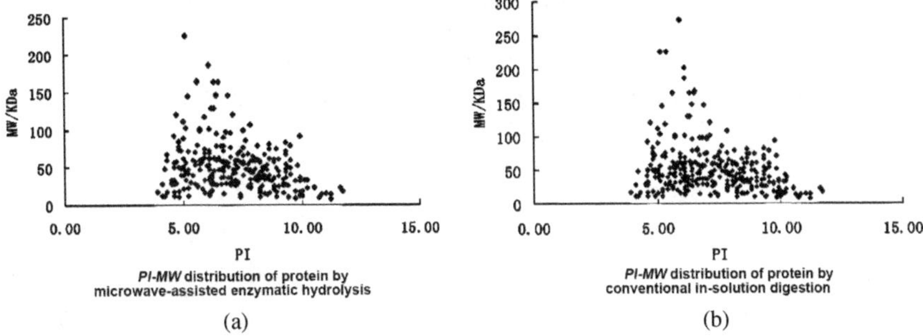

Figure 3.24: *pI-MW* distribution of protein obtained by microwave-assisted enzymatic hydrolysis of trypsin immobilized amino-magnetic nanoparticle and conventional in-solution digestion.

Table 3.2: The information of protein identified by combination of microwave-assisted trypsin immobilized amino-magnetic nanoparticle and LC-ESI MS/MS, and corresponding parameters ($p < 0.01$).

Accession	Coverage	Matching Peptide (Hits)	Molecule Weight (MW)	Isoelectric Point (pI)
IPI00210644.1	47.40	81 (810000)	164474.8	6.30
IPI00551812.1	63.70	34 (331000)	56318.6	5.10
IPI00201413.1	51.60	21 (201000)	41844.5	7.90
IPI00210444.5	30.30	16 (160000)	56875.7	8.80
IPI00189819.1	24.80	6 (60000)	41709.7	5.20
IPI00205036.1	37.30	5 (50000)	15274.8	8.40
IPI00191711.1	16.60	3 (30000)	20723.5	5.80
IPI00231742.5	44.20	21 (210000)	59719.6	7.20
IPI00326195.3	31.90	5 (50000)	32830.8	8.60
IPI00211225.1	20.20	7 (70000)	47842.5	7.60
IPI00388209.2	3.60	1 (10000)	59181.2	4.30
IPI00339148.2	45.70	23 (230000)	60917.5	5.80
IPI00475676.3	16.60	7 (70000)	59604.5	6.40
IPI00388302.2	11.50	1 (10000)	21884.3	10.20
IPI00327518.4	34.60	10 (100000)	36480.9	6.60
IPI00205157.1	29.00	6 (60000)	34425.9	9.20
IPI00206624.1	33.60	20 (200000)	72302.5	4.90

(*Continued*)

Table 3.2: (*Continued*)

Accession	Coverage	Matching Peptide (Hits)	Molecule Weight (MW)	Isoelectric Point (pI)
IPI00191737.6	42.10	23 (230000)	68686.2	6.10
IPI00364321.3	27.10	6 (60000)	27670.1	7.80
IPI00197696.2	46.20	11 (110000)	35660.8	8.80
IPI00213659.3	26.90	7 (70000)	36109.7	9.30
IPI00207010.2	21.70	6 (60000)	46435.1	7.10
IPI00195123.1	28.60	5 (50000)	23382.8	10.50
IPI00370596.2	17.20	4 (40000)	43792.7	8.20
IPI00198887.1	30.80	14 (140000)	56915.8	4.70
IPI00324302.3	21.50	8 (80000)	44666.5	9.00
IPI00471911.7	25.00	11 (110000)	39593.2	8.40
IPI00396910.1	31.10	18 (180000)	59716.7	9.50
IPI00205018.2	27.40	10 (100000)	57882.7	8.20
IPI00211127.1	25.50	12 (11 1000)	46466.9	7.70
IPI00231631.13	4.10	1 (10000)	46984.4	7.20
IPI00190790.1	54.30	9 (90000)	14263.3	8.50
IPI00231643.5	36.40	6 (60000)	15901.8	5.90
IPI00230788.6	46.20	13 (130000)	29412.7	7.00
IPI00369166.3	11.10	1 (10000)	20354.5	8.80
IPI00204316.1	17.60	1 (10000)	12486.6	9.90
IPI00480620.1	29.90	6 (60000)	38308.4	9.00
IPI00324633.2	26.90	15 (150000)	61377.4	8.00
IPI00214454.2	11.80	1 (10000)	18473.4	4.80
IPI00361944.2	5.70	2 (0 2000)	62113.6	5.90
IPI00560967.1	5.80	1 (10000)	35126.3	8.40
IPI00202549.1	4.00	1 (10000)	62161.7	6.50
IPI00358163.3	9.60	4 (40000)	74352.0	8.90
IPI00331983.5	12.50	6 (60000)	39619.6	8.10
IPI00480679.4	17.70	5 (50000)	47732.4	5.00
IPI00325135.3	7.50	1 (10000)	29155.4	4.50
IPI00231013.2	46.00	3 (30000)	11399.7	5.20
IPI00231365.5	20.50	5 (50000)	34911.9	8.20

Table 3.2: (*Continued*)

Accession	Coverage	Matching Peptide (Hits)	Molecule Weight (MW)	Isoelectric Point (pI)
IPI00230897.5	58.50	9 (90000)	15969.3	8.20
IPI00231292.6	52.50	5 (50000)	14479.7	6.70
IPI00231264.6	4.50	1 (10000)	39471.2	5.00
IPI00359623.3	21.60	10 (100000)	53387.3	6.80
IPI00231192.5	26.50	4 (40000)	15972.3	9.20
IPI00195109.3	3.60	1 (10000)	55729.8	8.20
IPI00365985.4	10.30	7 (70000)	92474.1	4.60
IPI00201262.1	8.30	9 (90000)	163669.2	5.60
IPI00209807.3	5.20	3 (30000)	101375.7	6.10
IPI00395281.1	13.80	5 (50000)	68154.8	7.30
IPI00197711.1	21.70	5 (50000)	36427.4	8.30
IPI00207941.1	6.80	4 (40000)	95987.5	7.00
IPI00210435.1	18.90	16 (160000)	129607.8	6.20
IPI00372498.1	15.70	5 (50000)	59406.4	7.30
IPI00332027.7	44.20	13 (130000)	44947.9	7.80
IPI00207217.1	21.40	5 (50000)	31496.2	8.10
IPI00199482.2	7.80	2 (20000)	25537.0	5.60
IPI00211510.1	5.60	2 (20000)	74631.4	8.60
IPI00371957.2	4.00	1 (10000)	45807.8	6.70
IPI00393599.3	13.60	2 (20000)	37766.4	8.30
IPI00471577.1	4.00	1 (10000)	52815.5	5.50
IPI00231139.6	5.50	3 (30000)	71113.5	7.40
IPI00326433.11	13.70	1 (10000)	10894.9	9.40
IPI00365929.1	16.40	5 (50000)	48729.7	4.90
IPI00358005.1	5.50	2 (20000)	66346.5	8.40
IPI00470304.1	16.40	3 (30000)	39433.3	5.40
IPI00389571.6	32.70	14 (13 1000)	53985.3	5.70
IPI00325136.5	3.50	2 (20000)	58775.5	6.60
IPI00231864.5	34.30	4 (40000)	11598.0	10.00
IPI00411230.3	7.30	1 (10000)	25686.0	7.20
IPI00763308.1	9.30	1 (10000)	23863.5	7.80

(Continued)

Table 3.2: (*Continued*)

Accession	Coverage	Matching Peptide (Hits)	Molecule Weight (MW)	Isoelectric Point (pI)
IPI00207601.1	13.80	5 (50000)	40942.3	7.10
IPI00188989.1	21.70	12 (120000)	78128.3	6.60
IPI00197770.1	40.10	16 (160000)	56452.7	6.70
IPI00389611.3	5.70	2 (20000)	33368.4	5.10
IPI00231028.2	9.40	1 (10000)	28154.9	9.20
IPI00209480.5	3.10	1 (10000)	58589.2	7.20
IPI00373331.2	2.70	1 (10000)	60410.1	9.00
IPI00201103.1	2.70	1 (10000)	72238.7	6.10
IPI00370348.3	25.40	4 (40000)	27731.1	4.60
IPI00189813.1	8.50	4 (3 1000)	42023.9	5.10
IPI00190348.1	19.80	3 (30000)	13881.6	10.80
IPI00558154.1	10.50	3 (2 1000)	59976.5	6.10
IPI00192409.1	3.20	1 (10000)	46677.8	7.80
IPI00205332.4	23.70	6 (60000)	34929.5	8.50
IPI00369093.1	20.20	1 (10000)	10417.0	4.80
IPI00389152.4	16.30	1 (10000)	10064.9	8.90
IPI00197579.2	5.90	1 (10000)	49639.0	4.60
IPI00200593.1	5.50	1 (10000)	46532.2	5.20
IPI00212014.2	9.20	5 (50000)	89478.0	5.00
IPI00210920.1	25.30	10 (100000)	47284.2	9.30
IPI00230835.5	7.70	1 (10000)	28284.9	4.70
IPI00194550.5	13.80	6 (60000)	43577.3	7.80
IPI00365545.1	11.20	4 (40000)	54004.2	7.70
IPI00205374.1	8.60	2 (20000)	34169.7	8.40
IPI00198966.3	3.40	1 (10000)	50222.7	8.30
IPI00231253.5	52.50	8 (80000)	27228.6	9.20
IPI00231767.5	20.50	3 (30000)	26831.8	7.00
IPI00421885.2	3.90	1 (10000)	50388.1	6.90
IPI00197555.6	4.60	1 (10000)	36125.0	9.90
IPI00195372.1	16.20	6 (60000)	50082.2	9.40
IPI00231955.6	10.70	1 (10000)	16826.8	3.90

Table 3.2: (*Continued*)

Accession	Coverage	Matching Peptide (Hits)	Molecule Weight (MW)	Isoelectric Point (pI)
IPI00196725.5	8.90	5 (50000)	99063.9	5.70
IPI00231426.6	11.00	4 (40000)	44510.0	7.90
IPI00231745.8	28.70	8 (80000)	39584.2	5.40
IPI00210139.1	23.40	7 (70000)	39860.8	9.50
IPI00208917.4	11.30	4 (40000)	58711.5	7.80
IPI00421711.1	18.40	1 (10000)	11453.1	9.90
IPI00188804.1	39.10	3 (30000)	11684.9	4.20
IPI00362070.5	3.90	1 (10000)	56987.9	7.40
IPI00208026.1	8.70	2 (20000)	52498.5	6.10
IPI00209082.1	1.30	1 (10000)	102895.6	5.10
IPI00214480.1	11.20	4 (40000)	45946.1	6.70
IPI00205561.1	3.80	2 (20000)	76750.1	7.50
IPI00212622.1	25.30	18 (180000)	82460.7	9.30
IPI00230939.4	8.30	1 (10000)	17767.9	11.70
IPI00213057.2	5.20	3 (30000)	71245.0	9.30
IPI00363265.3	17.70	8 (80000)	73811.9	5.90
IPI00551702.2	4.60	1 (10000)	48894.5	8.90
IPI00212015.1	9.30	3 (30000)	46525.8	8.50
IPI00200659.1	8.70	4 (40000)	71569.8	6.70
IPI00208215.2	10.10	2 (20000)	28277.5	7.30
IPI00231648.5	16.00	3 (30000)	32528.4	7.20
IPI00337168.5	2.40	1 (10000)	58062.9	8.00
IPI00476295.7	13.70	4 (40000)	47507.3	6.10
IPI00196656.2	10.30	9 (90000)	107383.3	7.80
IPI00464815.11	11.80	5 (3 2000)	47098.3	6.10
IPI00326667.1	2.90	1 (10000)	45983.9	6.30
IPI00366293.3	20.90	5 (50000)	33385.8	7.80
IPI00211593.1	6.30	1 (10000)	24658.6	9.00
IPI00230889.5	8.40	1 (10000)	17460.2	9.90
IPI00231150.4	19.50	2 (20000)	25989.0	9.30
IPI00205275.1	13.90	1 (10000)	11060.3	7.40

(*Continued*)

Table 3.2: (*Continued*)

Accession	Coverage	Matching Peptide (Hits)	Molecule Weight (MW)	Isoelectric Point (pI)
IPI00198467.1	10.10	4 (40000)	51381.6	9.80
IPI00231638.6	8.10	1 (10000)	25590.6	9.20
IPI00326948.2	13.40	7 (70000)	81038.0	8.60
IPI00421428.9	5.50	1 (10000)	28813.9	6.80
IPI00207146.1	9.50	2 (20000)	16012.3	8.60
IPI00366020.2	5.20	1 (10000)	27846.1	5.50
IPI00557879.2	10.80	2 (20000)	31758.8	9.70
IPI00231611.7	10.50	3 (30000)	54429.1	9.30
IPI00194974.2	5.00	1 (10000)	50996.4	5.00
IPI00361708.3	5.20	1 (10000)	35534.7	6.10
IPI00205745.3	32.40	5 (50000)	22164.6	8.80
IPI00231260.5	26.80	4 (40000)	24803.0	5.60
IPI00215574.5	23.50	6 (5 1000)	32234.0	9.90
IPI00212651.1	14.70	1 (10000)	10451.0	8.20
IPI00776522.1	7.30	1 (0 1000)	31415.6	6.20
IPI00202616.1	4.90	1 (10000)	30227.2	7.30
IPI00209828.5	7.40	1 (10000)	27579.5	6.30
IPI00211779.1	35.20	5 (4 1000)	22095.3	8.20
IPI00189795.1	7.80	2 (20000)	50103.7	4.80
IPI00471539.4	13.40	4 (40000)	46958.8	7.70
IPI00382226.1	1.00	1 (10000)	144618.5	5.20
IPI00190557.2	9.00	2 (20000)	33291.9	10.20
IPI00358033.1	7.40	3 (30000)	79361.7	5.60
IPI00360618.3	9.10	2 (20000)	33964.1	8.00
IPI00231639.7	12.80	2 (20000)	25897.1	8.30
IPI00326972.6	18.10	9 (8 1000)	62107.8	6.10
IPI00382191.1	3.10	2 (20000)	120448.8	4.70
IPI00231963.5	42.40	3 (30000)	13124.8	6.10
IPI00373418.3	2.90	1 (10000)	53274.1	8.50
IPI00191728.1	11.80	4 (40000)	47965.9	4.20
IPI00769236.1	2.90	1 (10000)	65489.7	7.50

Table 3.2: (*Continued*)

Accession	Coverage	Matching Peptide (Hits)	Molecule Weight (MW)	Isoelectric Point (pI)
IPI00200996.2	4.30	1 (10000)	42793.1	7.20
IPI00421539.3	7.40	3 (30000)	85380.1	7.60
IPI00232011.9	8.30	4 (40000)	78608.5	9.50
IPI00195673.1	6.50	2 (20000)	50027.2	4.60
IPI00230857.7	9.60	2 (20000)	26362.6	6.40
IPI00359732.2	9.30	1 (10000)	16480.1	4.60
IPI00203690.5	15.40	6 (60000)	56432.7	6.90
IPI00231148.8	18.10	5 (50000)	37428.3	6.20
IPI00324893.4	12.70	2 (20000)	27753.7	4.60
IPI00564133.1	15.20	4 (40000)	51011.4	8.40
IPI00200466.3	8.40	2 (20000)	32880.2	10.10
IPI00199636.1	2.70	1 (10000)	67212.7	4.30
IPI00553996.3	3.50	2 (20000)	78483.4	6.40
IPI00193153.1	8.10	3 (30000)	63575.3	7.00
IPI00212220.2	8.60	4 (3 1000)	72674.8	4.80
IPI00231359.3	14.70	3 (30000)	44939.1	8.30
IPI00196751.2	2.00	1 (10000)	70142.2	5.50
IPI00369635.3	2.60	1 (10000)	68501.4	5.90
IPI00208288.4	8.90	2 (20000)	31808.9	8.70
IPI00210566.3	3.50	2 (1 1000)	84761.8	4.80
IPI00207038.3	2.20	1 (10000)	58060.1	7.90
IPI00198717.8	8.70	2 (20000)	36460.1	6.20
IPI00362058.3	8.00	1 (10000)	22204.8	10.50
IPI00400573.1	3.40	1 (10000)	49769.0	4.60
IPI00372520.1	5.90	1 (10000)	27952.4	4.10
IPI00205906.2	6.20	3 (30000)	56778.7	6.80
IPI00192301.1	13.90	2 (20000)	22244.4	7.80
IPI00195160.1	1.80	1 (10000)	61083.7	5.00
IPI00555299.1	6.80	1 (10000)	39333.1	4.90
IPI00231694.7	0.90	1 (10000)	146148.7	6.90
IPI00326561.3	3.40	1 (10000)	36172.5	8.00

(*Continued*)

Table 3.2: (*Continued*)

Accession	Coverage	Matching Peptide (Hits)	Molecule Weight (MW)	Isoelectric Point (pI)
IPI00207980.1	14.30	1 (10000)	14856.1	11.00
IPI00324741.2	10.20	5 (50000)	57043.1	5.80
IPI00325599.4	5.00	2 (20000)	51517.3	6.00
IPI00361193.5	4.40	3 (30000)	101674.9	5.70
IPI00202658.1	13.10	3 (30000)	35279.6	8.50
IPI00365944.5	19.20	2 (20000)	16964.2	4.30
IPI00362927.1	3.30	1 (10000)	49892.4	4.80
IPI00208205.1	4.20	2 (20000)	70827.3	5.20
IPI00359978.3	17.40	1 (10000)	7836.2	11.20
IPI00190179.3	1.90	1 (10000)	97421.1	6.80
IPI00188924.4	5.50	2 (20000)	48366.2	9.50
IPI00566764.1	1.50	1 (10000)	129694.9	6.30
IPI00211897.4	4.10	1 (10000)	25331.2	8.80
IPI00231368.5	22.90	2 (20000)	11665.7	4.60
IPI00360056.1	5.00	1 (10000)	37869.2	6.20
IPI00471540.1	4.40	1 (10000)	39903.2	6.20
IPI00231358.6	21.40	2 (20000)	14947.5	8.40
IPI00367152.1	9.30	2 (20000)	27360.9	6.20
IPI00209045.2	3.80	2 (20000)	75810.4	7.00
IPI00480639.3	1.70	2 (20000)	186460.2	6.10
IPI00188158.1	2.10	1 (10000)	57397.5	5.50
IPI00195860.1	15.70	1 (10000)	9346.9	10.70
IPI00475946.2	2.80	1 (10000)	54488.9	7.50
IPI00210823.1	8.00	3 (30000)	37354.1	6.20
IPI00387771.6	17.10	3 (30000)	17862.8	8.20
IPI00767154.1	3.60	1 (10000)	35784.8	7.00
IPI00231650.7	5.90	1 (10000)	21974.0	11.60
IPI00230838.5	21.10	2 (20000)	18751.6	6.20
IPI00212731.1	4.40	1 (10000)	44651.9	6.70
IPI00201561.3	8.60	1 (10000)	21770.1	5.20
IPI00204774.1	4.60	2 (20000)	62454.9	6.30

Table 3.2: (*Continued*)

Accession	Coverage	Matching Peptide (Hits)	Molecule Weight (MW)	Isoelectric Point (pI)
IPI00360930.2	4.60	1 (10000)	28282.2	7.00
IPI00471530.2	10.60	3 (30000)	56114.8	6.80
IPI00207184.1	3.50	1 (10000)	28557.0	6.20
IPI00370752.3	3.60	1 (10000)	30681.7	7.10
IPI00188688.1	21.50	2 (20000)	14179.9	11.20
IPI00327781.1	2.60	1 (10000)	57144.5	7.50
IPI00339123.2	2.10	1 (10000)	78396.1	8.10
IPI00364948.3	6.00	2 (20000)	54047.4	7.50
IPI00365813.3	3.10	1 (10000)	35018.0	8.80
IPI00208203.2	9.00	3 (30000)	42994.0	9.30
IPI00212316.1	1.20	1 (10000)	99124.8	7.50
IPI00421857.1	1.60	1 (0 1000)	64791.0	7.90
IPI00214373.1	7.10	2 (20000)	47753.7	8.00
IPI00471584.7	1.80	1 (0 1000)	83229.2	4.80
IPI00211392.1	2.40	1 (10000)	58876.7	5.70
IPI00368708.2	13.60	2 (20000)	24464.6	7.70
IPI00200145.1	14.00	1 (10000)	11490.7	4.10
IPI00358226.2	11.10	1 (10000)	13767.2	7.40
IPI00194324.2	4.50	1 (10000)	38957.1	6.20
IPI00470301.1	3.10	1 (10000)	59498.9	5.40
IPI00194045.1	4.10	1 (10000)	46704.6	6.60
IPI00556987.2	2.70	1 (10000)	49867.6	6.70
IPI00231641.5	3.20	1 (10000)	61364.7	6.30
IPI00211989.2	5.00	1 (10000)	32781.6	6.10
IPI00211100.1	11.20	2 (20000)	37004.0	6.70
IPI00205389.6	2.90	1 (10000)	54200.7	7.70
IPI00197553.1	7.20	1 (10000)	32539.8	4.50
IPI00205560.2	1.20	1 (10000)	146656.3	6.40
IPI00231245.5	4.50	1 (10000)	39176.4	7.50
IPI00195423.1	4.30	2 (20000)	60553.3	8.80
IPI00231106.5	3.40	1 (10000)	32919.5	5.90

(*Continued*)

Table 3.2: (*Continued*)

Accession	Coverage	Matching Peptide (Hits)	Molecule Weight (MW)	Isoelectric Point (pI)
IPI00197900.5	4.70	1 (10000)	28748.5	4.70
IPI00371043.2	8.40	1 (10000)	20374.6	8.20
IPI00212666.2	0.70	1 (10000)	165220.9	5.60
IPI00192246.1	10.30	1 (10000)	16119.3	6.10
IPI00203558.1	4.40	1 (10000)	27776.3	6.60
IPI00204634.4	1.90	1 (10000)	47018.8	6.50
IPI00454288.1	2.00	1 (10000)	67677.3	7.70
IPI00208209.1	8.10	2 (1 1000)	30988.1	6.20
IPI00230859.5	5.80	1 (10000)	36482.9	6.90
IPI00327079.5	4.90	1 (10000)	25476.3	9.40
IPI00231069.5	20.70	1 (10000)	10021.2	9.40
IPI00231978.5	16.90	1 (10000)	8249.5	9.70
IPI00208185.3	1.20	2 (20000)	147390.0	6.40
IPI00205135.6	2.20	1 (10000)	76887.2	4.80
IPI00203214.6	1.50	1 (10000)	95223.0	6.40
IPI00196661.1	4.10	1 (10000)	27760.8	4.50
IPI00212478.1	6.80	1 (10000)	16094.5	8.60
IPI00324019.1	2.40	1 (10000)	46106.6	5.70
IPI00203647.1	3.10	1 (0 1000)	55697.3	4.60
IPI00207014.2	2.20	1 (10000)	44608.7	5.10
IPI00210975.1	2.70	1 (10000)	111220.1	5.00
IPI00210280.1	6.10	1 (10000)	29578.3	5.30
IPI00361686.5	7.20	1 (10000)	30977.5	4.60
IPI00210164.3	18.80	2 (20000)	17835.5	9.70
IPI00204118.1	5.10	2 (20000)	66680.9	9.30
IPI00199203.1	4.50	1 (10000)	35832.8	6.20
IPI00190240.1	10.30	1 (10000)	17939.5	10.00
IPI00212110.1	3.40	1 (10000)	60945.7	8.60
IPI00215107.3	4.40	1 (10000)	32803.4	4.60
IPI00400739.1	3.30	1 (10000)	47342.9	4.90
IPI00214665.2	1.40	1 (10000)	120704.0	7.10

Table 3.2: (*Continued*)

Accession	Coverage	Matching Peptide (Hits)	Molecule Weight (MW)	Isoelectric Point (pI)
IPI00188304.3	10.90	1 (10000)	20806.4	5.00
IPI00371236.3	3.10	1 (10000)	49491.0	7.30
IPI00368110.4	1.10	1 (10000)	226131.6	5.10
IPI00382228.1	4.10	1 (10000)	39427.9	8.60
IPI00231756.7	3.40	1 (10000)	54525.0	7.20
IPI00769270.1	2.70	1 (10000)	60852.9	8.10
IPI00196457.1	5.30	1 (10000)	35740.7	6.10
IPI00367281.2	3.10	1 (10000)	45980.9	8.30
IPI00567668.2	1.70	1 (10000)	91496.1	9.90
IPI00206780.1	1.70	1 (10000)	90477.0	6.70
IPI00371036.1	7.20	1 (10000)	16984.0	8.90
IPI00471872.1	2.30	1 (0 1000)	57588.5	6.70
IPI00371562.2	0.90	1 (0 1000)	164152.1	6.50
IPI00364927.1	2.70	1 (0 1000)	58693.9	6.70
IPI00190082.4	1.20	1 (0 1000)	118070.3	5.90
IPI00324041.1	1.70	1 (0 1000)	78130.0	6.70
IPI00200100.1	2.80	1 (0 1000)	54599.2	7.50

(2) Combination of enzymatic hydrolysis technology and "shot-gun" technology

Mouse liver protein is used as a sample to verify the feasibility of the microwave-assisted enzymatic hydrolysis technology of trypsin immobilized amino-magnetic nanoparticle for the practical complex sample, and the obtained products are analyzed by LC-ESI MS/MS. Figure 3.23 shows the TIC chromatogram of the peptides generated by the enzymatic hydrolysis of a trypsin immobilized amino-magnetic nanoparticle. The rapid and efficient enzymatic hydrolysis greatly improves the analysis and identification of mouse liver proteome. Figure 3.24 shows the isoelectric point-molecular weight (*pI-MW*) distribution of protein obtained by the microwave-assisted enzymatic hydrolysis and the traditional in-solution digestion. By comparison, the *pI-MW* distribution of the protein identified by the two methods is very similar. The analysis and identification process of the whole mouse liver proteome is less than one hour. Data search adopts SEQUEST algorithm model, and a total of 313 proteins are identified under credibility standard of 99%, as shown in Table 3.2. For the 16 h of in-solution digestion, 350 proteins are identified. The molecular weight and isoelectric point of the proteins identified by the two methods are statistically analyzed, but no specific distribution is found.

3.2.4. Conclusion

This chapter mainly introduces a new method of microwave-assisted rapid enzymatic hydrolysis based on enzyme-immobilzied functional magnetic nanomaterials. In this section, a series of new simple methods for immobilization of enzyme on the surface of magnetic microspheres are introduced, and the protease is immobilized on the microsphere to improve the efficiency of enzymatic hydrolysis under microwave irradiation. Compared with the traditional in-solution enzymatic hydrolysis, the new technology reduces the time consumption from 12 h to 15 s, and the enzymatic hydrolysis effect is better than the traditional method. With the help of microwave radiation, even micrograms of proteins can be efficiently digested and identified. In order to further verify the practicability of this method, the complex biological sample — whole protein extracts of mouse liver — is enzymatically hydrolyzed in 15 s. In addition, with an external magnet, the enzyme-immobilized microsphere can be recycled.

3.3. Chip enzymatic hydrolysis based on enzyme-immobilized magnetic material

3.3.1. Preparation and basic principle of capillary/ chip enzyme reactor

3.3.1.1. Capillary/chip enzyme reactor

Capillary/chip enzyme reactors have attracted more and more attention in recent years.[3] Compared with conventional in-solution enzymatic hydrolysis, capillary/chip enzymatic reactors can reduce the amount of samples by several orders of magnitude. Because there is no dilution process, the detection sensitivity of the sample can be greatly improved when using small volumes. At the same time, the small space size shortens the mass transfer time and is conducive to accelerating the enzymatic hydrolysis.[40]

3.3.1.2. *Common method for the preparation of capillary/chip enzyme reactor*

Currently, the reported preparation methods include physical adsorption, sol–gel embedding, covalent bonding, etc.[41–45] However, if the protease is directly immobilized in the capillary or chip inwall by adsorption or covalent bonding, the immobilized amount of enzyme is bound to be limited due to the limited surface area of the inner wall of the microchannel. In addition, the enzyme reactor prepared by this method is not renewable. Once the protease loses its activity, the whole enzyme reactor can only be discarded. To overcome this shortcoming, protease can be immobilized on the surface of nano/micromaterial,[46,47] and then the material with immobilized protease can be placed in a specific position in the capillary/chip channel through a plug or membrane.[48] There is also a major challenge of using this method, that is, the repeated filling of the microchannel. Porous polymer capillary/chip monolithic column[49] is one of the effective methods to solve this problem at present, but unfortunately, the preparation of monolithic column requires more complex steps and is not easy to repeat, and the resulting enzyme reactor is not renewable. Therefore, in order to meet the needs of current proteome analysis, it is particularly important to develop a new type of capillary/chip enzyme reactor with simple preparation, high enzymatic hydrolysis efficiency, and renewable capacity.

3.3.1.3. *Novel enzyme-immobilzied magnetic micro-/nanomaterial for preparation of capillary/chip enzyme reactor*

The properties of magnetic carrier are very important for the application of immobilized enzymes. Novel magnetic micro-/nanomaterials for preparing capillary/chip enzyme reactor need to meet certain conditions: (i) good magnetic responsiveness; (ii) good biocompatibility; (iii) large surface area for guaranteeing smooth enzyme reaction and reducing the dispersion restriction of reaction matrix and product; (iv) a certain mechanical strength; (v) regeneration and reuse ability; and (vi) possessing reactive functional groups.

The magnetic silica with high saturation magnetic value of 68.2 emu·g^{-1} is first synthesized, and then two technical routes of metal ion chelation and covalent bonding are adopted to immobilize trypsin on the surface of the magnetic silica. The capillary/chip enzyme reactor is prepared by immobilizing the magnetic silica in the capillary or microfluidic chip channels with the help of an external magnetic field. The protein solution is pumped through the capillary or microfluidic chip channel for rapid enzymatic hydrolysis. MALDI is used to analyze and identify the collected products. Thanks to the magnetism of the material, once the immobilized enzyme loses its activity, the magnetic microsphere can be pushed out with the liquid through the removal of the magnet outside the capillary/chip channel, and then new enzyme-immobilized magnetic microspheres can realize the regeneration of the capillary/chip enzyme reactor. This method is easy to operate and low-cost, and it solves the problem that the capillary/chip enzyme reactor can only be used once as in the previous reports. When adopting the metal ion chelation method, copper ions on the surface of magnetic microspheres can be removed by EDTA, and new copper ion and protease can be reintroduced, so as to realize regeneration of capillary/microfluidic chip enzyme reactor.

3.3.2. *Magnetic silica for enzyme immobilization*

3.3.2.1. *Synthesis of magnetic silica*

Magnetic silica is synthesized for subsequent enzyme immobilization as described in Section 2.4 of Chapter 2.

3.3.2.2. *Immobilization of enzyme on the surface of magnetic silica*

(1) Physical adsorption (metal ion chelation) for enzyme immobilization

Metal ion chelation is a kind of physical adsorption method that mainly uses ion chelating reagent such as IDA bonded on the surface of the solid-phase carrier to immobilize metal ion, and then realizes the immobilization of enzyme through the Lewis acid–base interaction between metal ion and protease. Therefore, the metal ion with its chelated protease can be easily eluted from the surface of the solid-phase carrier through the washing of a strong chelating agent (such as EDTA), and the regeneration of the solid-phase carrier can be achieved by introducing new protease.[50] Zou *et al.*[51] used copper ion chelation method to immobilize trypsin in the capillary wall, and successfully applied it in the enzymatic hydrolysis of protein and MALDI-TOF-MS analysis.

In order to immobilize the protease on magnetic silica, the synthesis route shown in Figure 3.25 is designed. First, silanizing reagent with chelating group is synthesized by using carboxymethylated ammonia derivative as chelating reagent and Glymo as coupling reagent. Then the silanating rea-

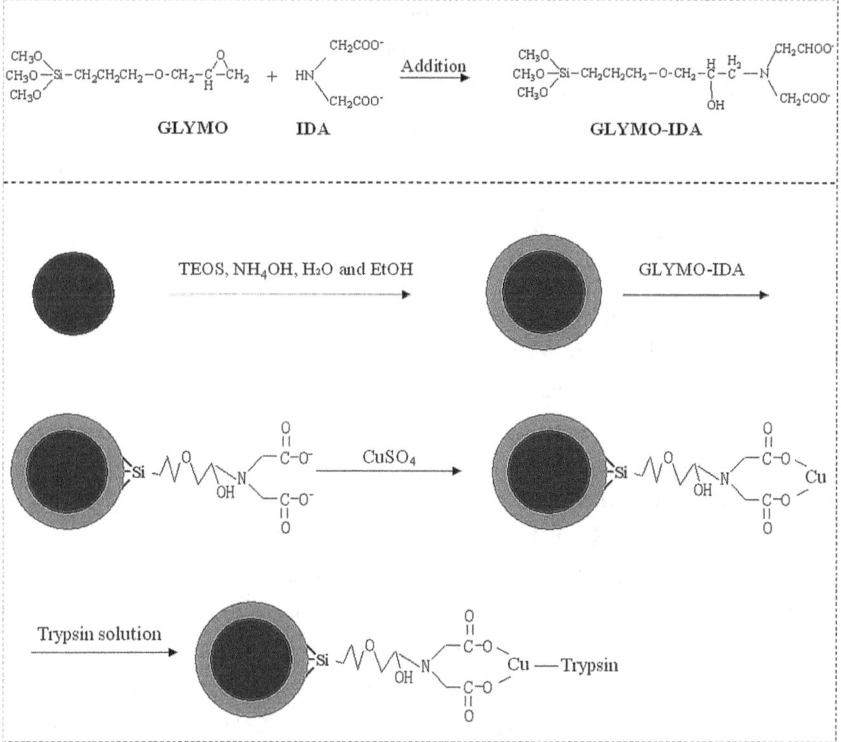

Figure 3.25: Schematic diagram of synthesis of Cu-IDA–Glymo-MS and metal ion chelation for trypsin immobilization.[52]

gent is reacted with magnetic silica before being mixed with $CuSO_4$ solution to make copper ions chelated on the surface of magnetic silica. Finally, trypsin is immobilized on magnetic silica by the interaction between copper ion and the histidine on the surface of protease. The detailed process is as follows:

- 4.20 g IDA and 50 mL 2 mol·L^{-1} Na_2CO_3 are added into a 100 mL round-bottomed flask, and then 10 mol·L^{-1} NaOH is used to adjust the pH value of solution to 11. The solution system is then placed in an ice water bath for stirring magnetically for 1 h, at the same time, 1.5 g Glymo is added in drops within 0.5 h. Afterwards, the mixture is heated to 65°C for 6 h and then cooled to 0°C. In order to ensure that the reaction occurs between IDA and Glymo and avoid cross-linking between Glymo molecules, the above reaction steps are repeated twice and Glymo is replenished at the same time. Finally, the pH value of the IDA derived silane coupling reagent solution is adjusted to 6 by concentrated hydrochloric acid for the next step.
- 0.02 g magnetic silica is dispersed in 50 mL anhydrous ethanol by ultrasonication, then 10.0 mL silane coupling reagent solution is added. The reaction system is kept in 40°C for 24 h. The obtained product is washed with ethanol.

The above obtained product is dispersed in 20 mL 0.2 mol·L^{-1} $CuSO_4$ solution for oscillating for 2 h. The new product is then repeatedly cleaned with deionized water to remove excess Cu^{2+}, and is kept in 60°C of vacuum overnight.

200 μL 2 mg·mL^{-1} trypsin solution is mixed with 3 mg Cu^{2+}-chelated magnetic silica in a mild shake for 2 h. The unchelated trypsin is washed with NH_4HCO_3 buffer (pH = 8.1). Trypsin-immobilized magnetic silica is kept in 20 mmol·L^{-1} NH_4HCO_3 buffer containing 0.02% sodium azide in 4°C.

In order to investigate the immobilized amount of trypsin on magnetic silica, the ultraviolet absorbancy degree of enzyme solution before and after reaction are measured and compared. The results show that the immobilized amount of enzyme on magnetic silica by metal ion chelation method is about 65 μg·mg^{-1}.

In order to investigate the stability of immobilized protein on the surface of magnetic silica, fluorescein isothiocyanate BSA (FITC-BSA) is selected as a fluorescence marker to be immobilized on the surface of the magnetic silica by metal ion chelation, and the silica is immobilized in the microchannel of the capillary under the force of the magnet. Magnetic silica is rinsed with H_2O and 20 mmol·L^{-1} NH_4HCO_3 buffer (pH = 8.0) 3 times, respectively, to remove non-specific adsorbed FITC-BSA. Next, on account that fluorescence intensity detected by laser-induced fluorescence (LIF) is proportional to the immobilized amount of FITC-BSA on magnetic silica, the immobilized stability of protein can be investigated by CE-LIF. With the external electric field, when NH_4HCO_3 buffer (pH = 8.0) is driven by electroosmosis to flow continuously in the capillary, the change of fluorescence intensity at a point in the capillary is also recorded. Finally, the spectrum of fluorescence intensity changing with the electrophoresis migration time is obtained (Figure 3.26), from which it can be seen that the fluorescence intensity does not decrease significantly within 20 min and maintains a good stability. This indicates that the protein immobilized on the surface of magnetic silica through metal ion chelation is relatively stable.

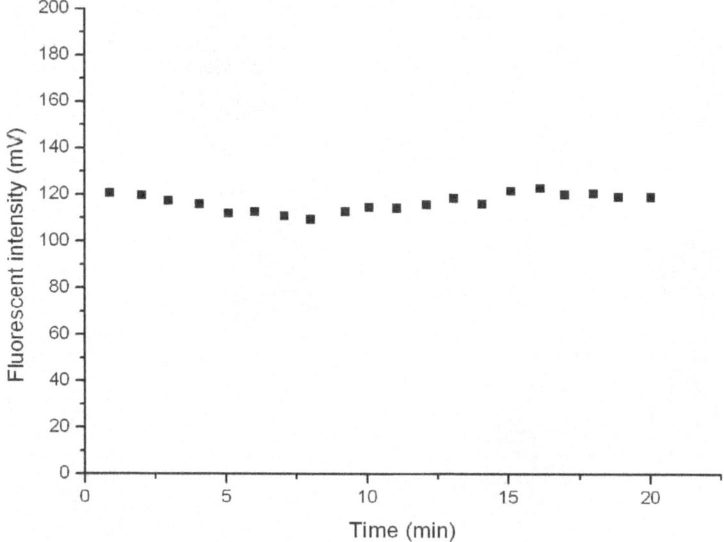

Figure 3.26: The change spectrum of fluorescence intensity of FITC-BSA marked capillary with electrophoretic migration time (emission wavelength is 473 nm, detection wavelength is 520 nm).

(2) Covalent bonding for enzyme immobilization

As shown in Figure 3.27, an amino silanating reagent (APTES) is first used to coat the magnetic silica with an amino group as follows: 3 g magnetic silica and 90 mL anhydrous toluene are put into 250 mL three-necked flask and mechanically stirred to disperse uniformity. Under the mechanical stirring, 8 mL APTES is added slowly into the dispersion, and the system is refluxed at 95°C for 4 h. After it is cooled to room temperature, the product is, respectively, washed with anhydrous toluene, water, and ethanol twice. Finally, the amino-modified magnetic silica is vacuum-dried at room temperature for 12 h.

It has been reported in the literature that glutaraldehyde (GA) can react with the primary amino group of protein quickly, which is beneficial to reduce the time required for immobilizing the protease,[54] moreover, the reaction usually occurs at a neutral pH.[55] Herein, the amino-modified magnetic silica is activated by glutaraldehyde. The reaction conditions are optimized through a series of experiments, and a buffer solution with neutral pH is used

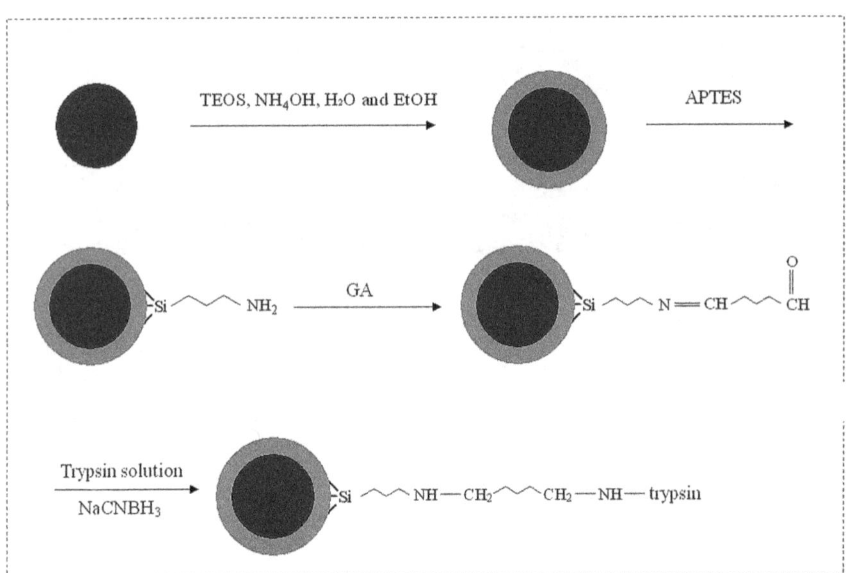

Figure 3.27: Schematic diagram of synthesis of magnetic microsphere and covalent bonding for trypsin immobilization.[53]

to activate the amino-magnetic silica as follows: 3 mg amino-modified magnetic silica is first washed by 25 mmol·L^{-1} 200 μL NH_4HCO_3 buffer (pH = 7.0) in a 1.5 mL centrifuge tube. The surface of magnetic silica is activated by aldehyde group via adding 200 μL 5% glutaraldehyde solution for 2 h under oscillation. The activated magnetic material is then washed 4 times with NH_4HCO_3 buffer (pH = 8.3). It is worth mentioning that the reason why pH value of buffer solution is increased to about 8.3. That's because the increased pH value is conducive to the nucleophilic reaction between the primary amino group of protease and the aldehyde group, according to the report of Walt and Agayn.[56] This means the increased pH can improve the immobilization efficiency of the enzyme.

Additionally, a H_2O molecule is eliminated and Schiff base is formed when the primary amino group of the protease reacts with the aldehyde group (Figure 3.27), and this reaction is reversible, resulting in a possibility of the enzyme shedding from the magnetic silica. The increased pH is conducive to the occurrence of reversible reaction, making it easier for the enzyme to fall off. In order to solve this problem, a reduction reagent ($NaBH_3CN$) is introduced to reduce the double-bond between carbon and nitrogen in the Schiff base to a stable single bond, thus inhibiting the occurrence of reversible reaction and stabilizing the enzyme on the surface of the magnetic silica within a certain pH range. The detailed process is as follows: 0.1 mg trypsin solution is dissolved in 100 μL NH_4HCO_3 buffer (pH = 8.3, containing 1% $NaBH_3CN$) and the activated magnetic silica is added, and kept under vibration for 3 h. Then the enzyme solution is removed and 200 μL NH_4HCO_3 buffer (pH = 8.3, containing 0.75% glycine and 1% $NaBH_3CN$) is added to maintain the reaction for 1 h under shaking to lock the aldehyde group which has not been bound to the protease. Finally, trypsin immobilized magnetic silica is treated with glycine solution to lock the aldehyde group that has not yet reacted with the protease, so as to avoid adsorption of targeted protein during the process of enzymatic hydrolysis.

In order to examine the immobilized amount of trypsin on the magnetic silica by covalent bonding, the ultraviolet absorbancy degree of the enzyme solution before and after the reaction is measured and compared. The results show that immobilized amount of enzyme by covalent bonding is about 70 μg·mg^{-1}, which is slightly higher than metal ion chelation (65 μg·mg^{-1}).

3.3.3. *Preparation of capillary/chip enzyme reactor based on enzyme-immobilized magnetic silica*

Magnetic silica is easy to be filled, positioned, rushed out, and refilled in capillary/chip channel due to its good dispersion in aqueous solution and strong magnetic responsiveness, and thereby greatly simplifying the preparation process of the capillary/chip enzyme reactor. The entire process is shown in Figure 3.28 (taking covalent bonding method as an example). Compared with the non-magnetic filler material, the application of magnetic silica skips the need to make any dam or plug in the microchannel to immobilize the material, and only a small magnet is needed to be placed outside the capillary/chip channel. The dispersion of enzyme-immobilized magnetic silica can flow through the capillary/chip channel in low flow rate using the impetus of the injection pump, resulting in the immobilization of enzyme-immobilized

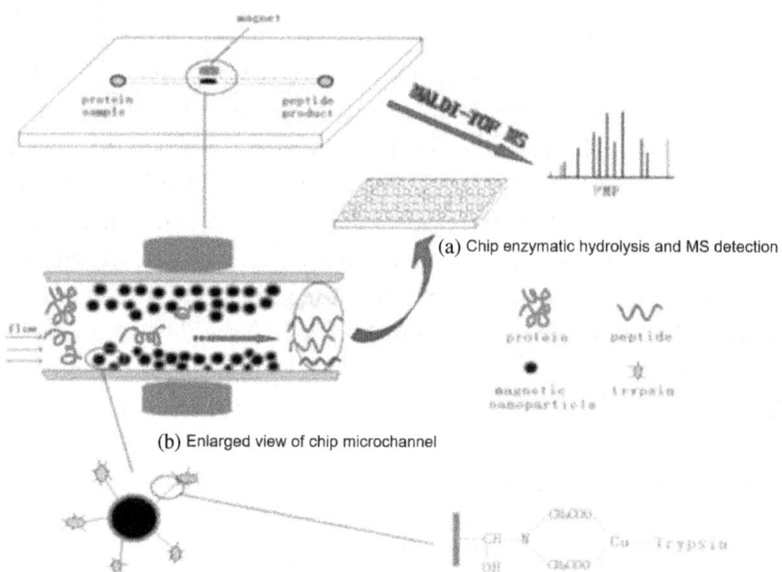

(c) Enlarged view of enzyme-immobilized magnetic silica   (d) Surface structure of enzyme-immobilized magnetic silica

Figure 3.28:   Schematic diagram of capillary/chip enzyme reactor.

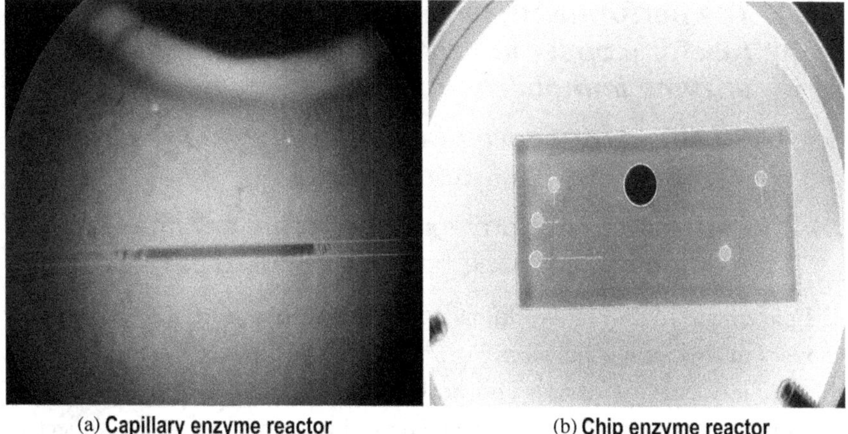

Figure 3.29: Photos of capillary/chip enzyme reactor under microscope.

magnetic silica for forming a packed bed of about 2–3 mm in the microchannel in less than 1 min. This is far less than the time in literature where commercial magnetic bead is used to form packed bed in the microchannel[4,5,57] (20–60 min). Figure 3.29 shows the packed bed photo in capillary/chip channel under a microscope. This method can be used to repeatedly fill magnetic silica in capillary/chip channel, and the packed bed is also very stable in the experiment of enzymatic hydrolysis. Once the experiment is completed, the magnetic sphere can be conveniently flushed out of the microchannel by removing the small magnet.

It is worth mentioning that the length of the entire packed bed is only 2–3 mm (Figure 3.29). This is because an excessive length will directly lead to the increase of column pressure, although the efficiency of enzymatic hydrolysis can be improved by increasing the length that increases the contact probability of protein and magnetic silica. If the resulting column pressure exceeds the magnetic attraction provided by the magnet, either the protein solution cannot flow through the packed bed, or the pump thrust pushes the entire packed bed out of the capillary/chip channel.

3.3.4. *The performance of capillary/chip enzyme reactor (copper ion chelation method for enzyme immobilization)*

3.3.4.1. *The performance of capillary enzyme reactor (copper ion chelation method)*

(1) The enzymolysis performance of capillary enzyme reactor (copper ion chelation method) for standard protein

The enzymatic hydrolysis workflow of capillary enzyme reactor prepared by copper ion chelation is shown in Figure 3.30. Protein solution is pumped into the capillary enzyme reactor and both ends are sealed for maintaining the reaction for 5 min at 50°C. The obtained products are pumped out, collected, and spotted on the MALDI plate. The acetonitrile solution containing alpha-cyano-4-hydroxyl sarcosilicic acid (CHCA) is then spotted on the sample and dried before being sent to the MALDI mass spectrometer for analysis. In order to examine the enzymolysis performance of the capillary enzyme reactor for the protein, 0.5 μL 0.2 μg·μL^{-1} BSA (a protein with multiple enzyme cutting sites, the molecular weight is about 66, 000) solution is injected into

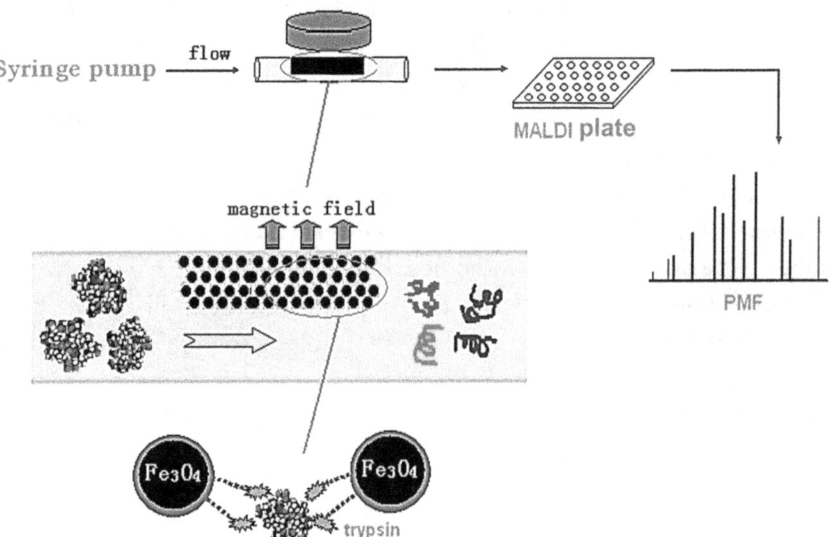

Figure 3.30: Schematic diagram of proteolysis by capillary enzyme reactor and MALDI-TOF MS detection.

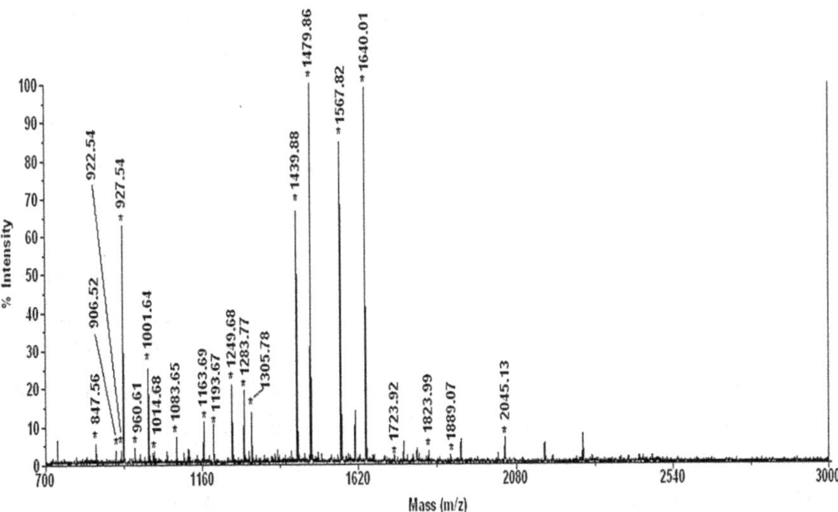

Figure 3.31: MALDI-TOF MS spectrum of peptides from BSA digested by capillary enzyme reactor. The BSA protein is dissolved in NH_4HCO_3 buffer (25 mmol·L^{-1}, pH=8.3) for forming a solution with final concentration of 0.2 $\mu g \cdot \mu L^{-1}$.

capillary enzyme reactor and digested under 50°C for a period of time, the peptides solution is collected and spotted on MALDI plate, followed by being covered by CHCA matrix and analyzed by MS. The obtained mass spectrum is shown in Figure 3.31. A total of 21 peptides matching BSA are identified, and the corresponding sequence coverage is up to 32%.

SDS gel electrophoresis is performed to test whether the protein has been adequately hydrolyzed within 5 min. The result shows that no obvious protein band can be observed on the SDS tape after 5 min of enzymatic hydrolysis, indicating that the protein has been completely hydrolyzed. In addition, this result is comparable to the traditional in-solution enzymatic hydrolysis (12 h, 22 peptides matching BSA are identified, corresponding sequence coverage is 35%).

(2) Reproducibility of capillary enzyme reactor

The capillary enzyme reactor is prepared by filling the capillary with enzyme-immobilized magnetic silica by copper ion chelation method, so it has the advantages that other enzyme reactors do not have, that is, the capillary enzyme reactor can be easily replaced and regenerated.

In terms of the replacement of capillary enzyme reactor, the magnetic silica can be flushed out by pumping NH_4HCO_3 buffer through capillary microchannel on the premise of removing the magnet. Then the capillary microchannel is washed with NH_4HCO_3 buffer and the magnet is placed outside the capillary microchannel, the new magnetic silica dispersion fluid can flow through the microchannel with the help of a pump, and thereby be immobilized in the microchannel to form 2–3 mm of packed bed.

In terms of the enzyme-immobilized magnetic silica by copper ion chelation method, the capillary enzyme reactor can also be regenerated. The flushed-out magnetic silica is soaked in EDTA solution for 2–4 h. Then $CuSO_4$ solution is used to soak magnetic silica under slight oscillation to reload copper ions, and the immersion time is 1–3 h. The Cu^{2+}-loaded magnetic silica is then immersed in trypsin solution to load the enzyme, the unchelated trypsin is washed with NH_4HCO_3 buffer (pH = 8.1), and the enzyme immobilized magnetic silica is prepared again. The above process of replacing the enzyme reactor is repeated to realize the regeneration of the capillary enzyme reactor with simple steps, short time, and low cost. Figure 3.32 shows the obtained MS spectra of BSA digest before and after regeneration of capillary enzyme reactor. It can be seen that the two MS spectra are very similar, indicating that the regeneration process does not affect the repeatability of proteolysis.

3.3.4.2. *The performance of chip enzyme reactor (copper ion chelation method)*

(1) The temperature and time of chip enzyme reactor (copper ion chelation method) for enzymatic hydrolysis

When digesting protein by trypsin, temperature generally is set at 37°C. Higher temperature will effectively destroy the structure of protein, increase the rate of enzymatic hydrolysis, and promote enzymatic hydrolysis. Cyc is enzymatically hydrolyzed in a chip enzyme reactor at different temperatures to investigate the effect of varying temperatures . As shown in Figure 3.33(a), when the temperature is lower than 50°C, the sequence coverage of Cyc slowly rises with the increase of temperature; while when the temperature is higher than 50°C, the sequence coverage has a downward trend. Therefore, 50°C is considered as the best temperature for chip enzyme reactor.

One of the problems in traditional in-solution enzymatic hydrolysis is that sufficient time is needed in order to obtain sufficient peptides. Generally,

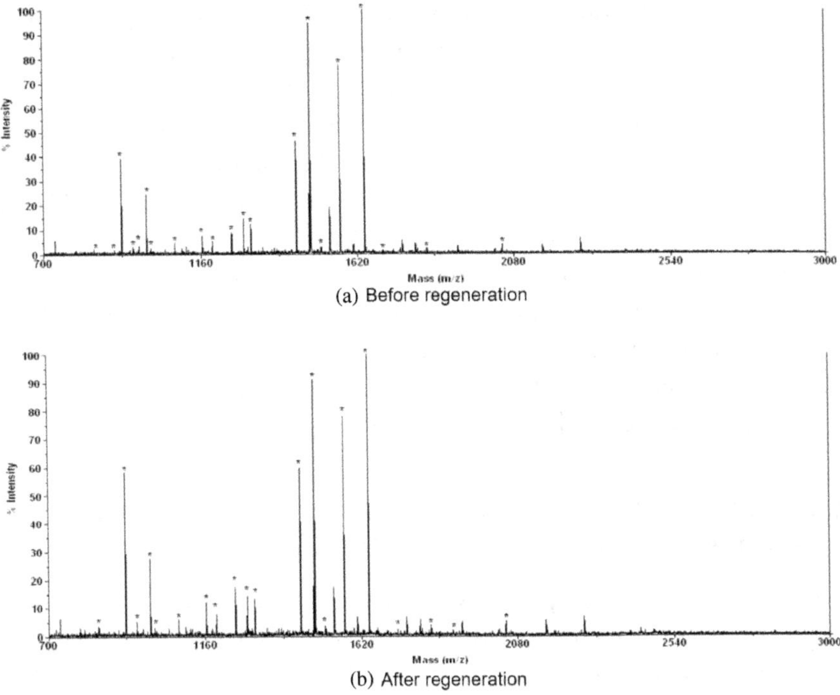

Figure 3.32: The MALDI-TOF MS spectra of BSA digest before and after regeneration of capillary enzyme reactor. (BSA is digested at 50°C for 5 min. "*" represents the peptides matching BSA.)

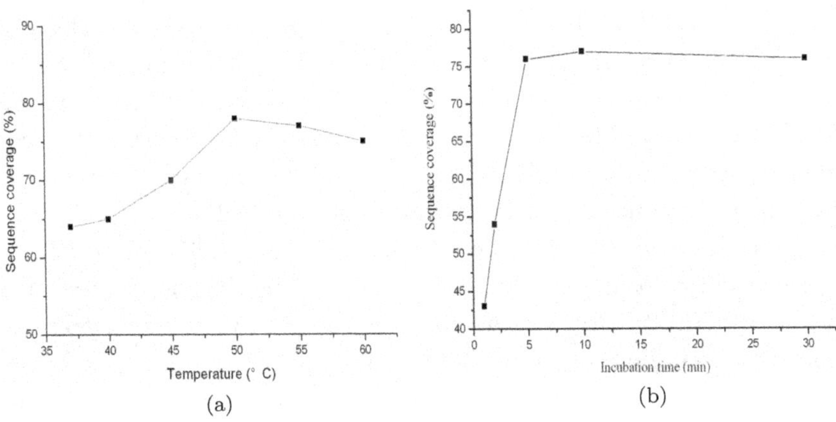

Figure 3.33: The effects of temperature and time on the enzymatic hydrolysis performance of chip enzyme reactor (copper ion chelation method). Cyc as sample and MALDI-TOF MS as detector.

the time for in-solution enzymatic hydrolysis is 4–24 h. The longer the time, the more sufficient the proteolysis, the more peptides are produced, and the stronger the MS signal. However, the decrease of time of enzymatic hydrolysis means more proteins can be identified in a shorter time, having great practical significance. Therefore, the effect of time on the efficiency of enzymatic hydrolysis is investigated further. Figure 3.33(b) shows the MALDI MS spectra of peptides from Cyc digested by chip enzyme reactor through different times of enzymatic hydrolysis at 50°C. When the time increases from 1 min to 5 min, the sequence coverage of Cyc increases from 43% to 77%; when the time continues to increase, the sequence coverage does not change significantly. It can be concluded that 5 min of enzymatic hydrolysis is sufficient when using chip enzyme reactor.

(2) The enzymolysis performance of chip enzyme reactor (copper ion chelation method) for standard protein

Two common standard proteins (Cyc, molecular weight 12384; BSA, molecular weight 66000) are used to investigate the efficiency of enzymatic hydrolysis of chip enzyme reactor prepared by copper ion chelation. For comparison, the two proteins are also enzymatically hydrolyzed by 12 h of traditional in-solution digestion. The enzymatic hydrolysates are collected and analyzed by MALDI-TOF/ TOF MS.

Figure 3.34 shows the obtained peptide mass fingerprinting (PMF) after enzymatic hydrolysis of Cyc and BSA in a chip enzyme reactor. Detailed results are listed in Tables 3.3 and 3.4. After enzymatically hydrolyzed by chip enzyme reactor and analyzed by MS, 81 amino acids in Cyc (104 in total) and 131 amino acids in BSA (583 in total) are identified. The sequence coverage of Cyc reaches 77%, while that of BSA reaches 21%. The results of enzymatic hydrolysis are comparable to the traditional method (Table 3.3).

In order to further test the stability and repeatability of enzymatic hydrolysis by chip enzyme reactor, one and the same chip enzyme reactor is used to conduct the enzymatic hydrolysis reaction of Cyc successively 6 times under the optimized conditions (50°C, 5 min). The enzymatic hydrolysates are collected for MALDI analysis (Figure 3.35). Between each two experiments of enzymatic hydrolysis, the chip channel is washed with buffer solution for more than 5 min to eliminate the memory effect. As shown in Figure 3.35,

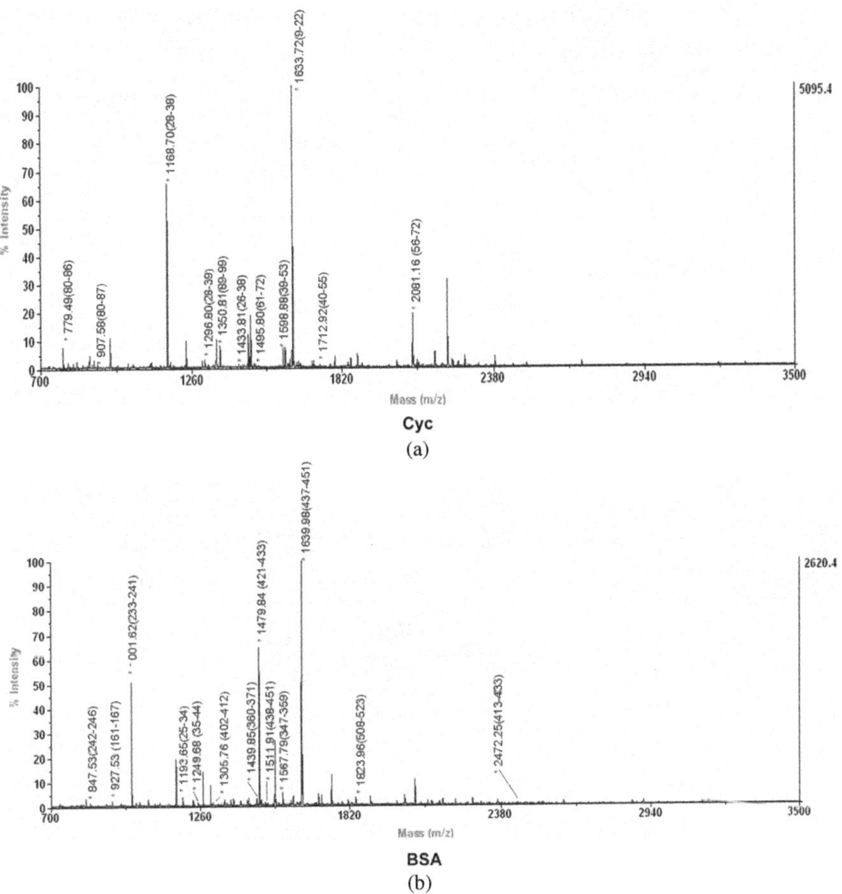

(a)

(b)

Figure 3.34: PMF of peptides from Cyc and BSA digested for 5 min by chip enzyme reactor at 50°C. The protein is dissolved in NH_4HCO_3 buffer (20 mmol·L^{-1}, pH = 8.0) for forming a solution with final concentration of 0.2 0 μg·μL^{-1}.

the first four experiments maintain good stability, and the obtained sequence coverage basically maintains at about 76% (RSD = 1.7%), while the sequence coverage is reduced to 59% at the fifth experiment, indicating that the protease has began to lose activity.

The chip enzyme reactor (copper ion chelation method) is the same as capillary enzyme reactor in this section. They are both prepared by filling enzyme-immobilized magnetic silica in the microchannel via copper ion

Table 3.3: MALDI-TOF MS results after enzymatic hydrolysis of Cyc and BSA in a chip enzyme reactor (copper ion chelation method).

	Cyc		BSA	
	Microreactor	**Solution**	**Microreactor**	**Solution**
Detected amino acid	81	80	130	253
Sequence coverage (%)	77	76	21	41
Time of enzymatic hydrolysis	5 min	≥6 h	5 min	≥6 h
Matched peptide	13	14	13	24
Accession	P00004	P00004	P02769	P02769
Molecule weight (MW)	11694.1	11694.1	69248.4	69248.4

Table 3.4: Fragment information detected by MALDI-TOF MS after enzymatic hydrolysis of Cyc and BSA in a chip enzyme reactor (copper ion chelation method).

Cyc		BSA	
9–22	IFVQKCAQCHTVEK	25–34	DTHKSEIAHR
26–38	HKTGPNLHGLFGR	35–44	FKDLGEEHFK
28–38	TGPNLHGLFGR	161–167	YLYEIAR
28–39	TGPNLHGLFGRK	233–241	ALKAWSVAR
39–53	KTGQAPGFTYTDANK	242–248	LSQKFPK
40–53	TGQAPGFTYTDANK	347–359	DAFLGSFLYEYSR
40–55	TGQAPGFTYTDANKNK	360–371	RHPEYAVSVLLR
56–72	GITWKEETLMEYLENPK	402–412	HLVDEPQNLIK
61–72	EETLMEYLENPK	413–433	QLINVCRDQFEKLGEYGFQNA
61–73	EETLMEYLENPKK	421–433	LGEYGFQNALIVR
80–86	MIFAGIK	436–451	VPQVSTPTLVEVSR
80–87	MIFAGIKK	437–451	KVPQVSTPTLVEVSR
89–99	TEREDLIAYLK	508–523	RPCFSALTPDETYVPK

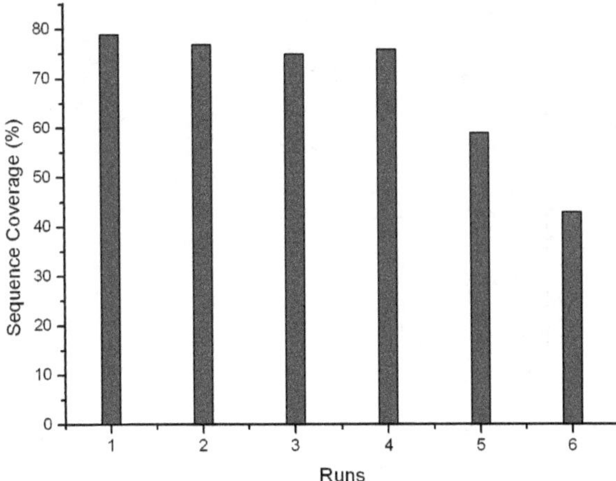

Figure 3.35: The obtained peptide coverage by MALDI-TOF MS after enzymatic hydroly-sis of Cyc 6 successive times in chip enzyme reactor.

chelation. As a result, the chip enzyme reactor can realize replacement and regeneration easily the same way described for capillary enzyme reactor.

3.3.4.3. *The performance of chip enzyme reactor (covalent bonding method)*

Figure 3.36 shows the sequence coverage after enzymatic hydrolysis of Cyc in chip enzyme reactor with different times of enzymatic hydrolysis when adopt-ing 50°C as the temperature of enzymatic hydrolysis. When the time increases from 1 min to 5 min, the sequence coverage increases from 51% to 77%; when the time continues to increase, the sequence coverage does not change signifi-cantly. It can be concluded that 5 min is sufficient for chip enzyme reactor.

Detailed results are listed in Table 3.5. After enzymatically hydrolyzed by chip enzyme reactor (covalent bonding method) and analyzed by MS, a total of 81 amino acids (104 in total) in Cyc are identified, and the sequence cover-age reaches 77%. The results of enzymatic hydrolysis is comparable to the traditional method (Table 3.5).

In order to test the repeatability of enzymatic hydrolysis by this chip enzyme reactor, one and the same chip enzyme reactor is used to conduct the enzymatic hydrolysis reaction of Cyc successively 5 times under the optimized conditions (50°C, 5 min). The enzymatic hydrolysates are collected for

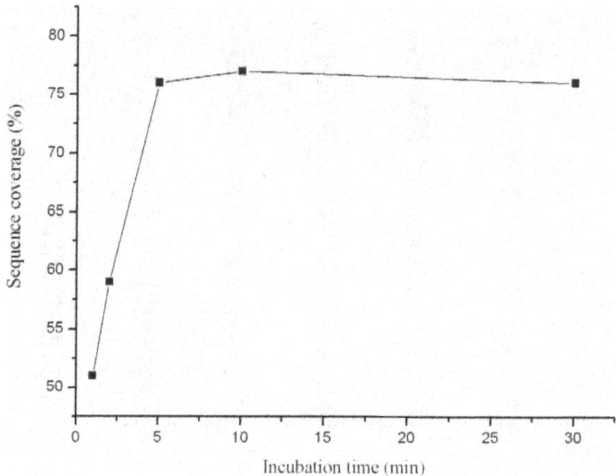

Figure 3.36: The effects of time on the enzymatic hydrolysis performance of chip enzyme reactor (covalent bonding method). Cyc as sample and MALDI-TOF MS as detector (5 times of repeated experiment for every incubation time).

MALDI analysis. Between each two experiments of enzymatic hydrolysis, the chip channel is washed with buffer solution for more than 5 min to eliminate the memory effect. The obtained sequence coverage basically maintains at about 76% (RSD = 1.4%).

In order to further test the repeatability of enzymatic hydrolysis of this chip enzyme reactor before and after replacement, three consecutive enzymatic hydrolysis experiments for Cyc are carried out by replacing the enzyme-immobilized magnetic silica. The resulting three MS spectra are very similar, indicating that the replacement of chip enzyme reactor does not affect repeatability of proteolysis. This is also confirmed by the results of library search, and the sequence coverage of Cyc remains basically around 77% (RSD = 2.1%).

The stability of magnetic silica immobilized by covalent bonding method is also tested. They are kept in 4°C refrigerator for 1 month and used for the preparation of chip enzyme reactor, and there is no significant difference in digestion efficiency compared to the new reactor. The enzyme activity is not significantly reduced.

Table 3.5: The obtained information after enzymatic hydrolysis of Cyc by chip enzyme reactor (covalent bonding method) and traditional method.

	Microreactor		Solution
Detected amino acids	81		80
Sequence coverage (%)	77		76
Time of enzymatic hydrolysis	5 min		12 h
Matched peptides	13		14
Accession	P00004		P00004
Molecule weight (MW)	11694.1		11694.1
Detected peptide	9–22	IFVQKCAQCHTVEK	
	26–38	HKTGPNLHGLFGR	
	28–38	TGPNLHGLFGR	
	28–39	TGPNLHGLFGRK	
	39–53	KTGQAPGFTYTDANK	
	40–53	TGQAPGFTYTDANK	
	40–55	TGQAPGFTYTDANKNK	
	56–72	GITWKEETLMEYLENPK	
	61–72	EETLMEYLENPK	
	61–73	EETLMEYLENPKK	
	80–86	MIFAGIK	
	80–87	MIFAGIKK	
	89–99	TEREDLIAYLK	

3.3.5. *The performance of chip enzyme reactor for enzymatic hydrolysis of practical bio-sample*

3.3.5.1. *The performance of chip enzyme reactor (copper ion chelation method) for enzymatic hydrolysis of practical bio-sample*

In order to investigate the application of chip enzyme reactor (copper ion chelation method) in proteome analysis based on "middle-down", the chip enzyme reactor is used to digest the proteins which are extracted from rat liver

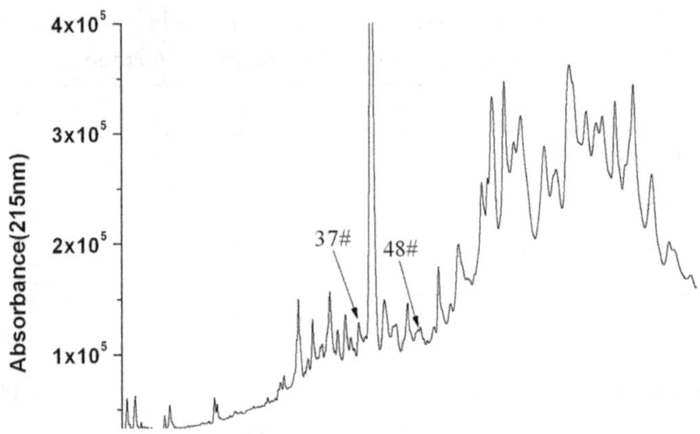

Figure 3.37: RPLC separation chromatogram of proteins extracted from rat liver tissue.

tissue and have been separated by RPLC. The RPLC spectrum of protein extraction (20 μg) is shown in Figure 3.37. One fraction is collected per minute among the 20–110 min of RPLC gradient. Every fraction is lyophilized and redissolved in 2 μL NH_4CO_3 buffer (pH = 8.0), and then the fraction solution is injected into chip enzyme reactor for 5 min of incubation at 50°C. The peptides are collected for MALDI-TOF MS/MS analysis. The collected fraction between 67 min and 68 min is taken as a sample (No. 48#, as indicated by arrows in Figure 3.37). The result shows that a total of seven proteins above score 59 are identified (Table 3.6).

3.3.5.2. *The performance of chip enzyme reactor (covalent bonding method) for enzymatic hydrolysis of practical bio-sample*

Similar to the test method for chip enzyme reactor (copper ion chelation method), the collected fraction between 56 min and 57 min is taken as a sample (No. 37#, as indicated by arrows in Figure 3.37) to test the performance of chip enzyme reactor (covalent bonding method). The ionic peaks of multiple peptides can be observed in the MALDI spectra of enzymatic hydrolysate (Figure 3.38), indicating that the protein mixture in the fraction has been effectively hydrolyzed in the chip enzyme reactor. The result shows that a total of five proteins above score 59 are identified (Table 3.7).

Table 3.6: Protein information of No. 48# fraction in Figure 3.37 (arrow) digested by chip enzyme reactor (copper ion chelation method) and identified by MALDI-TOF MS.

Name of Protein	Accession	Molecule Weight (MW)	Isoelectric Point (pI)	Peptide Counting	Score
3-hydroxy-3-methylglutaryl-Coenzyme A synthase 2 [Rattus norvegicus]	gi|54035469	56849.6	8.86	3	147
TH2A histone [Rattus norvegicus]	gi|57354	14275	11.02	4	114
3-hydroxyisobutyrate dehydrogenase [Rattus norvegicus]	gi|83977457	35279.6	8.73	3	112
sterol carrier protein — rat (fragment)	gi|2119443	10462.5	8.03	1	97
Fatty acid binding protein 1 [Rattus norvegicus]	gi|56541250	14263.3	7.79	3	86
gametogenetin-binding protein 1 [Rattus norvegicus]	gi|46410145	40828.3	5.78	7	65
alpha-globin [Rattus sp.]	gi|30027750	9345.7	6.49	2	64

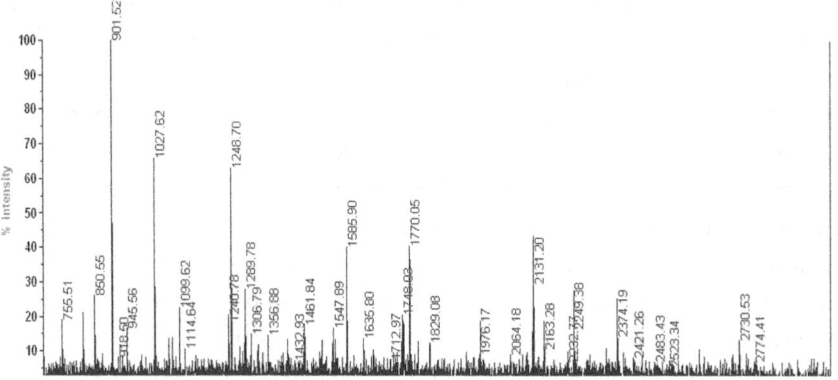

Figure 3.38: MALDI-TOF MS spectrum of the No. 37# fraction in Figure 3.37 digested by chip enzyme reactor (covalent bonding method).

Table 3.7: Protein information of No. 37# fraction in Figure 3.37 (arrow) digested by chip enzyme reactor (covalent bonding method) and identified by MALDI-TOF MS.

Name of Protein	Accession	Molecule Weight (MW)	Isoelectric Point (pI)	Peptide Counting	Score
unnamed protein product [Rattus norvegicus]	gi|55544	41844.4	8.09	3	83
ATP synthase D chain, mitochondrial	gi|1352051	18751.6	6.17	3	77
liver fatty acid binding protein p14	gi|204074	11410.8	6.74	1	76
Diazepam binding inhibitor [Rattus norvegicus]	gi|54261671	10021.2	8.78	2	69
Catalase [Rattus norvegicus]	gi|51980301	59719.5	7.07	3	67

3.3.6. *Conclusion*

This section introduces the application of enzyme-immobilized magnetic silica in capillary/chip enzymatic reactor. The enzyme-immobilized magnetic silica synthesized by metal ion chelation method or covalent bonding method is, respectively, immobilized inside the capillary or chip channel. The protein solution is pumped through the capillary or chip channel, resulting in rapid enzymatic hydrolysis. The collected enzymatic hydrolysates are analyzed and identified by MALDI MS. The reaction conditions such as time and temperature of enzymatic hydrolysis are optimized. The efficiency of enzymatic hydrolysis of capillary/chip enzyme reactor is investigated using standard proteins and compared with conventional in-solution enzymatic hydrolysis. The results show that the capillary/chip enzyme reactor prepared by two different methods of immobilizing enzyme could achieve the same or better enzymatic hydrolysis effect within 5 min than the 12-h conventional in-solution enzymatic hydrolysis. Due to the magnetic property of the material, only the magnet outside the capillary/chip channel needs to be removed once the immobilized enzyme loses its activity. The magnetic silica in the capillary/chip channel can be flushed out with liquid, and a new one can be introduced. The replacement of the capillary/chip enzyme reactor can be achieved easily. The method is easy to operate and low in cost, and well solves the shortcomings of the capillary/chip enzyme reactor which has only been used once in the previous literature. For metal ion chelation method, the regeneration of capillary/chip enzyme reactor can also be achieved by removing copper ions from the surface of magnetic silica with EDTA and reintroducing new copper ions and proteases.

3.4. On-plate enzymatic hydrolysis based on enzyme-immobilized magnetic material

3.4.1. Basic principle and development process

As one of the conceptual methods of great concern today, proteolytic technology on MS target plate has obvious advantages in accelerating enzymatic hydrolysis and reducing the transfer of samples before MS detection.[3,58–60] Nelson *et al.*[61,62] immobilized trypsin on a MALDI target plate and successfully identified 10^{-12} mole level of proteins with an enzymatic hydrolysis time of less than 30 min. Reilly's team[63 65] found that bacterial proteins were able to degrade rapidly and efficiently on the enzyme-attached MALDI target plate. Professor Lubman[66] reported a protein detection method combined with capillary reversed-phase liquid chromatography and proteolytic technology on MS target plate to study the proteins in human lung cancer cells. Compared to conventional method, their method is simple and effective in reducing sample loss. However, there are still some defects in immobilizing biological enzyme or other materials that can selectively adsorb proteins. The common on-plate enzymatic hydrolysis method in the literature is to directly add the free enzyme solution into the protein solution on the target plate,[67] or spot the enzyme solution on the target plate first and then add the protein solution.[63] After a period of enzymatic hydrolysis, the reaction is quenched by the addition of an acid such as 1% TFA. Since the volume of sample that can be accommodated on the plate is only about 1 μL, the mass transfer distance is small, so that both the enzyme and the protein solution can reach a higher concentration in a small space, thereby increasing the contact probability of the two. It speeds up the reaction and shortens the time of enzymatic hydrolysis (generally 15–30 min). However, this type of method still belongs to the scope of free in-solution enzymatic hydrolysis, and thus cannot overcome the problem of self-hydrolysis of the enzyme. In the obtained MS spectra by the above method, the peptide peaks are often interfered by the peaks produced by self-hydrolysis of enzyme, which is particularly evident when the concentration of protein sample itself is low. Immobilizing enzyme on the target plate to restrict its free migration can effectively solve this self-hydrolysis problem, and also using a higher enzyme/substrate ratio can further shorten the time of enzymatic hydrolysis.[58] Even so, since the enzyme is directly immobilized on the target plate, it is almost impossible to regenerate the enzymatic activity when the enzyme is inactivated, and the enzyme can only be immobilized on the new target plate. This not only greatly increases the experimental cost, but also reduces the repeatability of the enzymatic hydrolysis experiment. In addition, the introduction of solid particles on the MALDI target plate causes the substrate to be crystallized unevenly, leading to an unstable MS signal. Also, the nanomaterials on the target plate may also contaminate the ion source after entering the mass spectrometer. These are all issues that need to be resolved and improved.

In this section, the on-plate enzymatic hydrolysis technology based on enzyme-immobilized magnetic material will be described. The technology utilizes amino-magnetic microsphere and on-MALDI-plate enzymatic hydrolysis technology to identify proteins.

3.4.2. Preparation, characterization and enzyme immobilization of amino-magnetic microsphere

The amino-magnetic microsphere is synthesized by using $FeCl_3 \cdot 6H_2O$ and 1, 2-hexanediamine via one-step hydrothermal method, and then the enzyme is immobilized on its surface by activation of glutaraldehyde (as seen in Section 3.2).

3.4.3. Exploration of on-plate enzymatic hydrolysis method based on amino-magnetic microsphere

Figure 3.39 shows the schematic diagram of on-plate enzymatic hydrolysis process based on enzyme-immobilized amino-magnetic microsphere. The detailed process is as follows: the enzyme-immobilized amino-magnetic

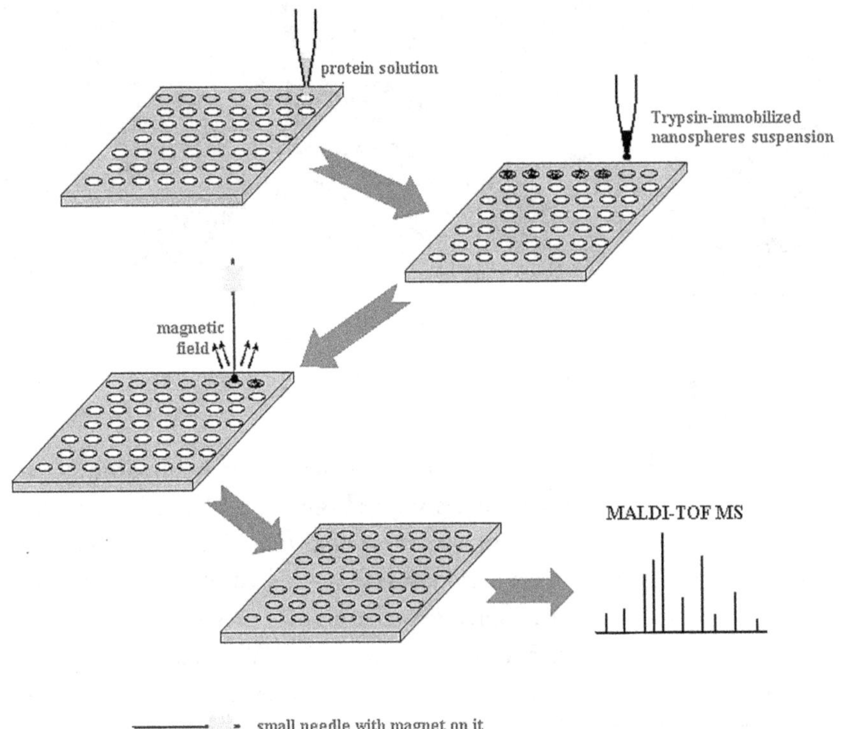

Figure 3.39: Schematic diagram of on-plate enzymatic hydrolysis process based on enzyme-immobilized amino-magnetic microsphere.

microsphere is dispersed in 25 mmol·L^{-1} NH$_4$HCO$_3$ solution (pH = 8.0) to achieve a dispersion solution with a final concentration of 1.5 μg·μL^{-1}. Thereafter, the proteolytic process is performed on a 192-well stainless steel MALDI target plate from Applied Biosystems. 0.8 μL 50 ng·μL^{-1} protein solution (25 mmol·L^{-1} NH$_4$HCO$_3$) is dropped into each well, and then 0.4 μL magnetic sphere suspension is added. Then the target plate is placed at 50°C for 5 min in a damp and closed container (such as enzyme analyzer). Thereafter, a magnetized iron needle is inserted into the solution of the well to separate the magnetic spheres sequentially from the respective well. Finally, 0.5 μL CHCA (solvent is the mixture solution of ACN/H$_2$O at 1:1, containing 0.1% TFA) solution is added to each sample. After crystallization in the air, MS analysis is performed.

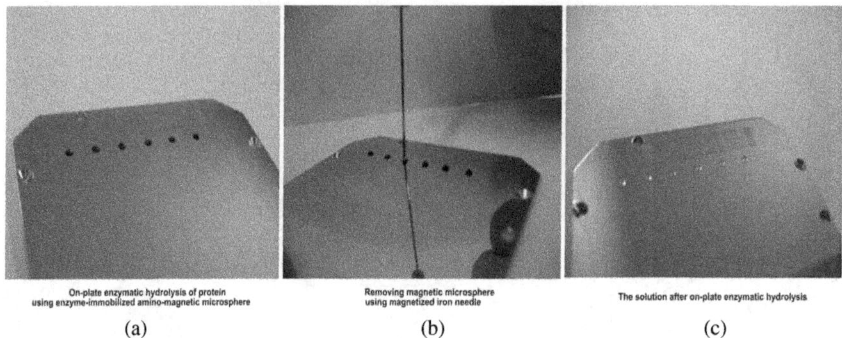

(a) (b) (c)

Figure 3.40: Photographs of the enzymatic hydrolysis process.

Figure 3.40 shows a photograph of the actual enzymatic hydrolysis process. Firstly, 1 μL protein solution is spotted onto the MALDI target plate (Figure 3.40(a)), followed by the dispersion of enzyme-immobilized amino-magnetic microsphere. After a period of enzymatic hydrolysis, the magnetic material is separated by a magnetized steel needle to stop the enzymatic hydrolysis reaction (Figure 3.40(b)). Figure 3.40(c) shows a photograph of the MALDI target after the magnetic material is removed, it can be seen from this that only a clear solution remains on the target plate, and no excess solid residue affects the crystallization and contaminates MS.

The on-plate enzymatic hydrolysis method based on amino-magnetic microsphere not only exerts the advantages of immobilized enzyme technology, improves the efficiency of enzymatic hydrolysis, accelerates the rate of enzymatic hydrolysis, and avoids the interference of the self-hydrolysis of enzyme, but also uses the magnetic property of the material itself to avoid contamination of the ion source grating of MS. In addition, the preparation of the amino-magnetic microsphere is inexpensive and simple, and the target plate can be reused after routine cleaning, thus greatly reducing experimental cost.

3.4.4. *Condition optimization of on-plate enzymatic hydrolysis based on amino-magnetic microsphere*

In order to achieve the best enzymatic hydrolysis, the temperature and time most likely to affect the enzymatic hydrolysis during the experiment are optimized. Two standard proteins are used in the optimization process: MYO and Cyc. Cyc is a protein that is relatively easily used to conduct in-solution enzymatic hydrolysis, while MTO is a globular protein, and cannot be easily used to conduct effective enzymatic hydrolysis without denaturation and structural change.

3.4.4.1. *Optimization of temperature*

The 3D structure of protein is one of the important factors that affects proteolysis.[68,69] Many proteins have a compact spherical structure and are difficult to be enzymatically hydrolyzed, so that efficient analysis cannot be performed. These proteins often play an important role in biological processes. The method of chemical denaturation can be used to change the 3D structure of the protein. In addition, high temperature can also effectively destroy protein structure and promote enzymatic hydrolysis.[70] The thermal denaturation process of the protein does not require additional sample processing and protein denaturants or solubilizers, so it is easy to handle. It is conceivable that high-temperature enzymatic hydrolysis can increase the rate of enzymatic hydrolysis, allowing complete proteolysis to produce sufficient peptides. Therefore, trypsin is required to exhibit stable activity at high temperature when considering the advantages of the protein thermal denaturation method to develop a new rapid enzymatic hydrolysis technique.

The proteolysis is carried out at different temperatures using MYO and Cyc as samples to investigate the effect of temperature on proteolysis. The proteolytic products on the plate are subjected to MALDI-TOF-MS analysis to obtain amino acid sequence coverage of the protein which is used as an evaluation index for the effect on proteolysis. 0.8 μL protein solution (25 mmol·L^{-1} NH$_4$HCO$_3$) and 0.4 μL enzyme-immobilized magnetic material are sequentially spotted on a MALDI stainless steel target plate, and kept at 18°C, 37°C, 50°C, and 60°C in a closed and humid environment for 5 min. After the enzymatic hydrolysis is completed, the MALDI target is taken out from the closed and humid environment, and the magnetic material is removed by a magnetic needle. After the sample is dried, 0.5 μL CHCA is covered on the proteolytic product to form a co-crystal for MALDI-TOF-MS analysis.

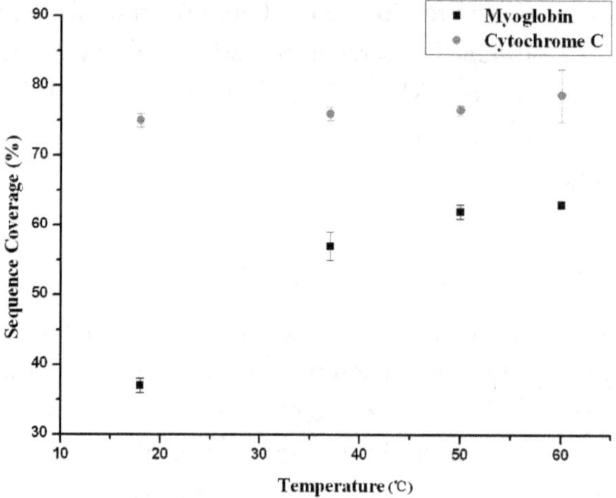

Figure 3.41: Changes in peptide coverage with temperature (18°C, 37°C, 50°C, 60°C) when adopting on-plate enzymatic hydrolysis based on amino-magnetic microsphere for MYO and Cyc. (3 times of repeated experiment for every temperature and 5 min of reaction time.)

The common temperature of adopting trypsin to enzymatic hydrolysis is 37°C. For the protein such as MYO which has a spherical structure and is difficult to be enzymatically digested in the conventional in-solution method, the efficiency of enzymatic hydrolysis significantly changes with temperature. When the temperature rises from 18°C to 50°C, the amino acid sequence coverage increases from 37% to 63%; when the temperature continues to increase to 60°C, the sequence coverage remains basically unchanged (Figure 3.41, black dot). For Cyc which is relatively easy to digest, the efficiency of enzymatic hydrolysis is basically not affected by temperature, and the amino acid sequence coverage is basically maintained at about 76% (Figure 3.41, red dot). The above experiments show that proteolysis on the MALDI target plate at high temperature improves the efficiency of enzymatic hydrolysis, which is beneficial to the reliable identification of the protein. Although proteolysis at high temperature may make trypsin unstable and increase the possibility of self-degradation, the effect of high temperature is alleviated by the use of immobilized trypsin. The results show that no self-degradation peak of trypsin is found in the MALDI spectrum. The enzymatic hydrolysis of trypsin at a high temperature for 5 min provides a higher efficiency of enzymatic hydrolysis compared to the proteolytic reaction at 37°C, so the temperature is preferably 50°C.

3.4.4.2. *Optimization of time*

The time optimization adopts the similar method described in Section 3.3.4.2. The temperature of enzymatic hydrolysis is set at 50°C, and the time is selected from several nodes between 0.5 min and 10 min. After the sample is naturally dried, 0.5 μL CHCA matrix is added for MALDI-TOF-MS analysis. The time control of all reactions is achieved by separating the magnetic spheres with a magnetic needle.

Figure 3.42 shows the effect of time on the proteolysis. When the time increses from 0.5 min to 5 min, the amino acid sequence coverage of MYO increases from 39% to 73%, and Cyc increases from 63% to 77%. However, when the time increases to 10 min, the efficiency of enzymatic hydrolysis of these two proteins do not change significantly compared to the results of 5 min, so it can be concluded that 5 min is enough for the on-plate enzymatic hydrolysis.

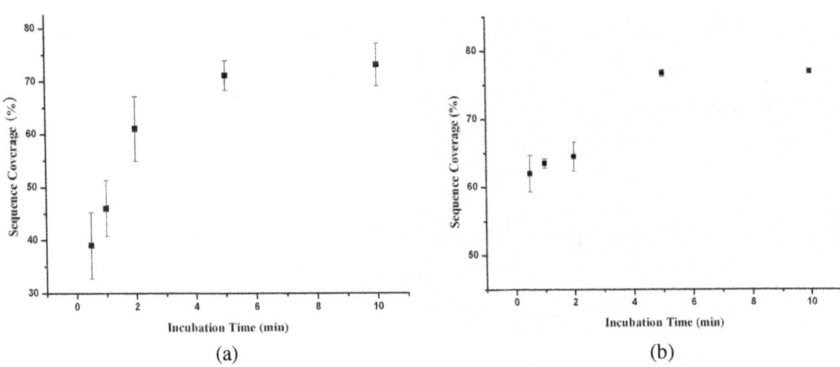

(a)   (b)

Figure 3.42: Changes in peptide coverage with time (30 s, 1 min, 2 min, 5 min, and 10 min) when adopting on-plate enzymatic hydrolysis based on amino-magnetic microsphere for MYO and Cyc. (4 times of repeated experiment for every incubation time and 50°C as reaction temperature).

3.4.5. *Performance of on-plate enzymatic hydrolysis based on enzyme-immobilized amino-magnetic microsphere*

Taking the two proteins of MYO and Cyc as samples, the proteolysis performance of on-plate method and in-solution method is compared. Figure 3.43 shows the obtained MS spectra of adopting on-plate method. The peptide peaks belonging to MYO and Cyc are indicated by "m" and "c", respectively. The detailed results are listed in Table 3.8. A total of four replicates are performed. The amino acid sequence coverage of MYO is in the range of 69–73%, and the Cyc is in the range of 77–83%. These results are similar to or even higher than that of the in-solution method.

The mixed protein solution containing Cyc and BSA with a concentration ratio of 2:1 is further used to study on-plate enzymatic hydrolysis of the mixed protein solution. Cyc is a protein that is easier to digest, and BSA is difficult to digest due to its large molecular weight and disulfide bond in its structure. In general, the more easily digested protein has a clear advantage under the same condition. The large number of produced peptides may interfere with another protein with lower intensity, and result in false positive result during MS library search. 0.8 μL mixed protein solution is spotted onto the MALDI target plate, 0.4 μL enzyme-immobilized magnetic suspension ($\sim$1.5 μg·μL^{-1}) is then added, and the temperature is kept at 50°C for 5 min. Next, the material is taken out by the magnetic needle. Finally, 0.5 μL CHCA matrix is added for MS identification. The obtained MS spectrum is shown in Figure 3.44, and the peaks of these two proteins are indicated by symbols. It can be seen that BSA and Cyc are efficiently hydrolyzed at the same tine on the target plate under this condition.

3.4.6. *Practical application of on-plate enzymatic hydrolysis based on enzyme-immobilized amino-magnetic microsphere*

Over the past decades, "middle-down" technology as an emerging proteomics research route has proven uniquely advantageous in studying some characteristics of protein,[71,72] such as the relative molecular weight, isoelectric point, hydrophobicity, and binding properties, since this technology is based on the separation of complete protein in a complex mixture. This also poses a serious problem, that is, how to realize a fast and efficient enzymatic hydrolysis

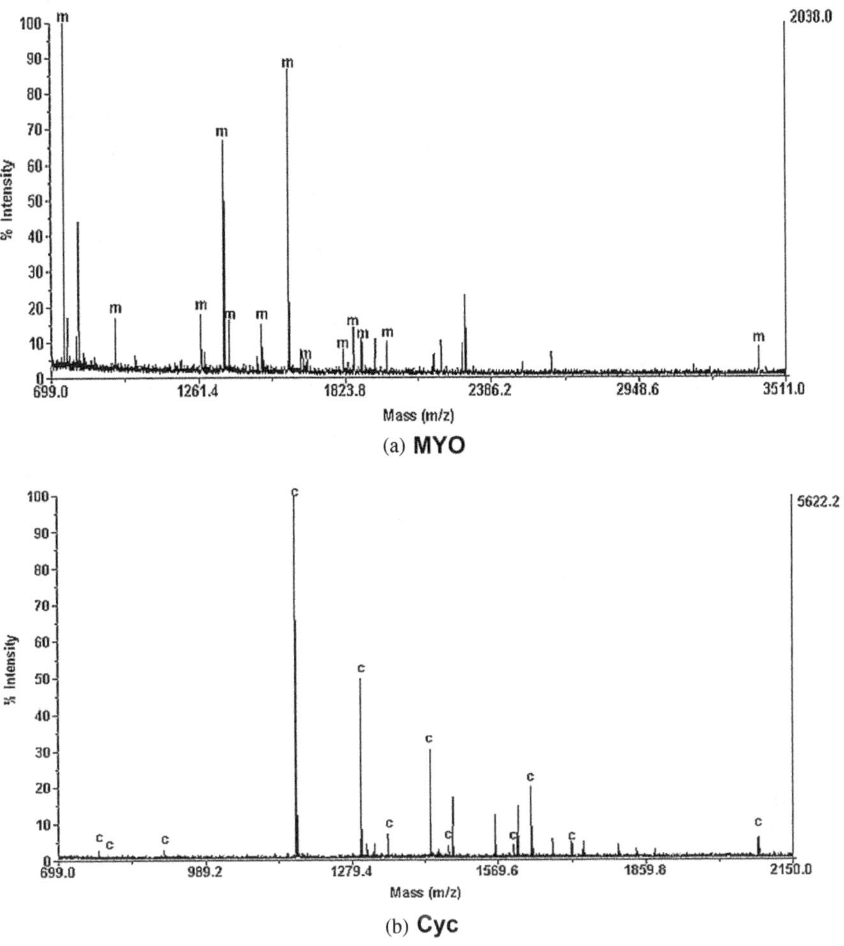

Figure 3.43: MALDI-TOF MS spectra of MYO and Cyc digested by on-plate enzymatic hydrolysis based on enzyme-immobilized amino-magnetic microsphere.

of the obtained protein before MS. The enzymatic hydrolysis on the plate is particularly suitable for use in the middle-down route because of its low sample consumption and the need of omitting sample transfer. Here, the rapid and simple separation and identification of protein based on the middle-down theory are realized by combining the on-plate enzymatic hydrolysis based on enzyme-immobilized amino-magnetic microsphere with chromatographic separation system of protein. The sample used throughout the experiment is the fraction of RPLC separated from rat liver.

Table 3.8: The obtained information after enzymatic hydrolysis of MYO and Cyc by on-plate enzymatic hydrolysis based on enzyme-immobilized amino-magnetic microsphere.

	MYO (AC P68082)		Cyc (AC P00004)	
	On-plate	Solution	On-plate	Solution
Detected amino acid	110	115	87	80
Sequence coverage (%)	69–73[a]	75	77–83	76
Time of enzymatic hydrolysis	5 min	12 h	5 min	12 h
Matched peptides	12	11	13	14

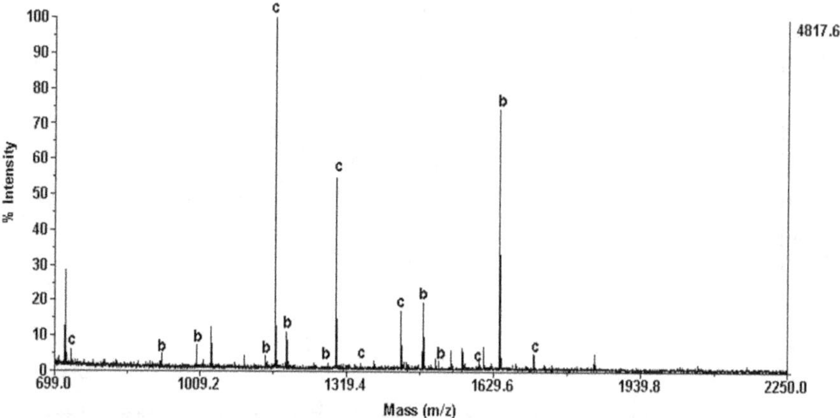

Figure 3.44: MALDI-TOF MS spectrum of mixed protein solution containing MYO and BSA (100 ng·μL^{-1} BSA + 50 ng·μL^{-1} Cyc) digested by on-plate enzymatic hydrolysis based on enzyme-immobilized amino-magnetic microsphere. b represents peptides from BSA, c represents peptides from Cyc.

During the RPLC separation of rat liver extracts, one fraction is collected in every minute. Then all the fractions are composed of two-part time period (5–12 min and 19.5–80.5 min). Each fraction is lyophilized and redissolved in 1 μL buffer, then spotted onto a MALDI target plate and enzymatically hydrolyzed at 50°C for 5 min by addition of enzyme-immobilized amino-magnetic microspheres. The magnetic microspheres are removed by a magnetic needle to stop the reaction, and the matrix is added for final analysis of MALDI-TOF MS/MS.

Figure 3.45(a) shows the MS spectrum of the 38# fraction (collected from 50.5 min to 51.5 min) after on-plate enzymatic hydrolysis, from which multiple peptide peaks after proteolysis in the fraction can be observed. However, many instances in the literature indicate that it is difficult to per-

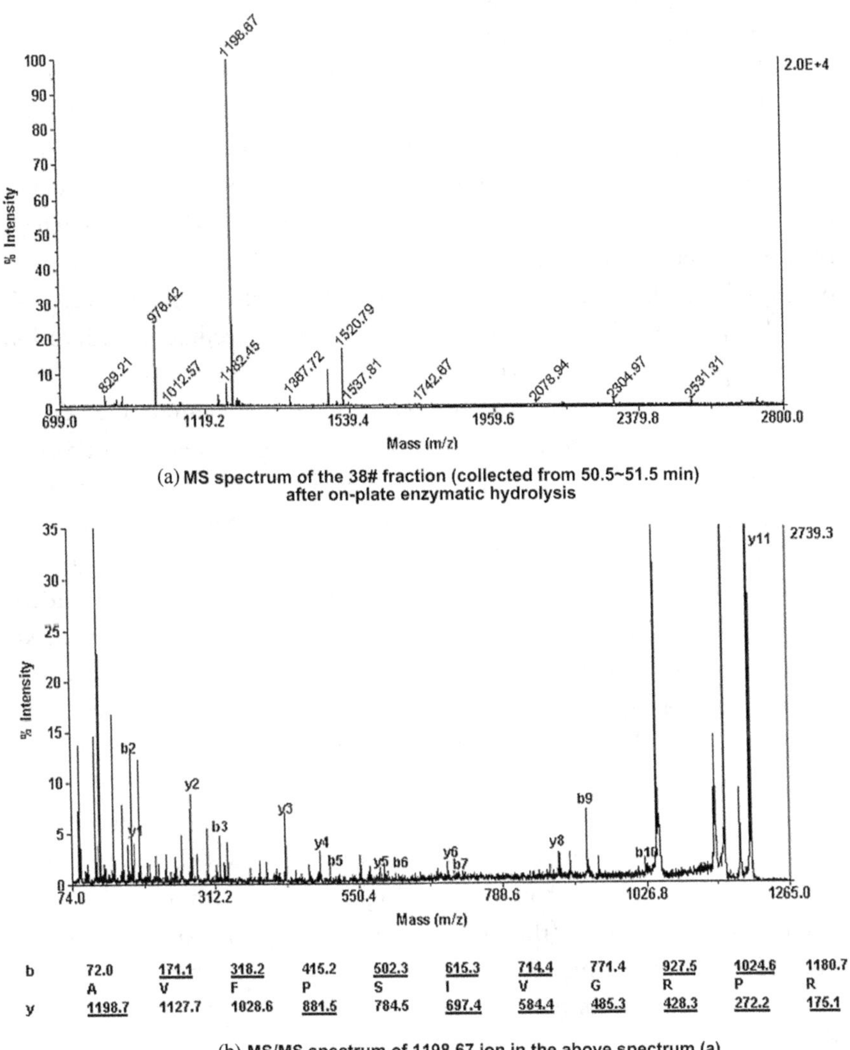

b	72.0	171.1	318.2	415.2	502.3	615.3	714.4	771.4	927.5	1024.6	1180.7
	A	V	F	P	S	I	V	G	R	P	R
y	1198.7	1127.7	1028.6	881.5	784.5	697.4	584.4	485.3	428.3	272.2	175.1

(b) MS/MS spectrum of 1198.67 ion in the above spectrum (a)

Figure 3.45: MS spectra of the No.38# fraction after on-plate enzymatic hydrolysis and fragment m/z=1198.67.

form accurate identification of protein when the number of peptides is not enough or the sequence coverage is not high enough, if only PMF is used for protein identification. This problem is particularly acute for the analysis of complex protein samples. This is because the content of most protein itself is low except for a small amount of high-abundance proteins in the practical biological samples, and the dilution during the separation process makes the content of protein lower, resulting in the difficulty to detect peptide by MS. In addition, there is a competitive effect in the process of both enzymatic hydrolysis and MS detection, and high-abundance protein is more easily identified. This is why researchers have been trying to separate proteins as completely as possible. However, unfortunately, no method has been developed to achieve satisfactory separating result. In the following experiment, a collected fraction through RPLC separation may contain several or even dozens of proteins, which will inevitably interfere with enzymatic hydrolysis and identification. Therefore, it is difficult to obtain accurate identification result simply by relying on PMF.

Tandem MS analysis of the obtained peptides is a common method for obtaining protein/peptide information and improving the identification credibility. Therefore, Tandem MS analysis is carried out during the experiment. PMF and tandem MS data are entered into the National Center of Biotechnology Information (NCBI, version of 070316) database for joint retrieval, allowing for a missed cutting site. Figure 3.45(b) shows the tandem MS spectrum of the peptide fragment of the parent-ion peak with the mass-to-charge ratio of 1986.67 in Figure 3.45(a). This spectrum gives the majority of the b- and y-daughter-ion peaks by br eaking the parent ion. The sequence of the peptide can be determined by the overall analysis of the b- and y-ions, and therefore the loss of individual ion peaks does not affect the determination of the entire peptide sequence. From Figure 3.45(b), the amino acid sequence of this peak can be obtained, which is AVFPSIVGRPR. The final identification result shows that there are eight proteins with score above 68, and after de-redundancy (the proteins identified by the same peptide are grouped together, the highest score is taken as the identified protein), 11 peptides are identified in one protein (gi|88953571). The specific results are shown in Table 3.9.

Table 3.9: Peptide fragment information of one protein (gi|88953571) in No. 38# fraction digested by on-plate enzymatic hydrolysis and identified by MALDI-TOF MS/MS.

Calc. Mass	Obsrv. Mass	± da	Start Seq.	End Seq.	Sequence
945.4271	945.5159	0.0888	510	518	GSENGQPEK
975.4377	975.417	−0.0207	392	400	GSENSQPEK
976.4482	976.4305	−0.0177	719	728	AGFAGDDAPR
1012.5019	1012.5684	0.0665	884	891	ELTDYLMK
1198.7054	1198.6785	−0.0269	729	739	AVFPSIVGRPR
1390.6882	1390.7312	0.043	2	15	VAEVDSMPAASSVK
1520.6282	1520.8127	0.1845	78	92	SNVGTSGDHDDSAMK
1534.8475	1534.7954	−0.0521	257	270	ALLLYGADIESKNK
1624.8615	1624.5433	−0.3182	884	896	ELTDYLMKILTER
2156.136	2156.1165	−0.0195	719	739	AGFAGDDAPRAVFPSIVGRPR
3239.3633	3238.3801	−0.9832	575	601	SRTPESQQFPDTENEEYHSDEQNDTQK

3.4.7. Conclusion

The magnetic microsphere with amino groups is directly synthesized by one-step hydrothermal method, and the reaction time is shortened from more than 3 days to less than 15 h. The synthesized amino-magnetic microsphere is further activated with glutaraldehyde, and finally the enzyme is immobilized on the surface of the magnetic microsphere by covalent reaction between aldehyde group and the primary amino group of the protease. The method for analyzing and identifying protein is developed by combining enzyme-immobilized amino-magnetic microsphere and enzymatic hydrolysis technique on the MALDI target plate. By taking advantage of the magnetic property of the material itself, the enzyme-immobilized amino-magnetic microsphere can be conveniently taken out from the target plate when the enzymatic hydrolysis process is stopped. The efficiency of enzymatic hydrolysis is improved by using the immobilized enzyme technology, and the contamination of the MS ion source is avoided. Finally, the liquid chromatographic fraction of the rat liver extract is subjected to enzymatic hydrolysis. This method is one of the research goals of future proteomics, and provides reference for the automation of protein identification in complex systems.

References

1. Kondo A., Kaneko T., Higashitani K. Development and Application of Thermosensitive Immunomicrospheres for Antibody Purification. Biotechnology and Bioengineering, 1994, 44: 1–6.

2. Li Y., Zhang X., Deng C. Functionalized Magnetic Nanoparticles for Sample Preparation in Proteomics and Peptidomics Analysis. *Chemical Society Reviews*, 2013, 42: 8517–39.

3. Li Y. Investigation of New Methodologies in Rapid Protein Digestion and Enrichment Based on Functionalized Magnetic Microspheres, Fudan University (2008).

4. Slovakova M., Minc N., Bilkova Z., Smadja C., Faigle W., Futterer C., Taverna M., Viovy J. L. Use of Self Assembled Magnetic Beads for on-Chip Protein Digestion. *Lab on a Chip*, 2005, 5: 935–42.

5. Bilkova Z., Slovakova M., Minc N., Futterer C., Cecal R., Horak D., Benes M., le Potier I., Krenkova J., Przybylski M., Viovy J.-L. Functionalized Magnetic Micro- and Nanoparticles: Optimization and Application to Mu-Chip Tryptic Digestion. *Electrophoresis*, 2006, 27: 1811–24.

6. Yan X., Zhu L., Dong J. Application of Polymeric Particles in Biochemistry and Biomedicine. *Journal of Functional Polymers*, 1997, 131–5.

7. Choi J. W., Oh K. W., Thomas J. H., Heineman W. R., Halsall H. B., Nevin J. H., Helmicki A. J., Henderson H. T., Ahn C. H. An Integrated Microfluidic Biochemical Detection System for Protein Analysis with Magnetic Bead-Based Sampling Capabilities. *Lab on a Chip*, 2002, 2: 27–30.

8. He J., Huang M., Wang D., Zhang Z., Li G. Magnetic Separation Techniques in Sample Preparation for Biological Analysis: A Review. *Journal of Pharmaceutical and Biomedical Analysis*, 2014, 101: 84–101.

9. Lin S. New Microwave-Assisted Protein Digestion Method Based on Functionalized Magnetic Nanoparticles, Fudan University (2008).

10. Krenkova J., Foret F. Immobilized Microfluidic Enzymatic Reactors. *Electrophoresis*, 2004, 25: 3550–63.

11. Deng H., Li X. L., Peng Q., Wang X., Chen J. P., Li Y. D. Monodisperse Magnetic Single-Crystal Ferrite Microspheres. *Angewandte Chemie-International Edition*, 2005, 44: 2782–5.

12. Qu Y., Moons L., Vandesande F. Determination of Serotonin, Catecholamines and Their Metabolites by Direct Injection of Supernatants from Chicken Brain Tissue Homogenate Using Liquid Chromatography with Electrochemical Detection. *Journal of Chromatography B*, 1997, 704: 351–8.

13. Zhu X.-r., Xu J. Preparation and Properties of β-Amylase Immobilized on Silk Fibroin. *Journal of Zhejiang University (Agric &Life Sci)*, 2002, 28: 64–9.

14. Yang C., Pan J., Zhong H., Hu N. The Synthesis of Monoglycerides by Cellulose Acetate Fixed Porcine Pancreatic Lipase–Catelyzed Glycerolysis of Lard. *Hubei Chemical Industry*, 2002, 19: 20–1.

15. Sakai-Kato K., Kato M., Toyo'oka T. On-Line Trypsin-Encapsulated Enzyme Reactor by the Sol-Gel Method Integrated into Capillary Electrophoresis. *Analytical Chemistry*, 2002, 74: 2943–9.

16. Kato M., Sakai-Kato K., Matsumoto N., Toyo'oka T. A Protein-Encapsulation Technique by the Sol-Gel Method for the Preparation of Monolithic Columns for Capillary Electrochromatography. *Analytical Chemistry*, 2002, 74: 1915–21.

17. Ma J., Duan J., Liang Z., Zhang L., Zhang W., Zhang Y. Immobilized Enzyme Reactor and Its Applications in Proteome Study. *Chinese Journal of Analytical Chemistry*, 2006, 34: 1649–55.

18. Guo Z., Zhang Q., Lei Z., Kong L., Mao X., Zou H. Studies on Rapid Micro-Scale Peptide Mapping Analysis Using a Capillary Micro-Reactor. *Chemical Journal of Chinese Universities*, 2002, 23: 1277–80.

19. Amankwa L. N., Kuhr W. G. Trypsin-Modified Fused-Silica Capillary Microreactor for Peptide-Mapping by Capillary Zone Electrophoresis. *Analytical Chemistry*, 1992, 64: 1610–3.

20. Deng H., Wu J., Xu Z., Xu Y., Patrick S. Recent Advances in Enzyme Immobilization on Membranes and Their Applications. *Membrane Science and Technology*, 2004, 24: 47–53.

21. Jiang H., Zou H., Wang H., Zhang Q., Ni J., Zhang Q., Guo Z., Chen X. The Combination of Immobilized Enzyme Microlift Reactor and MALDI-TOF MS for Study of Peptide Mapping of Protein. *Science in China (Series B)*, 2000, 30: 385–91.

22. Jiang H.-H., Zou H.-F., Wang H.-L., Ni J.-Y., ZHang Q. Covalent Immobilization of Trypsin on Glycidyl Methacrylate-Modified Cellulose Membrane as Enzyme Reactor. *Chemical Journal of Chinese Universities*, 2000, 21: 702–6.

23. Gao J., Xu J. D., Locascio L. E., Lee C. S. Integrated Microfluidic System Enabling Protein Digestion, Peptide Separation, and Protein Identification. *Analytical Chemistry*, 2001, 73: 2648–55.

24. Cooper J. W., Chen J. Z., Li Y., Lee C. S. Membrane-Based Nanoscale Proteolytic Reactor Enabling Protein Digestion, Peptide Separation, and Protein Identification Using Mass Spectrometry. *Analytical Chemistry*, 2003, 75: 1067–74.

25. Elander N., Jones J. R., Lu S. Y., Stone-Elander S. Microwave-Enhanced Radiochemistry. *Chemical Society Reviews*, 2000, 29: 239–49.

26. Pramanik B. N., Mirza U. A., Ing Y. H., Liu Y. H., Bartner P. L., Weber P. C., Bose M. K. Microwave-Enhanced Enzyme Reaction for Protein Mapping by Mass Spectrometry: A New Approach to Protein Digestion in Minutes. *Protein Science*, 2002, 11: 2676–87.

27. Juan H. F., Chang S. C., Huang H. C., Chen S. T. A New Application of Microwave Technology to Proteomics. *Proteomics*, 2005, 5: 840–2.

28. Lin S. S., Wu C. H., Sun M. C., Sun C. M., Ho Y. P. Microwalve-Assisted Enzyme-Catalyzed Reactions in Various Solvent Systems. *Journal of the American Society for Mass Spectrometry*, 2005, 16: 581–8.

29. Sun W., Gao S. J., Wang L. J., Chen Y., Wu S. Z., Wang X. R., Zheng D. X., Gao Y. H. Microwave-Assisted Protein Preparation and Enzymatic Digestion in Proteomics. *Molecular & Cellular Proteomics*, 2006, 5: 769–76.

30. Chen S. T., Chiou S. H., Wang K. T. Enhancement of Chemical-Reactions by Microwave Irradiation. *Journal of the Chinese Chemical Society*, 1991, 38: 85–91.

31. Zhong H. Y., Zhang Y., Wen Z. H., Li L. Protein Sequencing by Mass Analysis of Polypeptide Ladders after Controlled Protein Hydrolysis. *Nature Biotechnology*, 2004, 22: 1291–6.

32. Zhong H. Y., Marcus S. L., Li L. Microwave-Assisted Acid Hydrolysis of Proteins Combined with Liquid Chromatography MALDI MS/MS for Protein Identification. *Journal of the American Society for Mass Spectrometry*, 2005, 16: 471–81.

33. Kirschvink J. L. Microwave Absorption by Magnetite: A Possible Mechanism for Coupling Nonthermal Levels of Radiation to Biological Systems. *Bioelectromagnetics*, 1996, 17: 187–94.

34. Walkiewicz J. W., Clark A. E., McGill S. L. Microwave-Assisted Grinding. *IEEE Transactions on Industry Applications*, 1991, 27: 239–43.

35. Chen W.-Y., Chen Y.-C. Acceleration of Microwave-Assisted Enzymatic Digestion Reactions by Magnetite Beads. *Analytical Chemistry*, 2007, 79: 2394–401.

36. Wang L., Bao J., Wang L., Zhang F., Li Y. One-Pot Synthesis and Bioapplication of Amine-Functionalized Magnetite Nanoparticles and Hollow Nanospheres. *Chemistry-A European Journal*, 2006, 12: 6341–7.

37. Kelleher N. L., Lin H. Y., Valaskovic G. A., Aaserud D. J., Fridriksson E. K., McLafferty F. W. Top Down Versus Bottom up Protein Characterization by Tandem High-Resolution Mass Spectrometry. *Journal of the American Chemical Society*, 1999, 121: 806–12.

38. Meng F. Y., Cargile B. J., Miller L. M., Forbes A. J., Johnson J. R., Kelleher N. L. Informatics and Multiplexing of Intact Protein Identification in Bacteria and the Archaea. *Nature Biotechnology*, 2001, 19: 952–7.

39. Eng J. K., McCormack A. L., Yates J. R. An Approach to Correlate Tandem Mass-Spectral Data of Peptides with Amino-Acid-Sequences in a Protein Database. *Journal of the American Society for Mass Spectrometry*, 1994, 5: 976–89.

40. Kim J. S., Knapp D. R. Miniaturized Multichannel Electrospray Ionization Emitters on Poly(Dimethylsiloxane) Microfluidic Devices. *Electrophoresis*, 2001, 22: 3993–9.

41. Peterson D. S., Rohr T., Svec F., Frechet J. M. J. Enzymatic Microreactor-on-a-Chip: Protein Mapping Using Trypsin Immobilized on Porous Polymer Monoliths Molded in Channels of Microfluidic Devices. *Analytical Chemistry*, 2002, 74: 4081–8.

42. Sakai-Kato K., Kato M., Toyo'oka T. Creation of an on-Chip Enzyme Reactor by Encapsulating Trypsin in Sol-Gel on a Plastic Microchip. *Analytical Chemistry*, 2003, 75: 388–93.

43. Qu H. Y., Wang H. T., Huang Y., Zhong W., Lu H. J., Kong J. L., Yang P. Y., Liu B. H. Stable Microstructured Network for Protein Patterning on a Plastic Microfluidic Channel: Strategy and Characterization of on-Chip Enzyme Microreactors. *Analytical Chemistry*, 2004, 76: 6426–33.

44. Wu H. L., Tian Y. P., Liu B. H., Lu H. J., Wang X. Y., Zhai J. J., Jin H., Yang P. Y., Xu Y. M., Wang H. H. Titania and Alumina Sol-Gel-Derived Microfluidics Enzymatic-Reactors for Peptide Mapping: Design, Characterization, and Performance. *Journal of Proteome Research*, 2004, 3: 1201–9.

45. Liu Y., Lu H. J., Zhong W., Song P. Y., Kong J. L., Yang P. Y., Girault H. H., Liu B. H. Multilayer-Assembled Microchip for Enzyme Immobilization as Reactor toward Low-Level Protein Identification. *Analytical Chemistry*, 2006, 78: 801–8.

46. Wang C., Oleschuk R., Ouchen F., Li J. J., Thibault P., Harrison D. J. Integration of Immobilized Trypsin Bead Beds for Protein Digestion within a Microfluidic Chip Incorporating Capillary Electrophoresis Separations and an Electrospray Mass Spectrometry Interface. *Rapid Communications in Mass Spectrometry*, 2000, 14: 1377–83.

47. Sato K., Tokeshi M., Odake T., Kimura H., Ooi T., Nakao M., Kitamori T. Integration of an Immunosorbent Assay System: Analysis of Secretory Human Immunoglobulin a on Polystyrene Beads in a Microchip. *Analytical Chemistry*, 2000, 72: 1144–7.

48. Jemere A. B., Oleschuk R. D., Ouchen F., Fajuyigbe F., Harrison D. J. An Integrated Solid-Phase Extraction System for Sub-Picomolar Detection. *Electrophoresis*, 2002, 23: 3537–44.

49. Peterson D. S., Rohr T., Svec F., Frechet J. M. J. High-Throughput Peptide Mass Mapping Using a Microdevice Containing Trypsin Immobilized on a Porous Polymer Monolith Coupled to MALDI TOF and ESI TOF Mass Spectrometers. *Journal of Proteome Research*, 2002, 1: 563–8.

50. Kailasa S. K., Wu H.-F. Advances in Nanomaterial-Based Microwaves and Infrared Wave-Assisted Tryptic Digestion for Ultrafast Proteolysis and Rapid Detection by MALDI-MS. *Combinatorial Chemistry & High Throughput Screening*, 2014, 17: 68–79.

51. Guo Z., Xu S. Y., Lei Z. D., Zou H. F., Guo B. C. Immobilized Metal-Ion Chelating Capillary Microreactor for Peptide Mapping Analysis of Proteins by Matrix Assisted Laser Desorption/Ionization-Time of Flight-Mass Spectrometry. *Electrophoresis*, 2003, 24: 3633–9.

52. Li Y., Xu X., Yan B., Deng C., Yu W., Yang P., Zhang X. Microchip Reactor Packed with Metal-Ion Chelated Magnetic Silica Microspheres for Highly Efficient Proteolysis. *Journal of Proteome Research*, 2007, 6: 2367–75.

53. Li Y., Yan B., Deng C., Yu W., Xu X., Yang P., Zhang X. Efficient on-Chip Proteolysis System Based on Functionalized Magnetic Silica Microspheres. *Proteomics*, 2007, 7: 2330–9.

54. Jiang H. H., Zou H. F., Wang H. L., Ni J. Y., Zhang Q., Zhang Y. K. On-Line Characterization of the Activity and Reaction Kinetics of Immobilized Enzyme by High-Performance Frontal Analysis. *Journal of Chromatography A*, 2000, 903: 77–84.

55. Okuda K., Urabe I., Yamada Y., Okada H. Reaction of Glutaraldehyde with Amino and Thiol Compounds. *Journal of Fermentation and Bioengineering*, 1991, 71: 100–5.

56. Walt D. R., Agayn V. I. The Chemistry of Enzyme and Protein Immobilization with Glutaraldehyde. *Trac-Trends in Analytical Chemistry*, 1994, 13: 425–30.

57. Rashkovetsky L. G., Lyubarskaya Y. V., Foret F., Hughes D. E., Karger B. L. Automated Microanalysis Using Magnetic Beads with Commercial Capillary Electrophoretic Instrumentation. *Journal of Chromatography A*, 1997, 781: 197–204.

58. Stensballe A., Jensen O. N. Simplified Sample Preparation Method for Protein Identification by Matrix-Assisted Laser Desorption/Ionization Mass Spectrometry: In-Gel Digestion on the Probe Surface. *Proteomics*, 2001, 1: 955–66.

59. Ericsson D., Ekstrom S., Nilsson J., Bergquist J., Marko-Varga G., Laurell T. Downsizing Proteolytic Digestion and Analysis Using Dispenser-Aided Sample Handling and Nanovial Matrix-Assisted Laser/Desorption Ionization-Target Arrays. *Proteomics*, 2001, 1: 1072–81.

60. Warscheid B., Fenselau C. A Targeted Proteomics Approach to the Rapid Identification of Bacterial Cell Mixtures by Matrix-Assisted Laser Desorption/Ionization Mass Spectrometry. *Proteomics*, 2004, 4: 2877–92.

61. Dogruel D., Williams P., Nelson R. W. Rapid Tryptic Mapping Using Enzymatically Active Mass-Spectrometer Probe Tips. *Analytical Chemistry*, 1995, 67: 4343–8.

62. Nelson R. W., Dogruel D., Krone J. R., Williams P. Peptide Characterization Using Bioreactive Mass-Spectrometer Probe Tips. *Rapid Communications in Mass Spectrometry*, 1995, 9: 1380–5.

63. Harris W. A., Reilly J. P. On-Probe Digestion of Bacterial Proteins for MALDI-MS. *Analytical Chemistry*, 2002, 74: 4410–6.

64. Arnold R. J., Reilly J. P. Fingerprint Matching of E-Coli Strains with Matrix-Assisted Laser Desorption Ionization Time-of-Flight Mass Spectrometry of Whole Cells Using a Modified Correlation Approach. *Rapid Communications in Mass Spectrometry*, 1998, 12: 630–6.

65. Arnold R. J., Reilly J. P. Observation of Escherichia Coli Ribosomal Proteins and Their Posttranslational Modifications by Mass Spectrometry. *Analytical Biochemistry*, 1999, 269: 105–12.

66. Zheng S., Yoo C., Delmotte N., Miller F. R., Huber C. G., Lubman D. M. Monolithic Column HPLC Separation of Intact Proteins Analyzed by LC-MALDI Using on-Plate Digestion: An Approach to Integrate Protein Separation and Identification. *Analytical Chemistry*, 2006, 78: 5198–204.

67. Yu W.-j. Study on the Proteomic Strategy of Multi-Dimensional Liquid Chromatography Coupling with Fast Digestion and Its Applications, Fudan University (2006).

68. Fontana A., Fassina G., Vita C., Dalzoppo D., Zamai M., Zambonin M. Correlation between Sites of Limited Proteolysis and Segmental Mobility in Thermolysin. *Biochemistry*, 1986, 25: 1847–51.

69. Hubbard S. J., Eisenmenger F., Thornton J. M. Modeling Studies of the Change in Conformation Required for Cleavage of Limited Proteolytic Sites. *Protein Science*, 1994, 3: 757–68.

70. Park Z. Y., Russell D. H. Thermal Denaturation: A Useful Technique in Peptide Mass Mapping. *Analytical Chemistry*, 2000, 72: 2667–70.

71. Mao Y., Li Y., Zhang X. M. Array Based Capillary IEF with a Whole Column Image of Laser-Induced Fluorescence in Coupling to Capillary RPLC as a Comprehensive 2-D Separation System for Proteome Analysis. *Proteomics*, 2006, 6: 420–6.

72. Zhang J., Xu X., Gao M., Yang P., Zhang X. Comparison of 2-D LC and 3-D LC with Post- and Pre-Tryptic-Digestion Sec Fractionation for Proteome Analysis of Normal Human Liver Tissue. *Proteomics*, 2007, 7: 500–12.

Chapter 4

Enrichment Technology of Low-Abundance Proteomics Based on Magnetic Micro-/Nanomaterial

4.1. Basic enrichment principle of low-abundance proteome based on magnetic micro-/nanomaterial

Separation and enrichment of low-abundance protein/peptide are an important part of the development of proteomics and peptidomics. Magnetic microsphere has many advantages including large specific surface area, surface that can be easily modified, good dispersibility in solution, and sensitive magnetic responsiveness, providing a possibility for its application in the separation and enrichment of trace peptides. Like solid-phase microextraction, magnetic material for the separation and enrichment of low-abundance protein/peptide belongs the non-solvent selective extraction method, which is a kind of sample preparation and enrichment technology that can complete the extraction, separation, and enrichment at the same time. The technology key to the application of magnetic material in enrichment of low-abundance proteomics is the control of the surface property of the magnetic material. Normally, functionalized magnetic material is often used as solid-phase adsorbent for peptide/protein enrichment by using the special property on the surface of material.

4.1.1. *Reversed-phase magnetic nanomaterial*

It is well known that most amino acids contain hydrophobic groups such as phenyl, alkyl, etc. These amino acids are widely existent in the peptide/protein.

Therefore, reversed-phase liquid chromatography (RPLC), which has hydrophobic groups on the surface of the filler, has good retention for most peptides/proteins, and it is the most widely used liquid chromatographic separation mode. The method can remove salt and other impurities while effectively concentrating the sample. In order to prevent the salt and other impurities from entering the MS to affect the signal detection, a short reverse pre-column is connected in front of the separation column to achieve sample concentration and removal of salt effect, especially in the case of liquid chromatography coupled with MS. The commercially available products including Zip-tip and Zip-plate, which are widely used for sample desalination, are also based on the principle of chromatographic concentration[1] to fill a small amount of reversed-phase packing at the tip of the Zip-tip or the pore bottom of the 96-well plate. However, the operation is relatively cumbersome, and the amount of sample that can be concentrated is limited because of so little filler.

Currently, commercially available C8-magnetic bead is synthesized by coating organic polymer on the surface of magnetic microspheres, the particle size of which is about 1–10 nm. These materials often contain polymers on their surface such as polystyrene, which are not chemically stable and biocompatible, and the magnetic content in the magnetic bead is small, so the magnetic property is weak (< 30 emu·g^{-1}), with the result that long time needs to be taken to separate the magnetic bead from the sample (5 min for each separation). In addition, the C8-magnetic bead is generally on the order of micrometer scale, and specific surface area is relatively small.[2] In the latest researches, the magnetic nanomaterials generally have a particle size of less than 500 nm, which are strongly magnetic and often superparamagnetic. The reversed-phase magnetic nanomaterial modified with such as C1, C2, C3, C4, C8, C18 realizes the enrichment of the protein/peptide effectively by the hydrophobic interaction of the alkyl group on the surface and the peptide residue.

4.1.2. *Fullerene-modified magnetic nanomaterial*

Many other shapes of compounds are also hydrophobic and can act as reversed-phase affinity ligands apart from linear hydrophobic groups. As a method of purifying small molecules and hydrophilic peptides prior to MS analysis, carbon nanotubes have been the focus of the scientific community since it they were discovered in 1991.[3] The carbon nanotube is hollow, and the two ends are "hats" or hemispheres with a fullerene structure, and this structural feature binds its organic molecules closely.[4–7] The researchers also find that carbon nanotube has good biocompatibility.[8] It has been reported in the literature that carbon nanotubes are used in the enrichment and separation of peptides in human serum, with good results.[9] Fullerene (C60) is a spherical molecule with a very peculiar structure and a very stable nature. It contains a large amount of C=C, which is similar to the "hat" at both ends of the carbon nanotube. The application of C60 in proteomics has also attracted the attention of researchers at home and abroad. In 2007, Analytical Chemistry reported the research results of Austrian scientists Rania Bakry *et al.*, proving that C60-modified silica can be used for solid-phase microextraction of different molecules. For example, there is interaction between C60-modified silica and common peptide, protein and hydrophilic phosphopeptide, and the interaction is reversible.[10] Prof. Lu from Fudan University also added C60 to the polymer for the on-plate enrichment and desalination of proteomes.[11] Nowadays, many carbon nanotube/fullerene-modified magnetic nanomaterials have been used for the enrichment of low-abundance proteomics.

4.1.3. *Reversed-phase organic polymer-modified magnetic nanomaterial*

Porous polymer is the most widely used stationary phase in gas–solid chromatography, which is mainly made of styrene, divinylbenzene, or other monomers by proportional polymerization. The polarity of the stationary phase can vary depending on the nature of the added compound.[12] In recent years, a variety of reversed-phase polymer packings have been developed and applied to the separation of practical materials, such as C18 alkyl-derived cross-linked polystyrene, C18 alkyl polyacrylamide, polyalkyl methacrylate, vinyl acetate copolymer, and so on,[13] in the preparation of RPLC packings. Reversed-phase organic polymer is coated on the surface of magnetic nanoparticles via chemical modification, which can be used for efficient separation and enrichment of low-abundance protein/peptides.

4.1.4. *Metal ion-immobilized magnetic nanomaterial*

Amino acid contains both hydrophobic and hydrophilic groups, which is widely existent in peptides/proteins. The immobilized metal affinity chromatography (IMAC) technique for proteins and peptides is based on the reversible binding-dissociation mechanism between immobilized metal ions and target proteins/peptides. In particular, copper ions (Cu^{2+}) can be immobilized on the substrate relatively easily, and are bound to the amino acid chain of the proteins/peptides by covalent bonds, which can be destroyed to release peptides/proteins by using eluents. Metal ion immobilized magnetic nanomaterials can achieve the purpose of extracting proteins/peptides by affinity of metal ions. In recent years, IMAC has been newly developed and applied in protein purification.[14] The magnetic material is modified with a large amount of Cu^{2+} metal ions, which can be combined with the carboxyl group and the amino group in the peptide by covalent bond to achieve the purpose of effectively enriching the hydrophobic/hydrophilic peptide.

4.1.5. *Magnetic mesoporous material for peptidomics*

The research object of peptidomics is a biological sample, such as endogenous peptide and small molecular protein presented in cell lysate, tissue extract, and body fluid. The biological sample is generally complex, and contains a large amount of high-concentration macromolecular proteins apart from the impurities such as high-concentration salts. Most of the functional groups used for protein/peptide enrichment are modified on the outer surface of magnetic micro-/nanomaterial. These functional groups can not only bind endogenous peptides, but also bind macromolecular proteins, resulting in that most of the functional sites are occupied by proteins and the enrichment ability for low-concentration endogenous peptides is greatly impaired.[15] Therefore, it is necessary to develop novel functionalized magnetic micro-/ nanomaterials for selective separation and analysis of endogenous peptides.

Ordered mesoporous material has attracted great attention in catalytic science, adsorption science, and separation science since its first discovery in 1992.[16] In particular, silica particles with ordered mesoporous structure, such as M41s[16] and SBA-15,[17] have been widely used in the research field of separation and adsorption.[18,19] These silica materials have some unique properties, including large inner surface area of pore, fairly narrow pore size distribution, fully controllable pore size, and a large number of silanol and siloxane groups that are easily interconverted. Based on the principle of size exclusion,[20,21] these materials can selectively adsorb proteins/peptides with molecular weight within a certain range and exclude larger proteins. Professor Zou and co-workers reported a unique mesoporous silica, MCM-41, which was successfully applied to the enrichment of serum peptide. 988 peptides were identified by LC-MS/MS.[22] The outer layer of MCM-41 is coated by uniform ordered mesopore with a pore size of about 2.0 nm. The size-exclusion effect of mesopore can prevent macromolecular protein and allow only small peptides to enter the mesopore, so as to selectively capture them. In addition, the presence of mesopores allows the material to have a large surface area that enhances the capture efficiency of the material towards target molecules.

4.1.6. *Basic operation of separating and enriching low-abundance protein/peptide by magnetic nanomaterial*

The general operation of adopting functionalized magnetic micro-/nanomaterials to conduct low-abundance enrichment includes material activation and sample preparation.

(1) **Thorough exposure between material and sample:** The material is added into the sample solution, the enrichment time and the pH of the enriched solution are adjusted, the enrichment conditions are optimized, and the mixture is thoroughly mixed.
(2) **Separating material from the sample:** A magnet is placed closely on the sample container, and the superparamagnetic material for capturing the target protein/peptide is separated from the solution.
(3) **Cleaning material to remove impurity:** The separated material is washed 3 times with an appropriate buffer, and each time it is separated by a magnet to remove the washing buffer.
(4) **Eluting protein/peptide on the material:** The magnetic material for adsorbing peptide/protein is dispersed in a suitable eluent (for example, H_2O/ACN/TFA = 50/50/0.1, v/v/v), and the magnet separates the material to transfer the elute containing protein/ peptide.
(5) **MS analysis:** The magnetic material of adsorbing the peptide/protein can be directly transferred to the MALDI-TOF-MS target plate skipping the operation of the fifth step, and then MS analysis can be performed by

adding an appropriate amount of matrix solution (α-CHCA solution). Alternatively, after elution in the fifth step, the eluate containing peptide is transferred to a stainless steel target plate of MALDI-TOF-MS, and after it is dried, an appropriate amount of α-CHCA solution is added for MALDI-TOF-MS analysis, or LC-MS analysis.

(6) **Material recycling:** The magnetic material can be reused multiple times after being refreshed with a suitable solvent.

4.2. Magnetic micro-/nanomaterial with alkyl chain as reversed-phase affinity ligand

4.2.1. *Basic principle*

RPLC is the most widely used method in high performance liquid chromatography (HPLC). Statistics show that RPLC accounts for more than 80% of the totally applied number of liquid chromatographic separation modes, indicating that RPLC has strong separation ability and applicability for sample analysis. The stationary phase commonly used in RPLC is a non-polar bonded phase, abbreviated as C18 (ODS) and C8 column, and the alkyl chain is C1, C2, C3, C4, C6, C12, C16, or a longer carbon chain such as C22, C36, and so on.[12] As for the properties of the bonded phase of various chain lengths, the short-chain bonded phase has higher bonding density and surface coverage, is more suitable for separating highly polar samples, and can use a more acidic mobile phase. While the long-chain bonded phase has a higher carbon content and greater hydrophobicity, and exhibits greater separation ability and adaptability to various samples, especially for non-polar aromatic samples and polar drug, amino acid, and peptides. This section introduces the synthesis of octyl chain-modified magnetic materials by introducing alkyl group onto the surface of the magnetic micro-/nanomaterials, via the conventional silanization method with simplicity and effectiveness.

4.2.2. *C8-modified magnetic microsphere*

4.2.2.1. *Synthesis of C8-modified magnetic microsphere*

See Figure 4.1.[2,23,24]

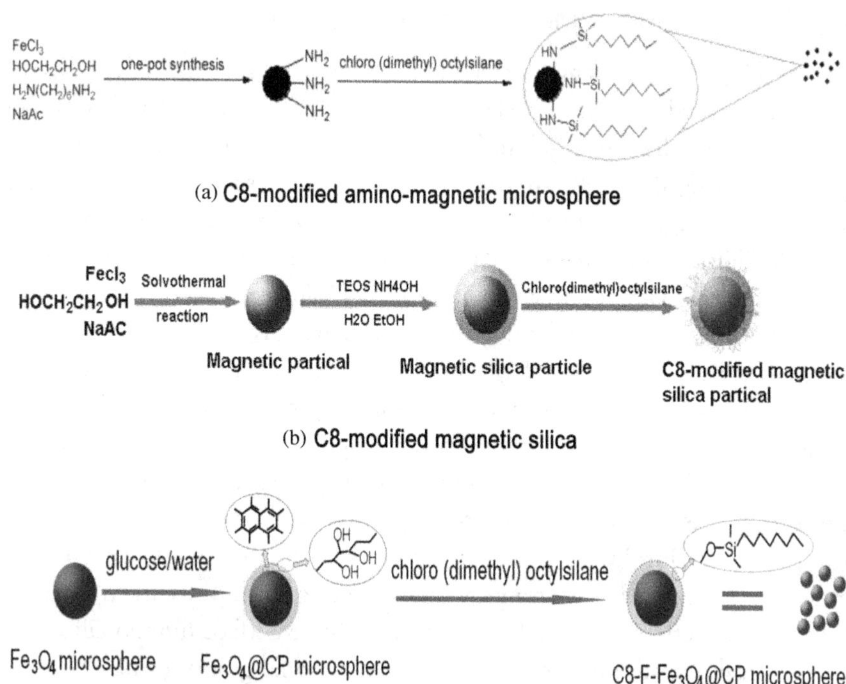

Figure 4.1: Schematic diagram of preparation of three C8-modified magnetic materials.

(1) C8-modified amino-magnetic microsphere

Magnetic nanoparticle chelated with hexamethylenediamine is synthesized via one-pot synthetic method (as seen in Section 2.3 of Chapter 2), the surface of which is covered with amino groups for subsequent modification. 0.01 g amino-magnetic nanoparticle is ultrasonically dispersed in 1.0 g anhydrous pyridine, and then 0.1 g chlorodimethyl-oetylsilane (C8) is added to be mechanically stirred at room temperature for 12 h. The synthesized C8-modified amino-magnetic nanoparticle is repeatedly washed with ethanol and water, and dried in vacuum for later use. The C8-modified amino-magnetic nanoparticle has strong magnetic response, and good effect on the enrichment of low-abundance peptide/protein in solution.

(2) C8-modified magnetic silica microsphere

The packing of RPLC is mainly based on porous silica gel. The preparation of silica gel-bonded phase is carried out by the reaction between silanol group on the surface of the silica gel and chlorosilane. Based on the preparation principle of liquid chromatography packing, the researchers developed a C8-bonded magnetic silica for the enrichment of low-abundance protein/peptide. Silica is used to coat magnetic microsphere for the following three reasons. First, it is well known that silica is a biologically inert material with good biocompatibility and can provide an inert surface for application of magnetic silica in a biological system. Therefore, magnetic silica can be used for the bio-medical field. Secondly, the silica layer can effectively shield the magnetic dipole attraction between the magnetic microspheres, which is beneficial for the dispersion of magnetic microspheres in liquid medium and protects magnetic microspheres from corrosion and dissolution in an acidic environment. Thirdly, the silica surface prepared by the sol–gel method has abundant silicon hydroxyl groups, which can not only improve the hydrophilicity of magnetic nanoparticles, but also facilitate surface functionalization by chemical modification so that different functional groups can be bonded to meet different practical application requirements.

The synthesis of $Fe_3O_4@SiO_2$ microsphere is carried out by hydrothermal method. The specific method is as described in Section 2.4 of Chapter 2. The resulting $Fe_3O_4@SiO_2$ microsphere is dried in vacuum at 60°C for 12 h. Thereafter, 0.01 g $Fe_3O_4@SiO_2$ microsphere is ultrasonically dispersed in 1.0 g anhydrous pyridine, and then 0.2 g chlorodimethyl-oetylsilane is added and kept under mechanically stirring at room temperature for 12 h. The obtained product is washed repeatedly with ethanol and water, and dried at 60°C for 24 h in vacuum.

(3) C8-modified magnetic polymeric carbon microsphere

Considering that the direct injection of magnetic material into the sample solution for low-abundance enrichment brings forward high request for the in-solution compatibility and biocompatibility of materials, the researchers improved the coating layer of Fe_3O_4 core, and synthesized a magnetic micro-/nanoparticle with both linear hydrophobic groups and hydrophilic groups on the surface, which is more suitable for enrichment of low-abundance peptides.

The Fe_3O_4 is added into the glucose aqueous solution to coat the polymerized carbon layer on the Fe_3O_4 core by hydrothermal reaction (as described in Section 2.4 of Chapter 2). 10 mg Fe_3O_4@CP microsphere is ultrasonically dispersed into 80 mL anhydrous toluene or anhydrous pyridine, and then 0.1 g silylating reagent — chloro(dimethyl)octylsilane — is added under mechanical stirring to maintain the reaction at room temperature for 12 h to obtain C8-modified magnetic polymeric carbon microsphere (Fe_3O_4@ CP-C8). The Fe_3O_4@CP-C8 is washed repeatedly with ethanol and water, and dried in vacuum at 60°C for 24 h. Since the polymeric carbon layer contains a large amount of hydroxyl groups, carboxyl groups, aldehyde groups, etc., a large amount of hydrophilic groups remain after modification of the C8 linear hydrophobic phase, which contributes to the dispersion of the material in the water-soluble biological sample. Based on the principle of RPLC concentration, this material is used for the enrichment and separation of low-abundance peptides in practical samples.

4.2.2.2. *Characterization of C8-modified magnetic microsphere*

(1) Characterization of C8-modified amino-magnetic microsphere

The FTIR spectrum (Figure 4.2) shows that the C8-modified amino-magnetic microsphere has strong absorption near 2,900 cm^{-1} compared to

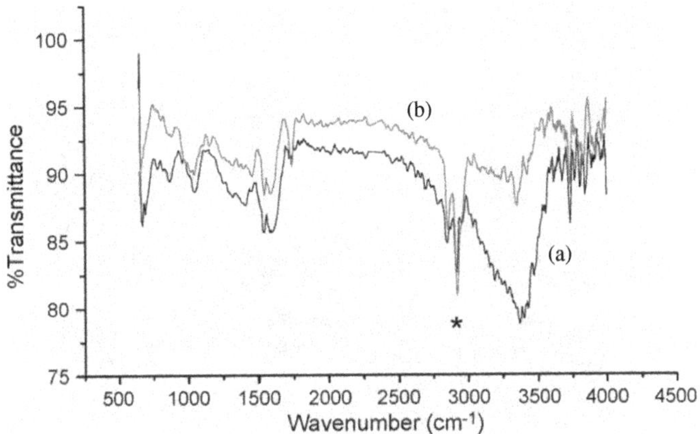

Figure 4.2: FTIR spectrum (curve (a) amino-magnetic microsphere; curve (b) C8-modified amino-magnetic microsphere).

the amino-magnetic microsphere, indicating that the modification of C8 introduces a large number of alkyl chains. [23]

(2) Characterization of Fe_3O_4@SiO_2-C8

Figure 4.3 shows the SEM image of Fe_3O_4@SiO_2-C8 microsphere, from which it can be seen that the magnetic material has good uniformity and dispersibility.[2] Figure 4.4 shows the TEM images of Fe_3O_4@SiO_2 and Fe_3O_4@SiO_2-C8, from which it can be seen that the magnetic microsphere with a core–shell structure is about 300 nm, and the shell layer is about 60 nm. The saturation magnetic value of the material is 65.5 emu·g^{-1}. Figure 4.5 shows the FTIR spectra of Fe_3O_4@SiO_2 and Fe_3O_4@SiO_2-C8. The peak of about 580 cm^{-1} corresponds to the vibration absorption of Fe–O–Fe, and the peak of about 1,095 cm^{-1} corresponds to the stretching vibration absorption of Si–O–Si. After silanization modification, an absorption peak of about 2,900 cm^{-1} corresponds to CH_2 of the silane coupling reagent, confirming that the alkyl chain is successfully bonded to the surface of the magnetic silica. In order to further determine the amount of alkyl groups, Fe_3O_4@SiO_2 and Fe_3O_4@SiO_2-C8 are characterized by TGA analysis. Figure 4.6 shows the weight loss curves of Fe_3O_4@SiO_2 and Fe_3O_4@SiO_2-C8, respectively, from which their weight loss is 25.7% and 19.2%,

Figure 4.3:   SEM image of Fe_3O_4@SiO_2-C8.

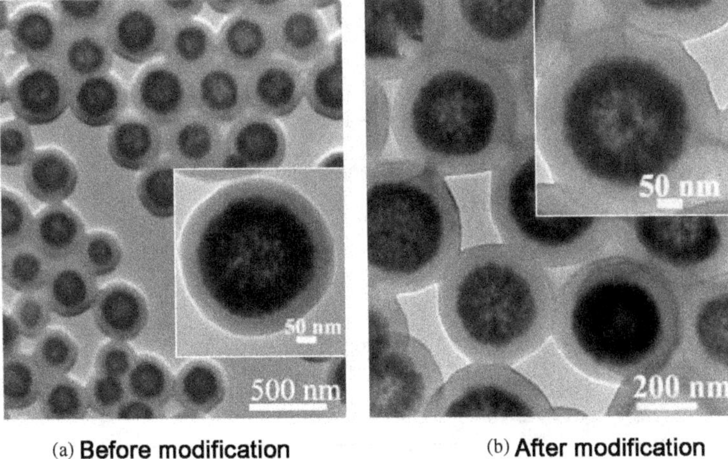

Figure 4.4: TEM images of magnetic silica before and after C8 modification (the inset is an enlarged view).

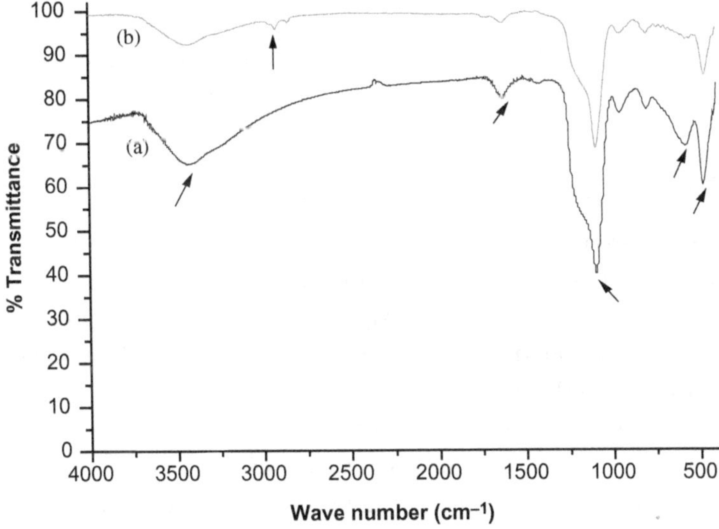

Figure 4.5: FTIR spectra (curve (a) $Fe_3O_4@SiO_2$; curve (b) $Fe_3O_4@SiO_2$-C8).

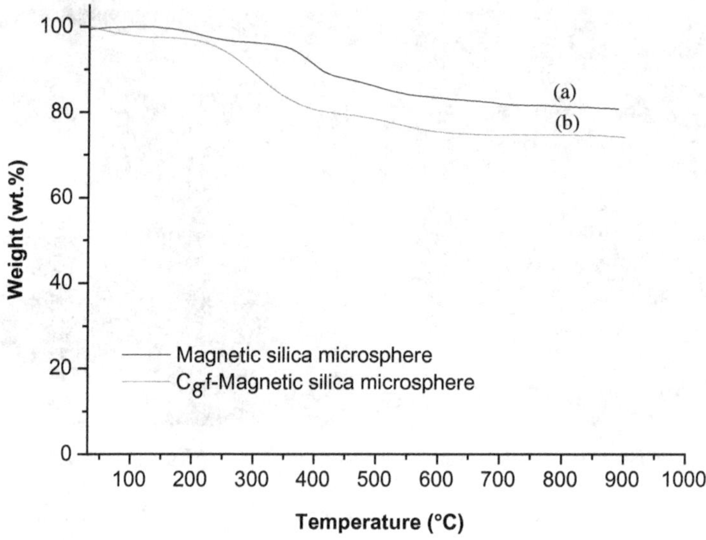

Figure 4.6: TGA analysis (curve (a) $Fe_3O_4@SiO_2$; curve (b) $Fe_3O_4@SiO_2$-C8; test temperature: 25–900°C, heating rate: 5°C·min^{-1}).[2]

respectively, indicating that the content of the alkyl groups in $Fe_3O_4@SiO_2$-C8 is about 6.5 wt.%.

(3) Characterization of $Fe_3O_4@CP$-C8

Figure 4.7 shows that monodisperse $Fe_3O_4@CP$-C8 has a smooth surface and an average particle size of about 300 nm.[24] Figure 4.8 shows the FTIR spectrum of $Fe_3O_4@CP$-C8 microsphere. The peaks at ~3,500 cm^{-1} and ~1,630 cm^{-1} correspond to the absorption of water adsorbed on the particle. The peaks at ~1,700 cm^{-1} and ~1,625 cm^{-1} correspond to the swing absorption of C=O and C=C, respectively, which are from the dehydrated aromatization structure of glucose on the carbon sphere during hydrothermal reaction. The peaks in the range of 1,200–1,400 cm^{-1} can be identified as the stretching and bending vibrations of –OH, indicating that glucose still remains in a large amount of hydroxyl groups during carbonization and dehydration. These large amounts of –OH and –C(H)=O groups covalently bonded to the carbon framework endow the microsphere with both good

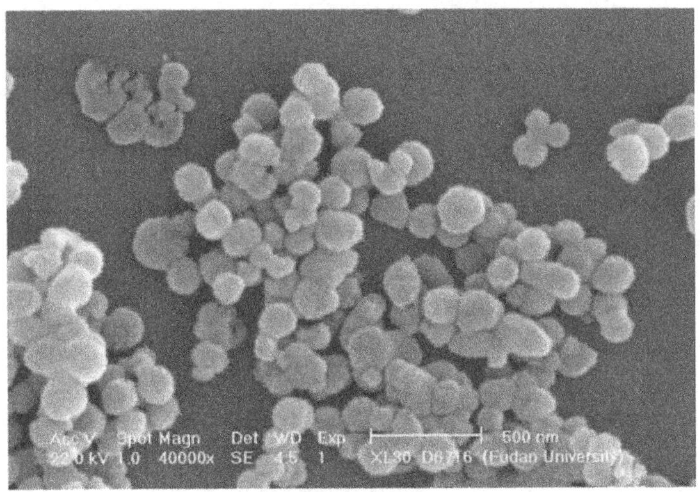

Figure 4.7: SEM image of Fe_3O_4@CP-C8.

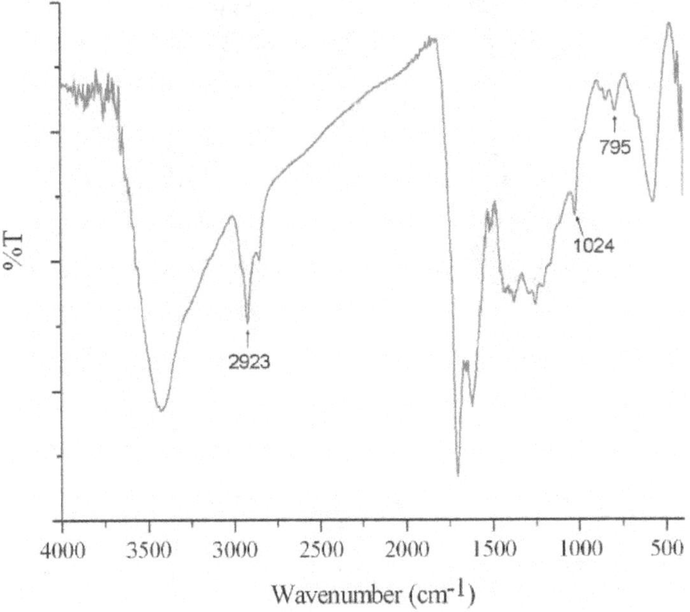

Figure 4.8: FT-IR spectrum of Fe_3O_4@CP-C8.

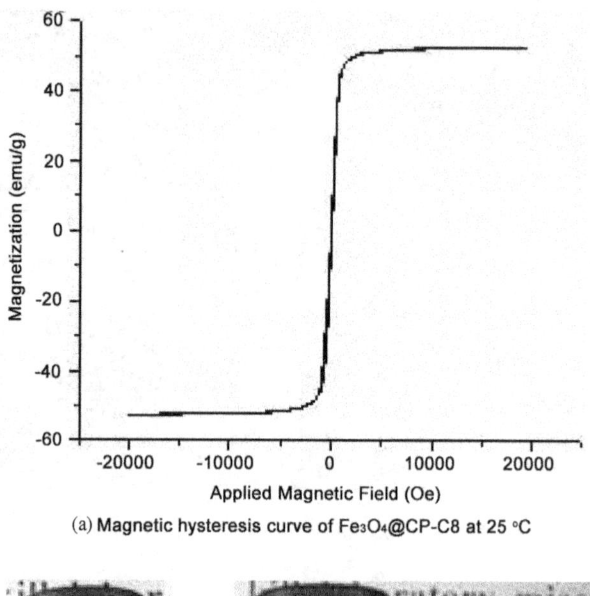

(a) Magnetic hysteresis curve of Fe₃O₄@CP-C8 at 25 °C

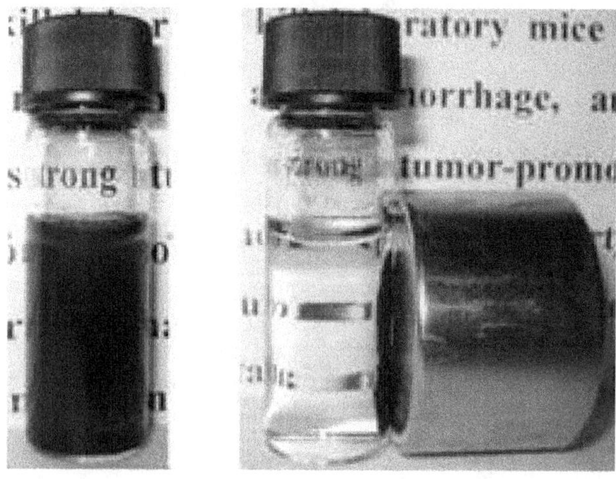

(b) Dispersion photograph of microsphere in water

(c) Separation photograph of microsphere from water by a magnet

Figure 4.9: Magnetic analysis of Fe_3O_4@CP-C8.

hydrophilicity and stability in aqueous systems. The peaks at ~2,923 cm⁻¹, ~1,024 cm⁻¹, and ~795 cm⁻¹ correspond to the absorption of CH_2 and O–Si–C in the silane coupling reagent, indicating that the C8 alkyl chain is successfully modified on the microsphere. The magnetic characterization (Figure 4.9(a)) shows that the Fe_3O_4@CP-C8 microsphere has excellent superparamagnetism and high magnetization ability, and its saturation magnetic value is about 52.0 emu·g^{-1}. The high magnetic property and superparamagnetic property make the Fe_3O_4@CP-C8 microsphere have a very fast response in the magnetic field and have good dispersion after removing the magnetic field. To prove this, the Fe_3O_4@CP-C8 microsphere is first dispersed in water, and then a magnet (2,000 Oe) is placed outside the centrifuge tube. The microsphere can all be adsorbed on the tube wall within 2 s as seen in Figures 4.9(b) and 4.9(c).

4.2.2.3. *C8-modified magnetic microsphere for the enrichment of low-abundance peptides/proteins*

(1) C8-modified amino-magnetic microsphere for the enrichment of low-abundance peptides/proteins

5 nmol·L^{-1} MYO digest is used as the enrichment sample to investigate the enrichment effect of C8-modified amino-magnetic microsphere on peptide.[23]

As seen from Figure 4.10, there is no signal in the direct MS detection of 5 nmol·L^{-1} MYO digest (Figure 4.10(a)). While after enrichment for 1 h by C8-modified amino-magnetic microsphere, the MS spectrum of 1 mL MYO digest shows the signal-to-noise (S/N) ratio of peptide with strongest signal (Mr = 1606.5) reaches 8,000 (Figure 4.10(b)), indicating that this material has a good enrichment effect on the peptide. In addition, different enrichment time (30 s, 5 min, 1 h) is adopted for the same sample, the results are shown in Figure 4.11. It can be seen that MS signal is enhanced along with the prolonging of enrichment time, but 30 s is sufficient for peptide detection. And experiment indicates that this material has good reproducibility for the enrichment of standard protein digest.

(2) C8-modified magnetic silica for the enrichment of low-abundance peptide/protein

A standard peptide (MW = 1046.2, pI = 6.74) is used as sample first. The signal is weak and the S/N ratio is only 13.88 when detecting 5 nmol·L^{-1} standard peptide solution directly by MS. While the standard peptide solution (1 mL 5 nmol·L^{-1}) is enriched by C8-modified magnetic silica (20 μL 2 mg·mL^{-1}), the MS signal is enhanced significantly and the S/N ratio reaches

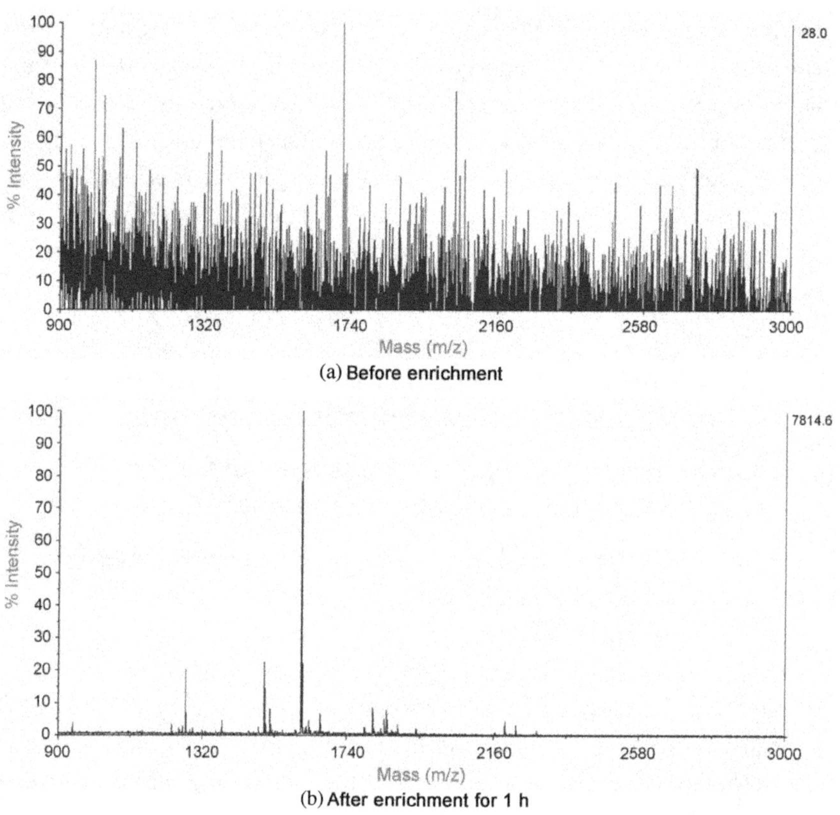

Figure 4.10: MS spectra of MYO (1 mL 5 nmol·L^{-1}) digest before and after enrichment by C8-modified amino-magnetic microsphere.

1326.04 (Figure 4.12). This experiment initially proves that this material has a good enrichment effect on the peptide.

In order to further verify the universality of this material for enriching peptides, 5 nmol·L^{-1} of BSA digest and MYO digest are used as samples. As seen in Figure 4.13, 7 peptides of BSA digest and none of the peptides in MYO digest are detected before enrichment, but after enrichment by C8-modified magnetic silica, 26 peptides of BSA digest are identified and the peptide coverage is 42%, 11 peptides of MYO digest are detected and the peptide coverage is 79%. The S/N ratio of the enriched peptide is significantly improved. The above results show that this magnetic material can be applied to the enrichment of peptides in complex mixtures.

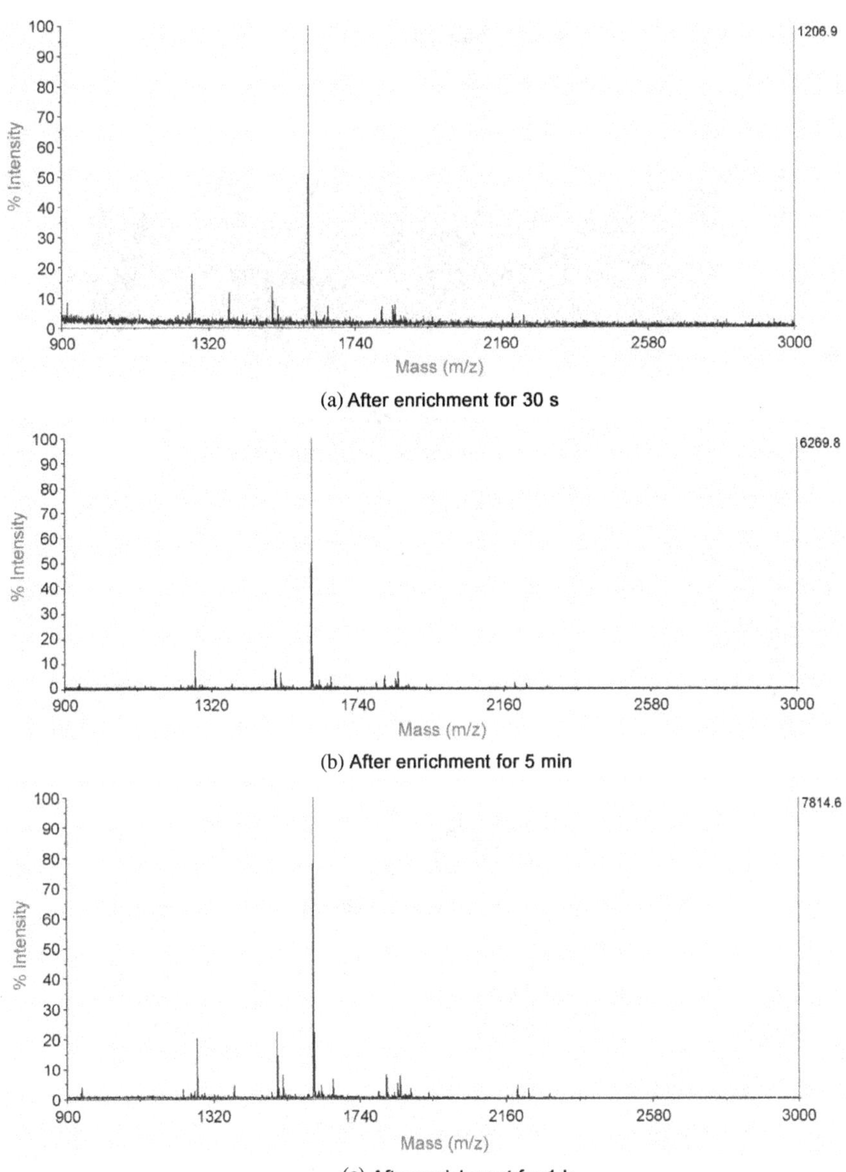

Figure 4.11: MS spectra of MYO (1 mL 5 nmol·L^{-1}) digest after enrichment by C8-modified amino-magnetic microsphere.

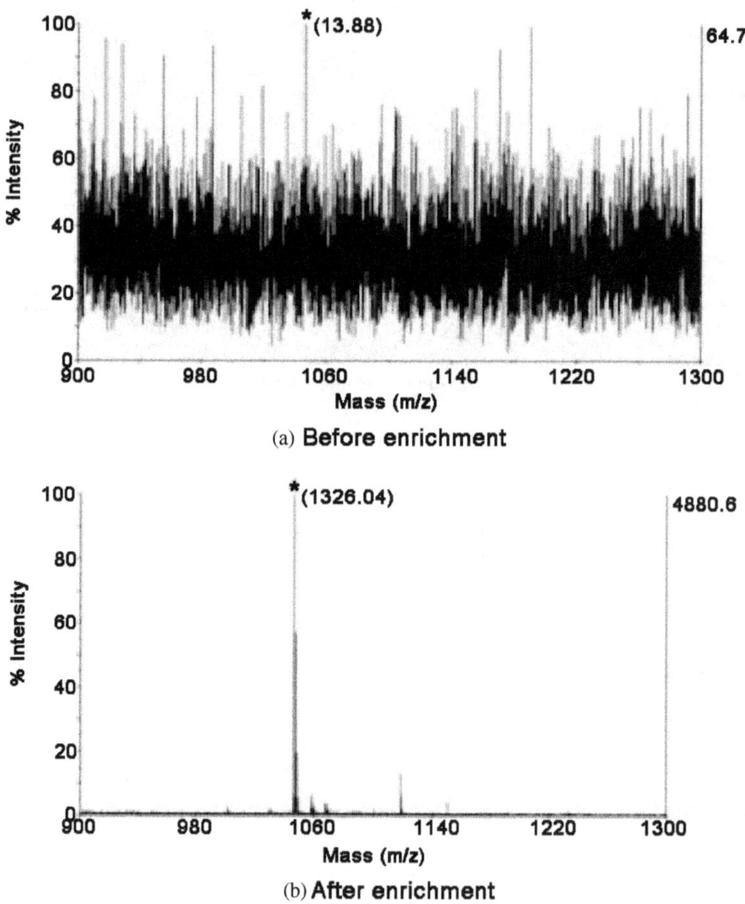

Figure 4.12 MALDI-TOF-MS spectra of standard peptide (MW = 1046.2, pI = 6.74, 1 mL 5 nmol·L^{-1}) before and after enrichment by C8-modified magnetic silica. "*" represents the signal peak of the standard peptide, the number in parenthesis represents S/N ratio.

In addition, the standard protein of Cyc is used to investigate the enrichment ability of Fe_3O_4@SiO_2-C8 for low-abundance protein. The MS signal of 0.4 ng·μL^{-1} Cyc solution (1 mL) is weak and the S/N ratio is only 16.97. After enrichment with Fe_3O_4@SiO_2-C8 (0.04 mg, dispersed in 20 μL H_2O), no Cyc signal is detected in the supernatant, while the signal of the eluent is significantly enhanced, and the S/N ratio is 351.38 (Figure 4.14), proving that this material has a good enrichment ability for low-abundance protein.

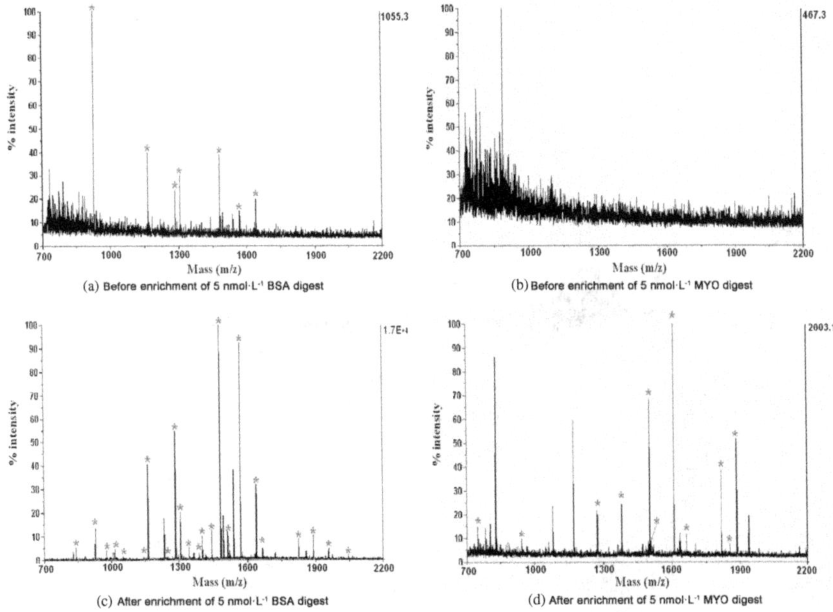

Figure 4.13: MALDI-TOF-MS spectra of BSA digest and MYO digest (5 nmol·L^{-1}) before and after enrichment by C8-modified magnetic silica. "*" represents the target peptide.

(3) C8-modified magnetic polymeric carbon microsphere for the enrichment of low-abundance peptide/protein

The standard peptide of Angiotensin II (amino acid sequence DRVYIHPF, MW = 1046.2, *pI* = 6.74) at a concentration of 5 nmol·L^{-1} is used to investigate the enrichment performance of Fe$_3$O$_4$@CP-C8 microsphere for peptide.[24] Figure 4.15 shows that the S/N ratio of this standard peptide increases to 1103.82 after enrichment by Fe$_3$O$_4$@CP-C8 microsphere, indicating that Fe$_3$O$_4$@CP-C8 microsphere has better ability to enrich low-abundance peptide. On the one hand, the good ability is attributed to the small particle size and large specific surface area of Fe$_3$O$_4$@CP-C8 microsphere which improves the enrichment efficiency of this microsphere. On the other hand, the good aqueous dispersion offered by polymeric carbon layer of Fe$_3$O$_4$@CP-C8 is beneficial to their interaction with peptide in the sample.

Subsequently, BSA digest is selected to investigate whether Fe$_3$O$_4$@CP-C8 microsphere has universal enrichment ability for most peptides.

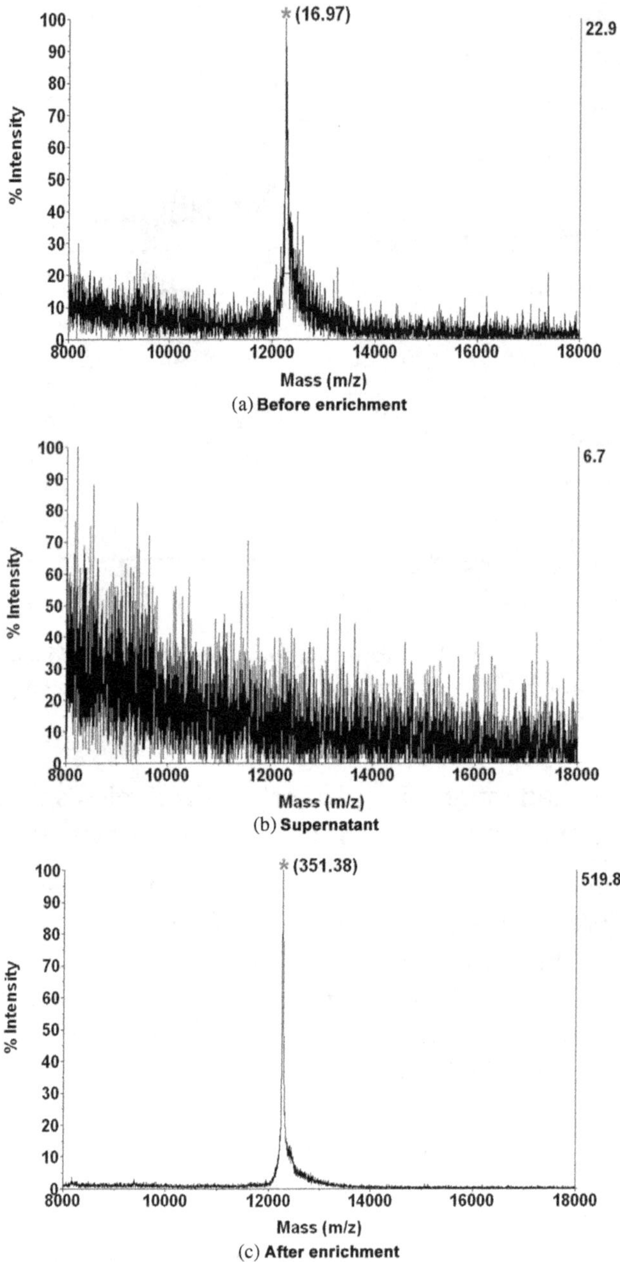

Figure 4.14: MALDI-TOF-MS spectra of Cyc (0.4 ng·μL^{-1}) before and after enrichment by C8-modified magnetic silica. "*" represents Cyc, the number in parenthesis represents S/N ratio.

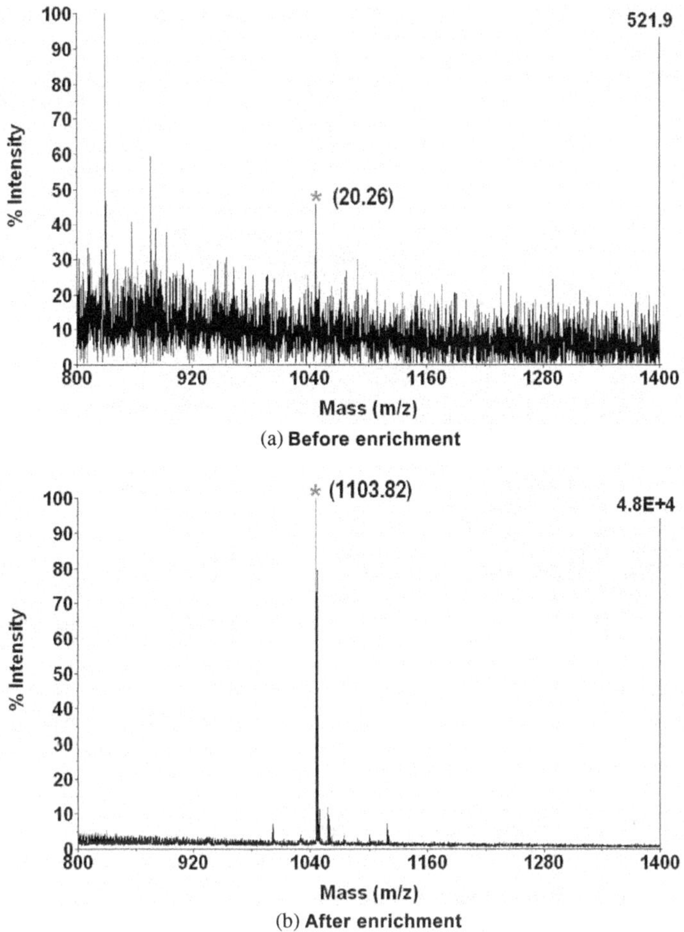

Figure 4.15: MALDI-TOF-MS spectra of Angiotensin II (0.4 mL 5 nmol·L^{-1}) before and after enrichment by Fe$_3$O$_4$@CP-C8. "*" represents Angiotensin II, the number in parenthesis represents S/N ratio.

The practical sample generally contains some salt, which may be contained in the sample itself or may have to be added during sample processing. The existence of salt can severely interfere with the crystallization of the analyte and matrix, resulting in a poor MALDI MS spectrum. Therefore, a one-step desalting process is usually required after protein concentration for obtaining a satisfactory result. Urea is a commonly used

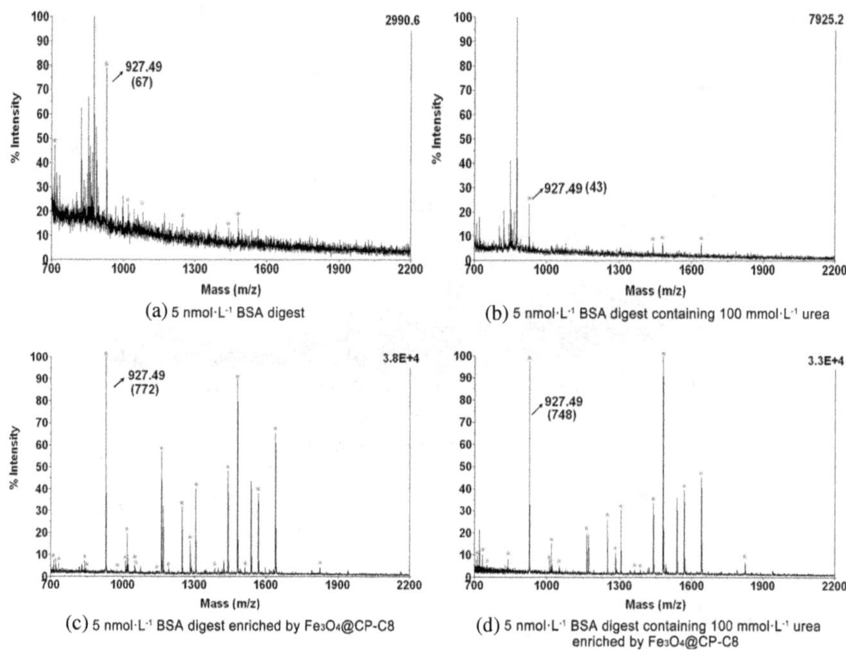

Figure 4.16: MALDI-TOF-MS spectra of sample without and with urea after enrichment by Fe_3O_4@CP-C8. "*" represents target peptide, the number in parenthesis represents S/N ratio.

reagent in sample processing. Therefore, urea as an interference is added into BSA hydrolysate. The sample with and without urea is enriched by Fe_3O_4@CP-C8 microsphere. The results of MALDI-TOF-MS analysis are shown in Figure 4.16. For observation and comparison, the peaks of the BSA digest are marked with "*", and the molecular weight and S/N ratio of the peak with the highest intensity are also marked. As shown in Figures 4.16(a) and 4.16(b), few peptides are identified before enrichment. In contrast, the S/N ratio of sample containing urea is worse and the number of detected peptides is less. After enrichment by Fe_3O_4@CP-C8 microsphere, 23 and 22 peptides are detected, respectively, from the samples without and with urea (Figures 4.16(c) and 4.16(d)). It is worth noting that the Fe_3O_4@CP-C8 microsphere exhibits a good enrichment ability for peptide of BSA hydrolysate samples with or without urea, and the enrichment results for these two samples are basically same. This result indicates that this material can eliminate the interference of salt when enriching peptides, and thus the sample treated by this method can be exempted from the subsequent desalting process.

4.2.2.4. *C8-modified magnetic microsphere for the enrichment of peptide in practical biological sample*

Blood is an ideal source of disease diagnosis, and it is convenient and quick to diagnose the disease by discovering biomarker in the blood. The identification of free peptide in plasma and serum is important for the discovery of biomarker for disease identification. However, it is difficult to analyze the blood sample due to the extreme complexity of itself and the high dynamic range of variation. Therefore, the development of extraction method of free peptide in serum is a hot topic of current research. There are two main methods currently used to separate peptide in plasma and serum. High-speed centrifugation with ability of precise relative molecular weight cutoff is currently widely used, which extracts peptide by removing large molecular weight protein based on the principle of size exclusion. However, this method is time-consuming and laborious, and some low molecular weight impurities (such as salt) are also concentrated at the same time as enriching the peptide. Another method is to use a material with adsorption property to enrich the peptide from plasma or serum sample. This method can remove the interference of impurity in the sample when enriching the peptide, and the research in this aspect has been extensive. Based on this, the authors' group initially applied C8-material to the enrichment of free peptide in human serum, and explored its potential in this field.

(1) C8-modified amino-magnetic microsphere for the enrichment of free peptides in serum

10 μL healthy serum is diluted with 20 μL ultrapure water, then the suspension of C8-modified amino-magnetic microsphere is added for intensive mixing for 30 s.[23] After magnetic separation, the material is washed twice with 20 μL H$_2$O. Finally, 10 μL CHCA matrix solution is added for MS detection. Figure 4.17 shows that 16 peptides are detected.

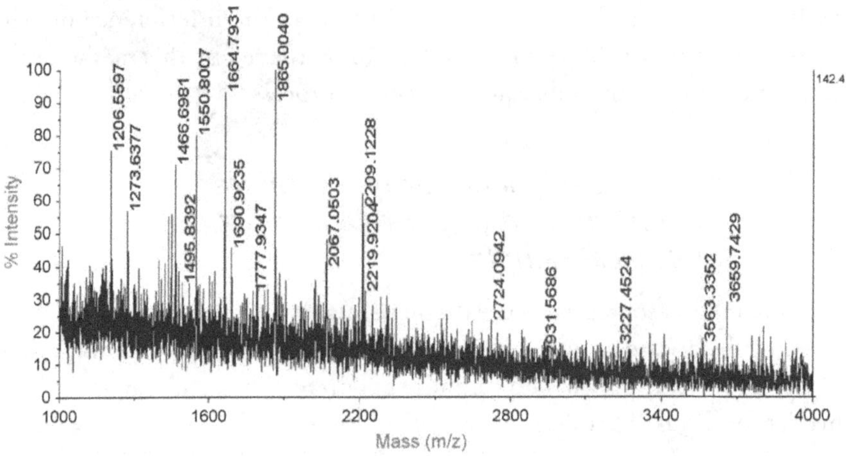

Figure 4.17: MALDI-TOF-MS spectrum of serum after enrichment by C8-modified amino-magnetic microsphere.

(2) C8-modified magnetic silica microsphere for the enrichment of free peptides in serum

Figure 4.18 shows the results of MS identification of peptides in a healthy human serum sample using C8-modified magnetic silica microsphere, from which it can be seen that more than 10 peptides in the serum have been successfully enriched.[2] The detailed information is listed in Table 4.1.

(3) Fe_3O_4@CP-C8 microsphere for the enrichment of protein dot hydrolysate from on-gel separation and enzymatic hydrolysis

The two-dimensional gel electrophoresis (2DE) technology with high reproducibility, operability, high resolution and other properties is still a powerful tool in proteomic analysis.[24] The traditional combination of 2DE with mass spectrometry usually involves the following steps. First, the biological sample is separated by 2DE. Then the protein dots separated by 2D gel are subjected to enzymatic hydrolysis on the gel and peptide extraction. Afterwards, the peptide extracts are lyophilized, re-dissolved, and then transferred to a target plate for analysis to obtaining protein information. This is the common

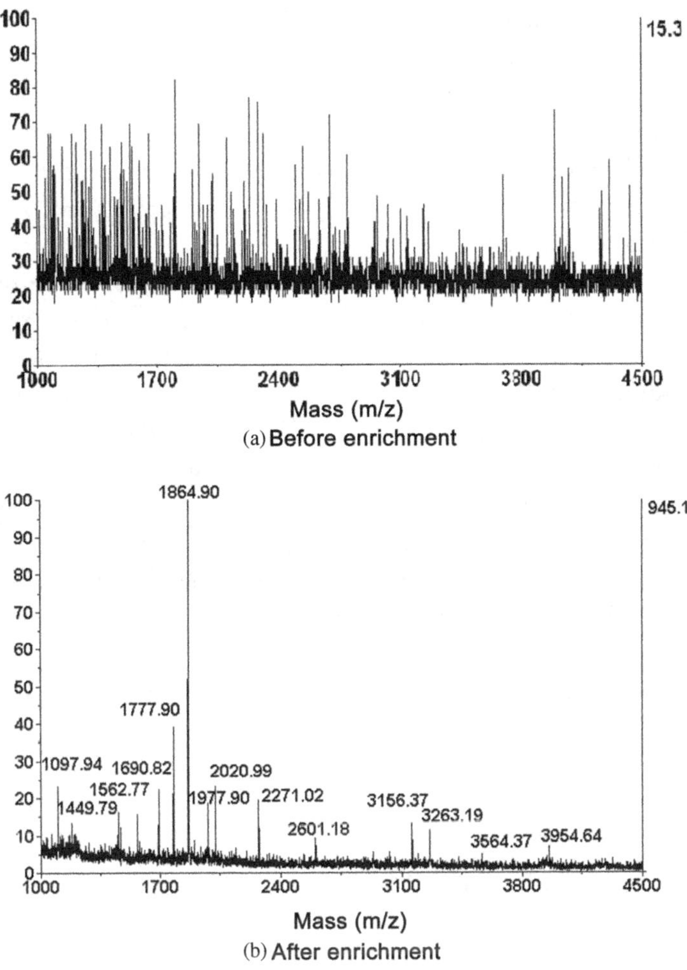

Figure 4.18: MALDI-TOF-MS spectra of serum peptide before and after enrichment by $Fe_3O_4@SiO_2$-C8 microsphere.

Table 4.1: Peptide information identified from human serum using $Fe_3O_4@SiO_2$-C8 microsphere.

Peptide Sequence	[M–H]$^+$	Score	Accession
A.GVSWRLPGKLVKKICFCS.P	2020.99	61	gi|47522368
L.VGDVAQAADVAIEHRNEA.E	1864.90	68	gi|17978483

"middle-down" technique in proteomics. In the lyophilization process, not only will some samples adhere to the tube wall but NH_4HCO_3 contained in the extract liquid (a solution for extracting peptide after enzymatic hydrolysis) is also concentrated, resulting in a low mass spectrometry success rate (about 50%). Based on this, the C8-material is applied to the enrichment of peptide obtained from on-gel enzymatic hydrolysis, and its application potential in this field has been studied.

Fe_3O_4@CP-C8 microsphere shows an advantage in the pretreatment of on-gel protein. After the separation of protein in the human lens by 2DE, a slightly lighter colored protein dot (Figure 4.19) is selected for on-gel enzymatic hydrolysis. The enzymatically decomposed peptide fragments are extracted 3 times with 50 μL extracting solution, then these extracting solutions are mixed and diluted with water to 400 μL to reduce the acetonitrile concentration. Then they are divided into two equal parts. Fe_3O_4@CP-C8 microsphere is added into one part for enrichment and the other part is not dealt with. No reliable proteins can be obtained in the untreated part (protein score greater than 64 is considered reliable). However, after enrichment with Fe_3O_4@CP-C8 microsphere, a protein with a score of up to 96 is identified. Figure 4.20 shows the MALDI-TOF-MS spectrum, and Table 4.2 shows the

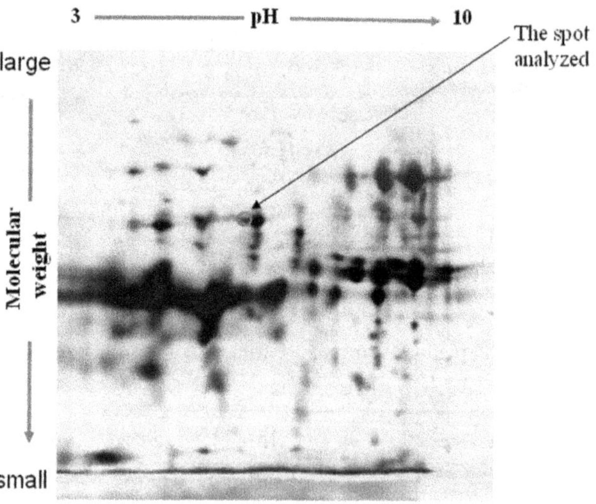

Figure 4.19: Separation diagram of human eye lens by 2DE. A lighter colored protein dot (shown by the arrow) on the gel for subsequent analysis.

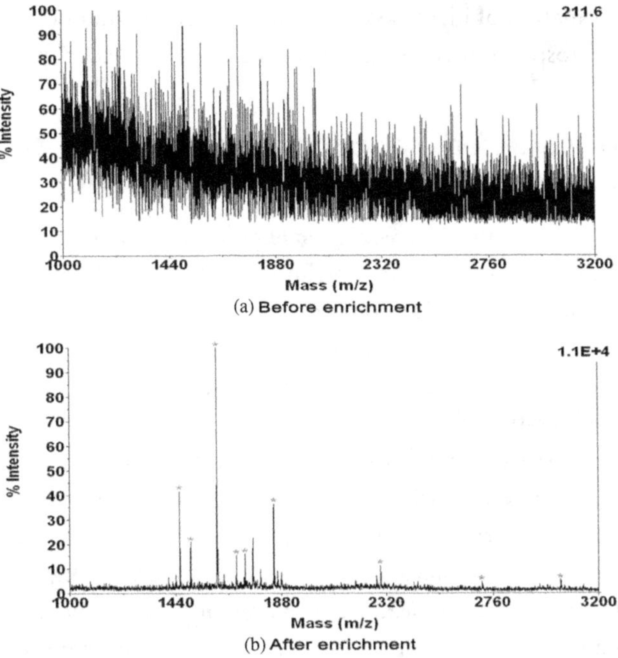

Figure 4.20: MALDI-TOF-MS spectra of extracting solution of human eye lens (separated by 2DE and digested by on-gel enzymatic hydrolysis) before and after enrichment by $Fe_3O_4@$ CP-C8 microsphere.

Table 4.2: Peptide information identified from human eye lens using $Fe_3O_4@SiO_2$-C8 microsphere.

Calculated m/z	Detected m/z	Sequence
1455.7114	1455.7206	ITIYDQENFQGK
1500.7515	1500.774	MTIFEKENFIGR
1611.8125	1611.8306	ITIYDQENFQGKR
1691.7561	1691.7736	WDAWSGSNAYHIER
1727.8459	1727.8635	EWGSHAQTSQIQSIR
1845.9077	1845.8767	ETQAEQQELETLPTTK
1847.9255	1847.8898	LMSFRPICSANHKESK
1863.9204	1863.9019	LMSFRPICSANHKESK
2294.0374	2294.0801	GEYPRWDAWSGSNAYHIER

Note: Protein name, crystalline, beta A3 [Homo sapiens]; Accession, gi|12056461; Score 96; sequence coverage, 42%.

peptide information of identified protein. This demonstrates the feasibility of magnetic microsphere-based enrichment method in practical proteomics.

4.2.3. *OA-modified magnetic nanoparticle for the enrichment of low-abundance peptide*

Oleic acid (OA) is a linear carboxylic acid commonly used in the laboratory, which has a carboxyl group at one end and a hydrophobic linear alkenyl group at the other end.[25] Oleic acid-modified nanoparticle is easy to be synthesized and is a kind of hydrophobic magnetic material which is commonly used in material synthesis science. The synthetic process of such particles is well known in the literature. Section 2.3 of Chapter 2 introduces the OA-modified magnetic nanoparticle which can be synthesized in one step by a one-pot method. This magnetic particle has small particle size, large specific surface area, and chelating end of carboxy group, which can utilize the linear hydrophobic end of the oleic acid molecule as an affinity ligand to achieve enrichment of the peptide, leading to its application in the enrichment of low-abundance peptide. Since the surface of OA-Fe_3O_4 nanoparticle has a large number of hydrophobic ends of oleic acid molecules, the particle can enrich the peptide in the solution by hydrophobic interaction, which is very similar to the C8 or C18-modified RPLC packing.

The OA-Fe3O4 nanoparticle has strong magnetic property, which enables the material to be gathered and dispersed flexibly in the presence of a magnetic field and a non-magnetic field. This improves the efficiency for its practical application. In the process of using it to separate the peptide in the solution, there is no tedious and lengthy centrifugation operation, which greatly shortens the time of sample processing.

4.2.3.1. *OA-modified magnetic nanoparticle for the enrichment of standard peptide*

The standard peptide of Angiotensin II (MW = 1046.2, pI = 6.74, amino acid sequence DRVYIHPF, 0.4 mL 5 nmol·L^{-1}) is first used as a sample to investigate the enrichment efficiency of OA-Fe_3O_4 nanoparticle for peptide. 10 min is used as the enrichment time. As seen from Figure 4.21(a), the S/N ratio of the 5 nmol·L^{-1} Angiotensin II before enrichment is 12.23 by MALDI-

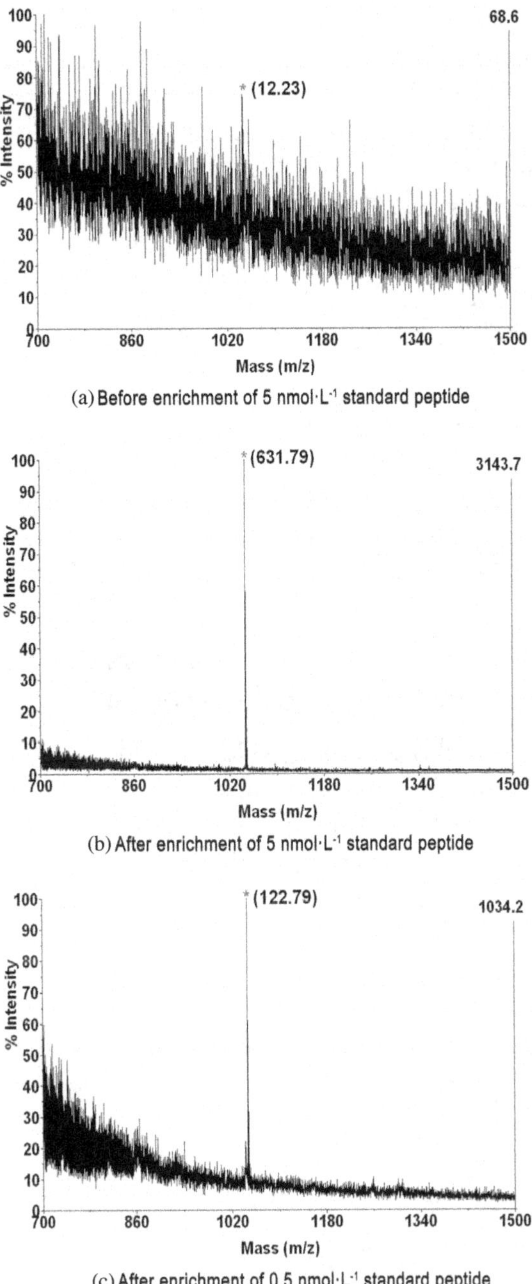

Figure 4.21: Standard peptide for testing the enrichment efficiency of OA-Fe$_3$O$_4$ nanoparticle. The number in parenthesis represents S/N ratio.

TOF-MS analysis. After 0.4 mL Angiotensin II is enriched by 0.05 mg OA-Fe_3O_4 nanoparticle, the S/N ratio is significantly enhanced to 631.79 (Figure 4.21(b)), which is more than 50 times of that before enrichment. Thereafter, OA-Fe_3O_4 nanoparticle is used to enrich the lower concentration of Angiotensin II aqueous solution (5 nmol·L^{-1}), the peptide peak is clearly visible, and the S/N ratio is 122.79 (Figure 4.21(c)).

4.2.3.2. *OA-modified magnetic nanoparticle for the enrichment of standard protein digest*

MYO hydrolysate is used to investigate the enrichment extensiveness of OA-Fe_3O_4 nanoparticle for peptide. As seen in Figure 4.22, 14 peptides are identified from 0.4 mL 5 nmol·L^{-1} MYO hydrolysate after enrichment by OA-Fe_3O_4 nanoparticle. Compared to that before enrichment, the S/N ratio is higher, indicating that OA-Fe_3O_4 nanoparticle has a good enrichment effect on complex peptide mixture.

4.2.3.3. *OA-modified magnetic nanoparticle for the enrichment of peptide in serum*

Human serum is readily available and is one of the most commonly used samples for clinical diagnosis. Human serum contains a large amount of protein and some inorganic salts and so forth, the composition of which is very complicated. The serum contains some endogenous peptides, which are now thought to be produced during the metabolism of human protein. Although these peptides are low in content, they may contain valuable metabolic or disease information. If the serum is not treated, it is impossible to analyze these small peptides by MALDI-TOF-MS as shown in Figure 4.23(a). While the 3 times of diluted serum is treated by 0.025 mg OA-Fe_3O_4 nanoparticle, more than a dozen peptides are clearly visible

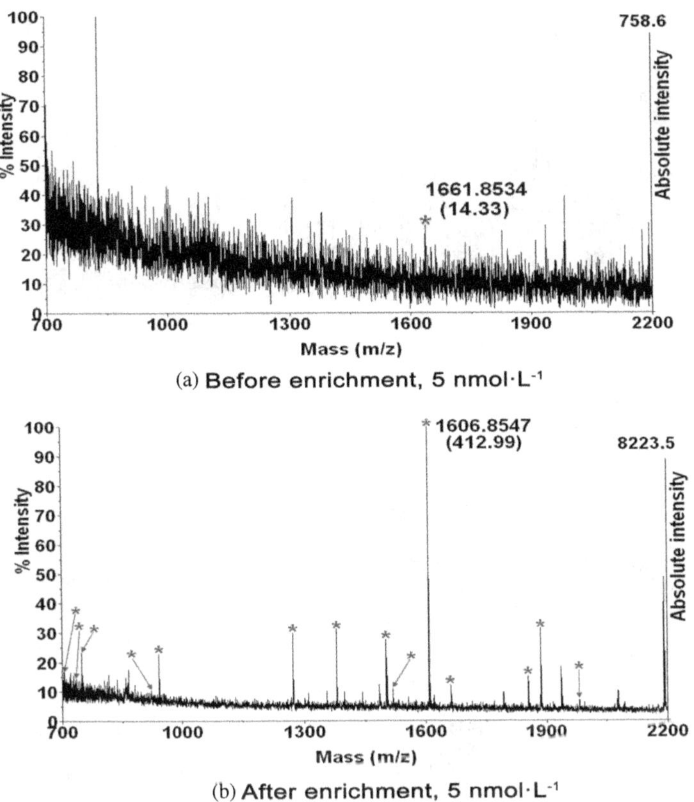

Figure 4.22: MYO hydrolysate for testing the enrichment efficiency of OA-Fe_3O_4 nanoparticle.

(Figure 4.23(b)). This is probably attributed to the small particle size, large specific surface area, and plenty of surface-modified hydrophobic oleic acid molecules, which provides a number of binding sites. And the above result indicates that this material has a good application potential in practical biological sample.

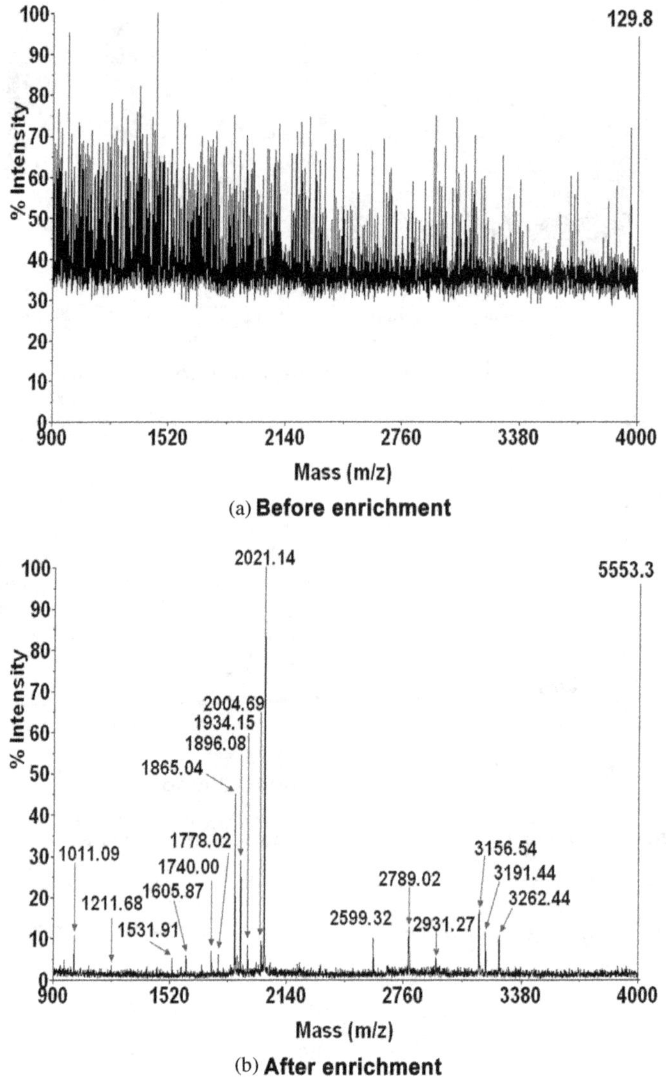

Figure 4.23: MS spectra of diluted serum before and after enrichment by OA-Fe_3O_4 nanoparticle.

4.2.4. *Conclusion*

This section introduces the magnetic micro-/nanomaterial with alkyl chains as reversed-phase affinity ligand for the enrichment of low-abundance peptide. C8-modified magnetic microsphere has good dispersibility in water and good biocompatibility, since the surface of the microsphere contains both linear hydrophobic C8 chain and a large number of unmodified hydrophilic functional groups. The hydrophilic groups are attributed to the presence of a large number of easily being modified hydroxyl, aldehyde, carboxyl and so forth on amino-magnetic microsphere, magnetic silica microsphere and polymeric carbon layer. After enrichment, the enriched peptide can be spotted together with the material on the target plate without the elution step. The OA-modified magnetic nanoparticle is synthesized via one step, the particle size of which is small, the specific surface area of which is large, the carboxyl end of which is chelated on the ferroferric oxide. And OA-modified magnetic nanoparticle utilizes the linear hydrophobic end of the oleic acid molecule as the affinity ligand to achieve the enrichment of peptide. These materials have good magnetic responsiveness, simple enrichment process and good enrichment effect, and all of them have been applied in practical proteomic analysis. The results show excellent feasibility of this magnetic enrichment method in practical samples.

4.3. Magnetic material with C60 as affinity ligand for the enrichment of low-abundance peptide/protein

4.3.1. *Basic principle*

This section describes a C60-functionalized magnetic silica ($Fe_3O_4@SiO_2$-C60) and its application in enrichment of low-abundance proteomics.[26] The $Fe_3O_4@SiO_2$-C60 microsphere is easy to prepare. The characterization of TEM and FTIR shows that the C60 content is high. C60 is modified tightly to the material by copolymerization. The material combines many advantages such as the superparamagnetism and strong magnetic responsiveness of Fe_3O_4 nanoparticle, the protection ability of silica layer to the internal magnetic nanoparticles and the surface of easily to being modified, the characteristics of C60 molecules. The microsphere is used to selectively enrich peptide/protein in solution by using C60 as a functional group for interaction with peptide/protein. At the same time, the microsphere has good loading property and desalination property, and can be used to enrich peptide/protein in human urine directly.

4.3.2. *Synthesis of C60-functionalized magnetic silica*

4.3.2.1. *Synthesis of $Fe_3O_4@SiO_2$*

The synthesis of $Fe_3O_4@SiO_2$ is described in Section 2.4 of Chapter 2. The synthesized $Fe_3O_4@SiO_2$ is dried in vacuum at 60°C for 12 h for later use.

4.3.2.2. *Modification of C60 on surface of $Fe_3O_4@SiO_2$*

As shown in Figure 4.24, the 3-methacryloxy propyltrimethoxysilane (MPS) is first modified on the surface of $Fe_3O_4@SiO_2$ as follows: 20 mg $Fe_3O_4@SiO_2$, 160 mL ethanol, 40 mL H_2O and 2.0 mL ammonia solution are added into a flask for ultrasonic dispersion. Then 2.0 mL MPS is injected with a syringe under mechanical stirring at 25°C. Next, after stirring at room temperature for 24 h, an MPS-modified $Fe_3O_4@SiO_2$ is formed. The MPS-modified $Fe_3O_4@SiO_2$ is separated by a magnet and washed several times with ethanol, and dried under vacuum at 60°C for 12 h for later use.

Secondly, the MPS-modified $Fe_3O_4@SiO_2$ and C60 are polymerized under the action of the catalyst. The operation is as follows: 10 mg MPS-modified $Fe_3O_4@SiO_2$, 5 mg C60 and 80 mL toluene solution containing 1 mg azodiisobutyronitrile (AIBN) are added into a three-necked flask. After the nitrogen gas is passed through the mixture for 30 min, the water bath is heated to 70°C for 6 h to polymerize C60 onto the magnetic microsphere. Finally, the C60-modified $Fe_3O_4@SiO_2$ is separated by a magnet, and the residual impurities are washed with ethanol. The C60-modified $Fe_3O_4@SiO_2$ is dried at 40°C.

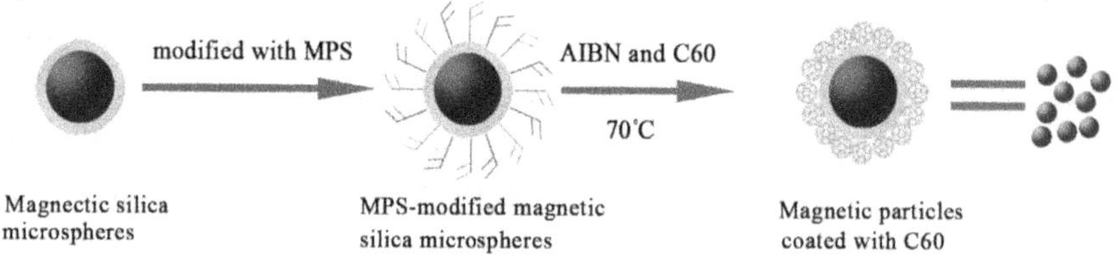

Figure 4.24: Schematic diagram of synthesis of C60-functionalized magnetic silica.

4.3.3. *Characterization of C60-functionalized magnetic silica*

It is observed by TEM that the magnetic silica has a Fe_3O_4 core with a particle diameter of about 250 nm and a SiO_2 shell with about 20 nm of thickness (Figure 4.25(a)). After modification of C60 by MPS coupling, the structure of original magnetic silica is maintained, but a thin layer is formed outside the silica (Figure 4.25(b)), indicating that C60 has been modified onto the magnetic microsphere. FTIR (Figure 4.2(6)) shows that an absorption peak at 1,730 cm^{-1} appears after the modification of MPS compared to magnetic silica, indicating that successful modification of MPS. After copolymerization with C60, the characteristic absorption peaks of C60 including 1,180 cm^{-1}, 1,143 cm^{-1}, and 530 cm^{-1} are shown in the FTIR spectrum, demonstrating the successful synthesis of C60-modified magnetic silica.

The saturation magnetic values of the Fe_3O_4 particle, magnetic silica, $Fe_3O_4@SiO_2$-MPS and $Fe_3O_4@SiO_2$-C60 are measured at 300 K (the applied magnetic field strength is $-20,000$ Oe to $20,000$ Oe) using the vibrating-sample magnetometer (VSM) (EG&G Princeton Applied Research Vibrating Sample Magnetometer, Model 155). The results show that these four

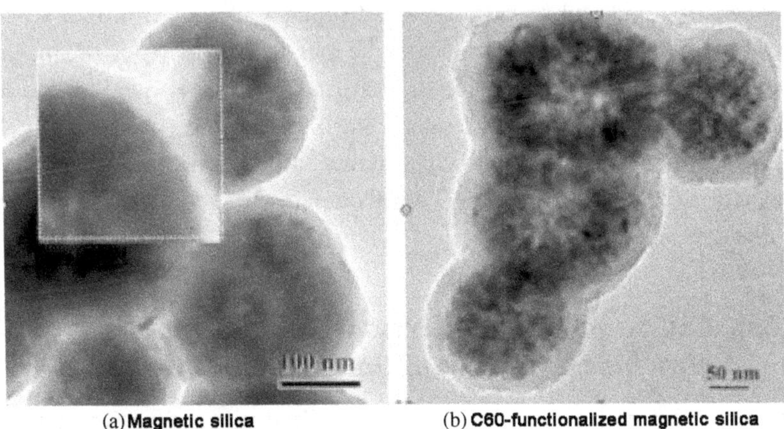

(a) **Magnetic silica**   (b) **C60-functionalized magnetic silica**

Figure 4.25:   TEM images of magnetic silica and C60-functionalized magnetic silica.

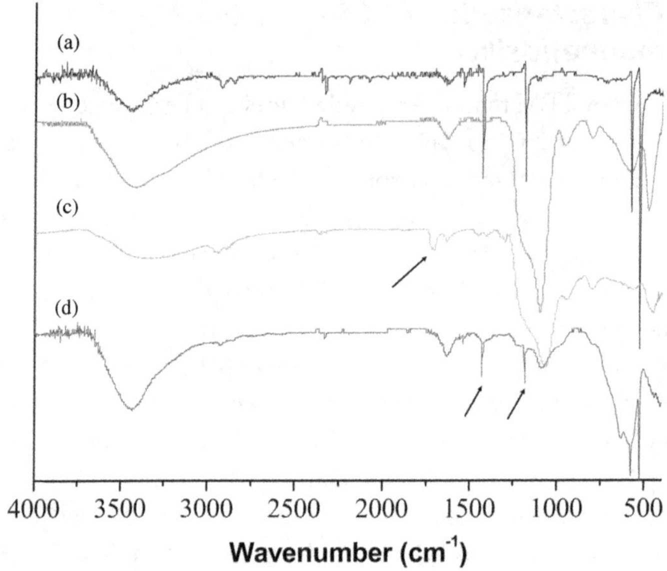

Figure 4.26: FTIR spectra of different materials (curve (a) C60; curve (b) $Fe_3O_4@SiO_2$; curve (c) $Fe_3O_4@SiO_2$-MPS; curve (d) $Fe_3O_4@SiO_2$-C60).

magnetic materials all possess very good superparamagnetism (that is, when the applied magnetic field disappears, the magnetic property of the magnetic sphere disappears with it), and the saturation magnetic values are 81.2 emu·g^{-1}, 76.5 emu·g^{-1}, 72.8 emu·g^{-1}, and 70.6 emu·g^{-1}, respectively. It is estimated that $Fe_3O_4@SiO_2$-MPS contains about 50 mg MPS per gram of magnetic sphere, and $Fe_3O_4@SiO_2$-C60 contains about 32 mg C60 per gram of magnetic sphere.

4.3.4. *C60-functionalized magnetic silica for the enrichment of low-abundance peptide*

The superparamagnetism of C60-functionalized magnetic silica makes the cumbersome centrifugation operation omitted in the enrichment process, and the material can be easily and quickly separated from the matrix solution by the magnet. The whole process is shown in Figure 4.27. The aqueous solution of $Fe_3O_4@SiO_2$-C60 (20 μL 5 mg·mL^{-1}) is added into 0.4 mL sample solution such as aqueous solution of standard peptide or standard proteolytic solution (diluted with water after enzymatic hydrolysis) or urine. The mixture is then shake at room temperature for 5 min. The magnet separates the magnetic microsphere to remove the supernatant. The separated material is washed 3 times with water, and each time it is separated by a magnet to remove the washing buffer. The protein/peptide fragment on the material is then eluted with 5 mL solvent (water/acetonitrile/TFA = 50/50/0.1, v/v/v) at 37°C. Finally, the stock solution and eluate are directly spotted onto a stainless-steel target plate of MALDI-TOF-MS for MS analysis. With the help of

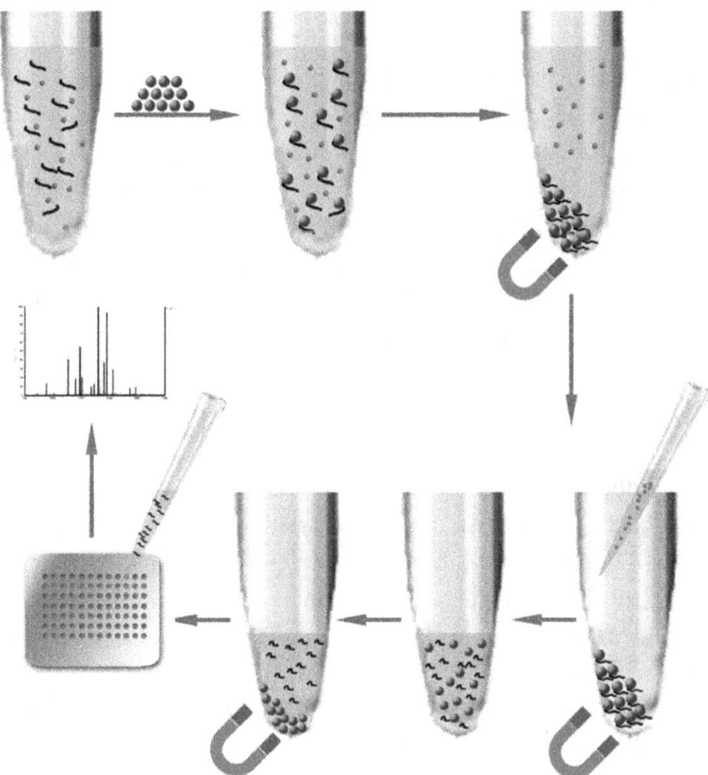

Figure 4.27: C60-functionalized magnetic silica for the fast enrichment of low abundance and MALDI-TOF-MS analysis.

magnet, the entire enrichment, washing and elution process is very convenient and fast.

4.3.4.1. *Study on loading capacity of Fe_3O_4@SiO_2-C60 towards peptide*

The standard peptide of Angiotensin II (amino acid sequence DRVYIHPF, MW = 1046.2, *pI* = 6.74) is used as a sample to investigate the loading capacity of Fe_3O_4@SiO_2-C60 towards peptide. On one hand, the amount of material is maintained, and the enrichment effect is observed by changing the content of Angiotensin II in the solution. On the other hand, the content of Angiotensin II in the solution is kept unchanged, and the enrichment effect is observed by changing the amount of the added material. The procedure is as follows: a set of aqueous solutions containing different amounts of standard peptide of Angiotensin II (0.2 ng, 2 ng, 4 ng, 8 ng, 12 ng, 24 ng, 40 ng, respectively, 400 μL). 20 μL Fe_3O_4@SiO_2-C60 (5 μg·μL^{-1}) is added into the above every solution. Or different volumes of Fe_3O_4@SiO_2-C60 (5 μg·μL^{-1}) are added into seven centrifuge tubes to make the microsphere have a mass of 2.5 μg, 5 μg, 10 μg, 50 μg, 100 μg, 150 μg, 200 μg, respectively. The supernatant is removed by a magnet, and then 400 μL Angiotensin II (aqueous solution, 5 μg·μL^{-1}) is added. After the mixture is shaken for 5 min, the material is separated, and the separated material is washed 3 times with water, and each time it is separated by a magnet to remove the washing buffer. The peptide on the material is then eluted with 5 μL solvent (water/acetonitrile/ TFA = 50/50/0.1, v/v/v) at 37°C. Finally, the stock solution and eluate are directly spotted onto a stainless-steel target plate of MALDI-TOF-MS for MS analysis. The S/N ratio of Angiotensin II after enrichment is analyzed according to MALDI-TOF-MS (Table 4.3), and the 2D maps of S/N ratio with two conditions are plotted (Figure 4.28). It can be seen from the results that when the amount of the material is kept constant, the S/N ratio is first increased, and then flattened because of saturation of material. When the sample amount is kept unchanged, the S/N ratio first increases, and then flattens, and finally decreases slightly, probably because too much material results in incomplete elution of peptide by a small amount of eluent. Based on these results, it is estimated that the loading capacity of the material towards the standard peptide of Angiotensin II is about 0.12 ng·mg^{-1}.

Table 4.3: The effect of dosage of Fe_3O_4@SiO_2-C60 and standard peptide on enrichment.

Effect of Dosage of Standard Peptide on Enrichment[a]		Effect of Dosage of Fe_3O_4@SiO_2-C60 on Enrichment[b]	
Dosage of Peptide (ng)	S/N	Dosage of Microsphere (μg)	S/N
0.2	176	2.5	354
2	766	5	1,791
4	1,495	10	2,500
8	2,643	50	2,622
12	3,006	100	2,643
24	3,105	150	2,612
40	3,111	200	2,371

Notes: [a]Adding 20 μL 5 μg$\cdot\mu$L^{-1} Fe_3O_4@SiO_2-C60 into each sample.
[b]Using 400 μL 0.02 ng$\cdot\mu$L^{-1} standard peptide solution for each sample.

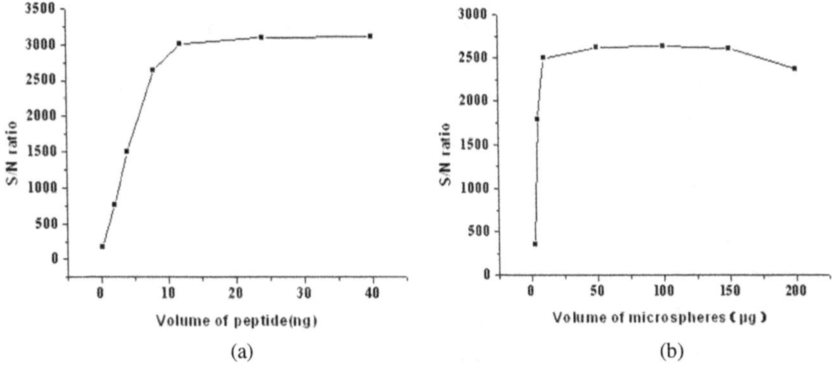

Figure 4.28: The effect of dosage of Fe_3O_4@SiO_2-C60 and standard peptide on enrichment.

4.3.4.2. *C60-functionalized magnetic silica for the enrichment of low-abundance standard peptide*

Angiotensin II is selected as the enrichment object. Figure 4.29 shows that the

S/N ratio of the 5 nmol·L^{-1} Angiotensin II before enrichment is 16.56 by MALDI-TOF-MS analysis (Figure 4.29(a)). The Angiotensin II is diluted with water to 0.5 nmol·L^{-1}. After enrichment with $Fe_3O_4@SiO_2$-C60 microsphere (0.4 mL, 0.1 mg), the signal peak of Angiotensin II is detected to be strong, and the S/N ratio is as high as 186.84 (Figure 4.29(b)). From this result, it can be estimated that the enrichment of $Fe_3O_4@SiO_2$-C60 can increase the signal of the peptide by more than 100 times.

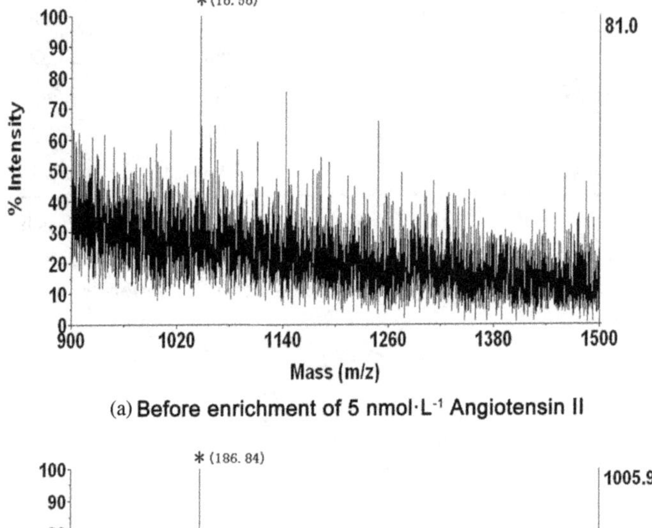

(a) Before enrichment of 5 nmol·L^{-1} Angiotensin II

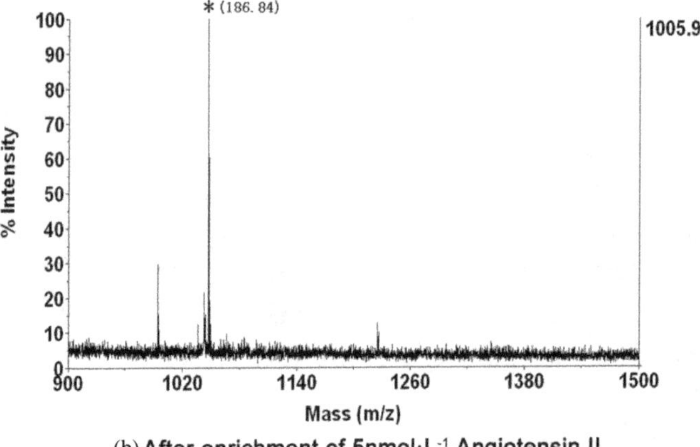

(b) After enrichment of 5nmol·L^{-1} Angiotensin II

Figure 4.29: MALDI-TOF-MS spectra of Angiotensin II (0.4 mL) before and after enrichment by $Fe_3O_4@SiO_2$-C60. The number in parenthesis represents S/N ratio.

4.3.4.3. *C60-functionalized magnetic silica for the enrichment of standard protein digest*

BSA digest is used as an experimental sample to investigate the enrichment extensiveness of the $Fe_3O_4@SiO_2$-C60 for peptide. As seen from Figure 30, eight peptides with weak signal are identified from 10 $nmol \cdot L^{-1}$ BSA digest before enrichment (Figure 30(a)), only 3 peptides with weaker signal are detected when the sample concentration is diluted to 1 $nmol \cdot L^{-1}$ (Figure 30(b)). However, after the 0.4 mL 1 $nmol \cdot L^{-1}$ BSA digest is enriched by $Fe_3O_4@SiO_2$-C60, 10 peptides with obviously increased S/N ratio are detected by MALDI-TOF-MS (Figure 30(c)). This result indicates that $Fe_3O_4@SiO_2$-C60 has a wide enrichment effect on mixed peptides.

4.3.4.4. *C60-functionalized magnetic silica for the enrichment of low-abundance peptide from salt-containing sample*

Biological sample usually contains a certain amount of inorganic salt. In the process of protein extraction and enzymatic hydrolysis, some inorganic salts are often added to maintain proper ionic strength in solution or to promote the dissolution of protein/peptide. Although MALDI MS has a certain salt tolerance theoretically, the presence of salt will inevitably affect the MALDI process. Especially when the salt is hard to volatilize or the concentration of salt is high, the quality of the spectrum will be seriously interfered, because these non-volatile low-molecular-weight contaminants can lead to the formation of complex adduct, increasing noise and causing significant signal suppression. However, impurities such as these non-volatile salts are also concentrated together with the peptide in the process of concentrating peptide by the conventional lyophilization method. In order to obtain satisfactory result in MS analysis, there is usually a one-step desalination process after enrichment of protein/peptide. Unlike the lyophilization method, the $Fe_3O_4@SiO_2$-C60 enriches peptide by the interaction between C60 and peptide. In order to investigate whether the $Fe_3O_4@SiO_2$-C60 will carry impurities such as inorganic salts in the solution when enriching peptide, BSA digest containing $CaCl_2$ (1 $nmol \cdot L^{-1}$ BSA digest containing 100 $mmol \cdot L^{-1}$ $CaCl_2$) is used as an experimental sample. At the same time, a

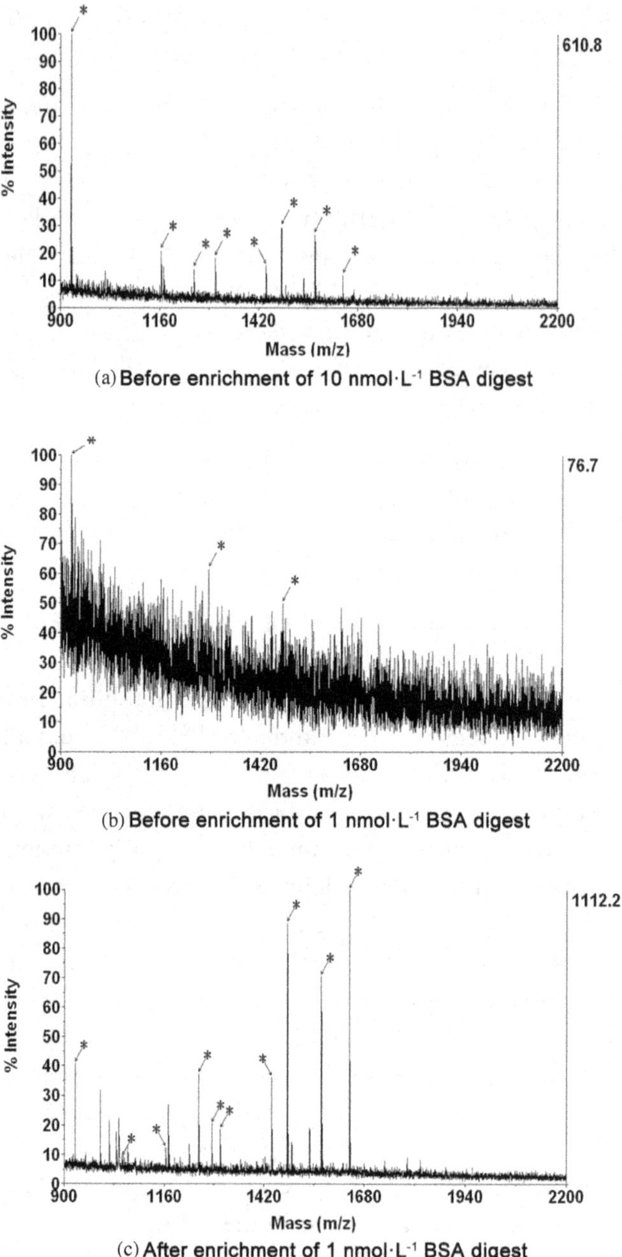

(a) Before enrichment of 10 nmol·L⁻¹ BSA digest

(b) Before enrichment of 1 nmol·L⁻¹ BSA digest

(c) After enrichment of 1 nmol·L⁻¹ BSA digest

Figure 4.30: MALDI-TOF-MS spectra of BSA digest before and after enrichment by $Fe_3O_4@SiO_2$-C60. "*" represents target peptide.

commercial ZipTipC18 is used to treat the same sample for comparison. Figure 4.31 and Table 4.4 show the results after enrichment with ZipTipC18 and $Fe_3O_4@SiO_2$-C60, respectively. It is obvious that 15 peptides can be detected after the BSA hydrolysate (0.4 mL 1 nmol·L^{-1}, containing 100 mmol·L^{-1} $CaCl_2$) is enriched by 0.1 mg $Fe_3O_4@SiO_2$-C60, while only three peptides can be detected when using ZipTipC18 (ZipTipC18 is used according to the standard procedure of MILLIPORE, the detailed process can be found at http://www.millipore.com/publications.nsf/docs/tn224). This

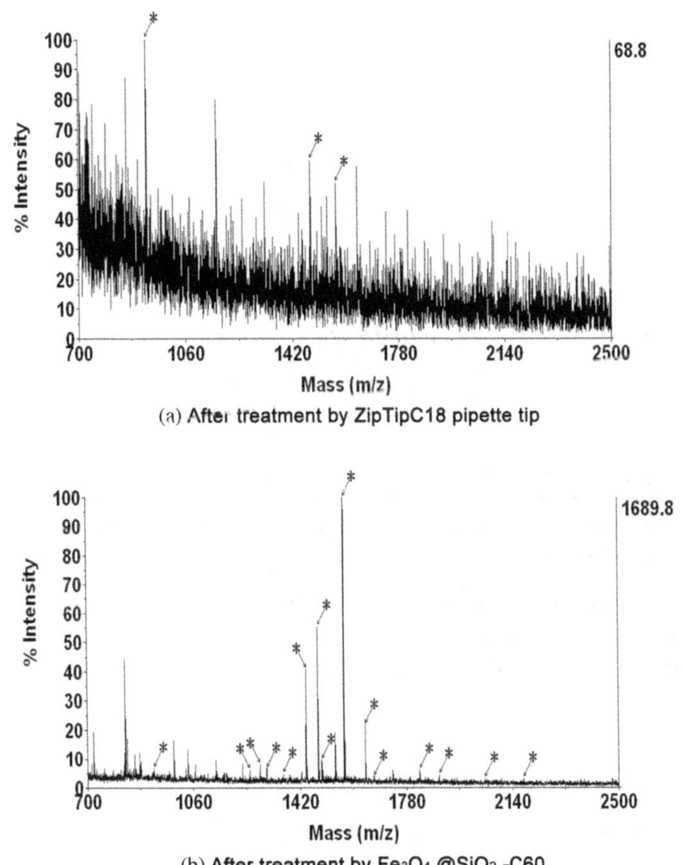

Figure 4.31: MALDI-TOF-MS spectra of 1 nmol·L^{-1} BSA hydrolysate containing 100 mmol·L^{-1} $CaCl_2$ after enrichment by two different methods. "*" represents target peptide.

Table 4.4: MS results of BSA digest after treatment by ZipTipC18 pipette tip and $Fe_3O_4@SiO_2$-C60.[a]

Calculated *m/z*	Database sequence	A[b]	B[b]
927.53	K.YLYEIAR.R	+	+
1249.68	R.FKDLGEEHFK.G		+
1283.76	R.HPEYAVSVLLR.L		+
1305.77	K.HLVDEPQNLIK.Q		+
1362.72	K.SLHTLFGDELCK.V		+
1439.87	R.RHPEYAVSVLLR.L		+
1479.87	K.LGEYGFQNALIVR.Y	+	+
1504.97	K.QTALVELLKHKPK.A		+
1567.83	K.DAFLGSFLYEYSR.R	+	+
1640.03	R.KVPQVSTPTLVEVSR.S		+
1667.95	R.MPCTEDYLSLILNR.L		+
1824	R.RPCFSALTPDETYVPK.A		+
1889.02	K.SLHTLFGDELCKVASLR.E		+
2045.12	R.RHPYFYAPELLYYANK.Y		+
2220.16	K.LFTFHADICTLPDTEKQIK.K		+
Matched peptide		3	15
Sequence coverage (%)		5	28

Notes: [a]The concentration of BSA digest is 1 nmol·L^{-1} (containing 100 mmol·L^{-1} CaCl$_2$).

[b]"A" refers to the result after treatment by ZipTipC18 pipette tip, "B" refers to the result after treatment by $Fe_3O_4@SiO_2$-C60, "+" refers to the target peptide detected by MALDI-TOF-MS.

indicates that $Fe_3O_4@SiO_2$-C60 do not enrich the salt in the solution while enriching the low-abundance peptide from very dilute and bulky sample, that is, it has an automatic desalting function, thereby avoiding the desalination process before MS analysis.

4.3.5. *C60-functionalized magnetic silica for the enrichment of low-abundance protein*

Because $Fe_3O_4@SiO_2$-C60 has small particle size and large specific surface area, and the surface of C60 has strong hydrophobicity, it can enrich low-abundance peptide and may also be used in low-abundance protein. Before enrichment, it is difficult to distinguish the MS peak of 0.2 ng·μL^{-1} Cyc by linear pattern of MALDI-TOF-MS (Figure 4.32(a)). After 0.4 mL Cyc solution is enriched by 0.1 mg $Fe_3O_4@SiO_2$-C60, there is no MS signal of Cyc in the supernatant (Figure 4.32(b)) and the MS signal peak of Cyc in eluent is high, and the S/N ratio reaches 78.68 (Figure 4.32(c)). The result indicates that $Fe_3O_4@SiO_2$-C60 can be used not only for the enrichment of low-abundance peptide, but also for the enrichment of low-abundance protein.

4.3.6. *C60-functionalized magnetic silica for the enrichment of peptide in human urine*

Human urine is a human body fluid sample that is easily obtained without any injury and is often used for clinical diagnosis. The $Fe_3O_4@SiO_2$-C60 is used for the enrichment and separation of peptide in human urine. Figure 4.33(a) shows that untreated urine has no discernible peptide signal by MALDI-TOF-MS analysis. This may be due to the complex composition of urine. There are many salts such as urea and NaCl in addition to the low-concentration peptide, and these impurities will inevitably interfere with high-resolution MS analysis. After enriched by, more than ten peptides are detected from urine (Figure 4.33(b)). This indicates that the $Fe_3O_4@SiO_2$-C60 combines the advantages of enrichment and desalination, and have good enrichment effect on peptide/protein even in complex biological sample.

4.3.7. *Conclusion*

Compared to the conventional C8 or C18 of reversed-phase magnetic microsphere, the surface modified by C60 functional group with hydrophobicity and a large number of π bonds shows more complex interaction with protein/peptide, therefore, the C60-modified material can realize the enrichment of target molecule with lower concentration. In addition, the C60 molecule does

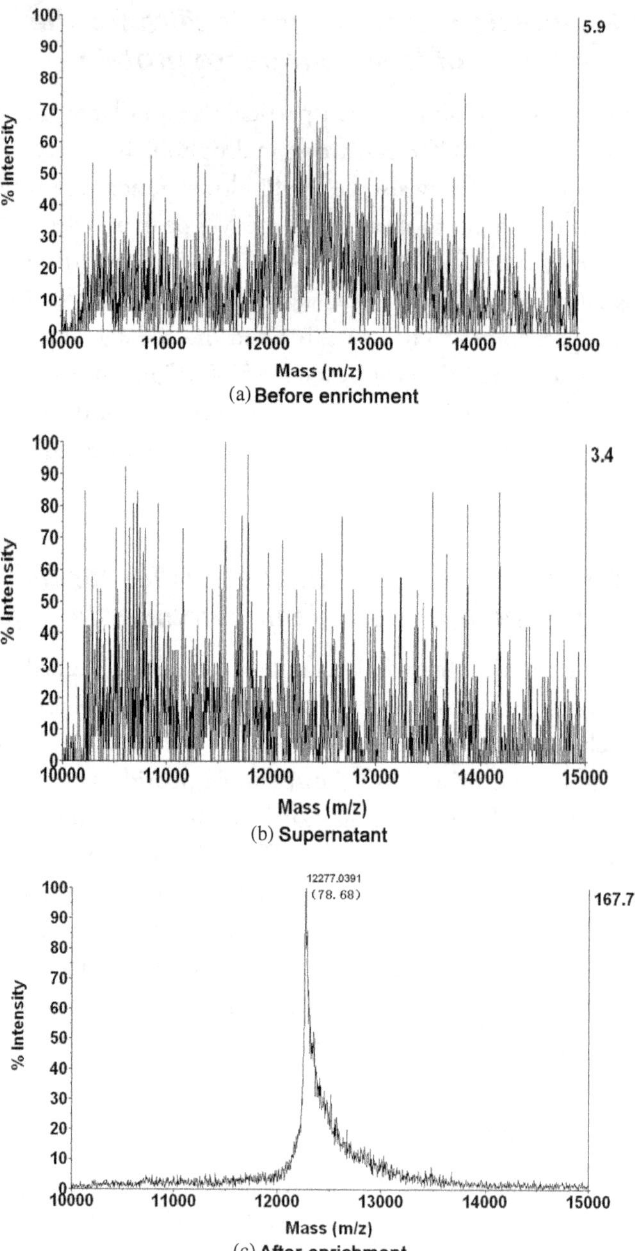

Figure 4.32: MALDI-TOF-MS spectra of Cyc (0.2 ng·μL^{-1}) before and after enrichment by Fe$_3$O$_4$@SiO$_2$-C60. The number in parenthesis represents S/N ratio.

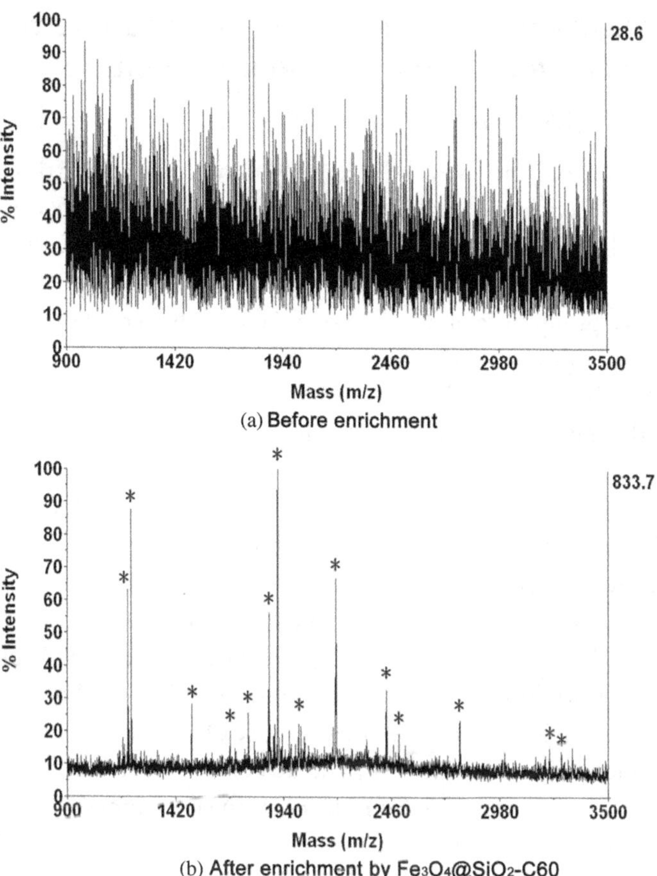

Figure 4.33: MS spectra of human urine. "*" represents target peptide.

not adsorb inorganic salt, so when the protein/peptide in the sample solution is enriched, the C60-modified magnetic silica can achieve the purpose of enrichment and desalination simultaneously. It can be said that the successful synthesis and application of this novel C60-modified magnetic silica provides new enlightenment for the application of magnetic material in proteomics research.

4.4. PMMA-modified magnetic microsphere for the enrichment of low-abundance peptide

4.4.1 *Basic principle*

In modern life, polymer is evolving at a rapid rate, and is inextricably linked to our life.[27] There are many kinds of polymers and their properties are very diverse. Both hydrophilic and hydrophobic polymer are very easy to obtain, and there are many kinds of polymers containing both hydrophilic group and hydrophobic group. Therefore, the potential of polymer-modified magnetic nanoparticle in proteomics is enormous.

Poly(methyl methacrylate) (PMMA) is an organic polymer commonly found in the industry. This polymer with hydrophobicity and good chemical stability is a linear molecule, non-toxic and environmentally friendly, and can be used for producing tableware, sanitary ware, etc. PMMA has good biocompatibility and is widely used in the biomedical industry, such as the manufacture of blood storage container. Since PMMA has good optical property and compatibility with human tissue, it is also used to manufacture human lens. According to the latest report, PMMA has a good function of enriching peptide/protein in aqueous solution.[28–30] Prof. Lu from Fudan University successively coated PMMA on $CaCO_3$, SnO_2 and other particles for enrichment of peptide/protein, and achieved good result.[28,29] However, these materials are inseparable from the centrifugation process due to lack of magnetism during the enrichment process. After the PMMA-coated $CaCO_3$ particle is used to enrich the peptide/protein, it is necessary to dissolve the $CaCO_3$ with acid, which is difficult to control.

This section introduces a PMMA-modified magnetic microsphere ($Fe_3O_4@SiO_2@PMMA$), which combines the advantages of convenient operation of magnetic material and enrichment ability of PMMA towards peptide/protein.

4.4.2. *Synthesis and characterization of $Fe_3O_4@SiO_2@$ PMMA microsphere*

The magnetic inorganic-organic composite of $Fe_3O_4@SiO_2@PMMA$ can be obtained by coating PMMA on magnetic silica via aqueous-phase free radical polymerization. The $Fe_3O_4@SiO_2@PMMA$ has uniform particle size, good magnetic responsiveness, and good dispersibility in water and good biocompatibility, the detailed synthetic process and characterization of $Fe_3O_4@SiO_2@PMMA$ are described in Section 2.4 of Chapter 2.

4.4.3. $Fe_3O_4@SiO_2@PMMA$ microsphere for the enrichment of low-abundance peptide/protein

4.4.3.1. Enrichment of standard peptide/protein

The standard peptide of Angiotensin II (DRVYIHPF, MW = 1046.2, pI = 6.74) and the standard protein of Cyc (0.5 mg·L^{-1}, MW = 12,384, pI = 9.59) are selected as enrichment targets for studying the enrichment effect of $Fe_3O_4@SiO_2@PMMA$ on peptide/protein. After the mixture of 0.05 mg $Fe_3O_4@SiO_2@PMMA$ and 0.4 mL sample is incubated for 10 min, the microsphere is washed with water 3 times. Then the MALDI matrix solution (CHCA) is added and directly spotted onto a MALDI target plate for analysis. The results are shown in Figure 4.34.

The main advantage of $Fe_3O_4@SiO_2@PMMA$ is that this microsphere can effectively concentrate the peptide/protein, and the microsphere can be directly spotted onto the target plate with the enriched molecule for direct analysis by MALDI-TOF-MS. The S/N ratio of 4 nmol·L^{-1} Angiotensin II by direct analysis is only 13.21, while the S/N ratio increases to 798.67 after enrichment by $Fe_3O_4@SiO_2@PMMA$. For 0.5 mg·L^{-1} Cyc solution, the S/N ratio increase to 95.59 after enrichment by $Fe_3O_4@SiO_2@PMMA$, which is much higher than that before enrichment. This may be related to the optimized structure of the microsphere. First, the porous PMMA on the surface of the microsphere is a linear hydrophobic chain, which makes the PMMA-modified microsphere have good hydrophobicity. Most amino acids contain hydrophobic groups such as alkyl group and phenyl group, etc., and these hydrophobic group-containing amino acids are widely distributed in proteins and peptides. Therefore, PMMA-modified magnetic microspheres can be used as hydrophobic probes for the enrichment of peptides/proteins. However, if the captured target molecule enters the depth of the microbead, it is difficult to be detected by MALDI-TOF-MS. Therefore, the sample is enriched with pure PMMA, resulting in low intensity and S/N ratio of the target molecule. The $Fe_3O_4@SiO_2$ core of this microsphere can prevent the

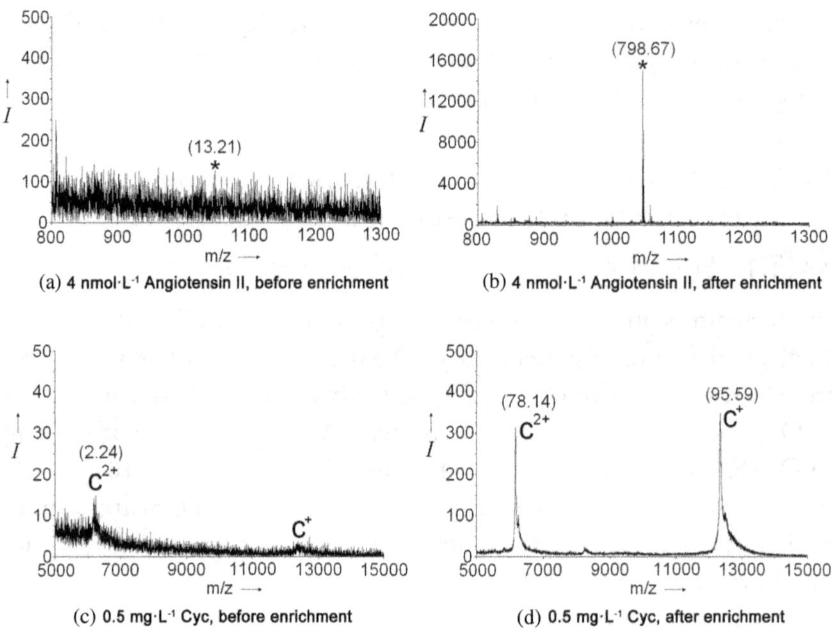

Figure 4.34: MALDI-TOF-MS spectra of Angiotensin II and Cyc before and after enrichment by $Fe_3O_4@SiO_2@PMMA$. C^+ and C^{2+} refer to the single and doubled charge peaks of Cyc, respectively. The number in parenthesis represents S/N ratio.

target molecule from going deep and retains the target molecule on the outermost layer of the microsphere. Therefore, high MS intensity can be detected after the sample is enriched by $Fe_3O_4@SiO_2@PMMA$.

4.4.3.2. *Enrichment of standard protein digest and investigation of salt tolerance*

Urea is a commonly used additive in the sample processing process, therefore, a part of BSA digest is mixed with urea to investigate the salt tolerance of the material. Figure 4.35 shows that there are only 3 peptides are detected from 2 nmol·L⁻¹ BSA digest, and none can be detected from 2 nmol·L⁻¹ BSA digest containing 100 mmol·L⁻¹ urea. When BSA digest (0.4 mL 2 nmol·L⁻¹,

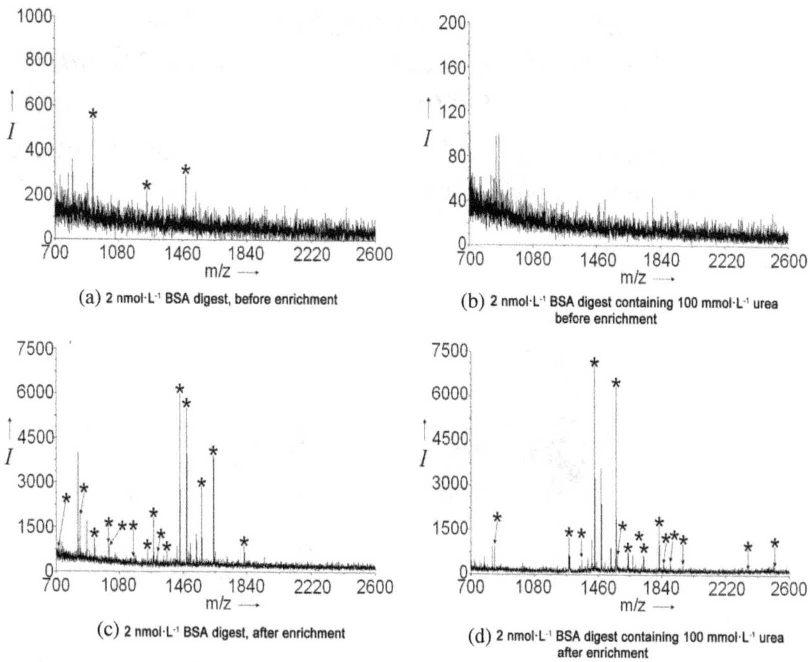

Figure 4.35. MALDI-TOF-MS spectra of BSA digest before and after enrichment by $Fe_3O_4@SiO_2@PMMA$. "*" represents target peptide, 0.4 mL solution is used for each experiment.

containing 100 mmol·L^{-1} urea) is enriched by ZipTipC18, 5 peptides with relatively low S/N ratio are detected. When BSA digest without or with urea is enriched by $Fe_3O_4@SiO_2@PMMA$, 15 peptides (sequence coverage reaches 24%) and 14 peptides (sequence coverage reaches 22%) are identified, respectively, indicating that effective enrichment can be obtained regardless of whether urea is or not contained in the solution. The good enrichment and desalination properties of $Fe_3O_4@SiO_2@PMMA$ are mainly attributed to the non-polarity of the hydrophobic polymer of PMMA, which exhibits strong hydrophobic interaction with peptide and weak interaction with hydrophilic molecules such as salts. Therefore, the $Fe_3O_4@SiO_2@$ PMMA microsphere can avoid concentration of the salt while effectively enriching the peptide in the solution.

4.4.4. Fe_3O_4@SiO_2@PMMA microsphere for the enrichment of peptide in protein extract from on-gel enzymatic hydrolysis

After the separation of protein in the human lens (Figure 4.36(a)) by 2DE, a slightly lighter colored protein dot (indicated by arrow in Figure 4.36(b)) is

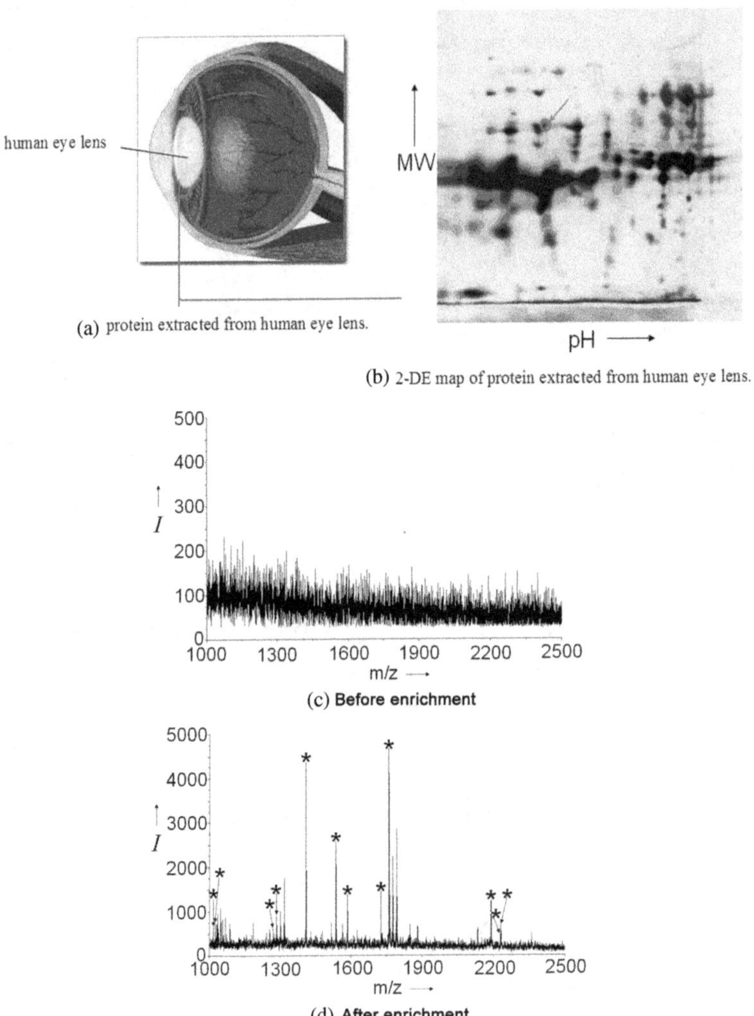

Figure 4.36: Separation diagram of human eye lens by 2DE, a protein dot (shown by the arrow) on the gel for subsequent experiment, and MALDI-TOF-MS spectra of selected protein dot digest before and after enrichment by Fe_3O_4@SiO_2@PMMA microsphere.

selected for on-gel enzymatic hydrolysis. The enzymatically decomposed peptide fragments are extracted 3 times with 50 μL extracting solution, then these extracting solutions are mixed and diluted with water to 400 μL to reduce the acetonitrile concentration. Then they are divided into two equal parts. Fe_3O_4@SiO_2@PMMA microsphere is added into one part for enrichment and the other part is not dealt with. None of reliable protein can be obtained in the untreated part (protein score greater than 64 is considered reliable) as seen in Figure 4.36(c). However, after enrichment with Fe_3O_4@SiO_2@PMMA microsphere, a protein with a score of up to 142 is identified. Figure 4.36d shows the MALDI-TOF-MS spectrum, and Table 4.5 shows the peptide information of identified protein. This further demonstrates the feasibility of magnetic microsphere-based enrichment method in practical proteomics.

Table 4.5: Peptide information identified from human eye lens using Fe_3O_4@SiO_2@PMMA.

Calculated m/z	Detected m/z	Sequence
1016.4366	1016.4633	DMQWHQR
1024.4484	1024.4702	WDSWTSSR
1269.627	1269.6583	IRDMQWHQR
1285.6219	1285.636	IRDMQWHQR
1408.7471	1408.8004	IILYENPNFTGK
1536.842	1536.896	IILYENPNFTGKK
1585.7645	1585.8148	GEQFVFEKGEYPR
1726.8143	1726.8778	DSSDFGAPHPQVQSVR
1760.8391	1760.9036	VQSGTWVGYQYPGYR
2190.021	2190.1189	GDYKDSSDFGAPHPQVQSVR
2231.0073	2231.1763	MEIIDDDVPSFHAHGYQEK
2232.1294	2232.3193	VDSQEHKIILYENPNFTGK

Note: Protein name, crystalline, beta B2 [Homo sapiens]; Accession, gi|4503063; Score 142.

4.4.5. *Conclusion*

In this section, the $Fe_3O_4@SiO_2@PMMA$ microsphere with core–shell–shell structure is synthesized by sol–gel method and aqueous phase rapid polymerization method. The $Fe_3O_4@SiO_2@PMMA$ microsphere has stable magnetic core. The intermediate SiO_2 layer protects magnetic core and provides an effective carrier for the modification of the outer PMMA. The organic hydrophobic PMMA shell has good dispersibility and affinity towards the peptide/protein. These microspheres can efficiently, quickly and conveniently enrich low-concentration peptide/protein and also have a desalting function. The microsphere can be analyzed by MALDI along with the captured target molecule, and will not interfere with the excitation and analysis of the target molecule. This work expands the design idea of using polymer to modify core–shell structured magnetic microsphere, and also paves the way for the application of magnetic polymer in biological separation.

4.5. Magnetic mesoporous nanomaterial for the enrichment of endogenous peptide

4.5.1. *Ordered mesoporous silica coated magnetic material for peptidomics*

4.5.1.1. *Development of ordered mesopore coated magnetic material*

Silica particles with ordered mesoporous structure, such as M41s and SBA-15, have been widely used in the field of separation and adsorption research.[15,31,32]

Carbon nanotube, also known as bucky tube, has typical layered hollow structural feature. Carbon nanotube can be classified into single-walled nanotube (SWCNT) and multi-walled nanotube (MWCNT) according to the layer number of graphene sheet. MWCNT is formed by crimping the hexagonal coaxial layer of SWCNT to form a hollow cylinder of having a polygonal conical cap. At the time of the initial formation of MWCNT, trap center is easily formed between layers to capture various defects, and thus the wall of MWCNT is usually covered with small hole-like defects. Compared with the MWCNT, the SWCNT is composed of a single-layer cylindrical graphite, which has a small distribution range of diameter, few defect, and higher uniformity. Carbon nanotube has good heat transfer performance and a very large aspect ratio, so its heat exchange performance along the length direction is high and the relative heat exchange performance in the vertical direction is low. The highly anisotropic heat conductive material can be synthesized using carbon nanotube by choosing appropriate direction. In addition, carbon nanotube has a high thermal conductivity, thus the adulteration of a small amount of carbon nanotubes may greatly improve the thermal conductivity of the composite material. Carbon nanotube has excellent physical and chemical properties, such as unique metal or semiconductor conductivity, high mechanical strength, hydrogen storage capacity, adsorption capacity and strong microwave absorption capacity. All countries in the world have invested a lot of researches in its preparation and application, with the hope to occupy the commanding height of this technology field.

Graphene is a novel material composed of carbon atoms, which has the single-layer planar film structure of the hexagonal honeycomb based on the sp^2 hybrid orbital. The basic structural unit of graphene is the most stable six-membered benzene ring in organic materials. Its theoretical thickness is only 0.35 nm, which is the thinnest 2D material that has been discovered so far.[33] The chemical bonds and electronic arrangements in graphene make it have special properties, such as large surface area ($2,630$ $m^2 \cdot g^{-1}$), tunable band gap; room temperature quantum Hall effect, high mechanical strength which is 200 times of steel, high elasticity, high electrical conductivity and thermal conductivity and so on. Designing and synthesizing graphene composite for separating and enriching target substance in complex system have become a research hotspot in analytical chemistry. All the advantages of graphene provides researchers with an excellent modification platform. In addition to the traditional advantages of graphene itself, it can be attached by relevant functional groups to make the material tailored according to the actual requirement. The application of graphene and its composite in peptidomics/proteomics pretreatment and MS identification is in the experimental stage. More and more related articles have been published. For example, Choi *et al.* used the high strength and ductility of graphene to prepare mesoporous graphene. On this basis, Co_3O_4 nanoparticles are bonded to extend the 2D graphene to the specific 3D graphene composite, which is successfully applied.[34]

For the demand of selectively enriching endogenous peptide in peptidomics analysis, mesoporous layer coated magnetic material has been developed to be a new method for the separation and analysis of endogenous peptide. Here are three kinds of magnetic materials coated with ordered mesoporous layer (the synthesis can be seen in Section 2.7 of Chapter 2). One is $Fe_3O_4@nSiO_2@mSiO_2$ microsphere with Fe_3O_4 particle as the core, a dense SiO_2 layer in the middle, and a mesoporous SiO_2 layer as outermost shell. This microsphere combines the advantages of easy handling superparamagnetic core and the mesoporous separation ability towards peptide. It is directly used in the study of selective isolation of endogenous peptide in rat brain protein extract, showing many advantages. The second is to coat the ordered mesoporous silica on the hybrid magnetic MWCNT to form a special hybrid material of having a hybrid magnetic MWCNT core and a mesoporous silica shell (c-MWCNTs/Fe_3O_4–@$mSiO_2$). After calcination under nitrogen protection, the surface silicon hydroxy group is converted into the silicon-oxygen bridge, which can be used to selective capture of peptide quickly and efficiently. The third is to coat the ordered mesoporous silica on surface of graphene-covered magnetic nanoparticle to form a magnetic double-sided mesoporous graphene composite material (Fe_3O_4–graphene@$mSiO_2$). The material of the double-sided structure has a well-arranged pore structure and a large pore volume, large surface area, high magnetic responsivity in the magnetic field, hydrophobic inner wall and so on, which are suitable for separating the target substance from the solution system and simplifying the enrichment process. The three kinds of mesoporous silica layer coated materials are calcined under nitrogen protection, and a large amount of "Si–OH" on the mesoporous surface is converted into a "Si–O–Si" bridge bond, which enhances hydrophobicity and the interaction between material surface and peptide/protein in the sample.

4.5.1.2. *Size-exclusion effect of ordered mesopore coated magnetic material*

The unique of the ordered mesoporous magnetic material is that its surface has an ordered mesoporous layer aligned perpendicular to the core. Taking $Fe_3O_4@nSiO_2@mSiO_2$ as an example to investigate the size-exclusion effect. Cyc with a molecular weight of ~12,300 Da and BSA with a molecular weight of approximately 69,248 Da are used for the study. As well known, the molecular weight of target peptidome is normally below 15 kDa,[35] Cyc is just below this value and BSA is greater than this value. Figure 4.37 shows MS

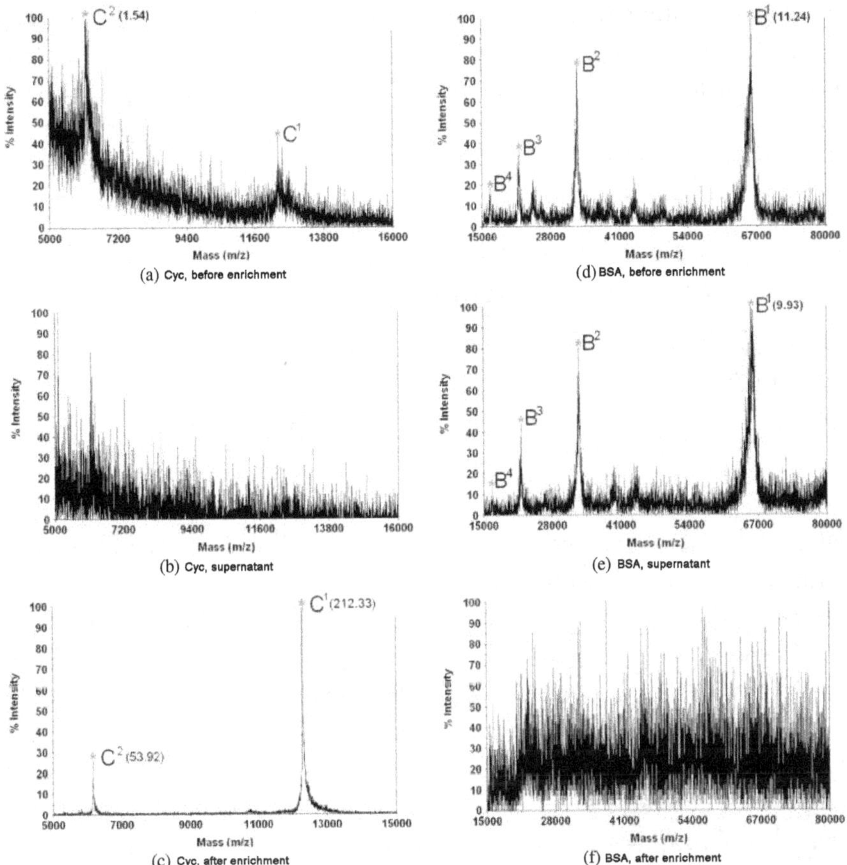

Figure 4.37: MS spectra of 0.5 mg·L^{-1} Cyc and 10 mg·L^{-1} BSA solution before and after enrichment by Fe$_3$O$_4$@nSiO$_2$@mSiO$_2$. The number in parenthesis represents S/N ratio, C^1, C^2, B^1, B^2 represents single and double charge peaks of Cyc and BSA, respectively.[31]

spectra of these two standard proteins before and after enrichment by using this Fe$_3$O$_4$@nSiO$_2$@mSiO$_2$, respectively.

As seen in Figures 4.37(a)–4.37(c), 0.5 mg·L^{-1} Cyc is barely visible before enrichment. After 0.4 mL of this Cyc solution is enriched by 0.1 mg Fe$_3$O$_4$@ nSiO$_2$@mSiO$_2$, there is no Cyc signal in the supernatant, however, this material is eluted with a small amount of solvent (water/acetonitrile/TFA = 50/50/0.1, v/v/v), and the Cyc signal is significantly enhanced, indicating that this mesoporous material has an enrichment effect on it. In contrast, the protein signal of 10 mg·L^{-1} BSA is substantially equal in the supernatant and the original solution (Figures 4.37(d) and 4.37(e)), while it is barely visible in eluent (Figure 4.37(f)). This indicates that the $Fe_3O_4@nSiO_2@mSiO_2$ can selectively allow peptide and protein with smaller molecular weight to enter the pore, and block protein with large molecular weight such as BSA outside the pore using good size-exclusion effect.

4.5.1.3. *Ordered mesopore coated magnetic material for the enrichment of peptide*

(1) $Fe_3O_4@nSiO_2@mSiO_2$ for the enrichment of standard peptide

The standard peptide of Angiotensin II (DRVYIHPF, MW = 1046.2, pI = 6.74, 0.4 mL, 5 nmol·L^{-1}) is used as an enrichment object to investigate the enrichment efficiency of $Fe_3O_4@nSiO_2@mSiO_2$.[31] In order to investigate the effect of the mesoporous layer on the enrichment efficiency, the $Fe_3O_4@nSiO_2$ with the non-mesoporous silica layer is also calcined under nitrogen for enriching the same Angiotensin II solution. The S/N ratio of 5 nmol·L^{-1} Angiotensin II is 25.58 before enrichment (Figure 4.38a). After 0.4 mL of this Angiotensin II solution is enriched by $Fe_3O_4@nSiO_2$, the S/N ratio increases to 75.02 (Figure 4.38(b)). After enrichment by $Fe_3O_4@nSiO_2@mSiO_2$, the S/N ratio increases to 3083.49 (Figure 4.38(c)). This result indicates that the enrichment effect of $Fe_3O_4@nSiO_2@mSiO_2$ on the peptide is much stronger than that of $Fe_3O_4@nSiO_2$ due to the presence of the mesoporous SiO_2 layer. This may be because the mesoporous layer has a high specific surface area and provides more available sites for peptide.

The 5 nmol·L^{-1} MYO hydrolysate is used to investigate the enrichment universality of $Fe_3O_4@nSiO_2@mSiO_2$. The results (Figure 4.39) shows that 15 peptides with strong S/N ratio are detected from 0.4 mL MYO hydrolysate after enrichment by $Fe_3O_4@nSiO_2@mSiO_2$, and the sequence coverage reaches 93%. While only five peptides with poor S/N ratio are identified by $Fe_3O_4@nSiO_2$ with 42% sequence coverage. The results show that the $Fe_3O_4@nSiO_2@mSiO_2$ has a wide enrichment effect on the peptide, which further indicates that the enrichment effect of $Fe_3O_4@nSiO_2@mSiO_2$ on the peptide can be significantly improved after the coating of mesoporous SiO_2 layer.

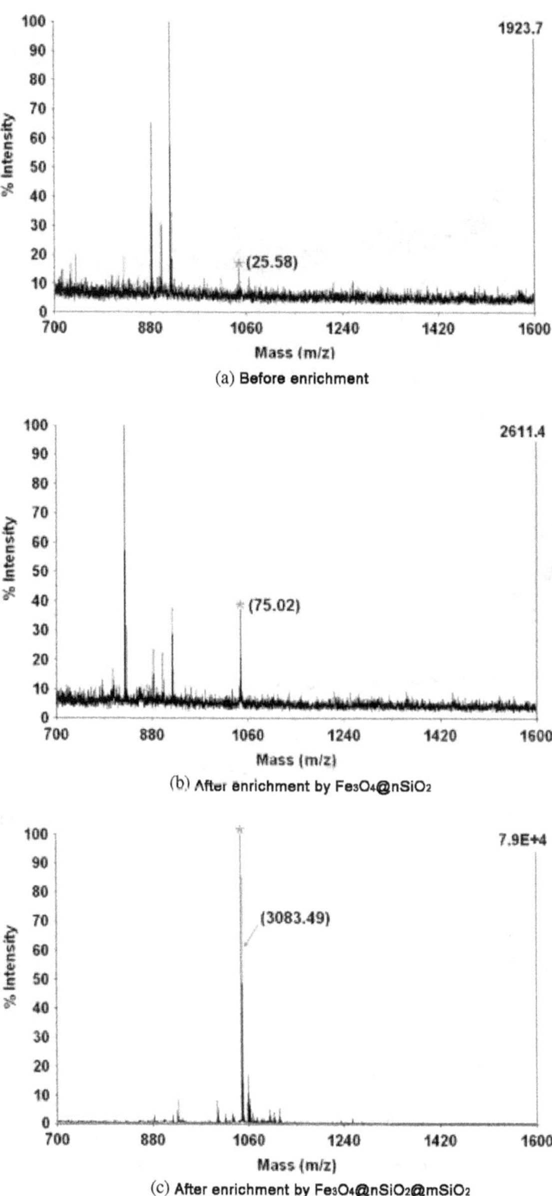

Figure 4.38: MALDI-TOF-MS spectra of 5 nmol·L^{-1} Angiotensin II before and after enrichment by two different microspheres. "*" represents Angiotensin II, the number in parenthesis represents S/N ratio.

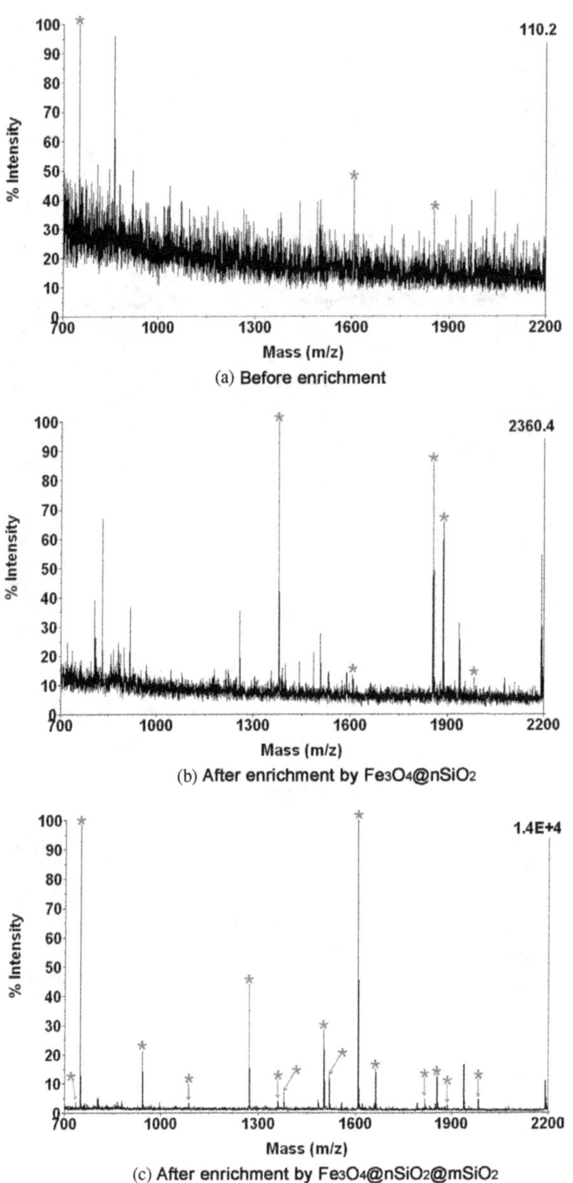

Figure 4.39: MALDI-TOF-MS spectra of 5 nmol·L^{-1} MYO hydrolysate before and after enrichment by two different microspheres. "*" represents target peptide.

(2) c-MWCNTs/Fe$_3$O$_4$-@mSiO$_2$ for the enrichment of low-abundance peptide

The most important structural feature of the c-MWCNTs/Fe$_3$O$_4$-@mSiO$_2$ is the mesoporous silica shell coated on the surface of MWCNTs/Fe$_3$O$_4$.[15] In order to confirm the enrichment ability of c-MWCNTs/Fe$_3$O$_4$-@mSiO$_2$ towards standard peptide is mainly from the weak hydrophobic surface of mesopore after calcination, MWCNTs/Fe$_3$O$_4$ and c-Fe$_3$O$_4$@mSiO$_2$ (c- refers to calcinated-) are also used for comparison. Figure 4.40 shows the MS spectra of the peptide enriched by these three different materials. Among them, c-MWCNTs/Fe$_3$O$_4$-@mSiO$_2$ shows the highest enrichment efficiency, and the S/N ratio of peptide signal is 1080.96. For MWCNTs/Fe$_3$O$_4$, the peptide signal is barely visible and is almost covered by noise (Figure 4.40(a)). c-Fe$_3$O$_4$@mSiO$_2$ has higher enrichment efficiency than MWCNTs/Fe$_3$O$_4$, but not the best one compared to c-MWCNTs/Fe$_3$O$_4$-@mSiO$_2$. As shown in Figure 4.40(b), the S/N ratio of the peptide after enrichment by c-Fe$_3$O$_4$@mSiO$_2$ is 531.63, which is about half of the peptide signal by c-MWCNTs/Fe$_3$O$_4$-@mSiO$_2$ (Figure 4.40(c)). This also indicates that the mesopores silica on the outer surface of c-MWCNTs/Fe$_3$O$_4$-@mSiO$_2$ replaces MWCNTs to become the effective part for enrichment, and the introduction of MWCNTs increases the surface area for effective enrichment.

Similarly, 0.4 mL 5 nmol·L^{-1} MYO hydrolysate is used to be enriched by c-MWCNTs/Fe$_3$O$_4$–@mSiO$_2$. As shown in Figure 4.41, seven peptides are detected with sequence coverage of about 54%, while only two peptides are detected before enrichment. These results indicate that the c-MWCNTs/Fe$_3$O$_4$–@mSiO$_2$ is capable of enriching peptide in relatively low concentration of solution, which can be used for the study of low-abundance peptide.

(3) Fe$_3$O$_4$–graphene@mSiO$_2$ for the enrichment of low-abundance peptide

The optimal conditions for the enrichment of peptide are studied using the standard peptide of Angiotensin II (5 nmol·L^{-1}).[32] The results show that 10 μL 10 mg·mL^{-1} as the dosage of Fe$_3$O$_4$–graphene@mSiO$_2$, 0.4 mol·L^{-1} NH$_3$·H$_2$O as eluent, the MS signal of 0.2 mL 5 nmol·L^{-1} Angiotensin II can achieves

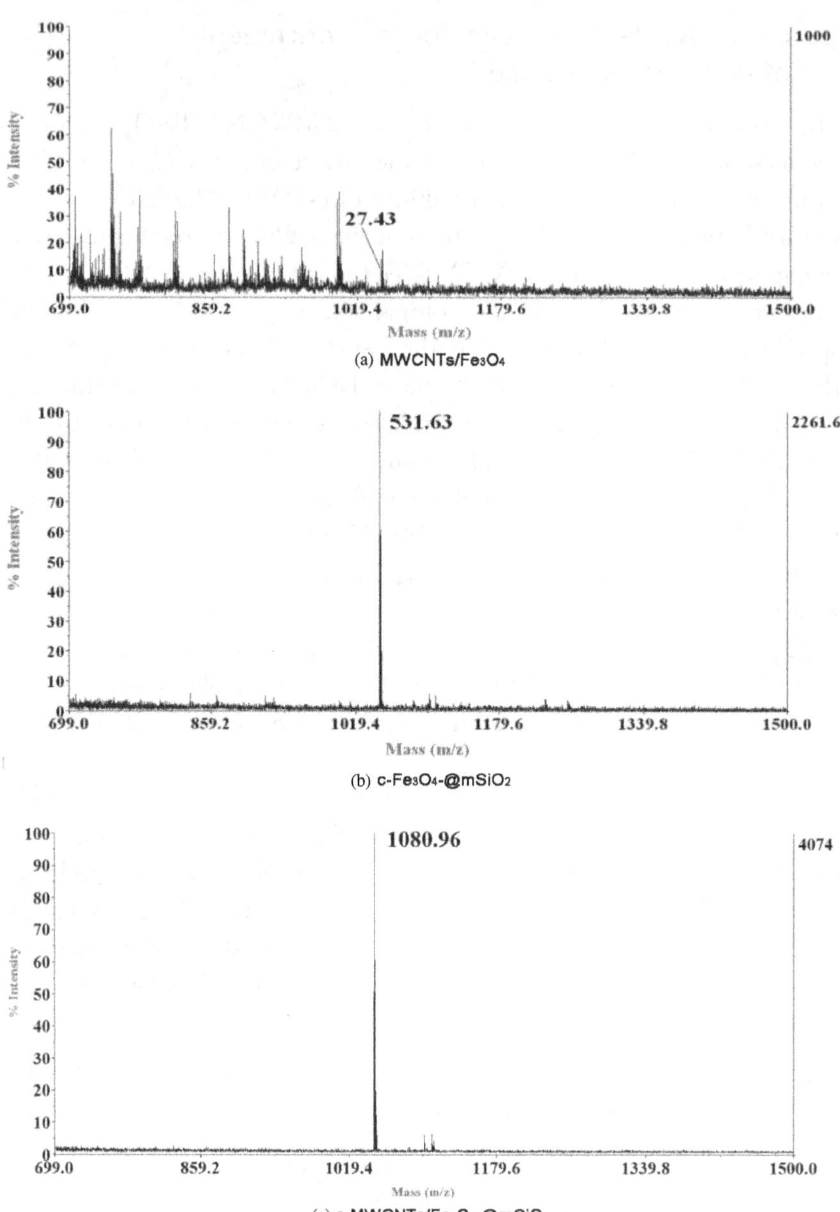

(a) MWCNTs/Fe₃O₄

(b) c-Fe₃O₄-@mSiO₂

(c) c-MWCNTs/Fe₃O₄-@mSiO₂

Figure 4.40: MS spectra of Angiotensin II (0.4 mL 5 nmol·L⁻¹) after enriched by three different materials.

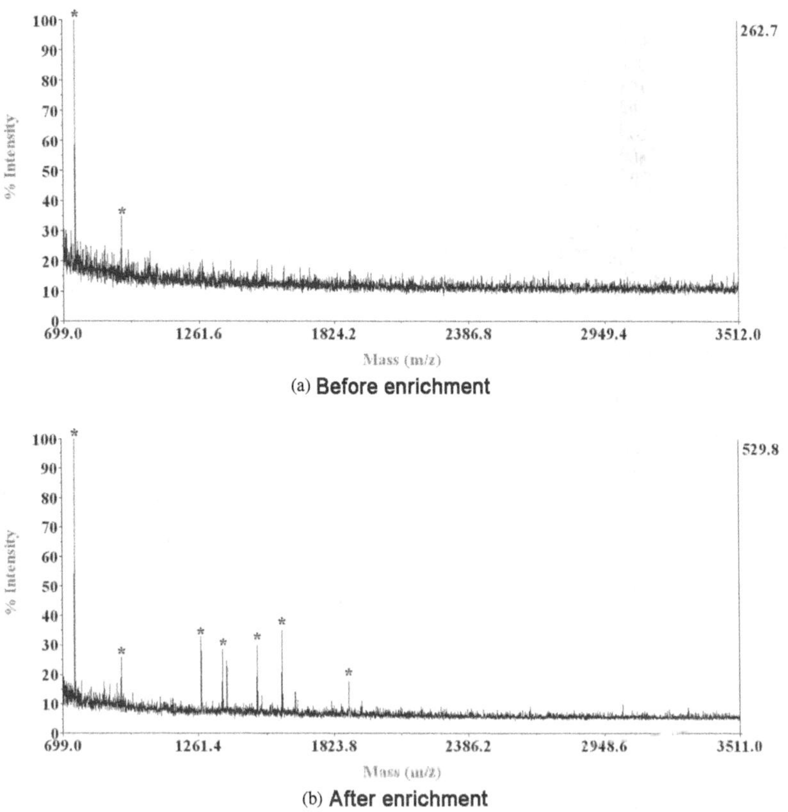

(a) **Before enrichment**

(b) **After enrichment**

Figure 4.41: MALDI-TOF-MS spectra of 5 nmol·L^{-1} MYO hydrolysate before and after enrichment by c-MWCNTs/Fe$_3$O$_4$–@mSiO$_2$. "*" represents target peptide.

best. The limit of detection (LOD) of Fe$_3$O$_4$–graphene@mSiO$_2$ for Angiotensin II is 0.1 nmol·L^{-1}, which is shown in Figure 4.42. The target peptide is indicated by the "*". The Fe$_3$O$_4$–graphene@mSiO$_2$ can be reused 5 times to maintain a good enrichment effect, with a maximum loading of 2.79 mg·mg^{-1}.

The MS results of 5 nmol·L^{-1} BSA digest and MYO digest are shown in Figure 4.43. Only one peptide is detected from BSA digest before enrichment, and 13 peptides are detected with sequence coverage of 22% after enrichment by Fe$_3$O$_4$–graphene@mSiO$_2$. Similarly, a total of five peptides are

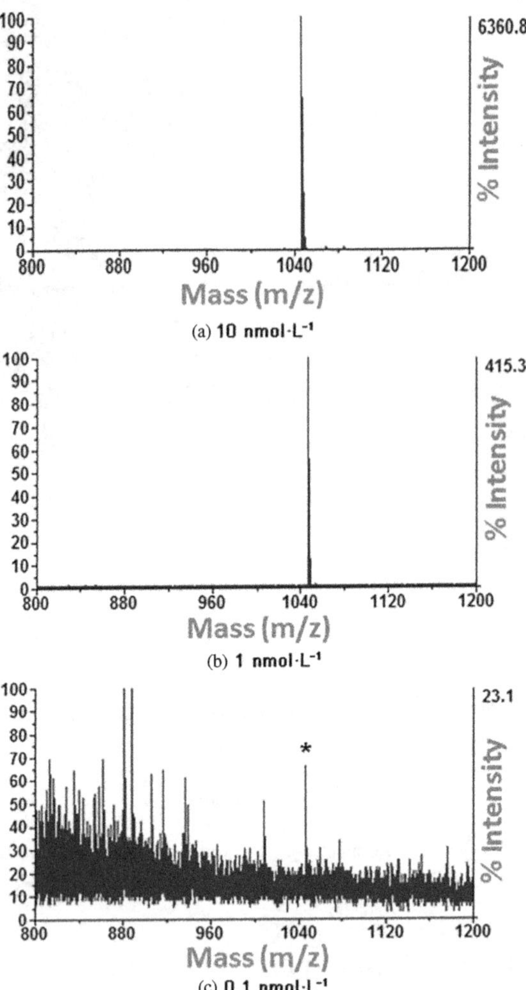

Figure 4.42: LOD examination of Fe_3O_4–graphene@mSiO$_2$ (10 μL 10 mg·mL^{-1}) using 0.2 mL Angiotensin II as sample.

found in MS spectrum of original MYO digest, and the sequence coverage is 42%. After enrichment by Fe_3O_4–graphene@mSiO$_2$, a total of eight peptides are detected with sequence coverage of 55%. These experimental results demonstrate that Fe_3O_4–graphene@mSiO$_2$ can be further applied in complex biological system.

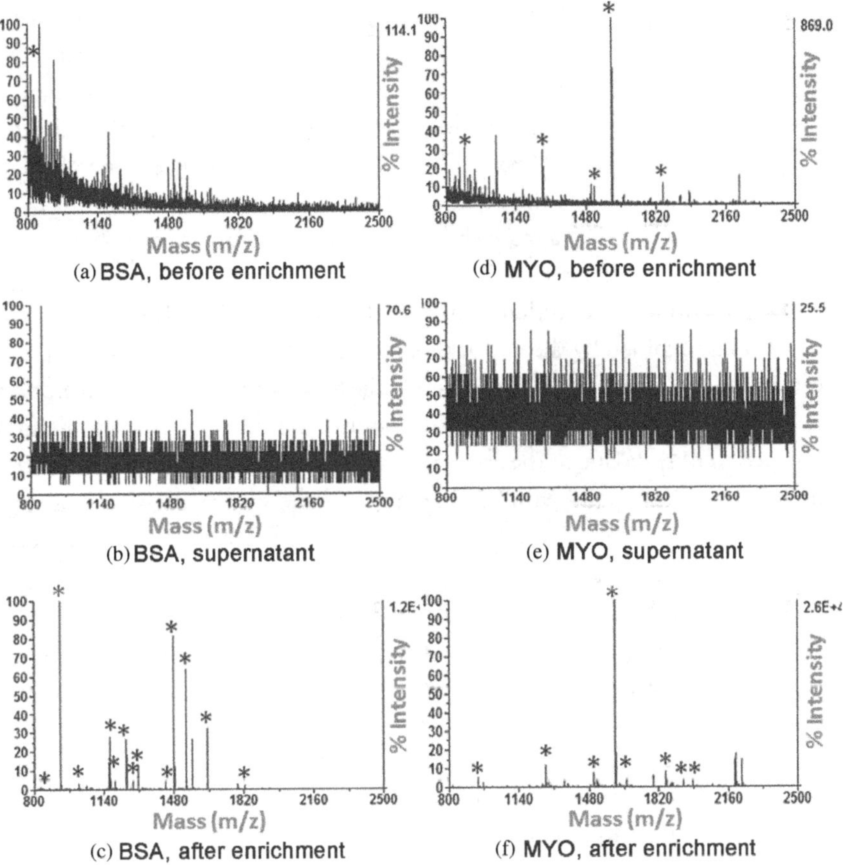

Figure 4.43: MALDI-TOF-MS spectra of BSA and MYO hydrolysate before and after enrichment by Fe_3O_4–graphene@mSiO$_2$.

4.5.1.1. *Laser elution of c-MWCNTs/Fe$_3$O$_4$–@mSiO$_2$ enrichment*

It has been reported that the combination of laser and carbon nanotube can cause tumor cell apoptosis, effectively inhibit tumor growth and enhance photothermal efficacy.[15,36] It has also been reported that MWCNTs can also absorb infrared light like SWCNTs. Therefore, the infrared laser-assisted elution effect of c-MWCNTs/Fe$_3$O$_4$–@mSiO$_2$ enrichment is investigated with the classical solution elution method as control.

The enrichment process is the same as the classical enrichment. Two standard peptide solutions (400 μL 5 nmol·L^{-1}) are prepared, and 3 μL 10 mg·mL^{-1} c-MWCNTs/Fe$_3$O$_4$–@mSiO$_2$ is added into the above two peptide solutions for oscillation for 5 min. Then the supernatant is discarded by a magnet and 50% ACN/50% H$_2$O–0.1% TFA is added. Then one of them is shaken for 20 min, and the other is eluted with a laser of 808 nm for 20 s. The eluate is separated by a magnet, lyophilized and re-dissolved in 2 μL eluate for MALDI-TOF-MS analysis. The results show that the laser-assisted elution effect is not significantly different from the classical elution. The S/N ratio of the peptides obtained by the two approaches is close. The S/N ratio of peptide by laser-assisted elution is 1083.18, while the S/N ratio of the classical elution is 1080.96, indicating that the infrared laser can indeed release peptide in a short period of time due to the absorption of infrared radiation by MWCNTs. The hybrid of Fe$_3$O$_4$ and MWCNTs maintains the surface physical property of MWCNTs at the same time of maintaining the sensitively magnetic response.

Since the c-MWCNTs/Fe$_3$O$_4$-@mSiO$_2$ is capable of absorbing infrared light and generating a thermal effect under laser irradiation, the eluent temperature and the peptide signal after laser irradiation are experimentally determined.

Figure 4.44(a) shows the changes in enrichment efficiency and eluent temperature with different material dosages and different irradiation times. The eluent temperature increases with the extension of laser irradiation time,

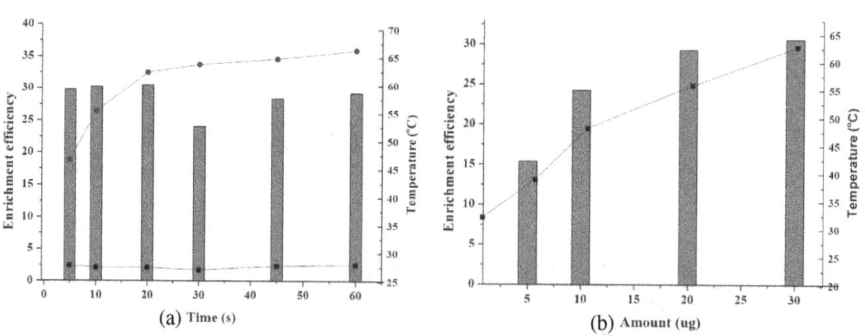

Figure 4.44: Change curve of enrichment factor and eluent temperature with laser irradiation time and c-MWCNTs/Fe$_3$O$_4$–@mSiO$_2$ dosage when using laser-assisted elution.

and it will reach a relative saturation value of 65°C after 30 s of laser irradiation. However, the enrichment efficiency is substantially unchanged since the dosage of material is same. Figure 4.44(b) shows the change trend of enrichment efficiency and eluent temperature along with the change of material dosage from 0 to 30 μg. Both the enrichment efficiency and eluent temperature increase with the increase of material dosage. These results further demonstrate that the c-MWCNTs/Fe_3O_4–@mSiO$_2$ is capable of absorbing infrared laser and is capable of delivering energy to the captured peptide for release. Therefore, the captured peptide can be successfully released under the auxiliary irradiation of the laser, just like a classical solution elution method, to achieve a certain enrichment efficiency.

4.5.1.2. *Ordered mesopore coated magnetic material for selective enrichment of endogenous peptide in practical biological sample*

Biological sample usually contains not only small peptide but also macromolecular protein, the complexity of which affects the applicability of peptidomics analysis technique, because high-abundance protein can also be adsorbed to interfere with peptide capture. Therefore, MS is difficult to directly detect and analyze the peptide therein. Human urine is a biological sample commonly used in scientific research because it is easily available and is rich in biomarkers that can be used for clinical examination and disease diagnosis. The rat brain is also a biological sample that is readily available and widely used in many laboratory. Compared with the general peptide-enriched adsorbent, the most important feature of the ordered mesopore coated magnetic material is the highly ordered mesoporous structure on the outer surface. Here, ordered mesopore coated magnetic material is used to selectively enrich endogenous peptide in human urine and rat brain sample.

(1) Fe_3O_4@nSiO$_2$@mSiO$_2$ for selective enrichment of endogenous peptide in rat brain

The whole protein extract of rat brain is used as a sample to investigate the selective enrichment performance of Fe_3O_4@nSiO$_2$@mSiO$_2$ for endogenous peptide in practical complex biological samples.[31] 1.0 mg Fe_3O_4@nSiO$_2$@

mSiO$_2$ is added into the diluted rat brain protein extract for 1 h at 4°C (the pH of mixture of 0.2 mL rat brain protein extract and 1.0 mL water is adjusted to 3 with hydrochloric acid). After that, the microsphere is separated by a magnet, and washed with deionized water, and eluted with 0.2 mL solvent (water/acetonitrile/TFA = 50/50/0.1, v/v/v). The eluate is lyophilized and re-dissolved for 1D nano-LC-ESI-MS/MS analysis.

As shown in Figure 4.45, the rat brain protein extract is directly enriched by Fe$_3$O$_4$@nSiO$_2$@mSiO$_2$ without the removing of high-abundance protein or desalting. The method relies on the large inner surface of the pore to capture the endogenous peptide in the sample through hydrophobic interaction. Calcination makes a large amount of silanol groups on the inner surface of the pore converted into siloxane bonds, and the surface hydrophobic property is enhanced. The mesoporous microsphere is superior to the centrifugal ultrafiltration in enriching endogenous peptide because the sample dosage is unconstrained. And low molecular weight contaminants, especially salts, are not simultaneously enriched. Therefore, this method is also superior to the usual magnetic bead with reversed-phase group, because the proper pore size of the ordered mesoporous layer allows the enrichment of endogenous peptide but

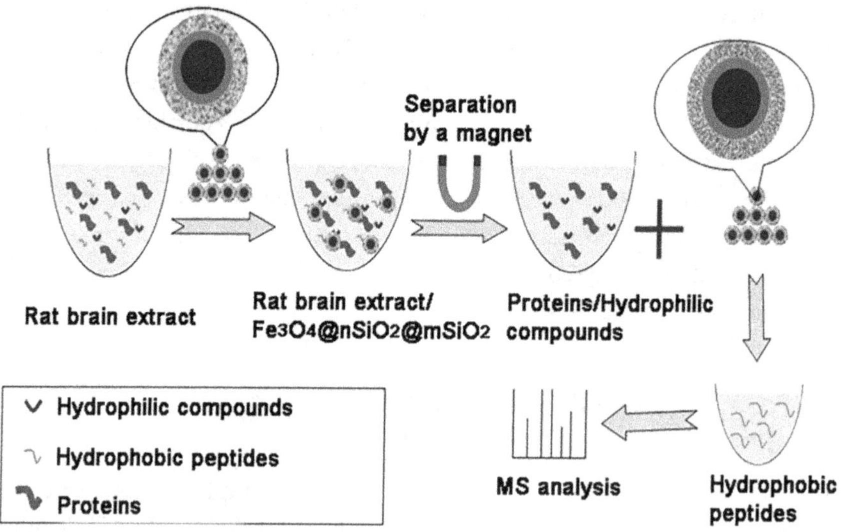

Figure 4.45: Schematic diagram of the rapid enrichment of endogenous peptide in biological sample by Fe$_3$O$_4$@nSiO$_2$@mSiO$_2$.

limits the entry of large protein. This enrichment method is very convenient and effective. The enriched peptides are eluted with an eluent and analyzed by 1D nano-LC-ESI-MS/MS. The positive library and the negative library search of obtained LTQ Orbitrap XL MS/MS spectra are carried out strictly according to the principle of SEQUEST algorithm. And the data is superimposed and simulated to obtain the Xcorr value under the standard of 99% confidence level. Finally, the peptides with an error rate of less than 5% are obtained by sieving.

After searching and screening the database, 60 different endogenous peptides of rat brain are obtained. Among these peptides, a regular peptide ladder as previously reported in the peptide analysis literature is found. Due to the limitation of LC-MS/MS condition, the molecular weight of the analyzed peptides is below 4.5 kDa. Figure 4.46 shows the molecular weight range distribution of the detected peptides. Most of the peptides are concentrated at 1,200–2,400 Da, which is in agreement with the best analytical result by LC-MS.

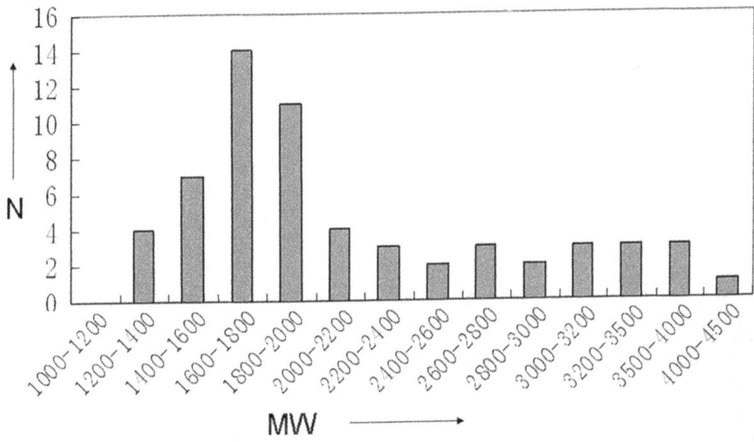

Figure 4.46: Molecular weight distribution of endogenous peptide from diluent of rat brain protein extract enriched by $Fe_3O_4@nSiO_2@mSiO_2$ and analyzed by 1D nano-LC/MS/MS. N is the peptide number.

(2) c-MWCNTs/Fe$_3$O$_4$-@mSiO$_2$ for selective enrichment of endogenous peptide in mouse brain

The pH of 1.2 mL diluent of mouse brain protein extract (1:2, v/v) is adjusted to 3, and then 1 mg c-MWCNTs/Fe$_3$O$_4$-@mSiO$_2$ is added for oscillation at 4°C for 1 h. [15] Then the material is separated and washed by 500 μL water for 3 times. Afterwards, 200 μL eluent (50% ACN/50% H$_2$O–0.1% TFA) is added for incubation for 1 h. Finally, the eluent is separated by a magnet, and lyophilized and re-dissolved in 2 μL mobile phase B for 1D nano-LC-MS/MS analysis. Figure 4.47 show the total ion chromatogram. The obtained LTQ Orbitrap XL MS/MS spectrum is set according to the principle of the SEQUEST algorithm with the following search parameters: non-enzymatic hydrolysis, no variable modification, peptide molecular weight deviation of 0.01%, and fragment ion deviation of 1 Da. In order to reduce the false positive results, the search of the positive and negative libraries is carried out, and

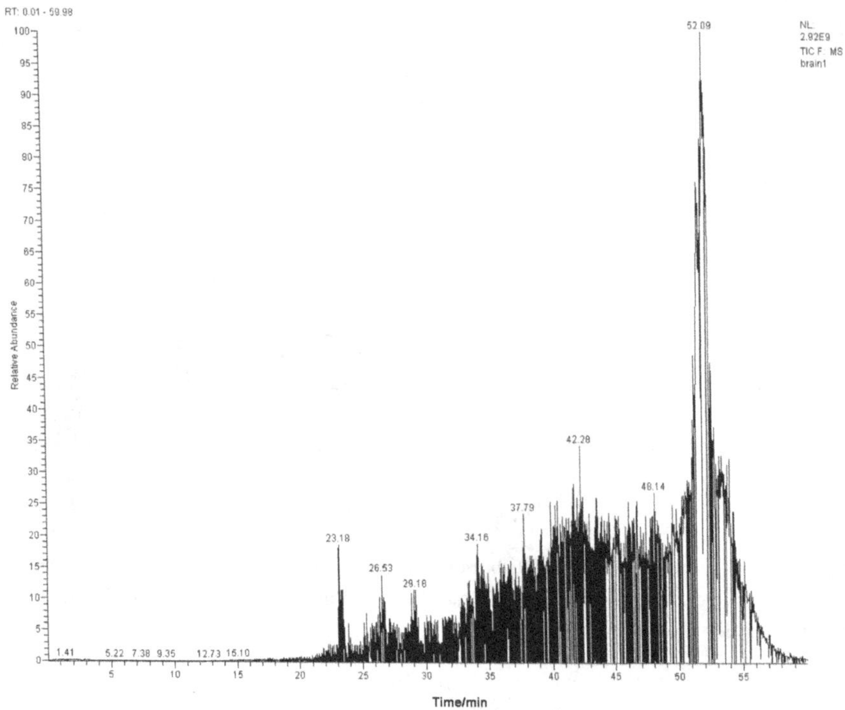

Figure 4.47: Total ion chromatogram of peptide from mouse brain extract enriched by c-MWCNTs/Fe$_3$O$_4$-@mSiO$_2$ and analyzed by 1D nano-LC-MS/MS.

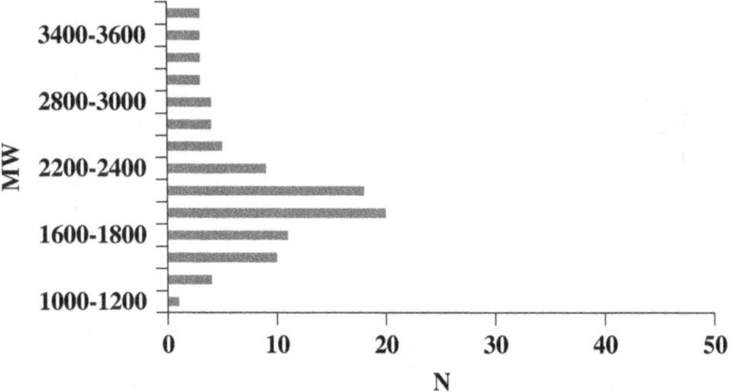

Figure 4.48: Molecular weight distribution of peptide from mouse brain protein extract enriched by c-MWCNTs/Fe_3O_4-@mSiO$_2$ and analyzed by 1D nano-LC-MS/MS. N is the peptide number.

the data is superimposed and simulated. The Xcorr values of the two-charged and three-charged ions are 3.16 and 3.34, respectively, under the criterion of 99% confidence level. Finally, the peptides with error rate less than 5% are obtained by sieving. After searching and screening the database, a total of 98 different mouse brain endogenous peptides are obtained.

Figure 4.48 shows the molecular weight range distribution of the detected peptides. Most of the peptides are concentrated at 1,000–2,600 Da. These results indicate the effectiveness of c-MWCNTs/Fe_3O_4-@mSiO$_2$ in selective enrichment of peptide. By taking advantage of the sensitively magnetic responsiveness, size-exclusion effect and infrared adsorption ability, it has been the new efficient material for peptide enrichment.

(3) Fe_3O_4–graphene@mSiO$_2$ for selective enrichment of endogenous peptide in practical biological sample

15 μL 10 mg·mL^{-1} Fe_3O_4–graphene@mSiO$_2$ and 200 μL diluted human urine (containing 10 μL raw human urine) are mixed for oscillation for 20 min.[32] Then the enriched peptides are eluted according to the previous elution step for MS detection. The results are shown in Figure 4.49, from which it can be seen that the difference of peptide number and the peak intensity between before and after enrichment, confirming that Fe_3O_4–graphene@mSiO$_2$ can successfully enrich the endogenous peptide in human urine.

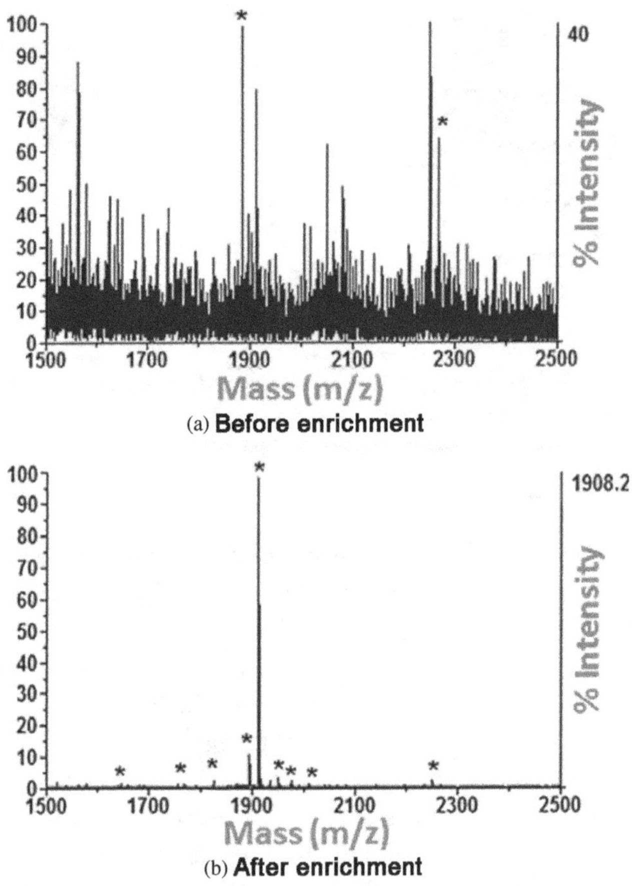

Figure 4.49: MALDI-TOF-MS spectra of human urine before and after enrichment by Fe_3O_4–graphene@mSiO$_2$.

Afterwards, mouse brain extract is selected as experimental object to investigate the enrichment performance of Fe_3O_4–graphene@mSiO$_2$ in more complex biological samples. 200 μL mouse brain extract is diluted 3 times and then is mixed with 0.2 mg Fe_3O_4–graphene@mSiO$_2$ for oscillation. The eluent is lyophilized and re-dissolved for nano-LC-MS/MS. A total of 409 peptides are identified by searching database, and the sequence information is shown in Table 4.6. The results show that Fe_3O_4–graphene@mSiO$_2$ can specifically enrich endogenous peptide from complex biological sample and exclude macromolecular protein and hydrophilic substance and so on by

Table 4.6: Endogenous peptide information identified from mouse brain using Fe_3O_4–graphene@mSiO$_2$.

MH$^+$	DeltaM	z	Peptide	XC
920.51999	0.00078	2	G.ILDSIGRF.F	2.538
973.5863	0.00086	2	E.AAVAIKAMAK.-	2.634
1005.5364	0.00135	2	F.VNDIFERI.A	2.542
1013.4435	0.00123	2	R.GGNFGFGDSR.G	2.525
1015.5935	0.00153	2	G.IAVAIKTYH.Q	2.504
1025.6102	0.00188	2	K.IGGIGTVPVGR.V	3.114
1029.5874	0.00492	2	-.MLRQILGQA.K	2.534
1043.49	0.00303	2	D.LMAYMASKE.-	2.55
1046.5477	0.00294	2	R.DTGILDSIGR.F	2.728
1047.5833	0.00014	2	E.FVDIINAKQ.-	2.613
1057.5888	0.00355	2	E.ANEALVKALE.-	2.662
1093.5524	0.00049	2	K.ALAAGGYDVEK.N	2.64
1094.4902	0.00158	2	E.YFSVAAGAGDH.-	2.606
1098.6881	0.00045	2	T.VAVGVIKAVDK.K	2.822
1107.5681	0.00051	2	K.ALAAAGYDVEK.N	2.872
1110.6055	−0.00202	2	G.VATALAHKYH.-	2.925
1114.6619	0.00085	2	E.AIAVQGPFIKA.-	3.314
1121.6466	0.0047	2	H.FFKNIVTPR.T	2.93
1136.6786	0.00438	2	Q.KVVAGVATALAH.K	2.7
1153.6576	−0.00204	2	S.VSTVLTSKYR.-	3.074
1160.7249	0.00207	2	E.VITISKSVKVS.L	2.632
1166.6164	−0.00058	2	K.GTGASGSFKLNK.K	2.762
1185.7929	−0.00315	2	R.IKLGLKSLVSK.G	3.018
1191.6078	0.00182	2	G.AGAPVYMAAVLE.Y	2.517
1199.7358	0.00138	2	Q.TVAVGVIKAVDK.K	3.176
1202.6813	0.00281	2	L.KQLLMLQSLE.-	2.685
1203.5827	0.00682	2	T.ITTAYYRGAMG.I	2.533
1224.7311	−0.00313	2	K.AVGKVIPELNGK.L	2.873
1226.7103	0.00336	2	K.APILIATDVASR.G	3.252
1229.6048	0.00531	2	E.KETAFEFLSSA.-	2.641
1238.664	0.00017	2	V.AGVATALAHKYH.-	2.515

(Continued)

Table 4.6: (*Continued*)

MH⁺	DeltaM	z	Peptide	XC
1240.5957	−0.00203	2	M.NSFVNDIFER.I	2.828
1250.5648	−0.00303	2	R.KQNDVFGEADQ.-	2.601
1250.6739	−0.00129	2	L.LPGELAKHAVSE.G	2.676
1256.6593	0.00634	2	E.ILELAGNAARDN.K	3.414
1258.7042	−0.00352	2	D.PLADLNIKDFL.-	2.508
1261.6821	0.00353	2	R.AITGASLADIMAK.R	3.551
1263.7307	0.00557	2	D.LIAYLKKATNE.-	3.08
1265.7688	0.0045	2	E.ALAAKAGLLGQPR.-	3.389
1268.7096	0.00244	2	F.LASVSTVLTSKY.R	2.651
1270.679	0.00379	2	A.KVIHDNFGIVE.G	2.598
1273.6383	−0.00155	2	F.KNLQTVNVDEN.-	3.04
1274.7289	0.00672	2	S.ILFMAKVNNPK.-	2.585
1283.6664	0.00351	2	S.LMNLEEKPAPAA.K	3.021
1285.7474	0.00788	2	S.ILKIDDVVNTR.-	2.658
1287.6943	0.00043	2	D.PKFEVIDKPQS.-	3.202
1288.7219	−0.03006	2	K.TKQGVAEAAGKTK.E	2.64
1300.7372	0.00473	2	T.FIAIKPDGVQRG.L	2.627
1308.6253	−0.00088	2	S.KYLATASTMDHA.R	2.848
1310.7355	−0.00018	2	F.LSIFSGILADFK.-	3.589
1313.6637	0.00423	2	M.VAYWRQAGLSY.I	2.645
1315.7104	−0.00293	2	T.LPTKETIEQEK.R	2.691
1317.8253	0.0048	2	S.VLISLKQAPLVH.-	2.734
1323.742	0.00691	2	T.KIYLPHSLPQQ.-	3.005
1325.579	−0.00035	2	E.DMQNAFRSLED.-	2.685
1326.7627	−0.00171	2	R.KASGPPVSELITK.A	3.596
1327.7944	−0.0001	2	R.QTVAVGVIKAVDK.K	3.354
1329.7447	−0.00113	2	R.VETGVLKPGMVVT.F	2.637
1331.7165	−0.00277	2	K.TSGVVQGVASVAEK.T	3.28
1332.7733	−0.00518	2	P.VSELITKAVSASK.E	2.765
1334.7467	0.00214	2	C.LLPAGWVLSHLE.S	2.807
1335.7631	0.00912	2	N.LISKLSAKFGGQS.-	3.205

Table 4.6: (*Continued*)

MH⁺	DeltaM	z	Peptide	XC
1336.6314	0.00124	2	K.YLATASTMDHAR.H	2.824
1339.7077	0.00044	3	R.HRDTGILDSIGR.F	3.86
1340.7784	−0.00559	2	R.KATGPPVSELITK.A	2.583
1342.7074	0.00308	2	N.LLAENGRLGNTQG.I	2.873
1353.6797	0.00281	2	M.NSFVNDIFERI.A	2.976
1356.7886	0.00092	2	E.IEAIAVQGPFIKA.-	2.6
1357.7798	0.00081	2	F.SGAGNIAAATGLVKK.E	2.94
1360.6777	0.00418	2	S.LVDAMNGKEGVVE.C	2.822
1366.7763	0.00268	2	-.MQIFVKTLTGKT.I	3.301
1367.6485	−0.00186	2	Y.LATASTMDHARHG.F	2.726
1369.7434	−0.00249	2	R.GAAQNIIPASTGAAK.A	2.67
1377.7849	−0.00533	2	N.IVTPRTPPPSQGK.G	3.362
1384.7318	0.00449	2	F.VDLEPTVIDEVR.T	2.68
1384.7543	−0.00136	2	E.ILELAGNAARDNK.K	3.492
1385.7019	−0.00368	2	F.SLSGAQIDDNIPR.R	3.063
1388.8148	0.0026	2	F.LGALALIYNEALK.-	2.767
1395.6573	−0.0012	2	R.SKYLATASTMDHA.R	2.809
1397.7787	0.00036	2	R.KAAFKELQSTFK.-	3.074
1400.8182	0.00536	2	E.LISVVKSMLKGPE.-	3.2
1402.7974	0.00183	2	Q.KEITALAPSTMKI.K	3.795
1403.6801	0.00345	2	M.LLADQGQSWKEE.V	3.253
1404.7886	−0.01678	2	R.NVFLLGFIPAKAD.S	2.527
1406.7122	0.00533	2	H.ASASSTNLKDVLSN.L	2.699
1414.6995	0.00169	2	E.QGQAIDDLMPAQK.-	2.742
1416.8685	0.01146	3	V.RLLLPGELAKHAV.S	3.24
1420.7682	0.00475	2	M.VKLIESKEAFQE.A	3.524
1422.8063	0.00404	2	R.IVAPPGGRANITSLG.-	2.61
1424.8108	0.00431	2	F.LASVSTVLTSKYR.-	3.554
1425.7332	−0.00107	2	E.AVAKADKLAEEHGS.-	3.516
1425.7332	0.00003	2	E.AVAKADKLAEEHGS.-	3.169
1434.8063	0.00124	2	N.IVTPRTPPPSQGKG.G	3.341

(*Continued*)

Table 4.6: (*Continued*)

MH⁺	DeltaM	z	Peptide	XC
1438.7723	0.00854	2	H.LIAKEEMIHNLQ.-	2.81
1444.7067	−0.00113	2	R.SLYSSSPGGAYVTR.S	2.691
1453.7508	0.00502	2	E.FMILPVGASSFRE.A	3.765
1456.8522	−0.0017	2	R.LSAKPAPPKPEPKP.K	2.685
1460.7169	−0.00597	2	R.TQDENPVVHFFK.N	2.864
1461.6176	0.00569	2	M.SSGAHGEEGSARMW.K	2.514
1467.7339	0.00387	2	E.NAFLSHVISQHQS.L	3.778
1468.8734	0.00392	2	E.KLGGSAVISLEGKPL.-	2.858
1470.805	0.00552	2	E.LGISTPEELGLDKV.-	3.282
1477.7937	−0.00126	2	R.KESYSVYVYKVL.K	2.844
1485.8285	0.00095	3	L.AARPNSGAIHLKPPG.-	3.171
1486.7748	−0.00275	2	V.SEGTKAVTKYTSSK.-	2.849
1488.7441	0.00385	2	E.NVIRDAVTYTEHA.K	2.744
1491.8278	−0.00232	2	F.KNIVTPRTPPPSQG.K	3.533
1492.7941	0.00716	3	Y.AKDVKFGADARALM.L	3.505
1499.739	0.00131	3	T.LASHHPADFTPAVH.A	4.626
1502.8577	0.00561	2	A.KVLTPELYAELRA.K	3.1
1515.7649	−0.00392	2	E.APAAAASSEQSVAVKE.-	2.57
1518.9036	0.0045	3	L.KSIKNIQKITKSM.K	3.257
1522.8588	0.00119	3	E.RLSVDYGKKSKLE.F	3.091
1524.8744	0.0034	2	N.AKVAVLGASGGIGQPLS.L	3.882
1530.8162	0.00418	2	E.AVYLITPSEKSVHS.L	2.503
1532.7704	−0.00363	2	G.FSLSGAQIDDNIPR.R	3.181
1537.8162	0.00639	3	L.LYRPGHYDILYK.-	3.826
1540.8457	0.00361	2	K.LAVNMVPFPRLHF.F	2.552
1542.801	−0.00132	2	F.VDLEPTVIDEVRTG.T	2.92
1551.7584	0.0003	2	R.SKYLATASTMDHAR.H	4.194
1551.7914	−0.00001	2	R.KGNYAERVGAGAPVY.M	3.583
1553.817	−0.00376	2	E.TNLESLPLVDTHSK.R	2.843
1555.7897	−0.00161	2	L.KQVHPDTGISSKAMG.I	2.739
1557.7332	0.01016	2	F.LVFHSFGGGTGSGFTS.L	2.85
1564.8958	−0.00169	2	Q.KVVAGVATALAHKYH.-	5.006

Table 4.6: (*Continued*)

MH$^+$	DeltaM	z	Peptide	XC
1567.7863	0.00027	2	R.KGNYSERVGAGAPVY.L	3.444
1571.8792	0.00388	2	K.FLASVSTVLTSKYR.-	3.872
1574.8173	0.00658	2	M.VLSGEDKSNIKAAWG.K	2.602
1575.9073	0.00578	2	S.KIQKLSKAMREML.-	3.029
1578.7871	−0.00022	2	K.ALAAAGYDVEKNNSR.I	4.293
1584.9472	−0.00017	3	R.LSAKPAPPKPEPKPK.K	3.787
1586.9741	0.00981	3	T.AVRLLLPGELAKHAV.S	3.135
1588.8251	−0.00602	2	L.QVEEEVDAMLAVKK.-	3.064
1594.8621	−0.00314	2	K.ASIKKGEDFVKNMK.-	3.623
1605.8959	0.00484	2	E.RADLIAYLKKATNE.-	3.767
1606.8145	0.00234	2	E.LLKQGQYSPMAIEE.Q	3.218
1607.8976	0.00952	3	E.TGRVLSIGDGIARVHG.L	3.02
1611.8701	0.00222	2	E.KVTSHAIVKEVTQGD.-	3.654
1616.8359	0.0107	2	E.VRECACAGLARLVQQ.R	2.764
1617.917	0.00371	3	P.VSELITKAVSASKER.G	3.993
1619.9228	−0.00067	3	F.KNIVTPRTPPPSQGK.G	3.461
1621.8288	−0.00488	2	L.SAMTEEAAVAIKAMAK.-	3.965
1622.8458	−0.00591	2	S.ISGEYNLKTLMSPLG.I	2.871
1623.9316	0.01004	2	E.LLEGKVLPGVDALSNV.-	2.788
1636.8952	0.01085	2	R.DGKLRHM*LARDLVP.G	2.562
1637.9585	0.00299	2	N.AKVAVLGASGGIGQPLSL.L	3.74
1638.8884	0.00636	3	S.LLMERLSVDYGKKS.K	3.154
1640.8854	0.00168	3	E.KNPLPSKETIEQEK.Q	3.22
1642.7595	−0.00922	2	V.FEYEAAGEDELTLR.L	2.574
1658.8068	−0.00262	2	S.KYLATASTMDHARHG.F	3.951
1659.8013	0.00305	2	G.FQQILAGEYDHLPE.Q	3.515
1660.7496	−0.00384	2	M.SSGAHGEEGSARMWKA.L	2.815
1661.719	−0.00049	2	Q.LTADSHPSYHTDGFN.-	3.201
1664.9013	−0.02646	2	H.GALVRKVHADPDCRK.T	2.504
1665.8153	−0.00267	2	R.FDEILEASDGIMVAR.G	4.208
1667.9877	0.00484	2	E.DLKLLAKKAQAQPIM.-	3.502
1668.9407	0.00385	2	R.KLAVNMVPFPRLHF.F	3.888

(*Continued*)

Table 4.6: (*Continued*)

MH$^+$	DeltaM	z	Peptide	XC
1669.8955	0.0068	3	F.KRISEQFTAMFRR.K	3.803
1670.9436	−0.00533	3	Q.VTNVGGAVVTGVTAVAQK.T	3.197
1676.9442	−0.00382	2	F.KNIVTPRTPPPSQGKG.G	2.76
1680.8704	0.00536	2	R.KGNYSERVGAGAPVYL.A	2.787
1682.8319	0.00446	2	R.KGNYAERVGAGAPVYM.A	3.503
1683.9105	0.00197	2	R.ALPFWNEEIVPQIK.E	2.724
1687.8216	0.00294	2	-.MQLKPMEINPEMLN.K	4.234
1696.9017	−0.02328	2	R.RLAVRFTALDLSYGD.A	2.705
1699.9741	−0.00356	3	D.KFLASVSTVLTSKYR.-	5.396
1701.8654	0.00595	2	E.ALARLEEEGQSLKEE.M	3.2
1704.8017	0.00558	2	G.FLVFHSFGGGTGSGFTS.L	4.128
1711.945	−0.00541	2	Y.AAQASAAPKAGTATGRIVA.V	2.99
1715.8058	−0.00302	2	Y.AEAAAAPAPAAGPGQMSFT.F	3.037
1718.98	0.00382	2	G.LISFIKQQRDTKLE.-	3.093
1720.9592	−0.00568	2	L.LLPGELAKHAVSEGTKA.V	3.19
1721.7725	0.00283	2	D.GQVINETSQHHDDLE.-	3.204
1723.8861	0.00263	2	E.LISNSSDALDKIRYE.S	3.256
1725.9745	−0.00562	2	K.ASGPPVSELITKAVAASK.E	3.813
1729.9483	0.00332	2	T.LWNGQKLVTTVTEIAG.-	2.526
1730.8708	0.00319	2	F.LENVIRDAVTYTEHA.K	3.223
1731.0527	−0.00592	2	T.IAQGGVLPNIQAVLLPK.K	3.265
1741.9807	0.00206	3	I.RNDEELNKLLGKVTI.A	3.693
1745.928	0.00727	2	E.SLATVEETVVRDKAVE.S	3.116
1747.8472	0.00073	2	F.LSQPFQVAEVFTGHMG.K	2.846
1749.9269	0.00572	2	E.ALELVKEAIDKAGYTE.K	4.43
1750.6787	−0.00043	2	T.LWTSDQQDDDGGEGNN.-	3.777
1751.8857	0.01239	2	D.SLKNCVNKDHLTAATH.W	2.593
1756.9704	0.00189	2	F.VNVVPTFGKKKGPNANS.-	2.503
1762.901	0.00132	2	P.LATYAPVISAEKAYHE.Q	2.87
1763.9327	0.00451	2	K.KSADTLWGIQKELQF.-	3.367
1764.9102	0.00227	2	N.KKAYSMAKYLRDSGF.-	3.086
1765.8828	−0.02342	2	A.GAAAGSAAAAAAAAAATAPGPAGA.A	2.892

Table 4.6: (*Continued*)

MH⁺	DeltaM	z	Peptide	XC
1768.944	−0.00463	3	E.KNPLPSKETIEQEKQ.A	3.572
1770.9207	0.00294	2	M.PMFIVNTNVPRASVPE.G	3.067
1772.9402	0.0056	3	E.NVIRDAVTYTEHAKR.K	5.094
1774.8139	0.01353	2	D.GQVNYEEFVQMMTAK.-	4.337
1777.9079	−0.00413	2	T.LTHGSVVSTRTYEKEA.-	3.442
1779.8773	0.01065	2	E.TAVNLAWTAGNSNTRFG.I	3.022
1780.0327	0.01009	3	M.VIRVYIASSSGSTAIKK.K	4.274
1787.9439	0.00599	3	R.KAGQVFLEELGNHKAF.K	4.302
1801.0364	−0.01063	2	I.ITLMIKSVQKNDVRVG.F	2.631
1804.8824	−0.00227	2	H.QSGFSLSGAQIDDNIPR.R	3.3
1806.8837	0.00572	3	-.MKGETPVNSTMSIGQAR.K	3.762
1815.8808	0.00041	2	E.MVPLKYFENWSASMI.-	3.458
1817.8857	0.00956	2	G.FLVFHSFGGGTGSGFTSL.L	2.629
1821.9164	−0.00006	3	R.RFDEILEASDGIMVAR.G	4.152
1823.0347	0.00705	2	-.MQIFVKTLTGKTITLE.V	3.101
1827.9422	0.00256	2	M.PMFIVNTNVPRASVPEG.F	3.813
1828.9651	−0.00103	2	E.GAELVDSVLDVVRKEAE.S	3.137
1832.8735	0.00296	2	G.IVMDSGDGVTHTVPIYE.G	2.978
1836.9755	0.0027	2	E.AFREANLAAAFGKPINF.-	3.513
1840.0228	0.00067	2	A.FQKVVAGVATALAHKYH.-	5.027
1843.9032	−0.0037	2	P.QAEAPAAAASSEQSVAVKE.-	5.58
1844.8966	0.00875	2	G.FSAKWDYDKNEWKK.-	3.155
1846.9518	−0.00769	2	F.LSSPEHVNRPINGNGKQ.-	3.106
1864.1266	0.00095	2	N.AKVAVLGASGGIGQPLSLLL.K	3.925
1865.8453	0.0081	2	L.FHNPHVNPLPTGYEDE.-	3.915
1868.9211	0.00944	2	R.AVNPSRGVFEEMKYLE.L	2.532
1869.8937	−0.0005	2	E.AASAAPGIPAEQTRDSPSGS.-	3.257
1871.0637	0.00578	3	R.KGLKEGIPALDNFLDKL.-	3.722
1874.8814	−0.00356	2	M.SSGAHGEEGSARMWKALT.Y	3.051
1877.8929	0.00684	3	A.SDLELHPPSYPWSHRG.L	3.971
1878.9668	0.00864	2	K.SLNLVSDRFSRTENIQ.K	2.528
1886.9351	−0.03424	2	P.GELLIALHNIDSVKCDM*.K	2.656

(*Continued*)

Table 4.6: (*Continued*)

MH⁺	DeltaM	z	Peptide	XC
1889.0015	0.00252	2	R.KSADTLWDIQKDLKDL.-	3.41
1892.8993	0.00402	2	E.AMRMLLADQGQSWKEE.V	3.297
1895.0134	−0.00328	2	G.VTSGNVGYLAHAIHQVTK.-	2.9
1896.0371	−0.0022	3	Y.KVLKQVHPDTGISSKAMG.I	3.995
1900.8415	0.00433	2	G.VMVGMGQKDSYVGDEAQS.K	3.985
1902.8676	−0.00188	2	M.ASGGGVPTDEEQATGLERE.I	3.883
1910.79	−0.00426	2	R.SSGSPYGGGYGSGGGSGGYGSR.R	4.233
1911.9851	0.00299	2	K.KLTGKDVNFEFPEFQL.-	2.533
1916.0389	−0.00036	3	R.KAGQVFLEELGNHKAFK.K	3.602
1920.0549	0.00714	3	K.KGERADLIAYLKKATNE.-	5.51
1922.0342	−0.00726	2	A.KHAVSEGTKAVTKYTSSK.-	5.223
1923.8277	−0.0021	2	K.VQEEFDIDMDAPETER.A	4.575
1926.9444	0.00521	2	E.KLTQDQDVDVKYFAQE.A	4.608
1928.0851	−0.00333	3	S.LDKFLASVSTVLTSKYR.-	3.753
1929.8672	−0.00941	2	K.SDAAPAASDSKPSSAEPAPSS.K	3.966
1930.048	0.00394	3	E.LFKRISEQFTAMFRR.K	4.186
1939.1045	0.00369	3	A.LVGKDMANQVKAPLVLKD.-	3.602
1940.96	0.0028	2	A.LLSPYSYSTTAVVSNPQN.-	2.736
1941.1055	0.00303	3	M.KTYVEKVDELKKKYGI.-	4.495
1946.8623	−0.00105	2	D.GDGQVNYEEFVQMMTAK.-	3.842
1952.9634	−0.00726	2	K.HLEGLSEEAIMELNLPTG.I	3.565
1958.0164	−0.02226	3	L.PKVAVPSTIHCDHLIEAQ.V	3.141
1964.0348	0.00587	3	E.ALEWLIRETEPVSRQH.-	3.129
1965.7945	0.00107	2	T.LWTSDTQGDEAEAGEGGEN.-	4.607
1978.9294	0.0061	2	T.LFHNPHVNPLPTGYEDE.-	3.775
1982.097	0.00216	3	Q.AAFQKVVAGVATALAHKYH.-	3.838
1984.9822	0.00296	2	K.NPLPSKETIEQEKQAGES.-	3.264
1989.9811	−0.04203	2	S.FGRDQDNKIAIKNCDGVP.A	2.537
1990.0572	0.0324	2	L.MQQASSGRQSLPIKAM*LK.S	2.658
1991.873	0.01001	2	E.TTQLTADSHPSYHTDGFN.-	3.736
1992.2216	0.00217	2	N.AKVAVLGASGGIGQPLSLLLK.N	5.889
1994.0203	−0.00714	2	A.FLSSPEHVNRPINGNGKQ.-	3.795

Table 4.6: (*Continued*)

MH⁺	DeltaM	z	Peptide	XC
1994.0454	0.00636	3	D.TGILDSIGRFFSGDRGAPK.R	4.421
2000.9892	0.001	2	L.KQVHPDTGISSKAMGIMNS.F	4.382
2007.78	0.00119	2	E.AYEMPSEEGYQDYEPEA.-	2.854
2009.11	0.00319	3	S.LLMERLSVDYGKKSKLE.F	3.871
2010.1416	0.00394	3	E.ALVGKDMANQVKAPLVLKD.-	4.006
2011.1182	−0.00766	3	K.ASGPPVSELITKAVAASKER.S	3.356
2013.04	−0.0034	2	A.AKAPAPAAPAAAEPQAEAPAAAA.S	3.69
2020.2026	−0.00829	2	Y.PLRRSQSLPTTLLSPVRV.V	2.676
2023.0529	0.04708	3	L.LEKIYRLEMEENQLKS.E	3.135
2024.0407	0.00441	2	L.KEARVAQGQGEGEVGPEVAL.-	3.341
2031.0717	−0.00247	2	L.LIKTVETRDGQVINETSQ.H	4.039
2037.9447	0.00239	2	M.SSGAHGEEGSARMWKALTY.F	3.135
2043.9844	0.00173	2	F.LVFHSFGGGTGSGFTSLLME.R	3.272
2049.0261	0.00924	3	R.HRDTGILDSIGRFFSGDR.G	3.131
2050.0451	0.00804	3	M.VGPIEEAVAKADKLAEEHGS.-	3.12
2057.9622	−0.0032	2	K.SDAAPAASDSKPSSAEPAPSSK.E	4.01
2062.1543	−0.00603	3	R.SGVSLAALKKALAAAGYDVEK.N	3.82
2068.9393	−0.02	2	K.VNSGPVTFSQASGICHSYGGT.L	2.523
2070.0475	−0.00115	3	S.LPQKSQHGRTQDENPVVH.F	3.096
2075.9	−0.00461	2	K.AGEASAESTGAADGAAPEEGEAK.K	4.267
2080.0445	0.00095	2	E.NSEALELVKEAIDKAGYTE.K	4.447
2082.1165	0.00972	2	E.AFEISKKEMQPTHPIRLG.L	3.628
2086.1543	0.00239	3	H.ASLDKFLASVSTVLTSKYR.-	5.384
2092.0909	0.00732	3	E.AWARLDHKFDLMYAKRA.F	3.153
2099.1062	−0.0062	2	-.MQLKPMEINPEMLNKVLA.K	2.939
2106.2645	0.00177	3	N.AKVAVLGASGGIGQPLSLLLKN.S	3.617
2109.0723	0.00224	3	R.DTGILDSIGRFFSGDRGAPK.R	4.027
2113.0772	−0.00497	3	E.KNPLPSKETIEQEKQAGES.-	3.811
2114.1829	−0.0061	3	L.KEVIRERKEREEWAKK.-	3.586
2122.1503	0.00397	2	E.KIQASVATNPIITPVAQENQ.-	5.696
2133.1761	0.00808	3	E.TLKTKAQSVIDKASETLTAQ.-	3.101
2139.2132	0.00741	3	R.KASGPPVSELITKAVAASKER.S	5.61

(*Continued*)

Table 4.6: (*Continued*)

MH+	DeltaM	z	Peptide	XC
2140.1617	−0.00302	3	Y.KVLKQVHPDTGISSKAMGIM.N	4.27
2141.1138	0.00117	3	R.TQDENPVVHFFKNIVTPR.T	4.359
2150.036	−0.00895	2	E.VAAGDDKKGIVDQSQQAYQE.A	5.005
2151.1451	0.0417	3	A.SSSQKLSFKERVRMASPRG.Q	3.254
2156.1573	0.00218	2	R.TLWTVLDAIDQMWLPVVR.T	3.314
2157.1874	−0.00047	3	K.TKEQVTNVGGAVVTGVTAVAQK.T	3.671
2162.894	−0.02086	2	N.QMQGFCPENEEKYRCVSD.S	2.571
2166.1012	0.00618	2	R.KGNYAERVGAGAPVYMAAVLE.Y	3.521
2169.2238	−0.00748	3	R.KATGPPVSELITKAVSASKER.G	4.197
2171.1819	0.00388	3	G.KKVITAFNDGLNHLDSLKGT.F	3.917
2175.0863	−0.03454	3	D.ADMPRPPETTTAVGAVVTAPHG.R	3.792
2181.0856	0.00653	3	Y.MVGPIEEAVAKADKLAEEHGS.-	3.049
2183.0761	−0.00724	3	G.MGQKDSYVGDEAQSKRGILT.L	4.635
2191.0529	0.00495	2	G.FLVFHSFGGGTGSGFTSLLME.R	3.959
2201.0689	0.01052	2	E.MEQRLEQGQAIDDLMPAQK.-	4.078
2218.0946	−0.00493	3	E.ALSVREAREEAEEKSEEKQ.-	3.067
2223.1616	−0.04496	3	E.AKSPAEAKSPAEAKSPAEAKSPA.E	3.395
2229.042	−0.01454	3	L.EEALVGNDSEGPLCELLFFF.L	3.313
2229.1245	−0.00438	3	K.TETQEKNPLPSKETIEQEK.Q	3.754
2230.2152	0.00139	3	E.SAMKTYVEKVDELKKKYGI.-	3.667
2237.1997	0.02111	3	L.ERLREESAAKDRLALELHT.A	3.179
2238.0536	−0.00042	2	A.SSWWTHVEMGPPDPILGVTE.A	3.984
2240.1095	−0.0052	3	E.SGYVPGYKHAGTYDQKVQGGK.-	3.615
2242.0591	−0.00034	3	G.VMVGMGQKDSYVGDEAQSKRG.I	5.071
2245.3465	0.00725	3	M.APIGLKAVVGEKIMHDVIKKV.K	4.134
2265.1735	0.00429	3	H.RDTGILDSIGRFFSGDRGAPK.R	3.091
2286.3658	0.05418	3	L.LSWMLSVVLLLFVDVRVALL.L	3.224
2293.0401	−0.0129	3	R.KAQPSQAAEEPAEKADEPMEH.-	4.671
2294.2602	−0.01174	3	E.LRPTLNELGISTPEELGLDKV.-	4.975
2322.1626	0.00291	3	S.FVNDIFERIAGEASRLAHYN.K	3.013
2341.2367	0.00163	3	Y.KVLKQVHPDTGISSKAMGIMNS.F	4.633
2345.2711	0.00795	3	G.LTTKNLDYVATSIHEAVTKIQ.-	4.024

Table 4.6: (*Continued*)

MH⁺	DeltaM	z	Peptide	XC
2352.1731	−0.00007	3	S.FVNDIFERIASEASRLAHYN.K	4.474
2357.2195	−0.00854	3	K.KTETQEKNPLPSKETIEQEK.Q	4.309
2373.4414	−0.00075	3	M.APIGLKAVVGEKIMHDVIKKVK.K	3.304
2377.2722	−0.00277	3	R.SGVSLAALKKALAAAGYDVEKNNS.R	3.881
2385.2483	0.03204	3	E.EKEEKECLSLNPLKPYHLSK.D	3.407
2402.2934	−0.00286	3	Y.VYKVLKQVHPDTGISSKAMGIM.N	4.008
2423.107	−0.00276	3	K.TVETRDGQVINETSQHHDDLE.-	5.277
2434.3314	0.00075	3	P.KIKAFLSSPEHVNRPINGNGKQ.-	3.685
2444.4309	0.00736	3	E.LLKMFGIDKDAIVQAVKGLVTKG.-	6.053
2445.2654	0.00337	3	E.IAQVATISANGDKDIGNIISDAMK.K	3.082
2468.198	0.0041	2	M.VNPTVFFDITADDEPLGRVSFE.L	3.827
2471.226	−0.002	3	E.TQEKNPLPSKETIEQEKQAGES.-	4.468
2481.4188	−0.00784	3	V.RLLLPGELAKHAVSEGTKAVTKY.T	5.199
2485.278	−0.00276	3	K.KTETQEKNPLPSKETIEQEKQ.A	3.081
2491.2682	−0.0033	3	E.KADSNKTRIDEANQRATKMLGSG.-	3.801
2502.3562	−0.00373	3	L.LPGELAKHAVSEGTKAVTKYTSSK.-	6.578
2504.1536	−0.00029	2	Q.LLRDNLTLWTSDQQDEEAGEGN.-	4.509
2510.3613	−0.00804	3	T.LSALVDGKSFNAGGHKLGLALELEA.-	4.053
2515.384	−0.00389	3	R.VETGVLKPGMVVTFAPVNVTTEVK.S	3.406
2522.3726	−0.00908	3	T.LSALLDGKNVNAGGHKLGLGLEFQA.-	3.854
2533.3733	0.00294	3	R.SGVSLAALKKALAAAGYDVEKNNSR.I	6.34
2535.4579	0.00525	3	A.LAIVEALVGKDMANQVKAPLVLKD.-	4.499
2538.316	0.0059	2	K.HLEGLSEEAIMELNLPTGIPIVY.E	4.173
2557.3409	0.00006	3	T.LSALIDGKNFNAGGHKVGLGFELEA.-	3.253
2558.3768	0.02757	3	L.ELGEKRKGMLEKSCKKFMLFR.E	3.08
2569.2749	0.01606	3	G.VMVGMGQKDSYVGDEAQSKRGILT.L	3.913
2576.1495	−0.0092	2	Q.LLRDNLTLWTSDQQDDDGGEGNN.-	4.785
2598.3985	−0.00047	3	K.LKKTETQEKNPLPSKETIEQEK.Q	3.393
2611.4395	−0.01268	3	P.FAKLVRPPVQVYGIEGRYATALY.S	3.507
2615.4403	−0.00176	3	L.LLPGELAKHAVSEGTKAVTKYTSSK.-	3.593
2618.2906	0.01552	3	E.KMDLTAKELTEEKETAFEFLSSA.-	3.76
2623.3475	0.00239	3	A.FAQFGSDLDAATQQLLSRGVRLTE.L	3.469

(*Continued*)

Table 4.6: (*Continued*)

MH$^+$	DeltaM	z	Peptide	XC
2624.4981	0.00153	3	E.KLVSTKKVEKVTSHAIVKEVTQGD.-	3.082
2631.4076	0.0031	3	R.KTVTAMDVVYALKRQGRTLYGFGG.-	3.621
2664.286	0.00347	3	L.IKTVETRDGQVINETSQHHDDLE.-	5.023
2672.3778	−0.0123	3	F.SGAGNIAAATGLVKKEEFPTDLKPEE.V	4.1
2701.3163	−0.00902	3	K.TETQEKNPLPSKETIEQEKQAGES.-	3.278
2706.2981	−0.00825	3	V.NPSDRYHLMPIITPAYPQQNSTY.N	3.396
2711.2808	0.00687	3	G.LAAQGRYEGSGDGGAAAQSLYIANHAY.-	4.589
2728.5244	0.00186	3	R.LLLPGELAKHAVSEGTKAVTKYTSSK.-	3.541
2738.4876	0.0014	3	D.FTPAVHASLDKFLASVSTVLTSKYR.-	5.068
2739.3294	0.01126	2	S.LYASGRTTGIVMDSGDGVTHTVPIYE.G	2.977
2744.3638	0.0018	3	M.VLSGEDKSNIKAAWGKIGGHGAEYGAE.A	5.447
2748.4904	0.059	3	F.NGLDRIISRRKNEKYLGFGTPSNL.G	3.443
2751.4572	−0.02525	3	Y.SVYVYKVLKQVHPDTGISSKAMGIM.N	3.368
2753.5634	0.0014	3	E.FALAIVEALVGKDMANQVKAPLVLKD.-	6.099
2777.37	0.01094	3	L.LIKTVETRDGQVINETSQHHDDLE.-	5.668
2788.5356	0.012	3	A.QHFRKVLGDTLLTKPVDIAADGLPSP.N	3.292
2791.2653	−0.01238	3	Q.LLRDNLTLWTSDTQGDEAEAGEGGEN.-	5.183
2805.4319	0.00067	3	R.TQDENPVVHFFKNIVTPRTPPPSQG.K	3.513
2818.5569	−0.00659	3	E.KMIAEAIPELKASIKKGEDFVKNMK.-	5.289
2829.4112	−0.00695	3	K.KTETQEKNPLPSKETIEQEKQAGES.-	4.682
2840.5206	−0.00274	3	G.KDFTPAAQAAFQKVVAGVATALAHKYH.-	5.993
2858.4167	0.01013	3	K.APAPAAPAAAEPQAEAPAAAASSEQSVAVKE.-	4.299
2864.4901	0.00282	3	E.VAAFAQFGSDLDAATQQLLSRGVRLTE.L	4.194
2884.6255	0.00977	3	V.RLLLPGELAKHAVSEGTKAVTKYTSSK.-	6.45
2900.4472	−0.01299	3	Q.LMRIEEELGDEARFAGHNFRNPSVL.-	3.837
2901.4094	−0.00089	3	S.IVGRPRHQGVMVGMGQKDSYVGDEAQS.K	3.924
2906.3902	0.00195	3	R.VPSPVSSEDDLQEEEQLEQAIKEHLG.P	5.087
2915.6036	−0.00066	3	A.KRKTVTAMDVVYALKRQGRTLYGFGG.-	3.755
2933.5268	0.00054	3	R.TQDENPVVHFFKNIVTPRTPPPSQGK.G	6.591
2935.1444	0.00108	2	E.DMPVDPGSEAYEMPSEEGYQDYEPEA.-	2.538
2957.5062	−0.01752	3	L.KKTETQEKNPLPSKETIEQEKQAGES.-	4.687
2959.7667	0.00887	3	N.AKVAVLGASGGIGQPLSLLLKNSPLVSRLT.L	5.773

Table 4.6: (*Continued*)

MH⁺	DeltaM	z	Peptide	XC
2986.5116	−0.00959	3	A.KAPAPAAPAAAEPQAEAPAAAASSEQSVAVKE.-	3.765
3010.6659	0.0068	3	F.KKGDVVIVLTGWRPGSGFTNTMRVVPVP.-	4.871
3025.4857	−0.01376	3	T.YVPM*TGGAPSM*VTVDGTDTETRLVKLTPG.V	3.051
3028.599	0.00818	3	E.LASQPDVDGFLVGGASLKPEFVDIINAKQ.-	3.798
3039.5608	0.04888	3	L.VNNAALVIMQPFLEVTKEAFDRSFSVN.L	3.295
3057.5276	−0.00466	3	M.VLSGEDKSNIKAAWGKIGGHGAEYGAEALE.R	4.811
3058.5983	−0.0454	3	T.WDEVVLIAEQLKDLEALDLSENKLQF.P	3.059
3105.6752	0.00321	3	-.MQIFVKTLTGKTITLEVEPSDTIENVKA.K	5.483
3117.4332	0.00443	3	K.LISWYDNEYGYSNRVVDLMAYMASKE.-	4.416
3185.5015	−0.0015	3	S.TATQRTAGEDCSSEDPPDGLGPSLAEQALRL.K	3.065
3202.6275	0.0023	3	E.SYSVYVYKVLKQVHPDTGISSKAMGIMNS.F	4.103
3222.4782	0.001	3	E.KDAVDEAKPKESARQDEGKEDPEADQEHA.-	3.295
3266.7123	0.0075	3	M.VVDGVKLLIEMEQRLEQGQAIDDLMPAQK.-	5.701
3344.6692	0.00175	3	M.VLSGEDKSNIKAAWGKIGGHGAEYGAEALER M.F	7.533
3350.941	0.00254	3	N.AKVAVLGASGGIGQPLSLLLKNSPLVSRLTLYD.I	7.081
3353.7899	−0.0093/	3	Y.TEHAKRKTVTAMDVVYALKRQGRTLYGFGG.-	4.808
3410.7714	0.02223	3	D.MLFYLVSVCLCVAVIGAFQLTAFTFRENLAA.T	3.462
3433.707	0.00852	3	A.KYNQLMRIEEELGDEARFAGHNFRNPSVL.-	3.691
3489.8588	0.00223	3	S.LLGNIRSDGKISEQSDAKLKEIVTNFLAGFEP.-	5.347
3491.7376	0.00351	3	M.VLSGEDKSNIKAAWGKIGGHGAEYGAEALERM F.A	6.075
3495.6373	0.07262	3	I.VISGHFDGVQDLMWDPEGEFIITTSTDQTTR.L	3.099
3526.6395	−0.00881	3	S.LLTTAEAVVTEIPKEEKDPGMGAMGGMGGGM GGGMF.-	5.682
3562.7747	−0.00064	3	M.VLSGEDKSNIKAAWGKIGGHGAEYGAEALERM FA.S	7.187
3566.8754	0.00985	3	T.LASHHPADFTPAVHASLDKFLASVSTVLTSKYR.-	6.332
3667.853	0.00881	3	M.VNPTVFFDITADDEPLGRVSFELFADKVPKTA E.N	3.114
3746.847	0.01543	3	E.GLPINDFSREKMDLTAKELTEEKETAFEFLSSA.-	3.277
3772.8177	0.01446	3	G.LLSDRLHISPDRVYINYYDMNAANVGWNGSTFA.-	5.112
3833.7407	−0.01323	3	Q.LLRDNLTLWTSDMQGDGEEQNKEALQDVEDENQ.-	8.051

taking advantage of the hydrophobicity of mesoporous inner wall of Fe_3O_4–graphene@mSiO$_2$ and the size-exclusion effect. It can be seen from the experimental results that the Fe_3O_4–graphene@mSiO$_2$ has great potential and prospect in the separation and enrichment of complex biological sample.

4.5.2. *Copper ion-immobilized magnetic mesoporous material*

4.5.2.1. *Basic principle*

IMAC has been newly developed and applied in protein purification.[37,38] The IMAC technique is based on the reversible binding-dissociation mechanism between immobilized metal ion and target protein/peptide. In particular, copper ion (Cu^{2+}) can be immobilized on the substrate relatively easily, and is bound to the amino acid chain of the protein/peptide by covalent bond, and the elution reagent can be used to release the protein by breaking the covalent bond. In view of the characteristics of magnetic material and ordered mesoporous material, and the advantages of IMAC technology, the design and synthesis of the nanomaterial that can be used in peptidomics research has become a hotspot in proteomics.

Copper ion-modified magnetic silica microsphere (Fe_3O_4@mSiO$_2$–Cu^{2+}) includes the above technology characteristics. The core of this microsphere is magnetic Fe_3O_4. The surface of Fe_3O_4 is coated with a layer of mesoporous silica to form a functional magnetic mesoporous material with the core–shell structure. The mesoporous silica layer with vertical pores on the outer surface is directly formed by the surfactant template method, and the inner wall of the mesoporous channel is modified with copper ions. The obtained Fe_3O_4@mSiO$_2$–Cu^{2+} possesses a high magnetic response strength (43.6 emu·g^{-1}), a uniform mesoporous pore size (3.3 nm), a large surface area (85 cm^3·g^{-1}), and a large amount of Cu^{2+} modified on the inner wall of the pore. Since this material has highly open mesopore with a large amount of copper ions, the Fe_3O_4@mSiO$_2$–Cu^{2+} is successfully used to selectively enrich the peptide in complex biological sample.

4.5.2.2. *Synthesis of Fe₃O₄@mSiO₂-Cu²⁺*

As shown in Figure 4.50, the $Fe_3O_4@mSiO_2$ is first synthesized according to Section 2.4 of Chapter 2. Then 30 mg $Fe_3O_4@mSiO_2$ is dispersed in 30 mL anhydrous chloroform and stirred for 30 min in an ice water bath. Afterwards, 0.2 mL adipyl chloride (AC) is added to react with the silanol for 4 h in an ice water bath. The obtained solid product is washed several times with anhydrous chloroform to remove excess adipyl chloride, followed by the addition of 30 mL anhydrous chloroform and 200 mg iminodiacetic acid (IDA). The mixture is ultrasonically dispersed and continuously stirred at room temperature overnight. The $Fe_3O_4@mSiO_2$–AC-IDA microsphere is first washed with acetone for 1–2 times, then washed with ethanol several times, and dried in vacuum at 50°C for later use. Finally, 10 mg $Fe_3O_4@mSiO_2$–AC-IDA is immersed in 1 mL $CuSO_4$ solution (the concentration of $CuSO_4$ solution in the experiment: 0.02 mol·L⁻¹, 0.1 mol·L⁻¹, and 0.2 mol·L⁻¹) and shaken overnight. The final product of $Fe_3O_4@mSiO_2$–Cu²⁺ is obtained by deionized water washing and vacuum drying for subsequent experiment.

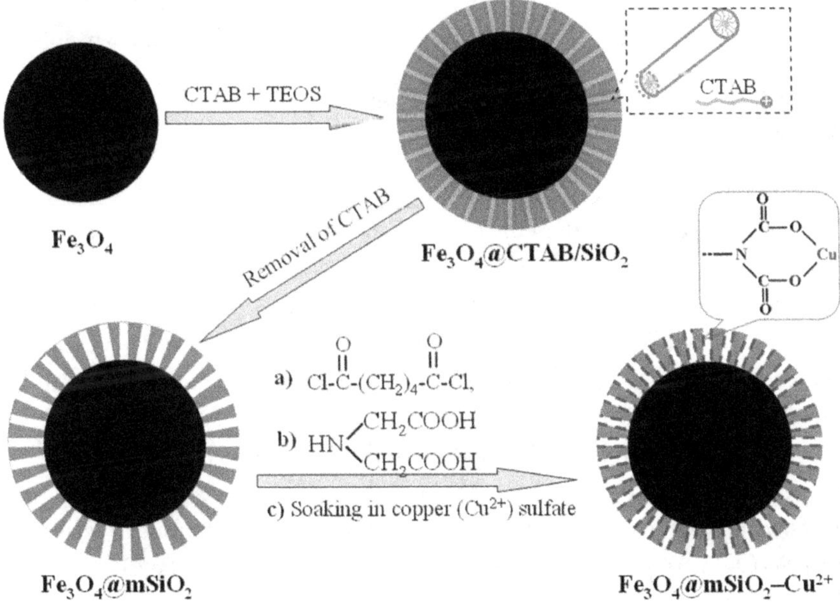

Figure 4.50: Schematic diagram of preparation of Fe₃O₄@mSiO₂–Cu²⁺.

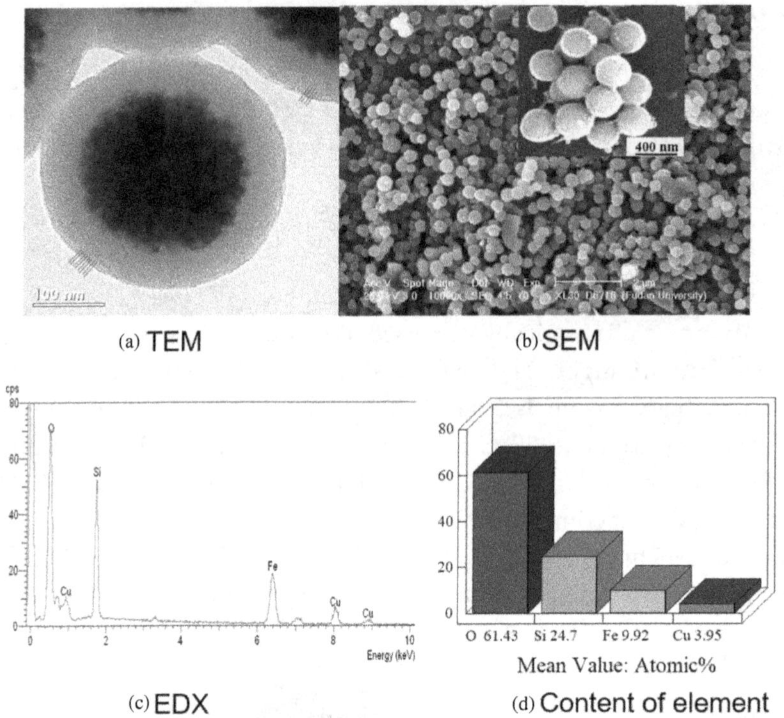

Figure 4.51: TEM image, SEM image, EDX spectrum and element content of $Fe_3O_4@mSiO_2$–Cu^{2+} using 0.2 mol·L^{-1} $CuSO_4$ solution as copper source.

In order to verify the effectiveness of the chemical modification, the chemical composition of $Fe_3O_4@mSiO_2$–Cu^{2+} is measured by SEM-EDX analysis (SEM-EDX) (as seen in Figures 4.51(c) and 4.51(d)). The amount of copper is 3.95% of the total amount of the main elements (Fe, O, Si, Cu, C), indicating that Cu^{2+} is effectively combined with the activating group by metal chelation.

Figure 4.52 shows the FTIR spectra of $Fe_3O_4@mSiO_2$ and $Fe_3O_4@mSiO_2$–Cu^{2+}. These two materials have obvious absorption peaks at 581 cm^{-1} and 1,085 cm^{-1}, which are corresponding to the stretching vibrations of Fe–O–Fe and Si–O–Si, respectively. Compared with $Fe_3O_4@mSiO_2$, $Fe_3O_4@mSiO_2$–Cu^{2+} shows strong absorption bands at 1,401 cm^{-1} and 1,629 cm^{-1}. These two absorption peaks belong to asymmetric stretching vibration and symmetric stretching vibration of C-O in the carboxylic acid. The comprehensive analysis

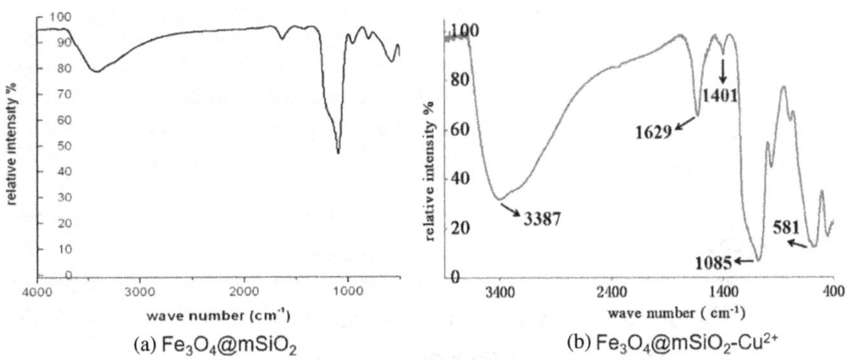

Figure 4.52: FTIR spectra of $Fe_3O_4@mSiO_2$ and $Fe_3O_4@mSiO_2-Cu^{2+}$.

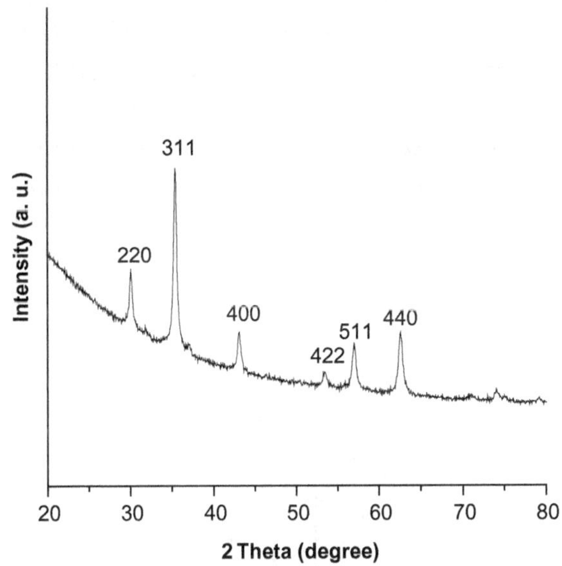

Figure 4.53: Wide-angle XRD pattern of $Fe_3O_4@mSiO_2-Cu^{2+}$.

of SEM-EDX and FTIR shows that Cu^{2+} is indeed immobilized on $Fe_3O_4@mSiO_2$ microsphere.

The wide-angle XRD pattern shows the apparent diffraction peaks of $Fe_3O_4@mSiO_2-Cu^{2+}$ (Figure 4.53). These diffraction peaks can accurately characterize the diffraction of magnetite, which indicates that the sol–gel reaction and surface chemical modification have no effect on the crystal property of magnetite Fe_3O_4. $Fe_3O_4@mSiO_2–Cu^{2+}$ still has the same crystal state as the Fe_3O_4 core. It is speculated that the $Fe_3O_4@mSiO_2–Cu^{2+}$ also has similar magnetic property to the Fe_3O_4 core, and the magnetic characterization of this material by SQUID at 25.8°C also shows that the $Fe_3O_4@mSiO_2–Cu^{2+}$ has very good superparamagnetism. The saturation magnetic value of $Fe_3O_4@mSiO_2–Cu^{2+}$ reaches 43.6 emu·g^{-1}, which facilitates the subsequent separation experiment.

The most important feature of $Fe_3O_4@mSiO_2–Cu^{2+}$ is its mesoporous structure. The small-angle XRD pattern (Figure 4.54(a)) shows a high-resolution diffraction peak at 2.38 corresponding to the 100 reflection of 2D regular hexagon of the mesoporous structure, and the two diffraction peaks at 4.58 may be attributed to the mesoporous positioning in a small range on the spherical surface. Next, the pore size distribution characteristics of $Fe_3O_4@mSiO_2–Cu^{2+}$ is studied by nitrogen adsorption. As shown in Figure 4.54(b), the nitrogen adsorption–desorption isotherms of $Fe_3O_4@mSiO_2–Cu^{2+}$ exhibit a type IV curve with a small hysteresis loop, indicating that some of the pore structure has been formed between the particles. According to Barrett–Joyner–Halenda (BJH) method, the data shows that $Fe_3O_4@mSiO_2–Cu^{2+}$ is porous, and its pore size distribution range is narrow, and its average pore

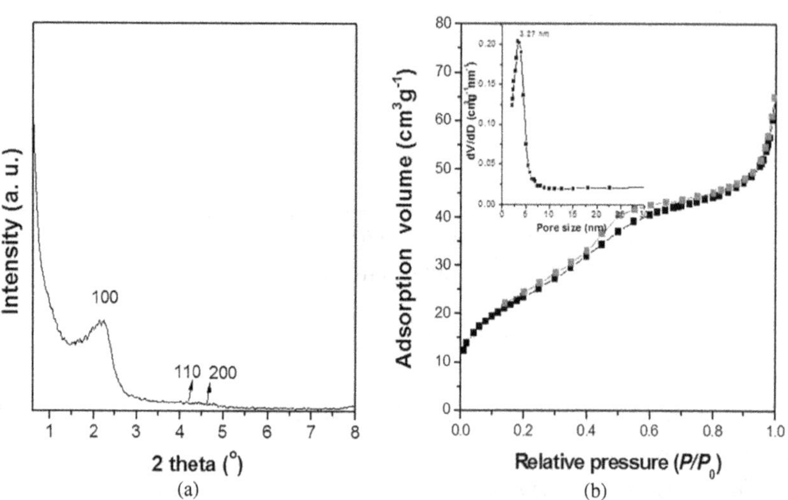

Figure 4.54: Small-angle XRD pattern, nitrogen adsorption–desorption isotherms and pore size distribution curve of $Fe_3O_4@mSiO_2–Cu^{2+}$.

diameter is about 3.3 nm. The calculated BET surface area and total pore volume are 85 $cm^3 \cdot g^{-1}$ and 0.1 $cm^3 \cdot g^{-1}$, respectively.

4.5.2.3. $Fe_3O_4@mSiO_2–Cu^{2+}$ for the enrichment of low-abundance peptide

Amino acid contains both hydrophilic group and hydrophobic group, and these amino acids are widely presented in peptide/protein. The large amount of metal ion Cu^{2+} modified on $Fe_3O_4@mSiO_2–Cu^{2+}$ can be used to effectively enrich hydrophobic/hydrophilic peptide by chelation and covalent bond with carboxyl group and amino group in the peptide. The $Fe_3O_4@mSiO_2–Cu^{2+}$ has good superparamagnetism and sensitive magnetic responsiveness to the external magnetic field, so that they can be separated from the dispersion quickly and conveniently, and avoid cumbersome centrifugation operation, which greatly shortens the experiment time.

(1) Enrichment of standard peptide

In the enrichment experiment of $Fe_3O_4@mSiO_2–Cu^{2+}$, two different hydrophobic peptides are used as target molecules to investigate the enrichment performance of Cu^{2+} for peptides. The enrichment process is shown in Figure 4.55. The two target peptides are hydrophobic Angiotensin II (linear peptide, amino acid sequence: DRVYIHPF) and hydrophilic microcystin (MC-LR, cyclic peptide, amino acid sequence see Figure 4.56).

5 $nmol \cdot L^{-1}$ Angiotensin II is difficult to be detected in MALDI-TOF-MS, the S/N ratio of which is only 25.18 as shown in Figure 4.57(a). After enrichment by $Fe_3O_4@mSiO_2–Cu^{2+}$, the S/N ratio is increased to 1017.33 (Figure 4.57(b)), and the enrichment factor (EF) is about 40. At the same time, 5 $nmol \cdot L^{-1}$ MC-LR solution is enriched by $Fe_3O_4@mSiO_2–Cu^{2+}$ for 2 min, the S/N ratio increases from 27.05 to 951.62, and the enrichment factor (EF) is about 35 (Figure 4.58).

The above experimental data demonstrates the excellent performance of $Fe_3O_4@mSiO_2–Cu^{2+}$ for the enrichment of hydrophobic and hydrophilic peptides. In order to study the effect of Cu^{2+} on the enrichment efficiency, $Fe_3O_4@mSiO_2$ and $Fe_3O_4@mSiO_2–AC\text{-}IDA$ are also used to enrich Angiotensin II (Figure 4.59). As shown in Figure 4.59(f), the enrichment

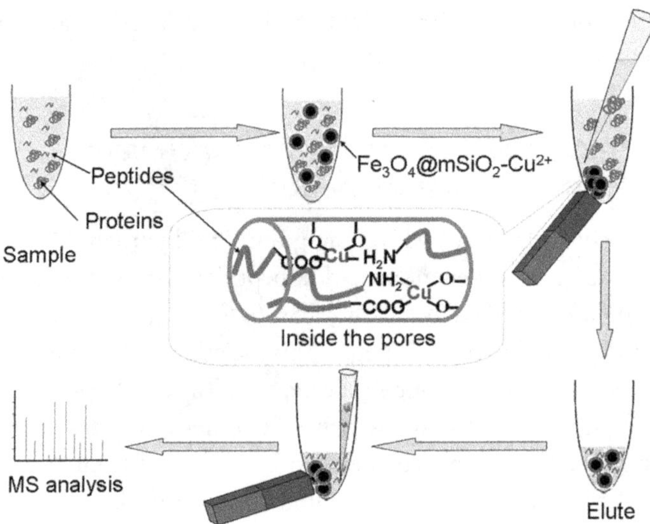

Figure 4.55: Schematic diagram of selective enrichment of peptide by Fe$_3$O$_4$@mSiO$_2$–Cu^{2+} and MALDI-TOF-MS analysis.

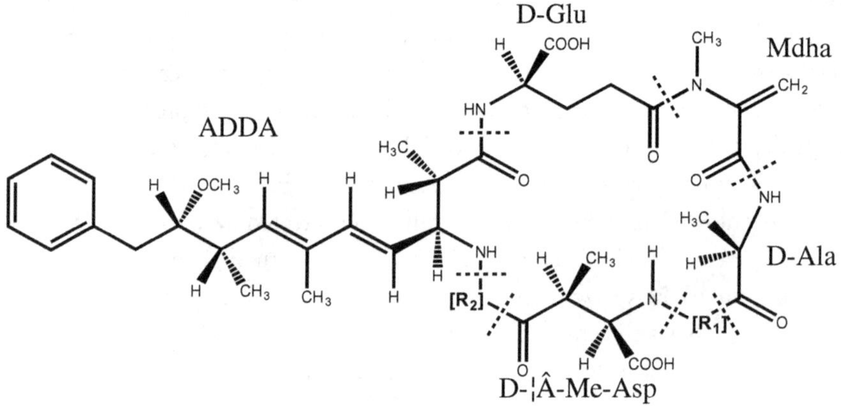

Figure 4.56: Schematic diagram of the circular peptide MC-LR (R$_1$ represents leucine and R$_2$ represents arginine).

factor of the unactivated Fe$_3$O$_4$@mSiO$_2$ as an adsorbent is 4.1, the activated Fe$_3$O$_4$@mSiO$_2$–AC-IDA shows a slightly higher enrichment factor of 6.9. The comparison of enrichment efficiency between these two materials indicates that the activated organic group HC-IA limits the enrichment of

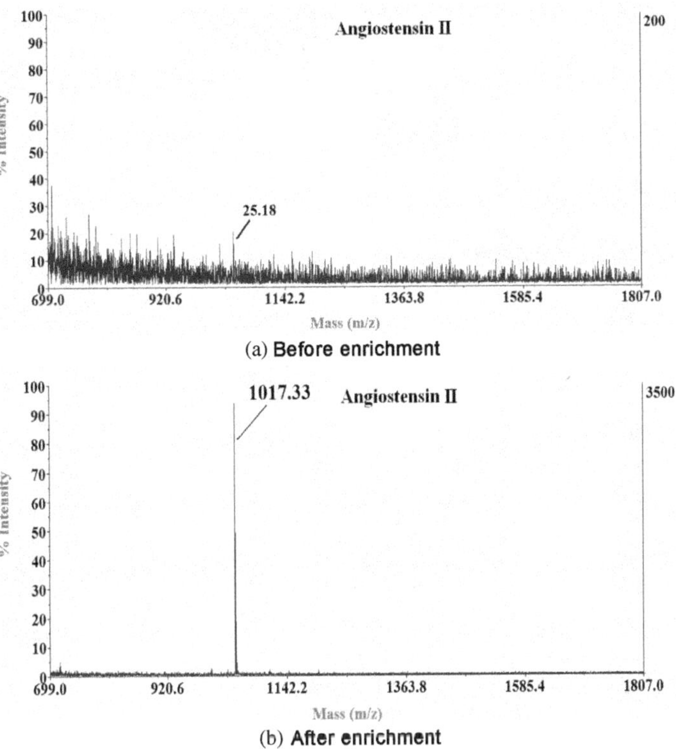

Figure 4.57: MALDI-TOF-MS spectra of Angiotensin II before and after enrichment by Fe$_3$O$_4$@mSiO$_2$–Cu^{2+} (25.18 and 1017.33 represents S/N ratio).

peptide. The enrichment factor is mainly derived from the mesoporous part. A small amount of peptides enter into the mesopore to combine with the surface –OH.

In addition, the effects of different concentrations of Cu^{2+} on the enrichment efficiency are compared. When applying Fe$_3$O$_4$@mSiO$_2$–Cu^{2+} (0.02 mol·L^{-1} CuSO$_4$ as copper source) to enrich Angiotensin II, the obtained enrichment factor is 16, which is about 4 times of Fe$_3$O$_4$@mSiO$_2$. This also indicates that Cu^{2+} immobilized on the inner wall of the mesopore can effectively enrich peptide. Next, 0.1 mol·L^{-1} and 0.2 mol·L^{-1} CuSO$_4$ are used to soak Fe$_3$O$_4$@mSiO$_2$, the Cu content of which is measured by SEM-EDX as shown in Figures 4.60 and 4.51. The enrichment factors also increase to 31.3 and 40.4 (Figure 4.59(f)).

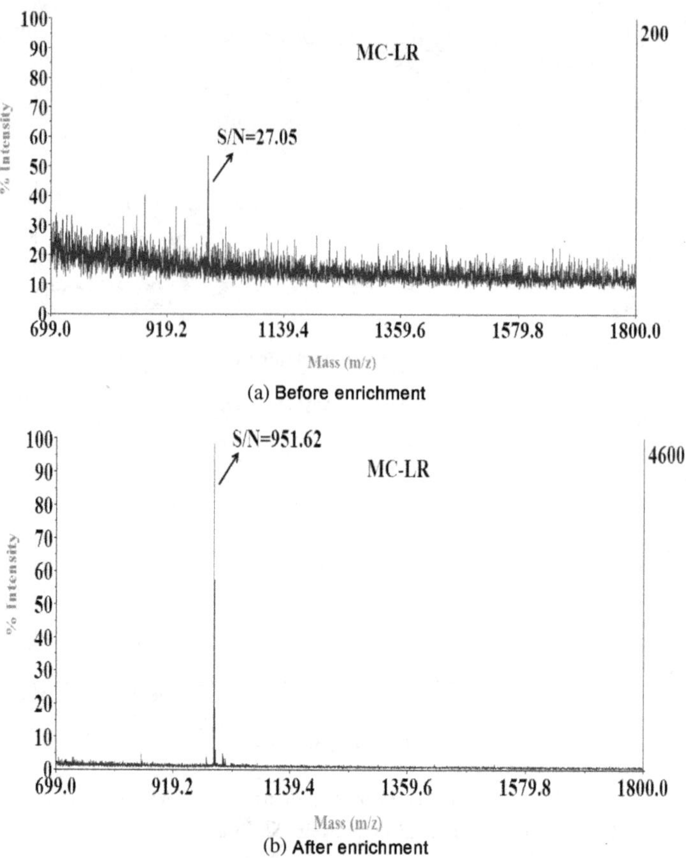

Figure 4.58: MALDI-TOF-MS spectra of MC-LR (MW = 995.5) before and after enrichment by $Fe_3O_4@mSiO_2–Cu^{2+}$.

From the above control experiments, it is concluded that the enrichment principle of $Fe_3O_4@mSiO_2\text{-}Cu^{2+}$ for peptides is mainly based on the covalent bonding of the Cu^{2+} on the inner wall of the mesopore with the carboxyl and amino groups of the peptide. It is also found that there is a necessary relationship between enrichment efficiency and Cu^{2+} content, the enrichment factor will increase significantly with the increase of Cu^{2+} content. And when the Cu^{2+} concentration is sufficiently high, a relatively high enrichment efficiency can be obtained, which lays a good foundation for future enrichment research

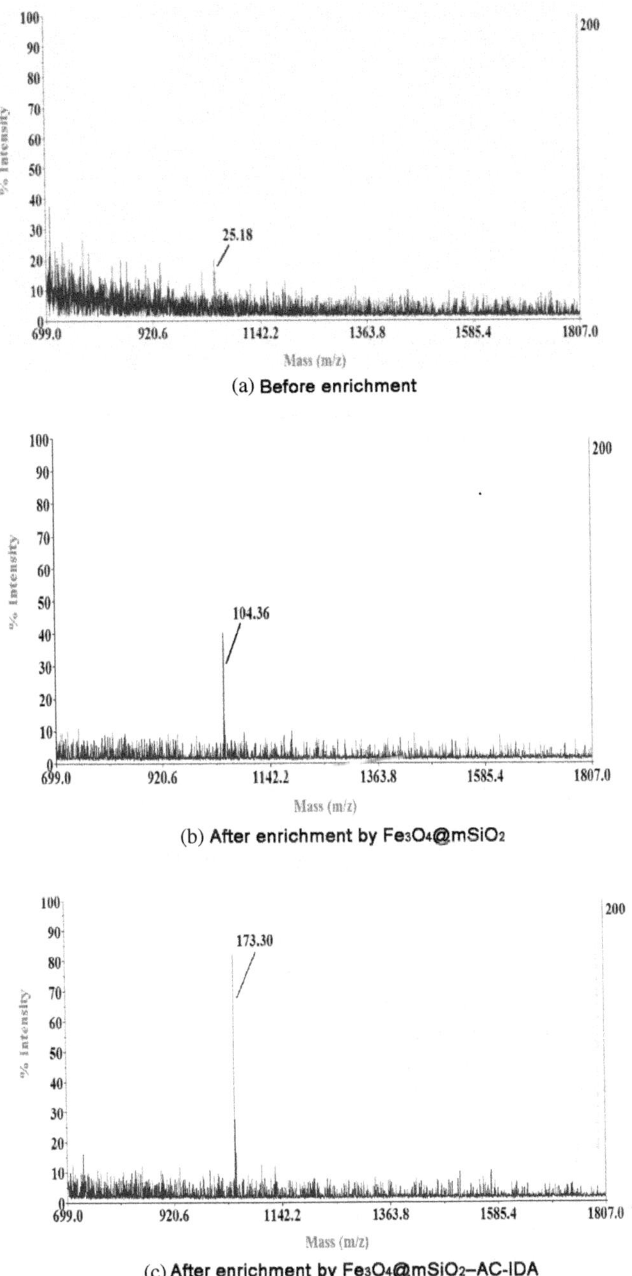

Figure 4.59: Analytical diagram of 5 nmol·L^{-1} Angiotensin II treated by different materials.

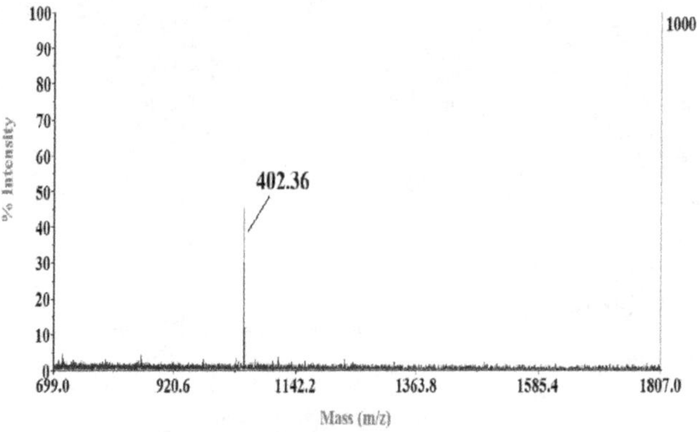

(d) After enrichment by Fe₃O₄@mSiO₂–Cu²⁺
(0.02 mol·L⁻¹ CuSO₄ as coper source)

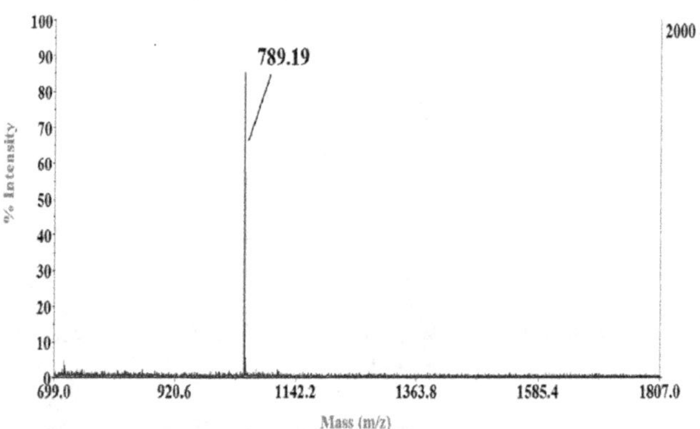

(e) After enrichment by Fe₃O₄@mSiO₂–Cu²⁺
(0.1 mol·L⁻¹ CuSO₄ as coper source)

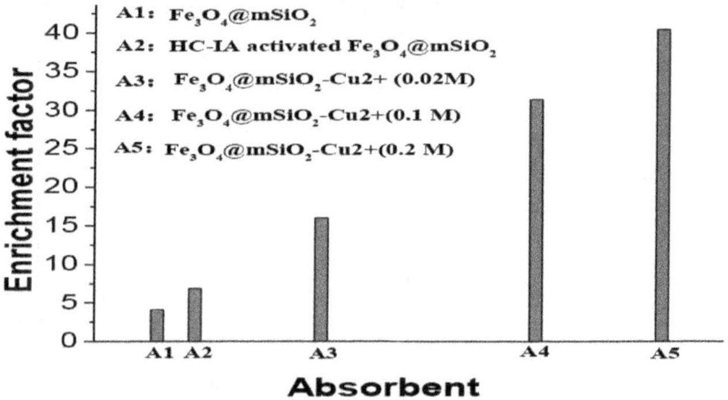

(f) Enrichment factor of different absorbents

Figure 4.59: *(Continued)*

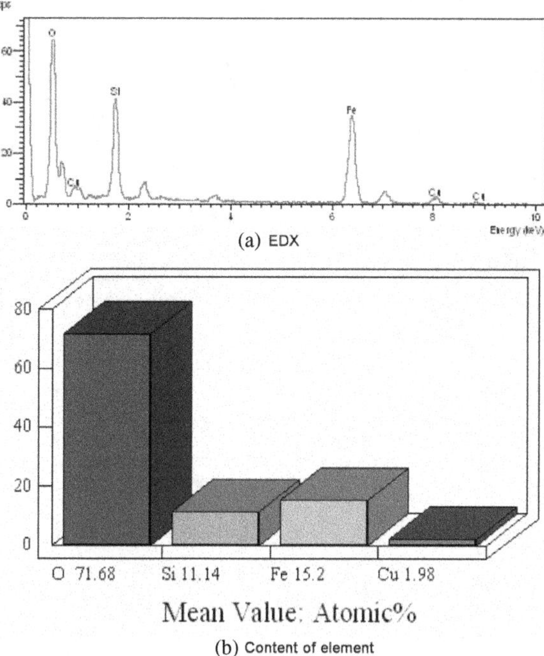

Figure 4.60: EDX spectrum and element content of Fe_3O_4@mSiO$_2$–Cu^{2+} using 0.1 mol·L^{-1} CuSO$_4$ solution as copper source.

(2) Enrichment of standard protein digest

BSA enzymatic hydrolysate is also used to investigate the enrichment universality of Fe_3O_4@mSiO$_2$–Cu^{2+} for peptide. From Figure 4.61, it can be seen that seven peptides with weak signal are detected from 5 nmol·L^{-1} BSA digest by direct analysis of MALDI-TOF-MS, and 18 peptides with obviously increased signal are identified after enrichment by Fe_3O_4@mSiO$_2$–Cu^{2+}, and the enriched peptides have an average hydrophobic value between −1.723 and 0.292. These data further indicate the enrichment universality of Fe_3O_4@ mSiO$_2$–Cu^{2+}, that is, it can be used as a general adsorbent for hydrophobic/hydrophilic peptide.

4.5.2.4. Fe_3O_4@mSiO$_2$-Cu^{2+} for the enrichment of peptide in human serum and urine

Biological sample such as serum, urine, tissue extract, etc., usually contain a large number of protein molecules, which interfere with the signal of the

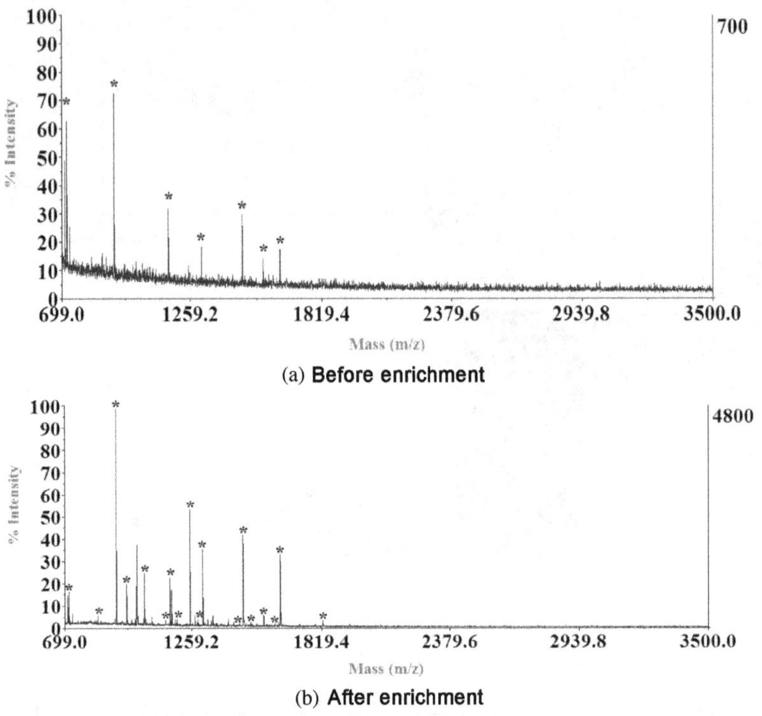

Figure 4.61: MALDI-TOF-MS spectra of BSA digest (0.4 mL, 5 nmol·L^{-1}) before and after enrichment by $Fe_3O_4@mSiO_2$–Cu^{2+}. "*" represents target peptide.

peptides in the original biological sample during MS identification, so that the peptide in the biological sample is difficult to be identified directly by MALDI-TOF-MS. In addition, the high-abundance proteins in biological sample have similar property to peptide and can be adsorbed on the surface of the adsorbent, therefore, the complexity of the biological sample also affects the current peptidomics analysis. However, $Fe_3O_4@mSiO_2$–Cu^{2+} can solve this problem well. $Fe_3O_4@mSiO_2$–Cu^{2+} has the unique advantages of size exclusion, magnetic separation and affinity of metal Cu^{2+} ion, which can be utilized in complex biological system.

Human serum contains a large number of proteolysis-derived peptides, which may also have biomarkers for clinical screening and disease diagnosis. Here, $Fe_3O_4@mSiO_2$–Cu^{2+} is used to enrich the peptide in the serum. First, 1 mg $Fe_3O_4@mSiO_2$–Cu^{2+} is dispersed in 1 mL deionized water. Then

100 μL fresh serum is diluted to 500 μL with water, and 20 μL 5 mg·mL^{-1} material dispersion is added into the diluted fresh serum solution for oscillation for 10 min. The finally obtained eluent is analyzed by MALDI-TOF-MS.

Figure 4.62 shows the MS spectra of serum before and after enrichment by $Fe_3O_4@mSiO_2$–Cu^{2+}. There is no signal detected in diluted serum sample before enrichment. However, after enrichment with $Fe_3O_4@mSiO_2$–Cu^{2+}, multiple peptides with molecular weight in the range of 800–3,500 Da are identified. Human urine from liver cancer patient is also used as a sample to examine enrichment performance of this material. As seen in Figure 4.63, multiple peptides with molecular weight in the range of 800–3,500 Da are also identified. These results demonstrate the enrichment potential of this material in complex biological samples.

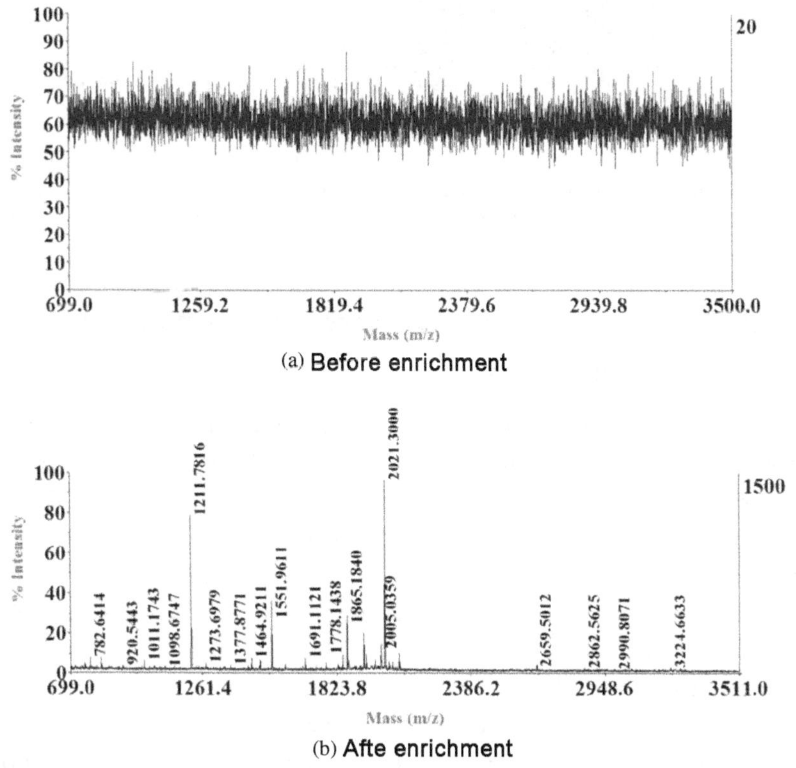

(a) **Before enrichment**

(b) **Afte enrichment**

Figure 4.62: MALDI-TOF-MS spectra of diluted human serum before and after enrichment by $Fe_3O_4@mSiO_2$-Cu^{2+}.

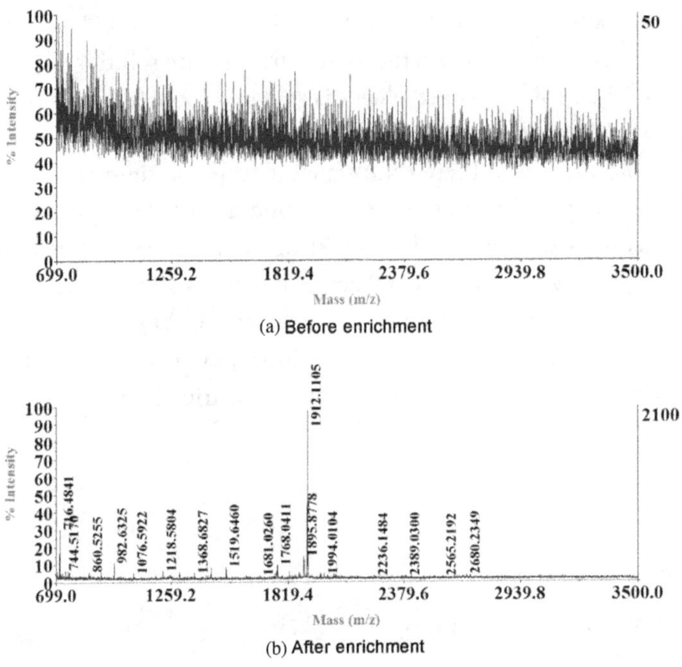

Figure 4.63: MALDI-TOF-MS spectra of diluted urine of liver cancer patient before and after enrichment by $Fe_3O_4@mSiO_2\text{-}Cu^{2+}$.

4.5.3. *C8-modified magnetic mesoporous material*

4.5.3.1. *Basic principle*

As well known, the functional group of reversed-phase filler in liquid chromatography, such as the long alkyl chains of C4, C8, C18, etc., can be modified on the magnetic bead to serve as solid phase extractant.[39] This solid-phase extractant can achieve high-throughput enrichment in peptidomics by hydrophobic–hydrophobic interaction, and has been widely applied. It has been found that functionalized octyl (abbreviated as C8) is superior to other alkyl groups such as C1, C2, C3, and C18 in peptide enrichment. C8-midified magnetic particle is ideal for enriching endogenous peptide in biological sample. Biological sample contains not only low concentration of endogenous peptides, but also high concentration of large proteins, therefore, the novel C8-modified magnetic material capable of selectively enriching low-abundance endogenous peptide in complex biological sample has practical implications for peptidomics research.

The authors of this book prepare C8-functionalized magnetic mesoporous silica (abbreviated as C8-Fe$_3$O$_4$@mSiO$_2$) and C8-modified mesoporous graphene composite (abbreviated as C8-graphene@mSiO$_2$). What is unique about them is that the C8 group is modified on the inner wall of the mesopores and can be applied to the selective enrichment of endogenous peptides in biological samples and exclude large proteins. One-pot method and sol–gel method are used in the synthetic process, CTAB is used as the surfactant template, TEOS and C8 silylating reagent (n-octyltriethoxysilane, C8TEOS) are used as the silica source and functional molecule. First, a layer of CTAB/C8 composite with mesoporous structure is uniformly coated on the surface of Fe$_3$O$_4$ or graphene, then the CTAB is removed by reflux to obtain C8-Fe$_3$O$_4$@mSiO$_2$ or C8-graphene@mSiO$_2$.

By this simple method, the C8 functional group and mesoporous silica can be directly introduced onto the magnetic material at the same time without undergoing surface chemical modification. C8 is modified on the inner wall of the mesopore and will not affect the silanol group of mesopore, so that the hydrophobic C8 and hydrophilic silanol group can coexist. This unique structure makes this C8-modified mesoporous material an amphoteric material, which has good aqueous dispersion and ability to enrich hydrophobic peptide/protein.

4.5.3.2. *Synthesis of C8-Fe$_3$O$_4$@mSiO$_2$*

The magnetic Fe$_3$O$_4$ with a diameter of about 250 nm is synthesized according to the high temperature hydrothermal reaction in Section 2.2 of Chapter 2.

The synthesis of C8-Fe$_3$O$_4$@mSiO$_2$ (Figure 4.64) employs a method of sol–gel reaction containing a surfactant. The Fe$_3$O$_4$ and the surfactant CTAB are mixed and dispersed in 50 mL water at a ratio of 50 mg/500 mg for ultrasonication for 30 min. Then 450 mL 1.1 nmol·L^{-1} NaOH solution is added for 5 min of ultrasonication to form a uniform and stable dispersion, which is then placed in a 60°C water bath for constant temperature reaction. After reacting for 30 min, a mixture of 2.55 mL TEOS/C8TEOS/ethanol (10/1/4, v/v/v) is slowly added under stirring, and the reaction is continued for 12 h in a 60°C water bath. The obtained solid product is separated by magnet, washed by ethanol and finally dispersed in 50 mL acetone for refluxing in a 60°C water bath to remove CTAB. In order to completely remove the surfactant CTAB, the refluxing process is repeated 5 times, and the finally

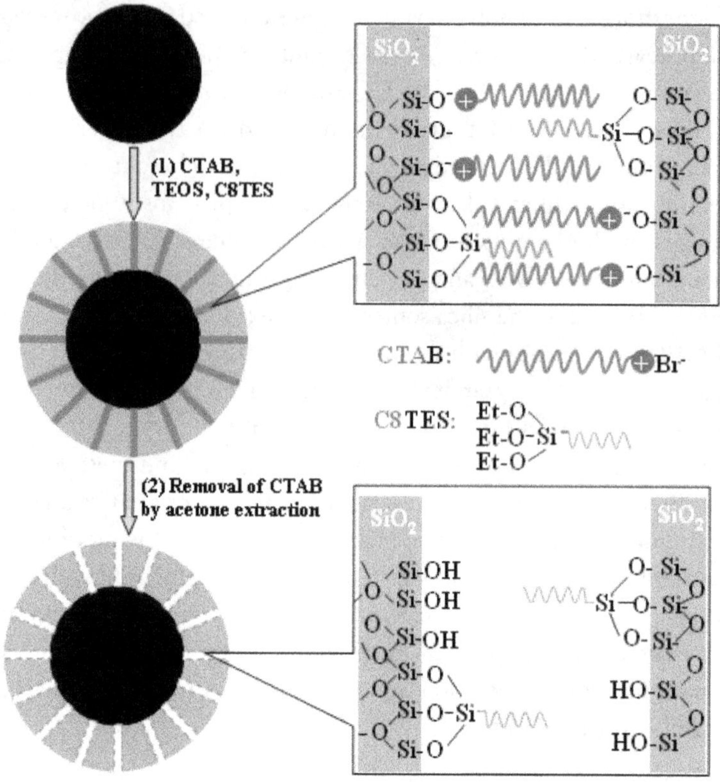

Figure 4.64: Schematic diagram of synthesis of C8-Fe$_3$O$_4$@mSiO$_2$.

obtained material is washed with ethanol and dried in vacuum at 50°C for later use.

Both the silica source of TEOS and n-octyltriethoxysilane can be hydrolyzed to react with the structure-directing reagent of CTAB to form the complex which is then coated outside the magnetic Fe$_3$O$_4$. Silicon hydroxyl and C8 group is randomly modified on the inner wall of the mesopore after the CTAB template is removed by solvent extraction, providing a bindable surface for the target molecule suitable for mesoporous size. The presence of porosity not only increases the surface area but also makes the mesoporous material an ultramicro-container with sufficient sites and spaces for adsorption of the target molecule. C8TEOS and TEOS are coated on the framework of mesoporous silica with the aid of structure-directing agent CTAB, so that the mesopore modified with C8 group can act as microreactor for capturing the target molecules by hydrophobic–hydrophobic interaction.

4.5.3.3. *Characterization of C8-Fe$_3$O$_4$@mSiO$_2$*

SEM is used to observe the morphology of C8-Fe$_3$O$_4$@mSiO$_2$. It can be seen that the C8-Fe$_3$O$_4$@mSiO$_2$ is a sphere with a diameter of about 300 nm (Figure 4.65(a)). TEM is used to observe the structural characteristic of C8-Fe$_3$O$_4$@mSiO$_2$. It can be clearly seen that the C8-Fe$_3$O$_4$@mSiO$_2$ has a core–shell structure, and the black core portion is Fe$_3$O$_4$ particle with a diameter of about 250 nm. The gray portion of the outer surface of the Fe$_3$O$_4$ particle is a porous silica layer of having a thickness of about 50 nm (Figure 4.65(b)). It can be more clearly seen from the high-power illustration in the upper right corner of Figure 4.65(b) that the mesopores are vertically aligned on the surface of the microsphere. This highly ordered open-type porous structure allows the target molecule to more easily enter the pore and simultaneously increase the surface area for efficient adsorption of guest molecule.

To verify the successful modification of C8 in the mesopore, structural analysis is performed using FTIR. As shown in Figure 4.66, the infrared absorption peaks at 582 cm^{-1} and 1,080 cm^{-1} are attributed to stretching vibrations of Fe–O–Fe and Si–O–Si, respectively. A broad absorption peak of O-H stretching vibration appears at 3,420 cm^{-1}, and a broad absorption peak at 962 cm^{-1} is attributed to stretching vibration of the –OH in the silica hydroxyl group. Compared with Fe$_3$O$_4$@mSiO$_2$, the most obvious difference

(a) **SEM** (b) **TEM**

Figure 4.65: SEM and TEM images of C8-Fe$_3$O$_4$@mSiO$_2$.

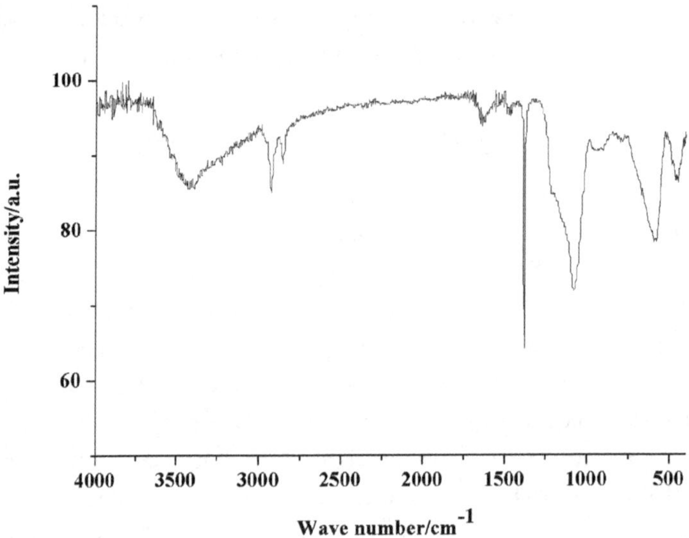

Figure 4.66: FTIR spectrum of C8-Fe$_3$O$_4$@mSiO$_2$.

is the stretching vibration of the alkyl CH$_2$ at 2,930 cm^{-1}, which can be considered as the vibration absorption peak of the alkyl chain in the C8TEOS, proving the existence of C8 group.

Next, the comparative analysis of C8-Fe$_3$O$_4$@mSiO$_2$ and Fe$_3$O$_4$ is carried out by XRD. The comparison of the characteristic diffraction peaks in Figure 4.67 shows that the peak positions are consistent, indicating that both are structures with inverse spinel magnetite. This result also indicates that the sol–gel process of coating the C8-functionalized mesoporous silica does not affect the property of the crystal and the phase state of the magnetite.

The magnetic property of C8-Fe$_3$O$_4$@mSiO$_2$ is detected by SQUID. At room temperature, C8-Fe$_3$O$_4$@mSiO$_2$ exhibits a high saturation magnetic value of 56.3 emu·g^{-1}, which is reflected in its fast response to the external magnetic field.

The porous nature of the C8-Fe$_3$O$_4$@mSiO$_2$ is determined by the nitrogen adsorption–desorption method at 77K. The type IV adsorption–desorption characteristic curve representing the cylindrical mesoporous pore structure is shown in Figure 6.48. The inset in Figure 4.68 shows the pore size distribution curve of the C8-Fe$_3$O$_4$@mSiO$_2$ according to the BJH method, from which it is seen that the average pore size of C8-Fe$_3$O$_4$@mSiO$_2$ is 3.4

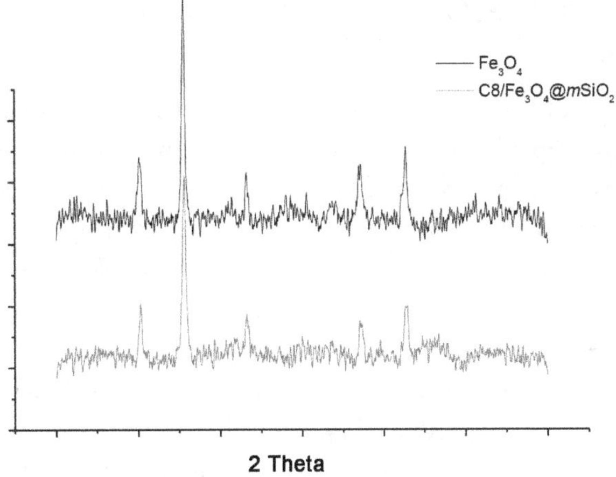

Figure 4.67: Wide-angle XRD pattern of C8-Fe$_3$O$_4$@mSiO$_2$.

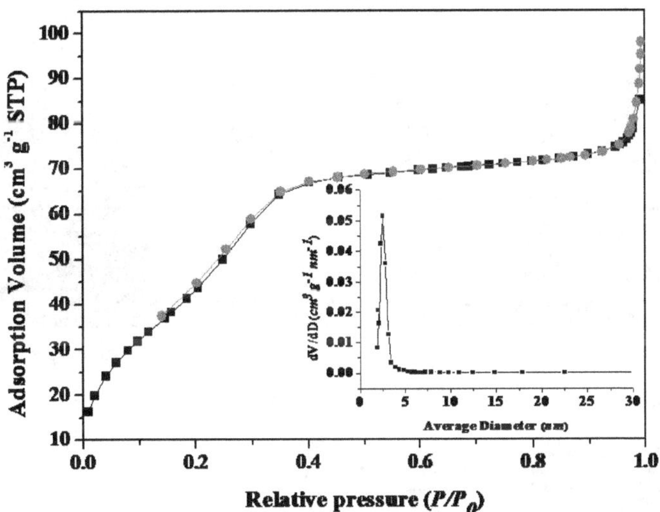

Figure 4.68: Nitrogen adsorption–desorption isotherms and pore size distribution curve of C8-Fe$_3$O$_4$@mSiO$_2$ at 77 K.

nm, and the distribution range is narrow. In addition, the BET surface area and total pore volume are calculated based on nitrogen adsorption data to be 162.5 m$^2 \cdot$g^{-1} and 0.17 cm$^3 \cdot$g^{-1}, respectively.

4.5.3.4. *Investigation on dispersion property of C8-Fe$_3$O$_4$@mSiO$_2$*

Studies have shown that some reversed-phase affinity materials can successfully perform peptide enrichment using the interaction of hydrophobic group, however, the long chain of the alkyl group has a strong hydrophobicity, resulting in that these materials exhibit a very weak hydrophilicity in aqueous solution, and a strong hydrophobicity on the contrary. While as we know that the general biological system is aqueous solution. Therefore, this is a difficult problem to be solved for peptide enrichment research.

For the control experiment about the dispersity of C8-Fe$_3$O$_4$@mSiO$_2$ synthesized by one-pot method, Fe$_3$O$_4$@mSiO$_2$ and C8-f-Fe$_3$O$_4$@mSiO$_2$ are synthesized by surface post-modification method. It is found that the three magnetic mesoporous materials show good dispersibility in ethanol, but there are significant differences in aqueous solution (Figure 4.69). The Fe$_3$O$_4$@mSiO$_2$ has good dispersibility in ethanol and water due to the large amount

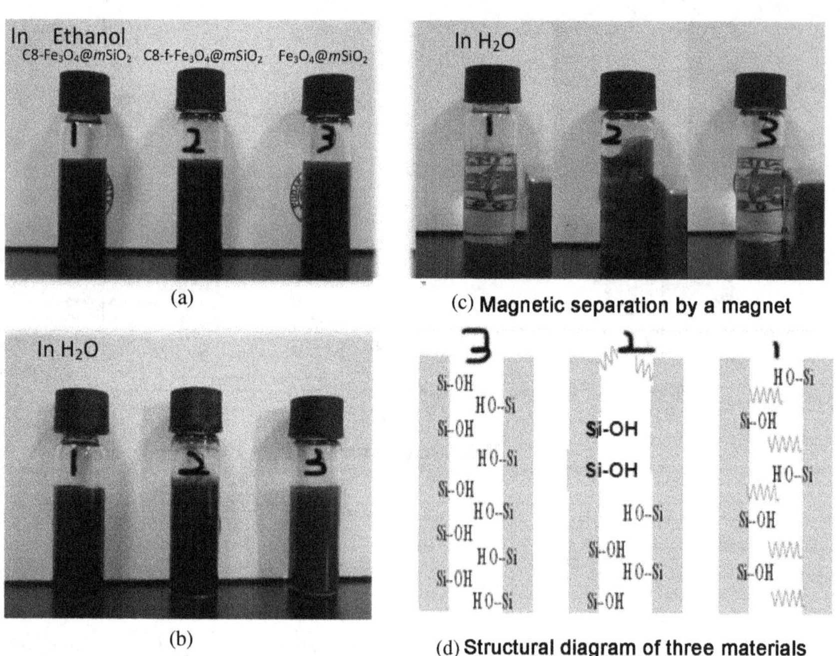

Figure 4.69: Dispersity comparison of C8-Fe$_3$O$_4$@mSiO$_2$ (1$^#$), C8-f-Fe$_3$O$_4$@mSiO$_2$ (2$^#$) and Fe$_3$O$_4$@mSiO$_2$ (3$^#$).

of silanol hydrophilic groups on its surface (Figure 4.69, 3$^\#$). C8-f-Fe$_3$O$_4$@mSiO$_2$ is well dispersed in ethanol, but when the solvent is changed to water, the dispersibility is extremely poor and the C8-f-Fe$_3$O$_4$@mSiO$_2$ floats on the water surface due to the presence of a large number of C8 chain hydrophobic groups on its surface. This results in that it is difficult to perform magnetic separation operation (Figure 4.69, 2$^\#$). And this phenomenon may be due to the fact that when C8 is modified by surface post-modification, the hydrophobic C8 group preferentially occupies the site of mesoporous entrance, and the outer surface of the material is modified with a large number of hydrophobic C8 groups, which hinders the solvation of silanol and reduces the dispersibility of the material. C8-Fe$_3$O$_4$@mSiO$_2$ shows good dispersibility in water compared to the other two materials because the silanol and C8 groups are randomly distributed on the surface of the mesopore during the synthetic process of mesopore (Figure 4.69, 1$^\#$). In addition, C8-Fe$_3$O$_4$@mSiO$_2$ and Fe$_3$O$_4$@mSiO$_2$ can be separated in the buffer by only 30 s under the aid of a magnet, but the C8-f-Fe$_3$O$_4$@mSiO$_2$ floats on the water surface due to the poor dispersibility and it is difficult to generate a magnetic response to the magnet, resulting in difficulty in separating from the solution.

The above study on the dispersibility of material shows that the C8-Fe$_3$O$_4$@mSiO$_2$ synthesized by one-pot method not only exhibits good dispersibility due to the presence of hydrophilic silicon hydroxyl groups on the surface, but also has fast magnetic response sensitivity to the external magnetic field. Moreover, the presence of the C8 group in its mesopore leads to its hydrophobic affinity towards guest molecules. The above characteristics of C8-Fe$_3$O$_4$@mSiO$_2$ make it an ideal adsorbent for peptide enrichment in biological samples.

4.5.3.5. *C8-Fe$_3$O$_4$@mSiO$_2$ for the enrichment of standard peptide*

The standard peptide of Angiotensin II (MW = 1046.2, sequence: DRVYIHPF, grand average of hydropathicity, GRAVY = −0.325) is used as hydrophobic guest molecule to investigate the hydrophobic affinity of C8 group of C8-Fe$_3$O$_4$@mSiO$_2$ towards target peptide. Through the optimization of experimental conditions, the best enrichment efficiency can be obtained when the enrichment time is 10 min, the elution is 30 min, the volume ratio of the eluent and the material dispersion ($10\ \mathrm{mg\cdot mL^{-1}}$) is 10:3, and the pH of sample aqueous solution is adjusted to be 7.

For the enrichment research, C8-f-Fe_3O_4@mSiO$_2$ and Fe_3O_4@mSiO$_2$ are used for comparison (Figure 4.70). The enrichment step is the same as follows: 3 μL the material dispersion is added into 400 μL 5 $\mathrm{nmol\cdot L^{-1}}$ Angiotensin II solution. The S/N ratio is 154.37 when 5 $\mathrm{nmol\cdot L^{-1}}$ Angiotensin II is directly analyzed by MALDI-TOF-MS. All the MS signals are enhanced after enrichment by the three materials. And the S/N ratio is 257.74, 706.29, and 7690.92 after enrichment by Fe_3O_4@mSiO$_2$, C8-f-Fe_3O_4@mSiO$_2$, and C8-Fe_3O_4@mSiO$_2$, respectively.

Compared with Fe_3O_4@mSiO$_2$ and C8-f-Fe_3O_4@mSiO$_2$, C8-Fe_3O_4@mSiO$_2$ is obviously superior in enrichment efficiency. The peptide signal after enrichment of C8-Fe_3O_4@mSiO$_2$ is about 30 times of Fe_3O_4@mSiO$_2$, which also indicates that the C8 group is modified on the inner wall of mesopore and the peptide capture mechanism is based on hydrophobic–hydrophobic interaction. Therefore, it can be considered that the enrichment efficiency of

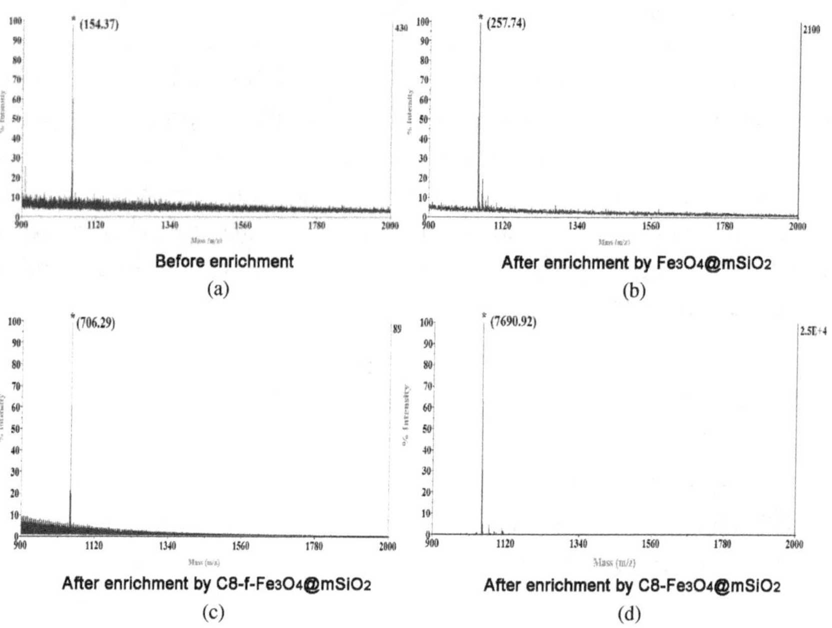

Figure 4.70: MALDI-TOF-MS spectra of Angiotensin II (0.4 mL 5 $\mathrm{nmol\cdot L^{-1}}$) before and after enrichment by different materials. The number in parenthesis represents S/N ratio.

the C8-Fe_3O_4@mSiO$_2$ on the peptide is mainly derived from the C8 group on the inner wall of the mesopore.

BSA and MYO hydrolysates are further used to study the enrichment performance of C8-Fe_3O_4@mSiO$_2$ for peptide. And C8-f-Fe_3O_4@mSiO$_2$ and Fe_3O_4@mSiO$_2$ are also used for comparison. The MS signal of a few detected peptides is too weak to search the library when 5 nmol·L^{-1} standard protein digest is directly spotted on targeted plate for MS analysis. Figure 4.71 and Table 4.7 show the MS spectra and peptide information of BSA hydrolysate after enriched by Fe_3O_4@mSiO$_2$, C8-f-Fe_3O_4@mSiO$_2$, and C8-Fe_3O_4@mSiO$_2$. As shown in Figure 4.71(a), only nine peptides with weak signal are detected when 5 nmol·L^{-1} BSA hydrolysate is analyzed directly by MALDI-TOF-MS. And 21 peptides with significantly enhanced S/N ratio are detected after enrichment by C8-Fe_3O_4@mSiO$_2$ (Figure 4.71(d)). It also can be obviously observed from the MS spectra that the enrichment performance of C8-Fe_3O_4@mSiO$_2$ is superior to the other two. Fe_3O_4@mSiO$_2$ can hardly enrich peptide from BSA hydrolysate (Figure 4.71(b)). The presence of C8 group results in a certain enrichment ability of C8-f-Fe_3O_4@mSiO$_2$ to peptide, however, the number of enriched peptides is less than C8-Fe_3O_4@mSiO$_2$ (Figure 4.71(c)). According to the peptide information in Table 4.7, the signal intensity of the same peptide enriched by C8-Fe_3O_4@mSiO$_2$ is much higher than the other two materials.

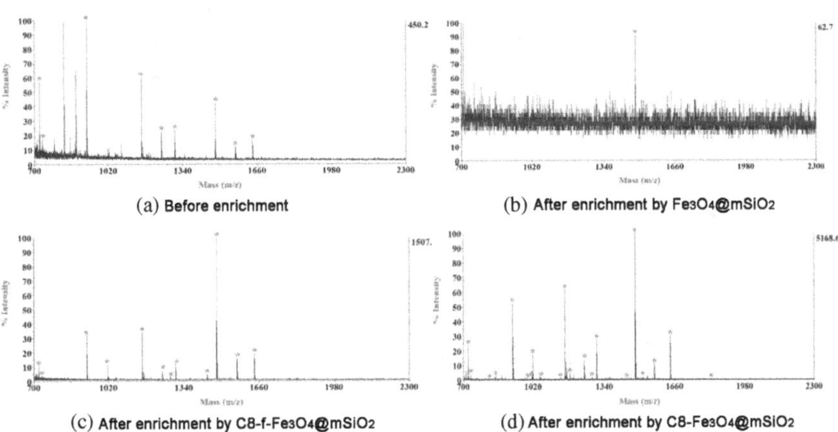

(a) Before enrichment

(b) After enrichment by Fe3O4@mSiO2

(c) After enrichment by C8-f-Fe3O4@mSiO2

(d) After enrichment by C8-Fe3O4@mSiO2

Figure 4.71: MALDI-TOF-MS spectra of BSA hydrolysate (0.4 mL 5 nmol·L^{-1}) before and after enrichment by different materials. "*" represents target peptide.

Table 4.7: The peptide information of BSA hydrolysate before (O) and after enrichment by $Fe_3O_4@mSiO_2$ (A), C8-f-$Fe_3O_4@mSiO_2$ (B) and C8-$Fe_3O_4@mSiO_2$ (C).

Calculated m/z	Peptide Sequence	Length	O	A	B	C
712.36	K.SEIAHR.F	6	20.41	—	28.59	123.02
733.39	K.VLTSSAR.Q	7	23.90	—	15.92	86.17
818.40	K.ATEEQLK.T	7	—	—	—	17.02
847.43	R.LSQKFPK.A	7	—	—	—	59.11
927.46	K.YLYEIAR.R	7	199.60	—	241.33	1204.62
990.48	R.EKVLTSSAR.Q	9	—	—	—	22.48
1014.53	K.QTALVELLK.H	9	—	—	—	43.54
1017.52	K.VLTSSARQR.L	9	—	—	82.52	418.61
1052.38	R.CCTKPESER.M	9	—	—	—	16.15
1142.63	K.KQTALVELLK.H	10	—	—	—	41.57
1163.54	K.LVNELTEFAK.T	10	149.12	—	302.82	1622.27
1193.51	R.DTHKSEIAHR.F	10	—	—	—	17.19
1249.57	R.FKDLGEEHFK.G	10	56.67	—	51.25	368.93
1283.63	R.HPEYAVSVLLR.L	11	—	—	19.34	51.92
1305.66	K.HLVDEPQNLIK.Q	11	53.19	—	103.61	785.04
1439.70	R.RHPEYAVSVLLR.L	12	—	—	39.95	22.59
1479.75	K.LGEYGFQNALIVR.Y	13	123.48	11.58	1012.00	3210.05
1511.74	K.VPQVSTPTLVEVSR.S	14	—	—	—	67.63
1567.70	K.DAFLGSFLYEYSR.R	13	29.66	—	152.32	350.83
1639.88	R.KVPQVSTPTLVEVSR.S	15	46.62	—	200.94	1018.87
1823.77	R.RPCFSALTPDETYVPK.A	16	—	—	—	24.70
Peptides matched			9	1	12	21

Similarly, for 0.4 mL 5 nmol·L^{-1} MYO enzymatic hydrolysate, only four peptides are detected before enrichment (Figure 4.72(a)), while 10 peptides are detected after enrichment by C8-$Fe_3O_4@mSiO_2$ and the sequence coverage reaches 62% (Figure 4.72(d)). The peptide information is shown in Table 4.8. Only one peptide with weak signal (S/N < 10) is enriched by $Fe_3O_4@mSiO_2$. Although the C8-f-$Fe_3O_4@mSiO_2$ can enrich a few peptides, but its enrichment efficiency is lower than C8-$Fe_3O_4@mSiO_2$

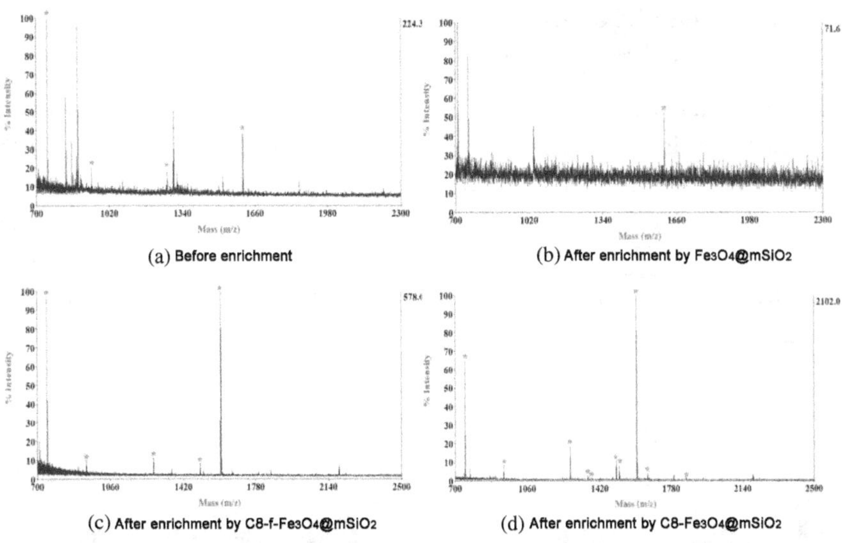

Figure 4.72: MALDI-TOF-MS spectra of MYO hydrolysate (0.4 mL 5 nmol·L^{-1}) before and after enrichment by different materials. "*" represents target peptide.

Table 4.8: The peptide information of MYO hydrolysate before (O) and after enrichment by Fe_3O_4@mSiO$_2$ (A), C8-f-Fe_3O_4@mSiO$_2$ (B) and C8-Fe_3O_4@mSiO$_2$ (C).

Calculated m/z	Peptide Sequence	Length	O	A	B	C
748.42	K.ALELFR.N	6	94.88	—	294.59	621.82
941.43	K.YKELGFQG.-	8	10.47	—	24.76	77.45
1271.62	R.LFTGHPETLEK.F	10	18.88	—	40.85	236.49
1360.72	K.ALELFRNDIAAK.Y	12	—	—	—	23.50
1378.78	K.HGTVVLTALGGILK.K	14	—	—	—	20.01
1502.61	K.HPGDFGADAQGAMTK.A	15	—	—	28.96	148.44
1518.60	K.HPGDFGADAQGAMTK.A (O)	15	—	—	—	100.55
1606.82	K.VEADIAGHGQEVLIR.L	15	53.11	< 10	439.79	1440.74
1661.78	R.LFTGHPETLEKFDK.F	14	—	—	—	40.63
1853.90	K.GHHEAELKPLAQSHATK.H	17	—	—	—	26.10
Peptides matched			4	1	5	10

4.5.3.6. *Investigation on the size-exclusion effect of C8-Fe$_3$O$_4$@mSiO$_2$*

Cyc (molecular weight 12,300 Da) and BSA (molecular weight 69,248 Da) are used in this experiment. Figure 4.73(a) shows the MALDI-TOF-MS spectrum of Cyc (2 mg·mL^{-1}) under linear mode before enrichment, there is no signal. When 0.5 mL of this Cyc solution is enriched by 0.3 mg C8-Fe$_3$O$_4$@mSiO$_2$ and C8-f-Fe$_3$O$_4$@mSiO$_2$, respectively, the MS signals are changed (Figures 4.73(b) and 4.73(c)), indicating that Cyc can be adsorbed by the C8

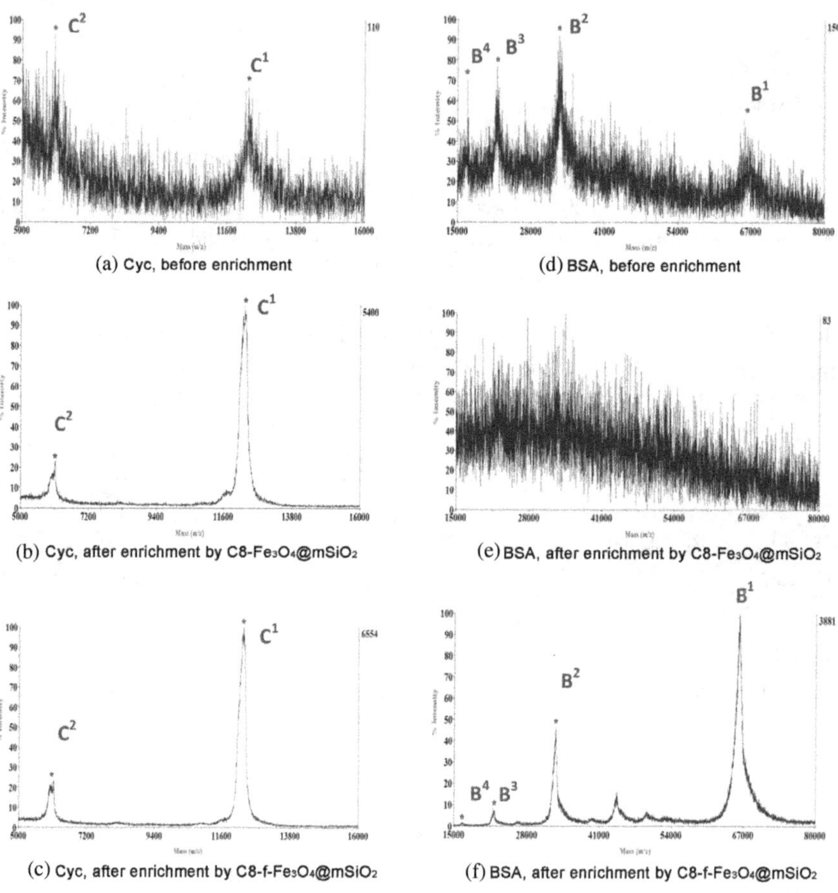

Figure 4.73: MALDI-TOF-MS spectra of Cyc solution (2 mg·mL^{-1}) and BSA solution (10 mg·mL^{-1}) before and after enrichment by different materials. "C^1, C^2" and "B^1, B^2, B^3, B^4" represent single-charge, two-charge, three-charge and four-charge ion peaks of Cyc and BSA, respectively.

group of the two materials. Then 10 mg·mL^{-1} BSA solution is treated in the same way. The single-charge and two-charge ion peaks of BSA can be seen in the original solution (Figure 4.73(d)). After treatment with C8-Fe$_3$O$_4$@mSiO$_2$, there is no ion peak signal in the MS spectrum (Figure 4.73(e)), but after treatment with C8-f-Fe$_3$O$_4$@mSiO$_2$, the ion peaks are shown in Figure 4.73(f). The results confirm that the C8-Fe$_3$O$_4$@mSiO$_2$ adsorb protein through the C8 group, and the mesopore could selectively extract peptide and small protein by size-exclusion effect.

4.5.3.7. *C8-Fe$_3$O$_4$@mSiO$_2$ for the enrichment of peptide in human serum*

Fresh serum sample is difficult to be analyzed directly by MALDI-TOF-MS, because large protein molecule and other inorganic salts in serum interfere with the peptide signal (Figure 4.74(a)). 50 mL serum is first diluted 5 times,

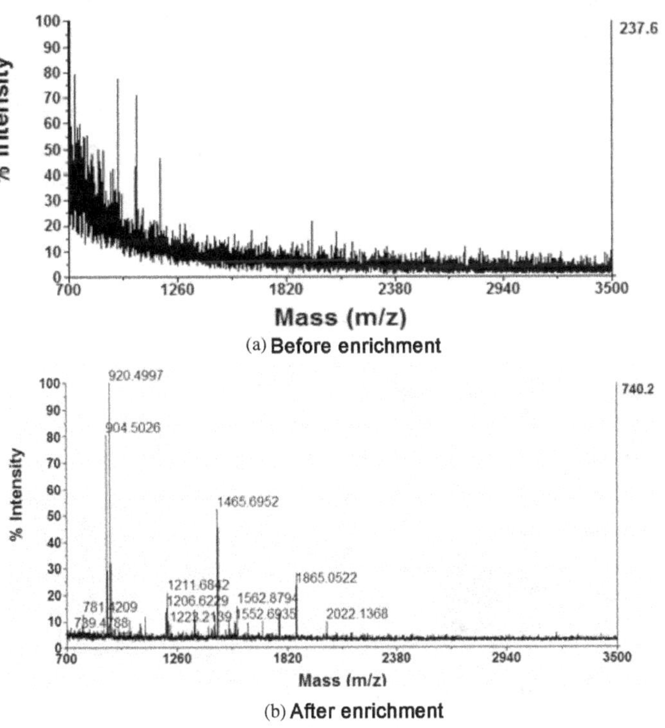

Figure 4.74: MALDI-TOF-MS spectra of diluted human serum before and after enrichment by C8-Fe$_3$O$_4$@mSiO$_2$.

and then part of the diluted serum is treated by C8-Fe$_3$O$_4$@mSiO$_2$. As seen from Figure 4.74(b), dozens of peptides are detected after enrichment by C8-Fe$_3$O$_4$@mSiO$_2$.

4.5.3.8. *C8-Fe$_3$O$_4$@mSiO$_2$ for the enrichment of peptide in mouse brain extract*

The enrichment workflow of C8-Fe$_3$O$_4$@mSiO$_2$ for endogenous peptide in mouse brain tissue extract is shown in Figure 4.75. 0.4 mL mouse brain tissue extract is diluted 3 times, then the diluent is treated by C8-Fe$_3$O$_4$@mSiO$_2$. Small peptides are induced into the functionalized mesopore to bind to the C8 group, while large protein molecules are excluded from the mesopore. The C8-Fe$_3$O$_4$@mSiO$_2$ of has adsorbed peptide is washed and eluted. The eluate is vacuum-lyophilized and re-dissolved in aqueous solution containing 0.1% FA and 5% CAN for being separated by nano-LC, and then analyzed by on-line electrospray tandem mass spectrometry. Finally, there are 267 peptides identified in mouse brain extract.

The total ion chromatogram of the peptide enriched from mouse brain extract is shown in Figure 4.76(a). MS/MS spectrum of MH$^+$ = 1,441 corresponding to the chromatographic peak at 22.54 min in the total ion chromatogram is shown in Figure 4.76(b), the corresponding peptide sequence is E.LNLPTGIPIVYEL.D, the b and y ions of which are highly matched.

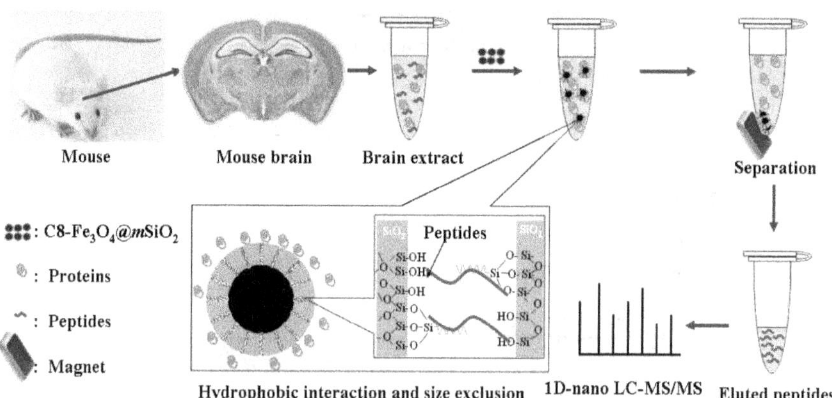

Figure 4.75: The enrichment workflow of C8-Fe$_3$O$_4$@mSiO$_2$ for endogenous peptide in mouse brain tissue extract.

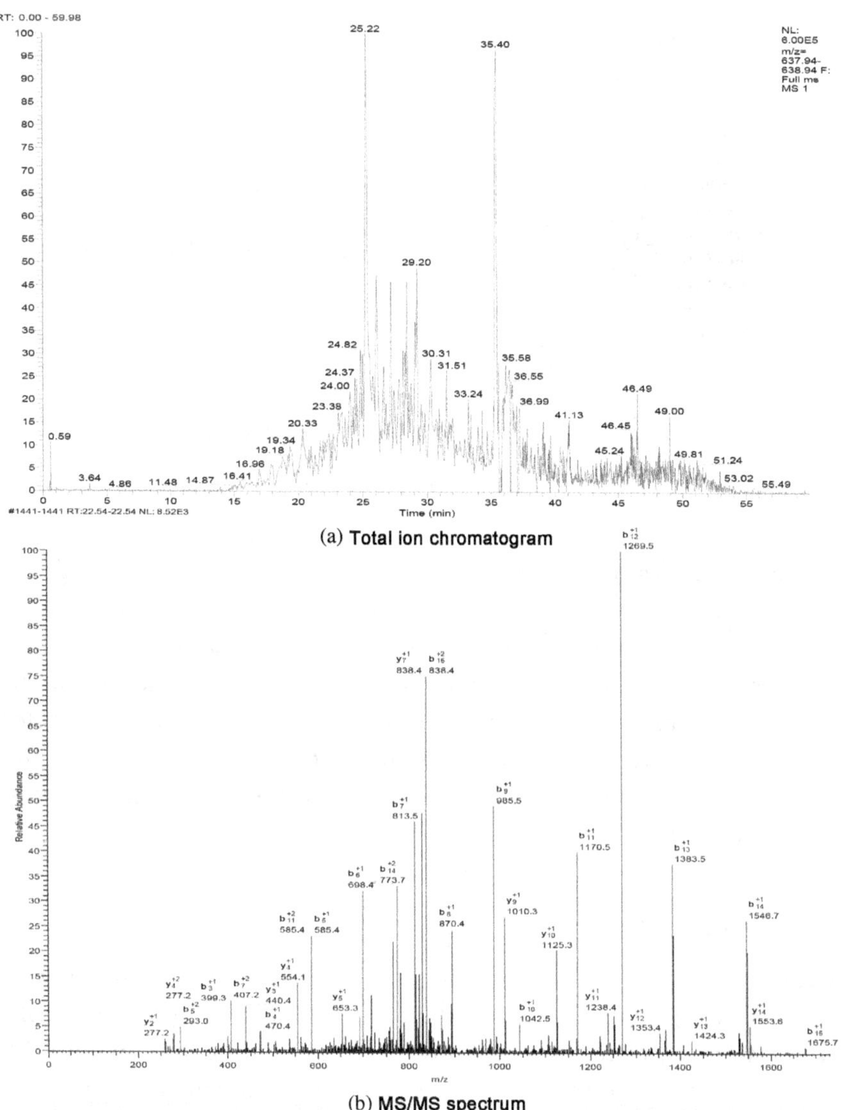

(a) Total ion chromatogram

(b) MS/MS spectrum

Figure 4.76: LC-ESI/MS analytical result of mouse brain extract after enrichment by C8-Fe$_3$O$_4$@mSiO$_2$ (N represents peptide number).

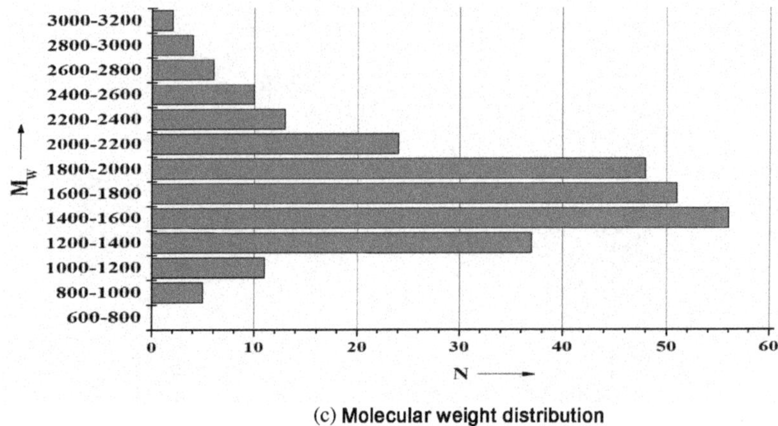

(c) **Molecular weight distribution**

Figure 4.76: (*Continued*)

Figure 4.76(c) clearly shows the distribution of the enriched peptides, mainly in the range of mass-to-charge ratio of 1,000–2,600 Da. Compared with the previous work, the peptides with molecular weight less than 1,200 Da also are enriched effectively. Moreover, it is found that about 85% of the peptides are less than 20 amino acids in length by counting the length of the detected peptides, which verify the size-exclusion effect of C8-Fe$_3$O$_4$@mSiO$_2$. This makes it an ideal material for the selective enrichment of endogenous peptide in complex biological sample.

4.5.4. *Conclusion*

Magnetic mesoporous nanomaterial synthesized by sol–gel reaction with surfactant template has sensitive magnetic responsiveness and large specific surface area. The material has a relatively narrow pore size distribution with an average pore diameter of about 3.3 nm. It is more meaningful to enrich the peptide/protein by calcining or modifying copper ion or introducing C8 group while synthesizing mesopore. The mesoporous material has a uniform ordered mesoporous surface, which can only allow small molecule to enter through the size-exclusion effect to achieve the purpose of removing large protein molecule, so it can also effectively enrich peptide in complex biological system (human serum, urine, mouse brain extract, etc.).

References

1. Chen H. Investigation of Functionalized Magnetic Nano-or Microspheres-Based Methodologies in Enrichment of Low-Abundance Proteins/Peptides for Proteome and Peptidome Analysis, Fudan University (2010).

2. Chen H., Xu X., Yao N., Deng C., Yang P., Zhang X. Facile Synthesis of C-8-Functionalized Magnetic Silica Microspheres for Enrichment of Low-Concentration Peptides for Direct MALDI-TOF MS Analysis. *Proteomics*, 2008, 8: 2778–84.

3. Iijima S. Helical Microtubules of Graphic Carbon. *Nature*, 1991, 354: 56–8.

4. Ichihashi T., Iijima S. Single-Shell Carbon Nanotubules of 1-nm Diameter. *Nature International Weekly Journal of Science*, 1993, 363: 603–5.

5. Cai Y. Q., Jiang G. B., Liu J. F., Zhou Q. X. Multi-Walled Carbon Nanotubes Packed Cartridge for the Solid-Phase Extraction of Several Phthalate Esters from Water Samples and Their Determination by High Performance Liquid Chromatography. *Analytica Chimica Acta*, 2003, 494: 149–56.

6. Li Q. L., Yuan D. X. Evaluation of Multi-Walled Carbon Nanotubes as Gas Chromatographic Column Packing. *Journal of Chromatography A*, 2003, 1003: 203–9.

7. Sang H. K., Park J. H. Intermolecular Interactions on Multiwalled Carbon Nanotubes in Reversed-Phase Liquid Chromatography. *Journal of Separation Science*, 2006, 29: 945–52.

8. Kam N. W. S., Dai H. J. Carbon Nanotubes as Intracellular Protein Transporters: Generality and Biological Functionality. *Journal of the American Chemical Society*, 2005, 127: 6021–6.

9. Xin L. S., Xu, Chensong, Pan, Houjiang, Zhou, Xiaogang, Jiang, Yu, Zhang, Mingliang, Ye, Hanfa, Zou. Enrichment of Peptides from Plasma for Peptidome Analysis Using Multiwalled Carbon Nanotubes. *Journal of Separation Science*, 2007, 6: 930–43.

10. Vallant R. M., Szabo Z., Bachmann S., Bakry R., Najam-ul-Haq M., Rainer M., Heigl N., Petter C., Huck C. W., Bonn G. K. Development and Application of C60-Fullerene Bound Silica for Solid-Phase Extraction of Biomolecules. *Analytical Chemistry*, 2007, 79: 8144–53.

11. Jia W., Wu H., Lui H., Li N., Zhang Y., Cai R., Yang P. Rapid and Automatic on-Plate Desalting Protocol for MALDI-MS: Using Imprinted Hydrophobic Polymer Template. *Proteomics*, 2007, 7: 2497–506.

12. Zhang X. *Modern Chromatography.* Fudan University Press, 2004.

13. Zou H. F., Zhang Y. K., Lu P. High Efficiency Liquid Chromatography. 1998.

14. Ren D. Y., Penner N. A., Slentz B. E., Regnier F. E. Histidine-Rich Peptide Selection and Quantification in Targeted Proteomics. *Journal of Proteome Research*, 2004, 3: 37–45.

15. Liu S. New Methods for Peptidomics Separation and Analysis Based on Functionalized Magnetic Mesoporous Materials, Fudan Univerity (2011).

16. Kresge C. T., Leonowicz M. E., Roth W. J., Vartuli J. C., Beck J. S. Ordered Mesoporous Molecular-Sieves Synthesized by a Liquid-Crystal Template Mechanism. *Nature*, 1992, 359: 710–2.

17. Zhao D. Y., Feng J. L., Huo Q. S., Melosh N., Fredrickson G. H., Chmelka B. F., Stucky G. D. Triblock Copolymer Syntheses of Mesoporous Silica with Periodic 50 to 300 Angstrom Pores. *Science*, 1998, 279: 548–52.

18. Hartmann M. Ordered Mesoporous Materials for Bioadsorption and Biocatalysis. *Chemistry of Materials*, 2005, 17: 4577–93.

19. Buchel G., Grun M., Unger K. K., Matsumoto A., Tsutsumi K. Tailored Syntheses of Nanostructured Silicas: Control of Particle Morphology, Particle Size and Pore Size. *Supramolecular Science*, 1998, 5: 253–9.

20. Han Y. J., Stucky G. D., Butler A. Mesoporous Silicate Sequestration and Release of Proteins. *Journal of the American Chemical Society*, 1999, 121: 9897–8.

21. Yiu H. H. P., Botting C. H., Botting N. P., Wright P. A. Size Selective Protein Adsorption on Thiol-Functionalised Sba-15 Mesoporous Molecular Sieve. *Physical Chemistry Chemical Physics*, 2001, 3: 2983–5.

22. Tian R., Zhang H., Ye M., Jiang X., Hu L., Li X., Bao X., Zou H. Selective Extraction of Peptides from Human Plasma by Highly Ordered Mesoporous Silica Particles for Peptidome Analysis. *Angewandte Chemie-International Edition*, 2007, 46: 962–5.

23. Yao N., Chen H., Lin H., Deng C., Zhang X. Enrichment of Peptides in Serum by C-8-Functionalized Magnetic Nanoparticles for Direct Matrix-Assisted Laser Desorption/Ionization Time-of-Flight Mass Spectrometry Analysis. *Journal of Chromatography A*, 2008, 1185: 93–101.

24. Chen H., Deng C., Li Y., Dai Y., Yang P., Zhang X. A Facile Synthesis Approach to C8-Functionalized Magnetic Carbonaceous Polysaccharide Microspheres for the Highly Efficient and Rapid Enrichment of Peptides and Direct MALDI-TOF-MS Analysis. 2009, 21: 2200–5.

25. Chen H., Liu S., Li Y., Deng C., Yang P. J. P. Development of Oleic Acid-Functionalized Magnetite Nanoparticles as Hydrophobic Probes for Concentrating Peptides with MALDI-TOF-MS Analysis. 2011, 11: 890–7.

26. Chen H., Qi D., Deng C., Professor, Yang P., Zhang X. J. P. Preparation of C60-Functionalized Magnetic Silica Microspheres for the Enrichment of

Low-Concentration Peptides and Proteins for MALDI-TOF-MS Analysis. 2010, 9: 380-7.

27. Chen H., Deng C., Zhang X. Synthesis of Fe_3O_4@SiO_2@PMMA Core-Shell-Shell Magnetic Microspheres for Highly Efficient Enrichment of Peptides and Proteins for MALDI-TOF-MS Analysis. *Angewandte Chemie-International Edition*, 2010, 49: 607–11.

28. Jia W., Chen X., Lu H., Yang P. J. A. C. $CaCO_3$–Poly(Methyl Methacrylate) Nanoparticles for Fast Enrichment of Low-Abundance Peptides Followed by $CaCO_3$-Core Removal for MALDI-TOF-MS Analysis. 2006, 118: 3423–7.

29. Shen W., Xiong H., Xu Y., Cai S., Lu H., Yang P. ZnO-Poly(Methyl Methacrylate) Nanobeads for Enriching and Desalting Low-Abundant Proteins Followed by Directly MALDI-TOF-MS Analysis. *Analytical Chemistry*, 2008, 80: 6758–63.

30. Xiong H.-M., Guan X.-Y., Jin L.-H., Shen W.-W., Lu H.-J., Xia Y.-Y. Surfactant-Free Synthesis of SnO_2@PMMA and TiO_2@PMMA Core-Shell Nanobeads Designed for Peptide/Protein Enrichment and MALDI-TOF-MS Analysis. *Angewandte Chemie-International Edition*, 2008, 47: 4204–7.

31. Chen H., Liu S., Yang H., Mao Y., Deng C., Zhang X., Yang P. Selective Separation and Enrichment of Peptides for MS Analysis Using the Microspheres Composed of Fe_3O_4@nSiO$_{(2)}$ Core and Perpendicularly Aligned Mesoporous SiO_2 Shell. *Proteomics*, 2010, 10: 930–9.

32. Yin P., Sun N., Deng C., Li Y., Zhang X., Yang P. Facile Preparation of Magnetic Graphene Double-Sided Mesoporous Composites for the Selective Enrichment and Analysis of Endogenous Peptides. *Proteomics*, 2013, 13: 2243–50.

33. Novoselov K. S., Geim A. K., Morozov S. V., Jiang D., Zhang Y., Dubonos S. V., Grigorieva I. V., Firsov A. A. Electric Field Effect in Atomically Thin Carbon Films. *Science*, 2004, 306: 666–9.

34. Choi B. G., Chang S.-J., Lee Y. B., Bae J. S., Kim H. J., Huh Y. S. 3D Heterostructured Architectures of Co_3O_4 Nanoparticles Deposited on Porous Graphene Surfaces for High Performance of Lithium Ion Batteries. *Nanoscale*, 2012, 4: 5924–30.

35. Tammen H., Schulte L., Hess R., Menzel C., Kellmann M., Mohring T., Schulz-Knappe P. Peptidomic Analysis of Human Blood Specimens: Comparison between Plasma Specimens and Serum by Differential Peptide Display. *Proteomics*, 2005, 5: 3414–22.

36. Zhou F., Wu S., Wu B., Chen W. R., Xing D. Mitochondria-Targeting Single-Walled Carbon Nanotubes for Cancer Photothermal Therapy. *Small*, 2011, 7: 2727–35.

37. Liu S., Chen H., Lu X., Deng C., Zhang X., Yang P. Facile Synthesis of Copper(ii) Immobilized on Magnetic Mesoporous Silica Microspheres for Selective Enrichment of Peptides for Mass Spectrometry Analysis. *Angewandte Chemie-International Edition*, 2010, 49: 7557–61.

38. Hu L., Zhou H., Li Y., Sun S., Guo L., Ye M., Tian X., Gu J., Yang S., Zou H. Profiling of Endogenous Serum Phosphorylated Peptides by Titanium (iv) Immobilized Mesoporous Silica Particles Enrichment and MALDI-TOF MS Detection. *Analytical Chemistry*, 2009, 81: 94–104.

39. Liu S., Li Y., Deng C., Mao Y., Zhang X., Yang P. Preparation of Magnetic Core-Mesoporous Shell Microspheres with C8-Modified Interior Pore-Walls and Their Application in Selective Enrichment and Analysis of Mouse Brain Peptidome. *Proteomics*, 2011, 11: 4503–13.

Chapter 5

Phosphoproteomics Analysis Technology Based on Magnetic Micro-/Nanomaterial

5.1. Basic analysis principle of phosphoproteomics based on magnetic micro-/nanomaterial

Protein phosphorylation is one of the most studied post-translational modifications (PTMs). The reversible phosphorylation plays a vital biological role in living organisms. Numerous studies have confirmed that abnormal phosphorylation is associated with many diseases, such as cancer, heart disease, Alzheimer's disease, etc. Generally, proteomics focuses on the study of proteins which are closely associated with specific physiological pathology, and their environment or the relationship between the specific protein and its surrounding. The great progress in proteomics research over the years has largely been due to the rapid development of biomass spectrometry with high-throughput and high-efficiency.

There are three different analytical strategies for proteomics research: bottom-up strategy, middle-down strategy, and top-down strategy (Figure 5.1). From the current research, the "top-down" strategy, which depends on MS technology to analyze complete proteins, has proved the most difficult in proteomics research. The "middle-down" strategy requires appropriate enzymatic hydrolysis method for limited hydrolysis of proteins. However, the current enzymatic hydrolysis methods, such as using specific enzyme[2,3] or microwave-assisted acidic hydrolysis method,[4] cannot be used in large-scale proteomics research. "Bottom-up" is the most widely used strategy in proteomics research. It directly digests all the proteins in bio-sample into peptides, without any pre-separation at the level of proteins. Studies have shown

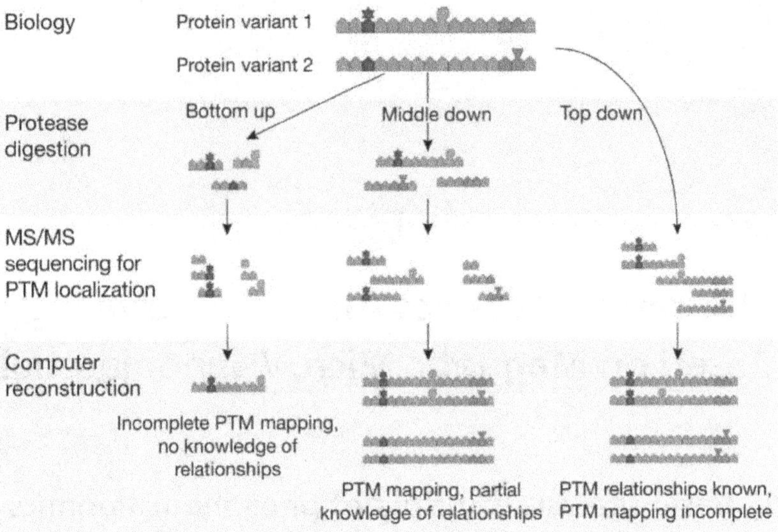

Figure 5.1: Common strategies for PTMs analysis.[1]

that this strategy is also very suitable for the study of phosphoproteomics. Nevertheless, the bottom-up strategy will enhance the complexity of the sample and make it more difficult to be detected by MS. Since there are a large number of non-phosphorylated peptides in bio-sample digest, which will inhibit the MS signal of phosphopeptide, it is necessary to extract and enrich the phosphorylated proteins/peptides in a biological sample before MS analysis to improve the relative concentration of target peptides. At present, the main enrichment methods for phosphoprotein/phosphopeptide include immunoprecipitation, affinity chromatography, chemical derivation, etc. Among them, magnetic microsphere, with the rapid separation ability and various functional groups modified on its surface for selective enrichment of peptide, are widely researched and used in phosphoproteomics analysis (Figure 5.2).

5.1.1. *IMAC-based magnetic micro-/nanomaterial*

IMAC is one of the most popular and mature methods for separation and enrichment of phosphorylated proteins/peptides. It is well known that IMAC stationary phase consists of three parts: carrier, chelating agent, and metal ion.

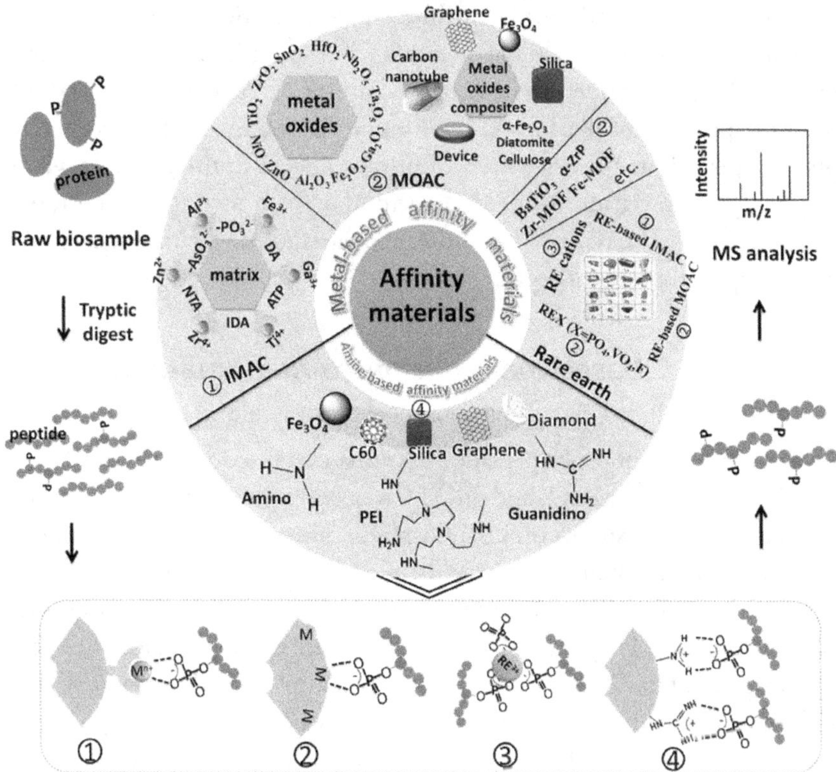

Figure 5.2: Different affinity materials for phosphopeptide enrichment. "①②③④" represent the enrichment mechanism of different affinity materials.[5]

Among the IMAC-based magnetic micro-/nanomaterials, magnetic microsphere or modified magnetic microsphere is usually used as carrier, and the carrier and metal ion can be closely linked by ideal chelating agent. However, one of the drawbacks of the IMAC method is that metal ions may be lost in the process of enrichment or elution, so it is very important to find suitable a chelating agent. An ideal chelating agent can not only tightly connect the carrier with metal ion, but also increase the biocompatibility of the magnetic micro-/nanomaterial. The common metal ions for phosphopeptide enrichment include Ti(IV), Fe(III), Al(III), Ga(III), Zr(IV), etc. The principle of phosphopeptide enrichment by IMAC-based magnetic micro-/nanomaterial is that the negatively-charged phosphorylated protein/peptide can chelate with positively-charged metal ion through electrostatic interaction under a certain acidic condition, and the interaction between metal ion and phospho-protein/phosphopeptide will be destroyed under a certain alkaline condition or in the presence of phosphate, resulting in the release of phosphoprotein/phosphopeptide for further identification. Since a multi-phosphorylated protein/peptide is more negative, and can form stronger interaction with metal ion, IMAC-based micro-/nanomaterials prefer to enrich multi-phosphorylated proteins/peptides.[6–9]

5.1.2. *MOAC-based magnetic micro-/nanomaterial*

MOAC is another widely used method for separation and enrichment of phosphoproteins/phosphopeptides. Various metal oxides, such as HfO_2, Nb_2O_5, Al_2O_3, ZrO_2, ZnO, and SnO_2, have been reported for extraction and enrichment of phosphoproteins/phosphopeptides. MOAC is reported to have higher enrichment selectivity for phosphoproteins/phosphopeptides, and has better tolerance to low pH, salt, and other low molecular impurities. The enrichment principle based on MOAC is that the metal oxide with posi-tive charge can bind with negatively-charged phosphorylated protein/peptide under a certain acidic condition. The metal oxide becomes negatively-charged and can bind with positive ion under a certain alkaline condition. Larsen[10] and Ficarro[11] report that MOAC prefers to enrich mono-phosphorylated peptides because they are more difficult to be eluted from the metal oxide, resulting in low enrichment efficiency.

5.1.3. *MOF-modified magnetic micro-/nanomaterial*

MOF is a new kind of inorganic–organic microporous crystal material, which is self-assembled by coordination bond between metal ion and organic ligand. A large number of available metal ions and organic ligands make it possible to prepare numerous novel MOF materials. MOF has attracted wide atten-tion because of its extraordinary properties, such as super-large specific surface area, excellent thermal stability, uniform micro-/nanoscale pore, the modifi-ability of its surface, and the modifiability of the inner wall of the pore. So far, MOF-modified magnetic micro-/nanomaterial, such as Er-MOF, Fe-MOF, Zr–MOF, has achieved excellent enrichment performance in phosphoprot-eomics analysis.

5.1.4 *Rare earth-modified magnetic micro-/nanomaterial*

Rare earth is a class of very important metal elements. Stable rare earth ion is usually trivalent or tetravalent cation. It is typical hard acid and can interact with hard base such as phosphate group. At present, rare earth element shows great prospect in the adsorption of phosphorylated protein/peptide. Free rare earth cation has great advantage in the separation of phosphoprotein/phosphopeptide, which can not only reduce the non-specific adsorption but also has high recovery of the target molecule. Rare earth ion-modified magnetic micro-/nanomaterials, such as Ce^{4+}, La^{3+}, Er^{3+} modified magnetic microspheres are based on IMAC mechanism to separate phosphoprotein/phosphopeptide. Similarly, the rare earth oxide can separate phosphoprotein/phosphopeptide based on MOAC mechanism.

5.1.5. *Amino-modified magnetic micro-/nanomaterial*

The phosphate group of phosphopeptide is negative and is able to interact with positively-charged group. Generally, the amino group is positive in a relatively wide pH range. For example, the guanido group has a pKa value of 13.6, which can not only form a hydrogen bond with the phosphate group, but also has an electron pair effect, resulting in stable and firm binding to the phosphate group. It has been reported that arginine has electrostatic interaction with phosphate group, which is similar to the covalent bond. Guanido or arginine modified composite has been widely used in enrichment of phosphoprotein/phosphopeptide. Amino-modified magnetic micro-/nanomaterial also shows outstanding performance in enrichment of phosphorylated protein/peptide.

5.2. IMAC-based magnetic micro-/nanomaterial for separation and analysis in phosphoproteomics

5.2.1. *Introduction*

Phosphorylation is one of the most important PTMs. Due to its relatively low abundance and poor ionization efficiency, it is difficult to directly analyze and identify phosphorylation by bio-mass spectrometry. Therefore, the sample needs to be pretreated to increase the relative content of phosphorylation. A number of strategies have been proposed for the separation and enrichment of phosphorylated proteins/peptides. Among them, IMAC technology is the most widely used method for capturing phosphorylated proteins/peptides from complex biological samples, and relies mainly on the affinity between metal ions and phosphate group. At present, IMAC technology mainly has the following problems: (i) the relatively small amount of immobilized metal ions limits the enrichment efficiency; (ii) metal ion has the possibility of falling off during the enrichment and elution processes; (iii) conventional IMAC material has small specific surface area and poor hydrophilicity. Therefore, seeking for novel IMAC materials with large specific surface area and good hydrophilicity has become imperative to further develop new methods for better analysis and identification of phosphoproteins. At present, there are many functionalization methods to immobilize metal ions on magnetic microspheres. For example, the carboxyl group of iminodiacetic acid (IDA),[12] phosphate group,[13] and polydopamine can be used as affinity ligand to immobilize metal ions.[14] This section will introduce the application of functionalized magnetic micro-/nanomaterial in phosphoproteomics research from the abovementioned aspects.

5.2.2. *IDA for immobilization of metal ion on magnetic microsphere*

5.2.2.1. *IDA-modified magnetic silica for immobilization of Fe^{3+} (Fe_3O_4@SiO_2@Glymo–IDA–Fe^{3+})*

(1) Preparation of Fe_3O_4@SiO_2@Glymo–IDA–Fe^{3+}

The synthesis of Fe_3O_4@SiO_2@Glymo–IDA–Fe^{3+} microsphere can be seen in Section 2.6 of Chapter 2.[12]

(2) Enrichment application of Fe_3O_4@SiO_2@Glymo–IDA–Fe^{3+} in phosphoproteomics

(a) *Preliminary study on the enrichment performance of Fe_3O_4@SiO_2@Glymo–IDA–Fe^{3+} for phosphopeptide in standard phosphorylated protein digest*

β-casein is a commonly used sample for phosphopeptide analysis when using MALDI-TOF-MS as detector. This protein contains two phosphorylation sites, and can be digested into two phosphopeptides by trypsin (Table 5.1). In this work, β-casein digest is selected to investigate the enrichment

Table 5.1: Detailed information of phosphopeptide identified from β-casein digest.

No.	AA	Phosphopeptide Sequence	$[M+H]^+$
1	33–48	FQ[pS]EEQQQTEDELQDK	2061.70
2	1–25	RELEELNVPGEIVE[pS]L[pS][pS][pS]EESITR	3121.98

performance of Fe_3O_4@SiO_2@Glymo–IDA–Fe^{3+} for phosphopeptides. Figure 5.3(a) shows the MALDI-TOF-MS spectrum of 2×10^{-7} mol·L^{-1} β-casein digest before enrichment. The detected multi-phosphorylated peptide has very weak signal except for one monophosphorylated peptide, and the non-phosphopeptide signals are very strong. As shown in Figure 5.3(b), the MS signals of these two phosphorylated peptides are significantly enhanced and the corresponding dephosphorylated segments can also be detected, after enrichment with Fe_3O_4@SiO_2@Glymo–IDA–Fe^{3+} microsphere. The m/z at 1967.3 is assigned to $m/z$$m/z$ at 2061.9, which means that the mono-phosphorylated peptide parent ion loses one phosphoric acid. Since the degraded parent ion is in the metastable state, the mass difference between m/z at 1967.3 and m/z at 2061.9 is 94.6 Da instead of 98 Da.[15] The peaks with a mass difference of 98 Da in Figure 5.3(b) correspond to the respective dephosphorylation of the tetra-phosphorylated peptide (m/z = 3122.3, M^+H). Compared with Figure 5.3(a), few non-phosphorylated peptides are identified in the elution, and the S/N ratio of the phosphopeptide is significantly improved. It indicates that Fe_3O_4@SiO_2@Glymo–IDA–Fe^{3+} microsphere has excellent enrichment ability for phosphopeptides. Moreover, for lower concentration of β-casein digest (2×10^{-8} mol·L^{-1}), Fe_3O_4@SiO_2@Glymo–IDA–Fe^{3+} microsphere still exhibits strong enrichment ability, as shown in Figures 5.3(d) and 5.3(e). The above results confirm that the Fe_3O_4@SiO_2@Glymo–IDA–Fe^{3+} microsphere has good enrichment ability and high enrichment sensitivity for phosphopeptides.

The phosphorylation sites of phosphopeptide enriched by Fe_3O_4@SiO_2@Glymo–IDA–Fe^{3+} microsphere are identified by MALDI-TOF-MS. Figure 5.3(c) is a MS/MS spectrum of the mono-phosphorylated peptide (m/z 2061.9). Considering the variable modifications such as phosphorylation of tyrosine, threonine, and serine, Mascot results indicate that the phosphopeptide sequence of 2061 is FQSEEQQQTEDELQDK and contains one phos-

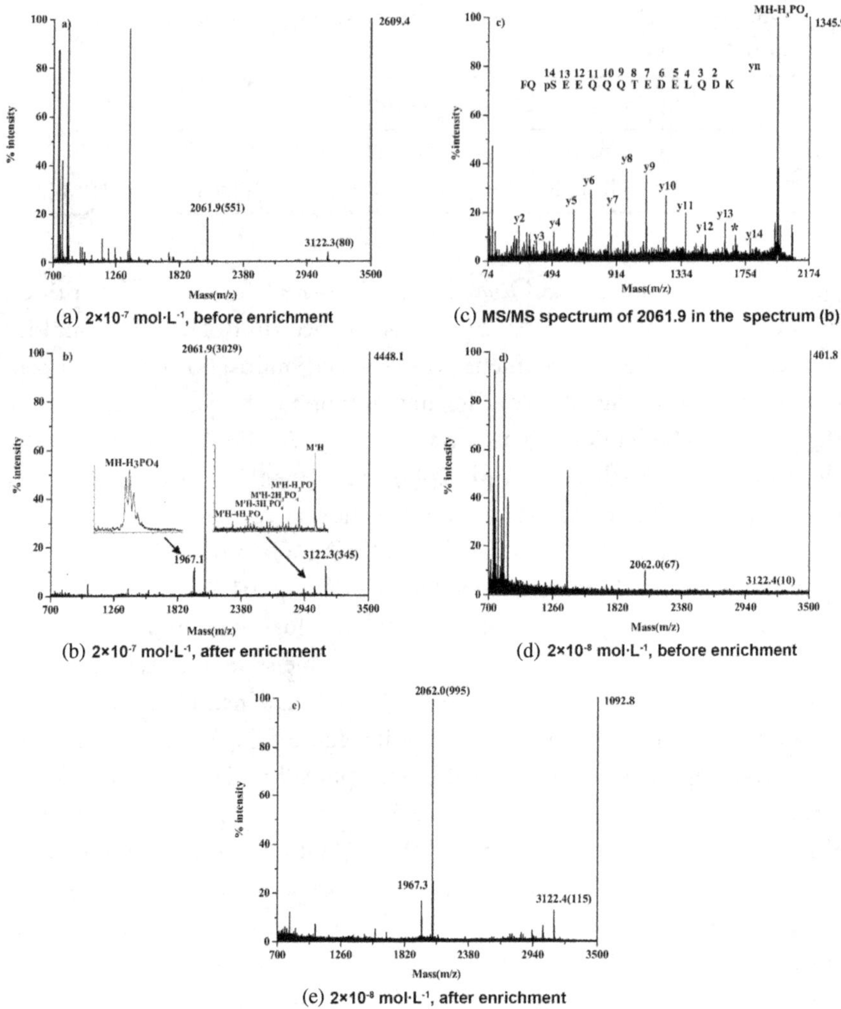

(a) 2×10^{-7} mol·L^{-1}, before enrichment

(c) MS/MS spectrum of 2061.9 in the spectrum (b)

(b) 2×10^{-7} mol·L^{-1}, after enrichment

(d) 2×10^{-8} mol·L^{-1}, before enrichment

(e) 2×10^{-8} mol·L^{-1}, after enrichment

Figure 5.3: MS spectra of β-casein digest (2×10^{-7} mol·L^{-1}, 2×10^{-8} mol·L^{-1}) before and after enrichment by Fe$_3$O$_4$@SiO$_2$@Glymo–IDA–Fe^{3+}. The number in parenthesis represents S/N ratio.

phorylation site. The MS/MS spectrum is analyzed by GPS software, and the information of the fragment ions is shown in Table 5.2. The MS/MS spectrum obtained shows co mplete y-series ions. The mass difference between y13 and y14 is 167, indicating that the phosphorylation site is on the serine residue, and the fact that y14 ion loses 98 Da further confirms the result.

Table 5.2: The information of fragment ions of phosphorylated peptide (m/z = 2,016).

M/Z	S/N	Ion Type	Matching Error (Da)
262.133	11.5	y2	0.007
390.170	12.3	y3	0.029
503.293	19.8	y4	0.010
632.218	47.9	y5	0.108
747.250	66.3	y6	0.121
876.504	54.3	y7	0.109
977.306	86.5	y8	0.137
1105.379	75.3	y9	0.122
1233.390	58.2	y10	0.170
1361.484	35.7	y11	0.134
1490.410	19.8	y12	0.251
1619.571	31.4	y13	0.133
1786.489	20.1	y14	0.213

(b) *Evaluation on enrichment condition for phosphopeptide in standard phosphoprotein digest using $Fe_3O_4@SiO_2@Glymo–IDA–Fe^{3+}$*

There are two general methods for enhancing the enrichment efficiency of IMAC technology. One is to use the dried methanol saturated with hydrochloric acid[16] to react with acidic side chain of aspartic acid and glutamic acid,[17] producing carboxymethyl ester via O-methyl esterification reaction. However, on the one hand, all the available carboxylic acid groups with 100% vacancy can be used for O-me thyl esterification reaction. On the other hand, the buffer used in the O-methyl esterification reaction usually leads to the partial removal of amide from aspartic acid and glutamic acid, and the amino acid residues can also be O-methylated, thus causing an increase in the sample complexity. The second way to enhance the enrichment efficiency is to adjust the pH of the solution. Acidification of the sample prior to IMAC enrichment can protonate the highly acidified carboxyl groups, thereby reducing non-specific adsorption. Therefore, the pH of the loading buffer is considered to be a key factor for enrichment selectivity. Most of the experiments and research works use 0.1–0.25 mol·L^{-1} acetic acid (pH = 2.7) as loading buffer. According to Saha *et al.*,[18] the pKa value of the phosphate will decrease to 1.1

when methylation happens, so the phosphopeptide has lower pKa value than phosphoric acid. Therefore, the selection of enrichment condition is crucial. In this work, the enrichment performance of microspheres is investigated in detail via β-casein digest, including the composition of the enrichment system, the pH condition of the enrichment system, the enrichment time, and the elution time.

(i) Optimization of the enrichment condition

To avoid the non-specific adsorption, four kinds of loading buffers (pH = 2): TFA–H$_2$O, 50% ACN–TFA, ACOOH–H$_2$O, 50% ACN–ACOOH, are selected for parallel experiments to investigate their effect on the enrichment performance of Fe$_3$O$_4$@SiO$_2$@Glymo–IDA–Fe^{3+} (Figure 5.4). It can be seen from Figure 5.4 that when 50% ACN is not added into the loading buffer, non-specific adsorption will interfere with the detection of phosphopeptides (Figures 5.4(b) and 5.4(d)). When 50% ACN is used, only the phosphopeptide is detected, indicating that acetonitrile can destroy the electrostatic adsorption between the acidic non-phosphopeptide with hydrophobic group and Fe$_3$O$_4$@SiO$_2$@Glymo–IDA–Fe^{3+} microsphere. Comparing Figures 5.4(c) and 5.4(e), it can be seen that both the loading systems have good selectivity for phosphopeptide, but considering that acetic acid is relatively mild, 50% ACN–ACOOH is selected as the final loading buffer.

(ii) Optimization of the concentration of ACOOH in loading buffer

Fe$_3$O$_4$@SiO$_2$@Glymo–IDA–Fe^{3+} microsphere enriches phosphopeptide via specific affinity interaction, the pH value of the loading buffer greatly affects the materials' enrichment performance. According to the literature, for the IMAC-based enrichment method, the adsorption of phosphate group will gradually decrease with the increase of pH value when the pH value is between 2 and 9.[19] In order to avoid the non-specific adsorption of peptide with acidic residue in the digest, it is necessary to ensure that the pH value of the loading buffer is about 2.5. Then the most acidic peptides containing carboxyl groups are positively charged, so it is difficult to produce adsorption with Fe$_3$O$_4$@SiO$_2$@Glymo–IDA–Fe^{3+} microspheres. Therefore, for the 50% ACN–ACOOH system, we examine its selectivity for phosphopeptides when the concentration of ACOOH is 1%, 5%, and 10%, respectively.

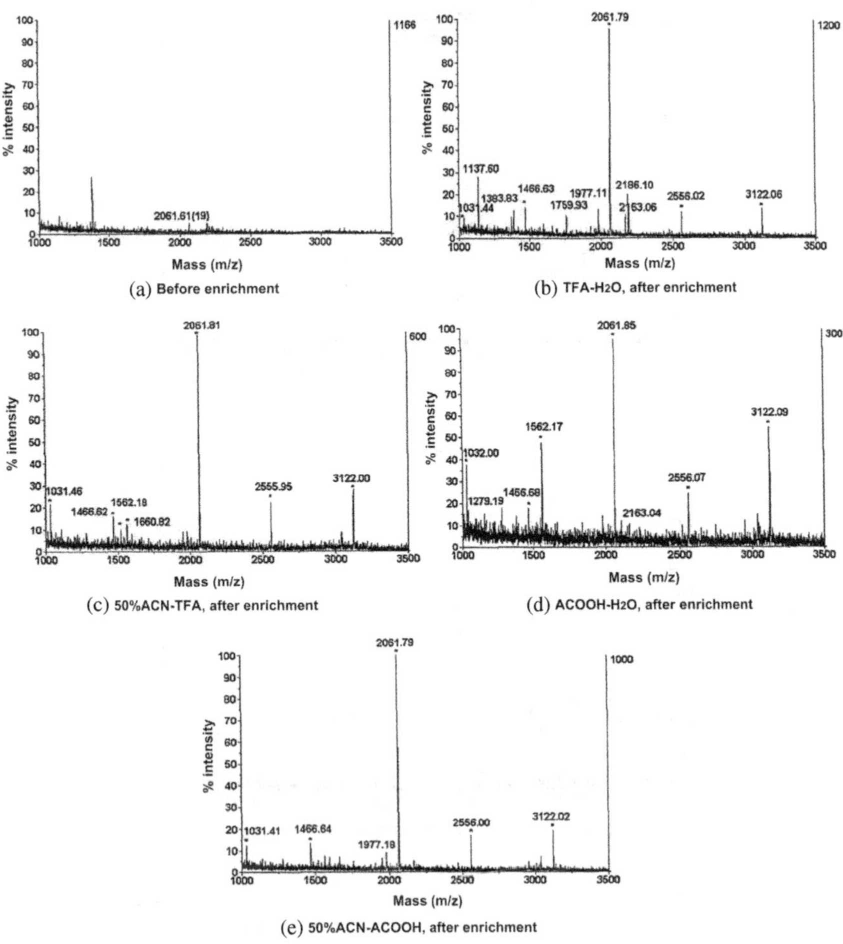

Figure 5.4: MS spectra of 2×10^{-8} mol·L^{-1} β-casein digest before and after enrichment by $Fe_3O_4@SiO_2@$Glymo–IDA–Fe^{3+} using different loading buffers.

As shown in Figure 5.5, when the acetic acid concentration is increased from 1% to 10%, the multi-phosphorylated peptides are gradually reduced, possibly due to the excessive acetic acid in the system. The coordination between the carboxylic acid group in the acetic acid and the Fe^{3+} ion will interfere with the interaction between Fe^{3+} ion and phosphate group in the phosphopeptide. Therefore, 1% ACOH is chosen for further experiment.

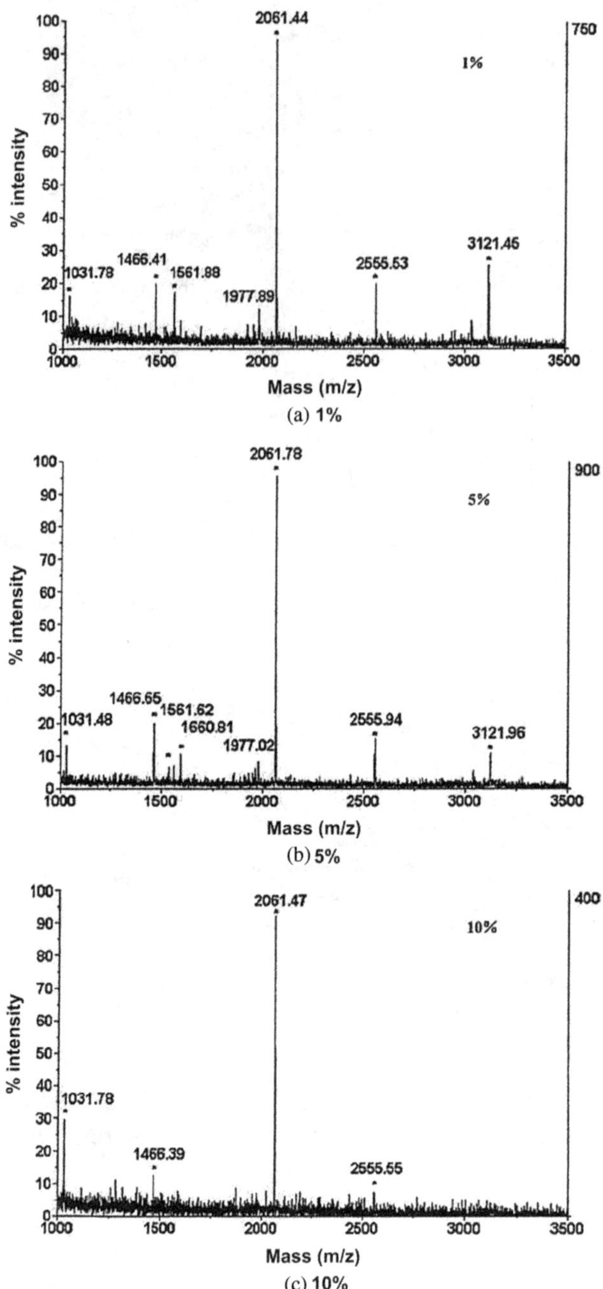

Figure 5.5: MS spectra of 2×10^{-8} mol·L^{-1} β-casein digest after enrichment by Fe$_3$O$_4$@ SiO$_2$@Glymo–IDA–Fe^{3+} under different concentrations of ACOOH.

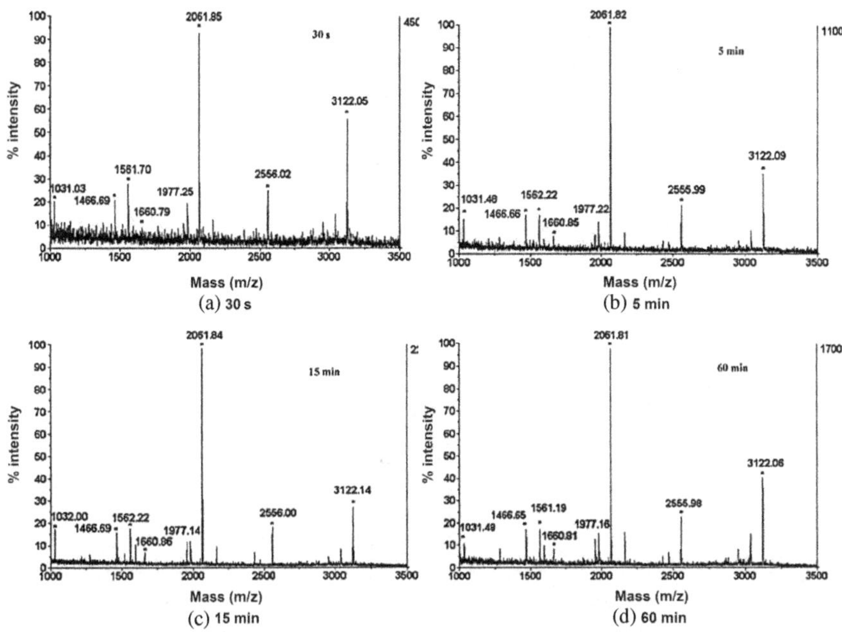

Figure 5.6: MS spectra of 2×10^{-8} mol·L^{-1} β-casein digest after enrichment by Fe$_3$O$_4$@SiO$_2$@Glymo–IDA–Fe^{3+} microspheres with different enrichment times.

(iii) **Optimization of enrichment time**

Taking 50% ACN–1% ACOOH as the loading buffer, the selective enrichment performance is evaluated at 30 s, 5 min, 15 min, and 60 min, respectively. As shown in Figure 5.6, the material shows efficient enrichment performance for phosphopeptides within 30 s. The reason is that Fe$_3$O$_4$@SiO$_2$@Glymo–IDA–Fe^{3+} microspheres have good dispersibility in the sample solution and large specific surface area, which provides a new method for simple, rapid, and effective enrichment of phosphopeptides. After comparing the figures, 10 min is chosen as the enrichment time.

(iv) **Optimization of elution time**

The enriched phosphopeptide needs to be eluted from the surface of the material. How to effectively elute phosphopeptides is necessary to be investigated. When IMAC technology is used, there are two ways for elution. One is to use a certain concentration of phosphate to elute the phosphopeptide according to the principle of displacement. But this method will introduce a large amount of phosphates, and the de-salt step is introduced before the

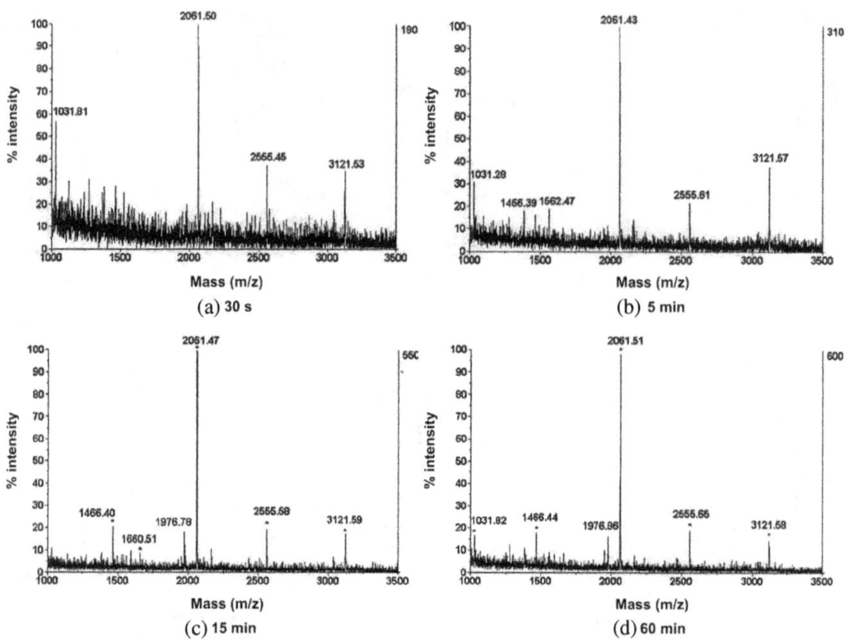

Figure 5.7: MS spectra of 2×10^{-8} mol·L^{-1} β-casein digest after enrichment by Fe$_3$O$_4$@ SiO$_2$@Glymo–IDA–Fe^{3+} microspheres with different elution times.

subsequent analysis and identification, which will also lead to sample loss. Another method is to use a certain concentration of ammonia to destroy the electrostatic interaction between phosphopeptide and the surface of the material, and a small amount of ammonia in the solution can be removed by evaporation. The operation is very simple and effective. Therefore, 0.5% ammonia solution is used as eluent in this work. Taking 50% ACN–1% ACOOH as loading buffer, 10 min as the enrichment time, the elution performance is investigated when elution time is 1 min, 5 min, 15 min, and 60 min. It can be seen from Figure 5.7 that the phosphopeptide can be eluted when the elution time is 1 min. When the elution time is increased to 15 min, the elution performance for phosphopeptides is stable. Therefore, the elution time is set to be 15 min for the subsequent experiment.

(v) **Direct analysis of the material with adsorbed phosphopeptide**

In order to simplify the traditional enrichment process of IMAC technique, Raska *et al.*[20] directly spot the Fe^{3+} stationary phase onto the MALDI plate,

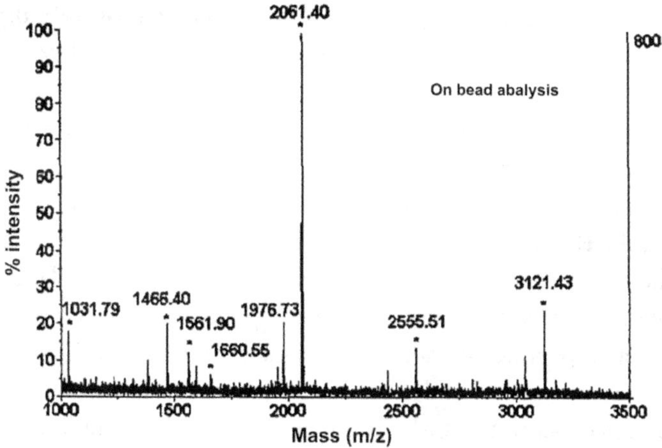

Figure 5.8: MS spectrum of direct analysis of phosphopeptide enriched by $Fe_3O_4@SiO_2@$ Glymo–IDA–Fe^{3+} without elution process.

and the enriched phosphopeptides are then detected by MS. The direct spotting method[20] can reduce the loss of phosphopeptides during the elution step. In this work, the obtained material after the enrichment is directly added into the matrix, without the elution process, for MALDI-TOF-MS analysis. As shown in Figure 5.8, the phosphopeptide signals by direct spotting have higher intensity and good S/N ratio. The mono- and multi-phosphorylated peptides can be detected, which confirms the efficiency of the direct spotting method. However, the direct spotting method will have certain adverse effects on the instrument during the ionization process, so the direct spotting method is not considered. In this work, the results of direct spotting method (Figure 5.8) are compared with the elution spotting method (Figure 5.7(c)). The signal intensity of the phosphopeptides is similar to each other, indicating that the enrichment method based on $Fe_3O_4@SiO_2@$Glymo–IDA–Fe^{3+} has a high recovery rate, and is more flexible in the operation.

(c) *Evaluation of the enrichment capability of $Fe_3O_4@SiO_2@$Glymo–IDA–Fe^{3+} microsphere for phosphopeptide in peptide mixture*

Taking 50% ACN–1% ACOOH as the loading buffer, 10 min as the enrichment time, and 15 min as the elution time, the enrichment selectivity of $Fe_3O_4@SiO_2@$Glymo–IDA–Fe^{3+} microsphere for phosphopeptide in the complex peptide mixture is investigated.

(i) Selective enrichment of phosphopeptide in casein protein digest

In order to further investigate the selective enrichment ability of $Fe_3O_4@$ $SiO_2@$Glymo–IDA–Fe^{3+} microsphere for phosphopeptide, the enzymatic hydrolysate of the complex commercial casein (including three phosphoproteins: α-casein-S1-Casein, α-casein-S2-Casein, and β-casein) is selected. Figures 5.9(a) and 5.9(b) show the MS spectra of direct analysis of casein digest and the eluent after enrichment with $Fe_3O_4@SiO_2@$Glymo–IDA–Fe^{3+} microsphere, respectively. By comparison, it can be seen that a large number of phosphopeptides are efficiently enriched by the microsphere. A total of 16 phosphorylated peptides are detected in Figure 5.9(b), including nine monophosphorylated peptides and seven multi-phosphorylated peptides. These phosphorylated peptides can be uniquely identified by the SWISS-PROT database. Details of their amino acid sequences and possible phosphorylation sites are listed in Table 5.3. This result indicates that the $Fe_3O_4@SiO_2@$ Glymo–IDA–Fe^{3+} microsphere can efficiently and selectively enrich the phosphopeptide in mixture of protein digest.

(ii) Enrichment of phosphopeptide in the enzymatic hydrolysate mixture of phosphorylated and non-phosphorylated protein

To further confirm the enrichment selectivity of $Fe_3O_4@SiO_2@$Glymo–IDA–Fe^{3+} microsphere for phosphopeptide, peptide mixture of two phosphorylated proteins (ovalbumin and β-casein) and three non-phosphorylated proteins (MYO, Cyc, and BSA) is selected. The addition of non-phosphorylated protein digest is used to increase the complexity of the sample and then investigate the selective

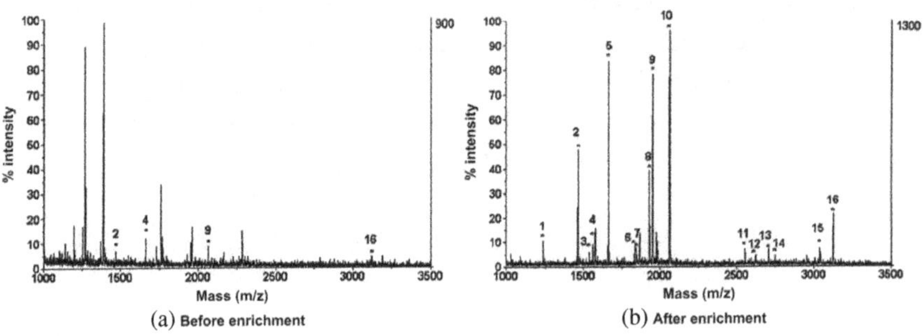

Figure 5.9: MS spectra of 2×10^{-7} mol·L^{-1} casein digest before and after enrichment with $Fe_3O_4@SiO_2@$Glymo–IDA–Fe^{3+} microsphere.

Table 5.3: Details of phosphopeptides identified from casein.

No.	AA	Phosphorylated Peptide Sequence	$[M+H]^+$
1	α-S2/138–147	TVDME[pS]TEVF	1237.4874
2	α-S2/138–149	TVDME[pS]TEVFTK	1466.5083
3	α-S2/126–137	EQL[pS]T[pS]EENSKK	1539.6237
4	α-S1/104–119	VPQLEIVPN[pS]AEER	1660.7974
5	α-S1/104–119	YKVPQLEIVPN[pS]AEER	1832.8342
6	α-S1/43–58	DIG[pS]ESTEDQAMEDIK	1847.7074
7	α-S1/43–58	DIG[pS]E[pS]TEDQAMEDIK	1927.4933
8	α-S1/104–119	YKVPQLEIVPN[pS]AEER	1951.9280
9	β/33–48	FQ[pS]EEQQQTEDELQDK	2061.7893
10	β/33–52	FQ[pS]EEQQQTEDELQDKIHPF	2555.9255
11	α-S2/2–21	NTMEHV[pS] [pS] [pS]EESII[pS]QETYK	2618.9685
12	α-S1/99–120	LRLKKYKVPQLEIVPN[pS]AEERL	2703.6973
13	α-S2/2–22	NTMEHV[pS] [pS] [pS]EESII[pS]QETYKQ	2746.7896
14	α-S2/62–85	NANEEEYSIG[pS][pS][pS]EE[pS]AEVATEEVK	3007.6777
15	α-S2/62–85	NANEEEY[pS]IG[pS][pS][pS]EE[pS]AEVATEEVK	3087.7534
16	β/1–25	RELEELNVPGEIVE[pS]L[pS][pS][pS]EESITR	3121.7006

enrichment ability of the microsphere for phosphopeptides. Figures 5.10(a) and 5.10(b) show the MS spectra of direct analysis of peptide mixture and eluent after enrichment with the microsphere, respectively. It can be observed from Figure 5.10(a) that a large number of non-phosphorylated peptides dominated the spectrum before enrichment, and their high intensity inhibits the detection of phosphopeptides. There are only weak phosphopeptide peaks at m/z 2088 and m/z 3121. However, the phosphopeptide signals in Figure 5.10(b) are significantly enhanced after enrichment. This result indicates that the microsphere has excellent selective enrichment ability for phosphorylated peptides, and can efficiently enrich phosphopeptides from complex peptide mixtures.

(d) *Evaluation on the enrichment ability of $Fe_3O_4@SiO_2@Glymo–IDA–Fe^{3+}$ microsphere for phosphopeptide in human serum*

Blood is a commonly used specimen in the clinic. Research on the composition of blood has been extensive, and researchers continue to hope to find

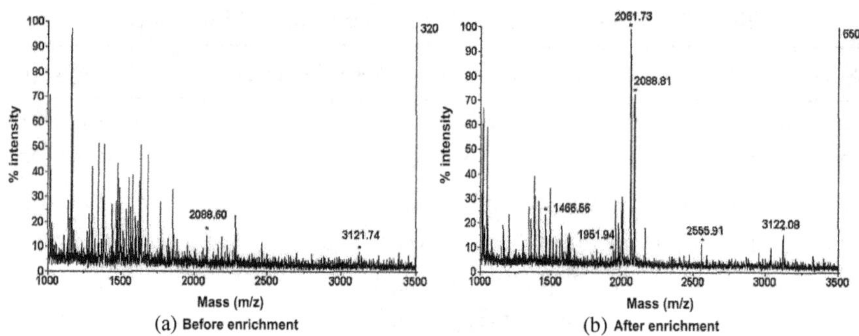

Figure 5.10: MALDI-TOF-MS spectra of the mixture of five protein digests before and after enrichment with Fe_3O_4@SiO_2@Glymo–IDA–Fe^{3+} microsphere.

biomarkers in blood so as to achieve a rapid and effective diagnosis of diseases. In this work, C8 modified hydrophobic magnetic silica microsphere and Fe_3O_4@SiO_2@Glymo–IDA–Fe^{3+} microsphere are used for enrichment of peptides in human serum, respectively. As shown in Figure 5.11, it is obvious that Fe_3O_4@SiO_2@Glymo–IDA–Fe^{3+} microsphere can selectively enrich the phosphopeptide in serum. Figures 5.11(b)–5.11(d) show the phosphopeptides enriched by Fe_3O_4@SiO_2@Glymo–IDA–Fe^{3+} microsphere from three human serum samples. The results show that the material has good enrichment reproducibility of phosphopeptides in human serum.

Tandem mass spectrometry is used to analyze the structure of enriched phosphopeptides (Figure 5.12). These phosphopeptides can be uniquely identified in NCBI database, and the detailed information of amino acid sequences are listed in Table 5.4.

(e) *Evaluation of the enrichment capability of Fe_3O_4@SiO_2@Glymo–IDA–Fe^{3+} microsphere for phosphopeptides in rat liver digest*

As shown in Figure 5.13, two routes are used to enrich the phosphopeptides in the rat liver digest. The first route is to enrich the total phosphopeptides in the rat liver digest and then they are identified by nano-LC-LTQ-MS. By manual confirmation of the spectra, 66 phosphopeptides and 91 different phosphorylation sites are identified by MS/MS, of which 24 are confirmed by MS/MS/MS. Figure 5.14 shows the results where 69% is mono-phosphorylated peptides, 23% is bi-phosphorylated

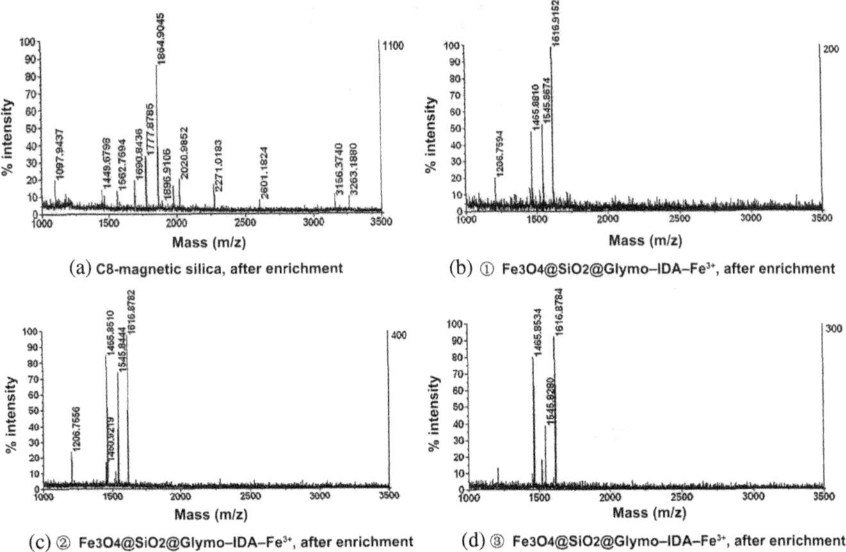

(a) C8-magnetic silica, after enrichment

(b) ① Fe3O4@SiO2@Glymo–IDA–Fe³⁺, after enrichment

(c) ② Fe3O4@SiO2@Glymo–IDA–Fe³⁺, after enrichment

(d) ③ Fe3O4@SiO2@Glymo–IDA–Fe³⁺, after enrichment

Figure 5.11: MS spectrum of peptide enriched by C8-modified hydrophobic magnetic silica from human serum and MS spectra of phosphopeptide enriched by Fe_3O_4@SiO_2@Glymo–IDA–Fe³⁺ from three healthy (①②③) human serum samples.

peptides, and 8% is tri-phosphorylated peptides among the identified 66 phosphopeptides. Moreover, 84% phosphorylation occurs on serine, 12% on threonine, and 4% on tyrosine. This result is basically consistent with the theoretical ratio.

The first analysis route is simple, but some low-abundance phosphopeptides are barely enriched due to the complexity of the system. When using the material to directly enrich phosphopeptide in the whole rat liver digest, a large number of non-phosphorylated peptides will cause severe interference. The second route can reduce the complexity of the system and increase the selective enrichment of phosphopeptides. As shown in Figure 5.13, the second route firstly uses reversed-phase chromatography to obtain 29 fractions, and then the material is used to, respectively, enrich phosphopeptide in each fraction. The enriched peptides are finally identified by MALDI-TOF-MS/MS. The library search results show a total of 107 possible phosphorylated peptides have been identified.

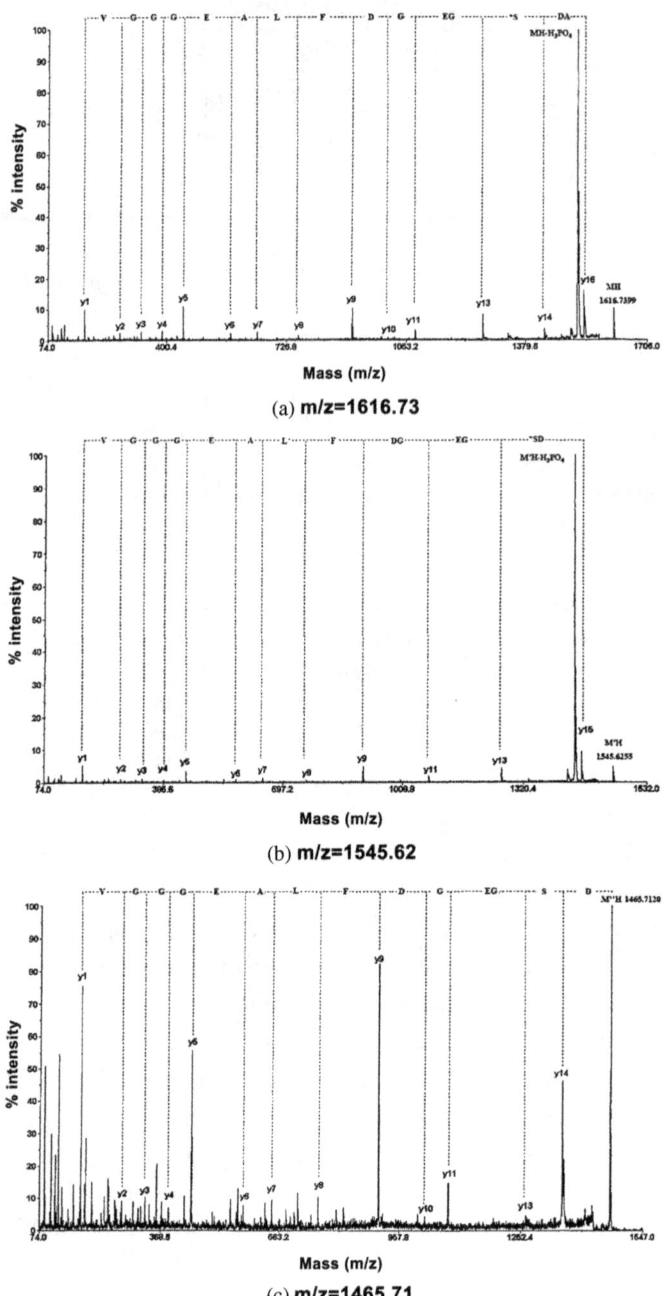

(a) m/z=1616.73

(b) m/z=1545.62

(c) m/z=1465.71

Figure 5.12: Tandem MS spectra of different parent ions after enrichment by $Fe_3O_4@SiO_2@Glymo$–IDA–Fe^{3+}.

Table 5.4: Detailed information of phosphopeptides identified from human serum.

AA	Peak	Sequence
1–16	1616.99	ADS*GEGDFLAEGGGVR (Fibrinopeptide A)
2–16	1545.94	DS*GEGDFLAEGGGVR
2–16	1465.96	DSGEGDFLAEGGGVR
5–16	1206.83	EGDFLAEGGGVR

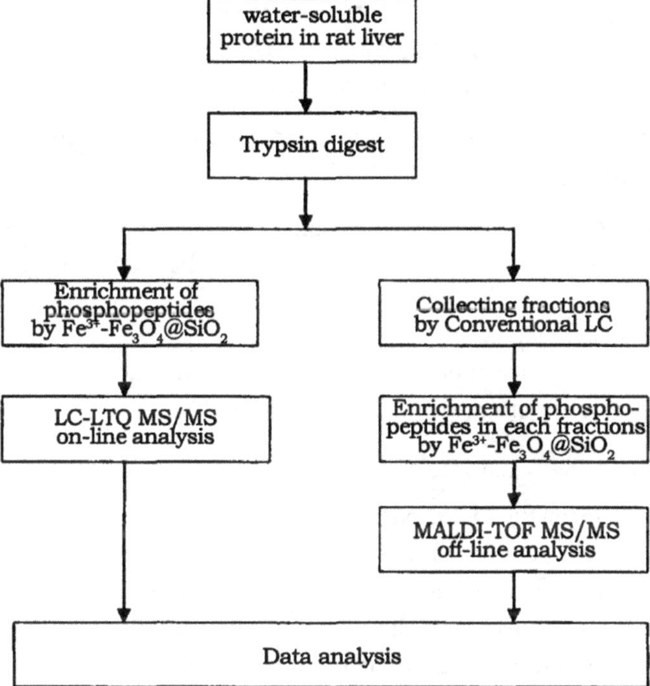

Figure 5.13: Flow chart of the enrichment procedure of Fe_3O_4@SiO_2@Glymo–IDA–Fe^{3+} for the enrichment of phosphopeptides in rat liver digest.

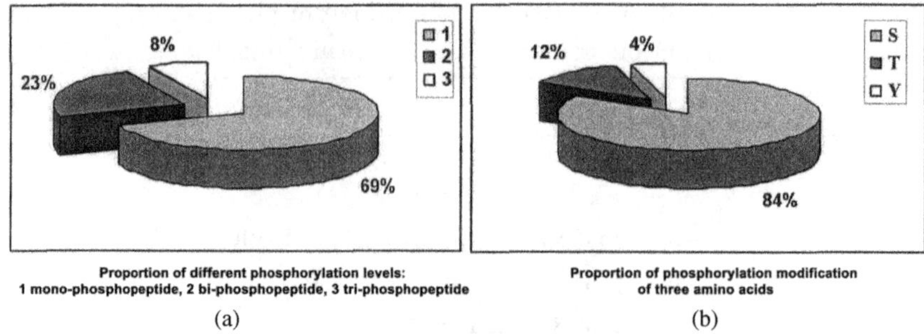

Figure 5.14: Types of the identified phosphorylated peptides.

5.2.2.2. *IDA-modified amino-magnetic microsphere for immobilization of different metal ions (M^{n+}-magnetic microsphere)*

(1) Synthesis and part characterization of M^{n+}-magnetic microsphere

The synthesis of Fe^{3+}-magnetic microsphere is taken as an example to introduce the synthesis of M^{n+}-magnetic microsphere (Figure 5.15).

(a) *Synthesis of amino-magnetic microsphere*
Amino-magnetic microsphere is synthesized by hydrothermal method.[21] The detailed synthesis steps are described in Section 2.3.

(b) *Synthesis of carboxyl-magnetic microsphere*
0.3 g amino-magnetic microsphere is ultrasonically dispersed in a mixture of 60 mL anhydrous toluene and 10 mL pyridine, and then 10 mL adipic acid chloride is added for 4-h reaction. After the reaction is finished, excess adipic acid chloride is removed, and the material is washed 6 times with anhydrous toluene. The resulting solid product is then dispersed in 40 mL anhydrous toluene, and 10 mL IDA is added for further reaction for 4 h. After thoroughly washing with anhydrous toluene, it is dried in vacuum for later use.

(c) *Synthesis of M^{n+}-magnetic microsphere*
The abovementioned product is dispersed in the aqueous solution of 20 mL 0.2 mol·L^{-1} $FeCl_3$, $Ga(NO_3)_3$, $Al(NO_3)_3$, $ZrOCl_2$, $In(NO_3)_3$, $Ce(NO_3)_3$ for oscillation for 2 h. The excess metal ions are removed by washing with deionized water, and the final product is dried in vacuum at 60°C for further experiment.

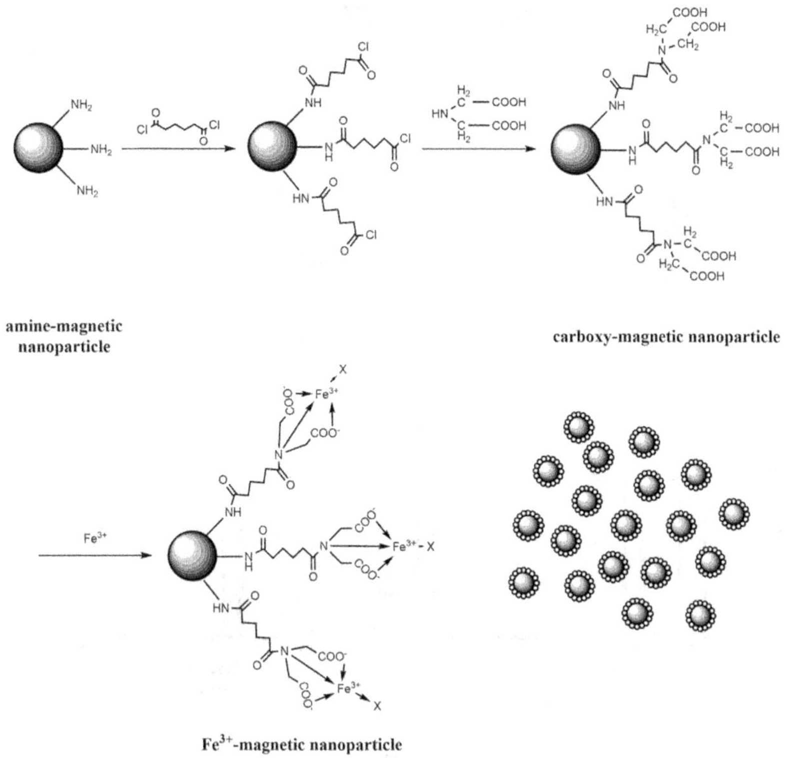

Figure 5.15: Schematic diagram of the synthesis of Fe^{3+}-magnetic microsphere.

As shown in Table 5.5, EDX analysis of different metal ions in M^{n+}-magnetic microsphere indicates that the metal ions are successfully immobilized on the surface of magnetic microsphere.

(d) *Comparison of the enrichment ability of different M^{n+}-magnetic microspheres for phosphopeptide*

Casein protein digest (including α-casein-S1-Casein, α-casein-S2-Casein, and β-casein) is selected for the parallel enrichment experiment. As shown in Figure 5.16, it shows that various M^{n+}-magnetic microspheres have good enrichment selectivity for phosphorylated peptide, among which Fe(III), Zr(III), Al(III), and Ga(III) show better performance than that of Ce(III) and In(III).

Table 5.5: EDX analysis of different metal ions in M^{n+}-magnetic microsphere.

Metal Ion	Percentage
Al(III)	1.51%
Ce(III)	2.02%
Ga(III)	1.88%
Zr(IV)	0.81%
In(III)	2.58%

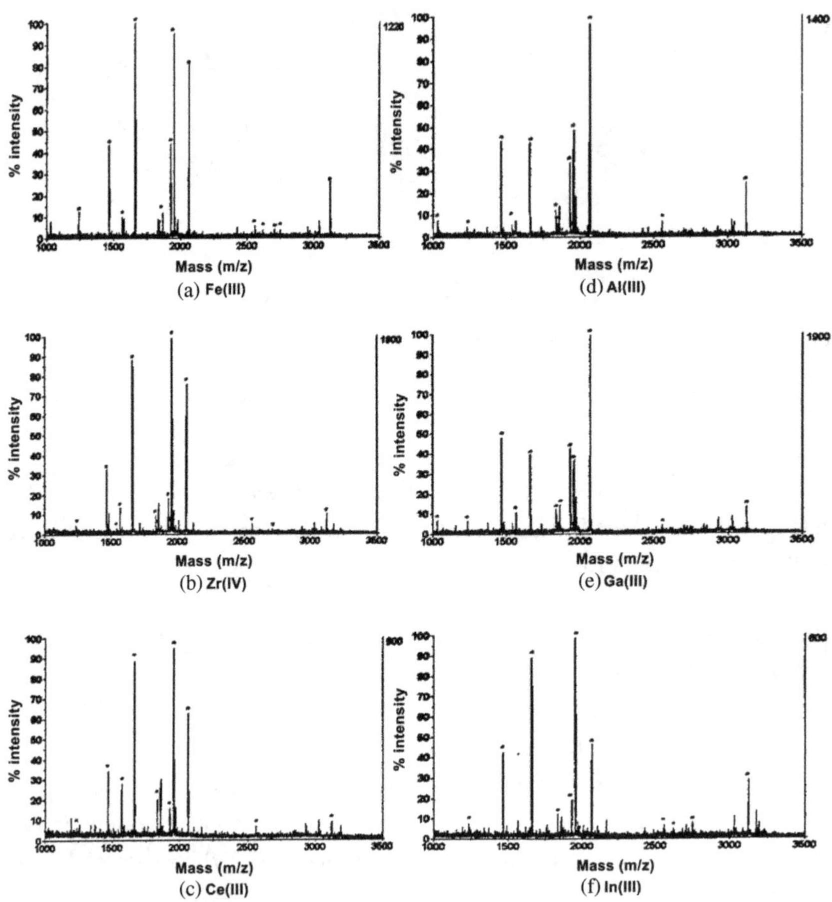

Figure 5.16: MS spectra of 2×10^{-7} mol·L^{-1} casein digest after enrichment with different M^{n+}-magnetic microspheres.

5.2.3. *Phosphate group-modified magnetic microsphere for immobilization of Zr^{4+} (Fe^3O^4@Phosph–Zr^{4+})*

5.2.3.1. *Preparation of Fe_3O_4@Phosph–Zr^{4+} microsphere*

The preparation of Fe_3O_4@Phosph–Zr^{4+} microsphere can be seen in Section 2.6 of Chapter 2.[13]

5.2.3.2. *Enrichment application of Fe_3O_4@Phosph–Zr^{4+} microsphere in phosphoproteomics*

(1) Evaluation of the enrichment capability of Fe_3O_4@Phosph–Zr^{4+} microsphere for standard phosphoprotein digest

First, the effect of different pH values on the enrichment selectivity of Fe_3O_4@Phosph–Zr^{4+} microsphere is investigated. Peptide mixture of β-casein (0.04 pmol) and BSA (1 pmol) is chosen. As shown in Figure 5.17(a), when the peptide mixture is directly analyzed by MS, only a large number of non-phosphorylated peptides are detected, and no signal of any phosphorylated peptide is detected. As shown in Figure 5.17(b), when 1.0% TFA is used as the loading buffer, a great many phosphopeptides and non-phosphopeptides are observed in the MS spectrum and the five peaks are identified to be derived from BSA digest (amino acid sequence is shown in the figure). All of these five non-phosphopeptides contain at least one acidic amino acid residue, and the acidic amino acid is considered to be the main cause of non-specific adsorption in IMAC technology. As shown in Figure 5.17(c), when 1.5% TFA is used as the loading buffer, all of the peaks are phosphopeptides, and there is no non-phosphopeptide in the spectrum. As shown in Figure 5.17(d), when a higher concentration of TFA is used as the loading buffer, the enrichment ability of Fe_3O_4@Phosph–Zr^{4+} is significantly reduced. Although no non-phosphorylated peptide is detected, the peak intensity of phosphopeptide is also significantly reduced. This is because when the pH is lowered to 0.68, the phosphate group is protonated, and thus the enrichment efficiency becomes poorer. In summary, 50% ACN+1.5% TFA is selected as the optimal loading buffer.

After the optimal enrichment condition is determined, the enrichment sensitivity of this new method is examined. Figure 5.18 shows the phosphopeptides enriched with different concentrations of β-casein digest by using Fe_3O_4@Phosph–Zr^{4+} microsphere. It can be seen from the figure that the

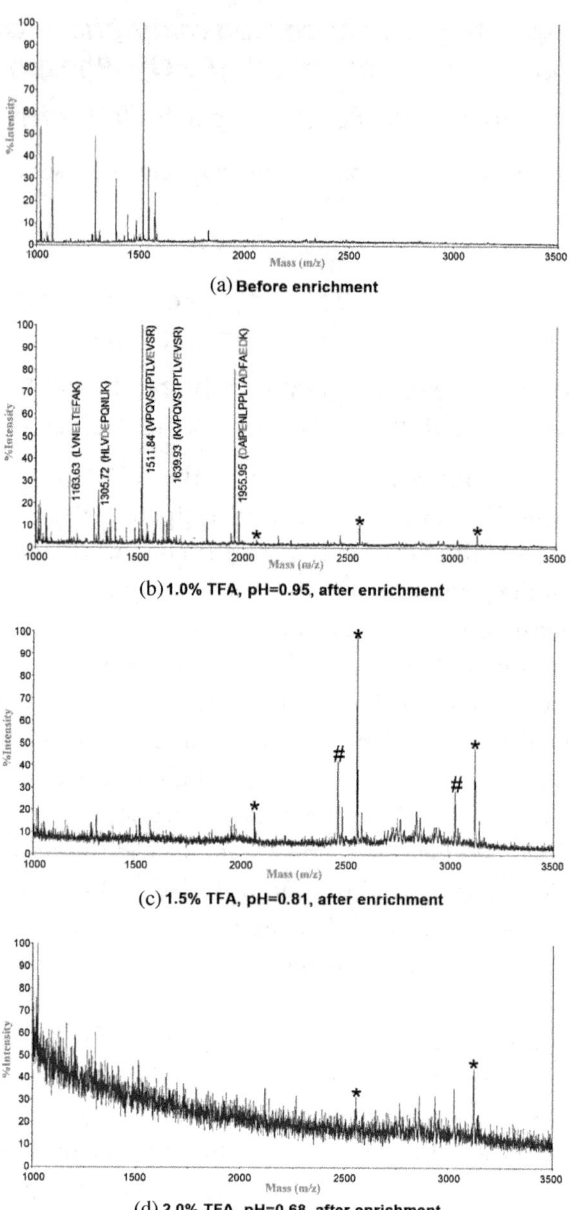

Figure 5.17: Effect of different TFA concentrations on the enrichment selectivity of Fe_3O_4@Phosph–Zr^{4+} microsphere. * represents phosphopeptide, # represents dephosphorylated fragment.

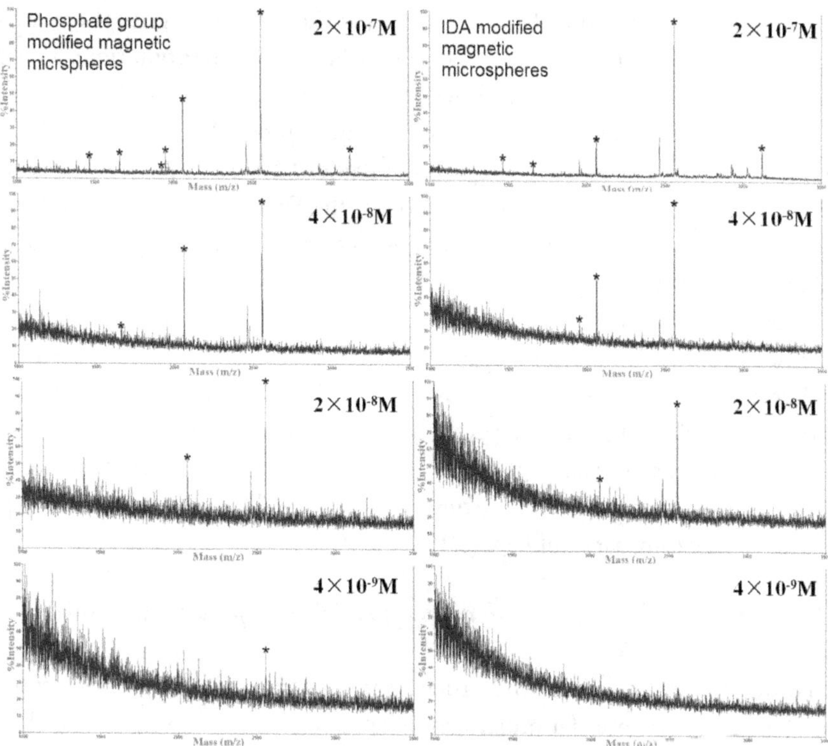

Figure 5.18: MS spectra of phosphopeptide enriched with different concentrations of β-casein digest by two different materials.

phosphopeptide can still be identified when the concentration of β-casein digest is as low as 4×10^{-9} mol·L^{-1}, indicating that the enrichment sensitivity of this new method is very high. In previous work, IDA-modified magnetic microsphere is used to immobilize metal ion for enrichment of phosphopeptide.[22] In this work, both materials are immobilized with Zr^{4+} ion and the enrichment sensitivity of the two methods is compared. As shown in Figure 5.18, when the concentration is as low as 4×10^{-9} mol·L^{-1}, the IDA-modified magnetic microsphere cannot enrich any phosphopeptide, which illustrates the advantage of phosphate group-modified magnetic microsphere for phosphopeptide enrichment.

RPLC is used to quantitatively analyze the recovery of phosphopeptide and non-phosphopeptide after enrichment with Fe$_3$O$_4$@Phosph–Zr^{4+} microsphere.

First, two standard non-phosphorylated peptides and two phosphorylated peptides are mixed for RPLC analysis. According to the retention time, a total of 4 fractions are collected and labeled as 1#, 2#, 3#, 4#. Then, the fractions are identified by MALDI-TOF-MS, respectively. Figure 5.19 shows the RPLC spectrum and the tandem mass spectra of the four fractions. Fractions 2# and 4# are identified as phosphopeptides. Next, the Fe_3O_4@Phosph–Zr^{4+} microsphere is applied to the peptide mixture. After enrichment, the supernatant and eluent are quantitatively analyzed by RPLC. As shown in Figure 5.19(a), the eluate contains two phosphorylated peptides and the non-phosphorylated peptide 3#. The tandem mass spectrum of the 3# shows that it contains four acidic amino acid residues. The recovery rates of the two phosphorylated peptides are 86.3% and 93.4%, respectively. The recovery of the 3# is about 0.6%. There is no phosphorylated peptide in the supernatant after enrichment compared to the original peptide mixture. These results illustrate the outstanding enrichment performance of Fe_3O_4@Phosph–Zr^{4+}.

(b) *Evaluation of the enrichment ability of Fe_3O_4@Phosph–Zr^{4+} microsphere for non-fat milk digestion*

In order to further evaluate the enrichment performance of Fe_3O_4@Phosph–Zr^{4+} microsphere, non-fat milk digest is selected as the complex sample. The main components of skim milk are the two most common phosphoproteins, α-casein and β-casein. Therefore, skim milk is commonly used in the evaluation of phosphopeptide enrichment. Figure 5.20 shows the MS and MS/MS spectra of phosphorylated peptides enriched by Fe_3O_4@Phosph–Zr^{4+} from skim milk digest. All results indicate that Fe_3O_4@Phosph–Zr^{4+} microsphere has excellent enrichment capacity for phosphopeptide. The detailed information of the phosphopeptides enriched from skim milk is listed in Table 5.6.

(c) *Evaluation of the enrichment performance of Fe_3O_4@Phosph–Zr^{4+} microsphere for mouse brain digest*

Mouse brain is commonly used as the practical sample, which is rich in biological information. In order to further evaluate the enrichment performance of Fe_3O_4@Phosph–Zr^{4+} microsphere, the material is used to enrich phosphopeptides from mouse brain digest. In total, 192 phosphorylation sites are identified, one of which occurs at the tyrosine residue (0.52%), 27 at the threonine residue (14.06%), and 164 at the serine residue (85.42%). Figure 5.21 shows the MS/MS and MS/MS/MS spectrum of the identified NLLEDDpSDEEEDFFLR phosphopeptide.

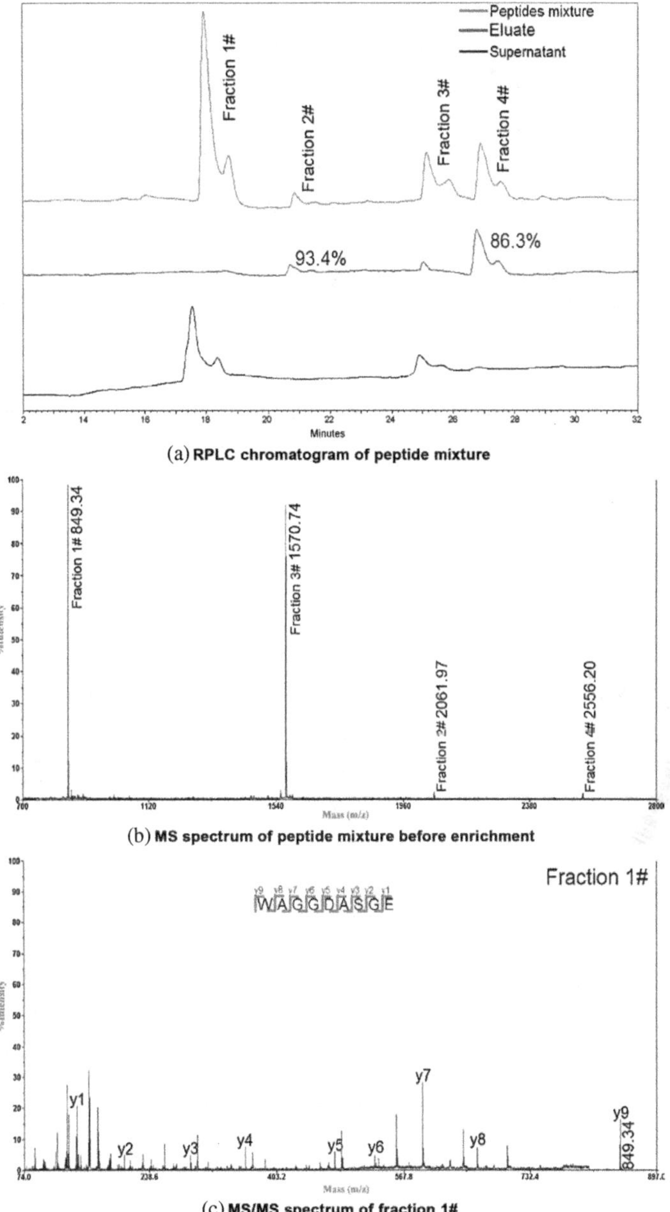

(a) **RPLC chromatogram of peptide mixture**

(b) **MS spectrum of peptide mixture before enrichment**

(c) **MS/MS spectrum of fraction 1#**

Figure 5.19: RPLC and MALDI-TOF-MS spectra of peptide mixture, MS/MS spectrum of fraction.

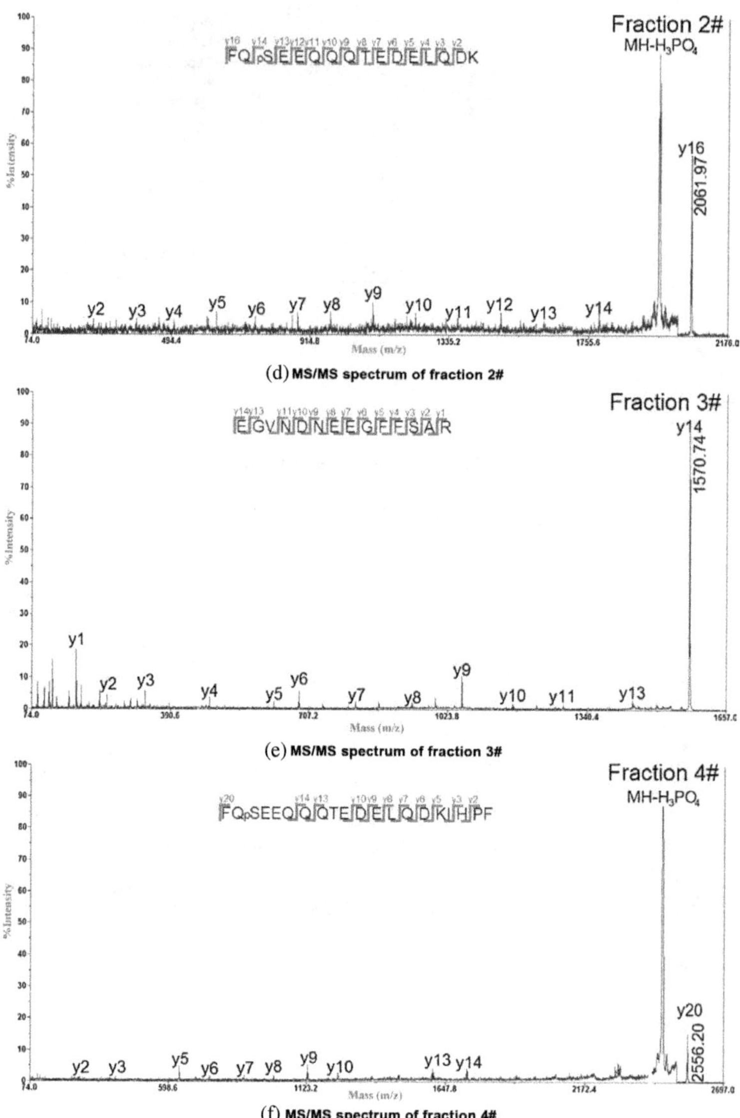

Figure 5.19: (*Continued*)

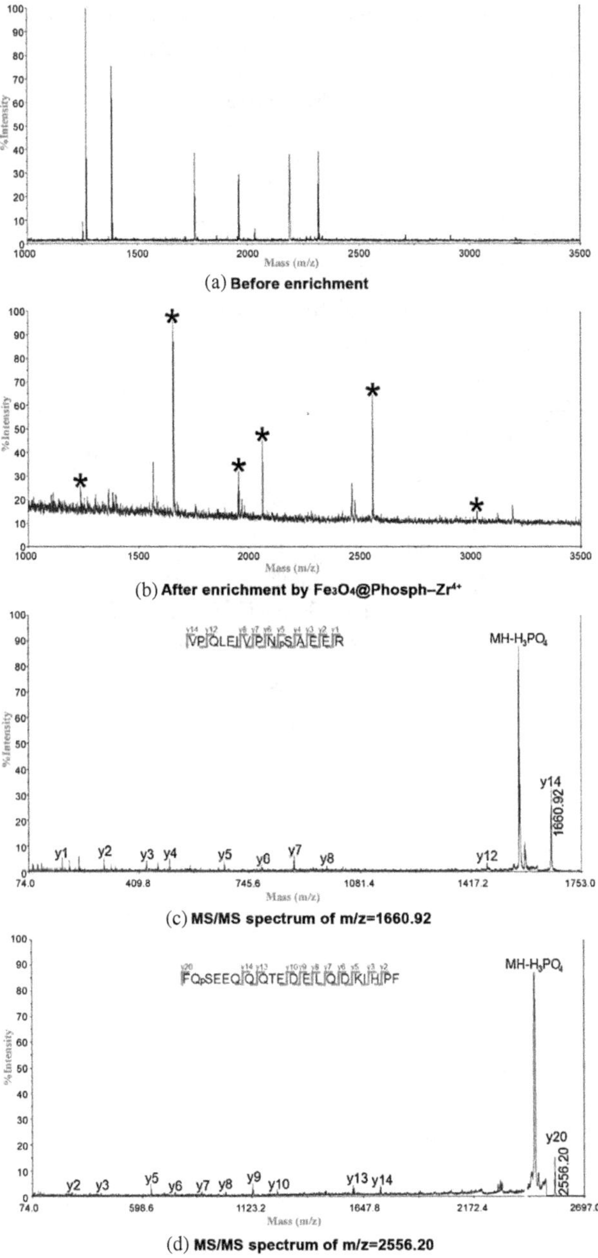

Figure 5.20: MS and MS/MS spectra of non-fat milk digest before and after enrichment with Fe$_3$O$_4$@Phosph–Zr^{4+}. * represents phosphopeptide, # represents dephosphorylated fragment.

Table 5.6: Detailed information of phosphopeptides enriched with Fe_3O_4@Phosph–Zr^{4+} microsphere from skim milk.

No.	Peptide Sequence	$(M + H)^+$
1	TVDMES#TEVFTK(α-S2-(153–164))	1466.73
2	VPQLEIVPNS#AEER(α-S1-(121–134))	1660.92
3	DIGS#ES#TEDQAMEDIK(α-S1-(58–73))	1927.83
4	YKVPQLEIVPNS#AEER(α-S1-(119–134))	1952.09
5	FQS#EEQQQTEDELQDK(β-c-(33–48))	2061.96
6	FQS#EEQQQTEDELQDKIHPF(β-c-(33–52))	2556.20
7	RELEELNVPGEIVES#LS#S#S#EESITR(β-c-(14–40))	3122.40

Note: #: Phosphorylation site.

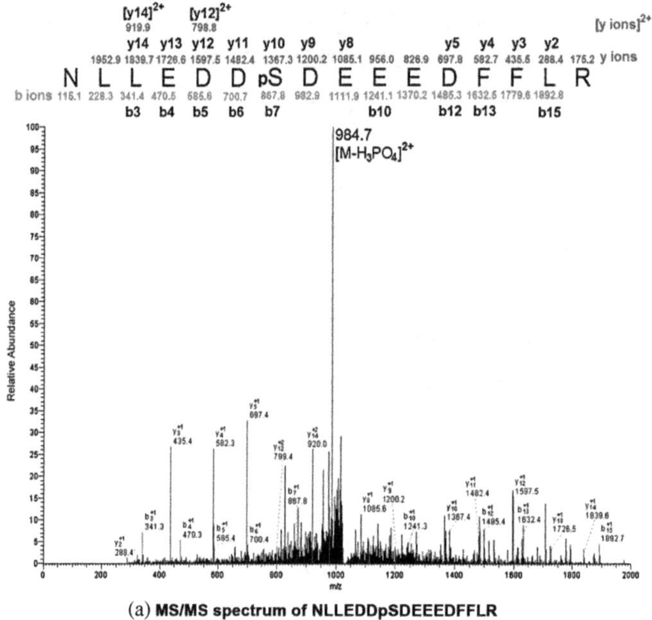

(a) **MS/MS spectrum of NLLEDDpSDEEEDFFLR**

Figure 5.21: MS/MS and MS/MS/MS spectra of the phosphopeptide.

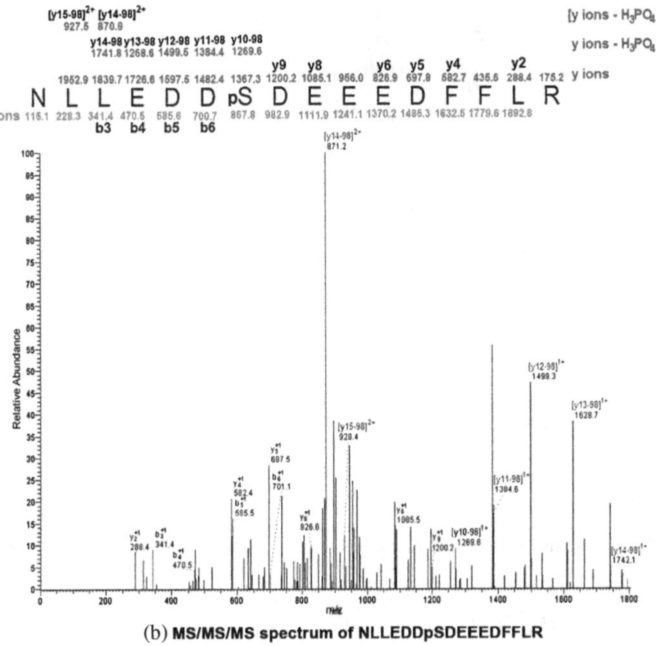

(b) **MS/MS/MS spectrum of NLLEDDpSDEEEDFFLR**

Figure 5.21: *(Continued)*

5.2.4. *Different methods for phosphate group modification — Phosphate group-modified magnetic microsphere for immobilization of metal ion*

5.2.4.1. *Polyethylene glycol as a linking agent to immobilize phosphate group (Fe₃O₄@SiO₂@ PEG–Ti⁴⁺)*

(1) Synthesis and part characterization of Fe₃O₄@SiO₂@PEG–Ti⁴⁺ microsphere

The synthesis of $Fe_3O_4@SiO_2@PEG–Ti^{4+}$ microsphere is shown in Figure 5.22.[23] Firstly, $Fe_3O_4@SiO_2$ microsphere is obtained by coating the silica shell on the surface of Fe_3O_4 microsphere via sol–gel method. Secondly, $Fe_3O_4@SiO_2$ microsphere reacts with 3-aminopropyltriethoxysilane (APTEOS) to obtain the $Fe_3O_4@SiO_2–NH_2$ microsphere, which then reacts with 2-bromoisobutyryl bromide to obtain $Fe_3O_4@SiO_2–Br$ microsphere. Afterwards, the PEG chain is attached to the surface of $Fe_3O_4@SiO_2–Br$ microsphere by atom transfer radical

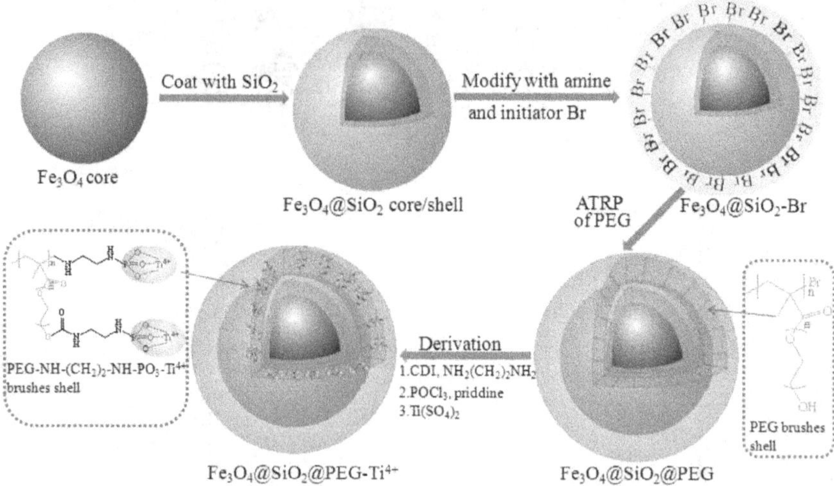

Figure 5.22: Schematic diagram of the synthesis of Fe$_3$O$_4$@SiO$_2$@PEG–Ti^{4+} microsphere.

polymerization. The hydroxyl group on PEG chains is converted to amino group by reaction with 1,1-carbonyldiimidazole and ethylenediamine. Finally, the product reacts with POCl$_3$ for immobilization of metal ion.

For comparison, two materials are synthesized by using the same procedure as above: Fe$_3$O$_4$@SiO$_2$–Ti^{4+} microsphere without PEG layer and Fe$_3$O$_4$@SiO$_2$@PEG–3h-Ti^{4+} microsphere with thin PEG layer. It can be seen from Figure 5.23 that the PEG layer of Fe$_3$O$_4$@SiO$_2$@PEG–Ti^{4+} microsphere is relatively thick (about 20 nm).

(2) The effect of the thickness of PEG on the enrichment performance of Fe$_3$O$_4$@SiO$_2$@PEG–Ti^{4+} microsphere for phosphopeptide

The enrichment protocol of Fe$_3$O$_4$@SiO$_2$@PEG–Ti^{4+} microsphere is shown in Figure 5.24.

Inductively coupled plasma atomic emission spectroscopy (ICP-AES) method[23] is used to determine that the content of titanium ion in Fe$_3$O$_4$@SiO$_2$@PEG–Ti^{4+} microsphere, Fe$_3$O$_4$@SiO$_2$@PEG–3h-Ti^{4+} microsphere, and Fe$_3$O$_4$@SiO$_2$–Ti^{4+} microsphere. And the corresponding content is 32.98 μg·mg^{-1}, 21.80 μg·mg^{-1}, and 9.97 μg·mg^{-1}, respectively. It indicates that when the PEG layer is thicker, the Ti^{4+} binding ability of the material is stronger.

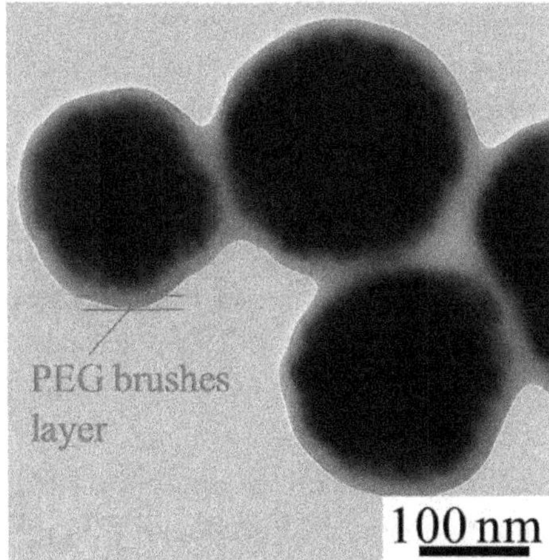

Figure 5.23: TEM image of $Fe_3O_4@SiO_2@PEG–Ti^{4+}$ microsphere.

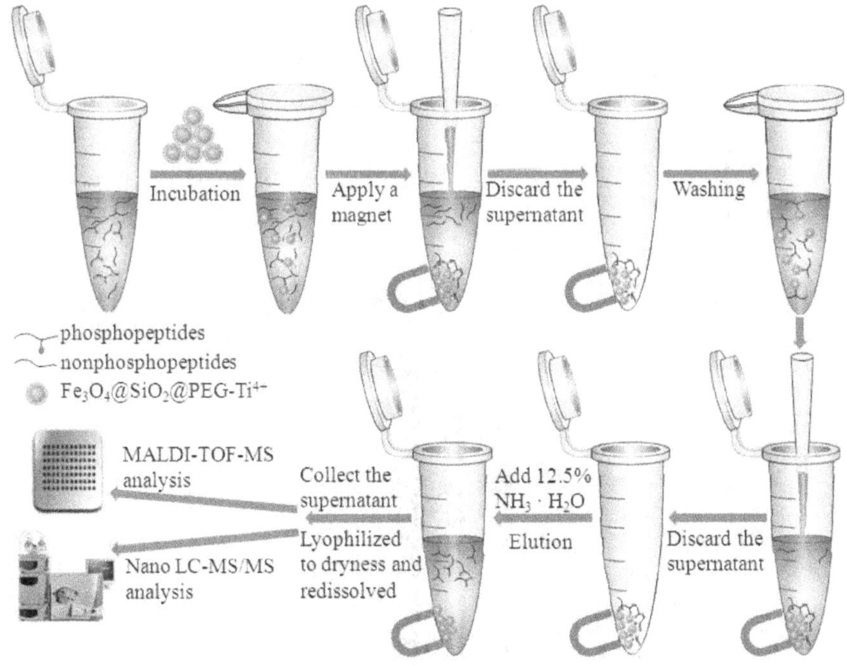

Figure 5.24: Flow chart of the enrichment process of $Fe_3O_4@SiO_2@PEG–Ti^{4+}$ microsphere.

These three materials are applied to the enrichment of phosphopeptides with various amounts of β-casein digest (0.1 to ~20 pmol, 200 μL). As shown in Figure 5.25, after enrichment with $Fe_3O_4@SiO_2@PEG–Ti^{4+}$ microsphere, the peak intensities of the two phosphopeptides, β1 and β2, are significantly higher than those enriched by $Fe_3O_4@SiO_2–Ti^{4+}$ microsphere. Especially when the β-casein content is 5 to 20 pmol, the intensities increase 2.5–3.8 times. As the thickness of the PEG layer decreases, the intensities of the enriched phosphopeptides also decrease. This may be because the phospho-peptides enrichment performance is directly proportional to the content of Ti^{4+}, indicating that the magnetic microspheres, coated with thicker PEG layer, have a stronger ability to enrich the phosphorylated peptides.

Also, four standard phosphorylated peptides (NVPL[pY]K, HLADL[pS]K, VNQIGTL[pS]E[pS]IK, VNQIG[pT]LSESIK) are used to test the enrichment recovery of the three different materials by means of stable isotope dimethyl labeling quantitative method.[23] The sample is the mixture of one kind of standard phosphorylated peptide (1 pmol) and BSA digest (1 pmol). The measured recovery rates are shown in Table 5.7. In addition, $Fe_3O_4@SiO_2@PEG-Ti^{4+}$ microsphere exhibits better performance than $Fe_3O_4@SiO_2-Ti^{4+}$ microsphere, both in terms of the limit of detection (Figure 5.26) and the

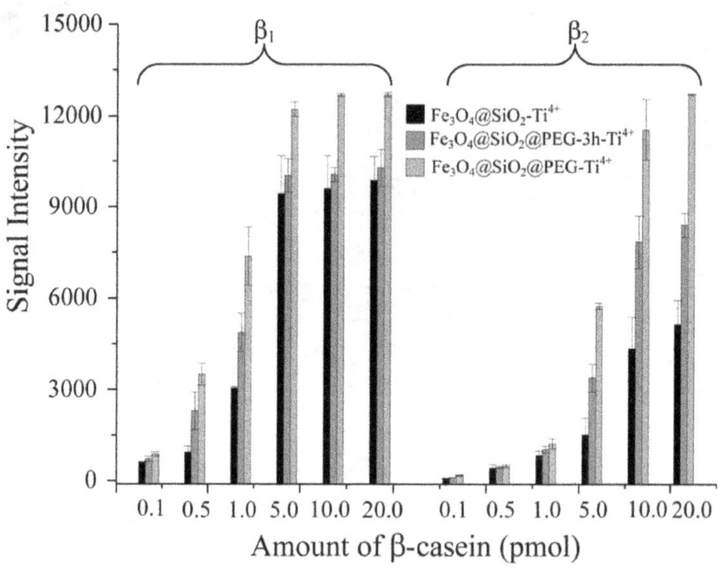

Figure 5.25: Comparison of the binding capacity of different materials for phosphopeptides (β-casein: 0.1–20 pmol, 200 μL).

Table 5.7: Enrichment recovery rates of four standard phosphorylated peptides (1 pmol) by using different materials.

Standard Phosphopeptide	Recovery ± S.D. (%, n = 3)		
	Fe_3O_4@ SiO_2–Ti^{4+}	Fe_3O_4@SiO_2@ PEG-3h–Ti^{4+}	Fe_3O_4@SiO_2@ PEG–Ti^{4+}
NVPL[pY]K	24.4 ± 0.7	51.9 ± 0.4	72.6 ± 1.0
HLADL[pS]K	39.2 ± 1.2	67.6 ± 1.4	79.8 ± 3.5
VNQIGTL[pS]E[pS]IK	67.4 ± 1.9	69.8 ± 2.7	71.8 ± 1.0
VNQIG[pT]LESEIK	45.2 ± 2.4	50.7 ± 1.3	80.8 ± 1.9

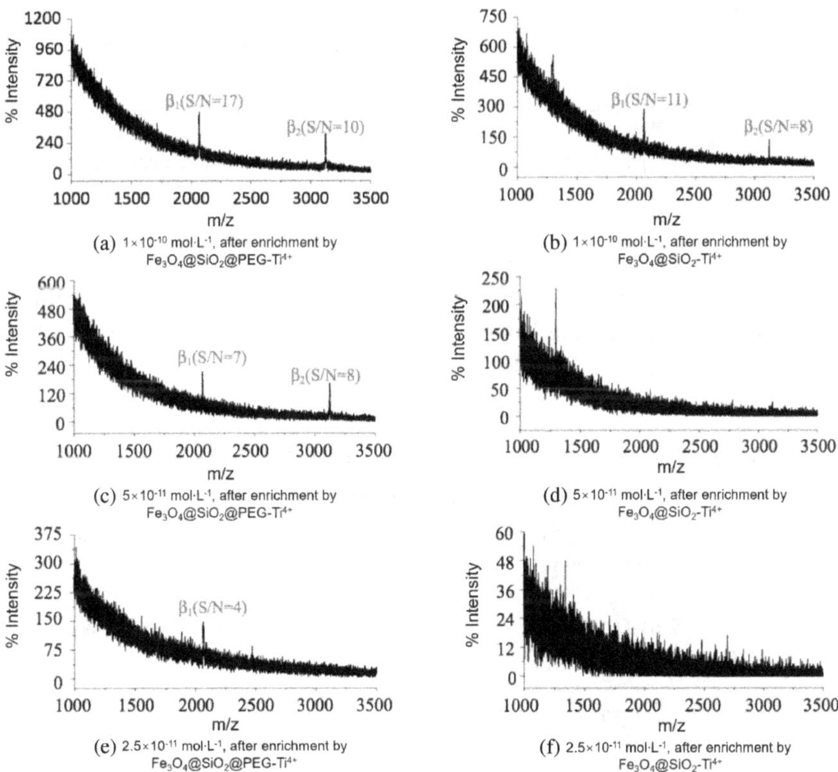

Figure 5.26: MS spectra of different concentrations of β-casein digest after enrichment with different materials.

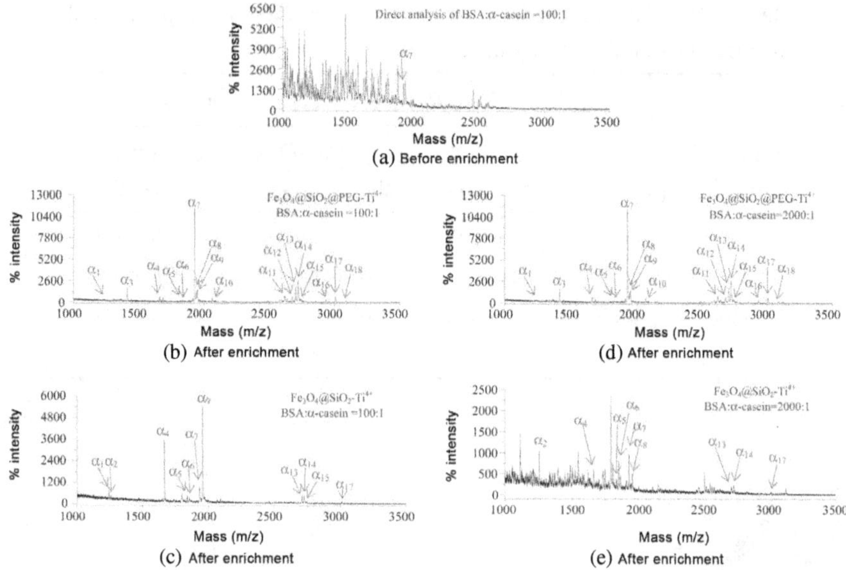

Figure 5.27: MS spectra of phosphopeptide enriched with different materials from the peptide mixture (200 μL) of BSA digest and α-casein (4 pmol) digest at different molar ratios.

enrichment selectivity (Figure 5.27). These results indicate that the coating of thick PEG layer for immobilization of Ti^{4+} can enhance the enrichment ability of the material for the phosphopeptide.

5.2.4.2. *Polyethylene glycol methacrylate phosphoric acid modified magnetic microsphere for immobilization of metal ion (Fe$_3$O$_4$@PMAA@ PEGMP-Ti^{4+})*

(1) Synthesis and part characterization of Fe$_3$O$_4$@PMAA@ PEGMP-Ti^{4+} microsphere

The schematic diagram of the synthesis of Fe_3O_4@PMAA@PEGMP–Ti^{4+} microsphere with Fe_3O_4 as core, polymethacrylic acid (PMAA) as the intermediate layer and polyethylene glycol methacrylate phosphoric acid (PEGMP) as the functional shell is shown in Figure 5.28.[24] In this case, the second route[24] is first used to directly coat the PEGMP shell on the surface of

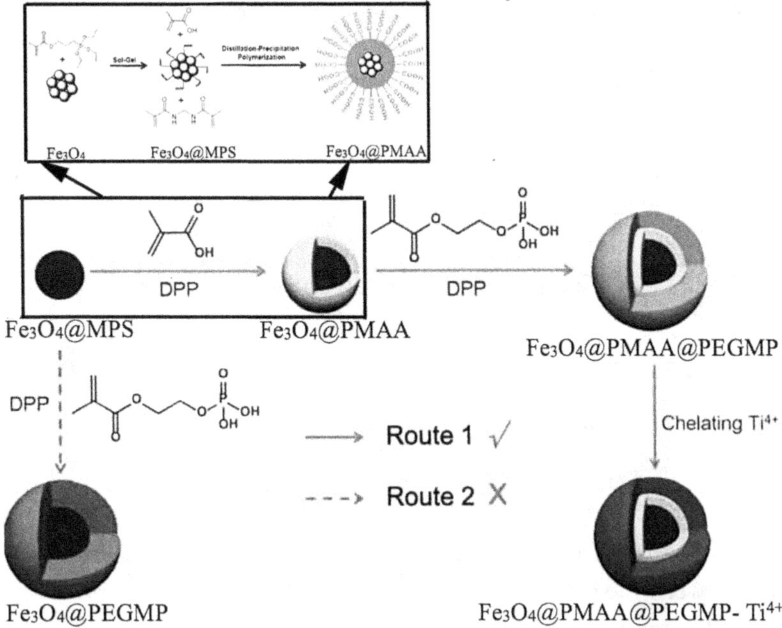

Figure 5.28: Schematic diagram of the synthesis of Fe_3O_4@PMAA@PEGMP–Ti^{4+} microsphere.

Fe_3O_4 microsphere, but unfortunately, the system becomes very unstable shortly after the reaction starts. It may be due to the fact that the monomer EGMP has a strong complexation with Fe in the Fe_3O_4 microsphere. When heated to a certain temperature, a large number of EGMP monomers will be attached to the surface of the Fe_3O_4 microsphere, which will seriously affect the dispersion stability of Fe_3O_4 microsphere. Thus, an intermediate layer needs to be carefully chosen and added between the Fe_3O_4 microsphere and the PEGMP shell. The selected intermediate layer needs to enable the Fe_3O_4 microsphere to disperse well in the acetonitrile, at the same time, there must be a strong interaction between the intermediate layer and the EGMP monomer to ensure the coating of the PEGMP shell. Based on these, PMAA is selected as the intermediate layer (see the first route).

The synthetic process is described as follows: Firstly, the synthesis of Fe_3O_4@PMAA composites can be found in Section 2.5 of Chapter 2.

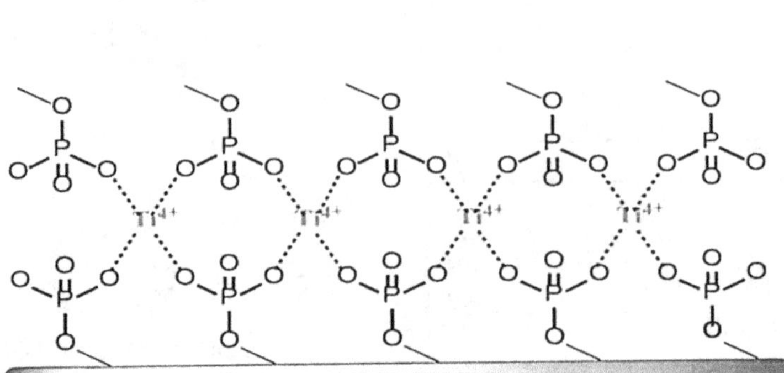

Figure 5.29: The interaction of Ti^{4+} with a phosphate group in the PEGMP shell or the phosphopeptide.

Secondly, Fe_3O_4@PMAA@PEGMP is prepared by hydrogen bonding interaction between the carboxyl group in PMAA shell and the phosphate group in EGMP monomer. Finally, the metal ion is immobilized.

Ti^{4+} is immobilized based on the metal phosphate chemistry between the Ti^{4+} and phosphate group in the PEGMP shell. As shown in Figure 5.29, each EGMP unit is considered a bidentate ligand that shares two oxygen atoms with two different metal atoms, respectively. Therefore, each metal ion does not interact with only one phosphate group. Likewise, each phosphate group does not interact with only one metal ion. This extremely strong combination provides a very stable Ti^{4+}-phosphate surface, allowing the immobilized Ti^{4+} on the surface to further interact with other phosphate groups, such as phosphate groups in the phosphopeptide, for selectively enriching the phosphopeptide. The loading amount of Ti^{4+} is determined by atomic absorption spectrophotometry, which shows that the loading amount of Ti^{4+} can be as high as 4.1%.

(2) Enrichment application of Fe_3O_4@PMAA@PEGMP–Ti^{4+} microsphere in phosphoproteomics

The peptide mixture of BSA and β-casein is selected to investigate the enrichment selectivity of the material. As shown in Figure 5.30, since the

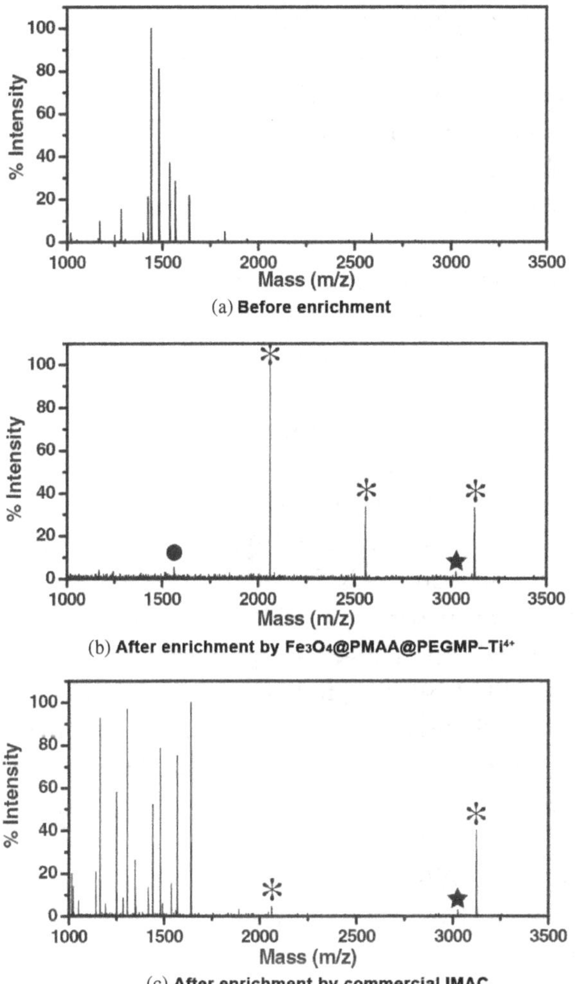

Figure 5.30: MS spectra of the peptide mixture of BSA and β-casein at molar ratio of 1/500 after enrichment with different materials. * and ★ represent phosphopeptide and dephosphorylated fragment, respectively.

mixture contains a large number of non-phosphorylated peptides, no signal of any phosphopeptide is detected with MALDI-TOF-MS before enrichment and the spectrum is occupied by the non-phosphopeptide peaks (Figure 5.30(a)). However, after enrichment with Fe_3O_4@PMAA@

PEGMP–Ti^{4+} microsphere, the signals of all phosphorylated peptides are detected very clearly and the background is very clean (Figure 5.30(b)). In contrast, when using commercial IMAC material to enrich the same peptide mixture, the spectrum is occupied by a large number of non-phosphorylated peptide peaks, and the signal of the phosphopeptide is too weak to be detected (Figure 5.30(c)).

In order to comprehensively evaluate the enrichment ability of Fe$_3$O$_4$@PMAA@PEGMP–Ti^{4+} microsphere for phosphopeptides, the enrichment capacity, sensitivity and recovery rate of the material are further investigated. The enrichment capacity of Fe$_3$O$_4$@PMAA@PEGMP–Ti^{4+} microsphere is about 75 mg·g^{-1}. When the concentration of β-casein digest is as low as 5 × 10^{-10} mol·L^{-1} (100 μL), Fe$_3$O$_4$@PMAA@PEGMP–Ti^{4+} microsphere can still enrich three phosphorylated peptides. The enrichment recovery of the material is determined by the ^{18}O labeling method. The principle is as follows: a certain amount of standard phosphorylated peptides (PSADGQHAGGLVK) are divided into two equal parts firstly. Then the ^{16}O of the C-terminus in one part is transferred to ^{18}O with self-made solid-phase trypsin to cause an increase in mass of 4 Da; the other part is used for enrichment experiment. The enriched peptides and the ^{18}O-labeled peptide are mixed and analyzed by mass spectrometry. As shown in Figure 5.31, the enrichment recovery rate can be estimated by comparing the relative intensities of different oxygen isotope peaks.[25] By calculation, the enrichment recovery of Fe$_3$O$_4$@PMAA@PEGMP–Ti^{4+} microsphere for phosphopeptide is as high as 87%. These results show that Fe$_3$O$_4$@PMAA@PEGMP–Ti^{4+} microsphere as an ideal magnetic IMAC material has excellent enrichment selectivity, high sensitivity, and recovery rate in phosphopeptide enrichment.

Fe$_3$O$_4$@PMAA@PEGMP–Ti^{4+} microsphere is further used for the enrichment of phosphopeptide from complex samples, such as milk protein digest and human serum. The results show that ten and four phosphorylated peptides can be successfully enriched from milk digest and human serum by using Fe$_3$O$_4$@PMAA@PEGMP–Ti^{4+} microsphere. This indicates that the material has outstanding enrichment capability for low-abundance phosphopeptides in practical samples. The detailed information of the phosphopeptide enriched from the milk digest is shown in Table 5.8.

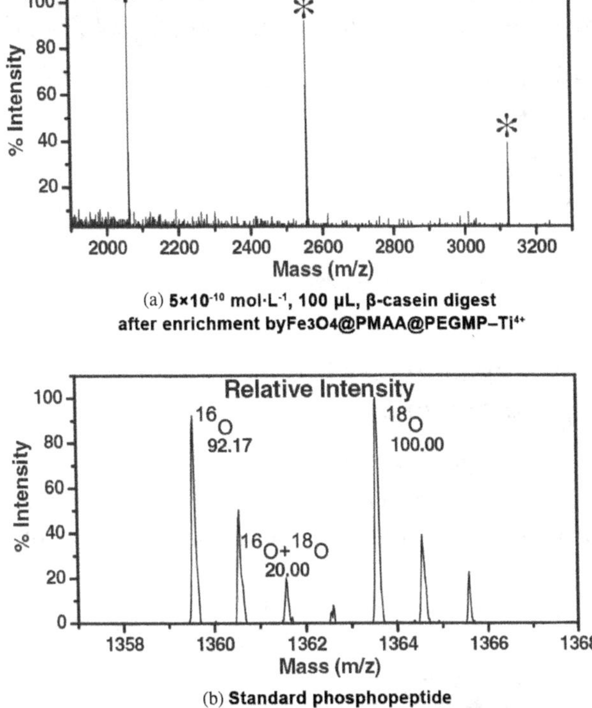

Figure 5.31: Evaluation on the ability of low abundance enrichment and recovery rate.

Table 5.8: Details of phosphopeptide enriched with Fe_3O_4@PMAA@PEGMP–Ti^{4+} from milk digest.

No	MH⁺	Peptide Sequence	Phosphorylation Site
1	1446.68	TVDME[pS]TEVFTK	1
2	1660.85	VPQLEIVPN[pS]AEER	1
3	1832.77	YLGEYLIVPN[pS]AEER	1
4	1927.74	DIG[pS]E[pS]TEDQAMEDIK	2
5	1952.01	YKVPQLEIVPN[pS]AEER	1
6	2061.83	FQ[pS]EEQQQTEDELQDK	1
7	2618.94	NTMEHV[pS][pS][pS]EE[pS]IISQETVYK	4
8	2703.93	Q*MEAE[pS]I[pS][pS] [pS] EEIVPN[pS]VEAQK	5
9	2925.57	NANEEEYSIG[pS][pS][pS]EEAEVATEEVK	3
10	3122.27	RELEELNVPGEIVE[pS]L[pS][pS]EESITR	4

5.2.4.3. *Adenosine-modified magnetic microsphere for immobilization of Ti^{4+} (Fe_3O_4@ATP–Ti^{4+})*

(1) Synthesis and part characterization of Fe_3O_4@ATP–Ti^{4+} microsphere

The synthesis of Fe_3O_4@ATP–Ti^{4+} microsphere is shown in Figure 5.32.[26] First, the amino-magnetic microsphere is synthesized according to Ref. 21. Then, the ATP is modified on the surface of amino-magnetic microsphere by glutaraldehyde. Finally, metal ion is immobilized via the phosphate group on ATP.

(2) Enrichment application of Fe_3O_4@ATP–Ti^{4+} microsphere in phosphoproteomics

ATP with three phosphate groups provides more active binding sites to enhance the ability of the material. In addition, the hydrophilic purine and the five-carbon sugar of adenosine enhance the hydrophilicity of the material

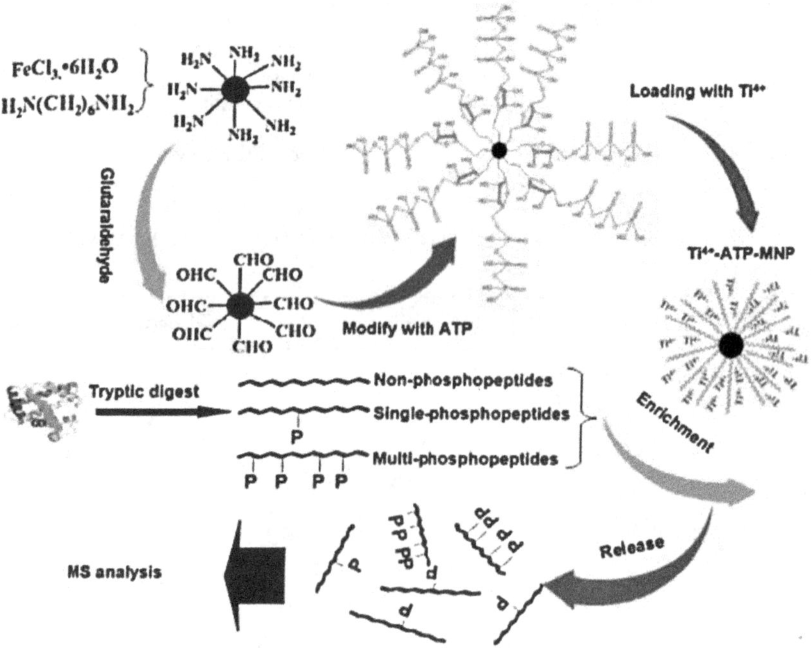

Figure 5.32: Schematic diagram of the synthesis of Fe_3O_4@ATP–Ti^{4+} microsphere.

to reduce non-specific adsorption. The peptide mixture of BSA digest and β-casein digest (4 pmol) is used to study the selective enrichment ability of Fe_3O_4@ATP-Ti^{4+} microsphere for phosphorylated peptides.[26] As shown in Figure 5.33, 7, 5, and 9 phosphopeptides can be detected after enrichment with Fe_3O_4@ATP–Ti^{4+} microsphere from peptide mixture of BSA and β-casein digest (4 pmol) at molar ratios of 1/500 (Figure 5.33(b)), 1/1,000

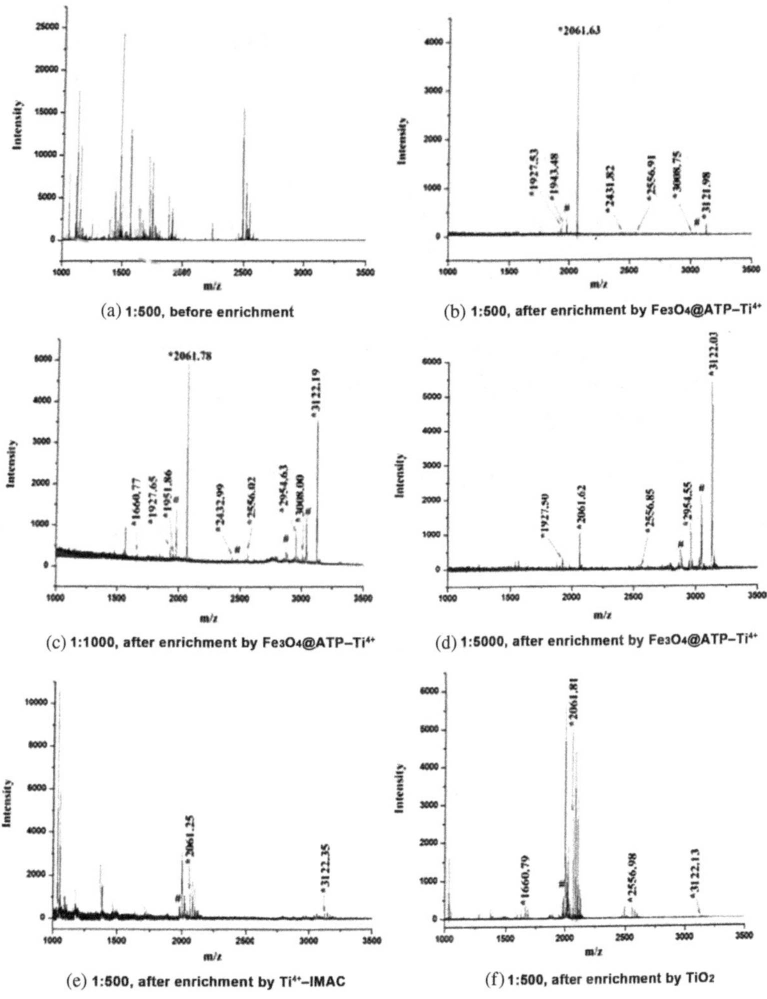

Figure 5.33: MS spectra of peptide mixture of BSA and β-casein (4 pmol) digest with different molar ratios before and after enrichment by different materials.

(Figure 5.33(c)), and 1/5,000 (Figure 5.33(d)), respectively. However, after enrichment with conventional Ti^{4+}-IMAC microsphere and TiO_2 from peptide mixture at molar ratio of 1/500, only 2 (Figure 5.33(e)) and 4 (Figure 5.33(f)) phosphopeptides are detected. This result indicates that the Fe_3O_4@ATP–Ti^{4+} microsphere has a better selective enrichment ability for phosphopeptides. Moreover, compared with other phosphate-based IMAC materials, the Fe_3O_4@ATP–Ti^{4+} microsphere with ATP as linker exhibits quite high enrichment selectivity (1/5,000), which is reported for the first time.

In addition, the enrichment sensitivity of Fe_3O_4@ATP–Ti^{4+} microsphere is also examined. As shown in Figure 5.34(b), even when the concentration of β-casein digest is as low as 3×10^{-12} mol·L^{-1}, two phosphorylated peptides are still detected. Finally, rat liver mitochondrial digest is used to further evaluate the enrichment ability of Fe_3O_4@ATP–Ti^{4+} microsphere. A total of 406 phosphopeptides are detected, including 538 phosphorylation sites corresponding to 313 phosphorylated proteins.

5.2.5. *Polydopamine-modified magnetic microsphere for immobilization of metal ion of Ti⁴⁺ (Fe₃O₄@PDA–Ti⁴⁺)*

5.2.5.1. *Preparation of Fe₃O₄@PDA–Ti⁴⁺ microsphere*

The synthesis protocol of Fe_3O_4@PDA–Ti^{4+} microsphere is shown in Figure 5.35.

(1) Synthesis of Fe₃O₄ microsphere

The detailed synthesis process is described in Section 2.2 of Chapter 2.

(2) Synthesis of Fe₃O₄@PDA microsphere

The detailed synthesis process is described in Section 2.5 of Chapter 2.

(3) Synthesis of Fe₃O₄@PDA–Ti⁴⁺ microsphere

10 mg Fe_3O_4@PDA microsphere is dispersed in 20 mL 0.1 mol·L^{-1} titanium sulfate aqueous solution, stirred at room temperature for 2 h, washed with water and ethanol, and finally dried for obtaining Fe_3O_4@PDA–Ti^{4+} microsphere.

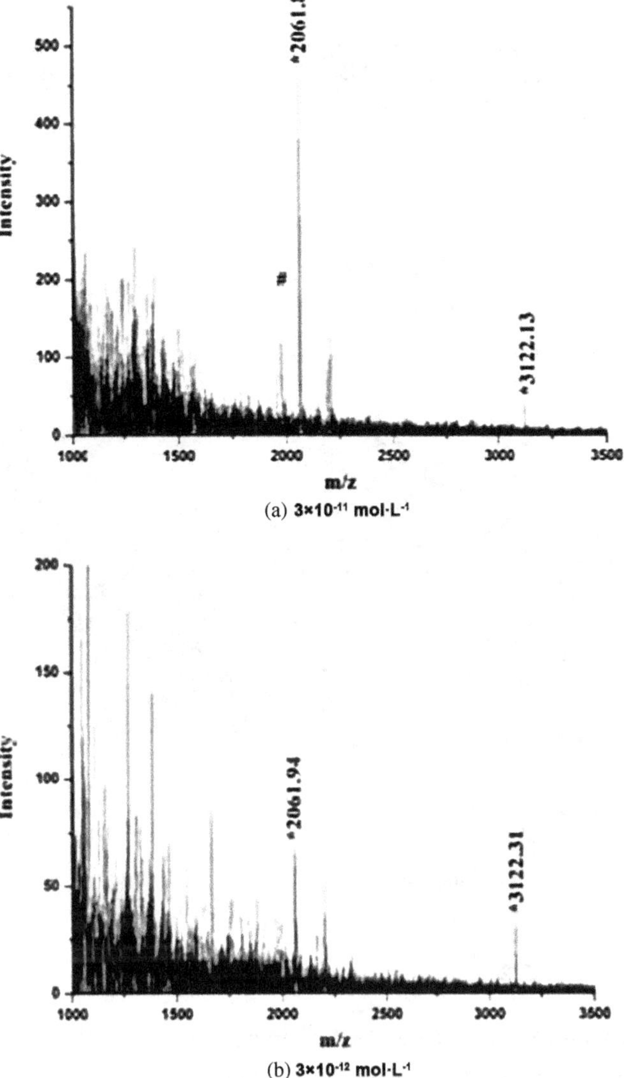

Figure 5.34: MS spectra of different concentrations of β-casein digests after enrichment with Fe$_3$O$_4$@ATP–Ti^{4+} microsphere. *: phosphorylated peptide, #: dephosphorylated fragment.

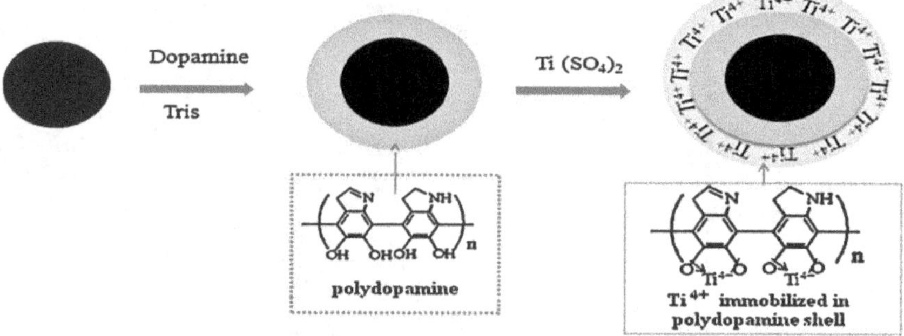

Figure 5.35: Schematic diagram of synthesis of Fe_3O_4@PDA–Ti^{4+} microsphere.

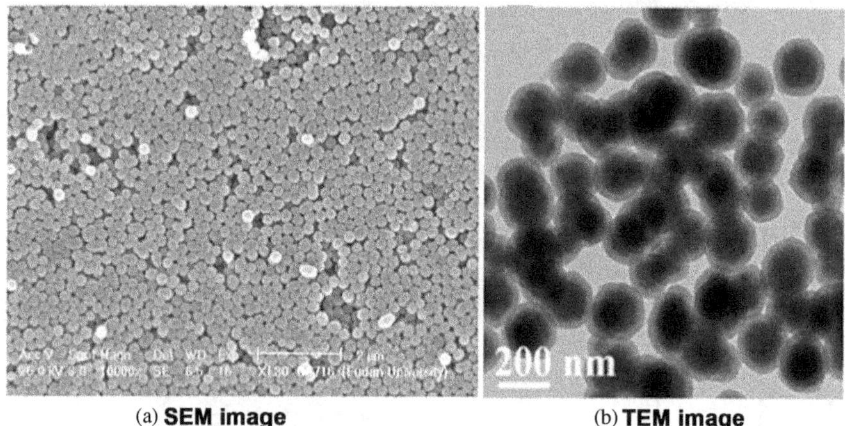

(a) **SEM image**   (b) **TEM image**

Figure 5.36: SEM image and TEM image of Fe_3O_4@PDA–Ti^{4+} microsphere.

5.2.5.2. *Characterization of Fe_3O_4@PDA–Ti^{4+} microsphere*

Figure 5.36 shows the SEM and TEM images of Fe_3O_4@PDA–Ti^{4+} microsphere. It can be seen from Figure 5.36(a) that the size of Fe_3O_4@PDA–Ti^{4+} microsphere is basically the same, and its diameter is about 210 nm. No significant agglomeration is observed. From Figure 5.36(b), it can be clearly observed that the Fe_3O_4@PDA–Ti^{4+} microsphere has a core–shell structure. The core is Fe_3O_4 microsphere, and the PDA layer around the Fe_3O_4 core is about 20 nm thick. The presence of Ti^{4+} is proved by the EDX analysis (Figure 5.37).

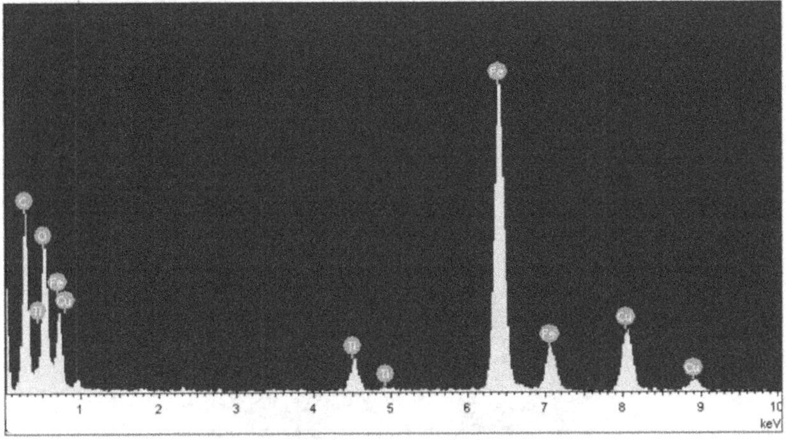

Figure 5.37: EDX analysis of Fe_3O_4@PDA–Ti^{4+} microsphere.

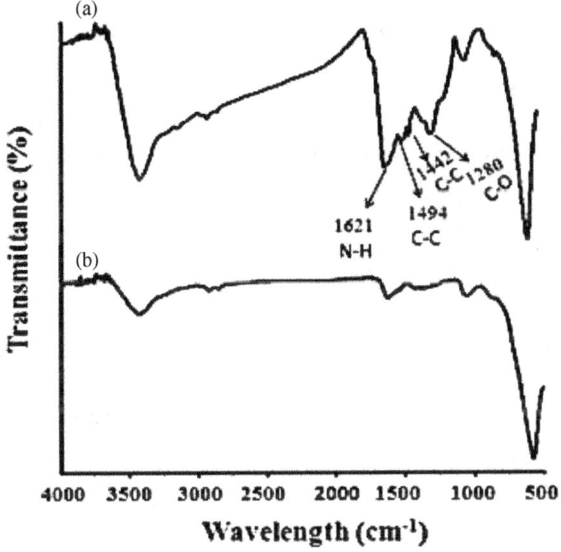

Figure 5.38: FTIR spectra of different materials (curve (a): Fe_3O_4@PDA–Ti^{4+} microsphere; curve (b): Fe_3O_4 microsphere).

Figure 5.38 shows the FTIR spectra of the Fe_3O_4@PDA–Ti^{4+} microsphere (Figure 5.38(a)) and the Fe_3O_4 microsphere (Figure 5.38(b)). Compared with Fe_3O_4, the FTIR spectrum of the Fe_3O_4@PDA–Ti^{4+} microsphere shows some

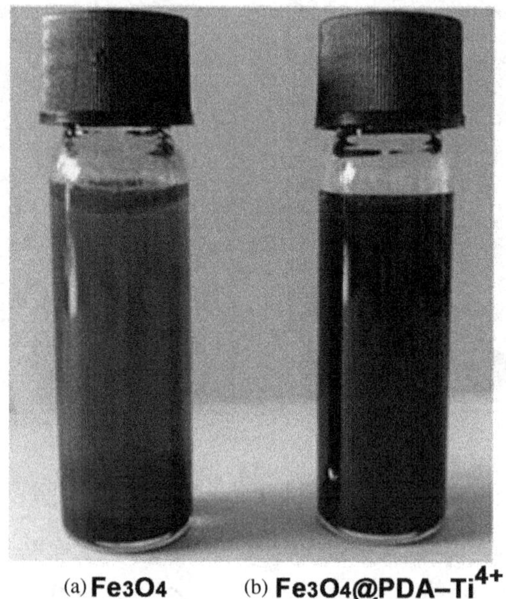

Figure 5.39: Photos of different materials after standing for 24 h.

new peaks: the absorption peak at 1,280 cm^{-1} is attributed to C-O vibration; 1,442 cm^{-1} and 1,494 cm^{-1} are attributed to C–C vibration; 1,621 cm^{-1} belongs to N–H stretching vibration. The appearance of these new peaks further confirms the successful synthesis of Fe_3O_4@PDA–Ti^{4+} microsphere.

Figure 5.39 shows the dispersity of Fe_3O_4 microsphere and Fe_3O_4@PDA–Ti^{4+} microsphere after standing for 24 h. After being coated with PDA, which has some hydrophilic groups such as amino group and hydroxyl group, the dispersion of Fe_3O_4@PDA–Ti^{4+} microsphere in water is still good after 24 h. However, the Fe_3O_4 microsphere has been precipitated after 24 h. This shows that the use of PSA as a linker endows the Fe_3O_4@PDA–Ti^{4+} microsphere with good water solubility, making it an ideal IMAC material for phosphoproteomics research.

5.2.5.3. *Enrichment application of Fe_3O_4@PDA–Ti^{4+} microsphere in phosphoproteomics*

The selection of dopamine as a novel chelating ligand to immobilize metal ion not only makes the synthesis process simple and easy, it also helps to enhance the hydrophilicity of the material and the binding ability to metal ion. By

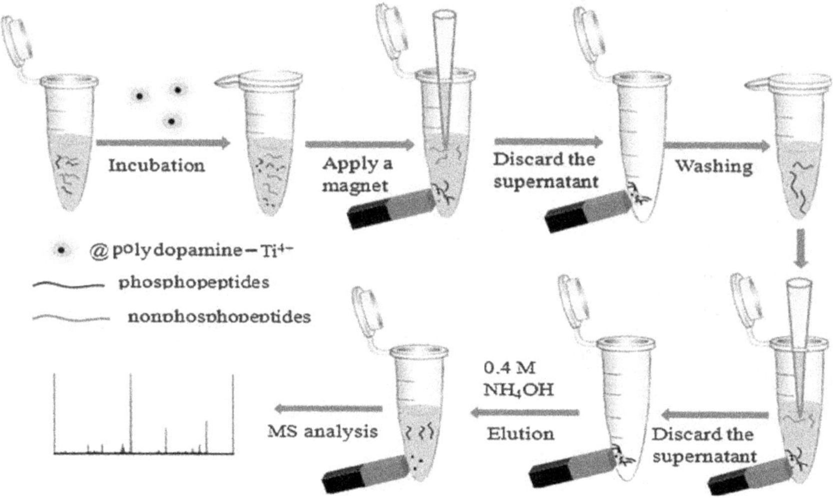

Figure 5.40: Flow chart of enrichment process of Fe_3O_4@PDA–Ti^{4+} microsphere for phosphopeptide.

self-polymerization of dopamine at room temperature and then immobilization of Ti^{4+} metal ion, the material has achieved great enrichment capacity in both standard phosphoprotein digest and mixed protein digest. The flow chart of the enrichment process is shown in Figure 5.40.

(1) Evaluation on the enrichment performance of Fe_3O_4@ PDA–Ti^{4+} microsphere for phosphopeptide in standard phosphoprotein digest

First, to test whether Fe_3O_4@PDA–Ti^{4+} microsphere has enrichment ability for phosphorylated peptides, β-casein (usually containing trace α-casein) is used as the standard phosphorylated protein to be digested for enrichment. Figure 5.41(a) is the MS spectrum of direct analysis of 1×10^{-6} mol·L^{-1} β-casein digest. Almost all the peaks are non-phosphorylated peptides and the signals of phosphorylated peptides are completely suppressed. After enrichment with Fe_3O_4@PDA–Ti^{4+} microsphere, the non-phosphopeptide peak disappears, and all the peaks are determined to be the phosphorylated peptides (marked with *, the dephosphorylated fragment peak is labeled with #, the detailed information of the detected phosphopeptide can be seen in Table 5.9).

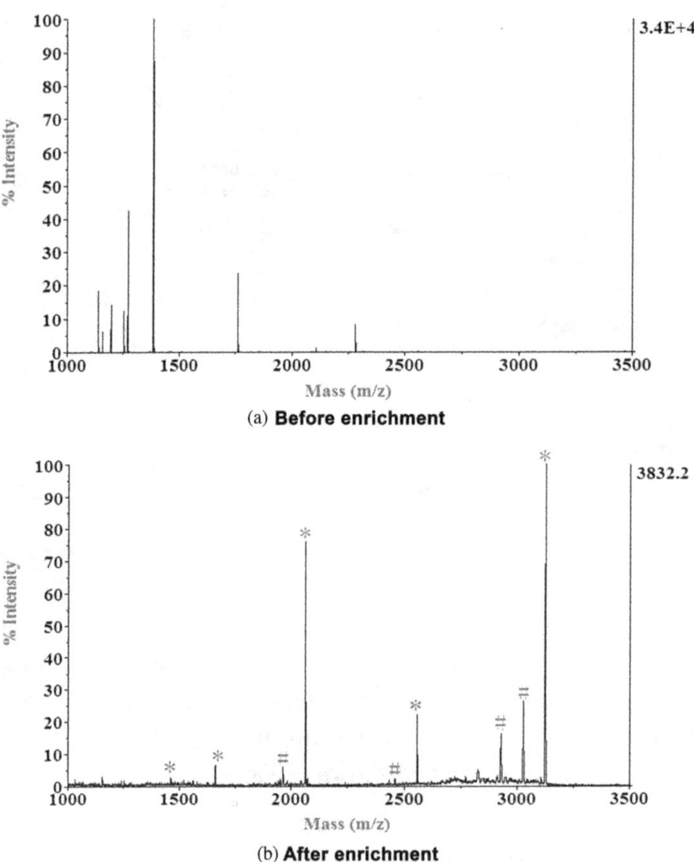

(a) **Before enrichment**

(b) **After enrichment**

Figure 5.41: MS spectra of 1×10^{-6} mol·L^{-1} β-casein digest before and after enrichment with Fe$_3$O$_4$@PDA–Ti^{4+} microsphere. *: phosphorylated peptide, #: dephosphorylated fragment.

Table 5.9: Detailed information of phosphopeptides detected from α-casein and β-casein digest.

m/z(Da)	Phospho-Sites	Peptide Sequence
1466.6	1	TVDMESTEVFTK
1660.7	1	VPQLEIVPNSAEER
2061.7	1	FQSEEQQQTEDELQDK
2556.2	1	FQSEEQQQTEDELQDKIHPF
3122.2	4	RELEELNVPGEIVESLSSSEESITR

It indicates that the newly designed Fe_3O_4@PDA–Ti^{4+} microsphere has outstanding enrichment ability for phosphorylated peptides.

To further investigate the enrichment ability of Fe_3O_4@PDA–Ti^{4+} microsphere, the material is applied to different concentrations of β-casein digests for phosphopeptide enrichment (Figure 5.42). As shown in Figure 5.42(b), when the concentration of β-casein digest is as low as 2×10^{-10} mol·L^{-1}, three phosphorylated peptides are still detected after enrichment with Fe_3O_4@PDA–Ti^{4+} microsphere. This indicates that the material has high enrichment sensitivity.

Then the enrichment selectivity (Figure 5.43) and repeatability of Fe_3O_4@PDA–Ti^{4+} microsphere (Figure 5.44) is verified. The peptide mixture of phosphoprotein (β-casein) and non-phosphoprotein (BSA) is selected to

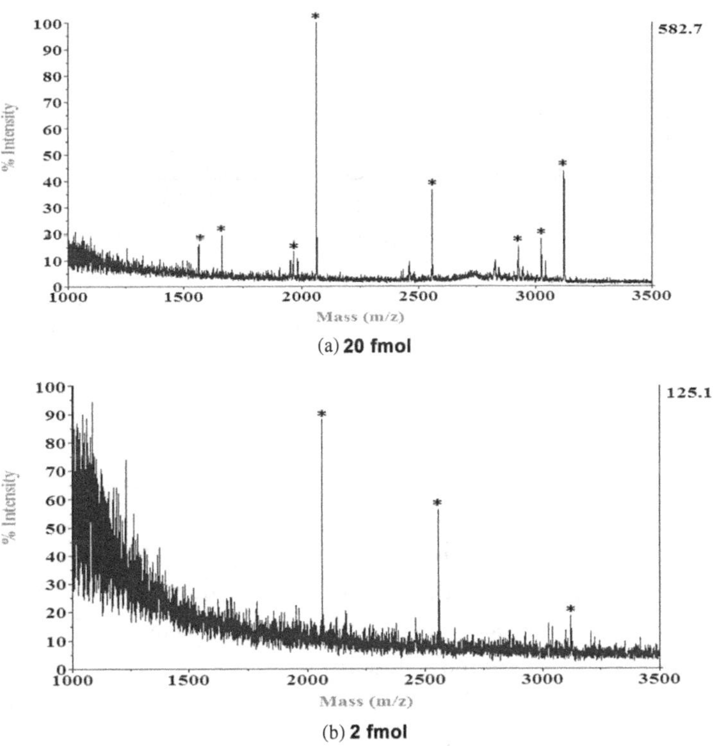

Figure 5.42: MS spectra of different concentrations of β-casein digests after enrichment with Fe_3O_4@PDA–Ti^{4+} microsphere.

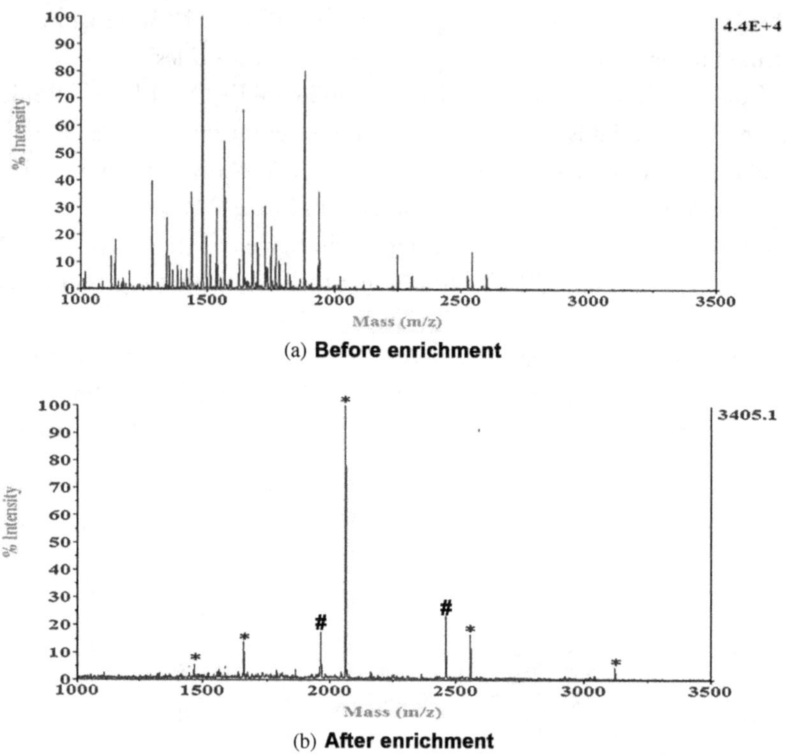

Figure 5.43: MS spectra of peptide mixture of β-casein and BSA at molar ratio of 1/500 before and after enrichment with Fe_3O_4@PDA–Ti^{4+} microsphere. *: phosphopeptide; #: dephosphorylated fragment.

evaluate the enrichment selectivity of Fe_3O_4@PDA–Ti^{4+} microsphere. As can be seen from the Figure 5.43, only non-phosphopeptides are detected before enrichment. But the phosphopeptide peak intensities are significantly enhanced after enrichment with Fe_3O_4@PDA–Ti^{4+} microsphere and few non-phosphopeptides are detected, demonstrating the synthesized Fe_3O_4@PDA–Ti^{4+} microsphere has good enrichment selectivity for phosphopeptides. As shown in Figure 5.44, after five cycles of enrichment and elution, the enrichment performance is still excellent, demonstrating the great regenerative capacity of the material and good enrichment reproducibility for phosphopeptides.

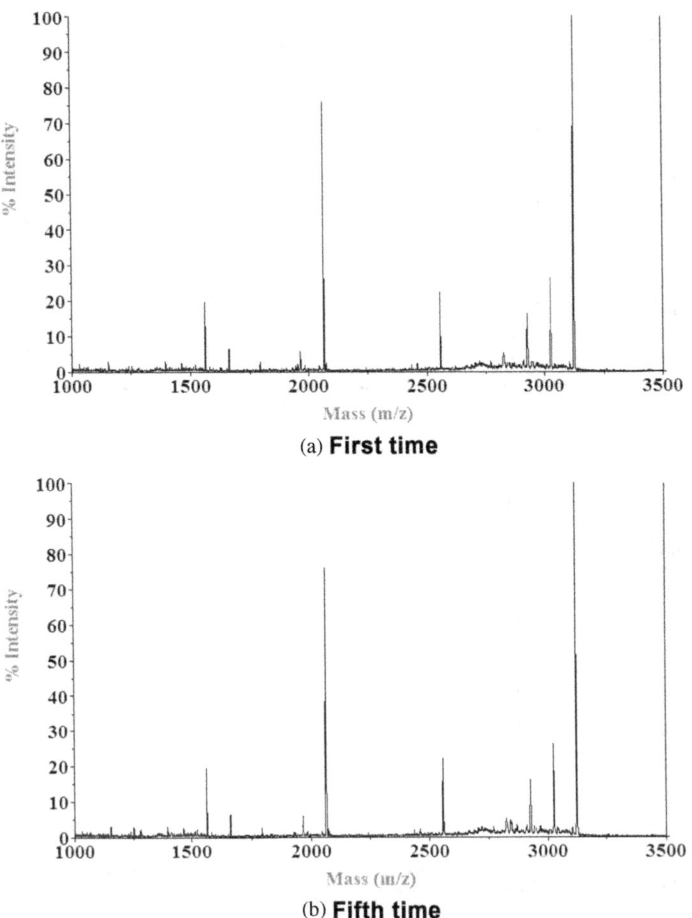

Figure 5.44: Repeatability of Fe_3O_4@PDA–Ti^{4+} microsphere for phosphopeptide enrichment.

(2) Evaluation of the enrichment of Fe_3O_4@PDA–Ti^{4+} microsphere for phosphopeptide in human serum

In order to investigate the enrichment selectivity of Fe_3O_4@PDA–Ti^{4+} microsphere for complex biological samples, the commonly used clinical sample, serum, is selected. As shown in Figure 5.45(a), no peak of any phosphorylated peptide is observed before enrichment. But after enrichment with Fe_3O_4@PDA–Ti^{4+} microsphere, four high-intensity phosphopeptides can be detected. The detailed information of phosphopeptides enriched from serum can be seen in Table 5.10.

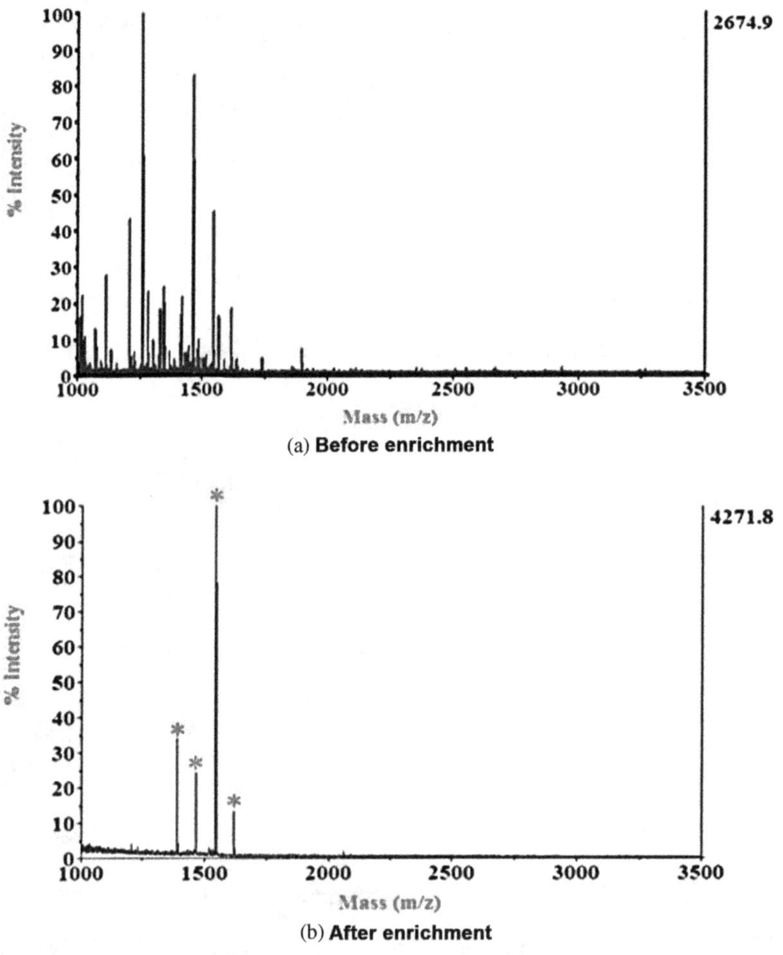

Figure 5.45: MS spectra of serum before and after enrichment with Fe_3O_4@PDA–Ti^{4+} material.

Table 5.10: Detailed information of phosphopeptides identified from human serum.

m/z(Da)	Peptide No.	Peptide Sequence
1389.5	1	ADpSGEGDFLAEGGGV
1460.6	2	DpSGEGDFLAEGGGV
1545.2	3	DpSGEGDFLAEGGGVR
1616.6	4	ADpSGEGDFLAEGGGVR

5.2.6. *Immobilization of metal ion on magnetic graphene (Mag GO)*

The synthesis protocol of Ti^{4+}-MGMSs composite is shown in Figure 5.46.[27]

(1) Preparation of Ti^{4+}-MGMSs composite

(a) *Synthesis of Mag GO*
The detailed synthesis process is described in Section 2.8 of Chapter 2.

(b) *Synthesis of Mag GO@mSiO$_2$*
The detailed synthesis process is described in Section 2.8 of Chapter 2.

(c) *Synthesis of Ti^{4+}-MGMSs composite*
10 mg Mag GO@mSiO$_2$ is dispersed in 10 mL 0.01 mol·L^{-1} Tris aqueous solution (pH = 8.5), and 20 mL ethanol is added for ultrasonication. 40 mg dopamine hydrochloride is dissolved in 15 mL deionized water, and then the dopamine hydrochloride aqueous solution is added to the previously described Tris solution, stirred at room temperature for 10 min. Afterwards, the product is washed with deionized water, and dispersed in

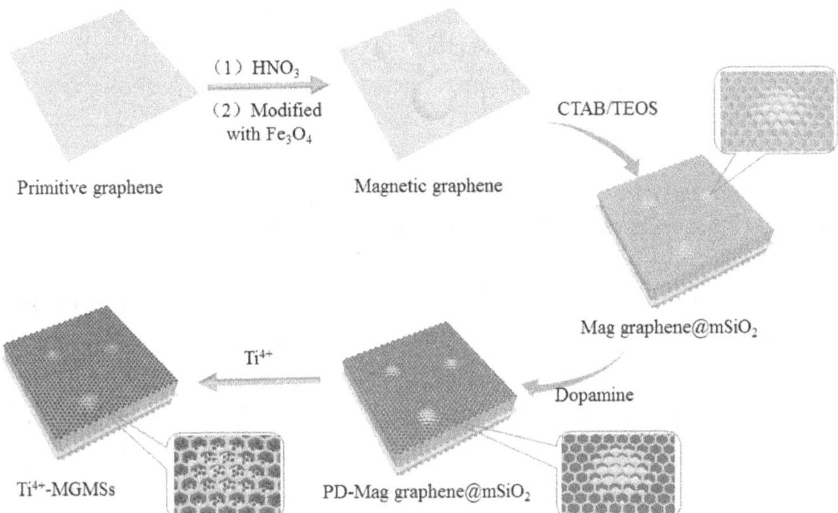

Figure 5.46: Schematic diagram of the synthesis of Ti^{4+}-MGMSs composite.

20 mL 0.1 mol·L^{-1} titanium sulfate solution, stirred at room temperature for 2 h, washed with deionized water and ethanol, and finally dried for further experiment.

(2) Characterization of Ti^{4+}-MGMSs composite

Figure 5.47 shows the nitrogen adsorption–desorption isotherms and the pore size distribution of Mag G@–SiO$_2$ and Ti^{4+}-MGMSs composites. As shown in Figure 5.47(a), the pore size of mesoporous silica is about 3.98 nm. After modification of PDA, the pore size changes to 2.17 nm. This pore size can still be considered to allow the entry of small molecular peptides and have size-exclusion effect for proteins. In addition, EDX analysis confirms the presence of titanium ion in Ti^{4+}-MGMSs composite (Figure 5.48).

(3) Enrichment application of Ti^{4+}-MGMSs composite in phosphoproteomics

(a) *Ti^{4+}-MGMSs composite for enrichment of standard phosphorylated peptide*
In this work, β-casein digest is first used to investigate the enrichment performance of the material. The Ti^{4+}-MGMSs composite is used for phosphopeptide enrichment with different concentrations of β-casein digests. When the concentration of β-casein is as low as 5×10^{-10} mol·L^{-1}, two phosphopeptides are still detected. This result indicates that the Ti^{4+}-MGMSs composite has good enrichment ability and high sensitivity for phosphopeptides.

(b) *Evaluation of the size-exclusion effect of Ti^{4+}-MGMSs composite*
To investigate whether Ti^{4+}-MGMSs composite has the size-exclusion ability for macromolecular protein, BSA (66 KDa), α-casein (23,690 Da), and β-casein digest are selected. As shown in Figure 5.50, when the ratio of BSA (66 KDa), α-casein (23,690 Da), and β-casein digest reaches 1:500:500, no phosphopeptide and non-phosphopeptide can be detected (Figure 5.50(b)), and only a large amount of protein is observed. After enrichment with Ti^{4+}-MGMSs composite, as shown in Figure 5.50(d), merely a few non-phosphorylated peptides are detected in the supernatant, and a large amount of protein is detected. In the eluent (Figure 5.50(f)), no protein signal is detected, and a

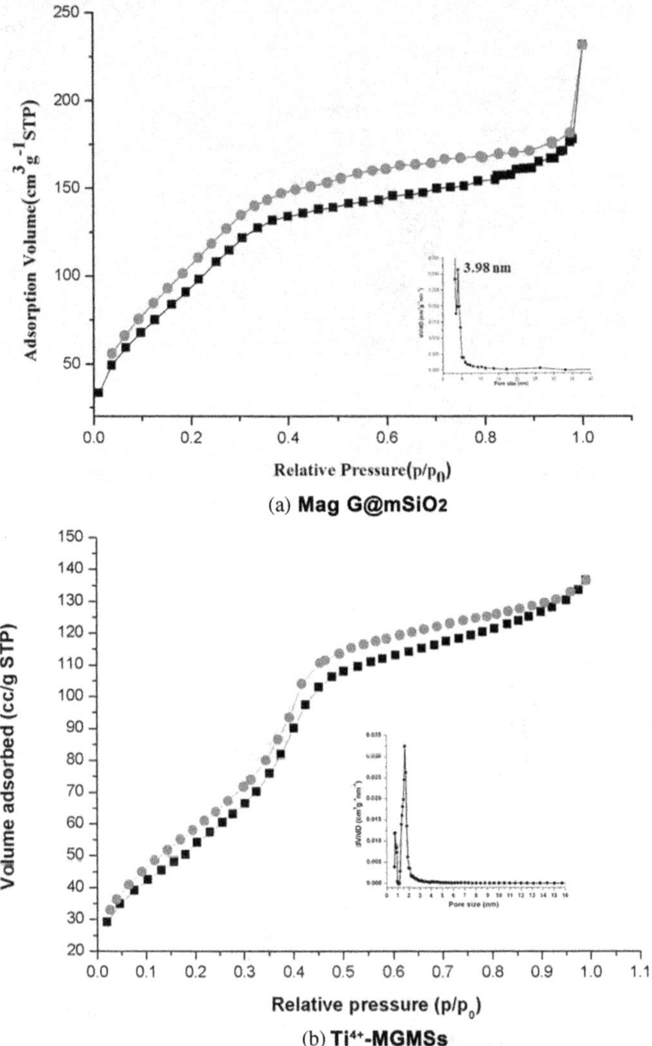

Figure 5.47: Nitrogen adsorption–desorption isotherms and pore size distribution of Mag G@–SiO$_2$ and Ti^{4+}-MGMSs composites.

large number of phosphopeptides are detected, and the background of the spectrum is clear. These results indicate that the Ti^{4+}-MGMSs composite indeed has size-exclusion effect on macromolecular proteins and selective enrichment ability for phosphorylated peptides.

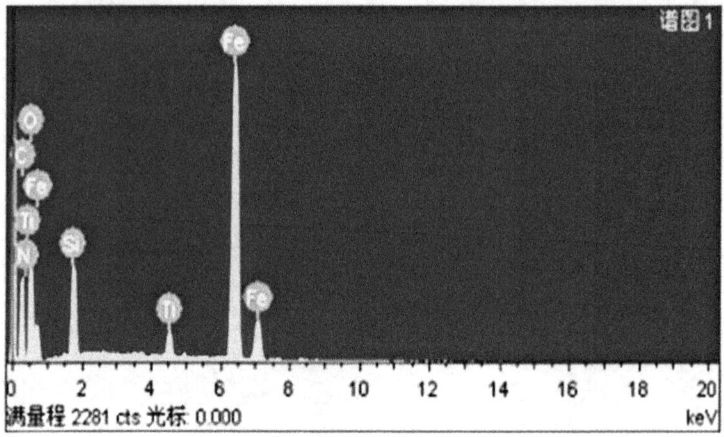

Figure 5.48: EDX analysis of Ti^{4+}-MGMSs composite.

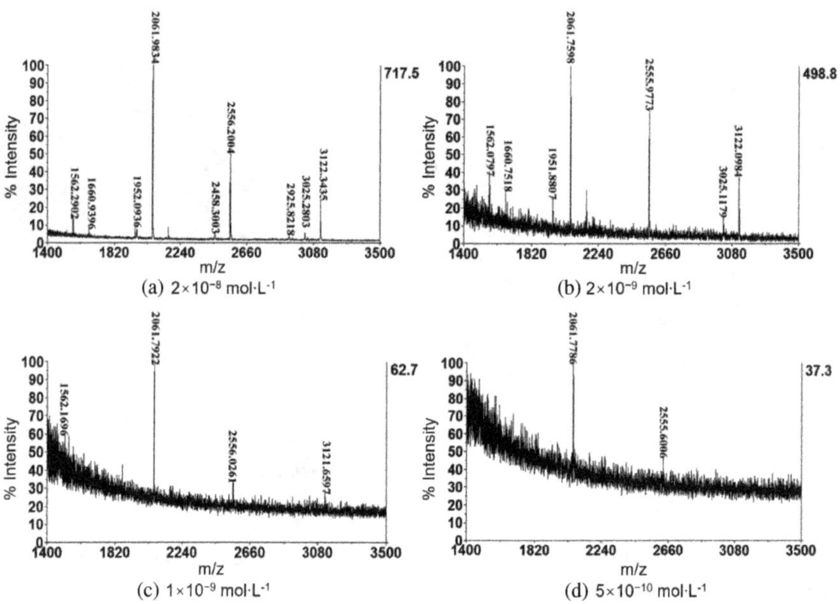

Figure 5.49: MS spectra of different concentrations of β-casein digests after enrichment withTi^{4+}-MGMSs composite.

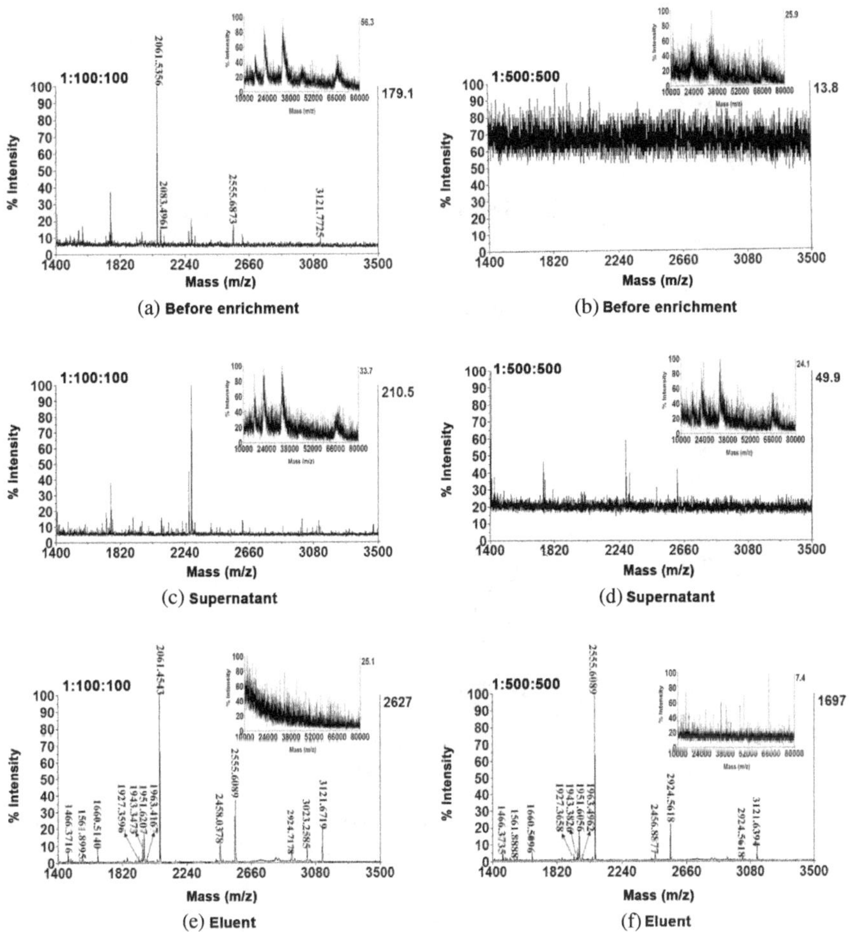

Figure 5.50: MS spectra of different mass ratios of BSA, α-casein and β-casein digest (5 μg) before and after enrichment with Ti^{4+}-MGMSs composite.

(c) *Evaluation of the enrichment capacity of Ti^{4+}-MGMSs for endogenous phosphopeptides in human saliva*

Human saliva is used to investigate the selective enrichment ability of Ti^{4+}-MGMSs composite for endogenous phosphorylated peptides. As shown in Figure 5.51, after enrichment with Ti^{4+}-MGMSs composite, a total of 14 endogenous phosphopeptides are detected.

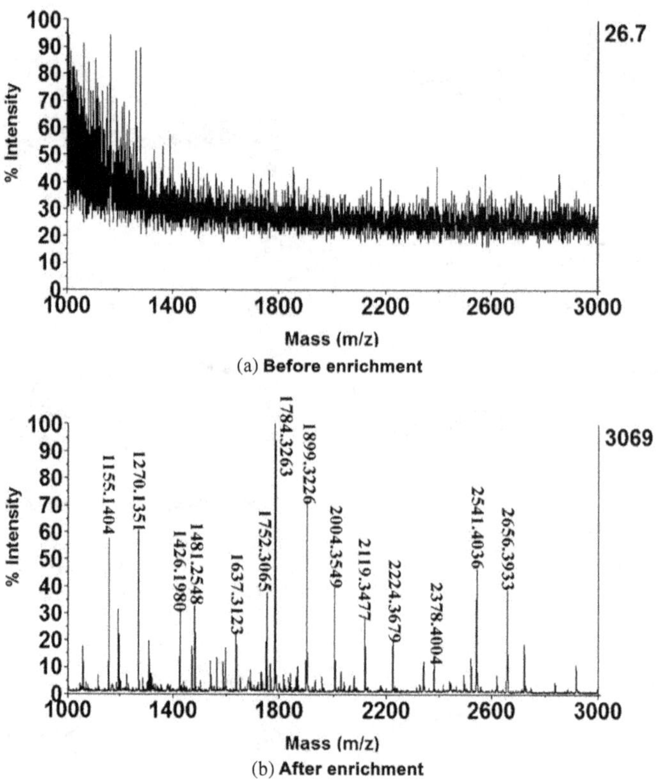

Figure 5.51: MS spectra of human saliva samples before and after enrichment with Ti^{4+}-MGMSs.

5.3. MOAC-based magnetic micro-/nanomaterial for separation and analysis in phosphoproteomics

5.3.1. *Introduction*

With the rapid development of the enrichment method for phosphorylated peptides, MOAC technology has gradually attracted widespread attention. Metal oxide plays an important role in the enrichment of phosphorylated peptides with high selectivity and good reproducibility. At present, TiO_2 is the main metal oxide for the separation of phosphopeptides, while there are also many reports on the enrichment of phosphopeptide by ZrO_2, Al_2O_3, and Ga_2O_3. Based on the magnetic material, the superparamagnetism of the material can solve the problem of centrifugation during separation process.

Chen[28–30] coated the metal oxide such as ZrO_2, Al_2O_3, and TiO_2 on the surface of commercial magnetic microsphere, which enables the simple enrichment of phosphopeptides. However, in their research, the dispersibility of the material is relatively poor, the structure is irregular, and the coating is also not tight enough, and may easily fall off from the surface of the magnetic microsphere. Therefore, it is necessary to find a suitable method to closely coat the metal oxide on the surface of the magnetic microsphere for the rapid separation of material from sample solution and selective enrichment of the phosphorylated peptide. This section describes several methods for coating metal oxide on magnetic microsphere and their application in phosphoproteomics.

5.3.2. Carbon layer-coated magnetic microsphere for modification of metal oxide ($Fe_3O_4@M_xO_y$)

5.3.2.1. Preparation of $Fe_3O_4@M_xO_y$ microsphere

The synthesis protocol of $Fe_3O_4@M_xO_y$ microsphere is shown in Figure 5.52. The Fe_3O_4 nanoparticle is first synthesized via the hydrothermal method, the surface of Fe_3O_4 nanoparticle is then coated with the carbon shell via the hydrothermal method, the carbon-coated magnetic nanoparticle is then immersed in the transition metal salt solution, and finally the $Fe_3O_4@M_xO_y$ microsphere is obtained by calcination.

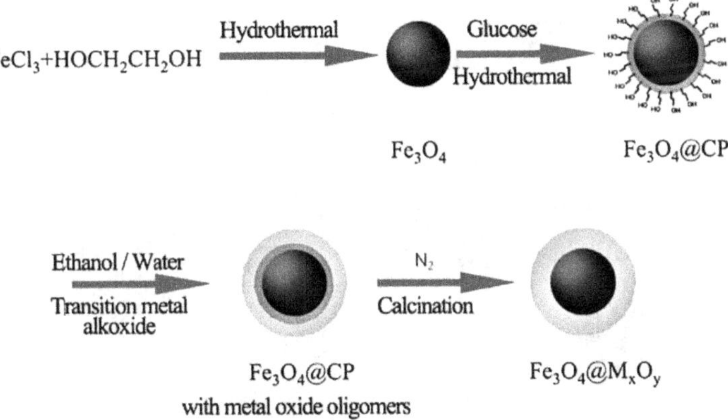

Figure 5.52: Schematic diagram of the synthesis of $Fe_3O_4@M_xO_y$ core–shell microsphere.

5.3.2.2. *Enrichment application of $Fe_3O_4@M_xO_y$ microsphere in Phosphoproteomics*

The enrichment process of $Fe_3O_4@M_xO_y$ microsphere is very simple. The detailed steps are shown in Figure 5.53. First, $Fe_3O_4@M_xO_y$ microsphere is added into the peptide mixture for incubation. Then, the $Fe_3O_4@M_xO_y$ microsphere is quickly separated from the solution by the external mag-

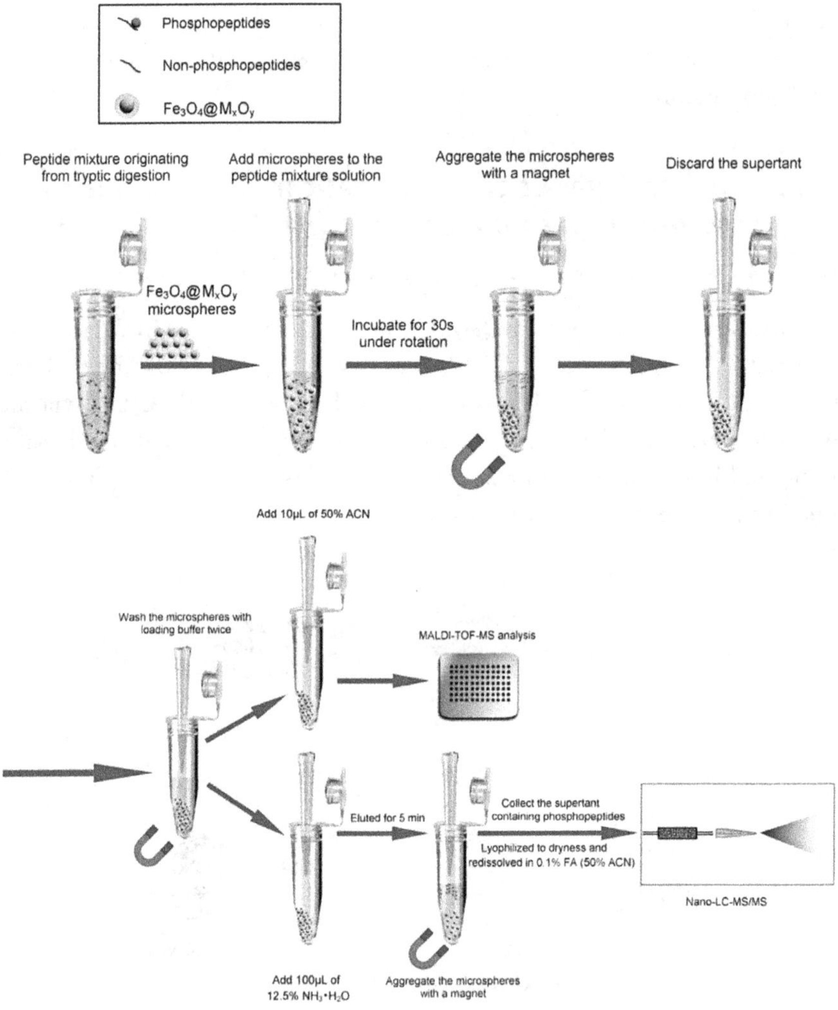

Figure 5.53: Flow chart of the phosphopeptide enrichment using $Fe_3O_4@M_xO_y$ microsphere.

netic field. Second, to remove the non-specific adsorption on the surface of the $Fe_3O_4@M_xO_y$ microsphere, the $Fe_3O_4@M_xO_y$ microsphere is washed 3 times with the loading buffer. Third, for standard protein sample, the phosphopeptides enriched with $Fe_3O_4@M_xO_y$ microsphere are eluted by 50% ACN, and directly analyzed by MALDI-TOF-MS. For complex biological samples, 12.5% $NH_3 \cdot H_2O$ is used to elute the phosphopeptides. The eluent is lyophilized and re-dissolved in 50% ACN+0.1% FA for LC-ESI-MS analysis.

(1) Enrichment application of $Fe_3O_4@TiO_2$ microsphere in phosphoproteomics

(a) *Evaluation of the enrichment conditions of $Fe_3O_4@TiO_2$ microsphere*
In this work, β-casein digest is selected to verify whether the $Fe_3O_4@TiO_2$ microsphere can selectively and efficiently enrich the phosphorylated peptide in a complex mixture.[31] Thus, the enrichment condition of $Fe_3O_4@TiO_2$ microsphere is explored to determine the optimal composition of the loading buffer, the pH condition, the enrichment time, and the elution time.

(i) **Optimization of the loading buffer**
Four loading buffers (pH = 2, 50% ACN–TFA, 50% ACN–ACOOH, TFA–H_2O, ACOOH–H_2O) are used to investigate their effects on enrichment selectivity. As shown in Figure 5.54, 50% ACN is more conducive to the enrichment selectivity of $Fe_3O_4@TiO_2$ microsphere. Comparing Figures 5.54(a) and 5.54(c), the background of the spectrum using TFA is clearer, so the final enrichment system is chosen to be 50% ACN–TFA.

(ii) **Optimization of the pH of the enrichment system**
For the selected 50% ACN–TFA enrichment system, the effect of pH value on the enrichment selectivity of phosphopeptide is investigated when the pH values are 2, 4, and 6, respectively. As shown in Figure 5.55, when the pH value is 2, the $Fe_3O_4@TiO_2$ microsphere has good selectivity for phosphorylated peptides. When the pH increases to 6, although $Fe_3O_4@TiO_2$ microsphere still has good selectivity for phosphopeptides, a few non-phosphorylated peptides can be seen in the spectrum. Therefore, the pH value of the enrichment system is finally selected to be about 2.

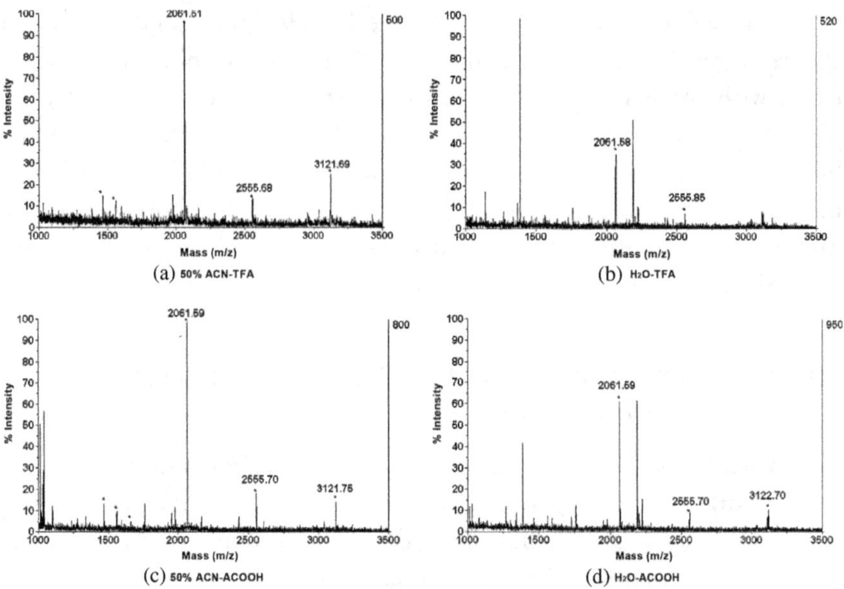

Figure 5.54: MS spectra of 2×10^{-8} mol·L^{-1} β-casein digest after enrichment with Fe$_3$O$_4$@TiO$_2$ microsphere in different loading buffers.

(iii) **Optimization of enrichment time**

The effect of enrichment time on the enrichment performance is investigated using 50% ACN–TFA and PH = 2 as enrichment system. As can be seen from Figure 5.56, the material achieves efficient selective enrichment of the phosphorylated peptides within 0.5 min. This may be due to the large specific surface area of the Fe$_3$O$_4$@TiO$_2$ microsphere and its good dispersibility in solution, which enables rapid selective enrichment of the phosphopeptides.

(iv) **Optimization of elution time**

The adsorption of phosphopeptide on the surface of the material is destroyed under alkaline condition. Therefore, 0.5% ammonia is selected as the eluent to investigate the effect of elution time on elution performance. It can be seen from Figure 5.57 that the phosphopeptide can be effectively detected at 1 min. When the elution time reaches 5 min, the elution performance is stable. Moreover, it can be seen that the Fe$_3$O$_4$@TiO$_2$ microsphere has better enrichment selectivity for the multi-phosphorylated peptide.

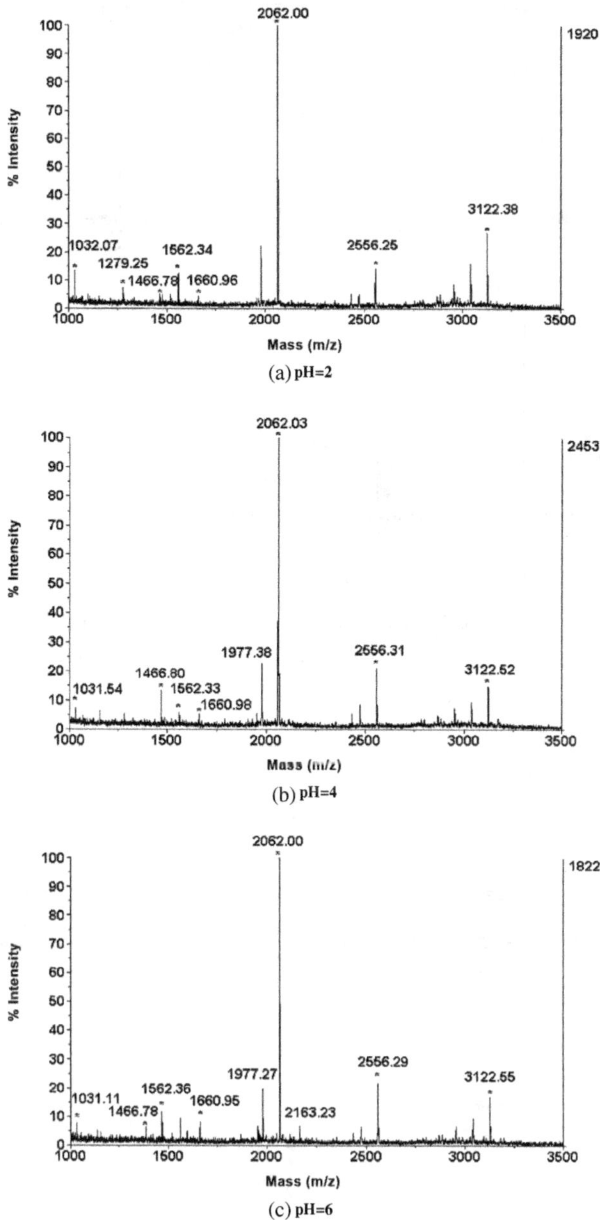

Figure 5.55: MS spectra of 2×10^{-8} mol·L^{-1} β-casein digest after enrichment with Fe$_3$O$_4$@ TiO$_2$ microsphere at different pH.

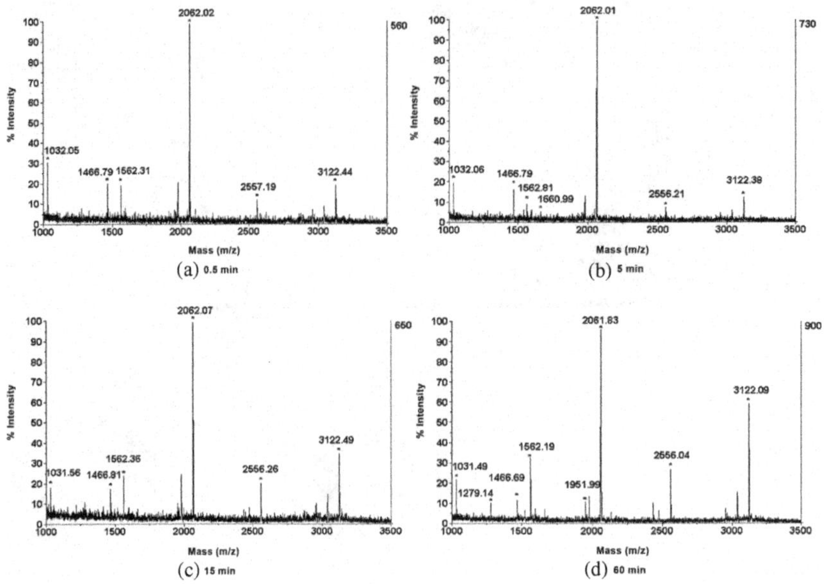

Figure 5.56: MS spectra of 2×10^{-8} mol·L^{-1} β-casein digest after incubation with Fe$_3$O$_4$@ TiO$_2$ microsphere at different enrichment times.

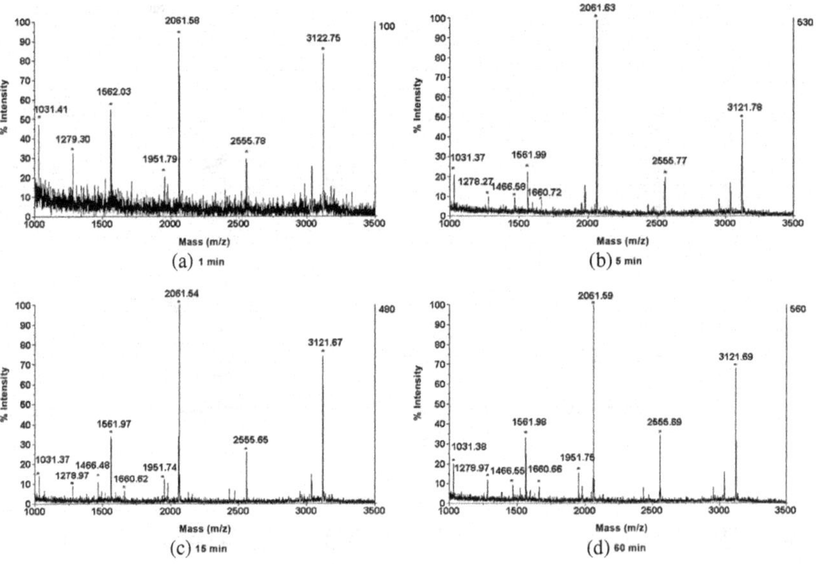

Figure 5.57: MS spectra of 2×10^{-8} mol·L^{-1} β-casein digest after enrichment with Fe$_3$O$_4$@ TiO$_2$ microspheres at different times.

(b) *Evaluation of the enrichment capability of $Fe_3O_4@TiO_2$ microsphere for phosphopeptide in standard phosphoprotein digest*

According to the optimal enrichment condition, 50% ACN–TFA, pH = 2, 5 min enrichment time and 5 min elution time are used to enrich phosphopeptides from 400 fmol β-casein digest. As shown in Figure 5.58(a), almost no signal of phosphopeptides can be detected before enrichment. However, after enrichment with $Fe_3O_4@TiO_2$, as shown in Figure 5.58(b), the monophosphorylated peptides of *m/z* 2061 and *m/z* 2556 and tetra-phosphorylated peptides of *m/z* 3122 are all detected, indicating that the $Fe_3O_4@TiO_2$ microsphere is a highly sensitive affinity extraction technique in the enrichment of phosphorylated peptides.

(c) *Evaluation of the enrichment capability of $Fe_3O_4@TiO_2$ microsphere for complex peptide mixture*

A more complex mixture of casein protein (including three phosphorylated proteins, α-S1-casein, α-S2-casein, and β-casein) is used to further investigate the selective enrichment ability of $Fe_3O_4@TiO_2$ microsphere (Figure 5.59). A large number of phosphorylated peptides can be efficiently enriched with the complex casein digest by $Fe_3O_4@TiO_2$ microsphere as seen in Figure 5.59(b). A total of 13 phosphorylated peptides are detected, including eight monophosphorylated peptides and 5 multi-phosphorylated peptides. These peptides can be identified by searching the Swiss-Prot database. The detailed information of amino acid sequence and possible phosphorylation sites is listed in Table 5.11. These results further demonstrate that $Fe_3O_4@TiO_2$ microsphere has selective enrichment ability for phosphorylated peptides.

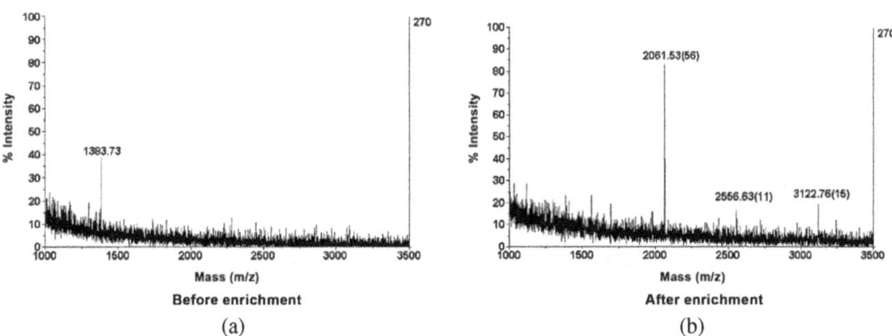

Figure 5.58: MS spectra of 2×10^{-9} mol·L^{-1} β-casein digest before and after enrichment with $Fe_3O_4@TiO_2$ microsphere.

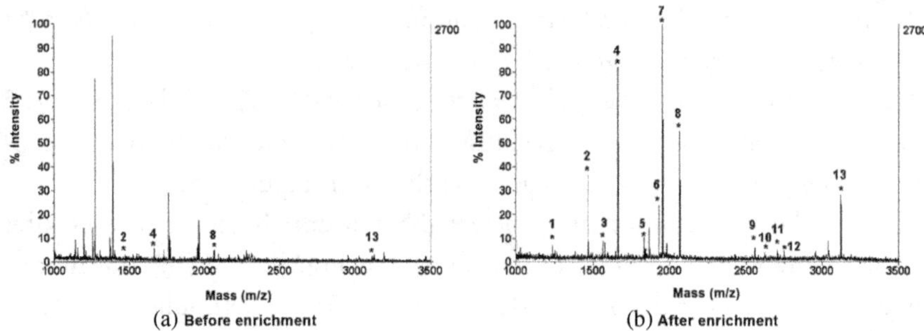

Figure 5.59: MS spectra of peptide mixture of casein digest before and after enrichment with $Fe_3O_4@TiO_2$ microsphere.

Table 5.11: The detailed information of amino acid sequence and possible phosphorylation site of the identified phosphorylated peptide after enrichment with $Fe_3O_4@TiO_2$ microsphere.

No.	AA	Peptide Sequence	m/z
1	α-S2/138–147	TVDME[pS]TEVF	1237.42
2	α-S2/138–149	TVDME[pS]TEVFTK	1466.51
3	α-S2/126–137	EQL[pS]T[pS]EENSKK	1562.04[a]
4	α-S1/106–119	VPQLEIVPN[pS]AEER	1660.661
5	α-S1/104–119	YKVPQLEIVPN[pS]AEER	1832.66
6	α-S1/43–58	DIG[pS]E[pS]TEDQAMTIK	1927.49
7	α-S1/104–119	YKVPQLEIVPN[pS]AEER	1951.76
8	β/33–48	FQ[pS]EEQQQTEDELQDK	2061.64
9	β/33–52	FQ[pS]EEQQQTEDELQDKIHPF	2555.86
10	α-S2/2–21	NTMEHV[pS] [pS] [pS]EESII[pS]QETYK	2618.84
11	α-S1/99–120	LRLKKYKVPQLEIVPN[pS]AEERL	2703.60
12	α–S2/2–22	NTMEHV[pS] [pS] [pS]EESII[pS]QETYKQ	2746.59
13	α/1–25	RELEELNVPGEIVE[pS]L[pS][pS][Ps]EESITR	3121.70

[a]Peaks of phosphopeptides refers to $[M + Na]^+$

(d) *Evaluation of the enrichment ability of $Fe_3O_4@TiO_2$ microsphere for phosphopeptide in rat liver digest*

The abovementioned results demonstrate the excellent enrichment ability of $Fe_3O_4@TiO_2$ microsphere, so a more complex biological sample is applied. 100 μg rat liver digest is enriched by $Fe_3O_4@TiO_2$ microsphere, and the eluent is lyophilized and then re-dissolved for C18 capillary column analysis. Finally, LTQ-Orbitrap mass spectrometry is performed. The obtained MS/MS data is searched by Sequest database, and the search results are screened by Xcorr value. For peptides with 1+, 2+, and 3+, the Xcorr values must be more than 2.5, 2.63, and 3.11, respectively. MS/MS/MS data is also searched by using Sequest database. MS/MS/MS data is mainly used to supplement the results of MS/MS. The Xcorr values of peptides with 1+, 2+, and 3+ must be more than 1.5, 2.0, and 2.5, respectively. Based on the results of MS/MS and MS/MS/MS, a total of 41 peptides are identified by both MS/MS and MS/MS/MS. The detailed information of identified peptide and the corresponding Xcorr values are shown in Table 5.12.

After manual confirmation, the MS/MS and MS/MS/MS spectra are considered to be of high quality and the results are credible. Taking one peptide as example, Figure 5.60 shows the MS/MS and MS/MS/MS spectra of a double-charged phosphorylated peptide (TPEELDDS*DFETEDFDVR). The b ion and y ion information in the MS/MS and MS/MS/MS spectra are consistent with the theoretically predicted peptide fragmentation information. In addition, from the MS/MS spectrum, the parent ion peak of m/z 1120.68 and the resulting fragment peak of m/z 1071.43 can be clearly recognized. Some peptides are identified by MS/MS but not by MS/MS/MS, so these results must be manually confirmed. In this work, only the mono-phosphorylated peptides identified in the MS/MS data are manually confirmed, and 15 mono-phosphorylated peptides are finally identified. The peptide sequence information and the corresponding Xcorr values identified only by the MS/MS spectrum are listed in Table 5.12. To sum up, a total of 56 phosphorylated peptides and 65 phosphorylation sites are identified in rat liver tissue, of which 85.7% are mono-phosphorylation sites, 12.5% are di-phosphorylation sites, and 1.8% are tri-phosphorylation sites (Figure 5.61(a)). Moreover, 90.8% identified phosphorylation sites occur on serine residues, 6.1% on threonine residues, and 3.1% on tyrosine residues (Figure 5.61(b)).

(2) Enrichment application of Fe$_3$O$_4$@ZrO$_2$ microsphere in phosphoproteomics

(a) Optimization of enrichment condition

In order to optimize the enrichment condition of Fe$_3$O$_4$@ZrO$_2$ microsphere for phosphopeptides, β-casein digest is selected.

Table 5.12: Detailed information of the identified phosphorylated peptide enriched with Fe_3O_4@TiO_2 microsphere from rat liver digest.

Protein No.	Peptide Sequence	Phospho-Sites	Charge	XC	MS2	MS3
IP100656420	K.DADEEDS*DEETSHLER.S	1	3	8.74	*	*
IP100767685	K.FNLFS*QEUDKK.S	1	2	2.93	*	*
IP100766722	R.WLDES'DAEMELR.A	1	2	3.73	*	*
IP100734740	K.KEES'EES*DEDMGFGLFD.-	2	2	4.88	*	*
IP100564566	R.T*LS*ElEUKVI1lA	2	2	2.9	*	*
IP100558327	R.TGDLGIPPNPEDRS*PS*PEPIYNSEGK.R	2	2	4.59	*	*
IP1005518l5	R.NYQQNYQNSESGEKNEGS*ES*APEGQAQQR.R	2	2	4.84	*	*
IP100480820	K.EGEEPTVYS*DDEEPK.D	I	2	3.44	*	*
IP100476899	K.YGPVSVADTIGSGAADAKDDDDIDLFGS*DDEEESEDAKR.L	11	2	8.74	*	*
IP100476698	R.MLPHAPGVQMQAIPEDAVHEDS*GDEDGEDPDKR.I	1	3	4.7	*	*
IP100476178	K.ESLKEEDES*DDDNM.-	1	2	4.6	*	*
IP100471911	K.GILAADES*VGTMG R.L	1	2	4.59	*	*
IP100471584	K.IEDVGS*DEEDDSGKDK.K	I	3	4.34	*	*
IP100393259	R. PT*AS*ISPGSPTSSMT.-	2	2	2.76	*	*
IP100382376	R.SSGS'PYGGGYGSGGGSGGYGSR.R	1	2	4.04	*	*
IP100382244	K.FHDS* EGDDTEETEDYR.Q	I	2	5.86	*	*
IP100373197	R.LLPGEEPSEYT*EEDTK.D	I	2	2.89	*	*
IP100370652	R.LGAS*PGGDAGTCPPVGR*GLK.T	2	3	5.32	*	*

IP100370209	K.VFDDS*DEKEDEEDTDVR.K	I	2	5.04	*	*
IP100369227	R.ESPRPPAAAEAPAGS*DGEDGGRR.D	I	3	3.63	*	*
IP100366370	K.TSI'DENDS*EELEDKDSK.S	1	2	4.18	*	*
IP100365935	K.DWEDDS*DEDMSNFDR.F	I	2	5.83	*	*
IP100365929	K.DGELPVEDDIDLS*DVELDDLEKDEL-	I	2	3.58	*	*
IP1100365864	K.VLHGAQTS*DEEKDF.-	1	2	4.04	*	*
IP100365663	R.S*VDEVNYWDK.Q	I	2	3.8	*	*
IP100365663	R.IGHHS*TSDDSSAYR.S	I	3	4.97	*	*
IP100365I 49	R.YTDQS*GEEEEDYESEEQIQHR.I	I	3	5.32	*	*
IP100359917	K.FIDKDQQPSGS'EGEDDDAEAALKK.E	I	3	4.43	*	*
IP100359I72	R.RES*GEGEEEVADSAR. L	I	2	2.76	*	*
IP100358406	R.TPEELDDS*DFETEDFDVR.S	I	2	5.42	*	*
IP100324618	K.LKDLGHPVEEEDES*GDQEDDDDELDDGDRDQDI.-	1	3	7.69	*	*
IP100210566	K.ESDOKPEIEOVGS*DEEEEEKK.D	I	3	5.78	*	*
IP100209277	K.LSSQLS*AGEEK.W	I	2	3.44	*	*
IP100208304	K.IYHLPDAES*DEDEDFKEQTR.L	1	3	5.28	*	*
IP100208277	K.SIDS*DES*EDEDDDYQQK.R	2	2	6.14	*	*
IP100200898	R.SAS*SDTSEELNAQDSPK.R	I	2	5.63	*	*
IP100197900	K.KGATPAEDDEDNDIDLFGS*DEEEEDKEAAR.L	I	3	7.96	*	*
IP100194102	R.FIIGSVSEDNS*EDEISNLVK.L	1	2	4.44	*	*
IP100193648	K.SAS*PAPADVAPAQEDUJ.T	I	2	4.93	*	*

(*Continued*)

Table 5.12: (*Continued*)

Protein No.	Peptide Sequence	Phospho-Sites	Charge	XC	MS2	MS3
IP100019I707	R.YHGHS*MS*DPGVS*YR.T	3	3	4.06	*	*
IP100189138	K.VVDYSQFQES*DDADEDYGR.D	I	2	4.56	*	*
IP100366533	K.NGIPYSFAFEUIDTGY*FGFLLPEMUK.P	I	3	3.25	*	
IP100364925	R.Y*PVAVSTLEEMAPGTAFK.P	1	2	2.66	*	
IP10036394 I	R.HSS*WGSVGLGGSLEASR.L	1	2	3.47	*	
IP100214258	K.EVEDKES*EGEEEDEDEDLSK.Y	I	2	3.17	*	
IP100210280	K.AIYQGPSS*PDKS.-	1	2	3.51	*	
IP100209618	R.APTAAPS*PEPR.D	1	2	3.21	*	
IP100208266	K.AALGLQDS*DDEDMVDIDEQIESMFNSK.K	1	3	3.94	*	
IP10020760I	K.NFETNDLAFS*PK.G	I	2	3.96	*	
IP100202703	R.TLSNAEDYLDDEDS*D.-	1	2	3.61	*	
IP100201103	R.PL.S*PTAFSLESLR.J.K	I	2	2.78	*	
IP100200145	K.KEES*EESEDDMGFGLFD.-	1	2	4.11	*	
IP100192480	R.VHGHS*DEEEEEEQPR.H	1	3	3.44	*	
IP100191707	R.YGMGTS*VER.A	I	2	3.I2	*	
IP100190024	R.MAMPINVS*DPDLLR.H	1	2	3.88	*	
IP100188053	R.GDS*ETDLEALFNAVMNPK.T	I	2	4.74	*	

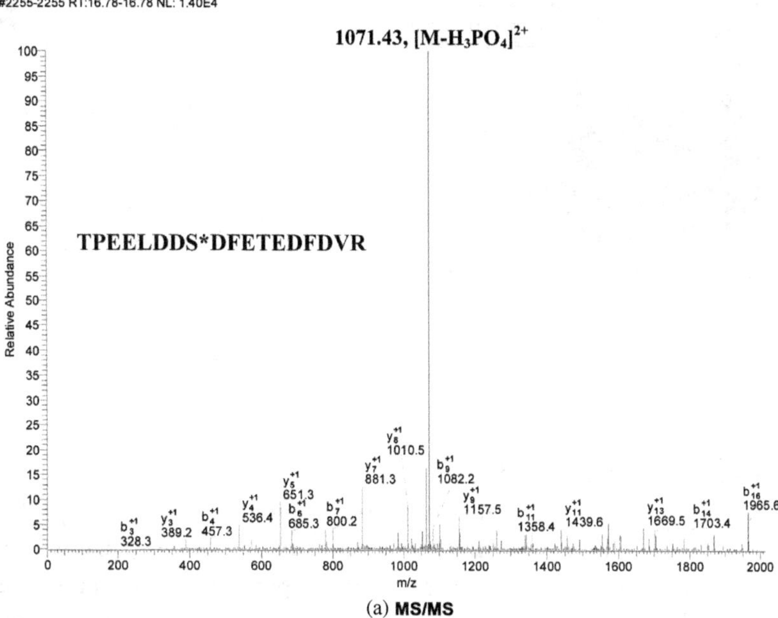

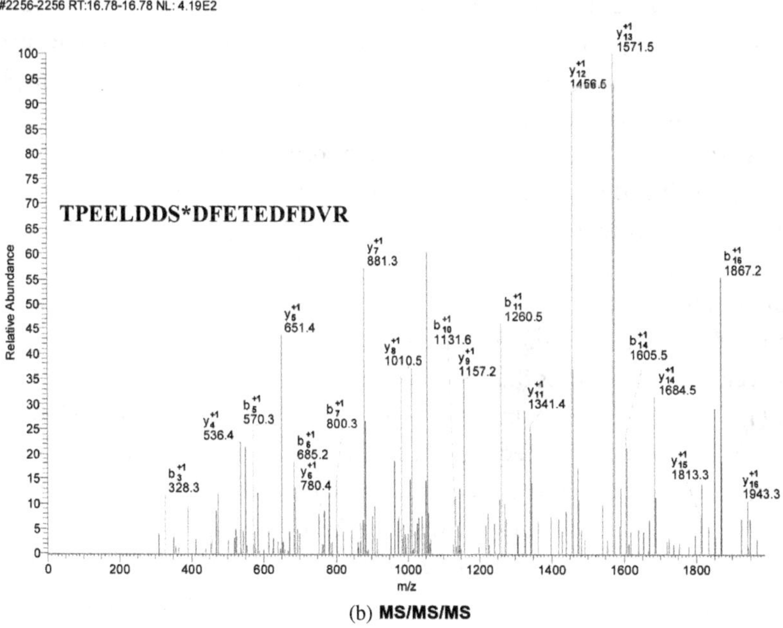

Figure 5.60: MS/MS and MS/MS/MS spectra of doubly charged phosphorylated peptide (TPEELDDS*DFETEDFDVR).

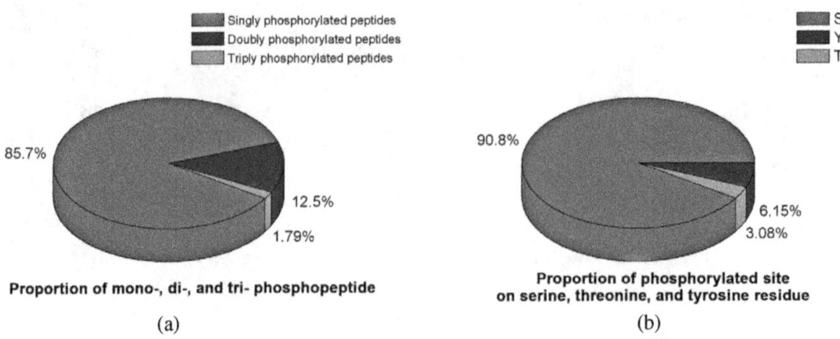

Figure 5.61: Types of the phosphorylated peptides identified in rat liver tissue.

(i) **Optimization of pH of the reaction solution**

The metal oxide of ZrO_2 is an amphoteric compound, that is, ZrO_2 behaves as Lewis acid or Lewis base under different pH conditions.[32] Under acidic condition, ZrO_2 behaves as a Lewis acid with a positive charge on the surface and ion exchange property. According to Refs. [33, 34], the bonding constant of the phosphate group and Lewis acid is significantly higher than other Lewis bases,[35,36] which provides the possibility of ZrO_2 for phosphopeptide enrichment: there should be a strong affinity between the phosphorylated peptide and ZrO_2 under suitable pH condition.

As shown in Figure 5.62, the three phosphorylated peptides and their corresponding dephosphorylated fragments in the β-casein digest are detected after enrichment with $Fe_3O_4@ZrO_2$ microsphere for 0.5 min at pH = 2. Among them, m/z 1978.99 is formed by dephosphorylation of the mono-phosphorylated peptide of m/z 2061.97. However, since the metastable ion is formed after the loss of the phosphate group, the mass difference between the two peaks is 82.98 Da instead of the theoretical difference of 98 Da (1 Da = 1.66×10^{-27} kg).[15] m/z 1031.57 is the doubly charged peak of the mono-phosphorylated peptide of m/z 2061.97. In addition, since β-casein usually contains trace amounts of α-casein, three phosphorylated peptides belonging to α-casein digest can be observed in the spectrum (m/z is 1466.76, 1660.98, 1952.08, respectively). Compared with Figure 5.62(a), there is almost no peak of the non-phosphorylated peptide in the spectrum

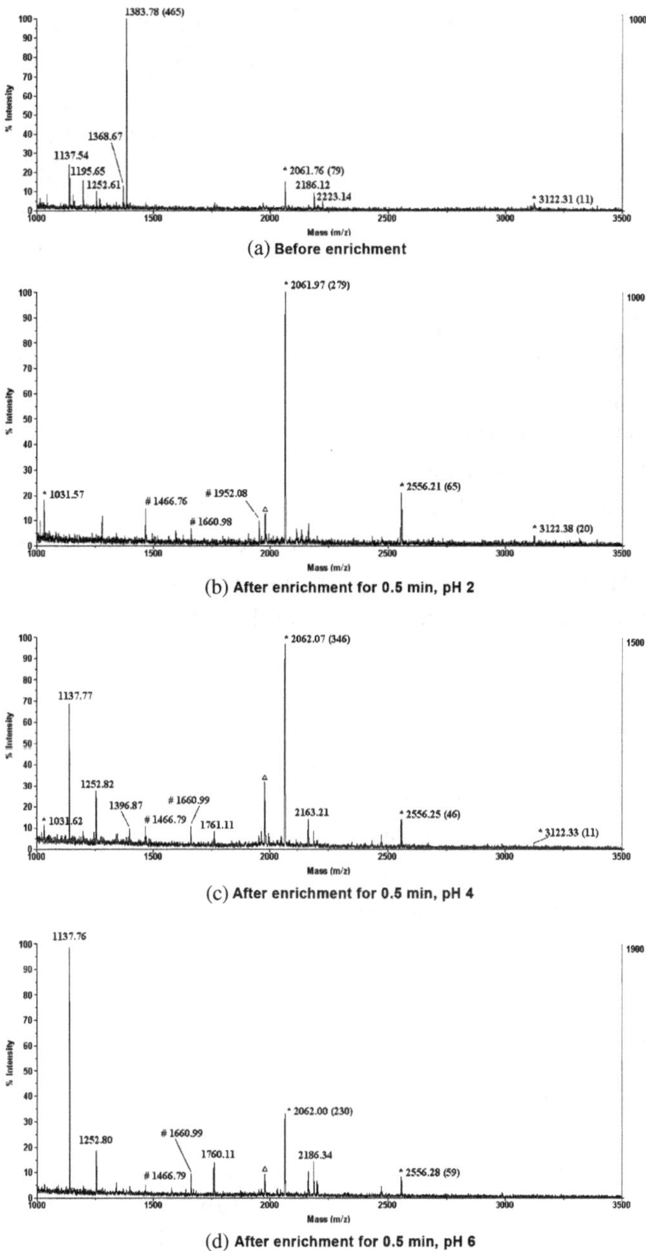

Figure 5.62: MS spectra of 2×10^{-7} mol·L^{-1} β-casein digest before and after enrichment with Fe$_3$O$_4$@ZrO$_2$ microsphere under different pH conditions. *: phosphorylated peptide in β-casein digest, #: phosphorylated peptide in α-casein digest, Δ: dephosphorylated fragment.

(Figure 5.62b) after enrichment with $Fe_3O_4@ZrO_2$ microsphere at pH 2, and the S/N ratio of all the phosphorylated peptide peaks is significantly improved.

When the pH is 4 as shown in Figure 5.62(c), although the signal of the phosphorylated peptide is the highest in the spectrum, the signal peaks of some non-phosphorylated peptides (*m/z* 1137.77, *m/z* 1252.82, *m/z* 1767.77, and *m/z* 2163.21) also appear in the spectrum, indicating that the increase in pH value of the loading buffer causes a decrease in the enrichment selectivity of the $Fe_3O_4@ZrO_2$ microsphere for the phosphorylated peptide. As shown in Figure 5.62(d), when the pH value of the solution rises to 6, the signal peak of the non-phosphorylated peptide is dominant in the spectrum due to the further weakening of the Lewis acidity of ZrO_2. Only two mono-phosphorylated peptides are detected and the peak of the tetra-phosphorylated peptide is absent in the spectrum. Based on these results, the pH value of the loading buffer has a great influence on the enrichment selectivity of the ZrO_2 for phosphorylated peptides. To increase the enrichment performance of $Fe_3O_4@ZrO_2$ microsphere for phosphorylated peptides, a lower pH value is required, so pH = 2 is selected.

(ii) Optimization of enrichment time

In order to optimize the enrichment time, 5 min, 15 min, and 60 min are tested for comparison with 0.5 min, respectively. It can be seen from Figure 5.63 that the enrichment time does not have much effect on the enrichment performance, and 0.5 min is sufficient for $Fe_3O_4@ZrO_2$ microsphere to complete the enrichment of phosphorylated peptides.

(b) *Evaluation of the enrichment capability of $Fe_3O_4@ZrO_2$ microsphere for phosphopeptide in standard phosphoprotein digest*

In biological samples, the abundance of phosphorylated proteins is often low, and the content of phosphorylated peptides in the digest is lower. Therefore, the enrichment sensitivity of the material is highly demanded. In this work, the concentration of β-casein digest is reduced to 2×10^{-8} mol·L^{-1} and 2×10^{-9} mol·L^{-1} to investigate the enrichment sensitivity of $Fe_3O_4@ZrO_2$ microsphere. As shown in Figures 5.64(a) and 5.46(c), phosphorylated peptides cannot be directly detected by MALDI-TOF-MS at such low concentrations. However, after enrichment with $Fe_3O_4@ZrO_2$ microsphere, as shown in

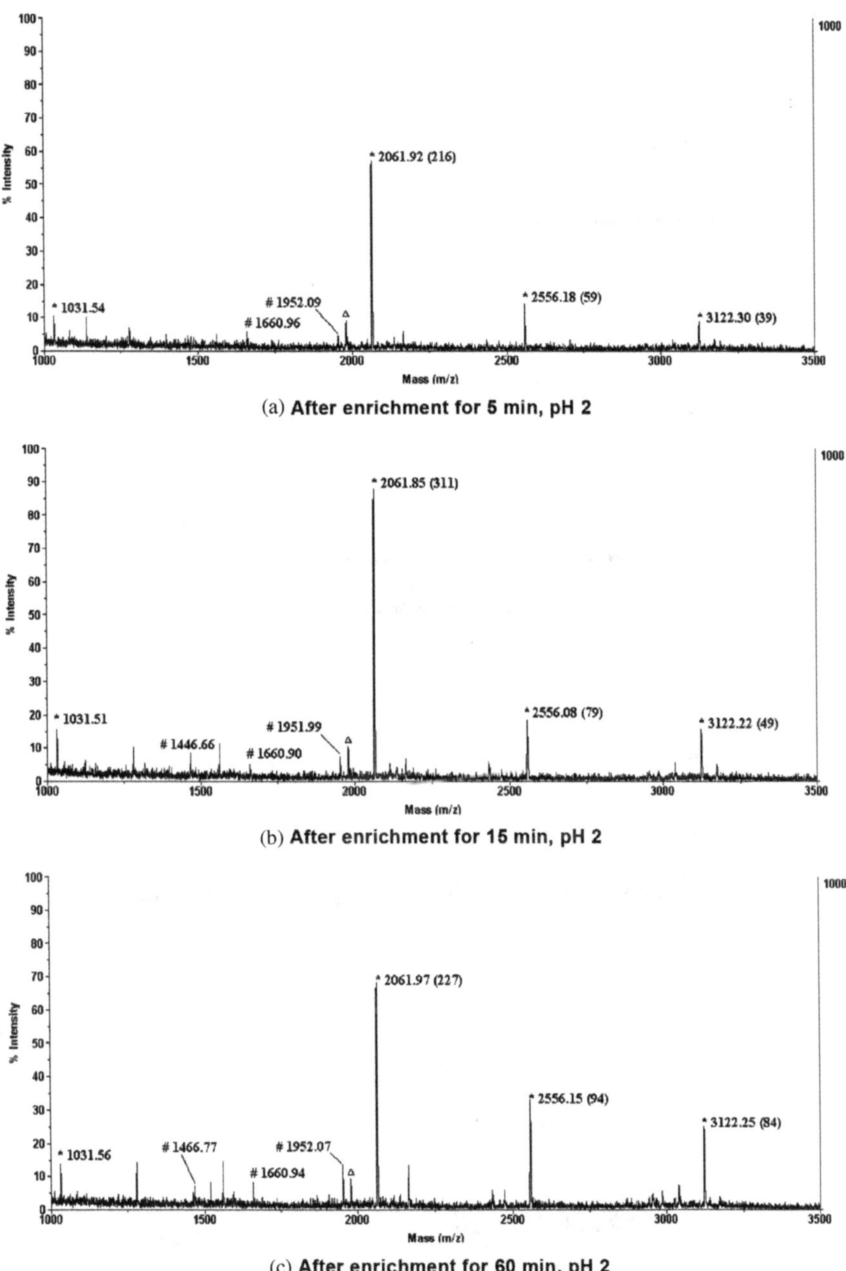

Figure 5.63: MS spectra of 2×10^{-7} mol·L^{-1} β-casein digest after enrichment with Fe$_3$O$_4$@ZrO$_2$ microsphere at pH = 2 using different incubation times. *: phosphopeptide in β-casein digest, #: phosphopeptide in α-casein digest, Δ: dephosphorylated fragment.

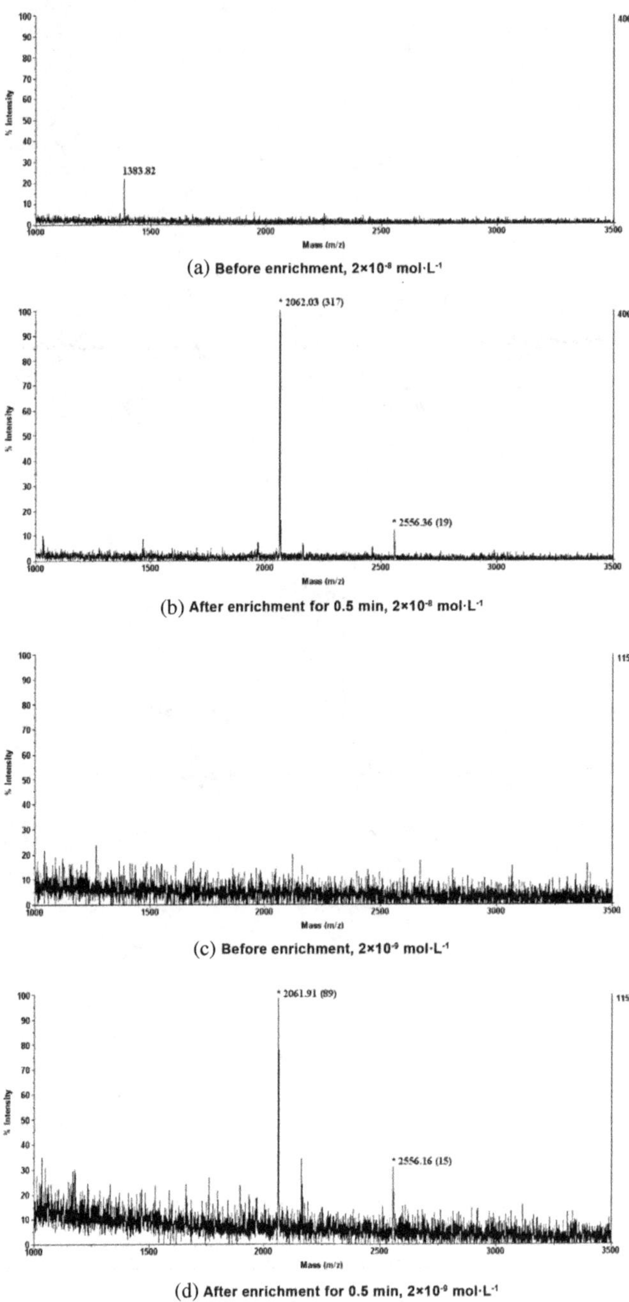

Figure 5.64: MS spectra of different concentrations of β-casein digest (200 μL) before and after enrichment with $Fe_3O_4@ZrO_2$ microsphere at pH = 2.

Figures 5.64(b) and 5.46(d), the two mono-phosphorylated peptides of m/z 2061.83 and m/z 2556.09 are detected. The reason why multi-phosphorylated peptide of m/z 3122.27 is not detected may be the poor ionization efficiency and the selectivity of ZrO_2. Håkansson et $al.$[34] compared the enrichment ability of ZrO_2 and TiO_2 in their research work and found that ZrO_2 is preferrable to enrich mono-phosphorylated peptides. In summary, $Fe_3O_4@ZrO_2$ microsphere still has strong enrichment ability for low abundance phosphorylated peptides using 0.5 min as enrichment time.

(c) *Evaluation of the enrichment ability of $Fe_3O_4@ZrO_2$ microsphere for phosphopeptide in complex peptide mixture*

(i) **Evaluation of the enrichment selectivity of $Fe_3O_4@ZrO_2$ microsphere for phosphorylated peptide in casein protein digest**

Casein protein digest (containing α-S1-casein, α-S2-casein and β-casein) is used to investigate the enrichment selectivity of $Fe_3O_4@ZrO_2$ microsphere. Figure 5.65(a) shows the MS spectrum of the casein peptide mixture directly analyzed by MALDI-TOF-MS. Only three weak phosphopeptide peaks are observed before enrichment. As shown in Figure 5.65(b), after enrichment with $Fe_3O_4@ZrO_2$ microsphere, the phosphorylated peptides in the casein digest are efficiently enriched and detected. The detailed information of amino acid sequence of the identified phosphorylated peptide is shown in Table 5.13. The peaks labeled 4, 5, 6, 7, and 11 in Figure 5.65(b) are assigned to α-S1-casein, the peaks labeled 1, 2, 3, and 10 are assigned to α-S2-casein, and the remaining peaks labeled 8, 9, and 12 are assigned to β-casein. It can be seen from Figure 5.65(b) that all the detected peaks in the spectrum are phosphorylated peptides in casein digest, indicating $Fe_3O_4@ZrO_2$ microsphere has a good enrichment selectivity for phosphopeptides in more complex protein samples.

(ii) **Evaluation of the selective enrichment ability for phosphorylated peptide in five kinds of phosphorylated and non-phosphorylated standard protein digests**

Five protein digests including two phosphorylated proteins (β-casein and ovalbumin) and three non-phosphorylated proteins (BSA, Cyc, MYO) are mixed in equal proportion for evaluation of the enrichment selectivity of $Fe_3O_4@ZrO_2$ microsphere. As shown in Figure 5.66(a), a large number of

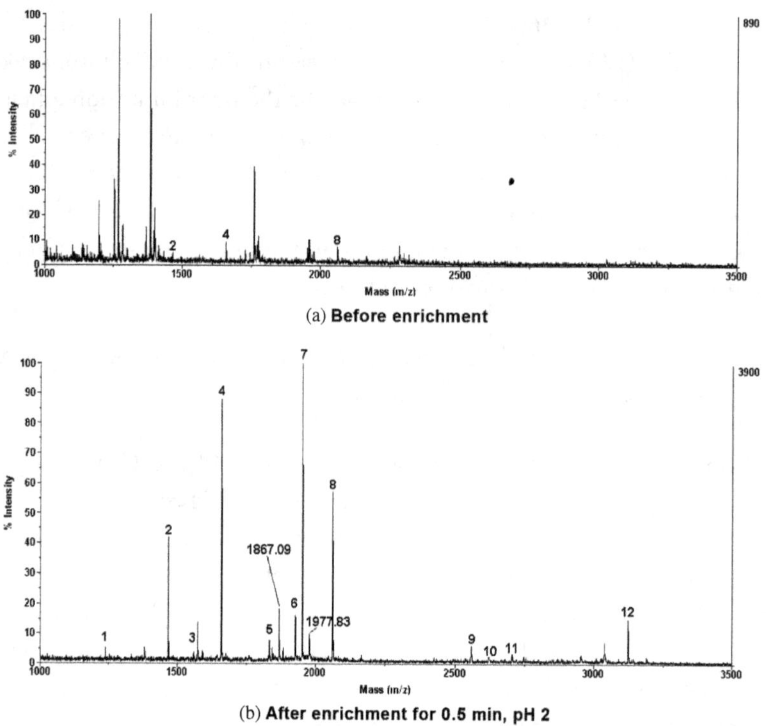

Figure 5.65: MS spectra of casein digest before and after enrichment with $Fe_3O_4@ZrO_2$ microsphere.

Table 5.13: Detailed information of the identified phosphorylated peptide from the casein digest enriched with $Fe_3O_4@ZrO_2$ microsphere.

No.	AA	Peptide Sequence	m/z
1	α-S2/138–147	TVDME[pS]TEVF	1237.36
2	α-S2/138–149	TVDME[pS]TEVFTK	1466.47
3	α-S2/126–137	EQL[pS]T[pS]EENSKK	1567.97*
4	α-S1/106–119	VPQLEIVPN[pS]AEER	1660.61
5	α-S1/104–119	YKVPQLEIVPN[pS]AEER	1832.62
6	α-S1/43–58	DIG[pS]E[pS]TEDQAMETIK	1927.46
7	α-S1/104–119	YKVPQLEIVPN[pS]AEER	1951.71
8	β33–48	FQ[pS]EEQQQTEDELQDK	2061.56
9	β33–52	FQ[pS]EEQQQTEDELQDKIHPF	2555.74

Table 5.13: *(Continued)*

No.	AA	Peptide Sequence	m/z
10	α-S2/2–21	NTMEHV[pS][pS][pS]EESII[pS]QETYK	2618.62
11	α-S2/99–120	LRLKKYKVPQLEIVPN[pS]AEERL	2703.56
12	β/1–25	RELEELNVPGEIVE[pS]L[pS][pS]EESITR	3121.94

*$[M + Na]^+$

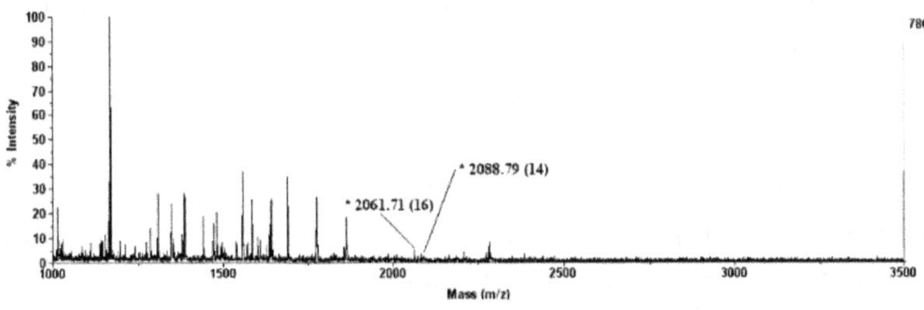

(a) **Before enrichment, 5 mixed proteins**

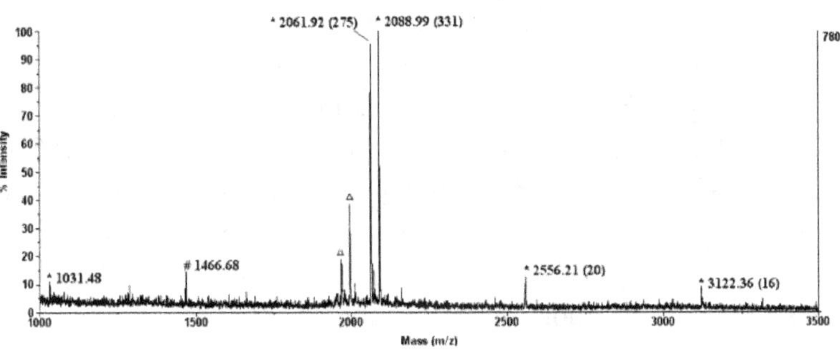

(b) **After enrichment of 5 mixed proteins for 0.5 min, pH 2**

Figure 5.66: MS spectra of peptide mixture of five kinds of proteins (2×10^{-8} mol·L^{-1}) before and after enrichment with $Fe_3O_4@ZrO_2$ microsphere. *: phosphopeptide in β-casein and ovalbumin digest, #: phosphopeptides in α-casein digest, Δ: dephosphorylated fragment.

non-phosphorylated peptides are detected before enrichment and the signals of the phosphorylated peptides are severely suppressed. Only two weak phosphorylated peptide peaks (*m/z* 2061.71 and *m/z* 2088.79) are observed. However, after enrichment with $Fe_3O_4@ZrO_2$ microsphere, eight phosphorylated peptides are detected and no non-phosphorylated peptide can be found in the spectrum. Among them, the peaks of m/z 2061.92, m/z 2556.21, and m/z 3122.36 belong to β-casein, and the peak of m/z 1031.57 is the double-charged peak of the mono-phosphorylated peptide of m/z 3122.36. The peak of m/z 2088.9 belongs to ovalbumin, while the peak of m/z 1466.68 belongs to α-casein. These results indicate that $Fe_3O_4@ZrO_2$ microsphere has good enrichment selectivity for phosphopeptides in complex biological samples.

(d) Evaluation of the enrichment ability of $Fe_3O_4@ZrO_2$ microsphere for phosphorylated peptides in human serum

The serum sample is diluted with an acidic buffer. After enrichment with $Fe_3O_4@ZrO_2$ microsphere, three peaks of m/z 1465.57, m/z 1545.56, and m/z 1616.60 are observed. The amino acid sequence of these three peaks are further identified by tandem mass spectrometry. All the tandem MS spectra give the relatively complete information on y ion fragmentation. In Figures 5.67(b) and 5.67(c), the mass difference between the fragment ions $y13$ (1263.60) and $y14$ (1430.77) is 167 Da, indicating the serine phosphorylation at Ser2 and Ser3. The MS and MS/MS data are searched via NCBI database, and the identified peptides are uniquely confirmed. The results show that all the three peptides identified in Figure 5.67(a) belong to the same fragment of fibrinopeptide A (gi|229185, ADSGEGDFLAEGGGVR). The peptides of m/z 1545.56 (D[pS]GEGDFLAEGGGVR) and m/z 1616.60 (AD[pS]GEGDFLAEGGGVR) are identified as phosphorylated peptides by tandem mass spectrometry and database search. The peptides of m/z 1465.57 and m/z 1545.56 have identical amino acid sequence without phosphorylation.

(3) Enrichment application of $Fe_3O_4@Al_2O_3$ microsphere in phosphoproteomics

(a) *Optimization of pH of the loading buffer*

In this section, the β-casein digest is selected to optimize the pH of the loading buffer. In Figure 5.68, three phosphopeptides of m/z 2061.94, m/z 2556.20, m/z 3122.43, and their corresponding dephosphorylated fragments are all efficiently enriched and identified under pH = 2.[37] Among them, m/z 1977.26 and m/z 2471.37, derived from mono-phosphorylated peptides of m/z 2061.94 and m/z 2556.20, are formed by dephosphorylation during ionization process. Peaks of m/z 1031.54 and m/z 1278.65 are the doubly charged peaks of the mono-phosphorylated peptides of m/z 2061.97 and m/z 2556.20,

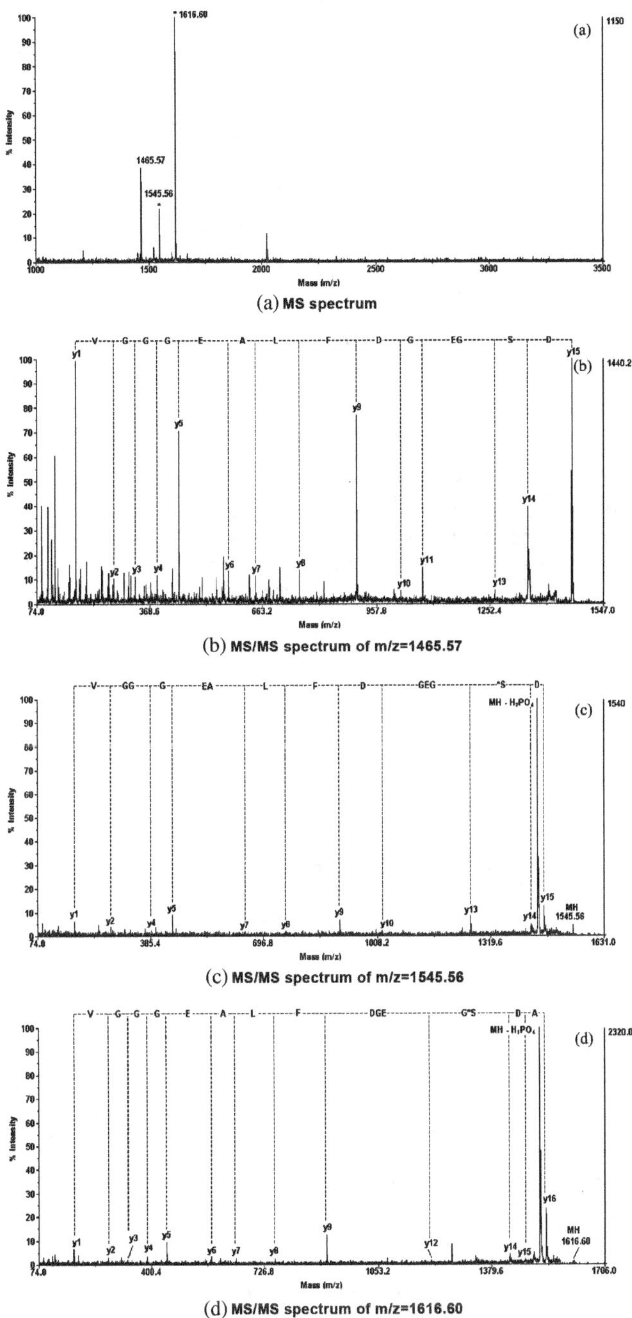

Figure 5.67: MS and MS/MS spectra of phosphopeptides enriched with $Fe_3O_4@ZrO_2$ microsphere from healthy human serum.

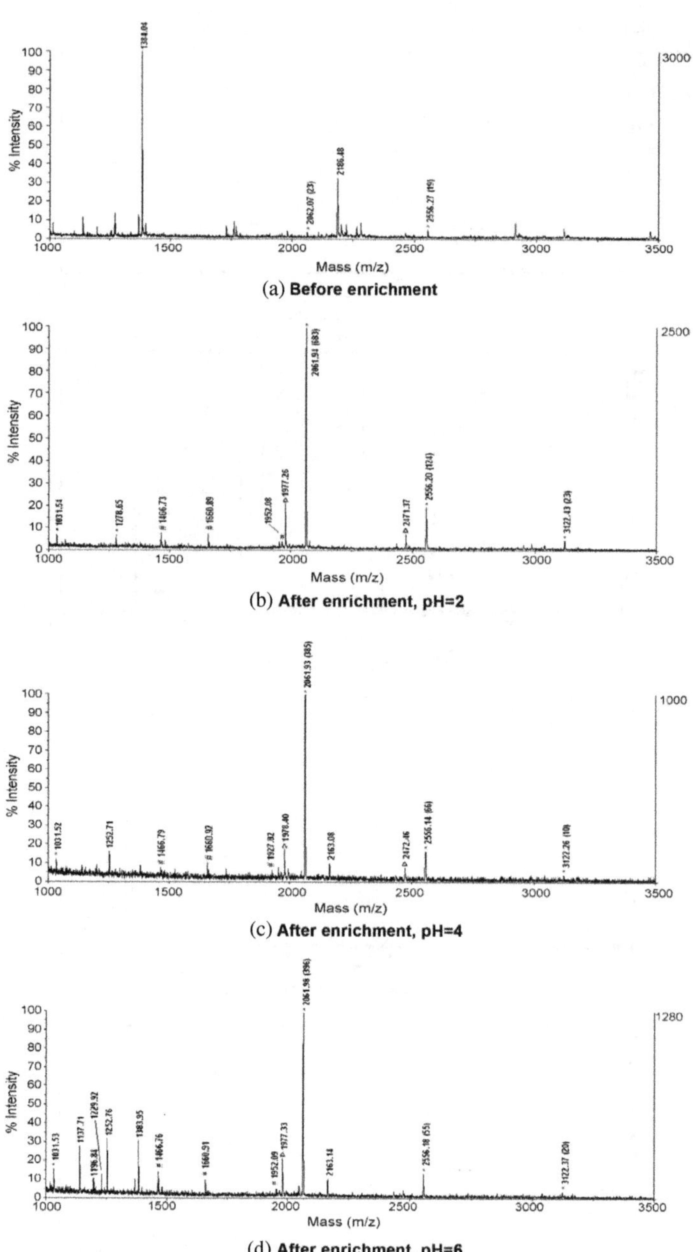

Figure 5.68: MS spectra of 2×10^{-7} mol·L^{-1} β-casein digest before and after enrichment with $Fe_3O_4@Al_2O_3$ microsphere at different pH conditions. *: phosphopeptide in β-casein digest, #: phosphopeptide in α-casein digest, Δ: dephosphorylated fragment.

respectively. In addition, there are three phosphorylated peptides of m/z 1466.73, m/z 1660.89, and m/z 1952.08 belonging to α-casein. Compared with the spectrum before enrichment in Figure 5.68(a), the S/N ratio of all the phosphorylated peptides is significantly improved after enrichment with $Fe_3O_4@Al_2O_3$ microsphere in Figure 5.68(b), and almost no non-phosphorylated peptide is observed.

At pH = 4, the MS spectrum after enrichment (Figure 5.68(c)) has no significant change when compared to the MS spectrum obtained at pH = 2 (Figure 5.68(b)). When the pH value of the loading buffer rises to 6, the phosphopeptides play a dominant role in the spectrum as shown in Figure 5.68(d), but the non-phosphorylated peptide peaks increase significantly. From these results, it can be concluded that although pH has little effect on the enrichment of phosphorylated peptides, the selection of the suitable pH is still beneficial to enhance the enrichment performance of $Fe_3O_4@Al_2O_3$ microspheres.

Similar to the $Fe_3O_4@ZrO_2$ microsphere, the $Fe_3O_4@Al_2O_3$ microsphere shows better enrichment performance at lower pH. The possible reason may be that the Al_2O_3 is also an amphoteric compound. In the acidic solution, Al_2O_3 has positively charged Al^{3+} on the surface, which is Lewis acid, and has ion exchange characteristic. The other reason may be that, many isothermal points of acidic non-phosphorylated peptides are relatively low, the net charge of which is negative at higher pH condition, so that they are easily adsorbed to the surface of the material by electrostatic action. When the pH value drops below 2, the carboxyl groups of most acidic non-phosphorylated peptides will not dissociate, the net charge of which is positive and does not cause non-specific adsorption to the material.

(b) *Evaluation of the enrichment ability of $Fe_3O_4@Al_2O_3$ microsphere for phosphopeptide in complex peptide mixture*

The five protein digests (two phosphorylated proteins, ovalbumin and β-casein, and three non-phosphorylated proteins, BSA, MYO, and Cyc) are selected to evaluate the enrichment selectivity of the material. As shown in Figures 5.69 and 5.70, the results show that $Fe_3O_4@Al_2O_3$ microsphere has excellent selective enrichment ability for phosphorylated peptides from complex samples.

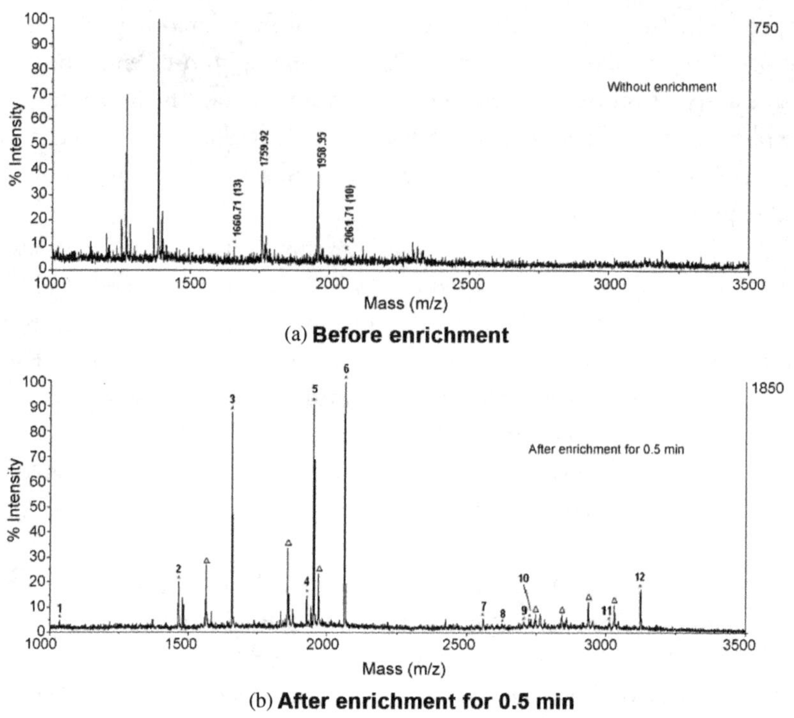

(a) **Before enrichment**

(b) **After enrichment for 0.5 min**

Figure 5.69: MS spectra of the peptide mixture before and after enrichment with $Fe_3O_4@$ Al_2O_3 microsphere. *: phosphopeptide in β-casein and ovalbumin digest; #: phosphopeptide in α-casein digest; Δ: dephosphorylated fragment.

(c) *Evaluation of the enrichment ability of $Fe_3O_4@Al_2O_3$ microsphere for phosphopeptide in rat liver digest*

Rat liver digest is selected as the biological sample to investigate the enrichment ability of $Fe_3O_4@Al_2O_3$ microsphere for phosphopeptides in practical samples. 100 μg rat liver digest is enriched with $Fe_3O_4@Al_2O_3$ microsphere. The eluent is lyophilized and then re-dissolved for capillary C18 column for separation and finally identified by LTQ-Orbitrap MS. The obtained MS/MS data is searched by Sequest database, and the results are screened according to the Xcorr value. Based on the results of MS/MS and MS/MS/MS, a total of 15 peptides are simultaneously identified by MS/MS and MS/MS/MS. The detailed information of the identified phosphopeptides and the corresponding Xcorr values are shown in Table 5.14.

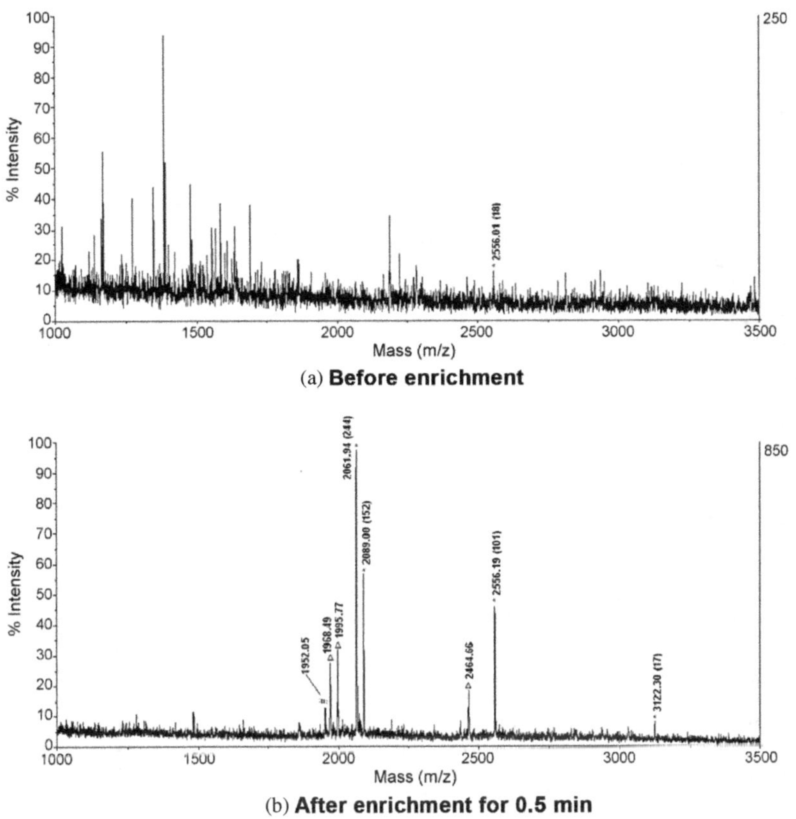

(a) **Before enrichment**

(b) **After enrichment for 0.5 min**

Figure 5.70: MS spectra of peptide mixture (2×10^{-8} mol·L^{-1}) before and after enrichment with $Fe_3O_4@Al_2O_3$ microsphere. *: phosphopeptide in β-casein and ovalbumin digest; #: phosphopeptide in α-casein digest; Δ: dephosphorylated fragment.

After manual confirmation, the MS/MS and MS/MS/MS spectra of these peptides are considered to be of hig h quality and the results are credible. Taking one peptide (IGHHS[pS]TSDDSSAYR) as an example, as shown in Figure 5.71, the *b*-ion and *y*-ion information in the MS/MS and MS/MS/MS spectra are consistent with the theoretically predicted peptide fragmentation information. In addition, from the MS/MS spectrum, the parent ion peak of *m/z* 537.55 and its fragment peak of *m/z* 505.78 can be clearly observed. In this work, the mono-phosphorylated peptides identified in the MS/MS are manually confirmed, and 19 mono-phosphorylated peptides are finally identified.

Table 5.14: Detailed information of the phosphorylated peptides enriched with $Fe_3O_4@Al_2O_3$ microsphere in rat liver digest.

No.	Protein No.	Peptide Sequence	Charge	XC	MS2	MS3
1	IP100476899	K.YGPVSVADTIGSGAADAKDDDDIDLFGS*DDEEESEDAK.R	3	8.772	*	*
2	IP100197900	K.KGATPAEDDEDNDIDLFGS*DEEEEDKEAAR.L	3	7.682	*	*
3	IP100361246	K.GDMS*DEDDENEFFDAPEIITMPENLGHK.R	3	7.298	*	*
4	IP100I97900	K.GATPAEDDEDNDIDLFGS*DEEEEDKEAAR.L	3	4.838	*	*
5	IP100208304	K.IYHLPDAES*DEDEDFKEQTR.L	3	4.150	*	*
6	IP100210566	K.ESDDKPEIEDVGS*DEEEEEKK.D	3	4.635	*	*
7	IP100476899	K.DDDDIDLFGS*DDEEESEDAKR.L	3	5.352	*	*
8	IP100195102	RKDEDS*DDESQSSHAGK.K	3	5.225	*	*
9	IP100476899	K.DDDDIDLFGS*DDEEESEDAK.R	2	5.343	*	*
10	IP100365935	K.DWEDDS*DEDMSNFDR.F	2	6.352	*	*
II	IP100471584	K.IEDVGS*DEEDDSGKDK.K	2	4.861	*	*
12	IP100365663	R.IGHHS*TSDDSSAYR.S	3	4.125	*	*
13	IP100471584	K.IEDVGS*DEEDDSGK.D	2	4.465	*	*
14	IP100192336	R.KRET*DDEGEDD	2	3.394	*	*
15	IP100188079	K.KSEGS*PNQGK.K	2	2.706	*	*
16	IP100187860	K.MESEAGADDS*AEEGDLLDDDDNEDRGDDQLELK.D	3	7.591	*	
17	IP100187860	K.MESEAGADDS*AEEGDLLDDDDNEDR.G	3	6.101	*	
18	IP100388302	K.VIHDNFGIVEGLMTIVHAITAT*QK.T	3	5.442	*	

19	IP100187860	K.DDEKEPEEGEDDRDS*ANGEDDS	3	4.369	*
20	IP100210566	K.ESDDKPEIEOVGS*DEEEEEK.K	3	4.403	*
21	IP100471584	K.IEDVGS*DEEDDSGKDKK.K	3	4.152	*
22	IP100362946	R.NRFTIDS*DAISASSPEK.E	2	2.984	*
23	IP100200898	R.SASS*DTSEELNAQDSPK.R	2	4.683	*
24	IP100I93648	K.S*ASPAPADVAPAQEDLR.T	3	4.298	*
25	IP100476178	K.ESLKEEDES*DDDNM	2	3.943	*
26	IP100231770	R.EDEIS*PPPPNPVVK.G	2	2.668	*
27	IP100766722	R.WLDES*DAEMELR.A	2	2.906	*
28	IP100201032	K.NEEDEGHSNSS*PR.H	2	4.141	*
29	IP10020760l	K.NFETNDLAFS*PK.G	2	3.178	*
30	IP100360386	R.S*GDETPGSEAPGDK.A	2	3.303	*
31	IP100209277	K.LSSQLS*AGEEK.W	2	3.238	*
32	IP100209618	R.APTAAPS*PEPR.D	2	2.640	*
33	IP100326566	R.AGDMLEDS*PK.R	2	3.065	*
34	IP100189138	K.TSAS*PPLEK.S	2	2.651	*

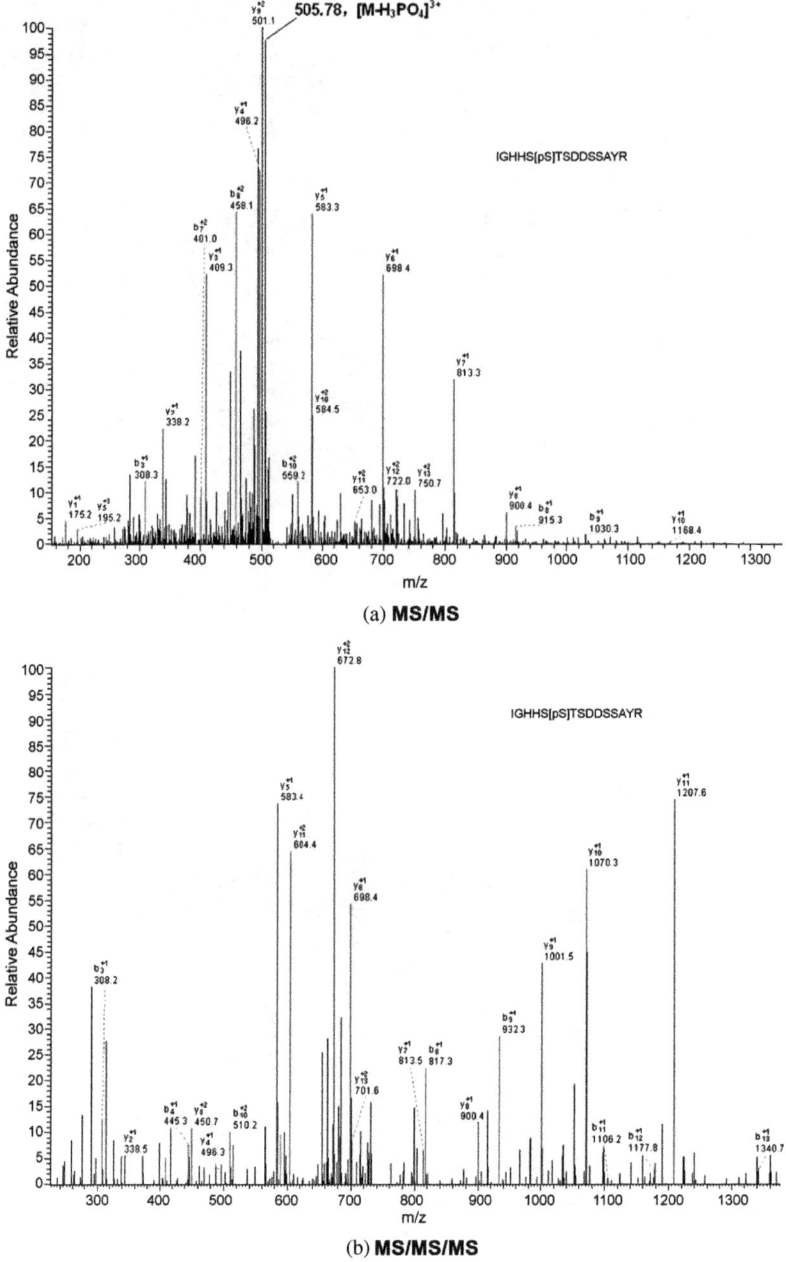

Figure 5.71: MS/MS and MS/MS/MS spectra of the three-charge phosphorylated peptides (IGHHS[pS]TSDDSSAYR).

The detailed information of the phosphopeptides identified only by the MS/MS and the corresponding Xcorr values are also listed in Table 5.14. In summary, a total of 34 phosphorylated peptides are identified in rat liver tissue.

(4) Enrichment application of $Fe_3O_4@Ga_2O_3$ microsphere in phosphoproteomics

The sensitivity of $Fe_3O_4@Ga_2O_3$ microsphere for β-casein digest is as low as 2×10^{-10} mol·L^{-1} (40 fmol).[38] It indicates that the $Fe_3O_4@Ga_2O_3$ microsphere also has excellent enrichment ability for phosphopeptide from low-concentration sample within 0.5 min.

In addition, five protein digests (two phosphorylated proteins, ovalbumin and β-casein and three non-phosphorylated proteins, BSA, MYO, Cyc) are selected to evaluate the enrichment selectivity of the material. The results show that $Fe_3O_4@Ga_2O_3$ microsphere has the ability to enrich phosphopeptide from complex sample.

The rat liver digest is enriched with $Fe_3O_4@Ga_2O_3$ microsphere and identified by LTQ-Orbitrap MS, MS/MS, and MS/MS/MS analysis. A total of 16 peptides are simultaneously identified by MS/MS and MS/MS/MS, and the detailed information and corresponding Xcorr values are shown in Table 5.15.

(5) Comparison of the enrichment performance of $Fe_3O_4@Ta_2O_5$ microsphere and $Fe_3O_4@TiO_2$ microsphere for phosphorylated peptides

The non-fat milk (α-casein and β-casein) is used for the comparison of the enrichment performance of $Fe_3O_4@Ta_2O_5$ microsphere and $Fe_3O_4@TiO_2$ microsphere.[39] 50 mg·mL^{-1} DHB, 50% ACN+0.1% TFA is selected as the loading buffer for enrichment of the phosphorylated peptides in the non-fat milk digest. As shown in Figure 5.72, the $Fe_3O_4@Ta_2O_5$ microsphere has similar enrichment performance and selectivity to $Fe_3O_4@TiO_2$ microsphere. To further explore whether there is difference in the phosphopeptide enrichment between $Fe_3O_4@Ta_2O_5$ microsphere and $Fe_3O_4@TiO_2$ microsphere. A more complex practical sample, rat liver digest, is selected. A total of 115 phosphorylation sites are identified when using $Fe_3O_4@Ta_2O_5$ microsphere,

Table 5.15: Detailed information of phosphopeptides enriched with $Fe_3O_4@Ga_2O_3$ microsphere in rat liver digest.

Protein No.	Peptide Sequence	Charge	X_{corr}	MS2	MS3
IP100231770	R.TDSREDEIS*PPPPNPWK.G	2	4.224	*	*
IP100326606	R.TAS*LTSAASIDGSR.S	2	4 001	*	*
IP100200898	R.SAS*SDTSEELNAQDSPK.R	2	4 527	*	*
IP100365663	R.S*VDEVNYWDKODHPISR.L	3	3.035	*	*
IP100190024	R.MAMPINVS*DPDLLR.H	2	3.704	*	*
IP100365663	R.IGHHS*TSDDSSAYR.S	3	4.956	*	*
IP10019402	R.FIIGSVSEDNS*EDEISNLVK.L	2	4.856	*	*
IP100209277	K.LSSOLS*AGEEK.W	2	3.708	*	*
IP100197900	K.KGATPAEDDEDNDIDLFGS*DE EEEDKEAAR.L	3	7.865	*	*
IP100208304	K.IYHLPDAES*DEDEDFKEOTR.L	3	4 655	*	*
IP100471584	K.IEDVGS*DEEDDSGKDK.K	2	5.028	*	*
IP100201032	K.IDASKNEEDEGHSNSS*PR.H	3	3.547	*	*
IP100210566	K.ESDDKPEIEDVGS*DEEEEEKK.D	3	5.044	*	*
IP100480820	K.EGEEPTVYS*DDEEPKDEAAR.K	3	3. 437	*	*
IP100365935	K.DWEDDS*DEDMSNFDR.F	2	6.301	*	*
IP100200661	K.AGS*DTELAAPK.S	2	3..794	*	*

of which 107 occur on serine residues (93.04%), five occur on threonine residues (4.35%), and three occur on tyrosine residues (2.61%). At the same time, 86 non-redundant phosphorylated peptides are identified, of which 63 phosphorylated peptides have one phosphorylation site, 18 phosphorylated peptides have two phosphorylation sites, four phosphorylated peptides have three phosphorylation sites, and two phosphorylated peptides have four phosphorylation sites. These results are shown in Figure 5.73. It shows that $Fe_3O_4@Ta_2O_5$ microsphere and $Fe_3O_4@TiO_2$ microsphere have similar enrichment performance, so the phosphorylation level and phosphorylation site cannot be used to reveal the difference between these two materials. Figure 5.73(c) shows that there is about 50% overlap of the phosphopeptide enriched by both the materials, that is, there is the enrichment difference between these two materials. At the same time, it also indicates that the

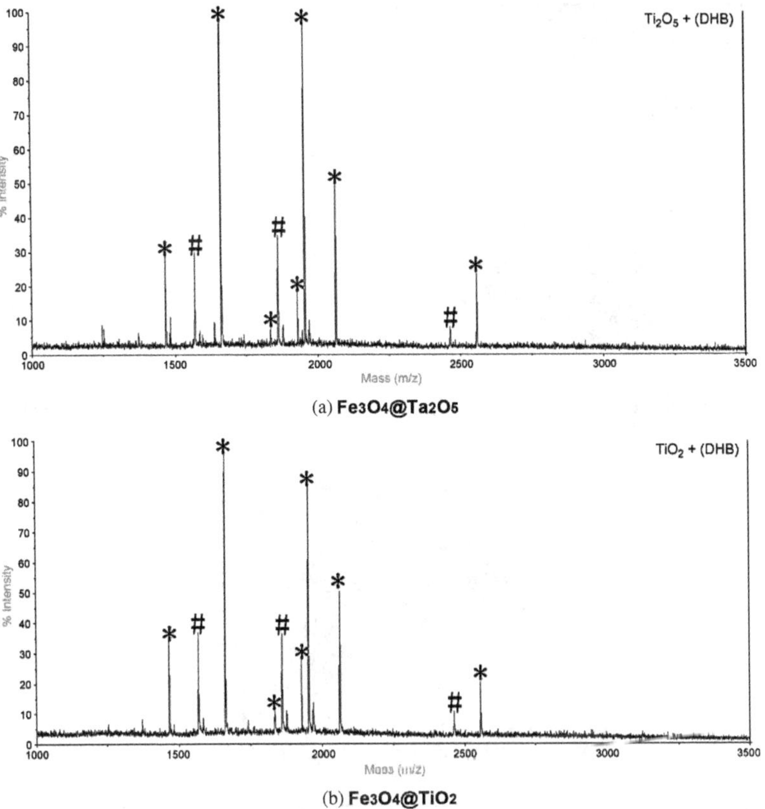

Figure 5.72: MS spectra of phosphopeptide in non-fat milk digest after enrichment with different materials. *: phosphopeptide, #: dephosphorylated fragment.

enrichment performance of $Fe_3O_4@Ta_2O_5$ microsphere and $Fe_3O_4@TiO_2$ microsphere is highly complementary.

(6) Comparison of the enrichment performance of $Fe_3O_4@SnO_2$ microsphere and $Fe_3O_4@TiO_2$ microsphere for phosphorylated peptide

The non-fat milk (containing α-casein and β-casein) is first used to compare the enrichment performance of $Fe_3O_4@SnO_2$ microsphere and $Fe_3O_4@TiO_2$ microsphere for phosphorylated peptides.[40] As shown in Figure 5.74, the $Fe_3O_4@SnO_2$ microsphere has excellent enrichment ability for phosphoryl-

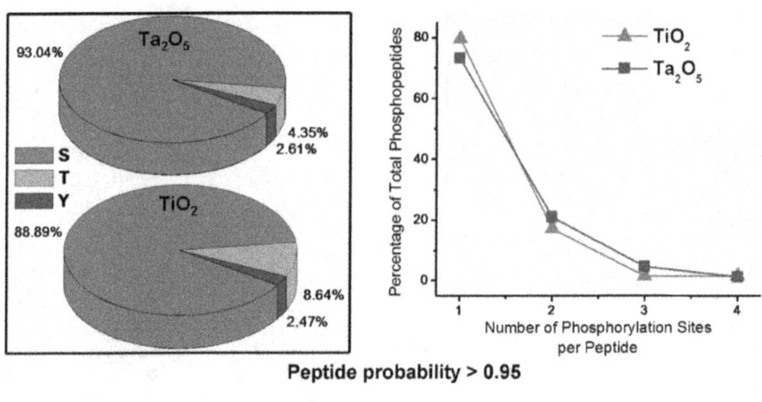

(a) Distribution comparison of phosphorylation site   (b) Comparison of phosphorylation level

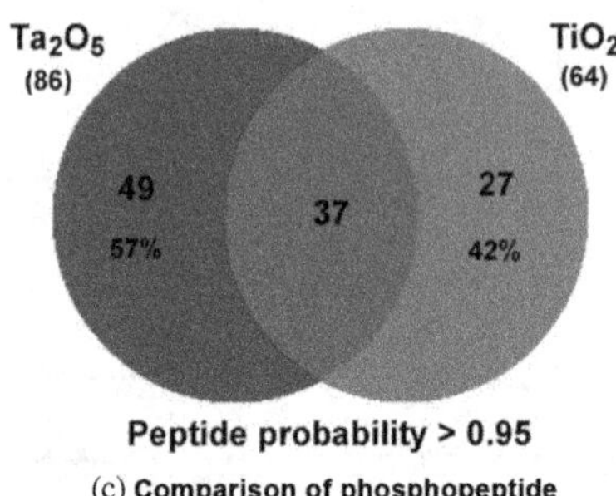

Peptide probability > 0.95

(c) Comparison of phosphopeptide

Figure 5.73 Comparison of the phosphorylation site distribution, phosphorylation level, and phosphorylated peptide identified by $Fe_3O_4@Ta_2O_5$ microsphere and $Fe_3O_4@TiO_2$ microsphere.

ated peptides. It is worth mentioning that, compared with $Fe_3O_4@SnO_2$ microsphere (Figure 5.74(b)), two peaks with significant difference are found at m/z 1244.73 and m/z 1635.86 after enrichment with $Fe_3O_4@TiO_2$ microsphere (Figure 5.74(c)), and their tandem MS spectra are listed in Figure 5.75. Since there is no significant mass loss of the phosphate group in the tandem MS spectra of the two peptides, it can be concluded t hat these two peptides are non-phosphorylated peptides. At the same time, the continuous b ion and y ion in the tandem MS spectra are detected, indicating that the two non-phosphorylated peptides contain at least five acidic amino acid residues, and thus affect the enrichment selectivity of metal oxide. As a Lewis acid, although the binding ability of TiO_2 to Lewis base is stronger than that of SnO_2, the enrichment selectivity is affected. This result demonstrates that $Fe_3O_4@SnO_2$

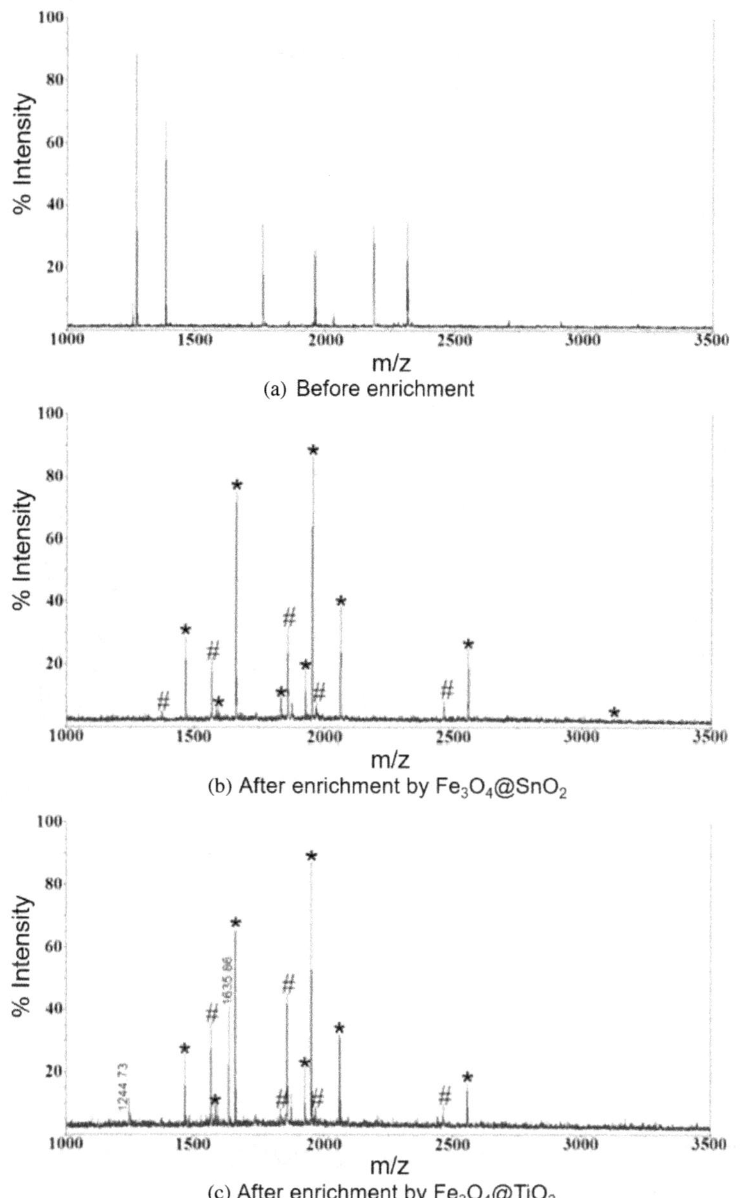

Figure 5.74: MS spectra of the non-fat milk digest before and after enrichment with two different materials *: phosphorylated peptide; #: dephosphorylated fragment.

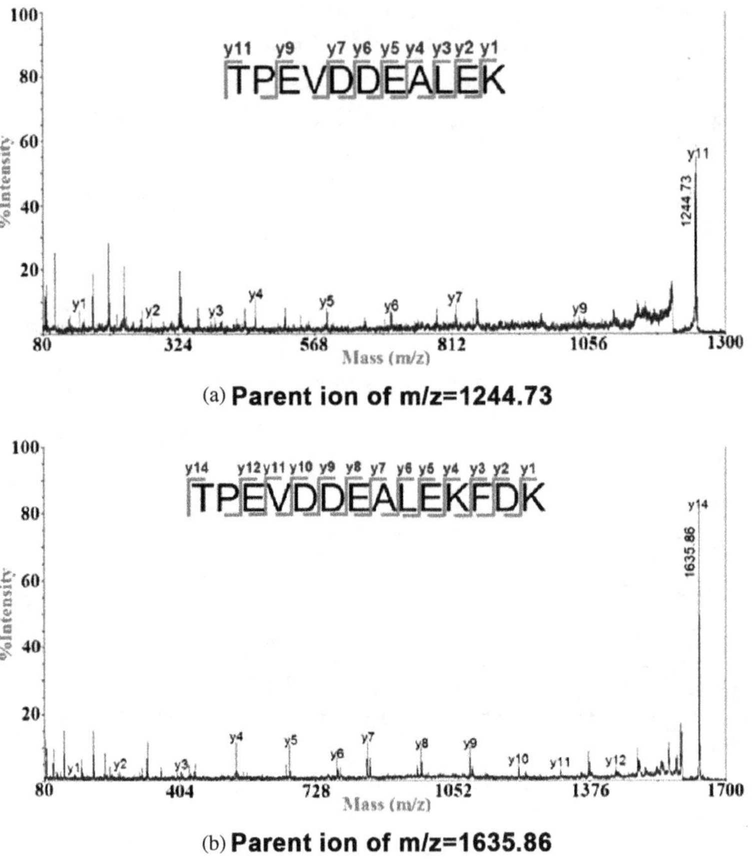

Figure 5.75: MS/MS spectra of non-phosphorylated peptide adsorbed by $Fe_3O_4@TiO_2$ microsphere.

microsphere has excellent enrichment ability and better enrichment selectivity for phosphopeptides than $Fe_3O_4@TiO_2$ microsphere.

5.3.3. *Mesoporous metal oxide-modified magnetic microsphere*

5.3.3.1. *$Fe_3O_4@mTiO_2$*

(1) Preparation of $Fe_3O_4@mTiO_2$ microsphere

The synthesis of $Fe_3O_4@mTiO_2$ microsphere can be seen in Section 2.7 of Chapter 2.[41]

(2) Enrichment application of $Fe_3O_4@mTiO_2$ microsphere in phosphoproteomics

(a) *Evaluation of the enrichment capability of $Fe_3O_4@mTiO_2$ microsphere for phosphopeptide in standard phosphoprotein digest*

The standard phosphoprotein of β-casein (containing trace amount of α-casein) is selected to investigate whether the $Fe_3O_4@mTiO_2$ microsphere has good enrichment ability for the phosphorylated peptide. As shown in Figure 5.76(a), the non-phosphorylated peptide peaks dominate the spectrum, and the phosphopeptide signals are severely suppressed when directly analyzing 4×10^{-7} mol·L^{-1} β-casein digest. However, after enrichment by $Fe_3O_4@mTiO_2$ microsphere, the signals of phosphorylated peptides are significantly enhanced. In addition, three phosphorylated peptide peaks from

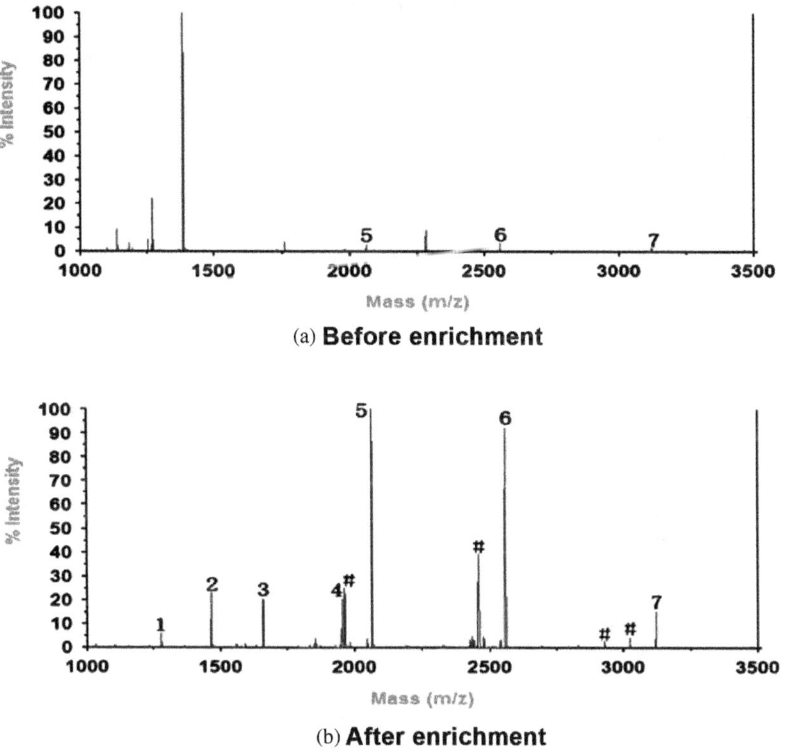

Figure 5.76: MS spectra of 4×10^{-7} mol·L^{-1} β-casein digest before and after enrichment with $Fe_3O_4@mTiO_2$ microsphere. Phosphopeptide labeled with number and # represents dephosphorylated fragment.

α-casein digest are also detected. This indicates that $Fe_3O_4@m\mathrm{TiO}_2$ microsphere has the enrichment ability for phosphopeptides.

As shown in Figure 5.77, when the concentration of β-casein digest is as low as 4×10^{-10} mol·L^{-1}, phosphopeptides with high intensity can still be

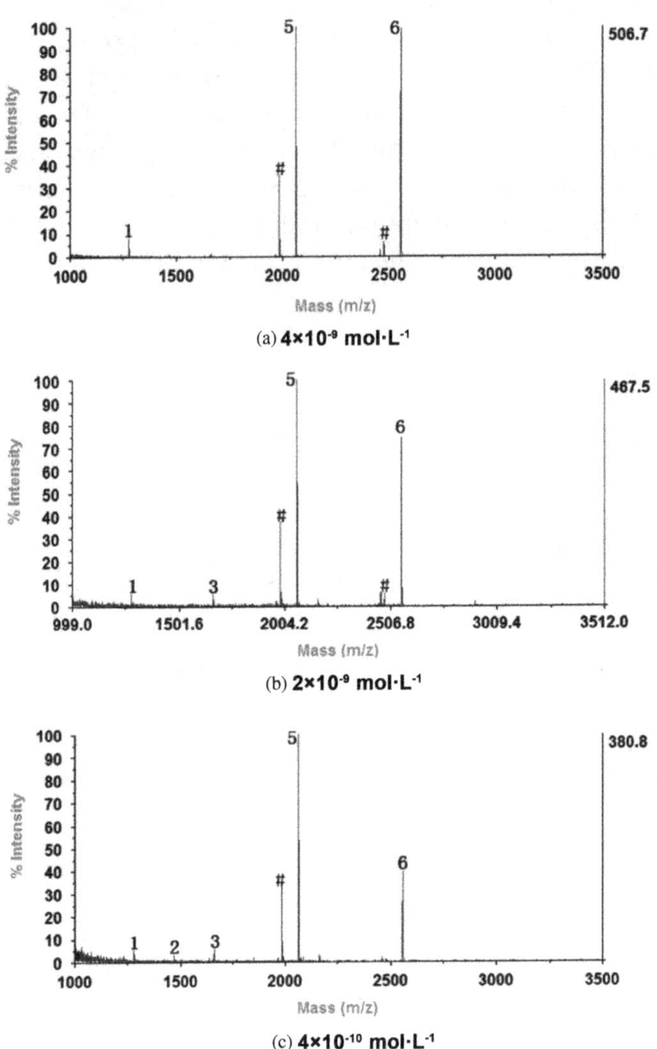

Figure 5.77: MS spectra of different concentrations of β-casein digests after enrichment with $Fe_3O_4@m\mathrm{TiO}_2$ microsphere. Phosphopeptide labeled with number and # represents dephosphorylated fragment.

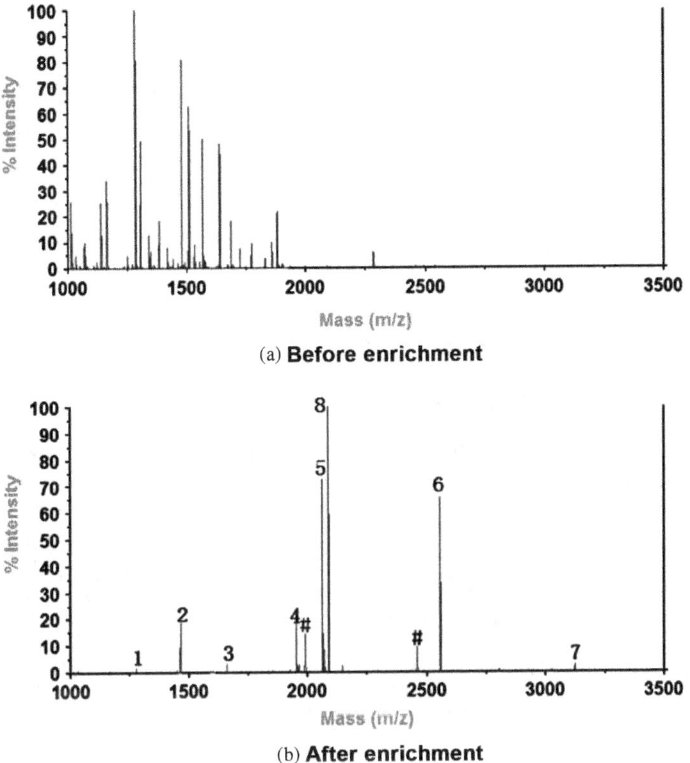

Figure 5.78: MS spectra of peptide mixture of β-casein/ovalbumin/BSA=1/1/50 before and after enrichment with Fe_3O_4@mTiO$_2$ microsphere. Phosphopeptide labeled with number and # represents dephosphorylated fragment.

observed after enrichment with Fe_3O_4@mTiO$_2$ microsphere. This result indicates that Fe_3O_4@mTiO$_2$ microsphere has higher enrichment sensitivity for phosphorylated peptides.

(b) *Evaluation of the enrichment ability of Fe_3O_4@mTiO$_2$ microsphere for phosphopeptide in complex peptide mixture*

The peptide mixture of β-casein/ovalbumin/BSA at molar ratio of 1/1/50 (4×10^{-8} mol·L^{-1} β-casein) is selected as sample. As shown in Figure 5.78(a), all the peaks are non-phosphorylated peptides when directly analyzing the peptide mixture. After enrichment with Fe_3O_4@mTiO$_2$ microsphere, almost no non-phosphorylated peptide is detected in Figure 5.78(b). The difference between the spectra indicates that the Fe_3O_4@mTiO$_2$ microsphere has good enrichment selectivity for the phosphopeptides.

(c) *Evaluation of the enrichment capability of $Fe_3O_4@mTiO_2$ microsphere for phosphopeptide in mouse brain digest*

Mouse brain digest is used to investigate the enrichment ability of $Fe_3O_4@mTiO_2$ microsphere for phosphopeptides. After enrichment with $Fe_3O_4@mTiO_2$ microsphere, 731 non-redundant phosphorylated peptides and 1774 phosphorylation sites are identified, of which 60.99% (1082) occur on serine residues, 28.75% (510) occur on threonine residues, and 10.26% (182) occur on tyrosine residues.

5.3.3.2. $Fe_3O_4@mTiO_2@mSiO_2$

(1) Preparation of $Fe_3O_4@mTiO_2@mSiO_2$ microsphere

The synthesis of $Fe_3O_4@mTiO_2@mSiO_2$ microsphere can be seen in Section 2.7 of Chapter 2.[42]

(2) Enrichment application of $Fe_3O_4@mTiO_2@mSiO_2$ microsphere in phosphoproteomics

$Fe_3O_4@mTiO_2@mSiO_2$ microsphere is considered to be suitable for enrichment of endogenous phosphopeptides. It is because the $mSiO_2$ shell has a size-exclusion effect for macromolecular proteins and the pore of the middle layer facilitates the diffusion of the endogenous phosphopeptides, increasing the interacting region between endogenous phosphopeptides and $Fe_3O_4@mTiO_2$. Besides, the $Fe_3O_4@mTiO_2$ with large specific surface area can provide a large number of affinity sites, which help to simplify the entire enrichment process. The selective enrichment process of the endogenous phosphopeptide by $Fe_3O_4@mTiO_2@mSiO_2$ microsphere is shown in Figure 5.79.

(a) *Evaluation of the enrichment selectivity of $Fe_3O_4@mTiO_2@mSiO_2$ microsphere for phosphopeptide*

First, peptide mixture of BSA digest/β-casein with different molar ratios is used to study the enrichment selectivity of $Fe_3O_4@mTiO_2@mSiO_2$ microsphere. As shown in Figure 5.80, when the molar ratio of BSA and β-casein increases from 10:1 to 1000:1, no significant change in phosphopeptide peak is observed, and all the three phosphopeptides in β-casein digest can be detected after enrichment with $Fe_3O_4@mTiO_2@mSiO_2$ microsphere. It indi-

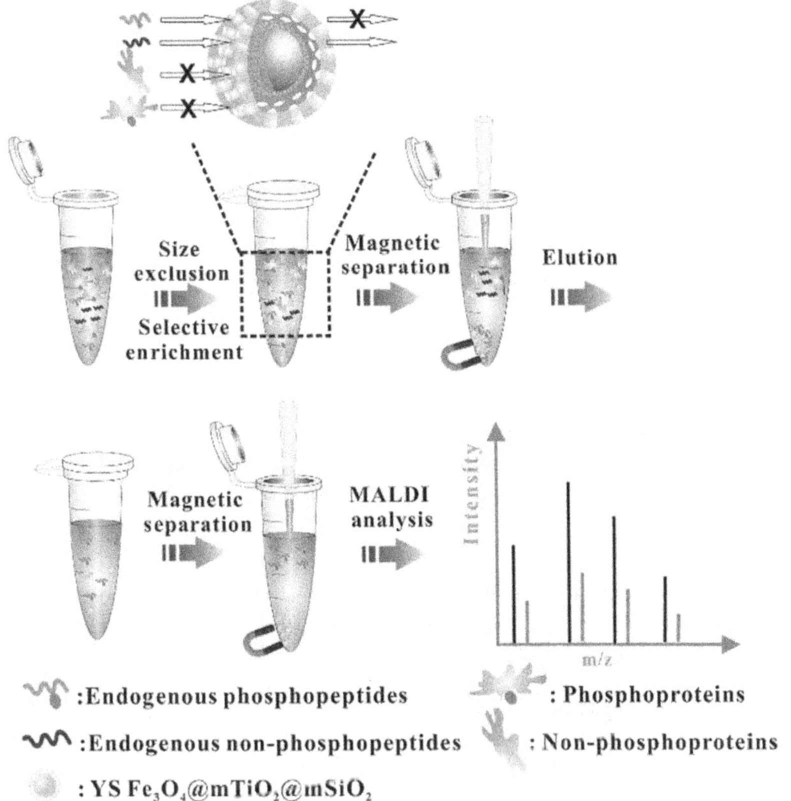

Figure 5.79: Schematic diagram of the selective enrichment process of endogenous phosphopeptide by $Fe_3O_4@m\text{TiO}_2@m\text{SiO}_2$ microsphere.

cates that $Fe_3O_4@m\text{TiO}_2@m\text{SiO}_2$ microsphere has good selective enrichment ability for phosphopeptides.

(b) *Evaluation of the size-exclusion effect of $Fe_3O_4@m\text{TiO}_2@m\text{SiO}_2$ microsphere* The mixture of α-casein protein (23,690 Da) and β-casein digest is used to investigate whether the $m\text{SiO}_2$ shell has the size-exclusion ability. At the same time, $Fe_3O_4@m\text{TiO}_2$ microsphere is also used to enrich the same sample for comparison.

As shown in Figures 5.81(c) and 5.81(d), it can be seen that both materials have enrichment ability for phosphopeptides, but due to the presence of a

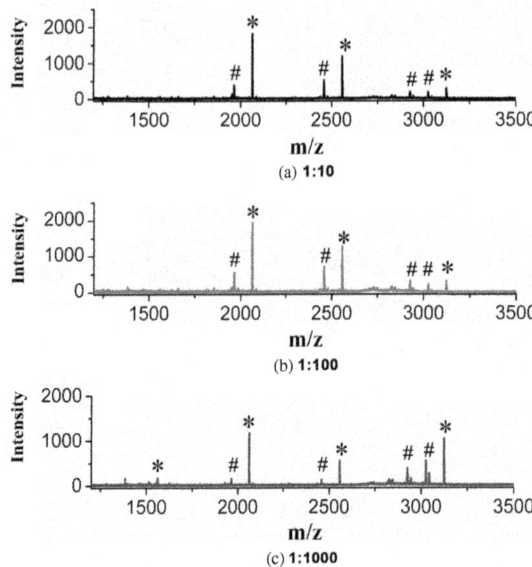

Figure 5.80: MS spectra of peptide mixture of BSA/β-casein at different molar ratios after enrichment with Fe_3O_4@mTiO_2@mSiO_2 microsphere. *: phosphorylated peptide, #: dephosphorylated fragment.

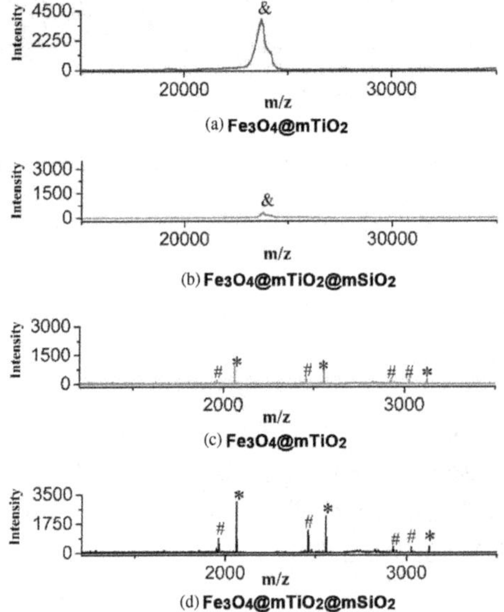

Figure 5.81: MS spectra of the mixture of α-casein protein and β-casein digest after enrichment with different materials. *: phosphorylated peptide, #: dephosphorylated fragment.

large amount of α-casein, the signals of phosphopeptides enriched with $Fe_3O_4@mTiO_2$ microsphere are very weak. As shown in Figure 5.81(a), a large amount of α-casein is detected in the eluent after enrichment with $Fe_3O_4@mTiO_2$ microsphere. In contrast, as shown in Figure 5.81(b), the signals of phosphorylated peptide enriched with $Fe_3O_4@mTiO_2@mSiO_2$ microsphere are stronger, and the signal of α-casein protein is weak, indicating the $mSiO_2$ layer has the size-exclusion effect on the α-casein protein.

(c) *Evaluation of the enrichment ability of $Fe_3O_4@mTiO_2@mSiO_2$ microsphere for phosphopeptide in human serum*

Human serum is used to investigate the enrichment selectivity of $Fe_3O_4@mTiO_2@mSiO_2$ microsphere for endogenous phosphopeptide in practical sample.

As shown in Figure 5.82, four phosphorylated peptides with high S/N ratio are detected in the eluent after enrichment with $Fe_3O_4@mTiO_2@mSiO_2$

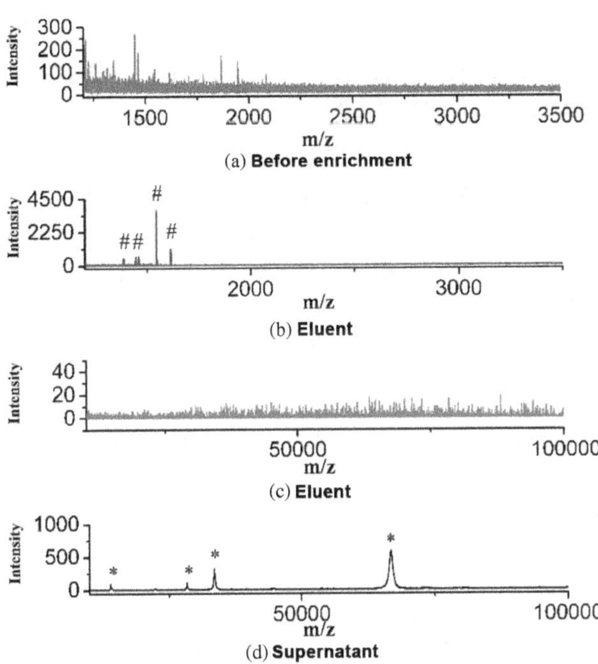

Figure 5.82: MS spectra of human serum before and after enrichment with $Fe_3O_4@mTiO_2@mSiO_2$ microsphere.

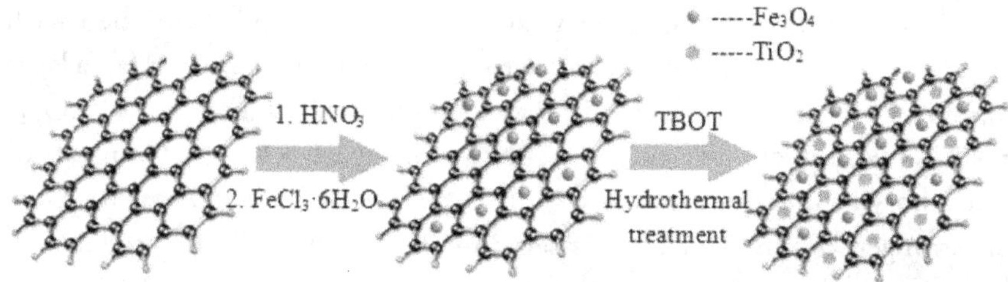

Figure 5.83: Schematic diagram of the synthesis of Mag G@TiO$_2$ composite.

microsphere. As shown in Figure 5.82(d), in the high molecular weight region, apparent protein peaks are observed (human serum albumin, 67 KDa, $5 \times 7 \times 7$ nm), but no protein is detected in the eluent (Figure 5.82(c)). This indicates that the Fe$_3$O$_4$@mTiO$_2$@mSiO$_2$ microsphere has excellent enrichment selectivity for endogenous phosphopeptide and has size-exclusion ability for the macromolecular protein.

5.3.4. *Immobilization of metal oxide on magnetic graphene (Mag GO)*

The Mag GO@TiO$_2$ composite with large specific surface area is successfully synthesized (Figure 5.83) and applied to the selective enrichment of phosphopeptide.[43] Mag GO@TiO$_2$ composite selectively enriches the phosphorylated peptide in human liver cancer digest, and more than 200 phosphopeptides are identified, indicating that this method is effective for enrichment of phosphopeptides in complex biological samples.

5.3.5. *Immobilization of hybrid metal oxide on magnetic graphene*

5.3.5.1. *Mag GO@(Ti-Sn)O$_4$ hybrid material*

(1) Preparation of Mag GO@(Ti-Sn)O$_4$ hybrid material

The synthesis of Mag GO@(Ti-Sn)O$_4$ hybrid material is shown in Figure 2.113 of Chapter 2.[44] The detailed synthesis process can be seen in Section 2.8 of Chapter 2.

(2) Enrichment application of Mag GO@(Ti-Sn)O$_4$ hybrid material in phosphoproteomics

(a) *Evaluation of the enrichment ability of Mag GO@(Ti-Sn)O$_4$ hybrid material for phosphopeptide in standard phosphoprotein digest*

β-casein digest is selected to investigate the enrichment ability of Mag GO@(Ti–Sn)O$_4$ hybrid material for phosphorylated peptides. At the same time, Mag GO@TiO$_2$ and Mag GO@SnO$_2$ are also used for comparison. As shown in Figure 5.84(a), when directly analyzing the 4×10^{-10} mol·L^{-1} β-casein

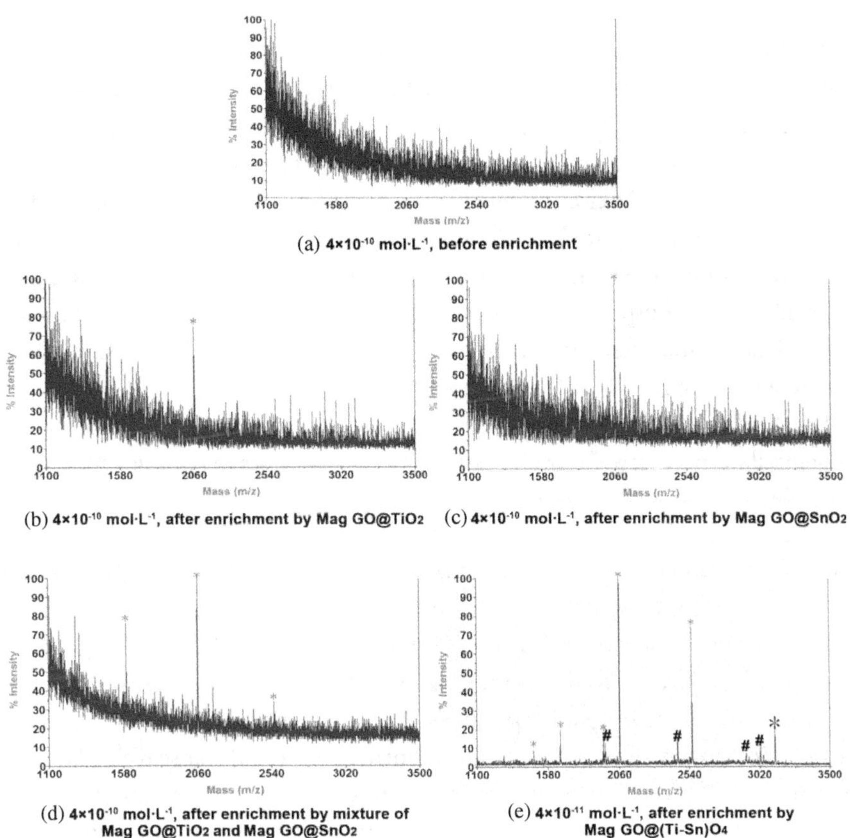

Figure 5.84: MS spectra of β-casein digest before and after enrichment with different materials. *: phosphopeptide; #: dephosphorylated fragment.

digest, no peptide is detected. After the sample is enriched with Mag GO@ TiO_2, Mag GO@SnO_2, and the mixture of Mag GO@TiO_2 and Mag GO@ SnO_2, as shown in Figures 5.84(b)–5.84(d), only one or two phosphorylated peptides can be detected. When the β-casein digest is as low as 4×10^{-11} $mol \cdot L^{-1}$, after enrichment with Mag GO@(Ti–Sn)O_4 hybrid material, six phosphopeptides and three dephosphorylated fragments are detected (Figure 5.84(e)), indicating that the Mag GO@(Ti–Sn)O_4 hybrid material has better enrichment ability and high enrichment sensitivity.

(b) *Evaluation of the enrichment selectivity of Mag GO@(Ti-Sn)O_4 hybrid material for phosphorylated peptide*

The digest mixture of β-casein and BSA with different mass ratios is selected to investigate the selectivity of Mag GO@(Ti–Sn)O_4 hybrid material. As shown in Figure 5.85(a), only a large number of non-phosphorylated peptides are detected when directly analyzing the peptide mixture (mass ratio of β-casein and BSA is 1/10). After enrichment with Mag GO@(Ti–Sn)O_4 hybrid material, seven phosphorylated peptides and four dephosphorylated fragments are detected as shown in Figure 5.85(b). When the ratio of β-casein to BSA reaches 1/1500, seven phosphorylated peptides and four dephosphorylated fragments are still detectable after enrichment with Mag GO@(Ti–Sn)O_4 hybrid material as shown in Figure 5.85(c). These results indicate that the Mag GO@(Ti–Sn) O_4 hybrid material has strong specificity and high selectivity, which has great potential in the selective enrichment of phosphopeptides in complex biological samples.

(c) *Evaluation of the enrichment ability of Mag GO@(Ti–Sn)O_4 hybrid material for phosphopeptide in mouse brain digest*

Mouse brain digest is used to investigate the enrichment ability of Mag GO@ (Ti–Sn)O_4 hybrid material for phosphopeptides in a practical sample. A total of 66 mono-phosphorylated peptides, 104 multi-phosphorylated peptides, and 349 phosphorylation sites are identified by database search. Among them, 79.37% (277) phosphorylation sites occur on the serine residues, 17.19% (60) on the threonine residues, and 3.44% (12) on the tyrosine residues.

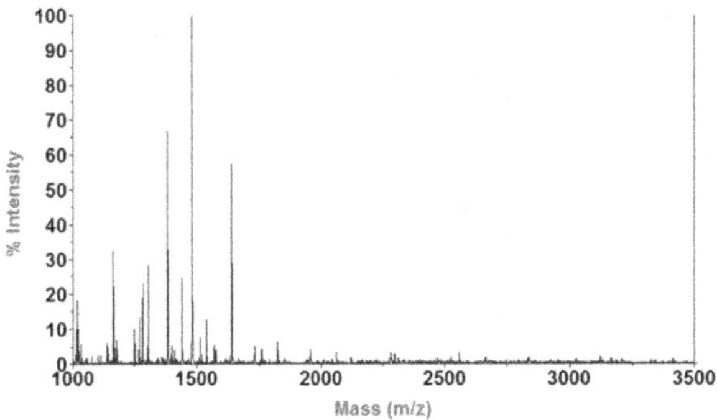

(a) **Mass ratio 1:10, before enrichment**

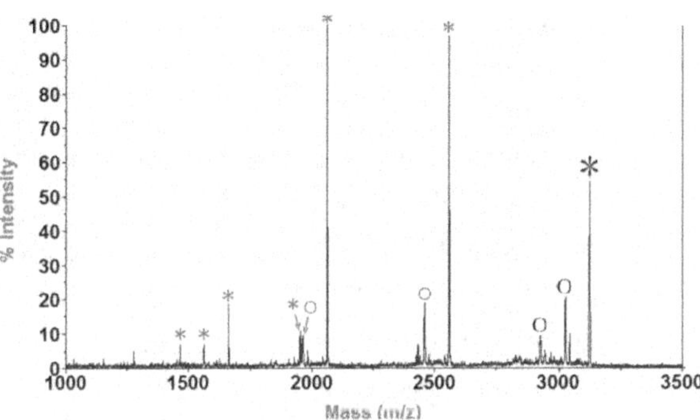

(b) **Mass ratio 1:10, after enrichment**

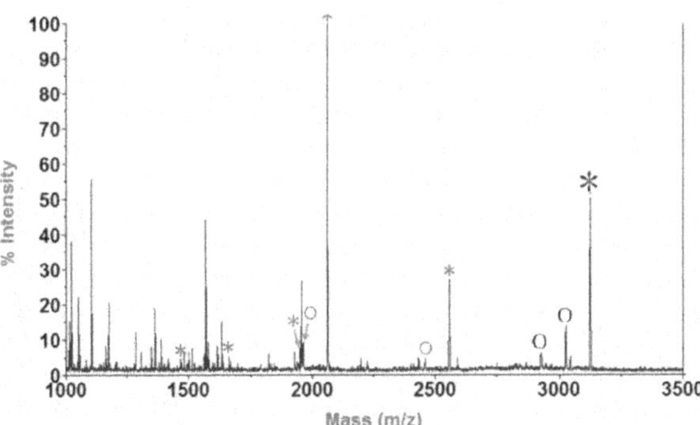

(c) **Mass ratio 1:500, after enrichment**

Figure 5.85 : MS spectra of the digest mixture of β-casein and BSA at different mass ratios before and after enrichment with Mag GO@(Ti–Sn)O$_4$ hybrid material.

5.3.5.2. *Mag GO@PDA@(Zr–Ti)O$_4$ hybrid material*

(1) Preparation of Mag GO@PDA@(Zr–Ti)O$_4$ hybrid material

Figure 5.86 shows the synthesis of Mag GO@PD@(Zr–Ti)O$_4$ hybrid material. The detailed synthesis protocol can be seen in Section 2.8 of Chapter 2.

(2) Enrichment application of Mag GO@PDA@(Zr–Ti)O$_4$ hybrid material in phosphoproteomics

(a) *Evaluation of the enrichment ability of Mag GO@PDA@(Zr–Ti)O$_4$ hybrid material for phosphopeptide in standard phosphoprotein digest*

β-casein digest is selected to investigate the enrichment ability of Mag GO@PDA@(Zr–Ti)O$_4$ hybrid material for phosphopeptides. At the same time, Mag GO@PDA@ZrO$_2$ and Mag GO@PDA@TiO$_2$ are also used to enrich the same sample, As shown in Figure 5.87(a), no peptide is detected for the direct analysis of 4×10^{-10} mol·L^{-1} β-casein digest. After enrichment with Mag GO@PDA@ZrO$_2$, Mag GO@PDA@TiO$_2$. and the mixture of Mag GO@PDA@ZrO$_2$ and Mag GO@PDA@TiO$_2$, as shown in Figures 5.87(b)–5.87(d), three phosphorylated peptides from β-casein can be detected, but a large number of non-phosphorylated peptides are also detected in the MS spectra. However, after enrichment with Mag GO@PDA@(Zr–Ti)O$_4$ hybrid material, as shown in Figure 5.87(e), not only the three phosphopeptides from β-casein digest are detected with high intensity, the corresponding dephosphorylated fragments and the two phosphopeptides from α-casein digest are also detected, but no non-phosphorylated peptide is detected with

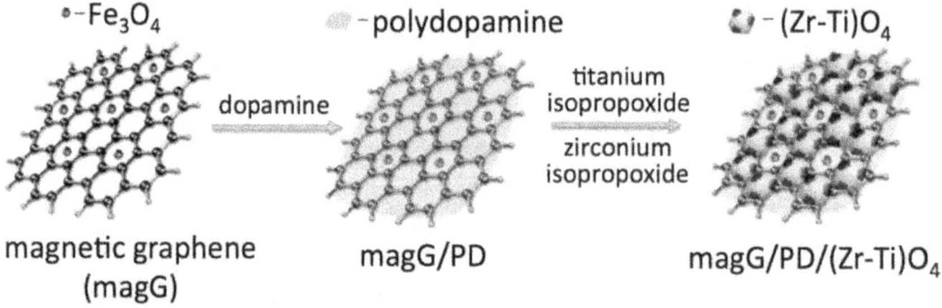

Figure 5.86: Schematic diagram of the synthesis of Mag GO@PDA@(Zr–Ti)O$_4$ hybrid material.

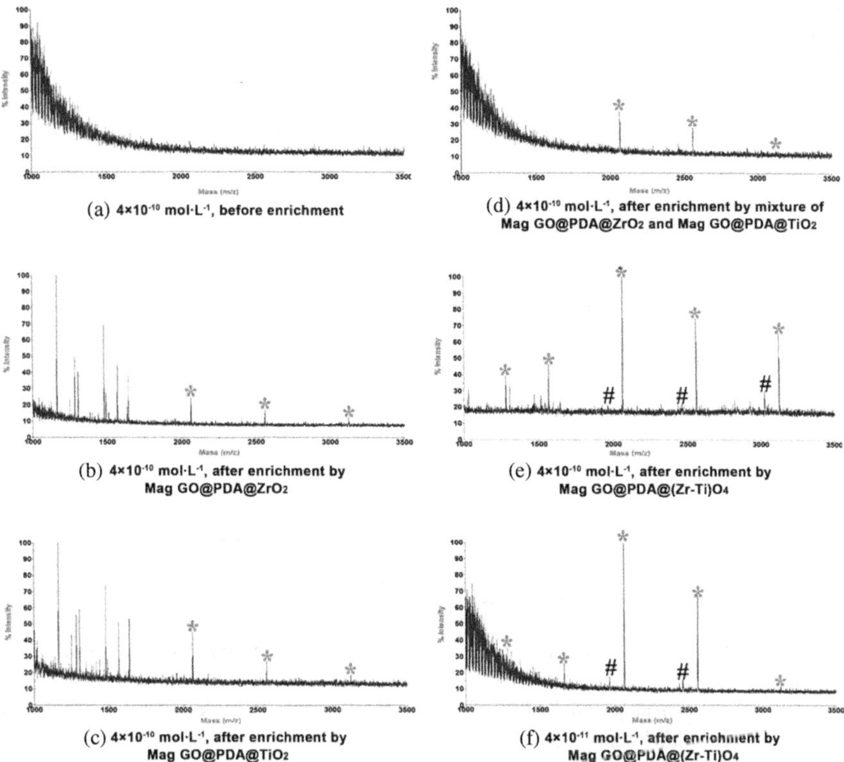

Figure 5.87: MS spectra of β-casein digest before and after enrichment with different materials. * represents phosphorylated peptide, # represents dephosphorylated fragment.

high intensity. Besides, when the concentration of β-casein digest is as low as 4×10^{-11} mol·L^{-1}, after enrichment with Mag GO@PDA@(Zr–Ti)O$_4$ hybrid material, five phosphorylated peptides and two dephosphorylated fragments are detected. All these results demonstrate the enrichment performance of Mag GO@PDA@(Zr–Ti)O$_4$ hybrid material for phosphopeptides.

(b) *Evaluation of the enrichment selectivity of Mag GO@PDA@(Zr–Ti)O$_4$ hybrid material for phosphopeptide*

The digest mixture of β-casein and BSA with different mass ratios are selected to investigate the enrichment selectivity of Mag GO@PDA@(Zr–Ti)O$_4$ hybrid material for phosphopeptide. It is worth mentioning that when the mass ratio of β-casein to BSA reaches 1:8000, eight phosphopeptides and

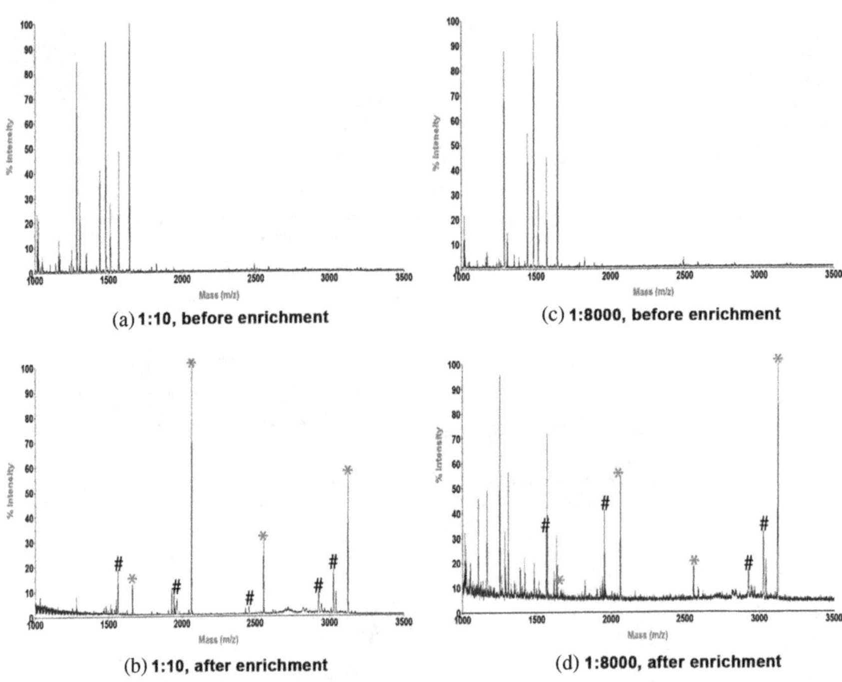

Figure 5.88 MS spectra of the digest mixture of β-casein and BSA with different mass ratios before and after enrichment with Mag GO@PDA@(Zr–Ti)O$_4$ hybrid material. *: phosphorylated peptide, #: dephosphorylated fragment.

their dephosphorylated fragments are observed in Figure 5.88(d). Although some non-phosphorylated peptides appear, the phosphorylated peptide peaks are absolutely prominent. Compared to Mag GO@(Ti–Sn)O$_4$ hybrid material, the Mag GO@PDA@(Zr–Ti)O$_4$ hybrid material exhibits a stronger specific adsorption capacity for phosphorylated peptides.

(c) *Evaluation of the enrichment ability of Mag GO@PDA@(Zr–Ti)O$_4$ hybrid material for phosphopeptide in mouse brain digest*

163 mono-phosphorylated peptides, 459 multi-phosphorylated peptides, and 1436 phosphorylation sites are successfully identified from the mouse brain digest, of which 87.53% occur on serine residues (1257), 11.98% occur on the threonine residues (172), and 0.49% occur on the tyrosine residues (7). The detected phosphorylation sites by Mag GO@PDA@(Zr–Ti)O$_4$ hybrid material are 4.11 times of that enriched with Mag GO@(Ti–Sn)O$_4$

hybrid material, while the phosphopeptides enriched with Mag GO@PDA@ $(Zr–Ti)O_4$ hybrid material are 3.66 times of that enriched with Mag GO@ $(Ti–Sn)O_4$ hybrid material.

5.4. MOF-modified magnetic micro-/nanomaterial for separation and analysis in phosphoproteomics

5.4.1. *Introduction*

MOF is self-assembled by coordination of central metal ion and organic ligand. It is another kind of important porous material after zeolite and carbon nanotube, which has been developed rapidly in recent years. Due to its large specific surface area, diverse structure, and tunable porous channel, it has a wide range of applications in gas adsorption, separation, catalysis, and biomedicine. In addition, because of its molecular sieve effect, it can effectively enrich peptides and block macromolecular proteins. Therefore, MOFs show great application prospects in proteomics. At present, many MOF-based materials have been synthesized for the extraction of low-abundance peptides and immobilization of enzymes. However, considering that MOFs are not magnetic in nature, direct application of them for enrichment in proteomics can lead to problems such as loss of sample and reagent, contamination, and low efficiency in operation. Therefore, many researchers are seeking reliable technology to modify MOF on the surface of magnetic microspheres. The successful synthesis of magnetic MOF will be the great advantageous means for sample separation. This section will describe two commonly used methods for modifying MOF on magnetic microspheres and the application of MOF-modified magnetic micro-/nanomaterial in phosphoproteomics.

5.4.2. *PDA-modified magnetic microsphere for immobilization of MOF (Fe₃O₄@PDA@Zr-MOF)*

5.4.2.1. *Preparation of Fe₃O₄@PDA@Zr-MOF composite*

The synthesis of Fe_3O_4@PDA@Zr–MOF composite can be seen in Section 2.6 of Chapter 2.[45] At the same time, the Fe_3O_4@PDA@Zr^{4+} microsphere is synthesized for comparison. The synthesis protocol is shown as follows: 100 mg Fe_3O_4@PDA microsphere is dispersed in 0.1 mol·L^{-1} $ZrOCl_2$ solution for reaction at 70°C for 2–3 h. The product is washed with ethanol.

5.4.2.2. *Enrichment application of Fe_3O_4@PDA@Zr-MOF composite in phosphoproteomics*

(1) Evaluation of the enrichment ability of Fe_3O_4@PDA@ Zr-MOF composite for phosphopeptide in standard phosphoprotein digest

Different concentrations of β-casein digests are selected to investigate the enrichment ability of Fe_3O_4@PDA@Zr-MOF composite. As shown in Figure 5.89(a), when the concentration is 5×10^{-11} mol·L^{-1}, six phosphorylated peptides can be detected by MALDI-TOF-MS. When the β-casein

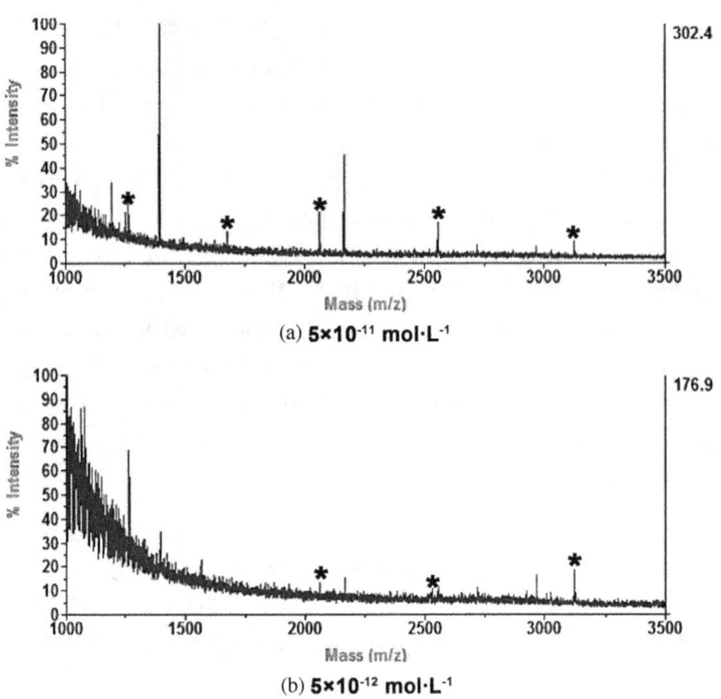

Figure 5.89: MS spectra of different concentrations of β-casein digests after enrichment with Fe_3O_4@PDA@Zr-MOF composite.

digest is as low as 5×10^{-12} mol·L^{-1}, three phosphorylated peptides can still be detected (Figure 5.89(b)), indicating that the Fe$_3$O$_4$@PDA@ Zr-MOF composite has good enrichment ability and high sensitivity for phosphopeptides.

(2) Evaluation of the enrichment selectivity of Fe$_3$O$_4$@PDA@ Zr-MOF composite for phosphopeptide

The digest mixture of β-casein and BSA at different molar ratios is used to investigate the enrichment selectivity of Fe$_3$O$_4$@PDA@Zr-MOF composite for phosphopeptide. As shown in Figure 5.90, the spectra have only non-phosphorylated peptides before enrichment, regardless if the molar ratio of β-casein to BSA is 1/200, 1/400, or 1/500. However, after enrichment with the Fe$_3$O$_4$@PDA@Zr-MOF composite, six to eight phosphorylated peptides

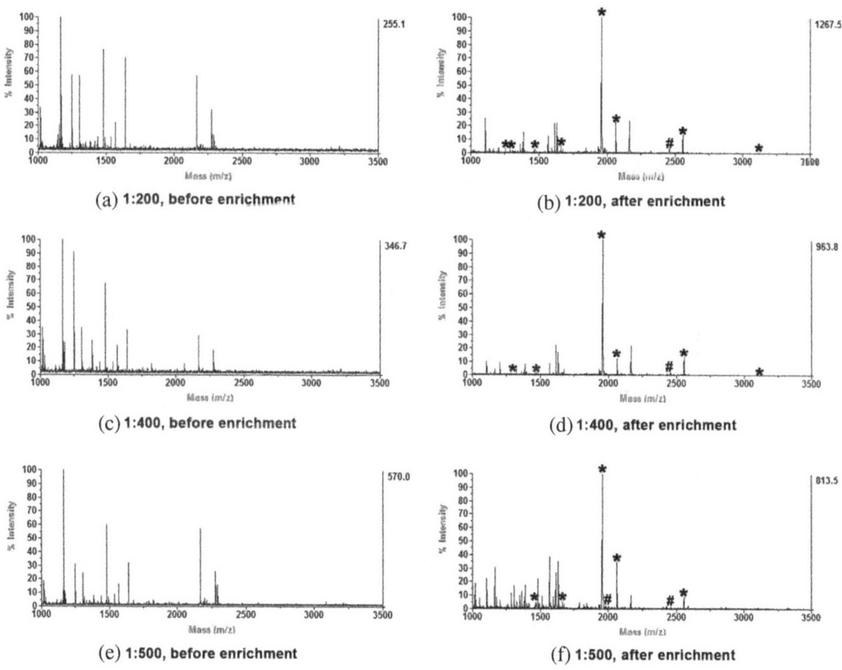

Figure 5.90: MS spectra of digest mixture of β-casein and BSA at different molar ratios before and after enrichment with Fe$_3$O$_4$@PDA@Zr-MOF composite. *: phosphorylated peptide, #: dephosphorylated fragment.

and their corresponding dephosphorylated fragments are detected. This indicates that Fe_3O_4@PDA@Zr-MOF composite has good selective enrichment ability for phosphorylated peptides.

(3) Evaluation of the enrichment ability of Fe_3O_4@PDA@Zr-MOF composite for phosphopeptide in human serum

Human serum is selected to further evaluate the selective enrichment ability of Fe_3O_4@PDA@Zr-MOF composite for phosphorylated peptide. As shown in Figure 5.91, four typical phosphorylated peptides are detected after enrichment with the Fe_3O_4@PDA@Zr-MOF composite. This indicates that the Fe_3O_4@PDA@Zr-MOF composite has good selective enrichment ability for the phosphopeptides in complex biological samples.

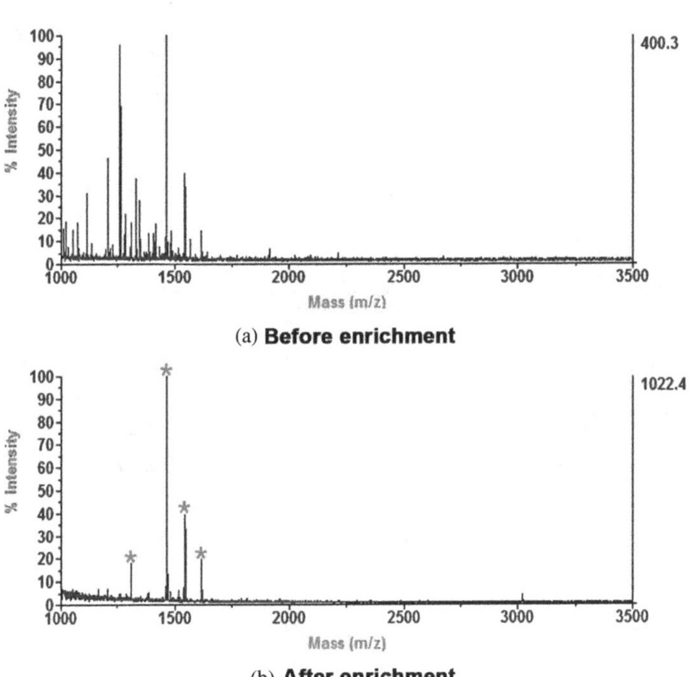

Figure 5.91: MS spectra of the human serum sample before and after enrichment with Fe_3O_4@PDA@Zr-MOF composite.

5.4.3. *Mercaptoacetic acid-modified magnetic microsphere for immobilization of MOF (Fe_3O_4@MIL-100(Fe))*

5.4.3.1. *Preparation of Fe_3O_4@MIL-100(Fe) composite*

The synthesis of Fe_3O_4@MIL-100(Fe) composite can be seen in Section 2.6 of Chapter 2.[46]

5.4.3.2. *Enrichment application of Fe_3O_4@MIL-100(Fe) composite in phosphoproteomics*

(1) Evaluation of the enrichment ability of Fe_3O_4@MIL-100(Fe) composite for phosphopeptide in standard phosphoprotein digest

The digests of β-casein and α-casein are used to investigate the enrichment performance of Fe_3O_4@MIL-100(Fe) composite for phosphorylated peptide. As shown in Figure 5.92, when the concentration of β-casein digest is as low as 1×10^{-12} mol·L^{-1}, still one phosphorylated peptide can be detected, confirming the enrichment ability and high sensitivity of Fe_3O_4@MIL-100(Fe) composite for phosphopeptides.

From Figure 5.93, the Fe_3O_4@MIL-100(Fe) composite also exhibits good enrichment performance for phosphopeptides in α-casein digest. The detailed information of the enriched phosphorylated peptides is shown in Table 5.16.

(2) Evaluation of the enrichment selectivity of Fe_3O_4@MIL-100(Fe) composite for phosphopeptide

The peptide mixture of β-casein (0.5 pmol) and BSA at different molar ratios is selected for evaluation of enrichment selectivity of Fe_3O_4@MIL-100(Fe) composite. As shown in Figure 5.94(a), when the molar ratio of β-casein and BSA is 1/100, only a large number of non-phosphorylated peptides are detected. But after enrichment with Fe_3O_4@MIL-100(Fe) composite, three phosphorylated peptides from β-casein and their corresponding dephosphorylated fragments are observed in Figure 5.94(b), and the background is clean. When the molar ratio of β-casein to BSA is 1/500 (Figure 5.94(c)), the spectrum is not significantly different from Figure 5.94(b), indicating that

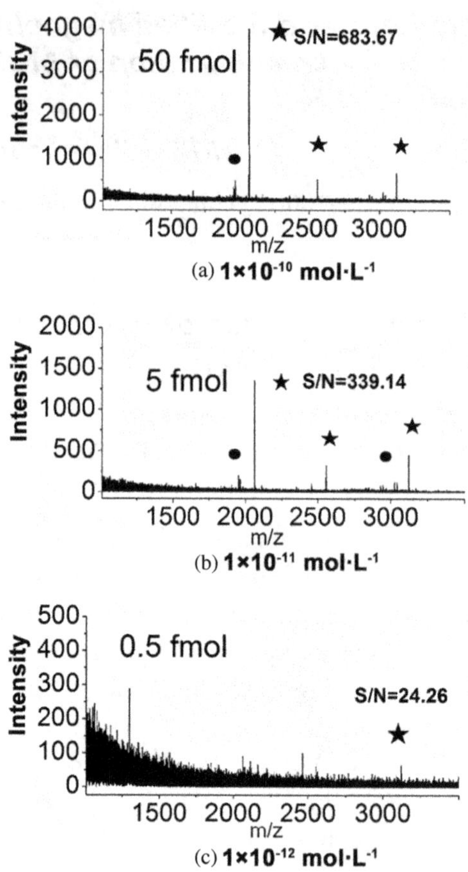

Figure 5.92: MS spectra of β-casein digest with different concentrations after enrichment with Fe_3O_4@MIL-100(Fe) composite. ★: phosphorylated peptide, ●: dephosphorylated fragment.

Fe_3O_4@MIL-100(Fe) composite has excellent enrichment selectivity for phosphorylated peptides.

(3) Evaluation of the size-exclusion effect of Fe_3O_4@ MIL-100(Fe) composite

According to the pore size distribution of Fe_3O_4@MIL-100(Fe) composite, the pore size is mainly concentrated at 1.93 nm and 3.91 nm. Therefore, BSA

Table 5.16: Detailed information of phosphopeptide enriched from α-casein digest by Fe₃O₄@MIL-100(Fe) composite.

No.	m/z	Peptide Sequence	Phosphor-Sites
α1	1237.07	TVDME[pS]TEVF	1
α2	1337.28	HIQKEDV[pS]ER	1
α3	1466.27	TVDME[pS]TEVFIK	1
α4	1842.27	TVD[Mo]E[pS]TEVFTK[b]	1
α5	1660.57	VPQLEIVPN[pS]AEER	1
α6	1847.50	DIGSE[pS]TEDQAMEDIK	1
α7	1927.28	DIG[pS]E[pS]TEDQAMEDIK	2
α8	1943.45	DIG[pS]E[pS]TEDQA[Mo]EDIK[a]	2
α9	1957.72	YEVPQLEIVPN[pS]AEER	1
α10	2061.61	FQ[pS]EEQQQTEDELQDK	1
α11	2618.69	NTMEHV[pS][pS][pS]EESII[pS]QETYK	4
α12	2677.47	VNEL[pS]KDIG[pS]E[pS]TEDQAMEDIK	3
α13	2703.71	Q*MEAE[pS]I[pS][pS][pS]EEIVPN[pS]VEAQ[b]	5
α14	2720.69	QMEAE[pS]I[pS][pS][pS]EEIVPNPN[pS]VEQK	5
α15	2934.98	KEKVNEL[pS]KDIG[pS]E[pS]TEDQAMEDIK	3
α16	3007.85	NANEEEYSIG[sP][sP][sP]EE[sP]AEVATEEVK	4

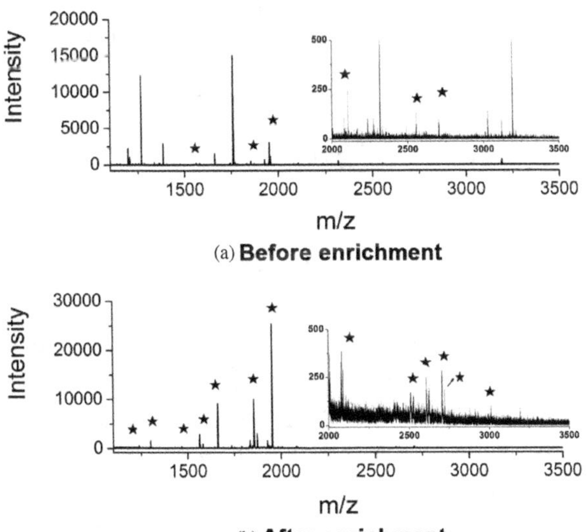

Figure 5.93: MS spectra of 1 × 10⁻⁹ mol·L⁻¹ α-casein digest before and after enrichment with Fe₃O₄@MIL-100(Fe) composite. ★: phosphorylated peptide.

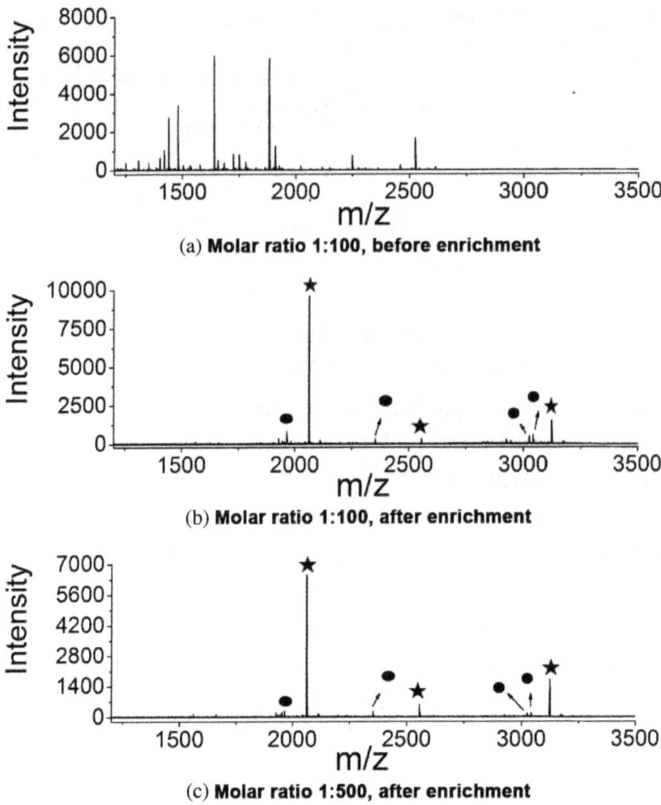

Figure 5.94: MS spectra of mixture peptide of β-casein (0.5 pmol) and BSA at different molar ratios before and after enrichment with Fe_3O_4@MIL-100(Fe) composite. ★: phosphorylated peptide, ●: dephosphorylated fragment.

protein is added into β-casein digest as interference to investigate whether the Fe_3O_4@MIL-100(Fe) composite has size-exclusion effect on macromolecular protein. As shown in Figures 5.95(a) and 5.95(b), no phosphorylated peptide is detected by MS due to the presence of high-abundance protein. After enrichment with Fe_3O_4@MIL-100(Fe) composite, as shown in Figures 5.95(c) and 5.95(d), the signal of BSA protein disappears, while three phosphorylated peptides from β-casein and their corresponding dephosphorylated fragments are detected. Moreover, the concentration of BSA protein in the supernatant is determined to be 0.030 mg·mL^{-1}, which is not much different from the original BSA concentration, indicating that the Fe_3O_4@MIL-100(Fe) composite has good size-exclusion effect.

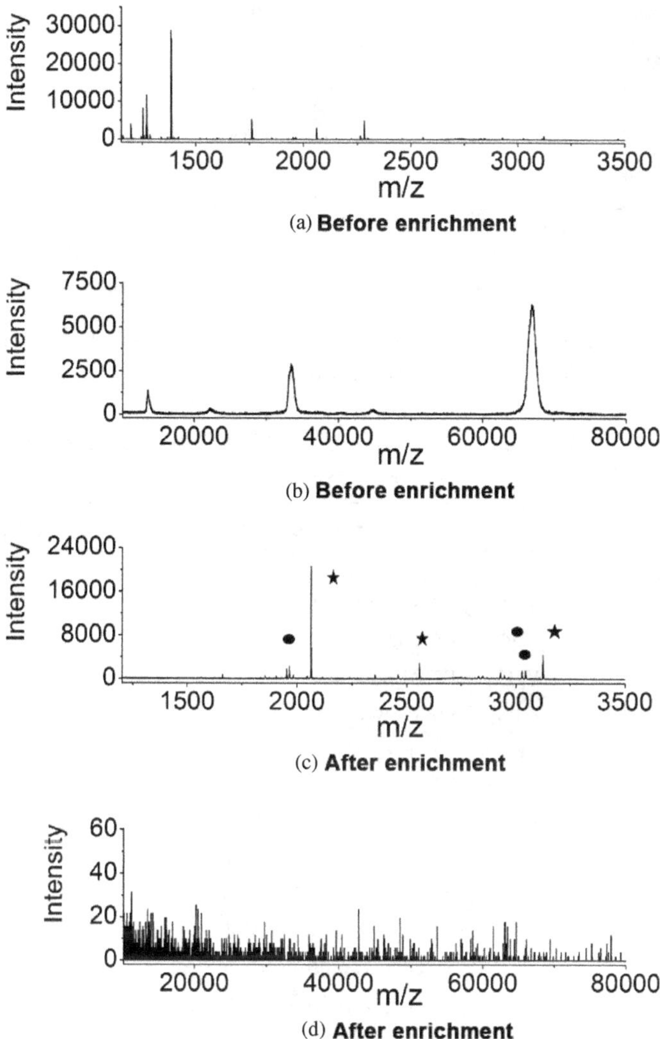

Figure 5.95: MS spectra of the mixture of BSA protein and β-casein digest before and after enrichment with Fe_3O_4@MIL-100(Fe) composite. ★: phosphorylated peptide, ●: dephosphorylated fragment.

(4) Evaluation of the enrichment ability of Fe_3O_4@MIL-100(Fe) composite for phosphopeptide in non-fat milk digest

Non-fat milk digest is selected to investigate the enrichment ability of Fe_3O_4@MIL-100(Fe) composite for phosphorylated peptide. As shown in Figure 5.96, 14 phosphopeptides from the non-fat milk digest are detected. The detailed

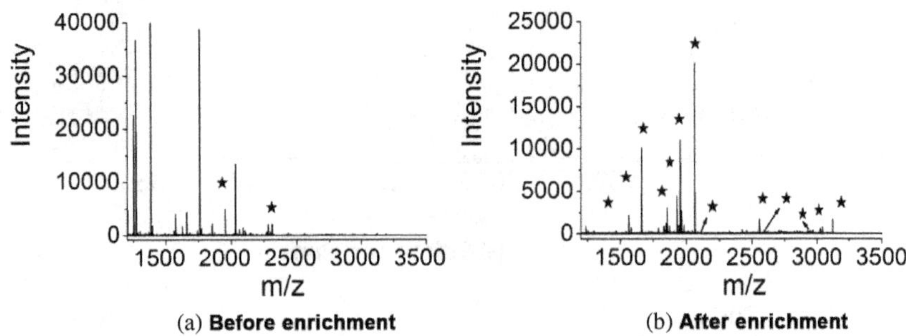

Figure 5.96: MS: spectra of non-fat milk digest before and after enrichment with Fe_3O_4@ MIL-100(Fe) composite. ★: phosphorylated peptide.

Table 5.17: The detailed information of phosphopeptide enriched from non-fat milk digest with Fe_3O_4@MIL-100(Fe) composite.

No.	m/z	Peptide Sequence	Phospho-Sites
1	1466.24	TVDME[sP]TEVFIK	1
2	1563.71	TVD[Mo]E[pS]TEVFTK[b]	1
3	1661.39	VPQLEIVPN[pS]AEER	1
4	1854.21	YLGEYLIVPN[pS]AEER	1
5	1927.28	DIG[pS]E[pS]TEDQAMEDIK	2
6	1952.01	YEVPQLEIVPN[pS]AEER	1
7	2062.61	FQ[sP]EEQQQTEDELQDK	1
8	2080.37	KKYKVPQLEIVPN[pS]AEREL	1
9	2555.65	FQ[pS]EEQQQTEDELQDKIHPF	1
10	2618.69	NTMEHV[pS][pS][pS]EESII[pS]QETYK	4
11	2720.69	QMEAE[pS]I[pS][pS][pS]EEIVPNPN[pS]VEQK	5
12	2965.46	ELEELNVPGEIVE[pS]l[pS][pS][pS]EESITR	4
13	3026.33	NANEEEYSIG[pS][pS][pS]EE[pS]AEVATEEVK	4
14	3122.98	RELEELNVPGEIVE[pS]L[pS][pS][pS]EESITR	4

information of phosphopeptide enriched from non-fat milk digest can be seen in Table 5.17.

(5) Evaluation of the enrichment ability of Fe_3O_4@MIL-100(Fe) composite for phosphopeptide in human serum

Serum of a uterine cancer patient is selected to investigate the selective enrichment ability of Fe_3O_4@MIL-100(Fe) composite for phosphopeptides in

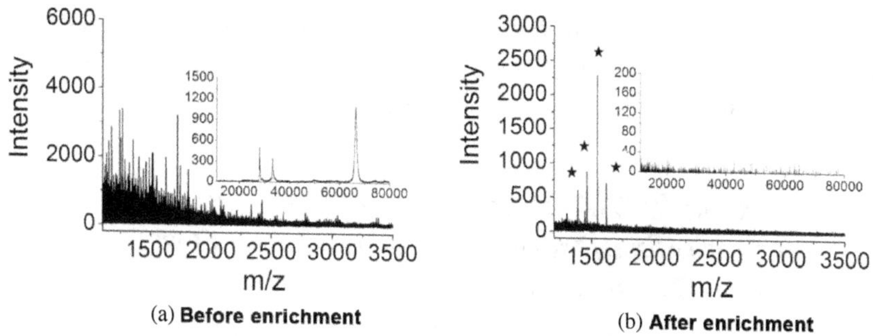

Figure 5.97: MS spectra of the serum of uterine cancer patient before and after enrichment with Fe_3O_4@MIL-100(Fe) composite. ★: phosphorylated peptide.

biological samples. As shown in Figure 5.97, no phosphorylated peptide can be detected due to the presence of a large amount of human serum albumin before enrichment. After enrichment with Fe_3O_4@MIL-100(Fe) composite, four phosphorylated peptides are detected, the m/z of which are similar to that of healthy human serum, indicating that the material also has selective enrichment ability for phosphorylated peptides in complex biological samples, and has the ability to exclude macromolecular proteins from the pore.

5.5. Rare earth-modified magnetic micro-/nanomaterial for separation and analysis in phosphoproteomics

5.5.1. Introduction

Rare earth elements include 17 kinds of metal elements with similar physical and chemical properties, including lanthanides, yttrium (Y), and strontium (Sc), which are similar in electronic structure and chemical properties. According to the Pearson's Hard and Soft of acids and bases reaction (HSAB) rule, rare earth ions are hard acids, so they easily form stable complexes with oxygen-containing ligands such as phosphine oxide extractant, β-diketone, and α-hydroxy acid. In addition, the organic ligands with both oxygen and nitrogen, such as the aminocarboxylate complexing agent, can form chelating ring with rare earth ions, resulting in stronger stability. Rare earth elements mainly exist in the ionic state in nature. The largest abundance in the earth's crust is Ce, which is higher than the common metal zinc. Rare earth elements are widely used in industrial production, agriculture and animal husbandry, and the medical field. In recent years, the biological effect of rare earth element has led to many applications in proteomics, and the number and variety of proteins identified by rare earth elements are very extensive.

5.5.2. *Cyclen (DOTA) modified magnetic silica for immobilization of rare earth ion ($Fe_3O_4@$ TCPP-DOTA-M^{3+})*

5.5.2.1. *Preparation of $Fe_3O_4@$TCPP-DOTA-M$_{3+}$ microsphere*

The schematic diagram of synthesis of $Fe_3O_4@$TCPP-DOTA-M^{3+} microsphere is shown in Figure 2.66 of Chapter 2.[47,48] The detailed synthesis process can be seen in Section 2.6 of Chapter 2. Herein, the M^{3+} represents the rare earth ions, such as Tb^{3+}, Tm^{3+}, Ho^{3+}, and Lu^{3+}.

5.5.2.2. *Enrichment application of $Fe_3O_4@$TCPP-DOTA-M$_{3+}$ microsphere in phosphoproteomics*

(1) Evaluation of the enrichment ability of $Fe_3O_4@$ TCPP-DOTA-M^{3+} microsphere for phosphopeptide in standard phosphoprotein digest

Lower pH can make the acidic residue protonate, and thus reducing the non-specific adsorption of the acidic non-phosphorylated peptide. But if the concentration of TFA is too high, the lanthanide ions easily fall off from the magnetic microsphere. Therefore, 1% TFA+50%ACN is selected as the loading buffer according to a series of optimization experiments.

In this work, the α-casein digest is used to investigate the enrichment performance of $Fe_3O_4@$TCPP-DOTA-M^{3+} microsphere for phosphopeptide. As shown in Figure 5.98(a), when directly analyzing 3.33×10^{-8} mol·L^{-1} α-casein digest, the signals of the phosphorylated peptides are severely inhibited by a large number of non-phosphorylated peptides. Only seven very weak phosphopeptide peaks are observed. After enrichment with $Fe_3O_4@$TCPP-DOTA-M^{3+} microsphere, seven mono-phosphorylated peptides and 12 multi-phosphorylated peptides are identified as shown in Figure 5.98(b). The detailed information of phosphorylated peptides is shown in Table 5.18.

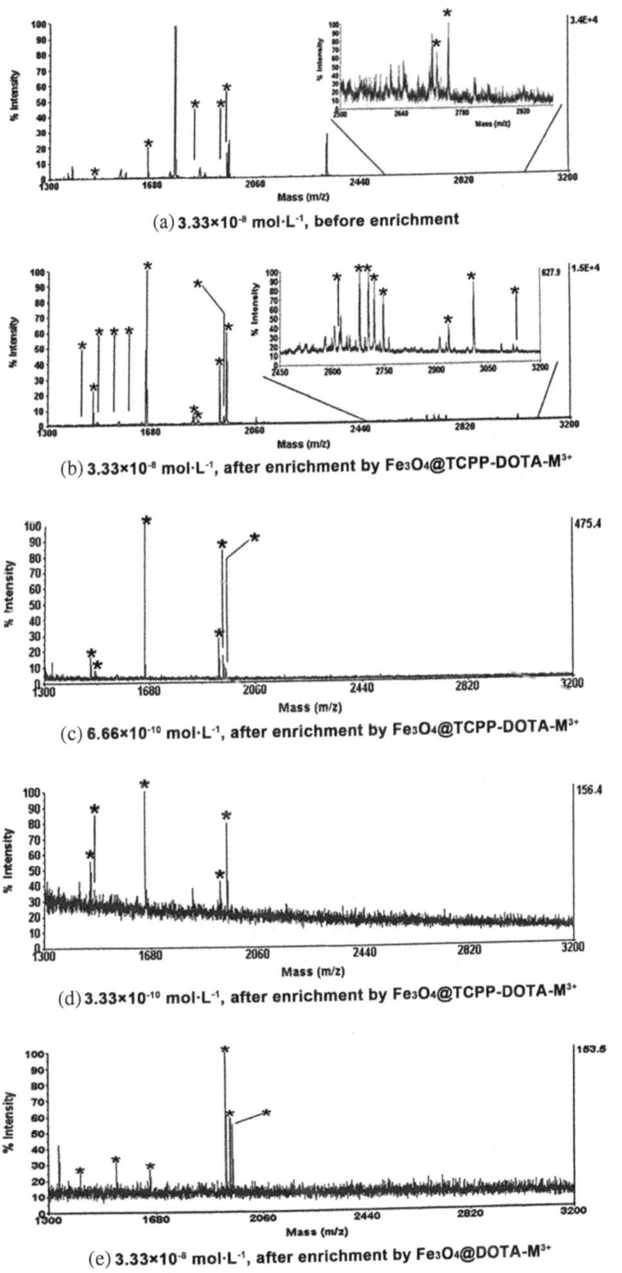

(a) 3.33×10^{-8} mol·L^{-1}, before enrichment

(b) 3.33×10^{-8} mol·L^{-1}, after enrichment by Fe$_3$O$_4$@TCPP-DOTA-M^{3+}

(c) 6.66×10^{-10} mol·L^{-1}, after enrichment by Fe$_3$O$_4$@TCPP-DOTA-M^{3+}

(d) 3.33×10^{-10} mol·L^{-1}, after enrichment by Fe$_3$O$_4$@TCPP-DOTA-M^{3+}

(e) 3.33×10^{-8} mol·L^{-1}, after enrichment by Fe$_3$O$_4$@DOTA-M^{3+}

Figure 5.98: MS spectra of α-casein digest before and after enrichment with different mate-
phosphorylated peptide.

Table 5.18. Detailed information of phosphopeptide enriched with $Fe_3O_4@$ TCPP-DOTA-M^{3+} microsphere from α-casein digest.

No.	Peptide Sequence	Phospho-Sites	(M + H)
1	EQL[pS]T[pS]EENSK S2-(141–151)	2	1411.59
2	TVDME[pS]TEVFTK S2-(153–164)	1	1466.72
3	TVD[Mo]E[pS]TEVFTK S2-(153–164)	1	1482.7
4	EQL[pS]T[pS]EENSKK S2-(141–152)	2	1539.7
5	TVDME[pS]TEVFKK S2-(153–165)	1	1594.82
6	VPQLEIVPN[pS]AEER S1-(121–134)	1	1660.92
7	YEGEYLIVPN[pS]AEER S1-(104–119)	1	1832.83
8	DIG[pS]E[pS]TEDQAMEDIK S1-(58–73)	1	1847.86
9	DIG[pS]E[pS]TEDQAMEDIK S1-(58–73)	2	1927.84
10	DIG[pS]E[pS]TEDQA[Mo]EDIK S1-(58–73)	2	1943.81
11	YKVPQLEIVPN[pS]AEER S1-(119–134)	1	1952.12
12	NTMEHV[pS][pS][pS]EE[pS]IISQETYK S2-(17–36)	4	2619.15
13	VNEL[pS]KIDG[pS]E[pS]TEDQAMEDIK S1-(52–73)	3	2678.24
14	Q*MEAE[pS]I[pS][pS][pS]EEIVPN[pS]VEAQK S1-(74–94)	5	2704.25
15	QMEAE[pS]I[pS][pS][pS]EEIVPN[pS]VEQAK S1-(74–94)	5	2721.16
16	NTMEHV[pS][pS][pS]EE[pS]IISQETYKQ S2-(17–37)	4	2747.23
17	EKVNEL[pS]KDIG[pS]E[pS]TEDQAMEDIK S1-(50–73)	3	2935.42
18	NANEEYSIG[pS][pS][pS]EE[pS]AEVATEEVK S2-(61–85)	4	3008.28
19	KNTMEHV[pS][pS][pS]EE[pS]IISQETYKQEK S2-(16–39)	4	3132.2

In order to further investigate the enrichment sensitivity of $Fe_3O_4@$ TCPP-DOTA-M^{3+} microsphere for phosphorylated peptide, different concentrations of α-casein digests are enriched with $Fe_3O_4@$TCPP-DOTA-M^{3+} microsphere. As shown in Figure 5.98(d), when the concentration of α-casein hydrolysate is as low as 3.33×10^{-10} mol·L^{-1}, five phosphorylated peptides can still be detected. In addition, the $Fe_3O_4@$TCPP-DOTA-M^{3+} microsphere is synthesized for comparison, which is used to enrich the α-casein digest of 3.33×10^{-8} mol·L^{-1}. In Figure 5.98(e), only six phosphorylated peptides and one non-phosphorylated peptide are detected, indicating that the intermediate linker of TCPP has a significant effect on the enrichment of phosphorylated peptides. The hydrophilic ligand of TCPP can increase the content of DOTA, and thus increasing the content of rare earth ion, plus its own hydrophilicity, the existence of TCPP can promote the adsorption of phosphopeptide by the material.

(2) Evaluation of the enrichment selectivity of $Fe_3O_4@$ TCPP-DOTA-M^{3+} microsphere for []phosphopeptide

The digest mixture of α-casein and BSA with molar ratio of 1/100 is selected to evaluate the selectivity of $Fe_3O_4@$TCPP-DOTA-M^{3+} microsphere for phosphopeptide. Before enrichment, no phosphorylated peptide is observed in Figure 5.99(a), and non-phosphorylated peptides dominate the spectrum. But after enrichment with $Fe_3O_4@$TCPP-DOTA-M^{3+} microsphere, the signals of the phosphorylated peptides are significantly enhanced. A total of 16 phosphorylated peptides are identified and the background of the spectrum is clean, indicating the excellent enrichment selectivity of the material.

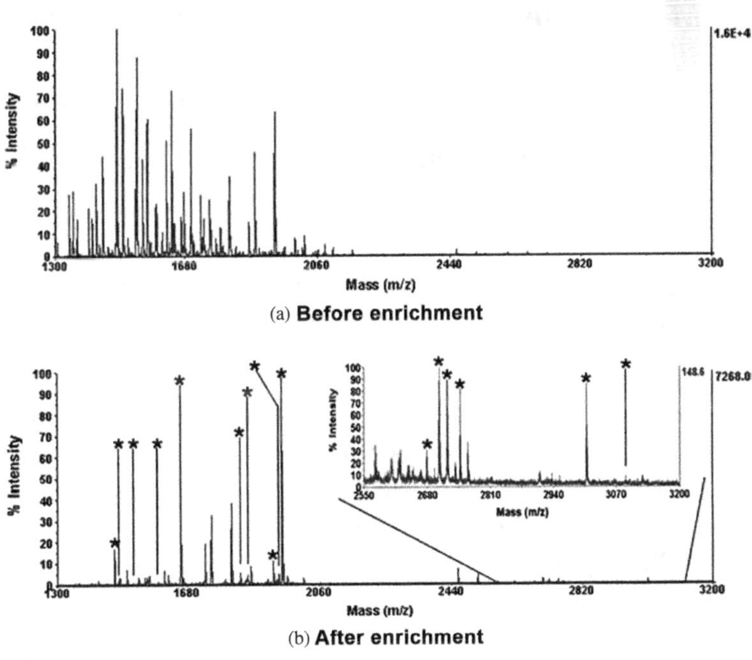

Figure 5.99: MS spectra of the peptide mixture of α-casein and BSA with molar ratio of 1/100 before and after enrichment with $Fe_3O_4@$TCPP-DOTA-M^{3+} microsphere. *: phosphorylated peptide.

(3) Evaluation of the enrichment ability of Fe_3O_4@TCPP-DOTA-M^{3+} microsphere for phosphorylated peptide in HeLa cell digest

To investigate the enrichment ability of Fe_3O_4@TCPP-DOTA-M^{3+} microsphere for phosphorylated peptides in biological samples, HeLa cell digest is selected. In a single MS analysis, 2103 phosphorylated proteins are detected, involving a total of 9048 phosphorylated peptides. Among these peptides, 3825 are uniquely identified, of which 73.85% are mono-phosphorylated peptides, 20.44% are bi-phosphorylated peptides, and 5.69% are tri-phosphorylated peptides. These results indicate that Fe_3O_4@TCPP-DOTA-M^{3+} microsphere has high selectivity and sensitivity for phosphorylated peptides in practical samples, and has great application prospect in the in-depth analysis of phosphoproteomics.

5.5.3. *Mesoporous rare earth oxide-modified magnetic silica (Fe_3O_4@SiO_2@$mCeO_2$)*

5.5.3.1. *Preparation of Fe_3O_4@SiO_2@$mCeO_2$ microsphere*

The synthesis of Fe_3O_4@SiO_2@$mCeO_2$ microsphere can be seen in Section 2.7 of Chapter 2.[49]

5.5.3.2. *Enrichment application of Fe_3O_4@SiO_2@$mCeO_2$ microsphere in phosphoproteomics*

The peptide mixture of β-casein and BSA, and non-fat milk digest are used to evaluate the enrichment ability of Fe_3O_4@SiO_2@$mCeO_2$ microsphere for phosphopeptide. Figure 5.100 shows the MS spectra of peptide mixture (Figure 5.100(a)) and non-fat milk digest (Figure 5.100(b)) enriched with Fe_3O_4@SiO_2@$mCeO_2$ microsphere. The phosphopeptide peaks dominate the spectra, indicating that Fe_3O_4@SiO_2@$mCeO_2$ microsphere has selective enrichment ability for phosphorylated peptides.

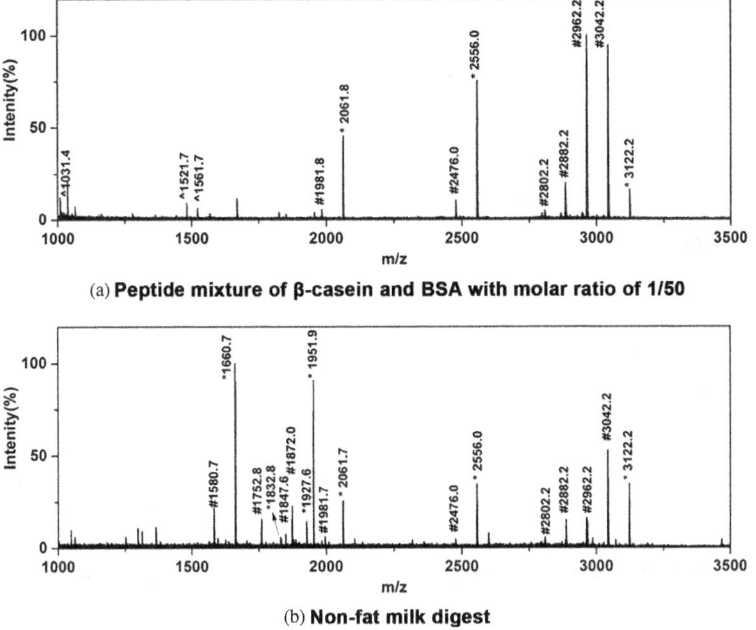

(a) **Peptide mixture of β-casein and BSA with molar ratio of 1/50**

(b) **Non-fat milk digest**

Figure 5.100: MS spectra of different samples after enrichment with $Fe_3O_4@SiO_2@mCeO_2$ microspheres.

5.5.4. *Lanthanum silicate-modified magnetic microsphere ($Fe_3O_4@La_xSi_yO_5$)*

5.5.4.1. *Preparation of $Fe_3O_4@La_xSi_yO_5$ microsphere*

The schematic diagram of the synthesis of $Fe_3O_4@La_xSi_yO_5$ microsphere is shown in Figure 2.97 of Chapter 2. The detailed synthesis process can be seen in Section 2.7 of Chapter 2.[50]

5.5.4.2. *Enrichment application of $Fe_3O_4@La_xSi_yO_5$ microsphere in phosphoproteomics*

The peptide mixture of β-casein and BSA with different molar ratios are selected to investigate the enrichment ability of $Fe_3O_4@La_xSi_yO_5$ microsphere for phosphorylated peptide. As shown in Figure 5.101, the phosphopeptide peaks are dominant in the spectrum after enrichment with $Fe_3O_4@La_xSi_yO_5$

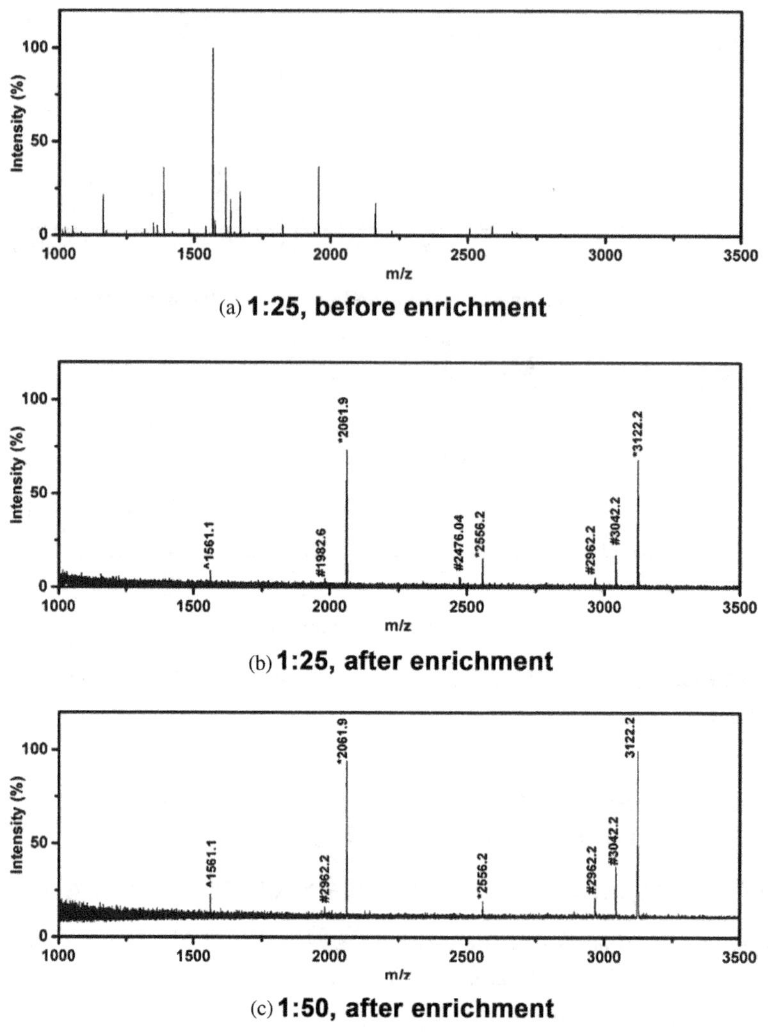

Figure 5.101: MS spectra of the mixture of β-casein and BSA at different molar ratios before and after enrichment with Fe_3O_4@LaxSiyO$_5$ microsphere.

microsphere, indicating that Fe_3O_4@La$_x$Si$_y$O$_5$ microsphere has selective enrichment ability for phosphorylated peptides. In addition, human serum is also selected to further investigate the enrichment selectivity of Fe_3O_4@La$_x$Si$_y$O$_5$ microsphere for phosphorylated peptides. Four typical phosphorylated peptides in human serum can be detected, indicating that Fe_3O_4@

$La_xSi_yO_5$ microsphere has selective enrichment ability for phosphopeptides in complex biological samples.

5.5.5. *Yttrium phosphate-modified magnetic microsphere (PA–Fe$_3$O$_4$@YPO$_4$)*

5.5.5.1. *Preparation of PA–Fe$_3$O$_4$@YPO$_4$ microsphere*

The preparation of PA–Fe$_3$O$_4$@YPO$_4$ microsphere is shown in Figure 5.102.[51]

(1) Synthesis of PA–Fe$_3$O$_4$ microsphere

0.540 g FeCl$_3$·6H$_2$O is dissolved in 20 mL ethylene glycol, and magnetically stirred to a transparent solution.[52] Then 1.2 g sodium acetate and 0.3 g sodium acrylate are added, and the mixture is stirred vigorously and transferred to 30 mL Teflon-lined stainless-steel reactor for being heated at 200°C for 10 h. After cooled to room temperature, the obtained PA–Fe$_3$O$_4$ microsphere is washed with deionized water and ethanol, and dried in vacuum.

(2) Synthesis of PA–Fe$_3$O$_4$@Y(OH)CO$_3$ microsphere

0.072 g Y(NO$_3$)$_3$·6H$_2$O and 0.375 g CO(NH$_2$)$_2$ are dissolved in 50 mL H$_2$O, and 0.020 g PA–Fe$_3$O$_4$ microsphere is added for ultrasonication for 20 min. Then the obtained mixture is reacted at 90°C with vigorous stirring for 2 h. The obtained PA–Fe$_3$O$_4$@Y(OH)CO$_3$ is washed with deionized water.

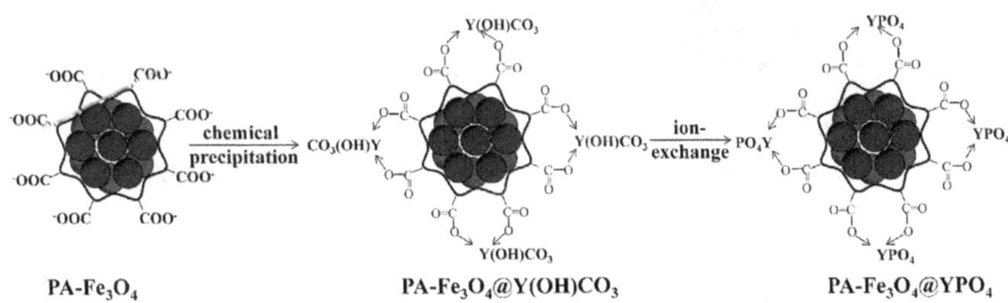

Figure 5.102: Schematic diagram of the synthesis of PA–Fe$_3$O$_4$@YPO$_4$ microsphere.

(3) Synthesis of PA–Fe$_3$O$_4$@YPO$_4$ Microsphere

0.045 g NH$_4$H$_2$PO$_4$ is dissolved in 20 mL H$_2$O, and then a certain amount of ammonia is added to adjust the pH of the solution to 11. Then, the obtained PA–Fe$_3$O$_4$@Y(OH)CO$_3$ is added for stirring at room temperature for 30 min. The dispersion is transferred to the reaction vessel, and reacted at 180°C for 12 h. The product is washed with deionized water and ethanol, and dried in vacuum.

5.5.5.2. *Characterization of PA–Fe$_3$O$_4$@YPO$_4$ microsphere*

From the TEM of the PA–Fe$_3$O$_4$@YPO$_4$ microsphere in Figure 5.103(a), the diameter of the microsphere is about 140 nm. The EDX analysis confirms the existence of P and Y (Figure 5.103(b)). The diffraction peaks of PA–Fe$_3$O$_4$@YPO$_4$ in the XRD pattern are also perfectly matched to Fe$_3$O$_4$ (JCPDS No. 19-0629) and YPO$_4$ (JCPDS No. 11-0264) (Figure 5.104(a)). These results illustrate the successful preparation of PA–Fe$_3$O$_4$@YPO$_4$ microsphere.

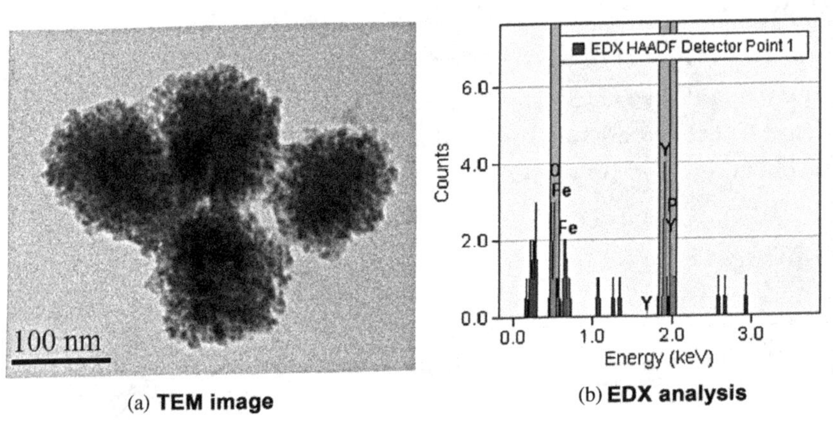

(a) **TEM image**　　(b) **EDX analysis**

Figure 5.103: TEM image and EDX analysis of PA–Fe$_3$O$_4$@YPO$_4$ microsphere.

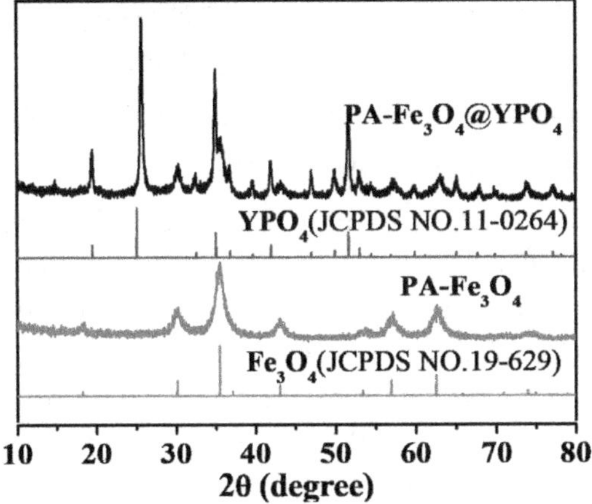

Figure 5.104: XRD pattern of different materials.

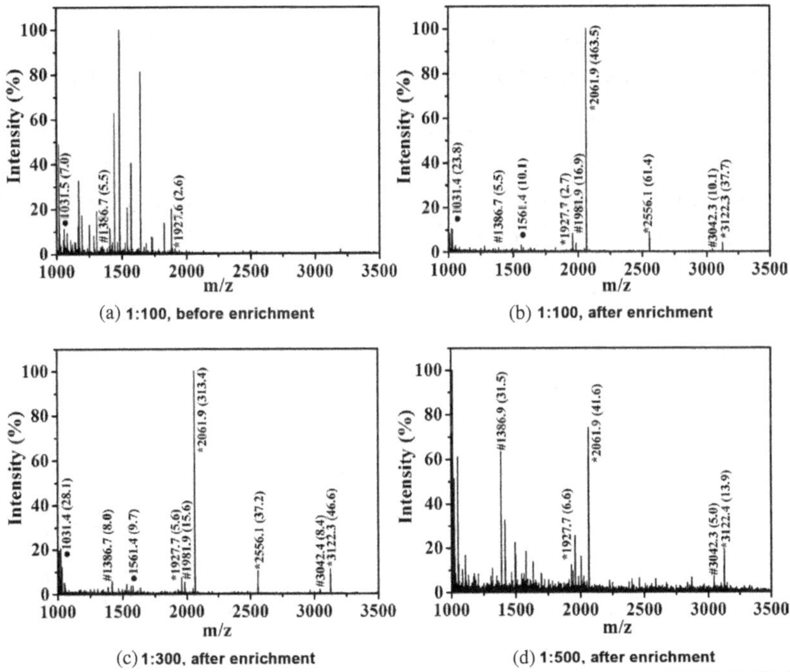

Figure 5.105: MS spectra of peptide mixture of β-casein and BSA at different molar ratios before and after enrichment with PA–Fe₃O₄@YPO₄ microsphere. *: phosphopeptide, #: dephosphorylated fragment, ●: double-charged phosphopeptide.

5.5.5.3. *Enrichment application of PA–Fe$_3$O$_4$@YPO$_4$ microsphere in phosphoproteomics*

(1) Evaluation of the enrichment selectivity of PA–Fe$_3$O$_4$@YPO$_4$ microsphere for phosphopeptide

The enrichment ability of PA–Fe$_3$O$_4$@YPO$_4$ microsphere is first investigated by using peptide mixture of β-casein and BSA at different molar ratios (Figure 5.105). As shown in Figure 5.105(c), after the digest mixture of β-casein and BSA with molar ratio of 1/300 is enriched with PA–Fe$_3$O$_4$@YPO$_4$ microsphere, the background is clear. When the molar ratio of β-casein to BSA reaches 1/500 (Figure 5.105(d)), although a few non-phosphopeptide peaks appear in the spectrum, the phosphopeptide peaks still dominate the spectrum. This indicates that the PA–Fe$_3$O$_4$@YPO$_4$ microsphere has excellent selective enrichment ability for phosphorylated peptides.

(2) Evaluation of the enrichment ability of PA–Fe$_3$O$_4$@YPO$_4$ microsphere for phosphopeptide in non-fat milk digest

As shown in Figure 5.106, 10 phosphorylated peptides and their corresponding dephosphorylated fragments are detected after enrichment with PA–Fe$_3$O$_4$@YPO$_4$ microsphere. This indicates that PA–Fe$_3$O$_4$@YPO$_4$ microsphere has selective enrichment ability for phosphorylated peptides in the practical samples.

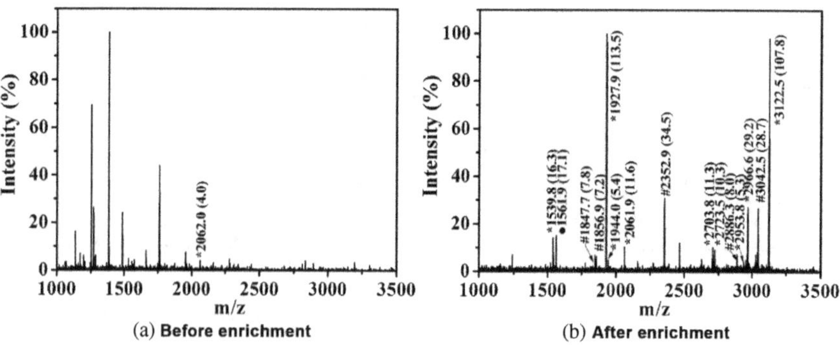

Figure 5.106: MS spectra of non-fat milk digest before and after enrichment with PA–Fe$_3$O$_4$@YPO$_4$ microsphere. *: phosphopeptide, #: dephosphorylated fragment, ●: double-charged phosphopeptide.

5.6. Amino-modified magnetic micro-/nanomaterial for separation and analysis in phosphoproteomics

5.6.1. *Introduction*

Amino group is a basic base in organic chemistry. It is positively charged and is an electron-repellent group. The phosphate group of the phosphorylated peptide is negatively charged, which provides possibility for amino-modified material to enrich the phosphopeptide. According to Refs. [53,54], the interaction between methyl guanidyl and dimethyl phosphate in aqueous solution is calculated by density functional theory and molecular dynamics, the positively charged arginine and phosphate group can form stable interaction similar to the covalent bond. The hydrogen bond can be formed between the phosphate group and the hydrogen atom of two nitrogen atoms in guanidine group, and thereby forming a double coordination complex. The structure of the polyethyleneimine also allows the amino group to form bi-ligand complex with the phosphate group. At present, the arginine and polyethyleneimine modified nanomaterials have been widely reported for the enrichment of phosphorylated peptide. This section will introduce two amino-modified magnetic microspheres in phosphoproteomics.

5.6.2. *Guanosilane-modified magnetic silica (Fe$_3$O$_4$@SiO$_2$@GDN)*

5.6.2.1. *Preparation of Fe$_3$O$_4$@SiO$_2$@GDN microsphere*

The synthesis of Fe$_3$O$_4$@SiO$_2$@GDN microsphere can be seen in Section 2.6 of Chapter 2.[55]

5.6.2.2. *Enrichment application of Fe$_3$O$_4$@SiO$_2$@GDN microsphere in phosphoproteomics*

(1) Evaluation on the enrichment selectivity of Fe$_3$O$_4$@SiO$_2$@GDN microsphere for phosphorylated protein

β-casein (phosphoprotein, MW = 24.0 KDa, pI = 4.6–5.1) and OVA (MW = 45.0 KDa, pI = 4.7) and six non-phosphorylated proteins with different sizes and pI, including BSA (MW = 67.0 KDa, pI = 4.8), hemoglobin (Hb, MW = 65.0 kDa, pI = 6.9), trypsin (Try, MW = 10.5 kDa, pI = 10.0), MYO

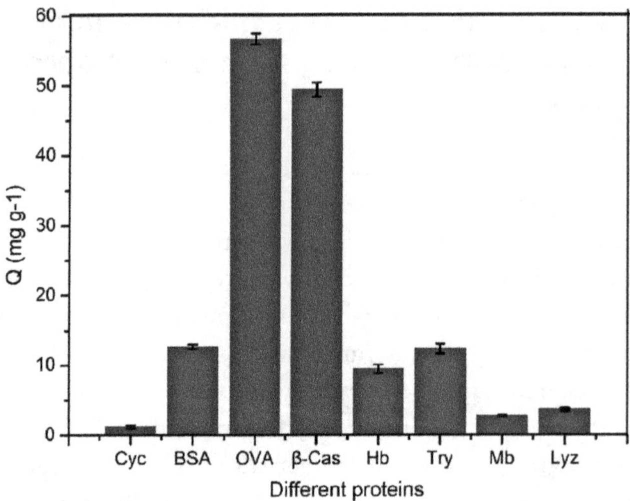

Figure 5.107: The recognition ability of $Fe_3O_4@SiO_2@GDN$ microsphere for different proteins.

(MW = 16.7, pI = 6.99), lysozyme (Lyz, MW = 14.0 kDa, pI = 11.0) and Cyc (MW = 12.4 kDa, pI = 9.8) are selected to investigate the selective enrichment ability of $Fe_3O_4@SiO_2@GDN$ microsphere. The concentration of each phosphorylated protein is 0.5 mg·mL^{-1}. In Figure 5.107, $Fe_3O_4@SiO_2@GDN$ microsphere shows significantly better recognition ability for β-casein (49.5 mg·g^{-1}) and OVA (56.7 mg·g^{-1}). The binding amount of $Fe_3O_4@SiO_2@GDN$ microsphere to non-phosphorylated protein ranges from 1.2 mg·g^{-1} (Cyc) to 12.7 mg·g^{-1} (BSA). Although BSA and OVA have similar isoelectric points, the binding ability of $Fe_3O_4@SiO_2@GDN$ microsphere to OVA is stronger than that of BSA. This indicates that $Fe_3O_4@SiO_2@GDN$ microsphere has selective enrichment ability for phosphorylated proteins. In addition, APTEOS-modified magnetic silica is also used to enrich β-casein and OVA, and the results show that the binding amount of APTEOS-modified magnetic silica to β-casein and OVA are 8.9 mg·g^{-1} and 5.7 mg·g^{-1}, respectively, indicating that the excellent binding ability of $Fe_3O_4@SiO_2@GDN$ microsphere to phosphorylated protein is due to the presence of guanidyl group.

Then, the mixture of β-casein and BSA and Cyc with different mass ratios is used to investigate the selectivity of $Fe_3O_4@SiO_2@GDN$ microsphere for phosphoproteins. The concentration of β-casein is 0.01 mg·mL^{-1}, and the concentrations of non-phosphoproteins BSA and Cyc range from 0.01 mg·mL^{-1} to 1.20 mg·mL^{-1}. As shown in Figure 5.108(a), when the mixed protein ratio is

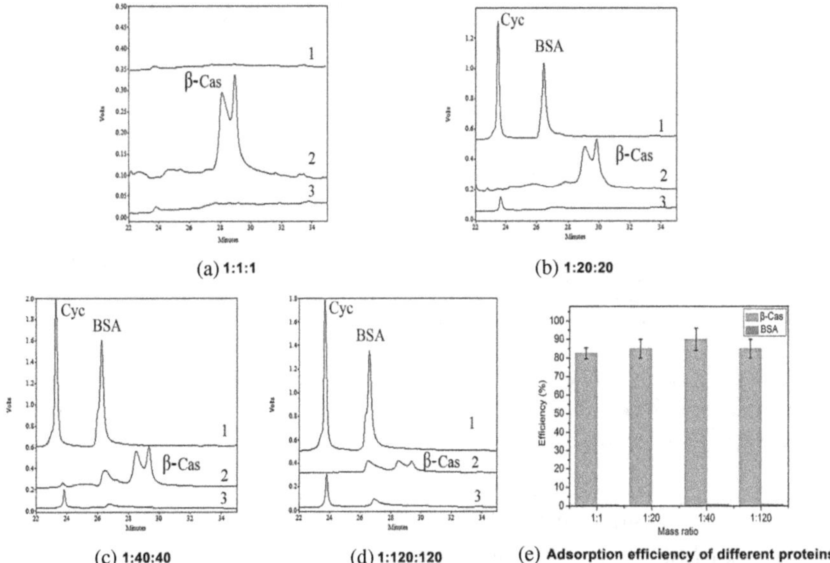

Figure 5.108: Chromatogram of the adsorption ability of $Fe_3O_4@SiO_2@GDN$ microsphere for protein mixtures with different mass ratios, and the corresponding adsorption efficiency for different proteins (curve 1: original solution; curve 2: eluent; curve 3: washing solution).

1:1:1, the concentration of the mixed protein is too low to be detected before enrichment. But after enrichment with $Fe_3O_4@SiO_2@GDN$ microsphere, β-casein peak appears in the chromatogram, indicating that $Fe_3O_4@SiO_2@GDN$ microsphere also has good ability to capture low concentration of phosphoprotein. As shown in Figure 5.108(c), when the mixed protein ratio is 1:40:40, a small number of characteristic peaks of BSA are also observed in the chromatogram of the eluent, but the adsorption efficiency of $Fe_3O_4@SiO_2@GDN$ microsphere to BSA is less than 5% (Figure 5.108(e)), while their adsorption efficiency to β-casein exceeds 80% at different ratios, indicating that $Fe_3O_4@SiO_2@GDN$ microsphere has good selective capture ability for phosphorylated protein in complex protein mixture.

(2) Evaluation of the tenrichment selectivity of $Fe_3O_4@SiO_2@$ GDN microsphere for phosphopeptide

The peptide mixture of β-casein and Cyc with molar ratio of 1/100 is used to investigate the enrichment selectivity of $Fe_3O_4@SiO_2@GDN$ microsphere for phosphopeptides. As shown in Figure 5.109, no phosphopeptide peak is

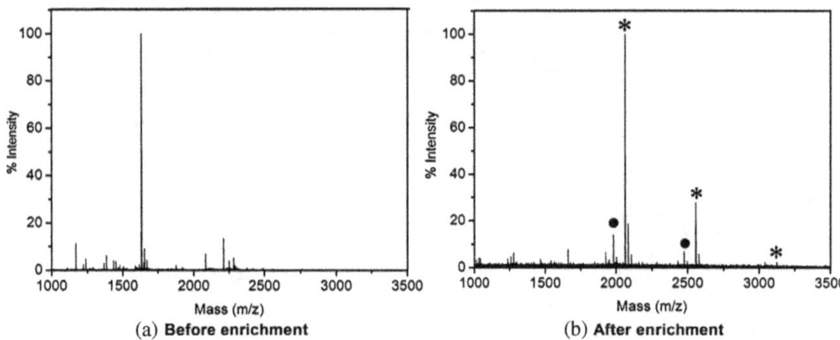

Figure 5.109:   MS spectra of the peptide mixture of β-casein and Cyc at the molar ratio of 1/100 before and after enrichment with $Fe_3O_4@SiO_2@GDN$ microsphere. *: phosphorylated peptide, •: dephosphorylated fragment.

detected before enrichment. However, after enrichment by $Fe_3O_4@SiO_2@$ GDN microsphere, three phosphorylated peptides and the corresponding dephosphorylated fragments belonging to β-casein digest are detected by MS, indicating that $Fe_3O_4@SiO_2@GDN$ microsphere has good enrichment selectivity for phosphopeptides in complex peptide mixtures.

5.6.3. *Polyethyleneimine-modified magnetic silica ($Fe_3O_4@SiO_2@PEI$)*

5.6.3.1. *Preparation of $Fe_3O_4@SiO_2@PEI$ microsphere*

The detailed synthesis process is described in Section 2.6 of Chapter 2.[56]

5.6.3.2. *Enrichment application of $Fe_3O_4@SiO_2@PEI$ microsphere in phosphoproteomics*

(1) Evaluation of enrichment condition

(a) *Optimization of enrichment system*

Two standard phosphorylated peptides (FQpSEEQQQTEDELQDK, m/z 2062.9 and RELEELNVPGEIVEpSLpSpSpSEESITR, m/z 3123.9) are selected to investigate the effect of enrichment system on the enrichment performance of phosphopeptide using $Fe_3O_4@SiO_2@PEI$ microsphere. As shown in Figure 5.110(b), only tetra-phosphorylated peptide is detected after enrichment with $Fe_3O_4@SiO_2@PEI$ microsphere under 50% ACN– 0.1% TFA. When using 100% ACN–0.1% TFA as loading buffer, both the

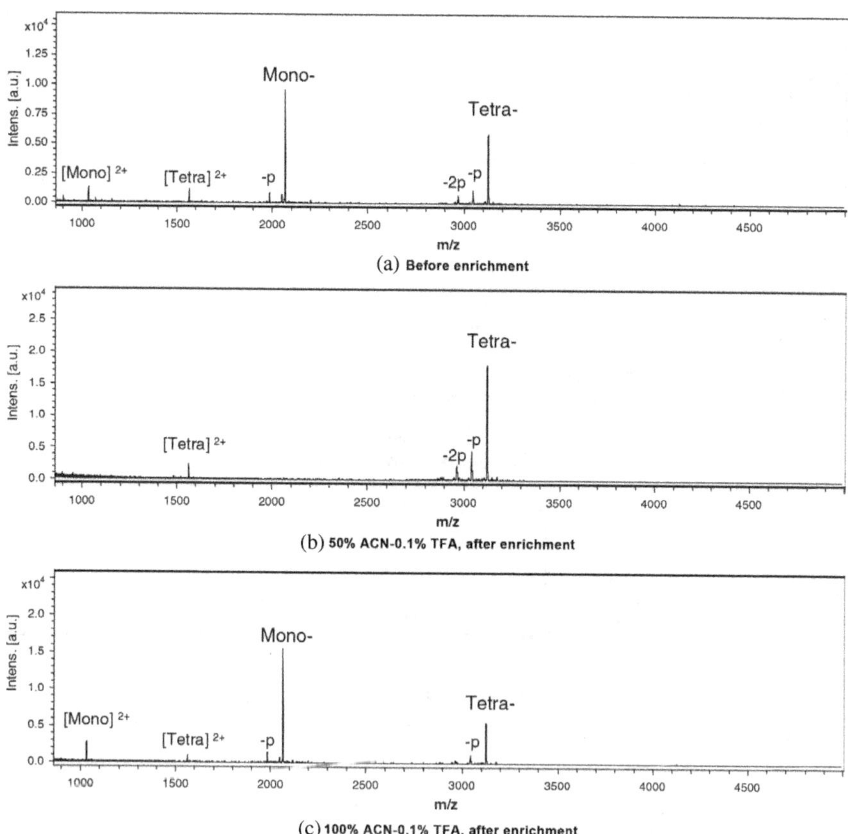

Figure 5.110: MS spectra of the mixture of standard phosphopeptide (100 μL 2 pmol·μL^{-1}) before and after enrichment with Fe$_3$O$_4$@SiO$_2$@PEI microsphere in different loading buffers. Mono-: mono-phosphopeptide; tetra-: tetra-phosphopeptide; -P: dephosphorylated fragment; -2P: dephosphorylated fragment.

mono-phosphorylated peptide and the tetra-phosphorylated peptide can be detected after enrichment with Fe$_3$O$_4$@SiO$_2$@PEI microsphere. This is because when the buffer system contains water, hydration enhances the stability between the multi-phosphorylated peptide and PEI, so the tetra-phosphorylated peptide can be observed after enrichment under 50% ACN–0.1% TFA. In the anhydrous buffer, there is no hydration, so both mono-phosphorylated peptide and the tetra-phosphorylated peptide are detected.

Next, 2 pmol·μL^{-1} α-casein is selected for further analysis. Figure 5.111(a) shows the direct analysis of α-casein digest, which is mainly occupied by the non-phosphorylated peptide peaks. Figure 5.111(b) shows the MS spectrum

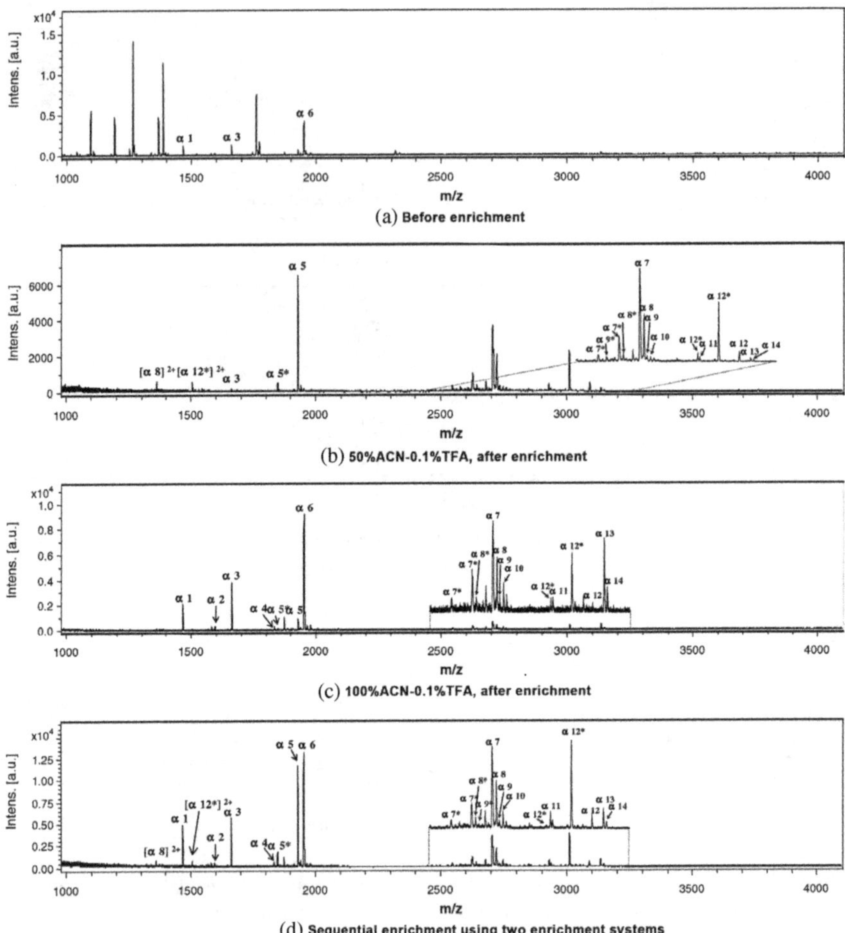

(a) Before enrichment

(b) 50%ACN-0.1%TFA, after enrichment

(c) 100%ACN-0.1%TFA, after enrichment

(d) Sequential enrichment using two enrichment systems

Figure 5.111: MS spectra of 2 pmol·μL^{-1} α-casein digest before and after enrichment with Fe$_3$O$_4$@SiO$_2$@PEI microsphere under different loading buffers.

of phosphorylated peptide captured by Fe$_3$O$_4$@SiO$_2$@PEI microsphere under 50% ACN–0.1% TFA. Multi-phosphorylated peptides dominate the spectrum, but the mono-phosphorylated peptides of α1, α2, α4, and α6 are not detected. However, when the loading buffer is 100% ACN–0.1% TFA, after enrichment with Fe$_3$O$_4$@SiO$_2$@PEI microsphere, mono-phosphorylated peptides of α1, α2, α4, and α6 and some multi-phosphorylated peptides can be observed as shown in Figure 5.111(c), but the signals of the mono-phosphorylated peptides are dominant.

Table 5.19: Detailed information of phosphopeptide enriched with $Fe_3O_4@SiO_2@PEI$ microsphere from α-casein digest.

No.	$[M+H]^+$ Da	Residues	Phospho-Sites	Peptide Sequence
α1	1466.6	S2-(153–164)	1	TVDMEpSTEVFTK
α2	1594.7	S2-(153–165)	1	TVDMEpSTEVFTKK
α3	1660.8	Sl-(121–134)	1	VPQLEIVPNpSAEER
α4	1832.8	S1-(104–119)	I	YLGEYLIVPNpSAEER
α5	1927.7	S1-(58–73)	2	DIGpSEpSTEDQAMEDIK
α6	1951.9	S1-(119–134)	1	YKVPQLEIVPNpSAEER
α7	2703.9	S1-pyro-(74–94)	5	pyroEMEAEpSIpSpSpSGEIVPNpSVEQK
α8	2720.9	S1-(74–94)	5	EMEAEpSIpSpSpSGEIVPNpSVEQK
α9	2736.9	S1-0-(74–94)	5	EoMEAEpSIpSpSpSGEIVP pSVEQK
α10	2747.1	S2-(17–37)	4	NTMEHVpSpSpSEEpS0SQETYKQ
α11	2935.1	S1-(50–73)	3	EKVNELpSKDIGpSEpSTEDQAMEDLK
α12	3087.9	S2-(61–85)	5	NANEEEYpS1GpSpSpSEEpSAEVATEEVK
α13	3132.2	S2-(16–39)	4	KNTMEHVpSpSpSEEpSIISQETYKQEK
α14	3148.2	S2-0-(16–39)	4	KNToMEHVpSpSpSEEpSDSQETYKQEK

Therefore, in order to enhance the signal of the multi-phosphorylated peptide, the 100% ACN–0.1% TFA buffer is firstly used as loading buffer for 30 s incubation with sample and $Fe_3O_4@SiO_2@PEI$ microsphere, then a certain amount of water is added to the above dispersion to form 50% ACN–0.1% TFA buffer, and then kept in incubation for another 30 s. In Figure 5.111(d), both the mono-phosphorylated peptides and the multi-phosphorylated peptides are detected, and the signals of the multi-phosphorylated peptides are significantly enhanced.

(2) Evaluation of the enrichment sensitivity of $Fe_3O_4@SiO_2@PEI$ microsphere for phosphopeptide in standard phosphoprotein digest

α-casein digest is used to further investigate the enrichment sensitivity of $Fe_3O_4@SiO_2@PEI$ microsphere for phosphopeptide.[56] As shown in Figure 5.112, when the concentration of α-casein digest is as low as 5×10^{-11} $mol \cdot L^{-1}$, three mono-phosphorylated peptides and one multi-phosphorylated

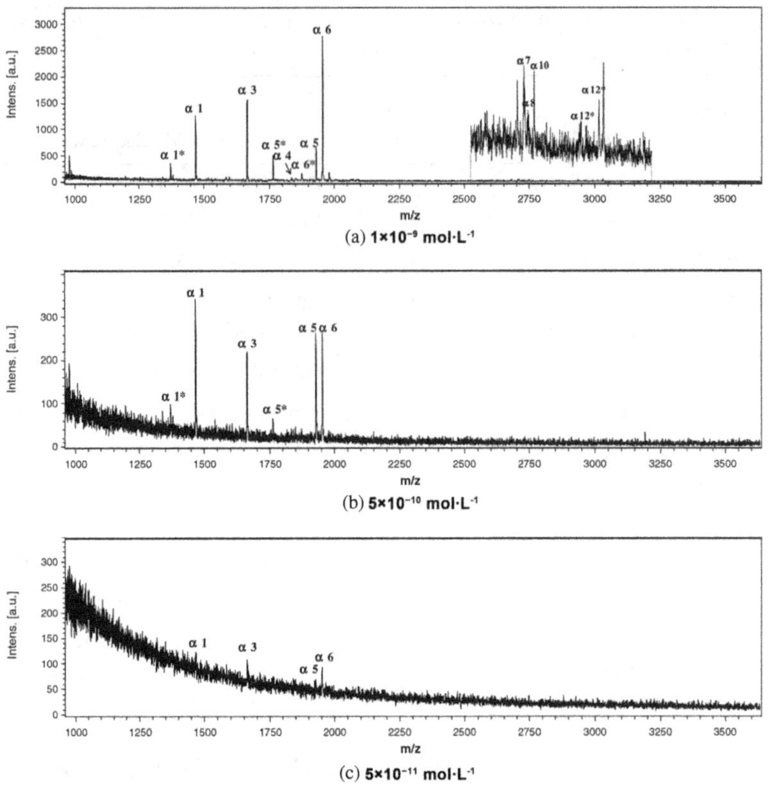

Figure 5.112: MS spectra of different concentrations of α-casein digests (100 μL) after enrichment with $Fe_3O_4@SiO_2@PEI$ microsphere.

peptide can be detected by MS after enrichment with $Fe_3O_4@SiO_2@PEI$ microsphere, indicating that $Fe_3O_4@SiO_2@PEI$ microsphere has high enrichment sensitivity.

(3) Evaluation of the enrichment selectivity of $Fe_3O_4@SiO_2@PEI$ microsphere for phosphopeptide

In order to investigate the enrichment selectivity of $Fe_3O_4@SiO_2@PEI$ microsphere for phosphopeptide, peptide mixture of α-casein (3×10^{-8} mol·L⁻¹), MYO, and Cyc with molar ratio of 1/1000/500 is selected. As shown in Figure 5.113(a), the spectrum is occupied by non-phosphorylated peptide

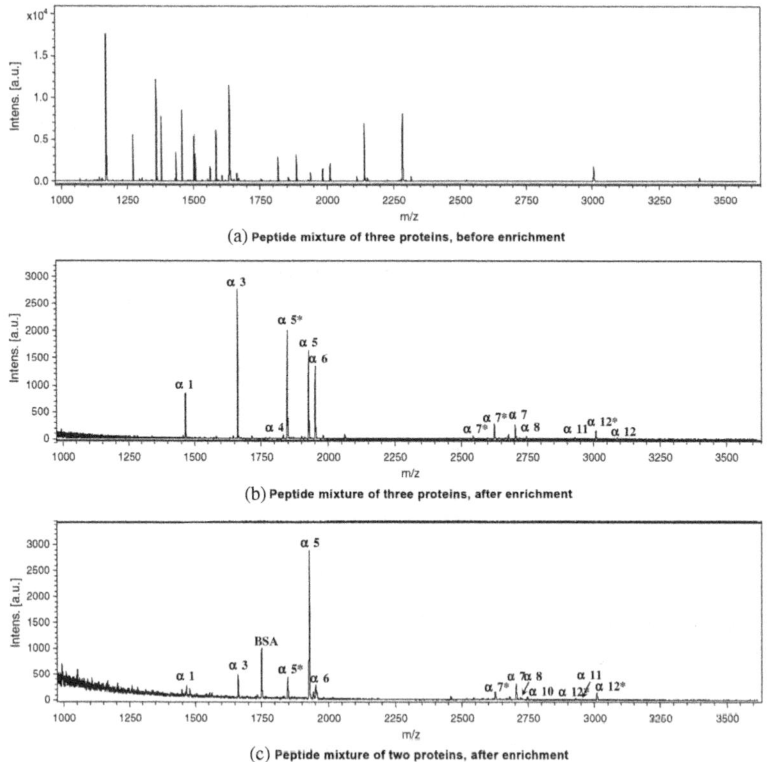

Figure 5.113: MS spectra of peptide mixture of α-casein, MYO, and Cyc with molar ratio of 1/1000/500 before and after enrichment with $Fe_3O_4@SiO_2@PEI$ microsphere and MS spectrum of peptide mixture of α-casein and BSA with molar ratio of 1/1,000 after enrichment with $Fe_3O_4@SiO_2@PEI$ microsphere.

before enrichment. After enrichment with $Fe_3O_4@SiO_2@PEI$ microsphere, the background of the spectrum is clean and more than a dozen phosphorylated peptides are detected, indicating that $Fe_3O_4@SiO_2@PEI$ microsphere has high enrichment selectivity for phosphopeptides.

Then the peptide mixture of α-casein and BSA with molar ratio of 1/1,000 is chosen to further investigate the enrichment selectivity of $Fe_3O_4@SiO_2@PEI$ microsphere for phosphorylated peptide. As shown in Figure 5.113(c), nine phosphorylated peptides and one non-phosphorylated peptide from BSA are detected, which further confirms the enrichment selectivity of $Fe_3O_4@SiO_2@PEI$ microsphere.

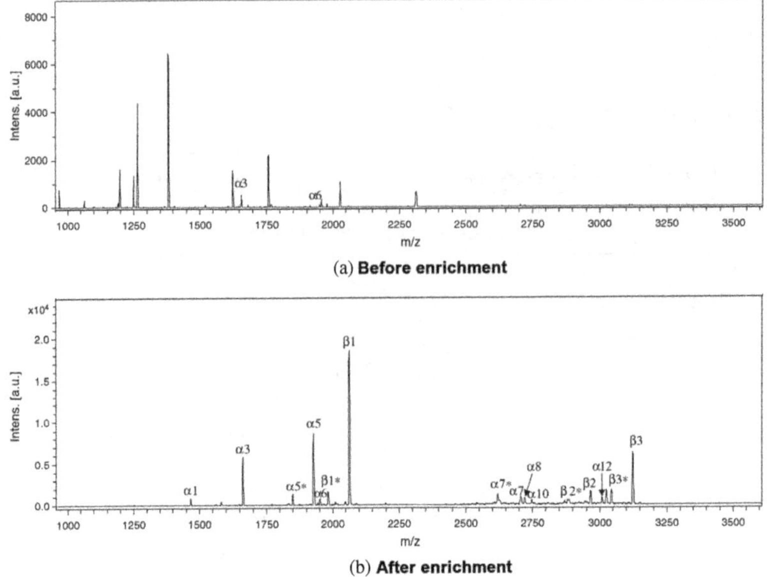

Figure 5.114: MS spectra of non-fat milk digest before and after enrichment with $Fe_3O_4@SiO_2@PEI$ microsphere.

(4) Evaluation of the enrichment ability of $Fe_3O_4@SiO_2@PEI$ microsphere for phosphopeptide in non-fat milk digest

The non-fat milk digest is selected to further investigate the enrichment ability of $Fe_3O_4@SiO_2@PEI$ microsphere for phosphopeptide in a complex sample. As shown in Figure 5.114(b), $Fe_3O_4@SiO_2@PEI$ microsphere exhibits excellent selective enrichment ability for the phosphorylated peptides in the non-fat milk digest.

References

1. Marx V. Making Sure Ptms Are Not Lost after Translation. *Nature Methods*, 2013, 10: 201–4.
2. Taouatas N, Drugan M. M., Heck A J., *et al.* Straightforward Ladder Sequencing of Peptides Using a Lys-N Metalloendopeptidase. *Nature Methods*, 2008, 5: 405–7.
3. Stensballe A, Andersen S, Jensen O. N. Characterization of Phosphoproteins from Electrophoretic Gels by Nanoscale Fe(III) Affinity Chromatography with Off-Line Mass Spectrometry Analysis. *Proteomics*, 2001, 1: 207–22.

4. Joe Cannon K. L., Colin Wynne., Yan Wang., Nathan Edwards., Catherine Fenselau. High-Throughput Middle-Down Analysis Using an Orbitrap. *Journal of Proteome Research*, 2010, 9: 3886–90.

5. Wang Z. G., Lv N., Bi W. Z., *et al.* Development of the Affinity Materials for Phosphorylated Proteins/Peptides Enrichment in Phosphoproteomics Analysis. *ACS Applied Materials and Interfaces*, 2015, 7: 8377–92.

6. Yang, C., Zhong, X., Li L. Recent Advances in Enrichment and Separation Strategies for Mass Spectrometry-based Phosphoproteomics. *Electrophoresis*, 2014, 35: 3418–29.

7. Nousiainen M., Sillje H. H., Sauer G., *et al.* Phosphoproteome Analysis of the Human Mitotic Spindle. *Proceedings of the National Academy of Sciences of the United States of America*, 2006, 103: 5391–96.

8. Nuhse T. S., Stensballe A., Jensen O. N., *et al.* Large-Scale Analysis of in Vivo Phosphorylated Membrane Proteins by Immobilized Metal Ion Affinity Chromatography and Mass Spectrometry. *Molecular & Cellular Proteomics*, 2003, 2: 1234–43.

9. Jensen S. S., Larsen M. R. Evaluation of the Impact of Some Experimental Procedures on Different Phosphopeptide Enrichment Techniques. *Rapid Communications in Mass Spectrometry*, 2007, 21: 3635–45.

10. Larsen M. R., Thingholm T. E., Jensen O. N., *et al.* Highly Selective Enrichment of Phosphorylated Peptides from Peptide Mixtures Using Titanium Dioxide Microcolumns. *Molecular & Cellular Proteomics*, 2005, 4: 873–86.

11. Ficarro S. B., Parikh J. R., Blank N. C., *et al.* Niobium(V) Oxide (Nb_2O_5): Application to Phosphoproteomics. *Analytical Chemisty*, 2008, 80: 4606–13.

12. Xu X., Deng C., Gao M., *et al.* Synthesis of Magnetic Microspheres with Immobilized Metal Ions for Enrichment and Direct Determination of Phosphopeptides by Matrix-Assisted Laser Desorption Ionization Mass Spectrometry. *Advanced Materials*, 2006, 18: 3289–93.

13. Qi D., Mao Y., Lu J., *et al.* Phosphate-Functionalized Magnetic Microspheres for Immobilization of Zr^{4+} Ions for Selective Enrichment of the Phosphopeptides. *Journal of Chromatography A*, 2010, 1217: 2606–17.

14. Yan Y., Zheng Z., Deng C., *et al.* Facile Synthesis of Ti^{4+}-Immobilized Fe_3O_4@Polydopamine Core-Shell Microspheres for Highly Selective Enrichment of Phosphopeptides. *Chemical Communications*, 2013, 49: 5055–57.

15. Muller D. R., Schindler P., Coulot H., *et al.* Mass Spectrometric Characterization of Stathmin Isoforms Separated by 2D Page. *Journal of Mass Spectrometry*, 1999, 34: 336–45.

16. Ficarro S. B., McCleland M. L., Stukenberg P. T., *et al.* Phosphoproteome Analysis by Mass Spectrometry and Its Application to Saccharomyces Cerevisiae. *Nature Biotechnology*, 2002, 20: 301–05.

17. He T., Alving K., Feild B., *et al.* Quantitation of Phosphopeptides Using Affinity Chromatography and Stable Isotope Labeling. *Journal of the American Society for Mass Spectrometry*, 2004, 15: 363–73.

18. Saha A., Saha N., Ji L. N., *et al.* Stability of Metal Ion Complexes Formed with Methyl Phosphate and Hydrogen Phosphate. *Journal of Biological Inorganic Chemistry*, 1996, 1: 231–8.

19. Wang J., Zhang Y., Cai Y., *et al.* Analysis of Protein Phosphorylation by Combination of IMAC,Phosphatase with Biological Mass Spectrometry. *Acta Biochimica et Biophysica Sinica*, 2003, 35: 459–66.

20. Raska C. S., Parker C. E., Dominski Z., *et al.* Direct MALDI-MS/MS of Phosphopeptides Affinity-Bound to Immobilized Metal Ion Affinity Chromatography Beads. *Analytical Chemistry*, 2002, 74: 3429–33.

21. Xu X.-Q. Development of Novel Isolation and Identification Approaches in Proteomics Analysis Using Functional Magnetic Technologies, Fudan University (2007).

22. Xu X. Q., Deng C. H., Gao M. X., *et al.* Synthesis of Magnetic Microspheres with Immobilized Metal Ions for Enrichment and Direct Determination of Phosphopeptides by Matrix-Assisted Laser Desorption Ionization Mass Spectrometry. *Advanced Materials*, 2006, 18: 3289–93.

23. Zhao L., Qin H. Q., Hu Z. Y., *et al.* A Poly(Ethylene Glycol)-Brush Decorated Magnetic Polymer for Highly Specific Enrichment of Phosphopeptides. *Chemical Science*, 2012, 3: 2828–38.

24. Ma W. F., Zhang Y., Li L. L., *et al.* Ti^{4+}-Immobilized Magnetic Composite Microspheres for Highly Selective Enrichment of Phosphopeptides. *Advanced Functional Materials*, 2013, 23: 107–15.

25. Yao X. D., Freas A., Ramirez J., *et al.* Proteolytic O-18 Labeling for Comparative Proteomics: Model Studies with Two Serotypes of Adenovirus. *Analytical Chemistry*, 2001, 73: 2836–42.

26. Zhang L. Y., Zhao Q., Liang Z., *et al.* Synthesis of Adenosine Functionalized Metal Immobilized Magnetic Nanoparticles for Highly Selective and Sensitive Enrichment of Phosphopeptides. *Chemical Communications*, 2012, 48: 6274–6.

27. Sun N., Deng C., Li Y., *et al.* Size-Exclusive Magnetic Graphene/Mesoporous Silica Composites with Titanium(IV)-Immobilized Pore Walls for Selective Enrichment of Endogenous Phosphorylated Peptides. *ACS Applied Materials and Interfaces*, 2014, 6: 11799–804.

28. Chen C-T., Chen Y-C. Fe_3O_4/TiO_2 Core/Shell Nanoparticles as Affinity Probes for the Analysis of Phosphopeptides Using TiO_2 Surface-Assisted Laser Desorption/Ionization Mass Spectrometry. *Analytical Chemistry*, 2005, 77: 5912–19.

29. Lo C-Y., Chen W-Y., Chen C-T., *et al.* Rapid Enrichment of Phosphopeptides from Tryptic Digests of Proteins Using Iron Oxide Nanocomposites of Magnetic Particles Coated with Zirconia as the Concentrating Probes. *Journal of Proteome Research*, 2007, 6: 887–93.

30. Chen, Chen W-Y., Tsai P-J., *et al.* Rapid Enrichment of Phosphopeptides and Phosphoproteins from Complex Samples Using Magnetic Particles Coated with Alumina as the Concentrating Probes for MALDI MS Analysis. *Journal of Proteome Research*, 2007, 6: 316–25.

31. Li Y., Xu X. Q., Qi D. W., *et al.* Novel Fe_3O_4@TiO_2 Core-Shell Microspheres for Selective Enrichment of Phosphopeptides in Phosphoproteome Analysis. *Journal of Proteome Research*, 2008, 7: 2526–38.

32. Li Y., Leng T. H., Lin H. Q., *et al.* Preparation of Fe_3O_4@ZrO_2 Core-Shell Microspheres as Affinity Probes for Selective Enrichment and Direct Determination of Phosphopeptides Using Matrix-Assisted Laser Desorption Ionization Mass Spectrometry. *Journal of Proteome Research*, 2007, 6: 4498–510.

33. Amphlett C. B. *Inorganic Ion Exchangers*. Elsevier Pub. Co., 1964.

34. Kweon H. K., Håkansson K. Selective Zirconium Dioxide-Based Enrichment of Phosphorylated Peptides for Mass Spectrometric Analysis. *Analytical Chemistry*, 2006, 78: 1743–49.

35. Blackwell J. A., Carr P. W. Fluoride-Modified Zirconium Oxide as a Biocompatible Stationary Phase for High-Performance Liquid Chromatography. *Journal of Chromatography A*, 1991, 549: 59–75.

36. Blackwell J. A., Carr P. W.. Study of the Fluoride Adsorption Characteristics of Porous Microparticulate Zirconium-Oxide. *Journal of Chromatography*, 1991, 549: 43–57.

37. Li Y., Liu Y., Tang J., *et al.* Fe_3O_4@Al_2O_3 Magnetic Core-Shell Microspheres for Rapid and Highly Specific Capture of Phosphopeptides with Mass Spectrometry Analysis. *Journal of Chromatography A*, 2007, 1172: 57–71.

38. Li Y., Lin H. Q., Deng C. H., *et al.* Highly Selective and Rapid Enrichment of Phosphorylated Peptides Using Gallium Oxide-Coated Magnetic Microspheres for MALDI-TOF-MS and Nano-LC-ESI-MS/MS/MS Analysis. *Proteomics*, 2008, 8: 238–49.

39. Qi D. W., Lu J., Deng C. H., *et al.* Development of Core-Shell Structure Fe_3O_4@Ta_2O_5 Microspheres for Selective Enrichment of Phosphopeptides for Mass Spectrometry Analysis. *Journal of Chromatography A*, 2009, 1216: 5533–9.

40. Qi D. W., Lu J., Deng C. H., *et al.* Magnetically Responsive Fe_3O_4@C@SnO_2 Core-Shell Microspheres: Synthesis, Characterization and Application in Phosphoproteomics. *Journal of Physical Chemistry C*, 2009, 113: 15854–61.

41. Lu J., Wang M. Y., Deng C. H., *et al.* Facile Synthesis of Fe_3O_4@Mesoporous TiO_2 Microspheres for Selective Enrichment of Phosphopeptides for Phospho-proteomics Analysis. *Talanta*, 2013, 105: 20–27.

42. Wan H., Li J., Yu W., *et al.* Fabrication of a Novel Magnetic Yolk-Shell Fe_3O_4@$mTiO_2$@$mSiO_2$ Nanocomposite for Selective Enrichment of Endogenous Phos-phopeptides from a Complex Sample. *RSC Advances*, 2014, 4: 45804–08.

43. Lu J., Deng C. H., Zhang X. M., *et al.* Synthesis of Fe_3O_4/Graphene/ TiO_2 Composites for the Highly Selective Enrichment of Phosphopeptides from Bio-logical Samples. *ACS Applied Materials & Interfaces*, 2013, 5: 7330–4.

44. Wang M., Deng C., Li Y., *et al.* Magnetic Binary Metal Oxides Affinity Probe for Highly Selective Enrichment of Phosphopeptides. *ACS Applied Materials & Interfaces*, 2014, 6: 11775–82.

45. Zhao M., Deng C., Zhang X. The Design and Synthesis of a Hydrophilic Core-Shell-Shell Structured Magnetic Metal-Organic Framework as a Novel Immo-bilized Metal Ion Affinity Platform for Phosphoproteome Research. *Chemical Communications*, 2014, 50: 6228–31.

46. Chen Y. J., Xiong Z. C., Peng L., *et al.* Facile Preparation of Core-Shell Magnetic Metal Organic Framework Nanoparticles for the Selective Capture of Phospho-peptides. *ACS Applied Materials & Interfaces*, 2015, 7: 16338–47.

47. Zhai R., Jiao F., Feng D., *et al.* Preparation of Mixed Lanthanides-Immobilized Magnetic Nanoparticles for Selective Enrichment and Identification of Phos-phopeptides by MS. *Electrophoresis*, 2014, 35: 3470–8.

48. Wei J. Y., Zhang Y. J., Wang J. L., *et al.* Highly Efficient Enrichment of Phos-phopeptides by Magnetic Nanoparticles Coated with Zirconium Phosphonate for Phosphoproteome Analysis. *Rapid Communications in Mass Spectrometry*, 2008, 22: 1069–80.

49. Cheng G., Zhang J. L., Liu Y. L., *et al.* Synthesis of Novel Fe_3O_4@SiO_2@CeO_2 Microspheres with Mesoporous Shell for Phosphopeptide Capturing and Label-ing. *Chemical Communications*, 2011, 47: 5732–4.

50. Cheng G., Liu Y. L., Zhang J. L., *et al.* Lanthanum Silicate Coated Magnetic Microspheres as a Promising Affinity Material for Phosphopeptide Enrichment and Identification. *Analytical and Bioanalytical Chemistry*, 2012, 404: 763–70.

51. Sun Y., Wang H-F. Ultrathin-Yttrium Phosphate-Shelled Polyacrylate-Ferrifer-rous Oxide Magnetic Microspheres for Rapid and Selective Enrichment of Phos-phopeptides. *Journal of Chromatography A*, 2013, 1316: 62–8.

52. Xuan S., Wang Y-X J., Yu J. C., *et al.* Tuning the Grain Size and Particle Size of Superparamagnetic Fe_3O_4 Microparticles. *Chemistry of Materials*, 2009, 21: 5079–87.

53. Woods A. S., Ferré S. Amazing Stability of the Arginine-Phosphate Electrostatic Interaction. *Journal of Proteome Research*, 2005, 4: 1397–402.

54. Frigyes D., Alber F., Pongor S., *et al.* Arginine-Phosphate Salt Bridges in Protein-DNA Complexes: A Car-Parrinello Study. *Journal of Molecular Structure-Theochem*, 2001, 574: 39–45.

55. Deng Q. L., Wu J. H., Chen Y., *et al.* Guanidinium Functionalized Superparamagnetic Silica Spheres for Selective Enrichment of Phosphopeptides and Intact Phosphoproteins from Complex Mixtures. *Journal of Materials Chemistry B*, 2014, 2: 1048–58.

56. Chen C. T., Wang L. Y., Ho Y. P. Use of Polyethylenimine-Modified Magnetic Nanoparticles for Highly Specific Enrichment of Phosphopeptides for Mass Spectrometric Analysis. *Analytical and Bioanalytical Chemistry*, 2011, 399: 2795–806.

Chapter 6

Glycoproteomics Analysis Technology Based on Magnetic Micro-/Nanomaterial

6.1. Basic analysis principle of glycoproteomics based on magnetic micro-/nanomaterial

Glycosylation is one of the most common, important and complex post-translational modifications, which has vital biological function as well as biological significance. Glycosylated proteins are abundant in tissues, cells and body fluids. They usually participate in a variety of life activities, such as the clearance of aging protein in plasma, immune response, cell adhesion, signal transduction, etc. Studies have reported that the occurrence of various diseases is usually accompanied by abnormal protein glycosylation. The occurrence of glycosylation, amount of glycosylated protein and the variation of glycan structure are connected directly with specific physiological and pathological states. Protein glycosylation, especially N-glycosylation, usually occurs on protein in extracellular environment, such as membrane protein, secreted protein and protein in body fluids. Generally, these proteins are readily available and can be served as a basis for disease diagnosis and treatment. A great many studies have shown that a lot of biomarkers and therapeutic targets are glycosylated proteins, such as CA125 for ovarian cancer, AFP for liver cancer and Her2/neu for breast cancer.

At present, the research on protein glycosylation is mainly based on biomass spectrometry. Biomass spectrometry, represented by MALDI-TOF-MS and ESI-Q-TOF, has greatly promoted the development of biomacromolecules by MS detection. The main topic of glycosylation research includes glycosylated protein/peptide analysis, glycosylation site analysis and glycan analysis. Hence, there are two main profiling strategies for glycoproteomics and glycomics. These two strategies are based on glycosylated protein/peptide or glycan, respectively, which is analyzed by a large-scale and high-throughput detection method. As shown in Figure 6.1, in order to achieve the analytical purpose, glycoproteomics usually has two routes: one is to enrich the glycosylated protein before enzymatic hydrolysis and then identify glycosylation site, and the other is to digest the protein before enrichment of glycopeptide. However, due to the relatively low abundance of glycoprotein/glycopeptide in total human protein/peptide, there is a great challenge for biomass spectrometry for the identification of glycosylated protein/peptide in a complex biosample. In the MS analysis process, the signal of glycopeptide is often obscured or suppressed by non-glycopeptide, so it is necessary to enrich the glycopeptide in a complex biological sample to improve its relative abundance for MS detection. At present, different methods, based on different mechanisms, are developed for enrichment of glycoprotein or glycopeptide, such as hydrophilic interaction chromatography (HILIC), lectin affinity, covalent binding and a combination of multiple methods. In recent years, various functionalized magnetic micro-/nanomaterials have been synthesized for separation and enrichment of glycoprotein/glycopeptide by utilizing their rapid separation capability and surface paintability.

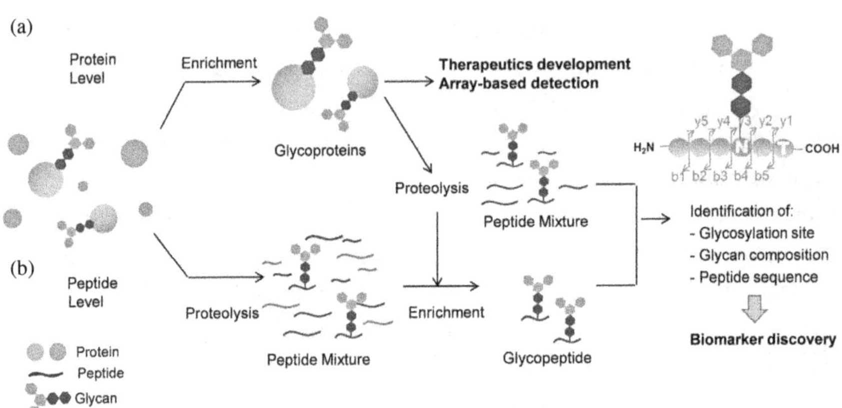

Figure 6.1: Two commonly used enrichment routes in glycoproteomics.[1]

6.1.1. *Hydrophilic functional group modified magnetic micro-/nanomaterial*

It is well known that the glycosylated protein/peptide contains a large amount of hydroxyl groups, so the glycan shows extreme hydrophilicity, while the non-glycosylated protein/peptide shows relatively strong hydrophobicity. Based on the difference between glycoprotein/glycopeptide and the non-glycosylated protein/peptide, hydrophilic material can be used to enrich the glycosylated protein/peptide by relatively strong hydrophilicity. At present, HILIC columns based on amino and agarose are widely used for separation and enrichment of glycosylated protein/peptide on the one hand. Among them, ion-pair HILIC has enhanced specific enrichment capability for glycopeptide due to the introduction of ion-pairing reagent. It is because ion-pairing reagent can effectively neutralize the charge on the peptide, and decrease the interaction with hydrophilic substrate, that t he glycopeptide can still combine with the hydrophilic substrate via hydrogen bonding. On the other hand, HILIC columns based on saccharide are also widely used for the separation of glycosylated protein/peptide. In such studies, chitosan, maltose, glucose, etc., with terminal alkynyl group, can be efficiently reacted with azide modified silica via click reaction to prepare the material with special surface structure and hydrophilicity. Such materials provide reliable methods for rapid and accurate analysis of glycosylation.

6.1.2. *Lectin-modified magnetic micro-/nanomaterial*

Lectin is a general term for protein that has selective recognition ability for soluble glycan, glycosylated protein, and glycolipid. As various types of glycans have their corresponding lectins, lectin affinity methods are widely used to separate and enrich glycoprotein. Different glycoproteins have specific affinity with different lectins. For example, wheat germ agglutinin (WGA) can selectively recognize acetylglucosamine (GlcNAc), while concanavalin A (Con A) has highly specific recognition of high mannose-type or hybrid *N*-glycan. At present, lectin immobilized silica material, gold foil and magnetic microsphere are widely used for the separation and analysis of glycosylated protein.

6.1.3. *Chelation-based magnetic micro-/nanomaterial*

The design purpose of such materials is to enrich sialylated glycoprotein. These materials are mainly metal oxides, such as TiO_2 and ZrO_2. The principle of enriching sialylated glycoprotein based on TiO_2 affinity method is that titanium ions have an electron-deficient structure, which can specifically bind to sialic acid residues with rich electrons. The highlight of this method is that it does not have any destructive effect in the enrichment process. Thus, complete glycan information can be obtained, which provides a strong basis for further identification of glycopeptide. Nowadays, TiO_2 modified magnetic microspheres have been widely used for large-scale enrichment of phosphopeptides, while they have little application in glycoproteomics. TiO_2 microsphere is widely used as chromatographic packing for separation and purification of sialylated glycoprotein.

6.1.4. *Covalent binding-based magnetic micro-/ nanomaterial*

Boronic acid chemistry and hydrazine chemistry are two widely used methods based on covalent binding for separation and enrichment of glycosylated protein/peptide. The boric acid molecule is hydroxylated in non-aqueous medium or alkaline solution, molecular configuration of which will change from planar triangle to tetrahedral. The tetrahedral anion can reversibly combine with cis-dihydroxy group of glycopeptide to form a cyclic diester for realizing glycopeptide enrichment. In acidic system, the above-described binding reaction will be reversed, and the enriched glycosylated protein/ peptide can be released. The boronic acid chemistry-based enrichment enables unbiased enrichment of glycosylated protein/peptide, and enrichment and elution of glycosylated protein/peptide can be simply realized by changing the pH of sample solution.

Hydrazine chemistry is a kind of solid phase enrichment method based on hydrazine reaction. The oxidized cis–diol on glycan of glycosylated protein/peptide will react with hydrazide to realize the capture of glycoprotein or glycopeptide. Currently, the reported hydrazine chemistry is to use periodate to oxidize the cis–diol on the glycan into the aldehyde group, then the aldehyde group will be covalently linked to the hydrazide group, and finally the glycoprotein or glycopeptide will be released via enzymatic hydrolysis.

6.2. Hydrophilic functional group-modified magnetic micro-/nano material for separation and analysis in glycoproteomics

6.2.1. *Introduction*

The hydrophilic material-based enrichment method for glycosylated protein/peptide is developped according to the structure and property of glycoprotein/glycopeptide. These methods with simple operations can remove most of the non-glycosylated peptides and salts, and enrich various types of glycosylated proteins/peptides with no bias, therefore, they are widely used in separation of glycan and glycopeptide. At present, different hydrophilic polymers are modified or co-modified on the surface of magnetic microsphere for the enrichment of glycoprotein and glycopeptide. In addition, the hydrophilicity of glucose, chitosan and galactose and so on is relatively strong due to the presence of a large amount of hydroxyl groups in their glycan structure. Different sugars are, respectively, modified or co-modified on the surface of magnetic microspheres for the separation and enrichment of glycosylated protein/peptide. Moreover, magnetic microsphere materials in which hydrophilic polymer and sugar are co-modified are also widely used. This section will introduce the application of hydrophilic micro-/nanomaterial in glycoproteomics from the above aspects.

6.2.2. *Sugar-modified magnetic microsphere*

6.2.2.1. *Glucose-modified magnetic mesoporous microsphere (Fe$_3$O$_4$@mSiO$_2$–glucose)*

(1) Preparation of Fe$_3$O$_4$@mSiO$_2$–glucose microsphere

The schematic diagram of the synthesis of Fe$_3$O$_4$@mSiO$_2$–glucose microsphere is shown in Figure 2.82, and the detailed synthesis process can be seen in Section 2.7 of Chapter 2.[2]

(2) Enrichment application of Fe$_3$O$_4$@mSiO$_2$–glucose in glycoproteomics

The enrichment flowchart of Fe$_3$O$_4$@mSiO$_2$–glucose microsphere is shown in Figure 6.2.

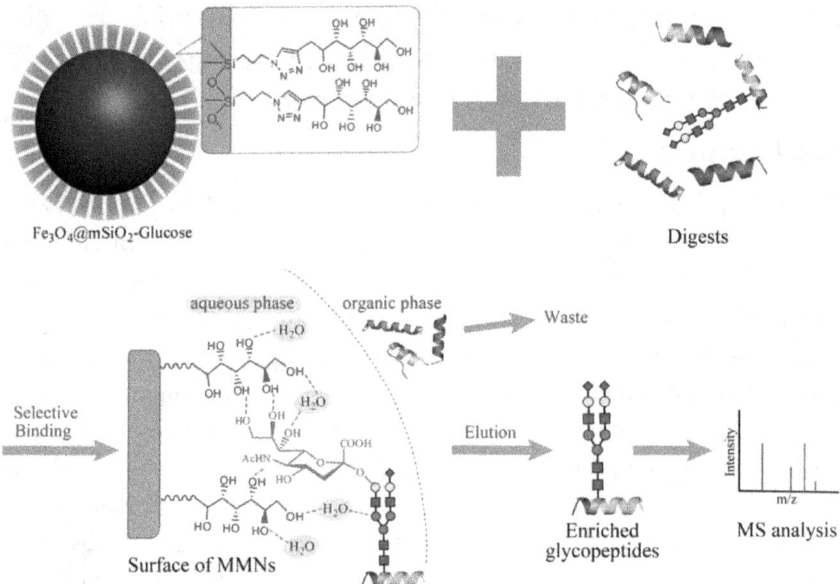

Figure 6.2: Enrichment flowchart of Fe$_3$O$_4$@mSiO$_2$–glucose microsphere.

(3) Evaluation of the enrichment capability of Fe$_3$O$_4$@ mSiO$_2$–glucose microsphere for glycopeptide in standard glycoprotein digest

Taking the digest of horseradish peroxidase (HRP) as the experimental object, ACN/H$_2$O/FA (v/v/v, 80/19.8/0.2) as loading buffer and 5 min as the loading time, the enrichment capability of Fe$_3$O$_4$@mSiO$_2$–glucose microsphere for glycopeptide is firstly evaluated. There are eight glycopeptides and nine glycosylation sites in HRP, but glycopeptides may be fragmented during the matrix-assisted laser desorption ionization process, so it is usual to detect more than eight glycopeptide signals in MS spectra. As shown in Figure 6.3(a), when 0.5 ng·μL^{-1} HRP digest is directly analyzed by ESI-Q-TOF tandem MS, only four glycopeptides are detected. But after enrichment by Fe$_3$O$_4$@mSiO$_2$–glucose microsphere, a total of 18 glycopeptides are detected and the spectrum background is clean. The detailed information of detected glycopeptides is shown in Table 6.1. When the concentration of HRP digest is reduced to 0.011 ng·μL^{-1}, still two glycopeptide peaks are detected after enrichment by Fe$_3$O$_4$@ mSiO$_2$–glucose microsphere. It demonstrates that Fe$_3$O$_4$@mSiO$_2$–glucose microsphere has excellent enrichment ability and sensitivity for glycopeptide.

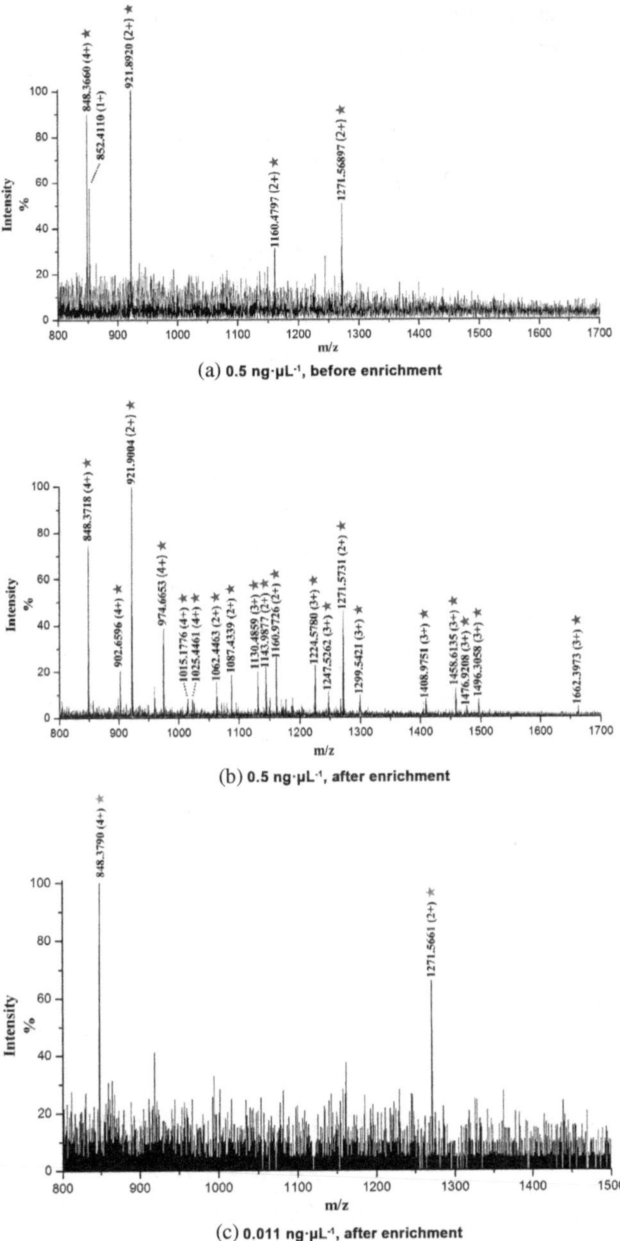

Figure 6.3: ESI-Q-TOF MS spectra of different concentrations of HRP digests before and after enrichment by $Fe_3O_4@mSiO_2$–glucose microsphere.

Table 6.1: Detailed information of glycopeptide enriched by $Fe_3O_4@mSiO_2$–glucose microsphere from HRP digest.

Amino Acid Sequence[a]	Glycan Composition	m/z with Charge States
QSDQELFSSPN#ATDTIPLVR	$Xyl_1 Man_2 GlcNA_{c2} Fuc_1$	848.3718(4+)
		1130.4859(3+)
GLC(NQCR)PLNN#LSALVDFDLR	$Xyl_1 Man_2 GlcNA_{c2} Fuc_1$	902.6596(4+)
		1203.2057(3+)
NVGLN#R	$Xyl_1 Man_2 GlcNA_{c2} Fuc_1$	921.9004(2+)
LHFHDCFVNGCDASILLDN#TISFR	$Xyl_1 Man_2 GlcNA_{c2} Fuc_1$	974.6653(4+)
		1299.5421(3+)
LHFHDCFVNGCDASILLDN#TISFR	$Xyl_1 Man_4 GlcNA_{c2} Fuc_1$	1015.1776(4+)
LDN#TISFR	$Xyl_1 Man_2 GlcNA_{c2} Fuc_1$	1062.4463(2+)
SFAN#STQTFF	$Xyl_1 Man_2 GlcNA_{c2}$	1087.4339(2+)
LDN#TTISFR	$Xyl_1 Man_2 GlcNA_{c2}$	1143.9877(2+)
GLIQSDQELFSSPN#ATDTIPLVR	$Xyl_1 Man_2 GlcNA_{c2} Fuc_1$	1160.9726(2+)
LHFHDCFVNGCDASILLDN#TISF	$Xyl_1 Man_2 GlcNA_{c2} Fuc_1$	1224.5780(3+)
SSPN#ATDTIPLVR	$Xyl_1 Man_2 GlcNA_{c2} Fuc_1$	1247.5262(3+)
QLTPTFYDN#SC(AAVESACPR) PNVSNIVR-H_2O)	$Xyl_1 Man_2 GlcNA_{c2} Fuc_1$	1271.5731(2+)
LYN#FSNTGLPDPTLN#TIY	$Xyl_1 Man_2 GlcNA_{c2} Fuc_1$	1408.6138(3+)
	$Xyl_1 Man_2 GlcNA_{c2} Fuc_1$	1458.6135(3+)
LYN#FSNTGLPDPTLN#TIYL	$Xyl_1 Man_2 GlcNA_{c2} Fuc_1$	1496.3058(3+)
	$Xyl_1 Man_2 GlcNA_{c2}$ Fuc_1	
LYN#FSNTGLPDPTLN#TIYLQTLR	$Xyl_1 Man_2 GlcNA_{c2} Fuc_1$	1662.4072(3+)
	$Xyl_1 Man_2 GlcNA_{c2} Fuc_1$	

[a] The N-glycosylation sites are marked with N#.GlcNAc: N-acetylglycosamine; fuc: fructose; man: mannose; xyl: xylose.

(4) Evaluation of the selective enrichment ability of $Fe_3O_4@$ mSiO$_2$–glucose microsphere for glycopeptide

In order to further investigate the selective enrichment ability of $Fe_3O_4@$ mSiO$_2$–glucose microsphere for glycopeptide in complex sample, the peptides mixture of HRP, MYO and β-casein with molar ratio of 1/1/10 is selected as the study object. At the same time, $Fe_3O_4@mSiO_2$–N_3 microsphere is selected for comparison. As shown in Figure 6.4(a), when the peptide mixture is directly analyzed by ESI-Q-TOF MS, only one glycopeptide is observed and a large number of non-glycopeptides occupy the spectrum. After enrichment by $Fe_3O_4@mSiO_2$–glucose microsphere, 19 glycopeptides dominate the spectrum, although some hydrophilic non-glycopeptides are still detected as shown in

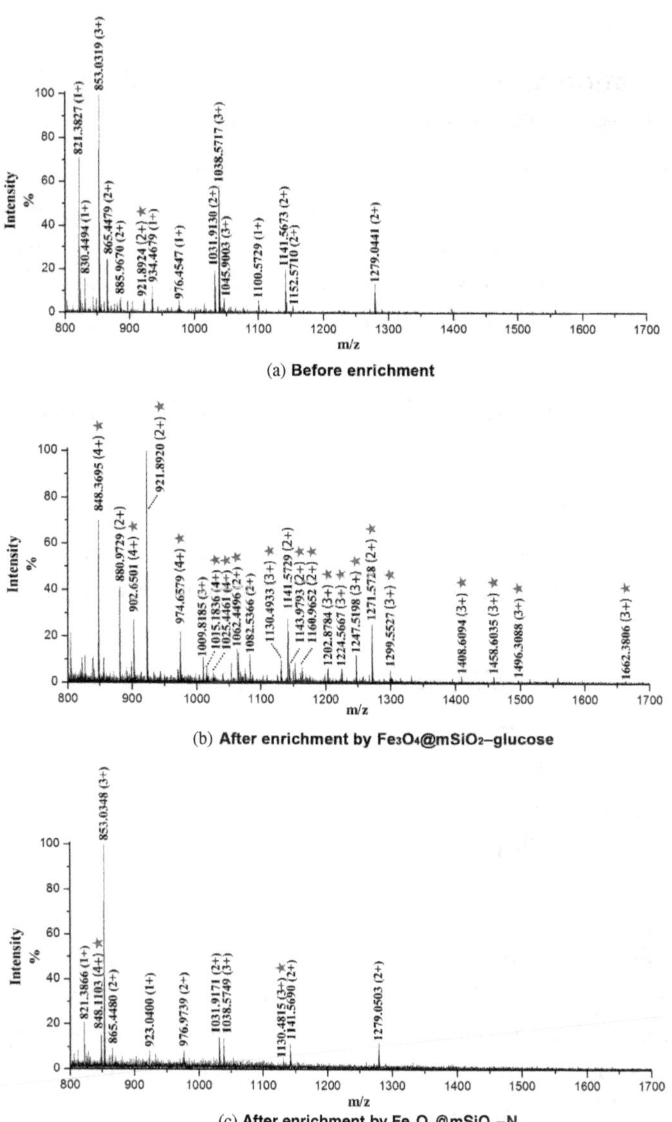

(a) **Before enrichment**

(b) **After enrichment by Fe₃O₄@mSiO₂–glucose**

(c) **After enrichment by Fe₃O₄@mSiO₂–N₃**

Figure 6.4: ESI-Q-TOF MS spectra of peptide mixture of HRP, MYO and β-casein with the molar ratio of 1/1/10 before and after being enriched by two different materials.

Figure 6.4(b). However, after enrichment by $Fe_3O_4@mSiO_2-N_3$ microsphere, only one glycopeptide is detected by MS as shown in Figure 6.4(c). It indicates that glucose increases the hydrophilicity of the material, which plays a vital role in glycopeptide enrichment. The selective enrichment ability of $Fe_3O_4@mSiO_2$–glucose microsphere in a complex sample is also demonstrated.

(5) Evaluation of the selective enrichment ability of $Fe_3O_4@$ $mSiO_2$–glucose microsphere for glycan in human serum

The human serum sample is also selected to further investigate the selective enrichment ability of $Fe_3O_4@mSiO_2$–glucose microsphere for glycan. As shown in Figure 6.5(a), only 12 glycans are detected after ultrafiltration treatment for 0.25 μL original human serum (treated by PNGase F) digest. While after enrichment by $Fe_3O_4@mSiO_2$–glucose microsphere, 42 glycans are identified.

6.2.2.2. *Chitosan-modified magnetic microsphere* *($F_{e3}O_4@CS$)*

(1) Preparation of $Fe_3O_4@CS$ microsphere

The schematic diagram of the synthesis of $Fe_3O_4@CS$ microsphere is shown in Figure 2.21, and the detailed synthesis process is shown in Section 2.5 of Chapter 2.[3]

(2) Enrichment application of $Fe_3O_4@CS$ microsphere in glycoproteomics

(a) *Evaluation of the enrichment ability of $Fe_3O_4@CS$ microsphere for* *glycopeptide in standard glycoprotein digest*

The enrichment ability of $Fe_3O_4@CS$ in human immunoglobulin G (IgG) digest is firstly studied. ACN/H_2O/TFA (v/v/v, 88/11.5/0.5, 200 μL) is selected as loading buffer and ACN/H_2O/TFA (v/v/v, 30/69.9/0.1, 10 μL) as eluent. Fe_3O_4 microsphere is utilized for comparison. As shown in Figure 6.6(a), 300 fmol human IgG digest is directly analyzed by MS, only two glycopeptides are detected and the interference from non-glycopeptide is very severe. But after being enriched by $Fe_3O_4@CS$ microsphere, 21 glyco-sylated peptides are detected by MS as shown in Figure 6.6(b). When IgG digest is enriched by Fe_3O_4 microsphere (Figure 6.6(c)), the spectrum does not change a lot with that before enrichment. The enriched glycosylated peptides

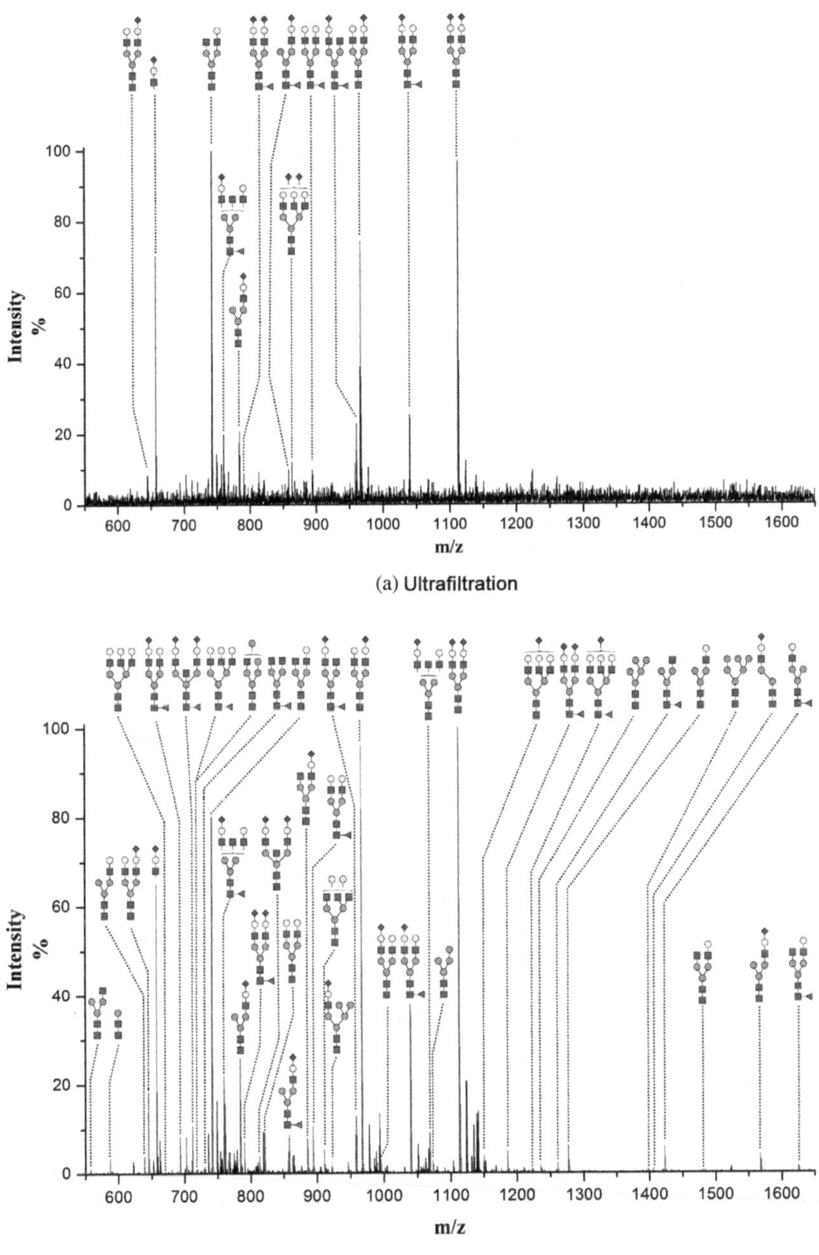

Figure 6.5: ESI-Q-TOF MS spectra of human serum (treated by PNGase F) after treated by different methods. Sialic acid (◆), galactose (●), mannose ( ), acetyl glucosamine (■), Trehalose (▲).

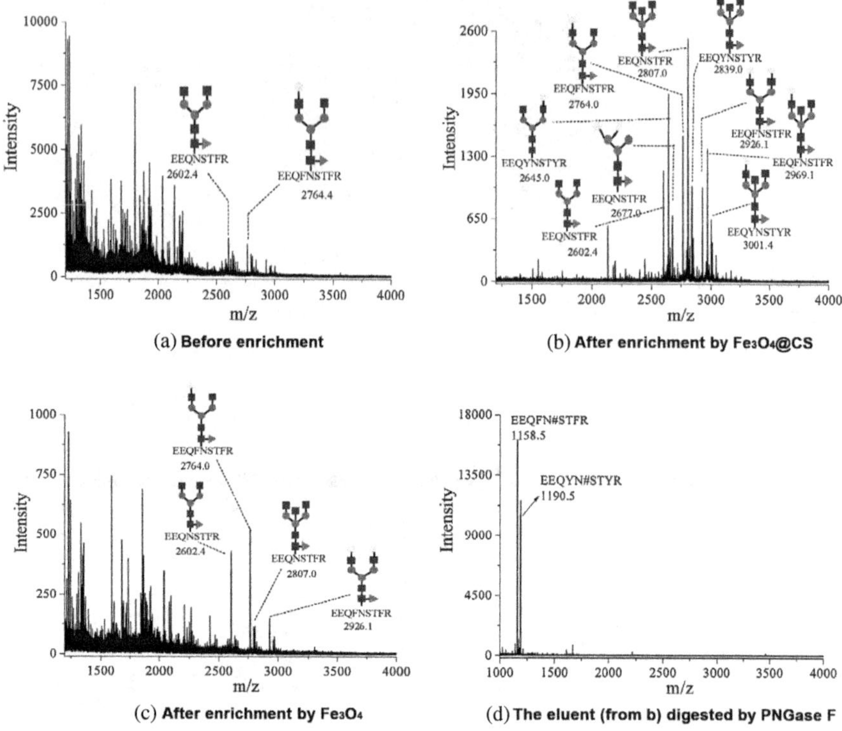

Figure 6.6: MS spectra of 300 fmol human IgG digest before and after being enriched by Fe$_3$O$_4$@CS microsphere. Galactose (●), mannose ( ), acetyl glucosamine (■), Trehalose (▲).

are then digested by PNGase F and analyzed by MS. As shown in Figure 6.6(d), the glycopeptide signals disappear in the spectrum, which confirms that the glycopeptides are indeed enriched by Fe$_3$O$_4$@CS microsphere.

In addition, Fe$_3$O$_4$@CS microsphere is used to enrich different amounts of human IgG digests to investigate their enrichment sensitivity. As shown in Figure 6.7, when human IgG digest decreases to merely 8 fmol, four glycopeptides are still detected.

(b) *Evaluation of the enrichment ability of Fe$_3$O$_4$@CS microsphere for glycopeptide in Hela cell protein digest*

The digest of Hela cell extract is selected to further investigate the selective enrichment ability of Fe$_3$O$_4$@CS microsphere. After 45 μg Hela cell digest is enriched by Fe$_3$O$_4$@CS, 273 glycopeptides and 283 *N*-glycosylation sites corresponding to 175 glycosylated proteins are identified.

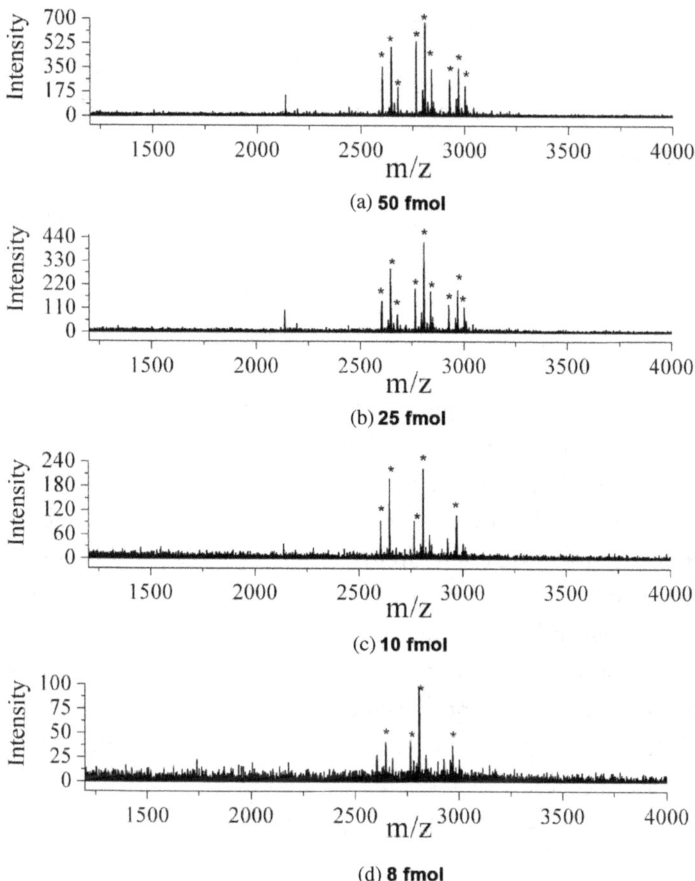

Figure 6.7: MS analysis of different amounts of human IgG digests after enrichment by $Fe_3O_4@CS$ microsphere.

6.2.3. Hydrophilic polymer modified magnetic micro-/nanomaterial

6.2.3.1. Amphoteric polymer-modified magnetic silica $(Fe_3O_4@SiO_2@PMSA)$ (PMSA, [2-(methacryloyloxy)ethyl]dimethyl-(3-sulfopropyl)ammonium hydroxide)

(1) Preparation of $Fe_3O_4@SiO_2@PMSA$ microsphere

The synthesis process of the $Fe_3O_4@SiO_2@PMSA$ microsphere is shown in Section 2.6 of Chapter 2.[4]

(2) Enrichment application of Fe$_3$O$_4$@SiO$_2$@PMSA in glycoproteomics

(a) *Evaluation of the enrichment ability of Fe$_3$O$_4$@SiO$_2$@PMSA microsphere for glycopeptide in standard glycoprotein digest*

ACN/H$_2$O/TFA (v/v/v, 86/13.9/0.1, 400 μL) is selected as loading buffer and ACN/H$_2$O/TFA (v/v/v, 30/69.9/0.1, 10 μL) as eluting buffer. Human IgG digest is selected as the standard sample to evaluate the enrichment ability of Fe$_3$O$_4$@SiO$_2$@PMSA microsphere for glycopeptide. As shown in Figure 6.8(a), merely four weak glycopeptide signals are detected when directly analyzing the human IgG digest. But after being enriched by Fe$_3$O$_4$@SiO$_2$@PMSA microsphere, as shown in Figure 6.8(b), 26 high-intensity N-glycopeptides are identified. For comparison, Fe$_3$O$_4$@SiO$_2$–MSA is also used to enrich glycopeptide in IgG digest. As shown in Figure 6.8(c), only eight glycopeptides are detected. This indicates that the reason why Fe$_3$O$_4$@SiO$_2$@PMSA microsphere has a better enrichment ability for glycopeptide is because the dense amphoteric polymer on the surface enhances the hydrophilic interaction between material and the glycan. The eluent is then digested by PNGase F to remove the glycan and then detected by mass spectrometry. As shown in Figure 6.8(d), the glycopeptide signals disappear, indicating that Fe$_3$O$_4$@SiO$_2$@PMSA microsphere indeed enriches the glycopeptide. The detailed information is listed in Table 6.2.

To further evaluate whether Fe$_3$O$_4$@SiO$_2$@PMSA microsphere has enrichment ability for different types of glycopeptides, HRP and OVA digests are selected. As shown in Figure 6.9, after enrichment by Fe$_3$O$_4$@SiO$_2$@PMSA microsphere, there are 14 glycosylated peptides (Figure 6.9(b)) and 15 glycosylated peptides (Figure 6.9(d)) detected, respectively. This result indicates that it is universal for Fe$_3$O$_4$@SiO$_2$@PMSA to enrich glycosylated peptides.

The enrichment sensitivity of Fe$_3$O$_4$@SiO$_2$@PMSA microsphere is then investigated using human IgG digest as sample. When the amount of IgG digest is as low as 0.1 fmol, the glycopeptide of *m/z* 2795.58 can still be observed, indicating that Fe$_3$O$_4$@SiO$_2$@PMSA microsphere has excellent enrichment sensitivity for glycopeptide (Figure 6.10).

(b) *Evaluation of the enrichment ability of Fe$_3$O$_4$@SiO$_2$@PMSA microsphere for glycopeptide in mouse liver protein digest*

ACN/H$_2$O/TFA (v/v/v, 86/13/1, 500 μL) is selected as the loading buffer and ACN/H$_2$O/TFA (v/v/v, 30/69.9/0.1, 3×100 μL) as the eluent to separate

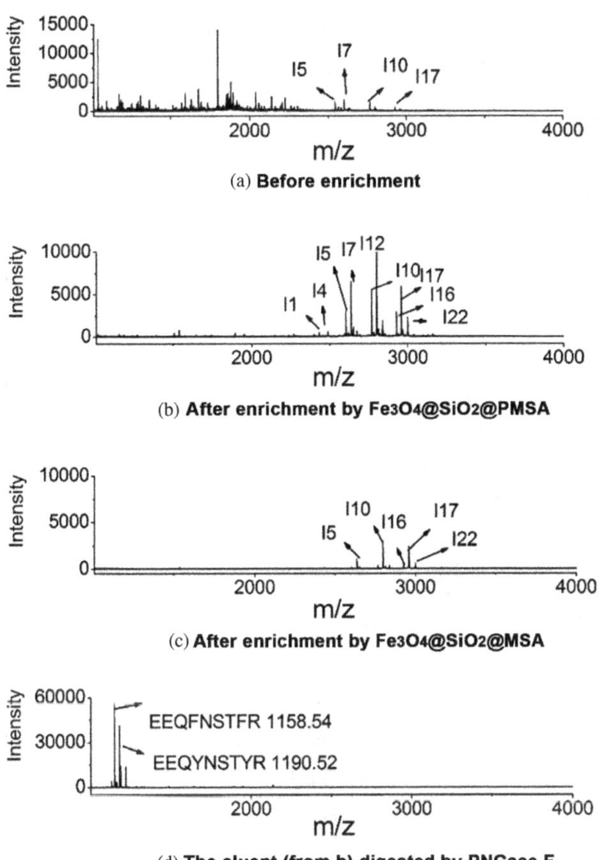

(a) **Before enrichment**

(b) **After enrichment by Fe3O4@SiO2@PMSA**

(c) **After enrichment by Fe3O4@SiO2@MSA**

(d) **The eluent (from b) digested by PNGase F**

Figure 6.8: MS spectra of 0.5 pmol human IgG digest before and after being enriched by different materials.

Table 6.2: The detailed information of N-glycopeptdie enriched by $Fe_3O_4@SiO_2@PMSA$ from human IgG digest.

No.	M/Z	Glycan Composition	Amino Acid Aequence
1	2286	[Hex]3[HexNAc]3	EEQYN#STYR
2	2432	[Hex]3[HexNAc]3[Fuc]1	EEQYN#STYR
3	2488	[Hex]3[HexNAc]4	EEQYN#STYR
4	2594	[Hex]4[HexNAc]3[Fuc]1	EEQYN#STYR
5	2603	[Hex]3[HexNAc]4[Fuc]1	EEQFN#STFR
6	2618	[Hex]4[HexNAc]4	EEQFN#STFR

(Continued)

Table 6.2: *(Continued)*

No.	M/Z	Glycan Composition	Amino Acid Aequence
7	2635	[Hex]3[HexNAc]4[Fuc]I	EEQYN#STYR
8	2650	[Hex]4[HexNAc]4	EEQYN#STYR
9	2658	[Hex]3[HexNAc]5	EEQFN#STYR
10	2764	[Hex]4[HexNAc]4[Fuc]1	EEQFN#STFR
11	2780	[Hex]5[HexNAc]4	EEQFN#STFR
12	2797	[Hex]4[HexNAc]4[Fuc]1	EEQYN#STYR
13	2806	[Hex]3[HexNAc]5[Fuc]I	EEQFN#STYR
14	2812	[Hex]5[HexNAc]4	EEQYN#STYR
15	2821	[Hex]4[HexNAc)5	EEQFN#STFR
16	2838	[Hex]3[HexNAc]5[Fuc]1	EEQYN#STYR
17	2853	[Hex]4[HexNAc]5	EEQYN#STYR
18	2926	[Hex)5[HexNAc]4[Fuc)1	EEQFN#STFR
19	2958	[Hex]5[HexNAc]4[Fuc]1	EEQYN#STYR
20	2968	[Hex]4[Hex7NAc]5[Fuc]1	EEQFN#STFR
21	2983	[Hex]5[HexNAc]5	EEQFN#STFR
22	3000	[Hex]4[HexNAc]5[Fuc]1	EEQYN#STYR
23	3087	[Hex)4[HexNAc)4[Fuc]1[NeuAc]1	EEQYN#STFR
24	3129	[Hex]5[HexNAc]5[Fuc]1	EEQFN#STFR
25	3161	[Hex]5[HexNAc]5[Fuc]1	EEQYN#STYR
26	3250	[Hex]5[HexNAc]4[Fuc]1(NeuAc]1	EEQYN#STYR

HexNAc = N-acetylglucosamine, Hex = mannose, Fuc = fuctose, NeuAc = sialic

Note: N#: glycosylation site.

glycopeptide from 65 μg mouse liver extract digest. As a result, a total of 905 *N*-glycosylation sites corresponding to 458 glycoproteins are identified by LC-MS/MS and database search.

6.2.3.2. *PEG-modified magnetic graphene oxide (GO/Fe$_3$O$_4$/Au/PEG)*

(1) Preparation of GO/Fe$_3$O$_4$/Au/PEG composite

The synthesis protocol of GO/Fe$_3$O$_4$/Au/PEG composite is shown in Section 2.8 of Chapter 2.[5]

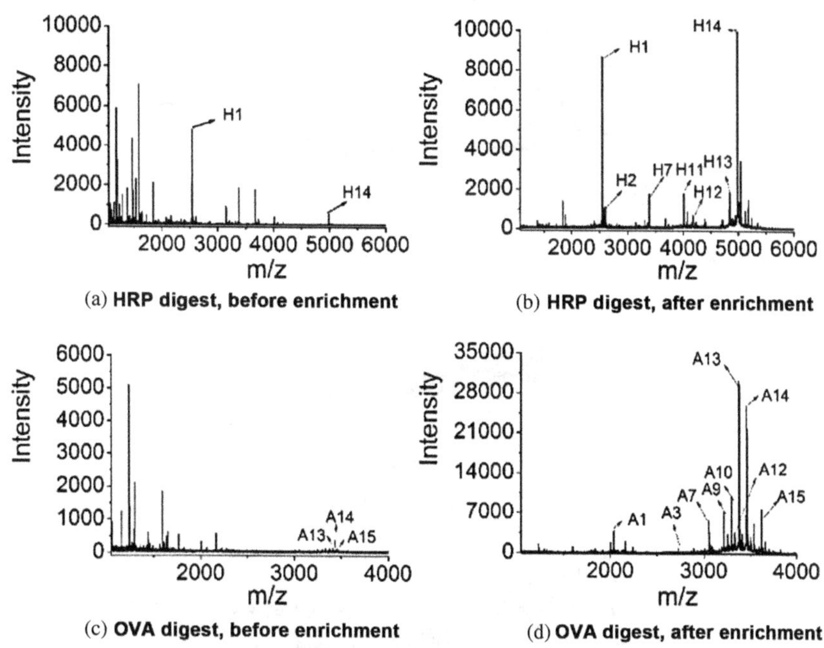

Figure 6.9:　MS spectra of HRP digest and OVA digest (1 pmol) before and after being enriched by $Fe_3O_4@SiO_2@PMSA$.

(2) Enrichment application of $GO/Fe_3O_4/Au/PEG$ in glycoproteomics

(a) *Evaluation of the enrichment ability of $GO/Fe_3O_4/Au/PEG$ for glycopeptide in standard glycoprotein digest*

$ACN/H_2O/FA$ (v/v/v, 80/20/0.1, containing 10 mmol·L^{-1} NH_4HCO_3, 50 μL) is selected as the loading buffer, and $ACN/H_2O/FA$ (v/v/v, 50/50/0.1, containing 10 mmol·L^{-1} NH_4HCO_3, 2 × 20 μL) as the eluent and 1 min as the enrichment time.

HRP digest is firstly used to investigate the enrichment ability of $GO/Fe_3O_4/Au/PEG$ composite for glycopeptide. As shown in Figure 6.11(a), when 2.5 pmol HRP digest is directly analyzed by MS, a great many non-glycopeptides dominate the spectrum, while after being enriched by $GO/Fe_3O_4/Au/PEG$ composite, 10 glycopeptides are detected (detailed information is listed in Table 6.3) and the non-glycopeptides disappear in Figure 6.11(b). For comparison, $GO/Fe_3O_4/Au$ and GO/Fe_3O_4 composite are also used to enrich glycopeptide from 2.5 pmol HRP digest. Only six and

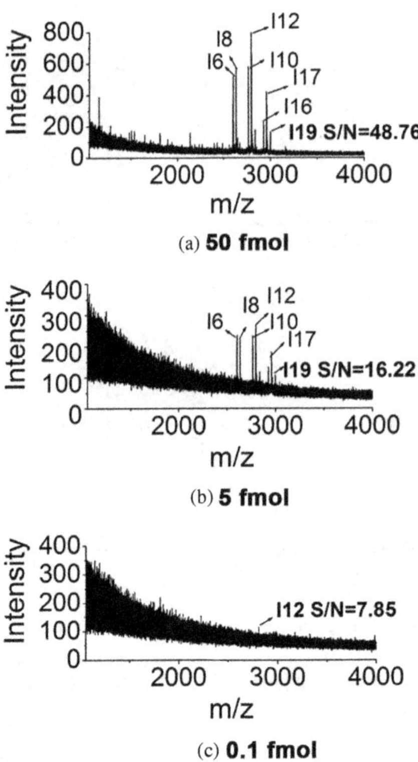

Figure 6.10: MS spectra of different amounts of human IgG digests after being enriched by $Fe_3O_4@SiO_2@PMSA$ microsphere.

three glycopeptides are detected, respectively, and non-glycopeptide can also be evidently observed in the spectra. It indicates that $GO/Fe_3O_4/Au/PEG$ composite has excellent enrichment ability for glycosylated peptide.

Similarly, human IgG digest which contains sialic acid is also applied to evaluate the enrichment ability of $GO/Fe_3O_4/Au/PEG$ for different types of glycopeptides. Since the human IgG digest contains sialic acid, under alkaline condition, the charge of the glycosylated peptide will be neutralized, thereby reducing the electrostatic interaction between the material and glycopeptide. Therefore, in this work, 75% ACN/20 mmol·L^{-1} NH$_4$HCO$_3$ is used as the loading buffer and 25 mmol·L^{-1} NH$_4$HCO$_3$ is used as the eluent. After 1 pmol human IgG digest is enriched by $GO/Fe_3O_4/Au/PEG$ composite, 20 glycopeptides are detected by MS (the detailed information of glycopeptide is shown in Table 6.4).

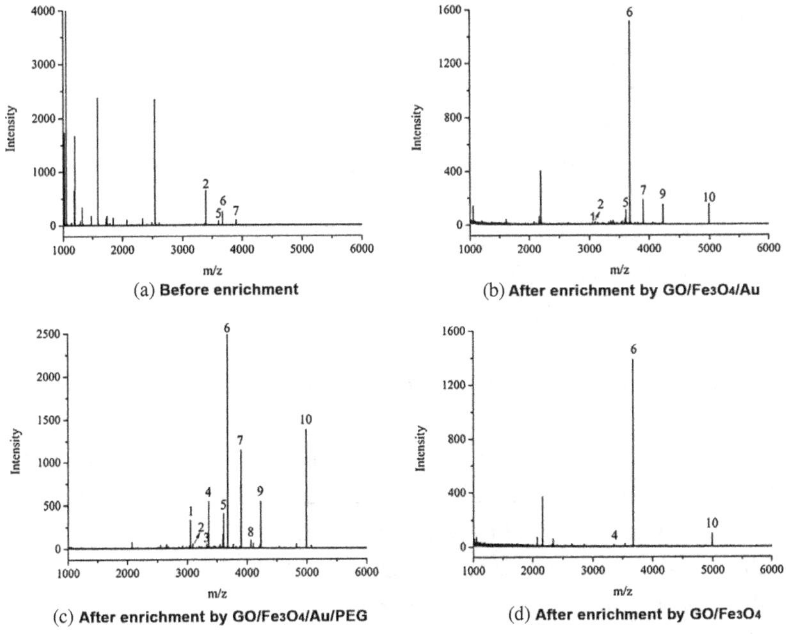

Figure 6.11: MS spectra of 2.5 pmol HRP digest before and after being enriched by different materials.

Table 6.3: Detailed information of glycopeptide enriched by GO/Fe$_3$O$_4$/Au/PEG composite from HRP digest.

Number	M/Z	Glycan Composite	Amino Acid Sequence
1	3050	[Hex]2[HexNAc][Xyl]1	SFAN#STQTFFNAFVEAMDR
2	3089	[Hex]3[HexNAc2][Fuc]1[Xyl]1	GLCPLNGN#LSALVDFDLR
3	3322	[Hex]3[HexNAc]2[Fuc]1[Xyl]1	QLTPTFYDNSCPN#VSNIVR
4	3354	[Hex]3[HexNAc]2[Fuc]1[Xyl]1	SFAN#STQTFFNAFVEAMDR
5	3607	[Hex]3[HexNAc]2[Fuc]1[Xyl]1	NQCRGLCPLNGN#LSALVDFDLR
6	3673	[Hex]3[HexNAc]2[Fuc]1[Xyl]1	GLIQSDQELFSSPN#ATDTIPLVR
7	3895	[Hex]3[HexNAc]2[Fuc]1[Xyl]1	LHFHDCFVNGCDASILLDN#TTSFR
8	4057	[Hex]3[HexNAc]2[Xyl]1	QLTPTFYDNSC(AAVESACPR)PN#VSNIVR-H$_2$O
9	4224	[Hex]3[HexNAc]2[Fuc]1[Xyl]1 [Hex]3[HexNAc]2[Fuc]1[Xyl]1	QLTPTFYDNSC(AAVESACPR)PN#VSNIVR
10	4985	[Hex]3[HexNAc]2[Fuc]1[Xyl]1	LYN#FSNTGLPDPTLN#TTYLQTLR

HexNAc = N-acetylglucosamine, Fuc = fructose, Hex = mannose, Xyl = xylose.

Note: N#: *N*-glycosylation site.

(b) *Evaluation of the selective enrichment ability of GO/Fe$_3$O$_4$/Au/PEG*
for glycopeptide

Peptides mixture of HRP and MYO with different mass ratios is selected to investigate the selective enrichment ability of GO/Fe$_3$O$_4$/Au/PEG composite for glycosylated peptide. As shown in Figure 6.12(a), when the mass ratio of HRP and MYO is at 1/10, 10 glycosylated peptides are detected. When the mass ratio reaches 1/100, there are still nine glycopeptides detected. It indicates that the GO/Fe$_3$O$_4$/Au/PEG composite has excellent selective enrichment ability for glycosylated peptides.

(c) *Evaluation of the enrichment ability of GO/Fe$_3$O$_4$/Au/PEG for glycopeptide*
in human serum digest

ACN/H$_2$O/FA (v/v/v, 75/25/0.1, containing 10 mmol·L^{-1} NH$_4$HCO$_3$, 100 μL) is selected as the loading buffer, ACN/H$_2$O/FA (v/v/v, 50/50/0.1, containing 10 mmol·L^{-1} NH$_4$HCO$_3$, 2×25 μL) as the eluent and 10 min as enrichment time to evaluate the enrichment ability of GO/Fe$_3$O$_4$/Au/PEG for glycopeptide in human serum digest. The maltose modified material is also chosen to enrich glycopeptide in human serum digest for comparison and the results are shown in Figure 6.13.

6.2.4. Sugar and hydrophilic polymer co-modified magnetic micro-/nanomaterial

6.2.4.1. Maltose and PEG co-modified magnetic silica (Fe$_3$O$_4$@SiO$_2$@PEG–maltose)

(1) Preparation of Fe$_3$O$_4$@SiO$_2$@PEG–maltose microsphere

The synthesis process of Fe$_3$O$_4$@SiO$_2$@PEG–maltose microsphere is shown in Figure 2.32, and can be seen in Section 2.6 of Chapter 2.[6]

(2) Enrichment application of Fe$_3$O$_4$@SiO$_2$@PEG–maltose in glycoproteomics

(a) *Evaluation of the enrichmentability of Fe$_3$O$_4$@SiO$_2$@PEG-maltose*
composite for glycopeptide in standard glycoprotein digest

ACN/H$_2$O/TFA (v/v/v, 88/11.9/0.1, 100 μL) is selected as the loading buffer, H$_2$O/TFA (v/v, 99.9/0.1, 2 × 10 μL) as the eluent and 10 min as the enrichment time. Two hydrophilic materials Fe$_3$O$_4$@SiO$_2$@PEG and Fe$_3$O$_4$@

Table 6.4: Detailed information of glycopeptide enriched by GO/Fe$_3$O$_4$/Au/PEG composite from human IgG digest.

Peak Number	Observed M/Z	Glycan Composition	Amino Acid Sequence
1	2400	[Hex]3[HexNAc]3[Fuc]1	EEQFN#STYR
2	2561	[Hex]4[HexNAc]3[Fuc]1	EEQFN#STYR
3	2603	[Hex]3[HexNAc]4[Fuc]1	EEQFN#STFR
4	2618	[Hex]4[HexNAc]4	EEQFN#STFR
5	2635	[Hex]3[HexNAc]4[Fuc]1	EEQYN#STYR
6	2650	[Hex]4[HexNAc]4	EEQYN#STYR
7	2764	[Hex]4[HexNAc]4[Fuc]1	EEQFN#STFR
8	2780	[Hex]5[HexNAc]4	EEQFN#STFR
9	2797	[Hex]4[HexNAc]4[Fuc]1	EEQYN#STYR
10	2806	[Hex]3[HexNAc]5[Fuc]1	EEQFN#STYR
11	2838	[Hex]3[HexNAc]5[Fuc]1	EEQYN#STYR
12	2926	[Hex]5[HexNAc]4[Fuc]1	EEQFN#STFR
13	2958	[Hex]5[HexNAc]4[Fuc]1	EEQYN#STYR
14	2968	[Hex]4[HexNAc]5[Fuc]1	EEQFN#STFR
15	3000	[Hex]4[HexNAc]5[Fuc]1	EEQYN#STYR
16	3087	[Hex]4[HexNAc]4[Fuc]1[NeuAc]1	EEQYN#STFR
17	3129	[Hex]5[HexNAc]5[Fuc]1	EEQFN#STFR
18	3161	[Hex]5[HexNAc]5[Fuc]1	EEQYN#STYR
19	3218	[Hex]5[HexNAc]4[Fuc]1[NeuAc1]	EEQFN#STFR
20	3250	[Hex]5[HexNAc]4[Fuc]1[NeuAc1]	EEQYN#STYR

HexNAc = N-acetylglucosamine, Fuc = fuctose, Hex = mannose, Xyl = xylose, NeuAc = Sialic

Note: N#: *N*-glycosylation site.

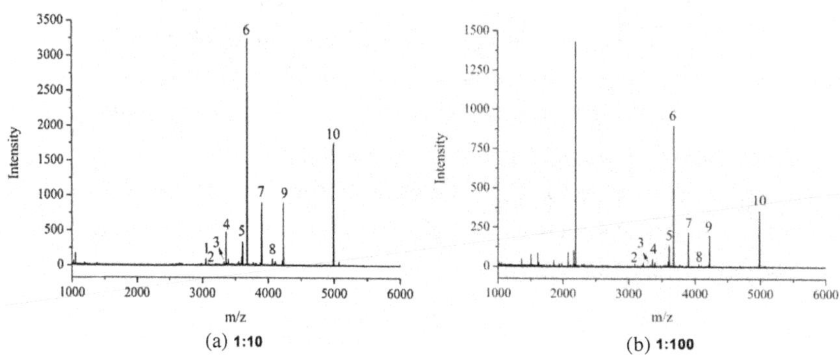

Figure 6.12: Peptides mixture of HRP and MYO with different mass ratios after being enriched by GO/Fe$_3$O$_4$/Au/PEG nanocomposite.

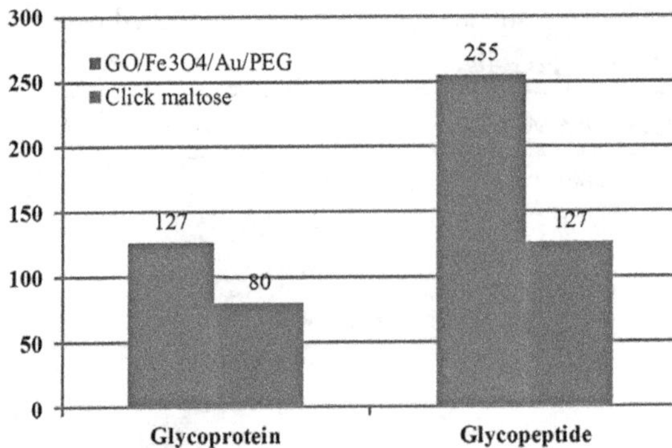

Figure 6.13: The number of glycoprotein and glycopeptide enriched by different materials from human serum.

SiO_2–maltose are used for comparison. The human IgG digest is used to investigate the enrichment ability of Fe_3O_4@SiO_2@PEG–maltose microsphere for glycopeptide. After enrichment with three kinds of materials, as shown in Figure 6.14, both of the S/N ratio and intensity of glycopeptide enriched by Fe_3O_4@SiO_2@PEG–maltose microsphere are stronger than those of the other two materials. As a result, a total of 27 glycosylated peptides assigned to human IgG digest are detected by MS. Moreover, it has been verified that they are indeed the glycosylated peptides. Detailed information can be seen in Table 6.5.

HRP digest is also selected to further investigate the enrichment universality of Fe_3O_4@SiO_2@PEG–maltose microsphere for different types of glycopeptides. The eluent is changed to ACN/H_2O/TFA (v/v/v, 30/69.9/0.1, 2×10 μL). Finally, a total of 19 glycosylated peptides are detected from 3 pmol HRP digests using Fe_3O_4@SiO_2@PEG–maltose microsphere. The detailed information of glycopeptide can be seen in Table 6.6.

(b) *Evaluation of the enrichment ability of Fe₃O₄@SiO₂@PEG–maltose microsphere for glycopeptide in human serum digest*

After enrichment by Fe_3O_4@SiO_2@PEG–maltose microsphere, 106 glycoproteins, and 204 *N*-glycosylation sites are identified from 15 μL human serum digest by 2D LC-MS/MS.

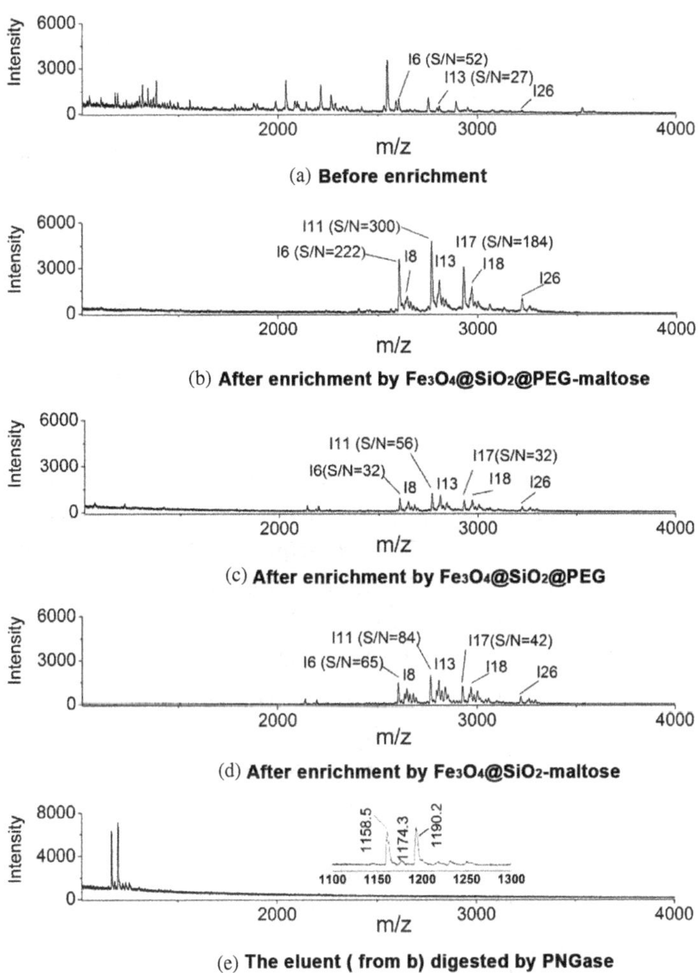

Figure 6.14: MS spectra of 5 pmol human IgG digest before and after being enriched by different materials.

Table 6.5: Detailed information of glycopeptide enriched by $Fe_3O_4@SiO_2@Peg$–maltose microsphere from human IgG digest.

Peak Number	Observed M/Z	Glycan Composition	Amino Acid Sequence
I1	2400.1	[Hex]3[HexNAc]3[Fuc]1	EEQFN#STFR
I2	2432.0	[Hex]3[HexNAc]3[Fuc]1	EEQYN#STYR
I3	2456.1	[Hex]3[HexNAc]4	EEQFN#STFR
I4	2488.1	[Hex]3[HexNAc]4	EEQYN#STYR

(Continued)

Table 6.5: (*Continued*)

Peak Number	Observed M/Z	Glycan Composition	Amino Acid Sequence
I5	2561.5	[Hex]4[HexNAc]3[Fuc]1	EEQFN#STFR
I6	2603.2	[Hex]3[HexNAc]4[Fuc]1	EEQFN#STFR
I7	2618.1	[Hex]4[HexNAc]4	EEQFN#STFR
I8	2635.6	[Hex]3[HexNAc]4[Fuc]1	EEQYN#STYR
I9	2674.4	[Hex]3[HexNAc]5	EEQFN#STYR
I10	2691.0	[Hex]3[HexNAc]5	EEQYN#STYR
I11	2764.6	[Hex]4[HexNAc]4[Fuc]1	EEQFN#STFR
I12	2780.1	[Hex]5[HexNAc]4	EEQFN#STFR
I13	2797.4	[Hex]4[HexNAc]4[Fuc]1	EEQYN#STYR
I14	2821.3	[Hex]4[HexNAc]5	EEQFN#STYR
I15	2838.1	[Hex]3[HexNAc]5[Fuc]1	EEQYN#STYR
I16	2853.3	[Hex]4[HexNAc]5	EEQYN#STYR
I17	2926.9	[Hex]5[HexNAc]4[Fuc]1	EEQFN#STFR
I18	2960.1	[Hex]5[HexNAc]4[Fuc]1	EEQYN#STYR
I19	2968.6	[Hex]4[HexNAc]5[Fuc]1	EEQFN#STFR
I20	2983.2	[Hex]5[HexNAc]5	EEQFN#STFR
I21	3000.6	[Hex]4[HexNAc]5[Fuc]1	EEQYN#STYR
I22	3015.1	[Hex]5[HexNAc]5	EEQFN#STYR
I23	3057.1	[Hex]4[HexNAc]4[Fuc]1[NeuAc]1	EEQYN#STFR
I24	3129.4	[Hex]5[HexNAc]5[Fuc]1	EEQFN#STFR
I25	3161.3	[Hex]5[HexNAc]5[Fuc]1	EEQYN#STYR
I26	3219.1	[Hex]5[HexNAc]4[Fuc]1[NeuAc]1	EEQFN#STFR
I27	3250.9	[Hex]5[HexNAc]4[Fuc]1[NeuAc]1	EEQYN#STYR

Note: N#: *N*-glycosylation site.

Table 6.6: The detailed information of *N*-glycosylated peptide enriched by Fe$_3$O$_4$@SiO$_2$@PEG–maltose microsphere from HRP digest.

	M/Z	Glycan Composition	Amino Acid Sequence
H1	1547.2	[Hex]2[HexNAc]1	PN#ATDTIPLVR
H2	1636.3	[Hex]2[HexNAc]1	SPN#ATDTIPLVR
H3	1844.1	[Hex]3[HexNAc]2[Fuc]1[Xy1]1	SPN#ATDTIPLVR
H4	2321.5	[Hex]2[HexNAc]2	MGN#ITPLTGTQGQIR

Table 6.6: *(Continued)*

	M/Z	Glycan Composition	Amino Acid Sequence
H5	2438.2	[Hex]3[HexNAc]2[Fuc]1[Xyl]1	SILLDN#TTSFR
H6	2509.5	[Hex]3[HexNAc]2[Fuc]1[Xyl]1	ASILLDN#TTSFR
H7	2543.2	[Hex]3[HexNAc]2[Fuc]1[Xyl]1	SSPN#ATDTIPLVR
H8	2612.6	[Hex]3[HexNAc]2[Xyl]1	MGN#ITPLTGTQGQIR
H9	2802.4	[Hex]3[HexNAc]2[Fuc]1[Xyl]1	LFSSPN#ATDTIPLVR
H10	2850.8	[HexNAc]1[Fuc]1	GLIQSDQELFSSPN#ATDTIPLVR
H11	3061.3	[Hex]3[HexNAc]2[Fuc]1[Xyl]1	QSDQELFSSPN#ATDTIPLVR
H12	3323.1	[Hex]3[HexNAc]2[Fuc]1[Xyl]1	QLTPTFYDNSCPN#VSNIVR
H13	3355.2	[Hex]2[HexNAc]2[Fuc]1[Xyl]1	SFAN#STQTFFNAFVEAMDR
H14	3674.0	[Hex]3[HexNAc]2[Fuc]1[Xyl]1	GLIQSDQELFSSPN#ATDTIPLVR
H15	3750.5	[Hex]3[HexNAc]2[Xyl]1	LHFHDCFVNGCDASILLDN#TTSFR
H16	3896.1	[Hex]3[HexNAc]2[Fuc]1[Xyl]1	LHFHDCFVNGCDASILLDN#TTSFR
H17	4059.0	[Hex]3[HexNAc]2[Xyl]1	QLTPTFYDNSC(AAVESACPR)PN#VSNIVR-H_2O
H18	4223.9	[Hex]3[HexNAc]2[Fuc]1[Xyl]1	QLTPTFYDNSC(AAVESACPR)PN#VSNIVR
H19	4986.2	[Hex]2[HexNAc]2[Fuc]1[Xyl]1 [Hex]2[HexNAc]2[Fuc]1[Xyl]1	LYN#FSNTGLPDPTLN#TTYLQTLR

Note: N#: Glycosylation site.

6.2.4.2. *Maltose and PAMAM co-modified magnetic graphene oxide (Fe_3O_4–GO@nSiO$_2$–PAMAM–Au–maltose)*

(1) Preparation of Fe_3O_4–GO@nSiO$_2$–PAMAM–Au–maltose composite

The synthesis process of the Fe_3O_4–GO@nSiO$_2$–PAMAM–Au–maltose composite is shown in Figure 2.120, and can be seen in Section 2.8 of Chapter 2.[7]

(2) Enrichment application of Fe_3O_4–GO@nSiO$_2$–PAMAM–Au–maltose composite in glycoproteomics

(a) *Evaluation of the enrichment ability of Fe_3O_4–GO@nSiO$_2$–PAMAM–Au–maltose composite for glycopeptide in standard glycoprotein digest*

ACN/H_2O/TFA (v/v/v, 88/11.7/0.3, 200 μL) is selected as the loading buffer, ACN/H_2O/TFA (v/v/v, 30/69.9/0.1, 2×100 μL) as the eluent (re-dissolved after lyophilization and then detected by MS). Human IgG digest is used to investigate the enrichment ability of Fe_3O_4–GO@nSiO$_2$–PAMAM–Au–maltose composite for glycopeptide. First, different amounts of human IgG digests are enriched by Fe_3O_4–GO@nSiO$_2$–PAMAM–Au–maltose composite, and the results are shown in Figure 6.15. When the human IgG digest is as low as 0.5 fmol, glycopeptide can still be detected, indicating that Fe_3O_4–GO@nSiO$_2$–PAMAM–Au–maltose composite has excellent enrichment ability and high enrichment sensitivity for glycosylated peptide. Besides, to prove that their enrichment ability is benefited from the co-modification of

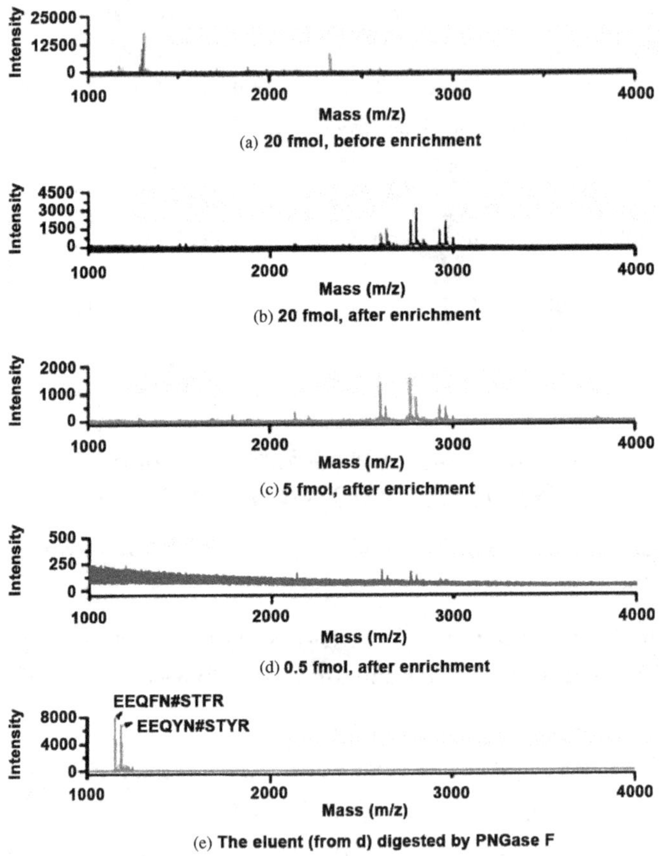

Figure 6.15: MS spectra of different amounts of human IgG digests after being enriched by Fe_3O_4–GO@nSiO$_2$–PAMAM–Au–maltose composite.

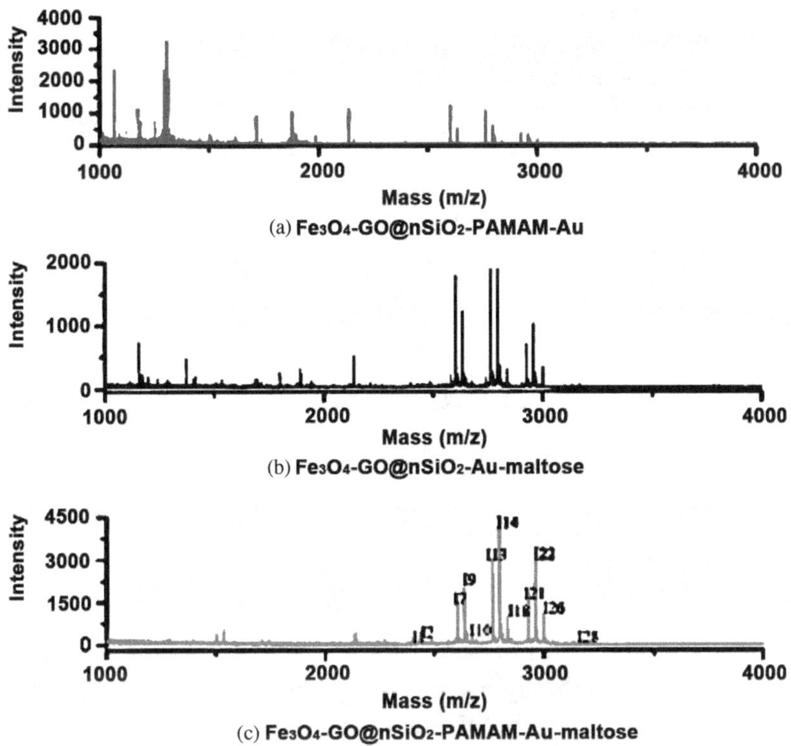

Figure 6.16: MS spectra of human IgG digest after enrichment by different materials.

polymer PAMAM and maltose, Fe_3O_4–GO@nSiO$_2$–PAMAM–Au composite and Fe_3O_4–GO@nSiO$_2$–Au–maltose composite are synthesized for enriching glycopeptide in IgG digest. As shown in Figure 6.16, although Fe_3O_4–GO@nSiO$_2$–PAMAM–Au composite and Fe_3O_4–GO@nSiO$_2$–Au–maltose composite also have certain enrichment ability for glycopeptide, there are a large number of non-glycosylated peptides in spectra, which seriously interfere with the detection of glycopeptide. However, after enrichment by Fe_3O_4–GO@nSiO$_2$–PAMAM–Au–maltose, the glycosylated peptides dominate the spectrum with a clean background. The above results indicate that it is the co-modification of PAMAM and maltose that plays a vital role in the efficient and selective enrichment of glycosylated peptide.

HRP digest is also selected to evaluate the enrichment universality of Fe_3O_4–GO@nSiO$_2$–PAMAM–Au–maltose composite for different types of

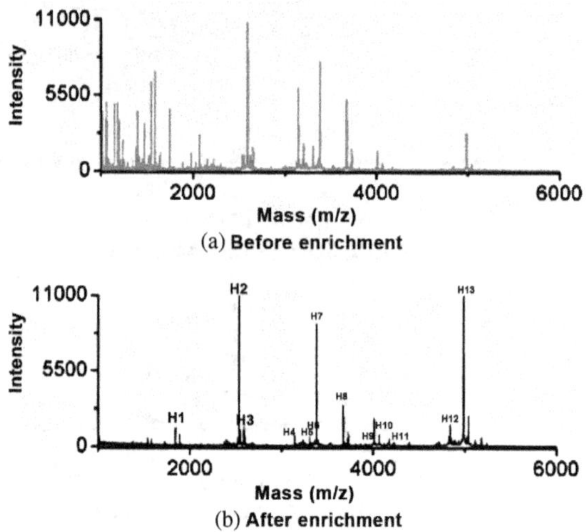

Figure 6.17: MS spectra of HRP digest before and after enrichment by Fe_3O_4–GO@nSiO$_2$–PAMAM–Au–maltose composite.

glycopeptides. As shown in Figure 6.17, non-glycosylated peptides severely suppress the signal of the glycopeptide before enrichment, but after enriched by Fe_3O_4–GO@nSiO$_2$–PAMAM–Au–maltose composite, the glycopeptides dominate the spectrum.

(b) *Evaluation of the enrichment ability of Fe_3O_4–GO@nSiO$_2$–PAMAM–Au–maltose for glycopeptide in mouse liver digest*

The 50-μg mouse liver digest is enriched by Fe_3O_4–GO@nSiO$_2$–PAMAM–Au–maltose composite. Finally, 1529 N-glycopeptides and 1254 N-glycosylation sites, corresponding to 760 glycoproteins are identified by MS.

6.2.5. Double sugar co-modified magnetic silica

In this section, the application of chitosan and hyaluronic acid co-modified magnetic silica will be introduced.

(1) Preparation of MNPs-(HA-CS)$_n$)

The synthesis process of MNPs-(HA-CS)$_n$ microsphere is shown in Section 2.6 of Chapter 2.[8]

(2) Enrichment application of MNPs-(HA-CS)$_{10}$ composite in glycoproteomics

(a) *Evaluation of the enrichment ability of MNPs-(HA-CS)$_{10}$ for glycopeptide in standard glycoprotein digest*

ACN/H$_2$O/TFA (v/v/v, 88/11.9/0.1, 400 μL) is selected as the loading buffer, ACN/H$_2$O/TFA (v/v/v, 30/69.9/0.1, 2×10 μL) as the eluent and 10 min as the enrichment time. Human IgG digest is used to investigate the enrichment ability of MNPs-(HA-CS)$_{10}$ microsphere for glycopeptide. For comparison, MNPs-HA microsphere is also used to enrich glycopeptide in IgG digest. As shown in Figure 6.18(a), when 5 pmol IgG digests are directly analyzed by MS, only two very weak glycopeptides are detected. While after being enriched by MNPs-(HA-CS)$_{10}$ microsphere, 24 enhanced glycosylated peptides are detected (Figure 6.18(b)), all of which are demonstrated to be glycopeptides through the enzymatic cutting of PNGase F. However, after being enriched by MNPs-HA (Figure 6.18(c)), only 14 glycosylated peptides are detected, their peak intensities are significantly lower than that in Figure 6.18(b).

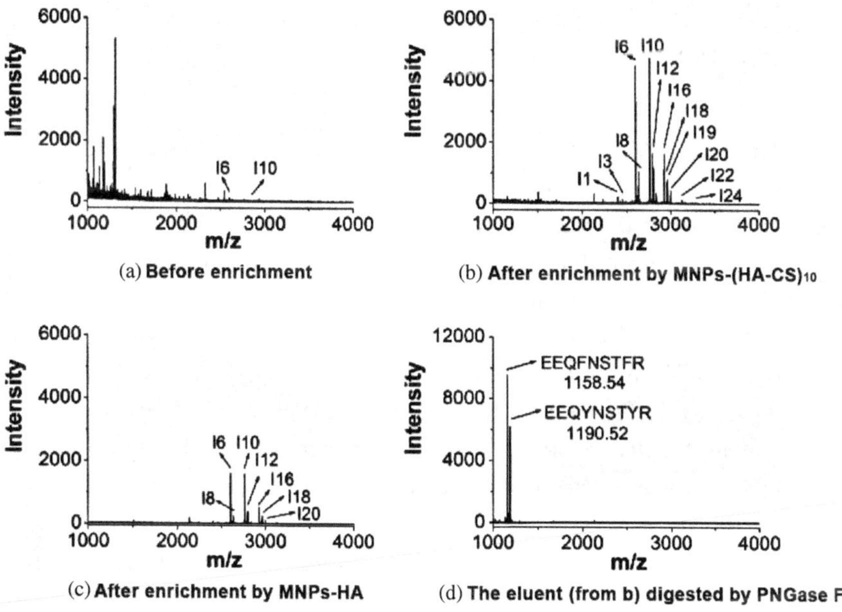

Figure 6.18: MS spectra of 5 pmol human IgG digest before and after being enriched by different materials.

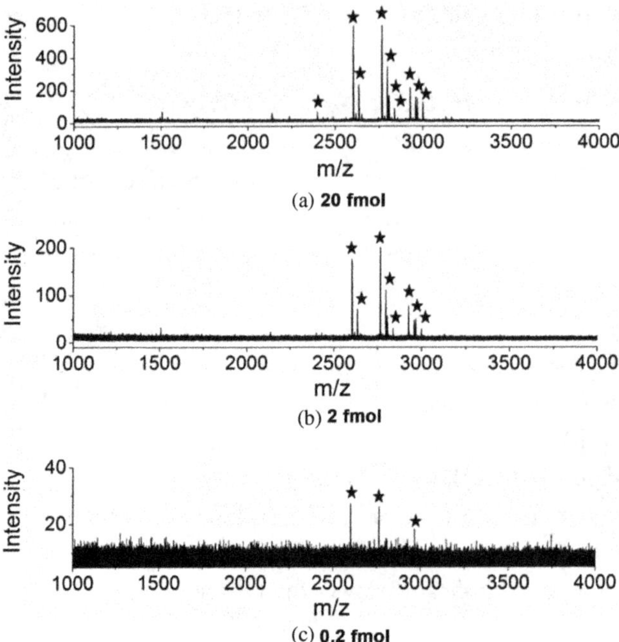

Figure 6.19: MS spectra of different amounts of human IgG digests (200 μL) after being enriched by MNPs-(HA-CS)$_{10}$ microsphere.

In addition, MNPs-(HA-CS)$_{10}$ microsphere is used to enrich different amounts of human IgG digests. As shown in Figure 6.19, when human IgG digest is as low as 0.2 fmol, three glycopeptides can still be detected. The above results indicate that MNPs-(HA-CS)$_{10}$ microsphere has great enrichment ability for glycopeptide.

OVA digest and HRP digest are selected to further investigate whether MNPs-(HA-CS)$_{10}$ microsphere can enrich different glycan types of glycopeptides. As shown in Figure 6.20(b), after being enriched by MNPs-(HA-CS)$_{10}$ microsphere, 16 glycopeptides from OVA digest and six glycosylation sites from HRP digest are, respectively, detected by MS.

(b) *Evaluation of the enrichment ability of MNPs-(HA-CS)$_{10}$ for glycopeptide in mouse liver digest*

With the same method, 20 μg mouse liver digest is treated by MNPs-(HA-CS)$_{10}$ microsphere and 616 glycopeptides and 605 glycosylation sites corresponding to 350 glycoproteins are identified by MS analysis.

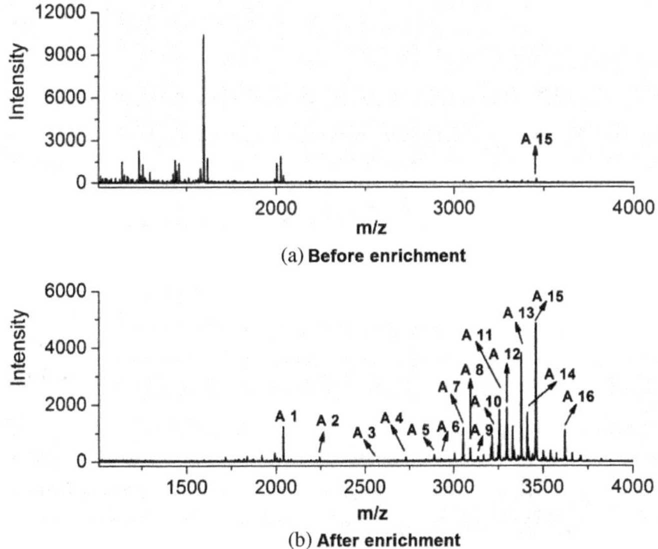

Figure 6.20: MS spectra of 5 pmol OVA digest before and after being enriched by MNPs-(HA-CS)$_{10}$.

6.2.6. *Glycopeptide dendrimer modified magnetic microsphere (dM-MNPs)*

6.2.6.1. *Preparation of dM-MNPs microsphere*

The synthesis of dM-MNPs microsphere is shown in Section 2.6 of Chapter 2.[9]

The M-MNPs microsphere is synthesized for comparison, and the difference of synthetic process of dM-MNPs from M-MNPs microsphere is summarized as follows. 20 mg MNPs-NH$_2$ microsphere is dispersed in anhydrous dichloromethane under nitrogen protection. Then 1.5 mL distilled triethylamine and 1 mL 2-bromoisobutyryl bromide are added in order under the protection of argon and ice bath, and then mechanically stirred for 2 h. Next, the reaction is continuously stirred at room temperature for 16 h. The obtained solid product is washed with dichloromethane, ethanol and deionized water and dried in vacuum. Afterwards, 50 mg product is dispersed in 30 mL DMF (containing 230 mg sodium azide and 167 mg ammonium chloride) for keeping reaction at 50°C for 26 h. The obtained MNPs-N$_3$ microsphere is washed with DMF, deionized water, methanol, and then dried in vacuum for further experiment.

6.2.6.2. *Enrichment application of dM-MNPs microsphere in glycoproteomics*

(1) Evaluation of the enrichment ability of dM-MNPs microsphere for glycopeptide in standard glycoprotein digest

ACN/H$_2$O/TFA (v/v/v, 88/7/5, 400 μL) is selected as the loading buffer, ACN/H$_2$O/TFA (v/v/v, 30/69.9/0.1, 2×10 μL) as the eluent, and 30 min as the enrichment time.

HRP is used to compare the enrichment ability of dM-MNPs and M-MNPs. As shown in Figure 6.21, UV–Vis spectroscopy shows that the binding ability of dM-MNPs microsphere to glycoprotein is significantly higher than that of M-MNPs microsphere.

Human IgG digest is also used to compare the enrichment ability of dM-MNPs and M-MNPs towards different types of glycopeptides. As shown in Figure 6.22, the dM-MNPs microsphere exhibits better enrichment ability. The above results indicate that the presence of glycopeptide polymer is critical for increasing the enrichment ability of the material.

Finally, different amounts of human IgG digests are treated with dM-MNPs and M-MNPs. As shown in Figure 6.23, when the human IgG digest is as low as 0.1 fmol, after enrichment by dM-MNPs microsphere, one glycopeptide can still be obtained. However, no apparent glycopeptide signal

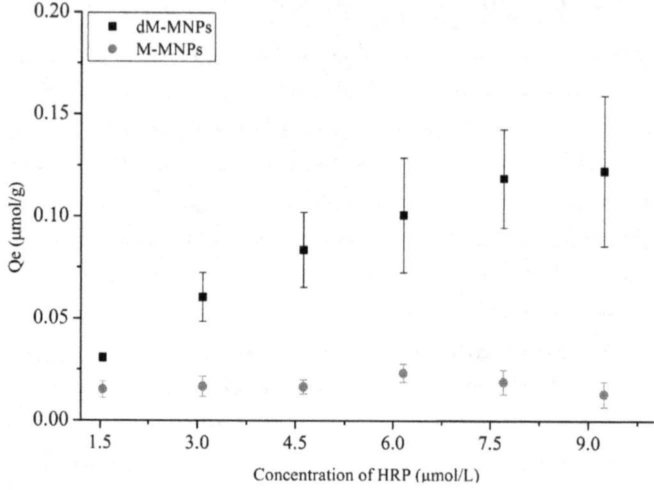

Figure 6.21: The binding ability of different materials for of HRP protein.

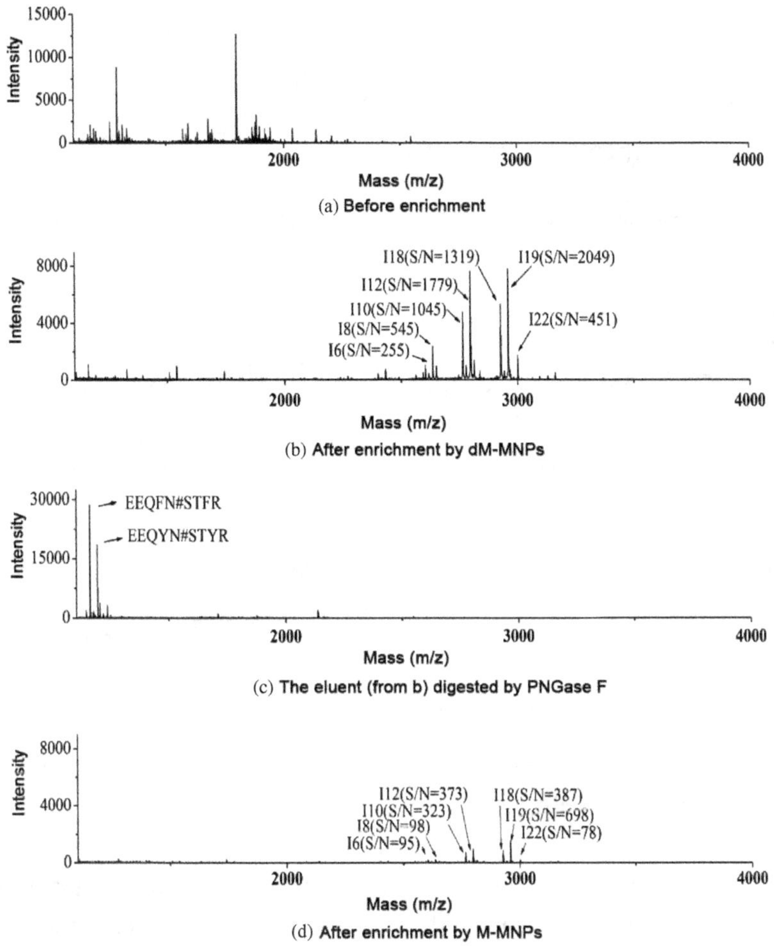

Figure 6.22: MS spectra of 0.5 pmol human IgG digest before and after being enriched by different materials.

can be found after enrichment by M-MNPs microsphere. This indicates that the dM-MNPs microsphere has better enrichment ability towards glycosylated peptide.

(2) Evaluation of the enrichment ability of dM-MNPs microsphere for glycopeptide in mouse liver digest

ACN/H_2O/TFA (v/v/v, 88/10/2, 400 μL) is selected as the loading buffer, ACN/H_2O/TFA (v/v/v, 30/69.9/0.1, 2×30 μL) as the eluent, and 30 min as

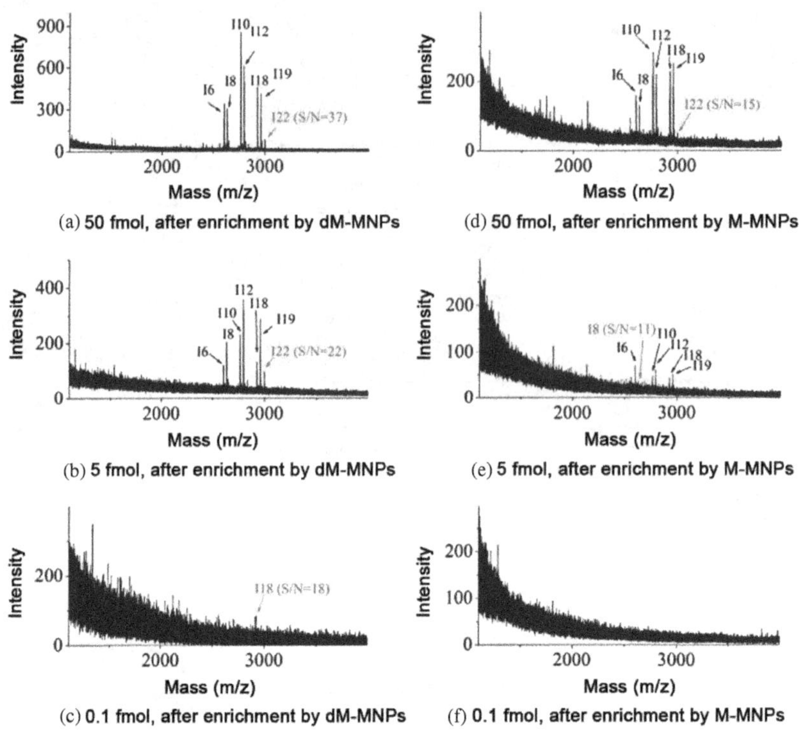

Figure 6.23: MS spectra of different amounts of human IgG digests before and after being enrichment by different materials.

the enrichment time. After 80 μg mouse liver digest is treated with dM-MNPs microsphere, 1009 *N*-glycopeptides and 1083 *N*-glycosylation sites, corresponding to 572 *N*-glycosylated proteins, are identified by MS.

6.3. Lectin-modified magnetic micro-/nanomaterial for separation and analysis in glycoproteomics

6.3.1. *Introduction*

Lectin is a kind of protein with specific affinity towards different types of glycans, which can specifically recognize and bind specific saccharide sequence. Common methods for immobilizing proteins are chemical bonding, physical adsorption, chemical derivatization, etc. Chemical bonding is to use target protein to form a covalent bond with the functional group on the substrate to realize the immobilization of protein. Physical adsorption immobilizes protein by hydrophobic interaction between the target protein and the substrate. Chemical derivatization usually needs to modify the target protein prior to its immobilization, which may result in protein inactivation. Therefore, the appropriate immobilization method needs to be selected according to the experimental need. Con A has a molecular weight of about 100 kDa and exists as a tetramer in aqueous solution. It has multiple binding sites for saccharide and high affinity for glycoprotein, which is one of the most widely used lectins. Currently, Con A is mostly used to prepare the lectin affinity chromatography column to separate glycosylated protein/peptide. But the lectin-modified magnetic microsphere has rarely been studied. In this section, we will describe the application of Con A-modified magnetic microsphere in glycoproteomics.

6.3.2. *Aminophenylboronic acid-magnetic microsphere for immobilization of Con A (Fe_3O_4@APBA–sugar–Con A)*

6.3.2.1. *Preparation of Fe_3O_4@APBA–sugar–Con A microsphere*

The synthesis process of Fe_3O_4@APBA–sugar–Con A microsphere is shown in Section 2.5 of Chapter 2.[10]

6.3.2.2. *Application of Fe_3O_4@APBA–sugar–Con A microsphere in glycoproteomics*

(1) Evaluation of the enrichment condition of Fe_3O_4@APBA–sugar–Con A microsphere for standard glycoprotein

When Con A is immobilized on the surface of Fe_3O_4@APBA microsphere, alkaline condition is adopted as the synthesis condition to immobilize the methyl alpha-D-Mannopyranoside on the Fe_3O_4@APBA microsphere, but Con A usually enriches glycoprotein under the neutral condition. To investigate the effect of different pH values on the glycoprotein enrichment, neutral buffer (PBS), and alkaline buffer (50 mmol·L^{-1} NH$_4$HCO$_3$) are respectively selected as the experimental conditions for immobilization of Con A and enrichment of standard glycosylated protein. As shown in Figure 6.24(a),

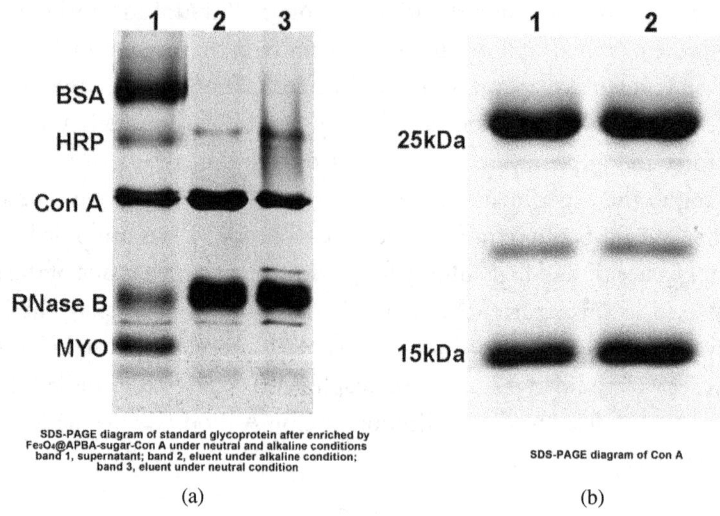

Figure 6.24: SDS-PAGE diagram of standard glycoproteins selectively enriched by Fe_3O_4@ APBA–sugar–Con A microsphere and Con A.

different experimental conditions have no significant impact in the enrichment performance of glycosylated protein. To protect the bonding interaction between boronic acid and monosaccharide, 50 mmol·L^{-1} NH_4HCO_3 solution is used as the loading buffer in the following experiment. It is worth noting that, although the Fe_3O_4@APBA–sugar–Con A microsphere has a better selective enrichment ability for glycosylated protein, the Con A band will always appear in the eluent. This may be because the Con A of the tetrameric structure will dissociate during the enrichment process. As shown in Figure 6.24(b), the molecular weight of Con A monomer is about 25 kDa, which is degraded to produce multiple bands during SDS-PAGE separation.

(2) Evaluation of the enrichment ability of Fe_3O_4@APBA–sugar–Con A microsphere for standard glycoprotein

Two standard glycoproteins of HRP and RNase B containing high mannose are selected to investigate the enrichment ability of Fe_3O_4@APBA–sugar–Con A microsphere. As shown in Figure 6.25(a), the bands of HRP (band 1)

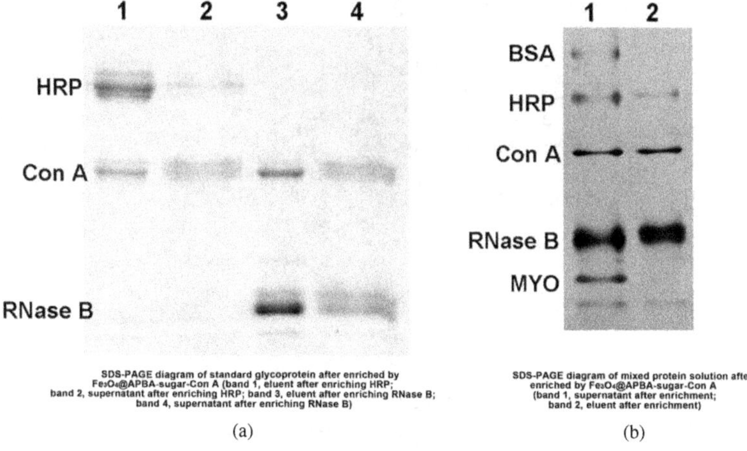

Figure 6.25: SDS-PAGE diagram of standard glycoprotein and mixed protein solution after enriched by Fe$_3$O$_4$@APBA–sugar–Con A microsphere.

and RNase B (band 3) appear in the eluate, indicating that both of the glycoproteins can be captured by Fe$_3$O$_4$@APBA–sugar–Con A microsphere.

Then, the mixture of HRP, RNase B, MYO, and BSA is used to further investigate the selective enrichment ability of Fe3O4@APBA–sugar–Con A microsphere. The MYO and BSA are non-glycosylated proteins so that they will not bind to Fe$_3$O$_4$@APBA–sugar–Con A microspheres. As shown in Figure 6.25(b), only the bands of HRP and RNase B appear in the eluent (band 2), and no MYO or BSA appears, which is consistent with the prediction. This result demonstrates that Fe$_3$O$_4$@APBA–sugar–Con A microsphere has excellent selective enrichment ability for glycoprotein.

(3) Evaluation of the enrichment ability of Fe$_3$O$_4$@APBA–sugar–Con A microsphere for glycoprotein in hepatoma cell line 7703

Human hepatocellular carcinoma cell line 7703, model of hepatocarcinoma, is used to investigate the enrichment ability of Fe$_3$O$_4$@APBA–sugar–Con A microsphere for glycoprotein. The enrichment conditions are as follows: 2 mg Fe$_3$O$_4$@APB–sugar–Con A microsphere is dispersed in 100 μL cell lysate

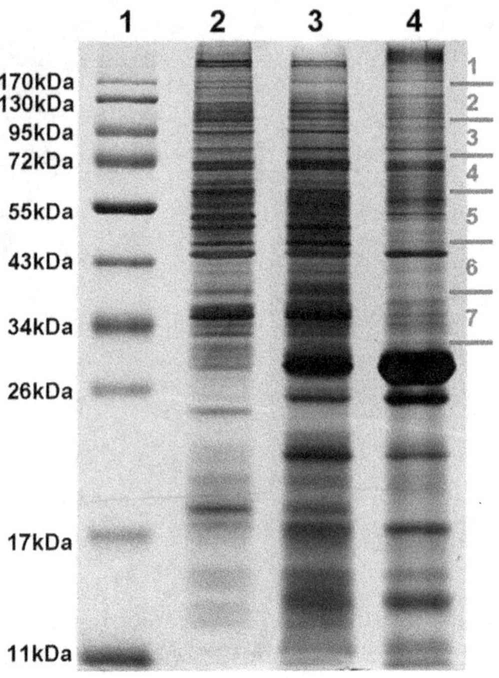

Figure 6.26: SDS-PAGE diagram of glycoprotein enriched by Fe_3O_4@APBA-sugar-Con A microsphere from human hepatoma cell line 7703 cell lysate. (band 1, standard molecular weight marker; band 2, original cell lysate; band 3, the supernatant; band 4, the eluent).

containing about 0.5 mg protein, and 1 mol·L^{-1} NH_4HCO_3 solution is added to make a final concentration of NH_4HCO_3 at 50 mmol·L^{-1}.

After enriching the lysate of cell line 7703 with Fe_3O_4@APBA–sugar–Con A microsphere, the cell lysate, the supernatant, and the eluent are analyzed by SDS-PAGE. As shown in Figure 6.26, the proteins below 25 kDa in the eluate are eliminated to avoid the interference from Con A in the subsequent MS analysis process, and the proteins above 25 kDa are divided into seven parts. These proteins are deglycosylated by PNGase F, digested by trypsin, and then analyzed by online nano-RPLC-ESI-MS/MS. The MS data are searched by a combination of positive and negative libraries, and the results are filtered by using PeptideProphet. Finally, only the p value of glycopeptide higher than or equal to 0.95 is accepted. The above experiments are repeated for 6 times. Based on the results, a total of 172 glycosylated peptides corresponding to 101 glycosylated proteins are identified. Except for the keratin, the total number of the identified proteins is 149.68 of the same glycopeptides are identified in all six replicates, more than 70% glycopeptides are identified in more than three replicates, and 21 glycosylated peptides are identified only in a single experiment.

6.4. Covalent bonding-based magnetic micro-/nanomaterial for separation and analysis in glycoproteomics

6.4.1. Introduction

The common glycosylated protein/peptide enrichment methods based on covalent bonding are boronic acid chemistry and hydrazine chemistry. Boronic acid chemistry is based on the stable and reversible covalent bond between boronic acid and cis–diol. In recent years, boronic acid and its derivatives functionalized magnetic micro-/nanomaterials have been developed rapidly in glycoprotein/glycopeptide enrichment. Among them, gold nanoparticle and silane reagent with unique physicochemical property are widely used as linking agent for immobilization of boric acid. In addition, azide-alkynyl-based click chemistry is also widely used to prepare boronic acid functionalized magnetic micro-/nanomaterial. The hydrazide reagent modified oxidized saccharide is a traditional research method. Its advantages are that it can enrich different types of glycoproteins/glycopeptides at one time, and the enrichment efficiency is high. At present, the functionalized magnetic micro-/nanomaterials derived from this traditional method have been used for glycoprotein/glycopeptide enrichment. But its application range is inferior to boronic acid chemistry because of the complex reaction and tedious operation. This section will describe different immobilization methods for boronic acid or hydrazide on magnetic microsphere in glycoproteomics.

6.4.2. Au nanoparticle for immobilization of phenylboronic acid on magnetic microsphere

6.4.2.1. Fe₃O₄@CP@Au–MPBA microspheres

(1) Preparation of Fe₃O₄@CP@Au–MPBA microsphere

The synthesis process of the Fe_3O_4@CP@Au–MPBA microsphere is shown in Figure 2.59, and can be seen in Section 2.4 of Chapter 2.[11]

(2) Application of Fe_3O_4@CP@Au–MPBA microsphere in glycoproteomics

(a) *Evaluation of the enrichment capability of Fe_3O_4@CP@Au–MPBA microsphere for glycopeptide in standard glycoprotein digest*

HRP digest is used to investigate the enrichment ability of Fe_3O_4@CP@Au–MPBA microsphere for glycopeptide. As shown in Figure 6.27(b), after being enriched by Fe_3O_4@CP@Au–MPBA microsphere, a total of 12 glycopeptide signals are detected. Compared with the supernatant in Figure 6.27(a), the Fe_3O_4@CP@Au–MPBA microsphere effectively enriches the glycosylated peptides from HRP digest.

The peptides mixture of HRP and β-casein with a molar ratio of 1/10 is then used to investigate the enrichment selectivity of Fe_3O_4@CP@Au–MPBA microsphere for glycosylated peptide. As shown in Figure 6.28(a), the identified glycosylated peptides reduce in the MS spectrum of direct analysis due to the addition of non-glycosylated peptides. However, after enrichment with Fe_3O_4@CP@Au–MPBA microsphere, as shown in Figure 6.28(b), seven glycosylated peptides are detected, indicating that Fe_3O_4@CP@Au–MPBA microsphere has outstanding enrichment selectivity for glycopeptide.

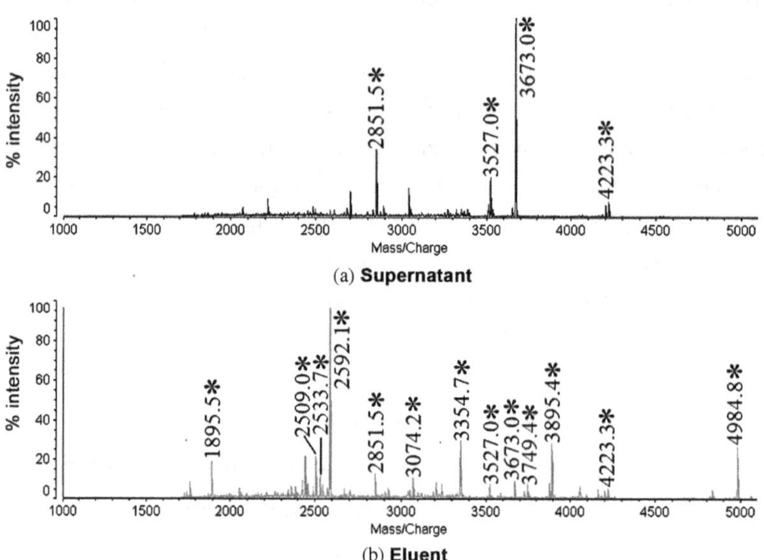

Figure 6.27: MS spectra of HRP digest (2.5 ng·μL^{-1}, 100 μL) after enrichment by Fe_3O_4@CP@Au–MPBA microsphere. *: glycosylated peptide.

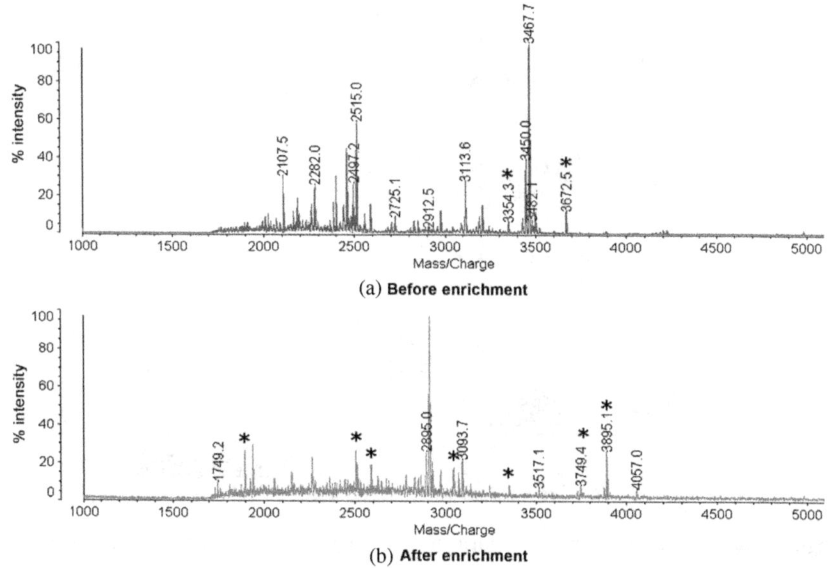

Figure 6.28: MS spectra of peptides mixture of HRP (2.5 ng·μL⁻¹) and β-casein with the molar ratio of 1/10 before enrichment by Fe_3O_4@CP@Au–MPBA microsphere.

(b) *Evaluation of the enrichment ability of Fe_3O_4@CP@Au–MPBA microsphere for standard glycoprotein*

The mixture of RNase B (glycoprotein) and MYO (non-glycosylated protein) is used to investigate the enrichment selectivity of Fe_3O_4@CP@Au–MPBA microsphere for glycoprotein. The results are analyzed by SDS-PAGE. As shown in Figure 6.29, two bands assigned to RNase B and MYO can be seen before enrichment. However, after being enriched by Fe_3O_4@CP@Au–MPBA microsphere, only RNB band can be observed. This result indicates that Fe_3O_4@CP@Au–MPBA microsphere can efficiently and selectively enrich glycoprotein.

6.4.2.2. Fe_3O_4@SiO$_2$@Au–APBA microsphere

(1) Preparation of Fe_3O_4@SiO$_2$@Au–APBA microsphere

The synthesis process of Fe_3O_4@SiO$_2$@Au–APBA microsphere is shown in Section 2.6 of Chapter 2.[12]

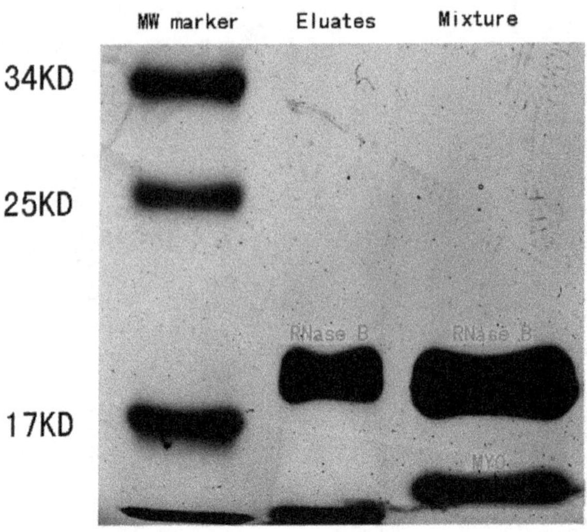

Figure 6.29: The enrichment performance of Fe_3O_4@C@Au–MPBA microsphere for glycoprotein.

(2) Application of Fe_3O_4@SiO$_2$@Au–APBA microsphere in glycoproteomics

(a) *Evaluation of the enrichment ability of Fe_3O_4@SiO$_2$@Au–APBA microsphere for glycopeptide in standard glycosylated protein digest*

HRP digest is used to investigate the enrichment ability of Fe_3O_4@SiO$_2$@ Au–APBA microsphere for glycopeptide. As shown in Figure 6.30(a), when 10 ng·μL^{-1} HRP digest are directly analyzed, only five weak glycopeptide peaks are detected. But after 2 ng·μL^{-1} HRP digest is enriched by Fe_3O_4@ SiO$_2$@Au–APBA microsphere, 17 glycopeptides are detected (Figure 6.30(b)). When the concentration of HRP digest further decreases to 0.1 ng·μL^{-1}, still 11 glycosylated peptides are detected (Figure 6.30(c)). It indicates that Fe_3O_4@SiO$_2$@Au–APBA microsphere has selective enrichment ability for glycopeptides.

Then, another commonly used glycoprotein, asialofetuin (AF), is used to investigate the enrichment universality of Fe_3O_4@SiO$_2$@Au–APBA microsphere for glycopeptide. The amino acid sequence length of AF is 359 amino acids, on which N99, N156, and N176 may be the *N*-glycosylation sites. As shown in Figure 6.30(d), when AF digest is 1 ng·μL^{-1}, 11 glycopeptides and

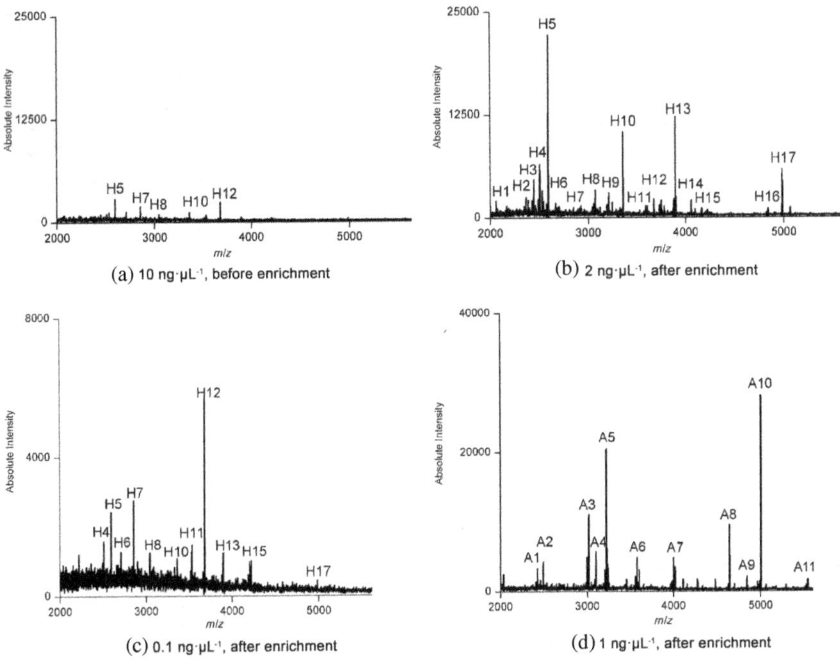

Figure 6.30: MS spectra of different concentration of HRP digests before and after enrichment by $Fe_3O_4@SiO_2@Au–APB$ and AF digest after enrichment by $Fe_3O_4@SiO_2@Au–APB$.

their fragments are detected by MS, and the detailed information of identified glycosylated peptides is shown in Table 6.7.

(b) *Evaluation of the enrichment ability of $Fe_3O_4@SiO_2@Au–APB$ microsphere for standard glycoprotein*

The protein mixture of non-glycosylated proteins including BSA, Cyt C, and glycosylated protein of HRP is used to investigate the enrichment ability of $Fe_3O_4@SiO_2@Au–APBA$ microsphere for glycoprotein. After enriched by $Fe_3O_4@SiO_2@Au–APBA$ microsphere, the original mixture and the eluent are compared through SDS-PAGE analysis. As shown in Figure 6.31, three standard proteins appear before enrichment. From the bottom to top, the bands are assigned to Cyt C, HRP, and BSA. However, after being enriched by $Fe_3O_4@SiO_2@Au–APBA$ microsphere, only HRP band appears. It indicates that the $Fe_3O_4@SiO_2@Au–APBA$ microsphere also has excellent enrichment ability for glycoprotein.

Table 6.7: Detailed information of glycoepeptides enriched by $Fe_3O_4@SiO_2@$ Au–APB microsphere from AF digest.

	M/Z	Glycon Composition	Amino Acid Sequence
A1	2428.3		b-NH$_3$ ion (+1) HAVEVALATFNAESNGSYLQLVE
A2	2494.2		y-NH$_3$ ion(+1) EVALATFNAESNGSYLQLVEISR
A3	3016.5		VVHAVEVALATFNAESN#GSYLQLVEISR
A4	3099.4	GllcNAc (Partial)	VVHAVEVALATFNAESN#GSYLQLVEISR
A5	3219.6	GlcNAc	VVHAVEVALATFNAESNGSYLQLVEISR
A6	3576.6	Man$_3$GlcNAc$_2$	b ion(+1) EVYDIEDTLETCHVLDPTPLAN#C
A7	4000.1	Man$_3$GlcNAc$_2$	b-H$_2$O ion (+1) PTGEVYDIEIDTLETCHVLDPTPLAN#CSV
A8	4638.9	Gal$_2$GlcNAc$_2$Man$_3$GlcNAc$_2$	VVHAVEVALATFNAESN#GSYLQLVEISR
A9	4842.4	Gal$_2$GlcNAc$_3$Man$_3$GlcNAc$_2$	VVHAVEVALATFNAESN#GSYLQLVEISR
A10	5004.1	Gal$_2$GlcNAc$_3$Man$_3$GlcNAc$_2$	VVHAVEVALATFNAESN#GSYLQLVEISR
A11	5545.7	Gal$_3$GlcNAc$_3$Man$_3$GlcNAc$_2$	RPTGEVYDIEIDTLETCHVLDPTPLAN#CSVR

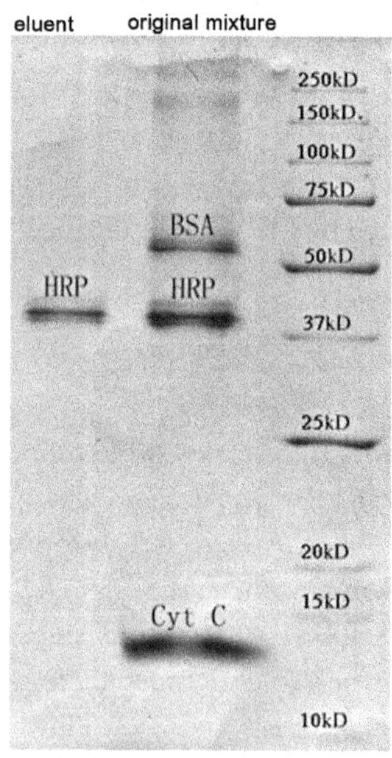

Figure 6.31: Enrichment performance of $Fe_3O_4@SiO_2@$Au–APB microsphere for glycoprotein.

(c) *Analysis of N-glycosylation site in human colorectal cancer tissue by $Fe_3O_4@SiO_2@Au–APB$ microsphere*

The third-stage clinical tissue sample of human colorectal cancer is used to evaluate the enrichment ability of $Fe_3O_4@SiO_2@Au–APBA$ microsphere for glycoprotein. The total amount of protein extracted from the tissue is first quantified, and then the protein is treated with excessive $Fe_3O_4@SiO_2@Au–APBA$ microsphere. The glycosylated protein in the eluate is digested with trypsin, and then deglycosylated by PNGase F. The peptide mixture is finally identified by 2D LC-MS/MS.

All secondary MS spectra are searched via MASCOT (Version 2.2) in the International Protein Index (IPI) Human Protein Non-Redundant Library (Version 3.35) to complete the protein matching work. The asparagine on the specific amino acid sequence Asn-Xaa-(Ser/Thr/Cys, Xaa is any other amino acid except proline) will cause a mass deviation of 0.98 Da due to deamidation, and *N*-glycosylation site can be identified with the mass deviation of 0.98 Da. In a single 2D LC-MS/MS analysis, a total of 190 glycopeptides and 155 glycoproteins are identified. As shown in Figure 6.32, 13 potential *N*-glycosylation sites, 16 reported *N*-glycosylation sites, and 165 new glycosylation sites are identified according to the Swiss-Prot database. It is worth mentioning that, in this case, the sample amount is only 270 μg, indicating that the $Fe_3O_4@SiO_2@$Au–APB microsphere has great enrichment performance.

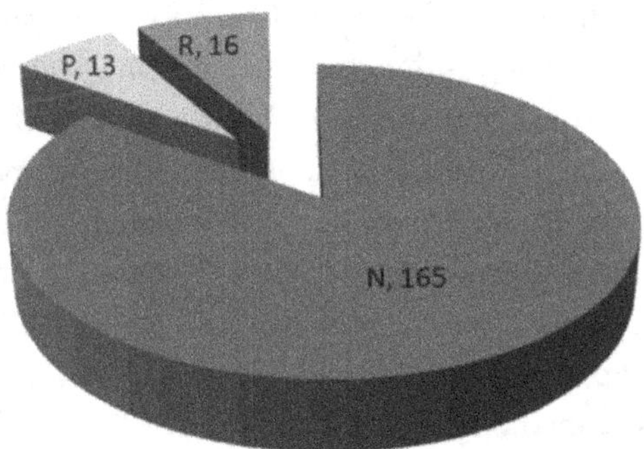

Figure 6.32: *N*-glycosylation sites identified from the clinical tissue of human colorectal cancer. R: the reported glycosylation sites; P: potential glycosylation sites via Swiss-Prot database; N: newly discovered glycosylation sites in this work.

6.4.3. *Silylated reagent for immobilization of boronic acid group on magnetic silica*

6.4.3.1. *Fe_3O_4@SiO_2@PSV microsphere (3-(methacryloyloxy)propyltrimethoxysilane, MPS as a linker)*

(1) Preparation of Fe_3O_4@SiO_2@PSV microsphere

The synthesis protocol of Fe_3O_4@SiO_2@PSV microsphere is shown in Section 2.6 of Chapter 2.[13]

(2) Application of Fe_3O_4@SiO_2@PSV microsphere in glycoproteomics

(a) *Evaluation of the enrichment ability of Fe_3O_4@SiO_2@PSV microsphere for glycopeptide in standard glycoprotein digest*

To investigate the enrichment ability and sensitivity of Fe_3O_4@SiO_2@PSV microsphere, 0.05 ng·μL^{-1} HRP digest (1.25×10^{-13} mol, 100 μL) is selected. As shown in Figure 6.33, when the concentration of HRP digest is as low as 0.05 ng·μL^{-1}, still four glycopeptides are detected after enrichment by Fe_3O_4@SiO_2@PSV microsphere.

In order to investigate the enrichment selectivity of Fe_3O_4@SiO_2@PSV microsphere, three non-glycosylated protein digests, 5 ng·μL^{-1} β-casein, BSA, and MYO, are used as negative controls. As shown in Figures 6.34(d)–6.34(f), after enrichment by Fe_3O_4@SiO_2@PSV microsphere, no peptide from β-casein or BSA digest is detected and only one peak from MYO digest is found. This indicates that the Fe_3O_4@SiO_2@PSV microsphere has no enrichment ability for non-glycosylated peptide. Then the peptide mixture of HRP and BSA at the mass ratio of 1/120 is enriched by Fe_3O_4@SiO_2@PSV microsphere. There is no glycopeptide detected before enrichment (Figure 6.34(g)), while after enrichment with Fe_3O_4@SiO_2@PSV microsphere, four glycopeptides are identified and their signals dominate the absolute advantage in the spectrum (Figure 6.34(h)). All the above data indicate that the Fe_3O_4@SiO_2@PSV microsphere has strongly specific enrichment ability, which can be applied to complex samples and exhibit high selectivity and high efficiency.

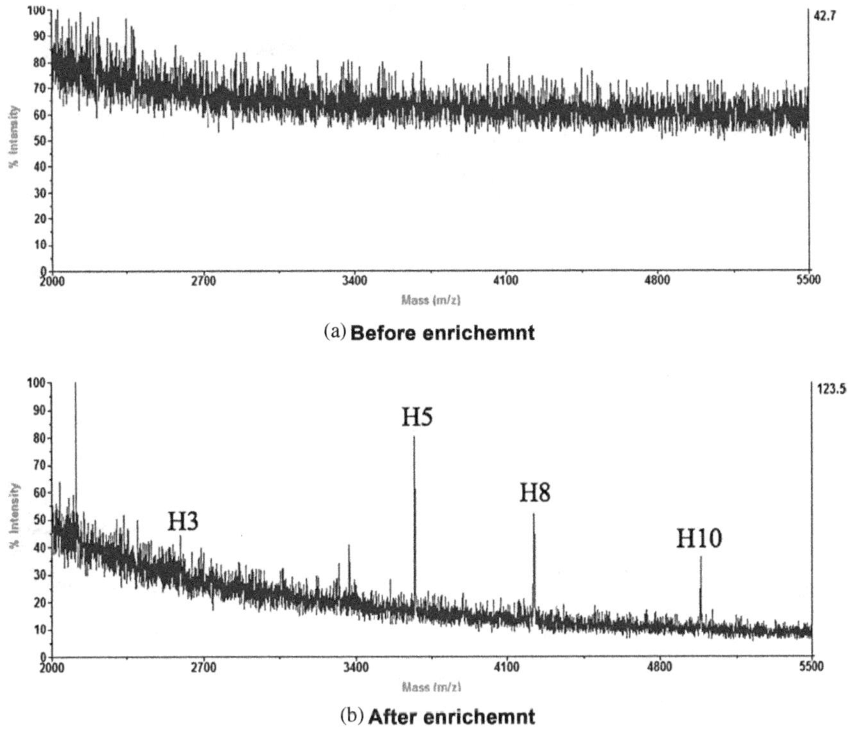

Figure 6.33: MS spectra of 0.05 ng·μL^{-1} HRP digest before and after enrichment by Fe$_3$O$_4$@SiO$_2$@PSV microsphere.

(b) *Analysis of N-glycosylation site in human serum sample using Fe$_3$O$_4$@SiO$_2$@PSV microsphere*

The human serum is used to further investigate the enrichment ability of Fe$_3$O$_4$@SiO$_2$@PSV microsphere for *N*-glycosylation. After LC-MS/MS analysis and database search, 103 glycosylation sites on 46 glycopeptides are successfully identified.

6.4.3.2. Fe$_3$O$_4$@SiO$_2$–APBA microsphere (GLYMO as a linker)

(1) Preparation of Fe$_3$O$_4$@SiO$_2$–APBA microsphere

The synthesis process of the Fe$_3$O$_4$@SiO$_2$–APBA microsphere is shown in Section 2.6 of Chapter 2.[14]

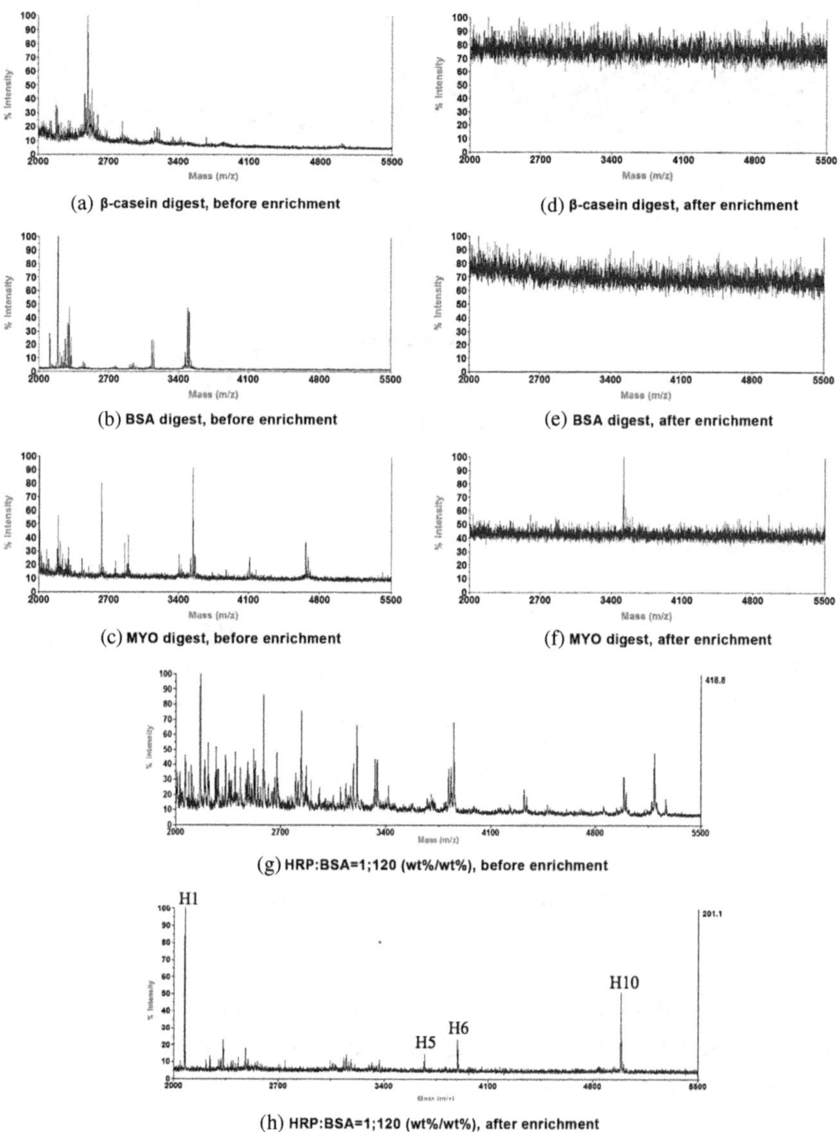

Figure 6.34: MS spectra of different peptide digests before and after enrichment by Fe3O4@SiO2@PSV microsphere.

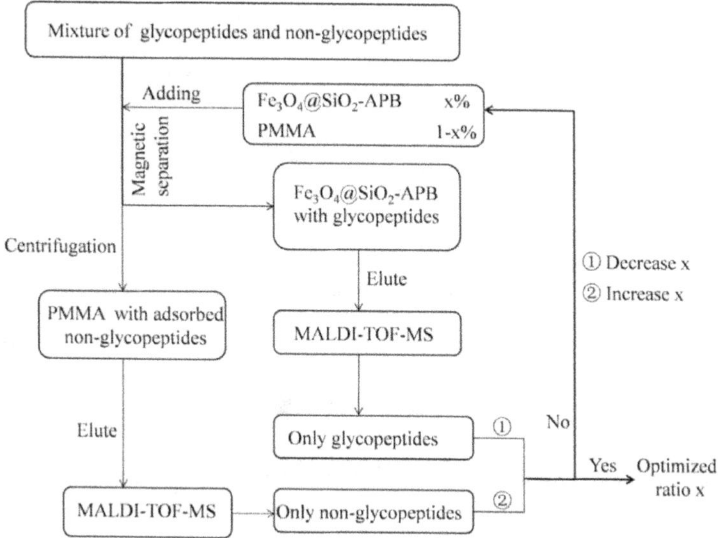

Figure 6.35: Synergistic enrichment mechanism of Fe_3O_4@SiO_2–APBA and PMMA.

(2) Application of Fe_3O_4@SiO_2–APBA microsphere in glycoproteomics

The synergistic enrichment mechanism of Fe_3O_4@SiO_2–APBA and PMMA is shown in Figure 6.35. The glycosylated peptide is captured by Fe_3O_4@SiO_2–APBA microsphere, and PMMA microsphere captures non-glycosylated peptide. The enrichment flowchart is shown in Figure 6.36. The enrichment conditions are as follows: 50 mmol·L^{-1} NH_4HCO_3 is the loading buffer and 20% ACN+1% TFA is the eluent.

(a) *Evaluation of the enrichment ability of Fe_3O_4@SiO_2–APBA and PMMA for glycopeptide in standard glycoprotein digest*

0.1 ng·μL^{-1} HRP digest is used to investigate the enrichment ability and sensitivity of Fe_3O_4@SiO_2–APBA and PMMA for glycopeptide. As shown in Figures 6.37(a)–6.37(c), after 0.1 ng·μL^{-1} HRP digest is enriched by Fe_3O_4@SiO_2–APBA (Figure 6.37(b)), only three glycopeptides are detected. However, after synergistically enriched by Fe_3O_4@SiO_2–APBA and PMMA (Figure 6.37(c)), 14 glycopeptides are detected, indicating the importance of synergistic enrichment.

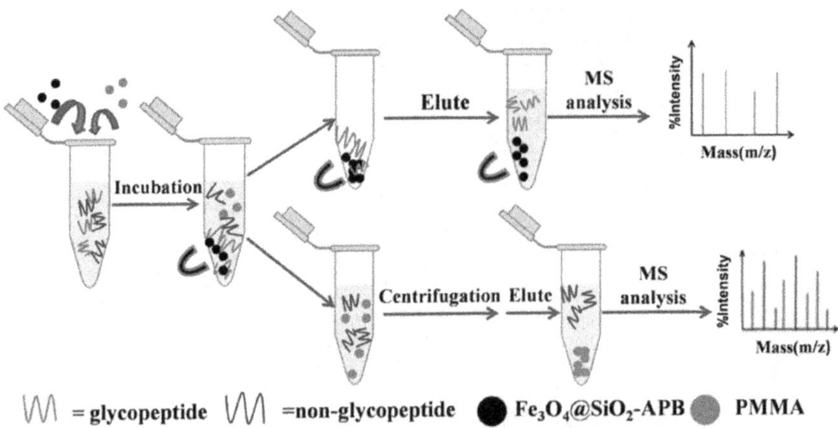

Figure 6.36: Synergistic enrichment flowchart of glycopeptide by $Fe_3O_4@SiO_2$–APBA and PMMA

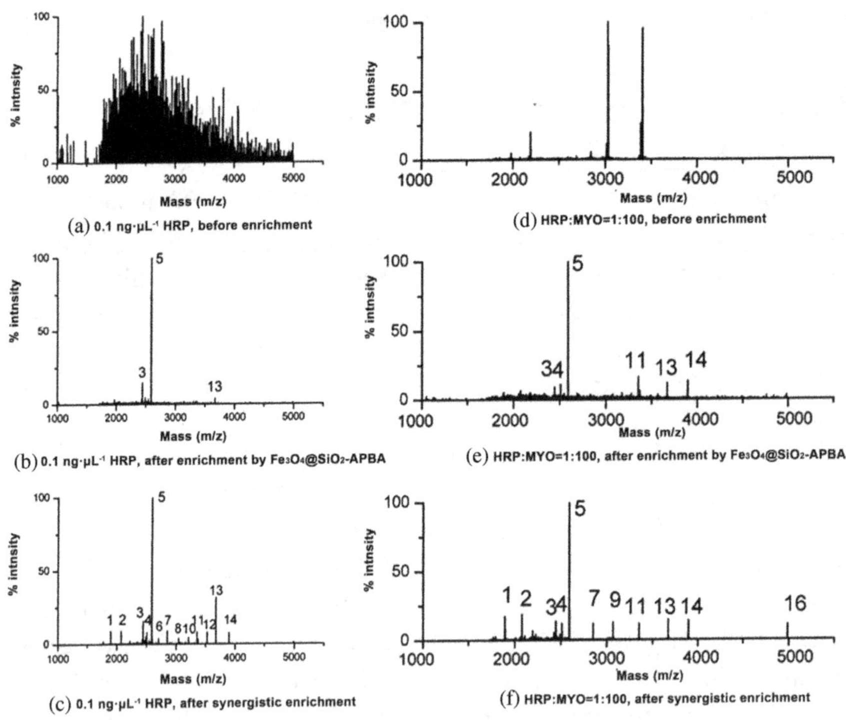

Figure 6.37: MS spectra of 0.1 ng·μL^{-1} HRP digest and peptide mixture of HRP and MYO at molar ratio of 1/100 before and after enrichment by $Fe_3O_4@SiO_2$–APBA or synergistic method.

Then, the selectivity of the synergistic enrichment method for glycopeptide is investigated using peptide mixture of HRP and MYO at molar ratio of 1/100. As shown in Figures 6.37(d) and 6.37(e), the synergistic enrichment of $Fe_3O_4@SiO_2$–APBA and PMMA shows great selectivity.

(b) *Analysis of N-glycosylation site in human serum sample using synergistic enrichment of $Fe_3O_4@SiO_2$–APB and PMMA*

1 μL human serum sample is applied to evaluate the synergistic enrichment performance of $Fe_3O_4@SiO_2$–APBA and PMMA for *N*-glycosylation. After LC-MS/MS analysis and database search, 153 glycosylation sites on 147 glycopeptides are successfully identified.

6.4.4. *Immobilization of boronic acid group on magnetic microsphere via click chemistry (Fe_3O_4@pVBC@APBA)*

(1) Preparation of Fe_3O_4@pVBC@APBA microsphere

The synthesis process of the Fe_3O_4@pVBC@APBA microsphere is shown in Figure 2.76 and can be seen in Section 2.6 of Chapter 2.[15]

(a) *Evaluation of the enrichment ability of Fe_3O_4@pVBC@APBA microsphere for standard glycoprotein*

Glucose, glycosylated proteins of OVA, transferrin (Trf) and HRP, non-glycosylated proteins of lysozyme (Lyz), Cyc, and BSA are selected to investigate the affinity between Fe_3O_4@pVBC@APBA microsphere and cis–diol molecule. Different concentrations of protein solutions (0.2–1.0 mg·mL^{-1}) are prepared with 0.02 mol·L^{-1} PBS buffer (containing 0.5 mol·L^{-1} NaCl) and then enriched by Fe_3O_4@pVBC@APBA microsphere. As shown in Figure 6.38(a), the binding capacity of Fe_3O_4@pVBC@APBA microsphere to glucose, OVA, Trf, and HRP is 912.6 mg·g^{-1}, 882.0 mg·g^{-1}, 596.1 mg·g^{-1}, and 263.2 mg·g^{-1} at pH = 9, while the binding ability towards BSA, Cyc, and Lyz is 39.7 mg·g^{-1}, 15 mg·g^{-1}, and 19.81 mg·g^{-1}, respectively. The results show that Fe_3O_4@pVBC@APBA microsphere has higher selective binding ability for glycosylated protein.

At the same time, the same experimental procedure is carried out under pH = 7.4, as shown in Figure 6.38(b), the binding ability of Fe_3O_4@pVBC@APBA microsphere for glycoprotein is slightly lower than that at pH = 9,

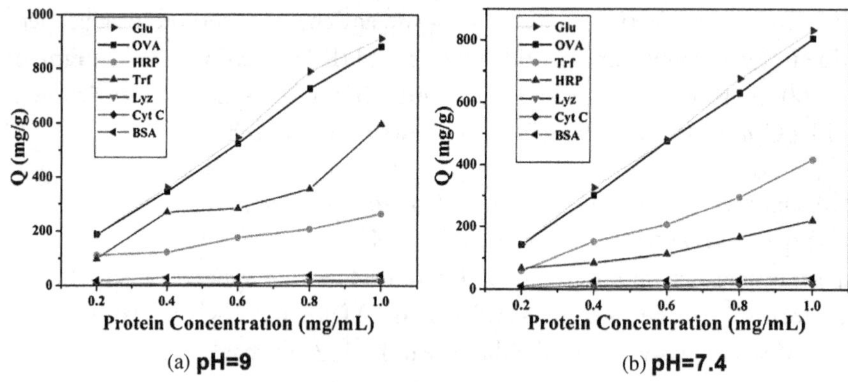

Figure 6.38. Adsorption isotherm of Fe_3O_4@pVBC@APBA for six proteins at different pH.

which is 806.1 mg·g^{-1}, 416.9 mg·g^{-1}, and 220.5 mg·g^{-1} for OVA, Trf, and HRP. But it still exhibits a higher selective binding ability towards glycoprotein.

(b) *Evaluation of the binding ability of Fe_3O_4@pVBC@APBA microsphere for glycoprotein in egg white*

The egg white sample is diluted into different concentrations and then used to further investigate the binding ability of Fe_3O_4@pVBC@APBA microsphere for glycoprotein. As shown in Figure 6.39, three high-abundance proteins, ovotransferrin (OVT, 76.7 kDa), OVA (46 kDa), and Lyz (14.4 kDa) can be observed when directly analyzing the 100-fold and 400-fold diluted samples (bands 2 and 5). After treatment with Fe_3O_4@pVBC@APBA microsphere, the OVT and OVA disappear on the band of the supernatant, and the bands of Lyz can still be observed (bands 3 and 6). The OVT and OVA appear on the band of the eluent, and no obvious Lyz can be observed (bands 4 and 7), indicating that the Fe_3O_4@pVBC@APBA microsphere also has selective binding ability for glycoprotein in complex samples.

6.4.5. *Hydrazine functionalized magnetic microsphere*

(1) Preparation of Fe_3O_4@PMAH microsphere

The synthesis process of Fe_3O_4@PMAH microsphere is shown in Figure 6.40, which is briefly described as follows[16]:

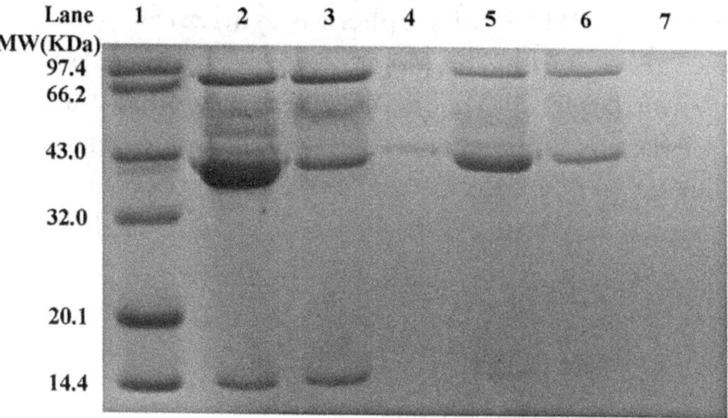

Figure 6.39: SDS-PAGE diagram of egg white samples with different concentrations. Protein marker: band 1. 100-fold diluted sample: band 2, direct analysis; band 3, supernatant after treatment by Fe_3O_4@pVBC@APBA; band 4, eluent after treatment by Fe_3O_4@pVBC@ APBA (HAc-NaAc, pH = 4). 400-fold diluted samples: band 5, direct analysis; band 6, after treatment by Fe_3O_4@pVBC@APBA; band 7, after treatment by Fe_3O_4@pVBC@APBA (HAc-NaAc, pH = 4).

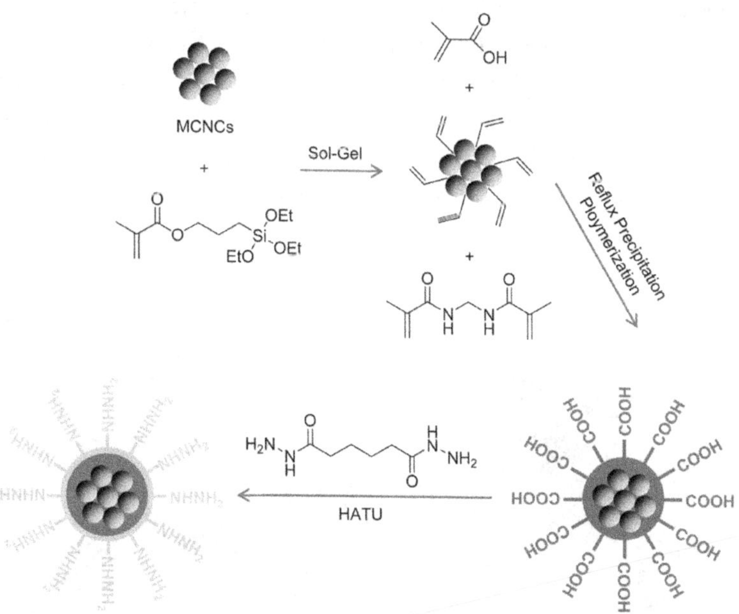

Figure 6.40: Schematic diagram of the synthesis of Fe_3O_4@PMAH microsphere.

First, Fe_3O_4@PMAA microsphere is synthesized according to the literature.[17] Then 50 mg Fe_3O_4@PMAA microsphere and 132.4 mg 2-(7-azobenzotriazole)-*N,N,N',N'*-tetramethyluron hexafluorophosphate (HATU) are dispersed in 1.16 mL DMF with ultrasonication under ice bath. After stirring for 10 min, 55.6 mg adipic acid dihydrazide (ADH) and 144 μL of *N,N*-diisopropylethylamine (DIPEA) are added. Afterwards, the whole reaction system is heated to room temperature and kept continuously stirred for 3 days. The obtained Fe_3O_4@PMAH microsphere is washed with DMF, deionized water, and ethanol for further experiment.

(2) Characterization of Fe_3O_4@PMAH microsphere

It can be observed from Figure 6.41(a) that the synthesized Fe_3O_4 microsphere is uniform and has a diameter of about 200 nm. After coating with the polymer layer, the surface is smoother (Figure 6.41(b)). The peaks at 1601 cm^{-1}, 1421 cm^{-1}, and 578 cm^{-1} observed in FTIR spectrum of Fe_3O_4 microsphere in Figure 6.41(a) are attributed to the asymmetric stretching vibration and symmetric stretching vibration of the carboxyl group and the stretching vibration of Fe–O, respectively. For the FTIR spectrum of Fe_3O_4@PMAA microsphere, the newly emerging peaks at 1716 cm^{-1} and 1533 cm^{-1} are attributed to the stretching vibration of the C=O and N–H of the MBA, indicating the successful coating of PMAA on the surface of magnetic microsphere. For the FTIR spectrum of Fe_3O_4@PMAH microsphere, the absorption peak at 1716 cm^{-1} disappears and a new peak appears at 1653 cm^{-1}, which is

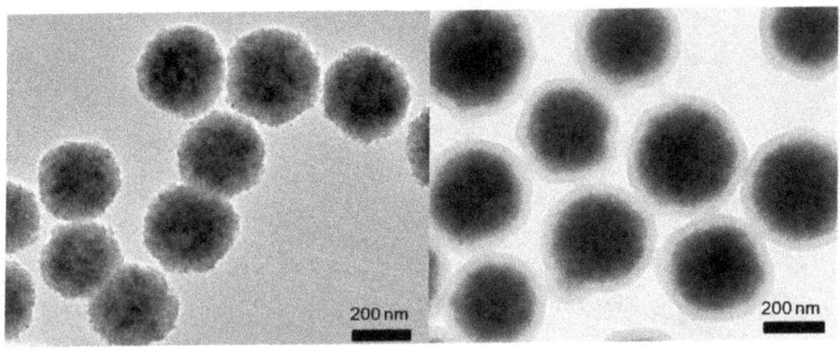

Figure 6.41. FE-TEM images.

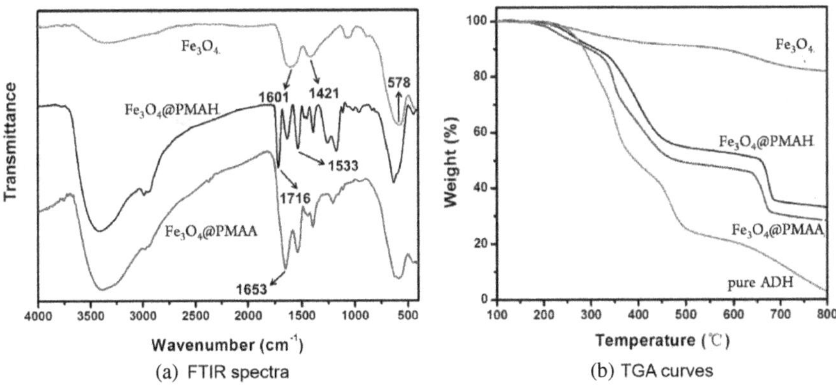

Figure 6.42: FTIR spectra and TGA curves of different materials.

attributed to C=O of ADH, indicating that the carbonyl group is modified with hydrazine group.

In Figure 6.42(b), it can be seen that the mass loss is 18 wt.% due to the presence of the citrate residue. The mass loss of Fe_3O_4@PMAA and Fe_3O_4@PMAH microsphere is 67 wt.% and 72 wt.%, respectively. Under the same condition, the mass loss of pure ADH is 97 wt.%, so that the content of polymer and ADH in Fe_3O_4@PMAH microsphere are estimated to be 49 wt.% and 5 wt.%, respectively.

(3) Application of Fe_3O_4@PMAH microsphere in glycoproteomics

The schematic diagram in Figure 6.43 illustrates the enrichment process of Fe_3O_4@PMAH microsphere. Firstly, the cis–diol group on the glycopeptide is oxidized to the aldehyde group by sodium periodate; secondly, the hydrazine group on the surface of Fe_3O_4@PMAH microsphere forms the hydrazone bond with the aldehyde group for completing the capture of glycopeptide; thirdly, the *N*-glycan is cleaved by PNGase F to release peptide for MS analysis.

(a) *Evaluation of the enrichment ability of Fe_3O_4@PMAH microsphere for glycopeptide in standard glycoprotein digest*

Two standard glycoprotein (fetuin and asialofetuin) digests are used to study the enrichment ability of Fe_3O_4@PMAH microsphere for glycopeptide. As

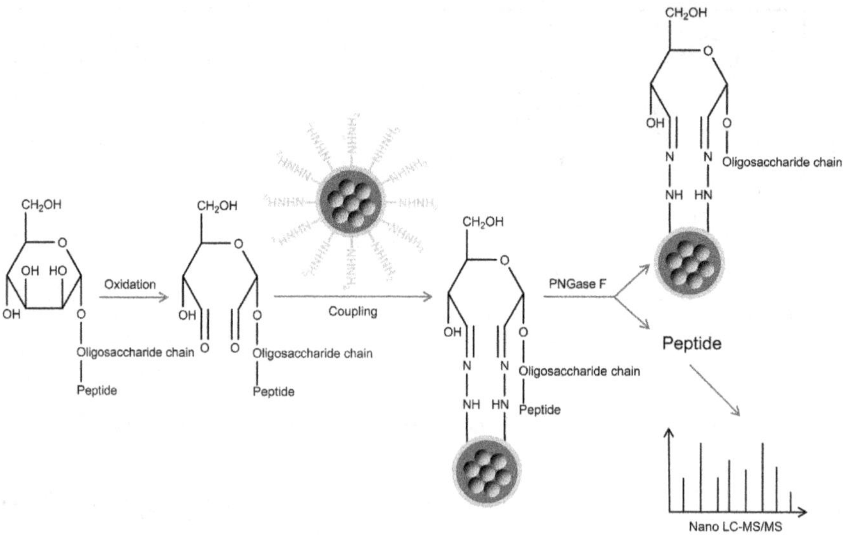

Figure 6.43: Schematic diagram of glycopeptide enrichment by Fe₃O₄@PMAH microsphere.

shown in Figure 6.44(a), when the fetuin digest is directly analyzed by mass spectrometry, a large number of non-glycosylated peptides are observed in the spectrum, and no glycosylated peptide is detected. After enriched by Fe₃O₄@ PMAH microsphere, as shown in Figure 6.44(b), four deglycosylated peptides from fetuin are detected, the *m/z* of which is 3556.2, 3017.1, 1753.6, and 1625.5, respectively and the corresponding amino acid sequences are RPTGEVYDIEIDTLETTCHVLDPTP LAN#CSVR (#: *N*-glycosylation site), VVHAVEVALATFNAESN#GSYLQLVEISR, KLCPDCPLLAPLN# DSR, and LCPDCPLLAPLN#DSR. Similarly, as shown in Figure 6.45, after the asialofetuin digest is enriched by Fe₃O₄@PMAH microsphere, four deglycosylated peptides from asialofetuin digest are detected.

(b) *Evaluation of the enrichment ability of Fe₃O₄@PMAH microsphere*
    for glycopeptide in serum digest from patient with intestinal cancer
In order to further investigate the selective enrichment ability of Fe₃O₄@PMAH microsphere for *N*-glycosylated peptide, serum digest of patient with intestinal cancer is selected. 8 mg Fe₃O₄@PMAH microsphere is applied to 20 μL serum digest for enrichment. Then, the enriched peptides are deglycosylated by

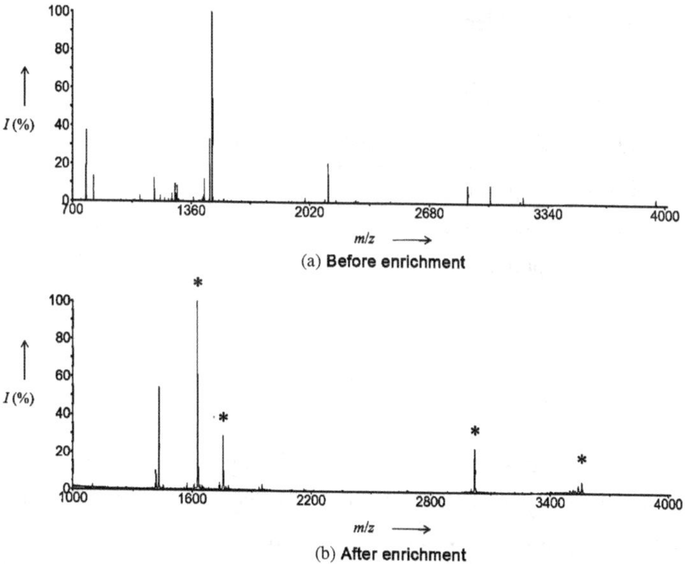

Figure 6.44: MS spectra of fetuin digest before and after enrichment by Fe$_3$O$_4$@PMAH microsphere. *: deglycosylated peptide.

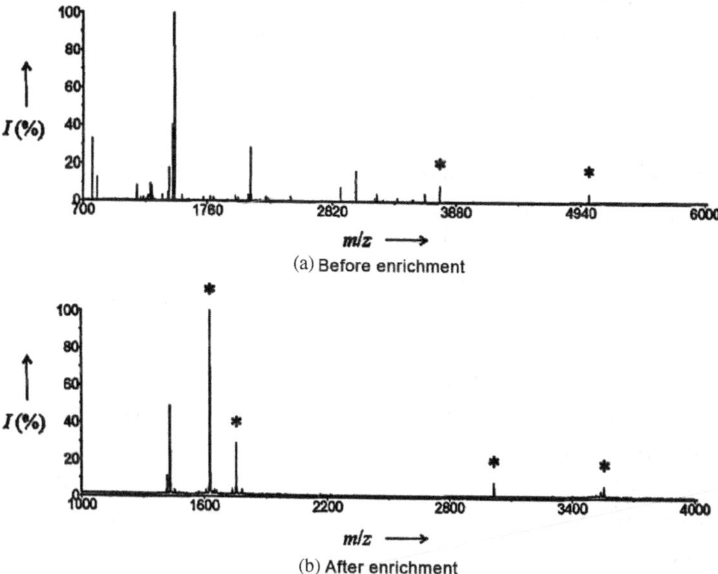

Figure 6.45: MS spectra of asialofetuin digest before and after being by Fe$_3$O$_4$@PMAH microsphere. *: deglycosylated peptide.

Table 6.8: Identification result of glycopeptide in serum digest from patient with intestinal cancer by Fe_3O_4@PMAH microsphere.

Identified	Enrich 1	Enrich 2	Enrich 3
Unique glysopeptides	135	135	139
Unique glycoproteins	55	56	54
Unique peptides	194	203	209
Unique proteins	68	70	69
Glycopeptide specificity	69.6%	66.5%	66.5%
Glycoprotein specificity	80.9%	80.0%	78.3%

PNGase F and fragmented by ESI-LTQ Orbitrap tandem mass spectrometry. The obtained CID spectra are searched by Swiss-Prot database and further analyzed by using the PeptideProphet software. The experiment is repeated 3 times and the results are shown in Table 6.8. It can be seen from the data that there is no significant difference between the results of the three replicates, indicating that Fe_3O_4@PMAH microsphere is feasible for complex bio-samples. Finally, a total of 181 glycosylation sites, corresponding to 175 glycopeptides and 63 glycoproteins, are identified in three replicates.

References

1. Chen C-C., Su W-C., Huang B-Y., *et al.* Interaction Modes and Approaches to Glycopeptide and Glycoprotein Enrichment. *Analyst*, 2014, 139: 688–704.

2. Zheng J. N., Xiao Y., Wang L., *et al.* Click Synthesis of Glucose-Functionalized Hydrophilic Magnetic Mesoporous Nanoparticles for Highly Selective, Enrichment of Glycopeptides and Glycans. *Journal of Chromatography A*, 2014, 1358: 29–38.

3. Fang C. L., Xiong Z. C., Qin H. Q., *et al.* One-Pot Synthesis of Magnetic Colloidal Nanocrystal Clusters Coated with Chitosan for Selective Enrichment of Glycopeptides. *Analytica Chimica Acta*, 2014, 841: 99–105.

4. Chen Y., Xiong Z., Zhang L., *et al.* Facile Synthesis of Zwitterionic Polymer-Coated Core-Shell Magnetic Nanoparticles for Highly Specific Capture of N-Linked Glycopeptides. *Nanoscale*, 2015, 7: 3100–08.

5. Jiang B., Wu Q., Deng N., *et al.* Hydrophilic Go/Fe_3O_4/Au/PEG Nanocomposites for Highly Selective Enrichment of Glycopeptides. *Nanoscale*, 2016, 8: 4894–7.

6. Xiong Z., Zhao L., Wang F., *et al.* Synthesis of Branched Peg Brushes Hybrid Hydrophilic Magnetic Nanoparticles for the Selective Enrichment of N-Linked Glycopeptides. *Chemical Communications*, 2012, 48: 8138–40.

7. Wan H., Huang J., Liu Z., *et al.* A Dendrimer-Assisted Magnetic Graphene-Silica Hydrophilic Composite for Efficient and Selective Enrichment of Glycopeptides from the Complex Sample. *Chemical Communications*, 2015, 51: 9391–94.

8. Xiong Z., Qin H., Wan H., *et al.* Layer-by-Layer Assembly of Multilayer Polysaccharide Coated Magnetic Nanoparticles for the Selective Enrichment of Glycopeptides. *Chemical Communications*, 2013, 49: 9284–6.

9. Li J., Wang F., Liu J., *et al.* Functionalizing with Glycopeptide Dendrimers Significantly Enhances the Hydrophilicity of the Magnetic Nanoparticles. *Chemical Communications*, 2015, 51: 4093–96.

10. Tang J., Liu Y., Yin P., *et al.* Concanavalin a-Immobilized Magnetic Nanoparticles for Selective Enrichment of Glycoproteins and Application to Glycoproteomics in Hepatocelluar Carcinoma Cell Line. *Proteomics*, 2010, 10: 2000–14.

11. Qi D. W., Zhang H. Y., Tang J, *et al.* Facile Synthesis of Mercaptophenylboronic Acid-Functionalized Core-Shell Structure Fe_3O_4@C@Au Magnetic Microspheres for Selective Enrichment of Glycopeptides and Glycoproteins. *Journal of Physical Chemistry C*, 2010, 114: 9221–26.

12. IZhang L., Xu Y., Yao H., *et al.* Boronic Acid Functionalized Core-Satellite Composite Nanoparticles for Advanced Enrichment of Glycopeptides and Glycoproteins. *Chemistry*, 2009, 15: 10158–66.

13. Wang M., Zhang X., Deng C. Facile Synthesis of Magnetic Poly(Styrene-Co-4-Vinylbenzene-Boronic Acid) Microspheres for Selective Enrichment of Glycopeptides. *Proteomics*, 2015, 15: 2158–65.

14. Wang Y., Liu M., Xie L., *et al.* Highly Efficient Enrichment Method for Glycopeptide Analyses: Using Specific and Nonspecific Nanoparticles Synergistically. *Analytical Chemisty*, 2014, 86: 2057–64.

15. Zhang X. H., He X. W., Chen L. X., *et al.* A Combination of Distillation-Precipitation Polymerization and Click Chemistry: Fabrication of Boronic Acid Functionalized Fe_3O_4 Hybrid Composites for Enrichment of Glycoproteins. *Journal of Materials Chemistry B*, 2014, 2: 3254–62.

16. Liu L. T., Yu M., Zhang Y., *et al.* Hydrazide Functionalized Core-Shell Magnetic Nanocomposites for Highly Specific Enrichment of N-Glycopeptides. *ACS Applied Materials & Interfaces*, 2014, 6: 7823–32.

17. Ma W. F., Zhang Y., Li L. L., *et al.* Ti^{4+}-Immobilized Magnetic Composite Microspheres for Highly Selective Enrichment of Phosphopeptides. *Advanced Functional Materials*, 2013, 23: 107–15.

Index

CPSIA information can be obtained
at www.ICGtesting.com
Printed in the USA
JSHW042047190621
16070JS00002B/12

9 789811 230547